Handbook of Brewing

Second Edition

FOOD SCIENCE AND TECHNOLOGY

A Series of Monographs, Textbooks, and Reference Books

Editorial Advisory Board

Gustavo V. Barbosa-Cánovas Washington State University–Pullman
P. Michael Davidson University of Tennessee–Knoxville
Mark Dreher McNeil Nutritionals, New Brunswick, NJ
Richard W. Hartel University of Wisconsin–Madison
Lekh R. Juneja Taiyo Kagaku Company, Japan
Marcus Karel Massachusetts Institute of Technology
Ronald G. Labbe University of Massachusetts–Amherst
Daryl B. Lund University of Wisconsin–Madison
David B. Min The Ohio State University
Leo M. L. Nollet Hogeschool Gent, Belgium
Seppo Salminen University of Turku, Finland
John H. Thorngate III Allied Domecq Technical Services, Napa, CA
Pieter Walstra Wageningen University, The Netherlands
John R. Whitaker University of California–Davis
Rickey Y. Yada University of Guelph, Canada

76. Food Chemistry: Third Edition, *edited by Owen R. Fennema*
77. Handbook of Food Analysis: Volumes 1 and 2, *edited by Leo M. L. Nollet*
78. Computerized Control Systems in the Food Industry, *edited by Gauri S. Mittal*
79. Techniques for Analyzing Food Aroma, *edited by Ray Marsili*
80. Food Proteins and Their Applications, *edited by Srinivasan Damodaran and Alain Paraf*
81. Food Emulsions: Third Edition, Revised and Expanded, *edited by Stig E. Friberg and Kåre Larsson*
82. Nonthermal Preservation of Foods, *Gustavo V. Barbosa-Cánovas, Usha R. Pothakamury, Enrique Palou, and Barry G. Swanson*
83. Milk and Dairy Product Technology, *Edgar Spreer*
84. Applied Dairy Microbiology, *edited by Elmer H. Marth and James L. Steele*
85. Lactic Acid Bacteria: Microbiology and Functional Aspects, Second Edition, Revised and Expanded, *edited by Seppo Salminen and Atte von Wright*
86. Handbook of Vegetable Science and Technology: Production, Composition, Storage, and Processing, *edited by D. K. Salunkhe and S. S. Kadam*
87. Polysaccharide Association Structures in Food, *edited by Reginald H. Walter*
88. Food Lipids: Chemistry, Nutrition, and Biotechnology, *edited by Casimir C. Akoh and David B. Min*
89. Spice Science and Technology, *Kenji Hirasa and Mitsuo Takemasa*
90. Dairy Technology: Principles of Milk Properties and Processes, *P. Walstra, T. J. Geurts, A. Noomen, A. Jellema, and M. A. J. S. van Boekel*
91. Coloring of Food, Drugs, and Cosmetics, *Gisbert Otterstätter*
92. *Listeria*, Listeriosis, and Food Safety: Second Edition, Revised and Expanded, *edited by Elliot T. Ryser and Elmer H. Marth*

93. Complex Carbohydrates in Foods, *edited by Susan Sungsoo Cho, Leon Prosky, and Mark Dreher*
94. Handbook of Food Preservation, *edited by M. Shafiur Rahman*
95. International Food Safety Handbook: Science, International Regulation, and Control, *edited by Kees van der Heijden, Maged Younes, Lawrence Fishbein, and Sanford Miller*
96. Fatty Acids in Foods and Their Health Implications: Second Edition, Revised and Expanded, *edited by Ching Kuang Chow*
97. Seafood Enzymes: Utilization and Influence on Postharvest Seafood Quality, *edited by Norman F. Haard and Benjamin K. Simpson*
98. Safe Handling of Foods, *edited by Jeffrey M. Farber and Ewen C. D. Todd*
99. Handbook of Cereal Science and Technology: Second Edition, Revised and Expanded, *edited by Karel Kulp and Joseph G. Ponte, Jr.*
100. Food Analysis by HPLC: Second Edition, Revised and Expanded, *edited by Leo M. L. Nollet*
101. Surimi and Surimi Seafood, *edited by Jae W. Park*
102. Drug Residues in Foods: Pharmacology, Food Safety, and Analysis, *Nickos A. Botsoglou and Dimitrios J. Fletouris*
103. Seafood and Freshwater Toxins: Pharmacology, Physiology, and Detection, *edited by Luis M. Botana*
104. Handbook of Nutrition and Diet, *Babasaheb B. Desai*
105. Nondestructive Food Evaluation: Techniques to Analyze Properties and Quality, *edited by Sundaram Gunasekaran*
106. Green Tea: Health Benefits and Applications, *Yukihiko Hara*
107. Food Processing Operations Modeling: Design and Analysis, *edited by Joseph Irudayaraj*
108. Wine Microbiology: Science and Technology, *Claudio Delfini and Joseph V. Formica*
109. Handbook of Microwave Technology for Food Applications, *edited by Ashim K. Datta and Ramaswamy C. Anantheswaran*
110. Applied Dairy Microbiology: Second Edition, Revised and Expanded, *edited by Elmer H. Marth and James L. Steele*
111. Transport Properties of Foods, *George D. Saravacos and Zacharias B. Maroulis*
112. Alternative Sweeteners: Third Edition, Revised and Expanded, *edited by Lyn O'Brien Nabors*
113. Handbook of Dietary Fiber, *edited by Susan Sungsoo Cho and Mark L. Dreher*
114. Control of Foodborne Microorganisms, *edited by Vijay K. Juneja and John N. Sofos*
115. Flavor, Fragrance, and Odor Analysis, *edited by Ray Marsili*
116. Food Additives: Second Edition, Revised and Expanded, *edited by A. Larry Branen, P. Michael Davidson, Seppo Salminen, and John H. Thorngate, III*
117. Food Lipids: Chemistry, Nutrition, and Biotechnology: Second Edition, Revised and Expanded, *edited by Casimir C. Akoh and David B. Min*
118. Food Protein Analysis: Quantitative Effects on Processing, *R. K. Owusu-Apenten*
119. Handbook of Food Toxicology, *S. S. Deshpande*
120. Food Plant Sanitation, *edited by Y. H. Hui, Bernard L. Bruinsma, J. Richard Gorham, Wai-Kit Nip, Phillip S. Tong, and Phil Ventresca*
121. Physical Chemistry of Foods, *Pieter Walstra*
122. Handbook of Food Enzymology, *edited by John R. Whitaker, Alphons G. J. Voragen, and Dominic W. S. Wong*

123. Postharvest Physiology and Pathology of Vegetables: Second Edition, Revised and Expanded, *edited by Jerry A. Bartz and Jeffrey K. Brecht*
124. Characterization of Cereals and Flours: Properties, Analysis, and Applications, *edited by Gönül Kaletunç and Kenneth J. Breslauer*
125. International Handbook of Foodborne Pathogens, *edited by Marianne D. Miliotis and Jeffrey W. Bier*
126. Food Process Design, *Zacharias B. Maroulis and George D. Saravacos*
127. Handbook of Dough Fermentations, *edited by Karel Kulp and Klaus Lorenz*
128. Extraction Optimization in Food Engineering, *edited by Constantina Tzia and George Liadakis*
129. Physical Properties of Food Preservation: Second Edition, Revised and Expanded, *Marcus Karel and Daryl B. Lund*
130. Handbook of Vegetable Preservation and Processing, *edited by Y. H. Hui, Sue Ghazala, Dee M. Graham, K. D. Murrell, and Wai-Kit Nip*
131. Handbook of Flavor Characterization: Sensory Analysis, Chemistry, and Physiology, *edited by Kathryn Deibler and Jeannine Delwiche*
132. Food Emulsions: Fourth Edition, Revised and Expanded, *edited by Stig E. Friberg, Kare Larsson, and Johan Sjoblom*
133. Handbook of Frozen Foods, *edited by Y. H. Hui, Paul Cornillon, Isabel Guerrero Legarret, Miang H. Lim, K. D. Murrell, and Wai-Kit Nip*
134. Handbook of Food and Beverage Fermentation Technology, *edited by Y. H. Hui, Lisbeth Meunier-Goddik, Ase Solvejg Hansen, Jytte Josephsen, Wai-Kit Nip, Peggy S. Stanfield, and Fidel Toldrá*
135. Genetic Variation in Taste Sensitivity, *edited by John Prescott and Beverly J. Tepper*
136. Industrialization of Indigenous Fermented Foods: Second Edition, Revised and Expanded, *edited by Keith H. Steinkraus*
137. Vitamin E: Food Chemistry, Composition, and Analysis, *Ronald Eitenmiller and Junsoo Lee*
138. Handbook of Food Analysis: Second Edition, Revised and Expanded, Volumes 1, 2, and 3, *edited by Leo M. L. Nollet*
139. Lactic Acid Bacteria: Microbiological and Functional Aspects: Third Edition, Revised and Expanded, *edited by Seppo Salminen, Atte von Wright, and Arthur Ouwehand*
140. Fat Crystal Networks, *Alejandro G. Marangoni*
141. Novel Food Processing Technologies, *edited by Gustavo V. Barbosa-Cánovas, M. Soledad Tapia, and M. Pilar Cano*
142. Surimi and Surimi Seafood: Second Edition, *edited by Jae W. Park*
143. Food Plant Design, *Antonio Lopez-Gomez; Gustavo V. Barbosa-Cánovas*
144. Engineering Properties of Foods: Third Edition, *edited by M. A. Rao, Syed S.H. Rizvi, and Ashim K. Datta*
145. Antimicrobials in Food: Third Edition, *edited by P. Michael Davidson, John N. Sofos, and A. L. Branen*
146. Encapsulated and Powdered Foods, *edited by Charles Onwulata*
147. Dairy Science and Technology: Second Edition, *Pieter Walstra, Jan T. M. Wouters and Tom J. Geurts*
148. Food Biotechnology, Second Edition, *edited by Kalidas Shetty, Gopinadhan Paliyath, Anthony Pometto and Robert E. Levin*
149. Handbook of Food Science, Technology, and Engineering - 4 Volume Set, *edited by Y. H. Hui*
150. Thermal Food Processing: New Technologies and Quality Issues, *edited by Da-Wen Sun*
151. Aflatoxin and Food Safety, *edited by Hamed K. Abbas*
152. Food Packaging: Principles and Practice, Second Edition, *Gordon L. Robertson*

153. Seafood Processing: Adding Value Through Quick Freezing, Retortable Packaging, Cook-Chilling and Other Methods, *Vazhiyil Venugopal*
154. Ingredient Interactions: Effects on Food Quality, Second Edition, *edited by Anilkumar G. Gaonkar and Andrew McPherson*
155. Handbook of Frozen Food Processing and Packaging, *edited by Da-Wen Sun*
156. Vitamins In Foods: Analysis, Bioavailability, and Stability, *George F. M. Ball*
157. Handbook of Brewing, Second Edition, *edited by Fergus G. Priest and Graham G. Stewart*

Handbook of Brewing

Second Edition

edited by
Fergus G. Priest
Graham G. Stewart

Taylor & Francis Group
Boca Raton   London   New York

A CRC title, part of the Taylor & Francis imprint, a member of the
Taylor & Francis Group, the academic division of T&F Informa plc.

Published in 2006 by
CRC Press
Taylor & Francis Group
6000 Broken Sound Parkway NW, Suite 300
Boca Raton, FL 33487-2742

© 2006 by Taylor & Francis Group, LLC
CRC Press is an imprint of Taylor & Francis Group

No claim to original U.S. Government works
Printed in the United States of America on acid-free paper
10 9 8 7 6 5 4 3 2 1

International Standard Book Number-10: 0-8247-2657-X (Hardcover)
International Standard Book Number-13: 978-0-8247-2657-7 (Hardcover)
Library of Congress Card Number 2005052938

This book contains information obtained from authentic and highly regarded sources. Reprinted material is quoted with permission, and sources are indicated. A wide variety of references are listed. Reasonable efforts have been made to publish reliable data and information, but the author and the publisher cannot assume responsibility for the validity of all materials or for the consequences of their use.

No part of this book may be reprinted, reproduced, transmitted, or utilized in any form by any electronic, mechanical, or other means, now known or hereafter invented, including photocopying, microfilming, and recording, or in any information storage or retrieval system, without written permission from the publishers.

For permission to photocopy or use material electronically from this work, please access www.copyright.com (http://www.copyright.com/) or contact the Copyright Clearance Center, Inc. (CCC) 222 Rosewood Drive, Danvers, MA 01923, 978-750-8400. CCC is a not-for-profit organization that provides licenses and registration for a variety of users. For organizations that have been granted a photocopy license by the CCC, a separate system of payment has been arranged.

Trademark Notice: Product or corporate names may be trademarks or registered trademarks, and are used only for identification and explanation without intent to infringe.

Library of Congress Cataloging-in-Publication Data

Handbook of brewing.--2nd ed. / [edited by] Fergus G. Priest, Graham G. Stewart.
    p. cm. -- (Food science and technology ; 157)
Includes bibliographical references and index.
ISBN 0-8247-2657-X
1 Brewing--Handbooks, manuals, etc. I. Priest, F.G. II. Stewart, Graham G., 1942- III. Food science and technology (Taylor & Francis) ; 157.

TP570.H23 2005
663'.3--dc22   2005052938

Taylor & Francis Group
is the Academic Division of Informa plc.

Visit the Taylor & Francis Web site at
http://www.taylorandfrancis.com

and the CRC Press Web site at
http://www.crcpress.com

Preface

The first edition of this handbook provided a groundbreaking coverage of the science and technology of brewing. In it, William Hardwick assembled a cast of highly experienced and authoritative authors to contribute on a comprehensive range of topics from commercial and economic aspects to the processes of manufacturing beer and included the economics and environmental aspects of brewery effluents. The result was a popular and highly valued handbook.

Much has changed in the 10 years since the publication of the first edition of this book. The industry has been transformed both commercially and technically. Many small (and not so small) companies have been subsumed into large multinationals, while at the other extreme, microbreweries have flourished in many parts of the world. Technology has been transformed and impinges on all aspects of the raw materials and production of beer. The massive improvements in computer power and automation have modernized the brewhouse, while developments in biotechnology including the sequencing of the yeast genome and the application of the polymerase chain reaction in molecular biology have steadily improved brewing efficiency, beer quality, and shelf life. For example, the range of hop extracts and products has grown almost exponentially. The composition and use of these products are covered in detail in Chapter 7. Finally, while traditional beers have their followers, markets for beer-like beverages are there to be explored in order to gain marketshare in a mature beer sector. Chapter 2 covers traditional beer styles as well as more obscure beverages such as chocolate- or coffee-flavored beers, while in the last chapter of the book, Inge Russell describes some fascinating new avenues to challenge the brewer's art of manufacturing a quality beverage from barley-based raw materials.

Just as the industry has changed, so has its personnel. While some of the original authors of the handbook were able to update their contributions, we have sought new authors for numerous areas where the original authors were no longer in a position to contribute. In so doing, we have been fortunate to enlist a team of international experts broadly recognized for their distinguished contributions to brewing science and technology. We are grateful to all these contributors for fitting in the writing around their "day jobs" and for responding, most of the time, as our deadlines passed. Finally, the publishers have changed, and Marcel Dekker which published the first edition of the Handbook has become part of Taylor & Francis. We thank Susan Lee and Kari Budyk of Taylor & Francis for their patience as we gathered the book together. It has been a long journey but the result makes it all worthwhile.

F.G. Priest

Editors

Fergus Priest is a professor of microbiology at Heriot-Watt University, Edinburgh, where he specializes in fermentation microbiology. His Ph.D. studies at the University of Birmingham focused on spoilage bacteria in brewery fermentations and their impact on beer flavor. More recently he has researched the role of lactic acid bacteria in Scotch whisky fermentations. He has published more than 70 research papers and several books. He was chief editor of *FEMS Microbiology Letters* and serves on the Editorial Board of the *Journal of the American Society of Brewing Chemists*. He was elected as a Fellow of the Institute of Brewing & Distilling in 1994.

Graham Stewart has been the director and a professor of the International Centre for Brewing and Distilling, Heriot-Watt University, Edinburgh, since 1994. He received his B.Sc. Hons in Microbiology and B.Sc. Hons in Biochemistry from the University of Wales at Cardiff, and Ph.D. and D.Sc. degrees from Bath University. He was a lecturer in biochemistry in the School of Pharmacy at Portsmouth College of Technology (now Portsmouth University) from 1967 until 1969. From 1969 to 1994, he held a number of technical/scientific positions with Labatt's in Canada and from 1986 to 1994, was its director of technical affairs. In addition to co-authoring and editing a number of books, he has published over 200 original papers, patents, and reviews.

Contributors

Raymond G. Anderson The Brewery History Society, High Ridge, Uttoxeter, Staffordshire, UK

Zane C. Barnes Coopers Brewery, Adelaide, Australia. *Present address:* Coors Brewing Company, Massanutten, Virginia, USA

Johannes Braun Otaru Beer, Zenibako Brewery, Otaru, Japan

Brian H. Dishman Otaru Beer, Zenibako Brewery, Otaru, Japan

Alexander R. Dunn International Centre for Brewing and Distilling, Heriot-Watt University, Edinburgh, Scotland, UK

Brian Eaton International Centre for Brewing and Distilling, Heriot-Watt University, Edinburgh, Scotland, UK

Tom Fetters Crown Cork and Seal, Alsip, Illinois, USA

Nick J. Huige Miller Brewing Company, Milwaukee, Wisconsin, USA

Kenneth A. Leiper International Centre for Brewing and Distilling, Heriot-Watt University, Edinburgh, Scotland, UK

Michaela Miedl International Centre for Brewing and Distilling, Heriot-Watt University, Edinburgh, Scotland, UK

James H. Munroe Anheuser-Busch, Inc., St. Louis, Missouri, USA

Geoffrey H. Palmer International Centre for Brewing and Distilling, Heriot-Watt University, Edinburgh, Scotland, UK

Charles Papazian Brewers Association, Boulder, Colorado, USA

George Philliskirk The Beer Academy, Camberley, Surrey, UK

Joseph Power Alltech Biotechnology Center, Nicholasville, Kentucky, USA

Fergus G. Priest International Centre for Brewing and Distilling, Heriot-Watt University, Edinburgh, Scotland, UK

Trevor R. Roberts Steiner Hops Ltd., Epping, Essex, UK

Inge Russell Alltech Biotechnology Center, Nicholasville, Kentucky, USA and International Centre for Brewing and Distilling, Heriot-Watt University, Edinburgh, Scotland, UK

David S. Ryder Miller Brewing Company, Milwaukee, Wisconsin, USA

Graham G. Stewart International Centre for Brewing and Distilling, Heriot-Watt University, Edinburgh, Scotland, UK

David G. Taylor D.G. Taylor Consultancy Associates Ltd., Corsham, Wiltshire, UK

Vernon E. Walter WAW Inc., Leakey, Texas, USA

Richard J.H. Wilson Steiner Hops Ltd., Epping, Essex, UK

Contents

Preface v

1. **History of Industrial Brewing** 1
   Raymond G. Anderson

2. **Beer Styles: Their Origins and Classification** 39
   Charles Papazian

3. **An Overview of Brewing** 77
   Brian Eaton

4. **Water** 91
   David G. Taylor

5. **Barley and Malt** 139
   Geoffrey H. Palmer

6. **Adjuncts** 161
   Graham G. Stewart

7. **Hops** 177
   Trevor R. Roberts and Richard J.H. Wilson

8. **Yeast** 281
   Inge Russell

9. **Miscellaneous Ingredients in Aid of the Process** 333
   David S. Ryder and Joseph Power

10. **Brewhouse Technology** 383
    Kenneth A. Leiper and Michaela Miedl

11. **Brewing Process Control** 447
    Zane C. Barnes

12. **Fermentation** 487
    James H. Munroe

13.	Aging and Finishing... *James H. Munroe*	525
14.	Packaging: A Historical Perspective........................ *Tom Fetters*	551
15.	Packaging Technology .... *Alexander R. Dunn*	563
16.	Microbiology and Microbiological Control in the Brewery .... *Fergus G. Priest*	607
17.	Sanitation and Pest Control.................................. *Vernon E. Walter*	629
18.	Brewery By-Products and Effluents .... *Nick J. Huige*	655
19.	Beer Stability.. *Graham G. Stewart*	715
20.	Quality... *George Philliskirk*	729
21.	Microbrewing .... *Johannes Braun and Brian H. Dishman*	771
22.	Innovation and Novel Products .... *Inge Russell*	817
Index		831

1
History of Industrial Brewing

Raymond G. Anderson

CONTENTS
Introduction 1
Brewing in an Agrarian World 2
The Eighteenth Century 4
   Porter: The First Industrial Beer 4
   Mechanization and Measurement 6
The Nineteenth Century 7
   Porter vs. Ale 8
   The Rush to Bottom Fermentation 10
   Science and Practice 12
The Twentieth Century 15
   Beer and Society 16
     Temperance and Prohibition 16
     Consumer Choice? 18
   Fewer and Bigger: The Path to Globalization 23
   Science Applied and Technology Transformed 27
References 33

Introduction

For most of its history, brewing was a domestic or small-scale commercial activity supplying an essential element of diet to a primarily agrarian population. It is now an industry increasingly dominated by a few large companies striving for global supremacy in the supply of branded recreational alcoholic beverages.[1] This chapter outlines the massive changes in the organization, economic importance, scale, scientific understanding and technology of brewing, and in the social function and nature of beer, that industrialization has engendered across the world.

Brewing in an Agrarian World

Brewing is generally considered to have originated as a by-product of the development of agriculture, although a minority opinion holds that the cultivation of cereals originated as a consequence of man's desire for alcohol rather than vice versa.[2] Whatever its exact origins, surviving historical artifacts allow us to trace brewing back to the Mesopotamians around 6000 or 7000 years ago. The Ancient Egyptians were brewers, and beer, brewed from the indigenous cereal sorghum, is still integral to the politics of African tribal life. The historical development of brewing and the brewing industry is, however, linked with northern Europe where cold conditions inhibited the development of viticulture.[3] The Romans commented in derogatory terms on the drinking of barley-based beverages by the Germans and the Britons.

From the tenth century, the use of hops in brewing spread from Germany across Europe to replace, or at least supplement, the plethora of plants, herbs, and spices popular at that time. The introduction of hops was met with resistance, but the pleasing flavor and aroma they provided and perhaps, more importantly, their action in protecting the beer from being spoiled by the then unknown microbes, eventually led to their widescale adoption. Brewers of unhopped beer depended upon high alcohol percentage to preserve their beers, but this was relatively inefficient and such beers generally had poor keeping qualities. Although brewing with hops was a more complicated operation, requiring extra equipment, it did allow the brewer to produce a weaker beer that was still resistant to spoilage and thus make a greater volume of product from the same quantity of raw material. Hops were introduced into Britain in the 15th century and reached North America in the early 17th century.

For a time, the terms ale and beer were being applied to distinct beverages made by separate communities of brewers. Ale described the drink made without hops, whereas the term beer was reserved for the hopped beverage. By the 16th century, ale brewers had also come to use some hops in their brews, but at a lower level than was usual for beer and an element of distinction remained. Ale would be recognized as a heavy, sweet, noticeably alcoholic drink characteristic of rural areas. Beer was bitter, often lighter in flavor and less alcoholic, but frequently darker brown in color than ale and was popular in towns.[3] Unhopped ale virtually disappeared from Europe during the 17th century, but there remained a vast variety of different beers available. Each region offered its own favorite brews influenced by availability and quality of raw materials and climate. The dominant cereal in use was barley, the easiest to malt, although it could be supplemented or even replaced by other cereals, particularly oats and wheat. In some regions, notably in parts of Germany and Belgium, wheat beers became a speciality. Taxes on beer became a growing feature along with a degree of consumer protection over serving measures and prices enforced by local authorities to regulate sales in taverns. In 1516, the Reinheitsgebot, literally "Commandment for Purity" was introduced in Bavaria. This early consumer or trade protection measure (outside Germany views differ)[4] decreed that only malt, hops, and water were to be used in brewing. Yeast was later added to the list when its necessity (if not its identity) became obvious and wheat allowed for speciality beers.

Beer was integral to the culture of the agrarian population of northern and central Europe in the medieval and early modern period. The weaker brews were accepted as an essential part of everyone's diet and the stronger beers as a necessary source of solace in all too brief periods of leisure in a harsh world. There is no reliable information on the level of consumption except for the frequent assertions that it was "massive" and "immense."[3] Nor can alcoholic strength be estimated with any accuracy without any data apart from general recipes. It was often the practice to carry out multiple extractions of the same grist to yield beers of different strengths. "Strong beer/ ale" was fermented using wort drawn from the first mash, with weaker beers derived from the second and third mashes. These latter brews (table or small beer) were everyday drinks consumed by people of all classes and ages at meals in preference to unreliable water and were an important source of nutrients in a frequently drab diet. The strong brews were particularly favored to celebrate church festivals and family celebrations. Only the elite ever saw wines or spirits. Brewing was restricted to the period roughly between October and March — attempts at summer brewing often led to spoilage.

The scale of brewing ranged from a few hectoliters annually in the average home to hundreds, or occasionally, thousands of hectoliters in the largest monasteries and country houses. Domestic brewing still accounted for well over half of the beer produced at the end of the 17th century. Commercial brewing was generally confined to taverns and small breweries. The latter produced a wide selection of beers of different strengths, light to dark brown in color, predominantly for local consumption. The biggest of these breweries could run to tens of thousands of hectoliters, but true industrialization of brewing did not begin until increased urbanization and concentrated population growth provided a ready market for beer produced on a massive scale.

The Eighteenth Century

In general, brewing changed little during the 18th century, with a mix of domestic and small-scale commercial production still the norm. Trade in beer remained predominantly local everywhere, whether by the tens of thousands of European brewers or the less than 150 breweries that existed in the fledgling United States by 1800. What little regional or national trade there was went by canal. International trade was an exception and confined to the most enterprising merchant brewers who could defray the cost of moving a bulky low-value product with reciprocal deals in other goods. Benjamin Wilson of Burton upon Trent, who traded extensively in the Baltic in the second half of the 18th century, is a prime example, but even here quantities were small at around a few thousand hectoliters per annum at best.[5]

The growth in population of Europe's cities was to prompt step changes in the scale of operation of breweries. London, capital of the first industrialized nation and the world's biggest and fastest growing city, provides the earliest example of this phenomenon.[6] Even at the beginning of the 18th century, beer production in London was dominated by "common brewers" who distributed beer to a number of public houses, many either owned or otherwise tied to them. Output from this source exceeded that of "brewing victuallers," who brewed only for sale in their own taverns, by a factor of over 100:1. In the country as a whole the output ratio at the time was 1:1. By 1750, the average output of London's top five common brewers was an impressive 80,000 hl; by 1799, it was 240,000 hl.[7] The breweries of Thrale/Barclay Perkins, Whitbread, Truman, Meux, and Calvert were wonders of the age. The product of these mammoth breweries, which far outstripped in size any others elsewhere, was a vinous, bitter tasting, inexpensive brown beer commonly known as porter.

Porter: The First Industrial Beer

The origins of porter, and indeed its very name, are unclear and controversial.[8] Based on the evidence of sparse contemporary reports, the name is most commonly taken to come from the drink's popularity amongst London's porters. First-hand evidence as to its origin comes down to little more than a pseudonymous letter published in the *London Chronicle* in 1760.[9] Embellishments on the story have it that porter was "invented" in 1722 at Ralph Harwood's Bell Brewhouse in Shoreditch in East London to provide a more convenient form of "three threads." This drink was a mix of three beers, most usually given as fresh brown ale (mild), matured pale ale (twopenny), and matured brown ale (stale). One explanation of why Harwood's beer had the contemporary name "entire butt" or just "entire," is because it was served as a single product from one cask rather than by the then practice of filling a glass from three separate casks containing different beers — a task that publicans found irksome.

The reality of porter's origin is almost certainly more complex than the mere result of a move to lighten the potman's workload. A scholarly "reconstruction from the fragments of contemporary testimony"[10] makes a persuasive case for the origins of porter lying in the reaction of London brewers to increased taxation on malt and the relative cheapness of hops as the 18th century dawned. Corran[11] builds on this and links the emergence of porter with the earlier tradition of brewing strong "October" beers. While these are important factors, to properly understand the appeal of entire butt/porter to both drinkers and brewers, a consideration of the strategies of beer production in the early 18th century and of the biochemistry and microbiology of the processes is necessary.

The practice at the time was for strong beers to be stored for many months in wooden casks or vats before consumption. It may now be recognized that storage in this way would promote secondary fermentation with strains of the yeast genus *Brettanomyces* and lead to the production of very high levels of fatty acids and their ethyl esters.[12,13] It was this extreme ester level — perhaps as much as ten times the taste threshold — rather than just the level of alcohol that gave the narcotic effect characteristic of English ales right up to the end of the 19th century.[14] The fresh brown beers produced in London were thin by comparison, hence the popularity of "three threads," which incorporated matured beers in the mix. In a similar manner to strong ales, some brown beers were also matured. But the lower alcohol content of the latter encouraged the growth of other organisms in addition to *Brettanomyces* and these beers developed a distinctly tart acidity in addition to an estery fullness. A third type of matured beer arose accidentally when, prompted by the relative cheapness of hops and increased tax on malt in the early 18th century, London brewers experimented with higher hop rates in their beers and ended up with what became known as porter. The breakthrough with porter was the discovery that a beer — even one made with cheap brown malt and significantly weaker than a strong ale — when brewed using a high enough level of hops became much less tart during storage than was usual for matured brown beer. Porter did, however, still develop the vinous, heavy, narcotic aroma, and flavor associated with expensive strong ales. We can deduce that this is what happened, because we now know that the high hop rate would have kept the lactic acid bacteria at bay through the antibacterial properties of hop bitter acids,[15] but would have no affect on yeasts such as *Brettanomyces*. Hence, Obadiah Poundage's observation[9] that porter "... well brewed, kept its proper time, became racy and mellow, that is, neither new nor stale..." The *Oxford English Dictionary* defines "racy" in this context as "having a characteristic (usually desirable) quality, especially in a high degree." In more modern brewers' parlance, porter "drank above its gravity" because of its ester content. The combination of relative cheapness and desirable flavor made porter irresistible to the urban laboring classes — a taste eagerly exploited by what were to become the behemoths of the brewing industry.

Even its dark brown color, typical of London-brewed beers, was to its advantage as it disguised any deterioration in clarity with age. This robustness and cheapness made it suitable for mass production and amenable to distribution far and wide. As sales took off, the need for long storage, often for a year or more, prompted the use of large vessels. Five thousand hectoliter capacity and greater storage vats eventually became commonplace and porter brewers could undercut on price their ale-brewing competitors who had none of the economies of scale or ability to use cheap materials in their more delicate products. Porter's retail price was 25% less than that of rival pale ales.[10] Porter was an entirely new beer that to an extent mimicked the attributes of mixed beers but also delivered an enhanced mellowness. It came in a convenient single serving, was easy to make, and sold at a competitive price. Thus, everybody was happy: the publican, the brewer, and the drinker. No wonder porter was a success.

Mechanization and Measurement

Mass-produced porter arrived on the scene prior to the mechanization of brewing; man and horsepower achieved large-scale output a generation before mechanization eased the burden. When Whitbreads became the second London brewery to install a steam engine in 1785, they were already producing 300,000 hl of beer per annum. Nonetheless, when it became available, the larger brewers were quick to make use of efficient steam power, purchasing the new improved engines of Boulton & Watt. It has been estimated that at least 26 steam engines were installed in breweries by the end of the 18th century, with use spreading to relatively small regional breweries thereafter.[16]

The first record of in-process quantitative measurement in brewing operations is the use of the thermometer by the London ale brewer Michael Combrune in the 1750s. Combrune experimented with drying temperatures required to give malts of different colors and recorded observations on mashing and fermentation temperatures. A big step forward came in 1784 when John Richardson, a Hull brewer, introduced his saccharometer for the measurement of the wort strength. For the first time, the relative value and efficiency of the use of extract-yielding materials could be quantitatively assessed with consequent economic benefit to the brewer.[17] By 1800, many of the larger common brewers had adopted the instruments that were promoted in treatises on brewing science and practice. From the writings of Richardson and his contemporaries,[18–20] which recorded original and sometimes present gravities, a rough calculation of the alcoholic strength

History of Industrial Brewing

of beers at the turn of the 18th century is possible. The data show wide variations but tend toward the following approximate bands for percent alcohol by volume (ABV): strong beer 7 to 9%, porter 6 to 7%, ale 5 to 7%, and small/table beer 2 to 3.5%.

The Nineteenth Century

Industrialization, population growth, urbanization, and increased consumption are the linked themes of 19th century brewing. In the leading European beer-drinking countries: the United Kingdom, Germany, and Belgium, there was a two- to four-fold increase in output between 1830 and 1900.[3] In 1800, the United States had a total commercial output of less than that of Whitbread's brewery in London; by 1900, it was the world's third largest producer of beer. The development of railway networks from the 1830s transformed distribution, prompting the larger-scale producers to make their regional beers available nationally to a growing population. The new breed of urban workers may have only exchanged the near serfdom of the countryside for the drudgery and grime of towns and cities but, with the novelty of relative prosperity in the blossoming industrial revolution, they drank heroic quantities of beer in the growing numbers of retail outlets. Levels of per capita consumption increased by up to 50% in some European countries. This rise in consumption was accompanied by the rise of commercial brewing and the decline of domestic brewing — indeed some have questioned the extent of the overall rise in consumption for this very reason; statistics on commercial production are liable to be more accurate than are those on domestic production.[3]

There was a vast range in the size of breweries, with outputs from a few thousand to millions of hectoliters per annum. Breweries became highly capitalized businesses and major employers of labor. As the economic importance of brewing increased, so did government interest in the industry, particularly as a generator of revenue. Brewery proprietors became more prominent socially and politically and welcomed the attention, stressing the importance of their industry to farming and the exchequer. In Britain, the industry's long-established links with agriculture gave brewers a head start over other industrialists on the social ladder and their growing wealth and widely heralded philanthropy boosted their position.[21] Adulteration of beer, both innocuous and harmful, which had been rife at the beginning of the century, petered out as breweries grew bigger and their owners aspired to join the gentry. With prosperity and social standing, the temptation to debase their products, as the brewing victualler and small producer of earlier generations had done, was easily resisted.

Between the 1870s and the 1890s, the majority of leading European brewing concerns became public companies. This rush to incorporation had little immediate influence, but over time a bureaucracy of salaried managers gradually replaced the original partners and took over the running of the companies.[22] As so often in brewing history, Bavaria differed from the norm. There, only 1% of breweries had gone public by the end of the century, but even in Bavaria these were the biggest companies comprising 17% of total production.

Porter vs. Ale

As a first approximation, at the beginning of the 19th century, the British beer market was characterized by the dominance of porter in London and the bigger cities and numerous strong, regionally variant ales in the rest of the country. In 1830, the 12 leading London brewers produced around 10% of England's beer and through the influence of the great metropolitan brewers Britain stood preeminent as a brewing nation. "We are the power loom-brewers" as Charles Barclay, one of the partners of the country's largest brewers, Barclay Perkins, boasted.[23] But 1830 was to be the peak of porter's popularity and over the next 50 years its position was to be usurped by the rise of mild and pale ale.[24] Why this change came about is impossible to say with any degree of certainty. One influence was the Beerhouse Act of 1830, which led to a great expansion in the number of outlets for beer and encouraged competition. Another important factor may well have been changes in porter itself. The ability to measure accurately the extract yield of malt with a saccharometer led to the discovery that porter brewed with pale malt and a small amount of very dark malt was actually cheaper to produce than porter brewed with traditional brown malt. By the 1820s, roasted barley was freely available and this also came to be used. Porter became ever darker, ultimately black. The consumer can hardly have failed to notice.

Victorian mild was well hopped, but noticeably sweet; a strong dark brown beer that was drunk young, that is, unaged. It grew in popularity with the laboring classes, first in London and then in the big industrial areas of the Midlands and Northwest England, at the expense of matured porter. With the joining of Burton upon Trent to the railway system in 1839, the London brewers also began to lose out to the now readily available quality pale ales brewed in the town. Although relatively expensive, these beers appealed to the aspirations of the growing band of lower middle class clerks and shopkeepers. Pale malt, dried over coke, rather than over wood or charcoal, had been available from the late 17th century, and pale ale was a favored premium-quality beer. The London brewer George Hodgson and his son Mark built up a respectable trade in the export of this type of beer for consumption by the British in India and Hodgson's lead was followed in the 1820s by the Burton brewers. The hard Burton water proved particularly suitable for this type of beer, and the Burton India Pale Ale (IPA) soon captured the export market and from the 1840s built up a considerable home trade. The major Burton brewers Bass, Ratcliff & Gretton and Samuel Allsopp & Sons led the field. Each decade their output trebled and, by the mid-1870s, Bass was briefly the biggest brewer in the world, with an output of nearly 1.5 million hl with Allsopps not far behind. At the turn of the century, Burton's 21 breweries (the peak number was 31 in 1888)[25] were producing around 10% of the beer in the United Kingdom.

Burton's star faded toward the end of the century, as its products became more difficult to sell. Rival brewers increasingly bought up or provided loans to pubs and so tied them to selling the brewery's own products to the exclusion of those of competitors. The market for Burton beers was also eroded by a shift in public taste away from matured, complex, stock winter-brewed beers like IPA to more easily produced lighter beers. Following advances in technology and technique these "running ales" could be brewed all the year around from the 1870s and required minimal maturation. Burton brewers could, of course, brew good examples of these beers, but so could many others. By the early 1900s, classic, double-fermented IPA had virtually disappeared. In Scotland, large brewing firms, notably William McEwan and William Younger, had developed their own version of pale ale, which, like the Burton brewers, they sold primarily through the wholesale market, as Scottish licensing laws did not permit brewers to acquire pubs directly. Because of this, Scotland itself, remained largely untouched by the surge in tied houses that distorted the English market (where 75% of outlets were tied to a brewer by 1900), but the Scottish brewers were heavily involved in exporting to England and suffered similarly to their Burton rivals.[26] Bass continued to prosper even in adversity through wise management and was producing 2.2 million hl by 1900; but the title of "world's largest brewer" had fallen to Guinness, the Dublin brewer, by the 1880s.

Arthur Guinness had started as an ale brewer in 1759, but in response to the success of imported London-produced porter in the Irish market, the company had switched entirely to porter by 1799, swiftly expanding its business through the new canal system.[27] The Irish had come to porter later than the British, but remained loyal to it longer. In common with other brewers, two strengths of porter were brewed by Guinness and, by 1840, the stronger version, known as stout, accounted for 80% of production. The brewer was the biggest in Ireland by 1833 and underwent massive expansion after the 1860s, brewing 3.8 million hl in 1900. By then, with an established export trade to Great Britain and the Empire, one third of Guinness's output went overseas. Total production represented 8.5% of all UK-produced beer; nearly twice as much as was brewed by all the Scottish brewers put together. This success was achieved without having to enter the increasingly expensive property market as Guinness, uniquely among major brewers in the British Isles, remained entirely as a wholesaler and not retailer of beer.

The Rush to Bottom Fermentation

Attracted by the scale and prosperity of its brewing industry, brewers from other countries came to Britain in the early 1800s to learn the latest practices. The historically most significant of these visits was that made by Gabriel Sedlmayr Jr. and Anton Dreher who traveled all around England and Scotland in 1833, picking up whatever information they could from breweries.[28] On return home, these men were quick to make the most of their experiences and instituted reformed practices in their breweries, including the use of the saccharometer. Sedlmayr at the Spaten brewery in Munich and Dreher at his eponymous brewery in Vienna and later in Michelob (Bohemia), Trieste, and Budapest were to build up important brewing empires and become instrumental in the spread of bottom-fermentation techniques across the continent.

Although little known outside the state, bottom-fermented beers had been brewed in Bavaria since at least the 1400s. Their defining characteristics were the utilization of yeasts that sank to the bottom of the vessel toward the end of fermentation and the use of low fermentation (4 to 10°C) and maturation (−2 to 4°C) temperatures. The rest of the world used yeasts that floated up to the surface of the fermenting wort and were accommodated to higher fermentation (15 to 25°C) and maturation (13°C) temperatures.[29] The adoption of the description "lager" (from the German verb lagern, to store) for bottom-fermented beer in anglophone countries has encouraged much misdirected comment. It is often stated, or at least tacitly accepted, that storage was a unique aspect of lager brewing. In reality, until the spread of artificial refrigeration from the 1870s made reliable summer brewing possible, it was necessary to store beers fermented in the cooler months for consumption in the warmer months, whether they were top-fermented "ales" or bottom fermented "lagers." That England and Bavaria adopted different techniques for preserving beer during storage was something generated by climate and geography. Conveniently for the Bavarian brewer monks, the foothills of the Alps provided cool caves for the storage of beer. When it was found that storage under these conditions led to the production of stable, bright, and sparkling products, commercial brewers mimicked the procedure, using ice taken from frozen lakes and rivers to cool the cellars of their breweries. In England, no such geographical advantage was available near brewing centers and heavy hopping and high alcohol was used as the preservative rather than cold storage. London porter and Munich lager were the result of these differences. Both were stored or vatted beers; porter was regularly stored for a year, lager rarely for more than 6 months.

Bavarian lager was brewed with malt dried at relatively high temperatures, leading to rather dark colored beers. The malt was also less well modified than the malt used in the production of top-fermented beers and thus required more intensive mashing to yield acceptable levels of extract. A "decoction" mashing system, involving extraction of the malt at three or

History of Industrial Brewing

so different temperatures by withdrawing and then heating a portion of the mash and adding it back to the bulk to give a step rise in temperature, came to replace the single-temperature "infusion" system used for ales. Later, "programmed upward infusion" mashing would achieve the same process more conveniently by gradually increasing the temperature of the bulk using steam-heated coils in the mash vessel.

Political changes in Germany, culminating in the eventual unification, were instrumental in fostering the opening of trade between the German states. Bavarian brewing practices became more widely known and some North German, Austrian, and Czech brewers adopted bottom fermentation in the late 1830s.[30] The first example of a straw colored lager produced using lightly dried, low color malt, seems to have been brewed with the soft water of Pilsen by a Bavarian-born brewer Josef Groll in October 1842. Both light colored, pilsner-style lager and dark lager based on the Bavarian münchner swept the world over the next 50 years, with the pilsner variety proving the most popular by the end of the century. Jacob Christian Jacobsen brewed the first Danish lager in 1847 using yeast he brought back from Munich. It was a dark lager. The pilsner style did not reach Denmark until brewed by Tuborg in the 1880s. Gerard A. Heineken switched from ale to lager brewing in Amsterdam in the 1870s, after seeing the demand for Anton Dreher's Vienna-brewed product at an international exhibition.[22]

The first lager brewed in the United States is credited to John Wagner in Philadelphia around 1840 using yeast from his native Bavaria, although it is Frederick Lauer, who set up a small commercial brewery in Pennsylvania in 1844 and was later called "the father of the American brewing industry," who was to be a more influential figure.[31] The wave of German immigration to the United States that came in the following 20 year brought with it such famous names as Bernard Stroh, Ebrehard Anheuser, Adolphus Busch, Frederick Pabst, Frederick Miller, Joseph Schlitz, and Adolph Coors. With immigration came a gradual switch in consumer preference from ale to lager and the drift westward of breweries, with Milwaukee and its plentiful supplies of ice from Lake Michigan and St Louis with its cool natural caverns as foci. For a time, these German–American brewers may have used the ingredients of their homeland — an early label for Budweiser, launched in 1876, notes the use of Saaz hops and Bohemian malt.[32] But during the 1880s, they developed a new style of lager brewed with readily available cereals, notably maize (corn) and, in Anheuser–Busch's, case, rice as diluents for the high-nitrogen six-rowed barleys grown in the United States. These adjuncts were used unmalted and were gelatinized before addition to the malt mash. Although banned in Bavaria, there was nothing new in the use of unmalted adjuncts, but what was different from mainstream European practice was the high level of use. This coupled with the development of an accelerated brewing process, where storage time was minimized and filtration used for clarification, led to the development of unique, very pale-colored beers of unrivaled blandness. Fueled by immigration, urbanization, industrialization, and the spread of the railroads brewing became a major U.S. industry — the latter being particularly exploited by the big vertically integrated "shipping brewers," Pabst and Anheuser-Busch, after the introduction of refrigerated rail cars in 1876. In 1850, there were 431 breweries operating in the United States producing 0.88 million hl of beer. By 1900, 1816 breweries made nearly 47 million hl and per capita annual consumption had gone from under 4 l to over 60 l.[33]

From the 1870s, the increasing availability of efficient artificial refrigeration freed lager brewers from the need for natural ice. Cold transport of beer became easier and lager and ale brewers alike adopted all-year-round brewing. Refrigeration was initially used to produce ice but was soon applied to direct cooling via expansion coils. Lager now became a world drink. The first successful commercial brewery in Japan, the Spring Valley Brewing Company, set up in Yokohama by an American W. Copeland in 1869, evolved into the Kirin Company and brewed lager.[34] German brewmasters brought brewing to China for the first time in the 1870s, Tsingtao lager being an early product.[35] Hampered by lack of suitable materials and the climate, local beer production in Australia only started to outstrip imports in the 1870s. Lager brewing reached the country in the 1880s, with the American émigré Foster Brothers beer, brewed using domestic cane sugar as an adjunct to malt, going on sale in Melbourne in 1889.[36]

But it was in its German homeland that lager prospered most. By the 1880s, Germany was the leading brewing nation with the greatest output of any country in the world. Bottom-fermented beer had triumphed with speciality wheat beers, the only significant top-fermented products still in production. Although predominantly a country of small brewers, Munich, Dortmund, and Berlin had become established as the main brewing centers with several large modern breweries.[30] Only the United Kingdom, with the exception of a few scattered attempts, and to a lesser extent Belgium, resisted the rush to bottom fermentation.

Science and Practice

Concomitant with the rise of lager and complementary to it was a new scientific approach to brewing. Increasing numbers of brewers looked to gain a greater understanding of their processes and thus improve efficiency. In 1843, in Prague, Carl Balling introduced his own version of the saccharometer. The instrument was quickly adopted in central Europe as were the teachings of his seminal work on fermentation chemistry published in the same year. The promotion of technical education in brewing followed.[37] Brewing courses began at Weihenstephan in 1865 with eight students taught by Carl Lintner.[38] In the United States, John Ewald Siebel, a German immigrant, founded a laboratory in Chicago in 1868, which became the Zymotechnic Institute in 1872, and began a school for brewers in 1882. In 1883, the Research and Teaching Institute for Brewing (VLB) was established in Berlin under Max Delbrück. Similar institutions appeared at this time in Austria and Switzerland. In Britain, education was on a more *ad hoc* basis with prospective brewers being taken on as pupils and receiving on-the-job tuition. There was no brewing school in England until 1900, when largely through the financial support of a local brewer, William Waters Butler, classes started at the newly formed University of Birmingham. Brewing tuition started at Heriot-Watt College in Edinburgh in 1904.[39] Trade/technical journals and societies proliferated from the 1860s. In England, the *Brewers' Journal* appeared in 1865 and the *Brewers' Guardian* in 1871; the *Transactions of the Laboratory Club*, which became *The Journal of the Institute of Brewing*, was first issued in 1886. Carl Lintner's journal *Bayerische Bierbrauer* came out in 1866. In 1882, *The Australian Brewers Journal* was launched and The Master Brewers Association of America was formed in 1887.

Until Louis Pasteur carried out his investigations on wine and beer fermentations in the 1860s and 1870s and showed the importance of eliminating deleterious bacteria, there was little meaningful scientific research in brewing. Curiously, although Pasteur's work on beer was carried out with the declared aim of redressing the balance in France's favor against the clearly superior German brewing industry, German breweries were amongst the first to employ heating of beer, that is, pasteurization, in order to preserve it.[40] Pasteur claimed the treatment as too severe for beer and did not recommend it in his famous book *Etudes sur la Bier* published in 1876. Rather, Pasteur devised a system of brewing that prevented ingress of bacteria in the first place, but his procedure found few users. Emil Christian Hansen, on the other hand, soon found his ideas finding application after he introduced the concept and the technology for achieving pure yeast culture at Carlsberg in 1883. Within 10 years, Hansen's yeast-propagation plants had been installed in 173 breweries in 23 different countries.[28]

The brewhouse also saw changes during the 19th century. Use of mashing rakes powered by horses and then steam were the norm by 1800 in large breweries. By then, the old technique of multiple extraction of a batch of malt to produce worts of decreasing strength and then fermenting them separately to produce different beers had largely been superseded. Worts were now blended prior to fermentation in order to produce a single beer or a range of beers. After 1800, the technique of "sparging" was increasingly introduced. This procedure, which seems to have originated in Scotland, was universal by the 1870s. James Steel introduced his mashing machine to give efficient mixing of ground malt with water on entry to the mash tun in 1853, which along with similar other devices soon found favor.

Bottling of beer, although probably started in earnest in the early 1700s, was of little importance until the 1860s. Bottles were originally corked; internal screw stoppers were patented in 1872, swing stoppers in

1875, and crown corks in 1892.[41] Hand bottling was the only option until the 1880s when bottling machinery was introduced, prompting a surge of interest. Lighter beers were produced specially for bottling using proportions of sugar and unmalted cereals. These adjunct beers tended to stay bright longer than all-malt beers. Narrow-mouthed bottles were patented in 1886. Multiple head fillers appeared in 1899 and fully automatic rotary fillers in 1903. "Naturally conditioned" bottled beer, packaged with a proportion of yeast and fermentable matter still present, allowing continued limited fermentation that generated carbon dioxide and gave it sparkle, remained the most usual product in many countries well into the 20th century. Gradually, however, it lost ground to "no deposit" bottled beer that was filtered and artificially carbonated using techniques introduced from the United States, where this type of beer was the norm by the 1890s. Bottled beer is estimated as having taken 20% of the U.S. market by 1900 and 10 year later one third of the beer sold in North Germany was bottled.

Appreciation of the importance of analytical data, particularly to evaluate water and malt quality, was a feature of the growing, if sometimes grudging, acceptance of the utility of scientific understanding in brewing in the 19th century.[42] Most breweries relied upon external consulting chemists for their analysis. There were around 12 specialists operating from London by the 1880s, major figures being Ralph Moritz (founder of the Laboratory Club), Alfred Chaston Chapman, and Lawrence Briant. Alfred Jorgensen founded his laboratory in Copenhagen in 1881, and the Danish-born Max Henius and the American Robert Wahl established a consultancy in Chicago in 1884. In Germany, the establishments at Weihenstephan, Berlin, Nuremberg, and elsewhere provided an analytical service to brewers. A number of the bigger breweries also provided laboratories for their scientifically trained brewers to carry out basic analysis, while a few of the biggest had specialist chemists on the payroll. The first brewery to appoint a qualified chemist in Britain (and perhaps the world) was the London firm of Truman, Hanbury, and Buxton in 1831. This was Robert Warington, destined to become the first secretary of the newly formed Chemical Society 10 years later. A group of talented scientists led by Cornelius O'Sullivan at Bass, Horace Brown at Worthington, Horace's half-brother Adrian at Salts, and Peter Griess at Allsopps advanced the cause of brewing science in Burton upon Trent from the 1860s to the turn of the century.[42] Carlsberg, with Hansen and the chemist Johan Kjeldahl, established what was to become a world famous laboratory in Copenhagen in 1876. Anheuser–Busch claims to have started the first brewery research laboratory in the United States in the 1870s.[43] The Pabst brewery in Milwaukee appointed a German Ph.D., Otto Mittenzurly, to its staff in 1886. The Edinburgh brewer William Younger had a full time chemist, John Ford, by 1889. Guinness started appointing Oxbridge chemistry graduates from 1893.[44] Many of these men took a full part in the mainstream science of the day and contributed presentations to learned societies and published in scholarly journals.

The Twentieth Century

As the new century dawned, the top three brewing countries, Germany, the United Kingdom, and the United States accounted for 68.5% of recorded beer production.[1] By 1906, the United States had overtaken the United Kingdom and by 1910, with production still rising, was ahead of Germany. German beer production, having doubled since the 1870s, peaked in 1909.[30] Production in the United Kingdom also stalled in the first decade of the century.[45] The First World War turned a slow down in the European industry into a collapse in output. Shortage of brewing materials and rises in prices and taxes were largely responsible. From 1915 to 1919, output in the United Kingdom and Germany fell by 37 and 52%, respectively, compared with the previous 5-year period. It was to be 40 years before world beer output returned to the levels reached in 1913. Local and regional breweries were still the major suppliers of beer. Only the biggest and most adventurous brewers had national and, in some cases, limited international distribution.

The brewing industry of continental Europe was again in turmoil between 1939 and 1945, although war actually boosted production in the United States, Canada, Australia, and the United Kingdom. In the postwar years, beer output remained flat or showed relatively modest growth until the mid- to late-1950s. Thereafter production soared. Between 1960 and 1990, world output increased threefold. Within this overall increase, national differences are evident.[46] After a fall in production in conditions of postwar austerity, output in the United Kingdom increased by nearly 70% after 1959 before peaking in 1979 — the year Margaret Thatcher came to power. In Germany, after the near collapse in 1945, volumes had recovered to prewar levels by 1960 and continued to grow before plateauing in the mid-1970s and then declining.[47] Helped to an extent by immigration, output in Australia, New Zealand, and Canada nearly trebled in the 35 years after the war before falling off. The United States saw a sluggish 1950s followed by 30 years of growth as output doubled by 1990. In general, increased demand in traditional western beer-drinking countries came primarily from young adults specifically targeted by marketing, resulting in per capita increases in consumption of around 40 to 80% between 1960 and 1980.[47] Although these rises in consumption were impressive, in the United Kingdom beer drinking never returned to pre-First World War levels. At its 20th century peak, per capita beer consumption in the United Kingdom was still only 70% of that in the 1870s. Furthermore, average beer strength was less than 4% ABV compared with nearly 6% ABV a century earlier.

The largest proportional increases in output came in Southern Europe and in southern hemisphere countries not previously noted for significant beer drinking. Brazil, for example, showed a sixfold increase in output between

1970 and 1990, with South Africa showing a tenfold increase in the same period. In Japan, there was a near 300% increase in per capita consumption between the 1950s and 1970s and then further spurts in the 1980s and 1990s as new products were introduced.

During the last quarter of the 20th century, China became the primary target for international companies looking for new markets. By the end of the century, output had topped 200 million hl, having been barely 1 million hl 30 years earlier. Although per capita consumption remained only a fraction of that in some other countries, with its enormous population, China was set to overtake the United States as the country with the biggest beer output early in the 21st century. At the end of the 20th century, the three countries, Germany, the United Kingdom and the United States, which had accounted for over two thirds of world output in 1901, now produced less than one third of the world's beer.[46] Even as the beer market in Western Europe and elsewhere stagnated, world output continued to rise primarily because of China. In 2002, overall world production of beer increased by 2.5% to an estimated 1400 million hl,[48] a volume over five times higher than it had been a century before.

Beer and Society

During the 20th century, temperance pressure, taxation, changes in public taste, and the power of marketing all had significant influences on beer and on drinking habits.

Temperance and Prohibition

The "demon drink" had two powerful interlinked institutional enemies — the church and the state. The increasingly powerful and freethinking urban workforce, unfettered by the dominance of the clergy, and the landowners of the countryside were perceived as a growing problem by the authorities. Their new freedom was associated, without any real proof, to rising drunkenness in the towns and cities.[3] In the forefront of the temperance movement were committed social reformers who in their work among the urban poor saw the dark side of drink — dependence and crime. The seamy nature of many British pubs and American saloons added to the unwholesome image of what in the United States the reformers called "the liquor traffic." It was not until teetotalism started to gather support in the mid-19th century that the brewers came under threat. Until then, temperance reformers had generally regarded beer as essentially harmless in relation to the cheap spirits on which they concentrated their attentions. The pressure on the brewers differed between countries. It was particularly significant in the United States, Britain, Scandinavia, Australia, and New Zealand.

During the decade prior to the First World War, American brewing technology, working practices, and marketing techniques were drawing admiration from European visitors. After the war, scope for further development of the U.S. industry seemed ripe, but other factors were at work. Prohibitionist sentiment had grown among the large and powerful nonconformist churches of the United States. Under the leadership of the Anti-Saloon League and the Women's Christian Temperance Union prohibitionist legislation swept through the south.[49] The adoption of the "local option" in many smaller towns and districts had led to them going dry, with small local breweries closing even while beer consumption nationally was increasing. The National Prohibition Enforcement or Volstead Act became law on January 16, 1920. Brewers turned to alternative products that ranged through yeast, malted milk, ice cream, malt syrup, ginger ale, corn syrup, soft drinks, etc. Some products were more ingenious than others. Anheuser–Busch marketed hopped malt syrup in 1921, ostensibly for use in baking. Much later, after prohibition was long over, it was admitted that cookies made with the syrup were inedible because of their bitterness.[50] By 1923, it was estimated that the equivalent of some 12 million hl of beer were being brewed at home. Breweries were allowed to produce legal "near beer" at 0.5% alcohol, but output fell from nearly 11 million hl in 1920 to just 3 million hl in 1932, as illicit alcohol proliferated.[50] Bootlegging was a lucrative enterprise for gangsters, with smuggled beer from Canada (which had its own rather haphazard form of prohibition) and Mexico playing its part. According to at least one source, however, most of the alcoholic drink consumed in the United States during prohibition was illegally produced within the country.[51]

The Volstead Act was modified with effect from April 7, 1933, with beer of 3.2% ABV declared to be a nonintoxicating beverage. This paved the way for a return to normality with the passing of the Twenty-First Amendment later in the year, which legalized alcoholic drink where not specifically prohibited by state law. Only a handful of states were still dry by the end of 1933, the last of them to hold out being Kansas, which relegalized alcohol on May 1, 1937.[52] The years of prohibition were undoubtedly difficult for U.S. brewers, but not perhaps as devastating as is sometimes supposed. Certainly the industry recovered from prohibition with great rapidity. Of the 1462 breweries authorized to operate in the United States in 1913, only 13% remained on May 1, 1933. Yet by mid-1934, numbers had risen to 48% of the 1913 level. To get this in perspective, the total number of breweries in the United Kingdom in 1934 was only 30% of its 1913 level.[53] Similarly, although beer production in the United States in 1934 was nearly 40% below that of 20 year earlier, output in the United Kingdom was nearer 45% down on prewar figures. By 1943, the United States had surpassed the output achieved in 1913; the United Kingdom did not do so until 1973.[46] With these comparisons in mind it is difficult to accept the view that prohibition had any lasting adverse influence on the U.S. brewing industry or indeed that it led to an unusual level of concentration. Paradoxically, prohibition may have had the unexpected benefit of allowing the already progressive U.S. industry the opportunity to digest the changes in consumer attitudes that occurred during the 13-year hiatus. The preference the U.S. public developed for buying prepackaged goods of all kinds is a prime example and the brewers were alert to this. Before prohibition, draught beer was easily the best seller. After prohibition, considerable effort was put into developing the beer can. First used for "Kruger Cream Ale," test marketed on January 24, 1935, the can was to become America's favorite beer package.

Iceland from 1915 and Finland from 1919 had preceded the United States in enforcing prohibition but had abandoned it as unworkable by the 1930s. Denmark, Norway, and Sweden came near and in the United Kingdom and Australia there was considerable prohibitionist fervor. New Zealand had the closest call of those countries that escaped prohibition.[54] Women's suffrage and temperance societies had much effect on the brewing industry in New Zealand through local option initiatives. Six o'clock closing became law in 1917. Votes for prohibition very nearly reached a majority in national referenda after the First World War; the measure only being defeated at the closest call by the votes of the returning troops. In 1923, in a defensive measure, the country's ten principal companies had amalgamated to form New Zealand Breweries Ltd. After 1925, sentiment swung against prohibition, and the danger passed. Australia also had a significant prohibitionist minority and, like New Zealand, introduced early closing during the First Word War leading to the "6 o'clock swill," as drinkers downed as much as possible in the limited hours available. Such measures had long-lasting effects. In Australia, opening was only extended to 10 p.m. in the 1960s. In the United Kingdom, restriction of opening hours introduced as a temporary measure in the First World War was only relaxed in 1988. Some Canadians still needed to buy drinking permits in the 1950s and similar attitudes prevailed in Norway, Sweden, and Finland. Even in the 21st century, the minimum drinking age in the United States is 21 and in Utah one needs to buy club membership to drink alcohol. American historians learn to couch their grant applications in terms of the social aspects of drink, rather than the industry as a whole, if they wish to have a chance of receiving research funding (D.W. Gutzke, personal communication, 2002).

Consumer Choice?

The strength of beer in the United Kingdom in 1919 was only 58% of the 1914 figure and it never returned to its earlier level. After 1914, for the first time, duty became the main determinant of beer prices in the United Kingdom, which rose inexorably to help finance the war. This may look like a recipe for disaster for the UK industry and, indeed for some it was, as many breweries went to the wall. But those who survived came out of the war selling weak beer with much lower raw material costs and an improved profit margin, with price increases hidden by the disproportionate duty. In the brief postwar boom, few breweries sold any more beer than they had 10 year earlier, but it was worth much more to them in profit terms. But this prosperity was not to last; by the mid-1920s, output of beer had stagnated across Europe and was to remain flat as high taxes, the depression, and increasing alternatives for leisure spending took effect. The latter had arguably the most complex influence. Working-class aspirations changed with the availability of mass-produced goods, the cinema, and better housing all conspiring to reduce the popularity of beer drinking. Attempts by the brewers to meet these changed priorities and circumstances, notably between the two world wars in the United Kingdom, with the reformed pub, which provided a range of attractions and greater comfort, were of limited success.[55] The working class felt uncomfortable in these often massive but cheerless pubs, and the middle class were not to be won over in sufficient numbers by buildings that aped their own mock Tudor homes. In the United States, the old time saloon did not survive prohibition and was replaced by the cocktail bar (successor to the speakeasy) for the better off and the tavern for the less aspirational working class.[50]

Porter disappeared from England in the 1930s. It hung on in Ireland until 1973 and saw a revival amongst microbrewers in the 1980s. Guinness had a virtual monopoly of bitter stout in Britain and weaker sweet stout emerged as an alternative among English and Scottish brewers. In Bavaria, the working classes maintained their enthusiasm for dark Münchner lagers, but even in their heartland the style lost out to lighter "helles" beers. Straw colored lagers swept the field around the world except in Britain. The most popular draught ale in England and Wales in the 1930s was mild, by then a weak (c. 3 to 3.5% ABV) beer. The alternative was bitter, a lineal descendent of the new running ales of the 1880s but again much weaker at c. 3.5 to 4.5% ABV. Scotland had its own distinctive reddish brown, rather sweet Scotch ales that were also of low strength compared with their postwar namesakes. The weakness and often uncertain condition of these draught products may have been a factor in the growth in demand for "light and bitter" in the wealthier South of England. This drink, the popularity of which had echoes of the 18th century's "three threads" about it, involved blending a bottle of pale or light ale with a half pint of draught bitter in an attempt to enliven the latter. Certainly the quality of draught beer in the United Kingdom left much to be desired in the interwar period. Complaints about it can even be found in the usually slavishly loyal publications of the Brewers' Society. Writing in 1933, the technical editor of *Bottling*[56] was moved to remark that: "Undoubtedly the brewer has a good deal to answer for in the diminution of his draught sales." In reference to poor cellar and dispense conditions he went on to observe how "The old beers could stand it. They were strong enough and matured enough to preserve their quality and condition in spite of the frightful conditions of draught. Not so the light beers of the present day." He attributed this sad state of affairs to "the futility of the brewer." Little wonder then that bottled beer appealed increasingly to drinkers up to mid century in the United Kingdom, rising from an estimated 4% in 1907, to 25 to 27% in 1935, and 33 to 36% in 1950. Although hard data are scant, even more marked shifts to small-pack beer may be traced in some other countries. In the United States, for example, draught beer sales were in the minority by 1941.

Prior to the First World War, the pint mug was the most usual measure in pubs in the United Kingdom, but after the war the half pint measure became the norm for increasingly expensive beer.[57] There was then a reversion to ordering in pints during the Second World War when fears of shortage and the need to make sure of one's "fair share," caused thirsty shipyard workers and miners to favor the larger measure in case the beer ran out (R. Anderson Sr., personal communication, 1965). Much of the European brewing industry entered the 1950s with largely antiquated run-down plants. This was certainly true of Britain, which had suffered from minimal investment for many years before the war. The more progressive British companies sought to rectify this situation first by investment in bottling and then in kegging of beer. During the 1960s, heavily advertised chilled and filtered keg ales such as Double Diamond and Red Barrel enjoyed enormous popularity. These rather anodyne beers have since been much derided[58] and their success attributed to expensive marketing, but they should be viewed within the context of the period. Opposition to chilled and filtered beer came not from the still largely working-class constituency of British beer drinkers, but rather from the sons of the middle class who had no memory of how bad interwar British beer could be and often was. Marketing undoubtedly played a big part in driving the sales of these products, but the attraction of a reliable sparkling draught beer to the pub consumer who for decades had been subjected to what was in many cases a mediocre, inconsistent, poorly presented pint should not be underestimated.

The industry had long realized the importance of brands and advertising. From the 1850s, Bass and Allsopp built their sales on easily recognized and memorable trademarks. Clear brand identification was a significant factor in the success of bottled Budweiser as was appreciated by Adolphus Busch on its launch in 1876. In February 1929, Guinness turned rather self-consciously to advertising in national newspapers for the first time and produced a long series of classic advertisements. But it was the spread of television after 1945 that really opened the doors to mass marketing of beer around the world.

The history of the Japanese brewing industry provides a useful insight into the interplay of consumer demand and marketing effort in the 20th century. Until their evacuation during the First World War, brewing in Japan was conducted by German brewmasters. The industry grew slowly, with beer a luxury item rather than a mass-market product. The country's total output in 1939 only amounted to 3 million hl, by which time many small-sized breweries had amalgamated with either Dai-Nippon or Kirin breweries. Dai-Nippon's share of the market was over 70% in 1949, and it was broken up under antimonopolies regulations to form what became Sapporo and Asahi Breweries.[59] After a period of rationing following the war, consumption increased rapidly in the 1950s as the economy boomed, reaching 40 million hl by the mid-1970s, representing a per capita increase of nearly 300%. The rate of increase then slowed, output reaching 50 million hl by the mid-1980s, with Kirin holding over 60% of what appeared to be a mature market. Then, a range of new products and packaging, particularly the heavily hyped "Super Dry" beer launched by Asahi Breweries in 1987 and soon copied by others, gave a boost to sales. Dry beer had a remarkable 34% of the market within 2 years and total beer consumption went up by 40% in 5 years, topping 70 million hl (per capita 57 l) by 1992, making Japan the world's fourth biggest producer of beer.[60] During the 1990s, a quick succession of other expensive marketing initiatives followed. These included the strategy of aiming at getting fresh beer to the customer rather than the more usual expedient of selling stabilized, pasteurized beer. Particularly notable was the introduction of "happoshu" beer by Suntory in 1994. This much-copied beer, which was low in malt, and hence in Japan low in tax, had captured around 30% of the market by 2001. In the same year Asahi, who were rather slow into the "happoshu" market, overtook Kirin as Japan's biggest brewer (further details in Chapter 22).

A similar story of the creation of a new category in the beer market and transformation of a company's fortunes had taken place in the United States somewhat earlier. The tobacco company Philip Morris completed its takeover of Miller in 1970 when the brewers were the seventh biggest in the United States. An eightfold increase in sales in 20 years saw them in second place as their "low calorie," "lite" beer, launched in 1975 and backed by an unprecedented level of advertising expenditure, proved a huge success. "Bud Lite" and "Coors Lite," among others, soon followed, and a new sector was established. Some have found the concept of these "lite" beers risible. As one widely published writer on the American brewing industry put it[50] "The irony of the 'lite' movement is that, mass-market beers in America are all so light that further dilution seems to push the very definition of beer." Whatever the truth of that, lite beers went on to take 40% of the U.S. market before fading in the 1990s and taking Miller on a downward trend with them. Ice-beer, introduced by Labatt of Canada in 1993, was aimed at a similar audience but never enjoyed the same success. Whether the "ice brewing process," which essentially involved passing beer through a bed of ice at low temperatures, actually changed the flavor of the beer remained a matter of faith rather than evidence, but was said to yield a product "rich in flavor and smoothness and yet uniquely easy to drink."[61]

The dominance of the world market by straw colored lagers was completed by the end of the 20th century. From a base of only 1% of the market in 1960 (20% in Scotland), and almost a century after the rest of the world, the British and the Irish turned increasingly to bottom-fermented beer. Some brands of lager sold in Britain were so low in alcohol that in an earlier age they would have been classified as small or table beer and indeed would have passed as nonintoxicating beverages under the 1933 modification of the Volstead Act in the United States.[62] But if the beers were feeble, then the advertising was not. "Probably the best lager in the world" and the beer "that refreshes the parts that other beers cannot reach" sold heavily and profitably. These "standard lagers," as they became known in the trade, had nearly 40% of the UK market by the end of the century.[46] By then, products closer in strength to their continental cousins had emerged and total lager sales came to over 62% of the market, having overtaken ales in 1989. British-brewed lager appealed particularly to the youth of the nation, to the extent that frequent overindulgence and associated rowdiness in public led to the creation of the judgmental neologism "lager lout" in the late 1980s.[63]

In a development that took major brewers by surprise, very small producers (craft or microbrewers), on a scale similar to the almost vanished licensed victuallers, staged a resurgence in the 1970s in the United Kingdom, with a similar trend in the United States, Canada, Australia, and elsewhere during the 1980s. These developments, while quantitatively small in terms of overall beer output (still only 2% of total UK sales in 2001) and of variable quality, have ensured that a wide range of beers remain available to the drinker even in the otherwise humdrum Australian and North American markets. In the United Kingdom, this increase in choice has been somewhat diluted by an accompanying decline in the number of operating regional brewers. Traditional British cask beer, the main product of these small and regional brewers in the United Kingdom, remains under threat. Despite the best efforts of the Campaign for Real Ale (CAMRA), which arguably saved this style of beer from extinction in the 1970s, cask beer suffered a severe downturn in the late 1990s, falling below 10% of total UK sales. Newly introduced "nitrokeg" beer proved a more serious rival for the attentions of the less-committed ale drinker than more highly carbonated keg bitters had been.

Only in the United Kingdom (c. 60%) and Ireland (c. 80%) did draught beer and drinking in the pub still have the largest share of the market at the end of the 20th century, reflecting the unusual preponderance of off-sales in these countries.[46] By the 1980s, the take-home trade comprised around 90% of the U.S. market. In Germany, bottled beer has the greatest sales; in the Czech Republic draught and bottle sell in similar volumes; in the Philippines, most of South America, South Africa, and Bulgaria sale of draught beer is only 1 to 2%; in Nigeria, it is unobtainable. In the United States, the can is easily the best seller; in Denmark for over 20 years, until January 2002, the sale of beer in cans was illegal.

During the 1960s, the old beer-drinking countries saw an increasing shift to wine drinking spreading down from their affluent middle classes. Food also received increasing attention in licensed premises. Many UK pubs changed their nature during the 1980s, as catering, for so long a secondary consideration to drinking, became increasingly important. This move was complemented by more liberal licensing laws that allowed children on the premises and brought longer opening hours. With more attention to drunken driving, alcohol-free (<0.05% ABV) and low alcohol (<1.2% ABV) beers were marketed during the 1980s and heady predictions made about their potential. The poor flavor of these products and uncertain market positioning confounded hopes of their success, and by the end of the century they held less than 1% of the world beer market.

During the 1990s, youth-oriented pub culture regularly turned city and town centers into flash points of drunken violence. Less-demanding alcoholic drinks such as alcopops and their successors, in the manufacture of which brewers played a full part, joined lager as the favorite alcoholic drink of the young. Brewers continued to advertise heavily and link themselves with sport. This often brought adverse images to public attention. As one writer[64] noted, shedding crocodile tears for the brewers, over hooliganism at a 1998 soccer tournament "... spare a thought for poor old Carlsberg and Budweiser, who have spent a fortune on TV advertising during the World Cup, only for the reputation of alcohol to be besmirched with every depressing news bulletin." In a counter to these negative images, the drinks industry has welcomed reports of the health benefits of moderate drinking. On the coattails of red wine, the brewers have been quick to point out that the epidemiological evidence of protection against coronary heart disease could also apply to beer[65] and promoted it as a natural drink to be treated with "reverence."[66] During the early years of the 21st century, beer and health became a regular focus for papers at congresses of the European Brewery Convention, as debate on the subject continued in the medical literature.

Disparity in taxation on beer between countries continued to cause problems. Post Second World War, taxes remained relatively stable in the United States and Germany while rising steeply in the United Kingdom. By the end of the century, U.S. taxes on beer were around double those in Germany but only a quarter of UK levels. During the 1990s, with excise duty in Britain seven times that in France, a lively trade developed in the channel ports as UK customers and entrepreneurs sought (sometimes illegally) to exploit the price differences. Despite attempts by Customs and Excise to restrict such imports, by 2003 it was estimated that 5 to 10% of beer drunk in the United Kingdom came from this source.[67] Alarmed by these developments, the British industry put pressure on government to follow the example of the Peruvian authorities, who were forced to reduce beer tax in 2002 in the face of a growing wave of contraband across the border from neighboring, lightly taxed Bolivia.

Fewer and Bigger: The Path to Globalization

Brewing of the same brand of beer in more than one plant, pioneered by Anton Dreher in Central Europe in the 19th century, was taken up by the

Pabst Brewery Company in the United States in the 1930s. In 1948, following acquisition of other breweries, Pabst became the first brewer with plants from the Atlantic to the Pacific. By the late 1940s, Schlitz and Anheuser–Busch had followed the trend. Multiplant brewing of major brands spread around the world during the following decades, providing a challenge in flavor matching that was not always satisfactorily met. Major national and international brewers erected plants of capacities up to 10 million hl per annum and occasionally exceeded that level in the United States. Regional brewers with 0.1 to 1.0 million hl per annum plants found themselves squeezed between these giants and craft or microbrewers producing 1000 to 10,000 hl per annum. Reductions in numbers and increases in size of breweries, which had long been a feature of the industry, have led to new levels of consolidation in the last 50 years. In many countries, by the end of the century, an oligopoly prevailed despite the efforts of government regulation, the resentment of smaller brewers, and the opposition of consumer groups.

Brewers in the UK were shaken during the 1960s by the intrusion of perceived outsiders attempting to gain a stake in the industry. A spate of defensive mergers followed with the formation of six major companies (Allied Breweries, Bass, Courage, Scottish & Newcastle, Watney, and Whitbread). By the 1970s, these companies controlled over 80% of beer production and half of full-on licensed outlets.[68] By 1990, only 65 of the 362 UK brewing companies active in 1950 remained, with an eightfold increase in average output per plant to 0.60 million hl per annum.

Changes in the United States during this period were even more dramatic. Of the 380 brewing companies active in the United States in 1950 only 30 remained in 1990, and the average output per plant had increased 28-fold to 3.76 million hl per annum.[47] U.S. brewers followed two methods of expansion. Some, notably Pabst and Anheuser–Busch, bought and built breweries in various parts of the country and had them brew the company's flagship brand. Others, such as Olympia and Falstaff, also bought breweries but continued to have them produce their own locally admired beers.[50] In time, it became clear that the national brand strategy was the most commercially successful. Not that all such attempts at national marketing worked. The Carling Brewing Company of Canada acquired numerous brewing companies over 20 years from the mid-1950s, with the declared aim of becoming the biggest in North America, spearheaded by its "Red Cap" ale and "Black Label" lager. It rose to become the 11th biggest brewer in the U.S. in 1975, but sold out to G. Heileman in 1979 without really getting near its goal.[50]

British brewing companies, already vertically integrated into material provision (malt and hops) and alcohol retailing, became horizontally integrated through takeover into spirit, wine, cider, soft drink, food, and leisure. For a time during the 1970s and 1980s, it appeared possible that the brewing industry would become subsumed by other businesses and lose its identity.

The takeover of Miller by Philip Morris in the United States and the creation of the B.S.N. group in France and Reemtsma in West Germany seemed to indicate a trend. But this proved to be the high water mark of such deals, as sentiment on stock exchanges turned against unwieldy conglomerates and toward more focused if still massive companies. By 1992, only five brewing companies remained in Japan operating 37 breweries with an average annual output of nearly 2 million hl per plant. At the end of the century, Carlsberg and Heineken dominated brewing in their home countries and had done for decades; South African Breweries brewed around 98% of that country's beer; 95% of Canada's beer was brewed by two companies; and 80% of Belgium's by three.

The Australian industry provides a good case study of how this type of consolidation can come about.[69,70] In Australia, 33 breweries in 1948 had been reduced to 22 by 1970, with an average output of 0.67 million hl. After years of relatively peaceful coexistence among the country's brewers, a shake up of the market led to the industry being a virtual duopoly by the mid-1980s. Until the late 1970s, each state had its own one or two breweries (Carlton and United in Victoria, Castlemaine-Perkins in Queensland, in New South Wales there was Tooths and also Tooheys, and in South Australia there was Swan). There was also tacit agreement that there would be no trespassing on each other's territory. Courage, the UK brewer, who had set up a brewery in Victoria in 1968, catalyzed a change in this cosy relationship when in 1976, in an attempt to revitalize its failing investment, the company began selling heavily discounted beer in New South Wales. This began a marketing war, which, exacerbated by the removal of the tied house system in 1979 and the involvement of egocentric entrepreneurs, resulted in the Australian market being dominated by just two companies. The largest and most aggressive of these, Elders IXL, which had acquired Carlton & United Breweries (CUB), then expanded out of Australia, taking over Courage in the United Kingdom in 1986 (having failed to get Allied Breweries) and Carling O'Keefe in Canada in 1987. The other giant Australian brewer, the Bond Corporation, took over the U.S. brewer Heileman in 1987 to become briefly the fourth biggest in the world, but floundered thereafter on a mountain of debt. Further changes followed during the 1990s and, by the end of the century, Fosters (the renamed CUB) and Lion Nathan of New Zealand (the inheritors of Bond) accounted for 97% of beer produced in the Antipodes.[54]

Consolidation in the United Kingdom has been less extreme than in some other countries, but has had a pronounced effect in placing the control of a major part of the industry in foreign hands. Although by the 1980s hardly the "picturesque dinosaur" of the 1950s, as so damningly tagged by *The Economist*,[71] the industry had fatally failed to make a significant impact beyond the Isle of Wight. It underwent traumatic upheaval after 1989 with the implementation of the Beer Orders, when the tie between the big brewer and the pub was severely weakened through government legislation.[72]

This greatly accelerated the shift away from manufacturing already being contemplated by some large brewing companies and resulted in a separation between production and retailing activities. Further changes followed, leading to a new dichotomy of pub-owning companies (pubcos) and wholesale brewers. In the decade following the implementation of the Beer Orders, the UK's national brewers continued to unravel, with parent companies withdrawing from brewing and then, as the pressure on profits continued, demerging or selling off their pub chains. Of the six vertically integrated national companies that had dominated the industry from the 1960s, the last to abandon a system that had for so long seemed immutable was Scottish & Newcastle. Trading as Scottish Courage after the takeover of the latter, and with 27% of the UK beer market, S&N sold its tied estate to a pubco in November 2003 in deference to pressure from the City. The remnants of the other indigenous national brewers' production capacity was by then owned by three foreign-controlled companies — Interbrew, Coors Brewers, and Carlsberg-Tetley — who produced over 50% of the UK's beer, and their tied estates had been ceded to the pubcos. But foreign stewardship failed to revitalize the UK brewing industry and the slide in output continued.

Concentration was also a feature of the German brewing industry but to a lesser extent than elsewhere.[73] In 1990, average output per brewery was only 0.09 million hl per annum[47] and the country still had an estimated 1150 brewing companies, about half the number in 1950. Even in Germany, large companies had emerged. In 1993, the top 13 brewers had 66% of the market[74] and as volumes continued to fall amalgamations gained pace. Also, as Gourvish points out,[47] the effective level of concentration in Germany may, in practice, be greater than raw statistics suggest because of interlocking shareholdings between major firms and the location of many small breweries in Bavaria.

While it was generally agreed that unprecedented consolidation within countries had taken place in the previous decade, it was still possible to argue at the end of the 20th century that the move to globalization in brewing was only in its infancy.[74] Certainly the 20% of the market held in 1999 by the top three world brewers was less than impressive when compared with the top three soft drinks companies who had 75% of world sales at that time. Local brands still had more than 70% of the market in all European countries except for Britain, France, Italy, and Greece. Foreign brands had made little impact in South America, North America, and Africa. In China, massive investment by international brewers in promoting expensive western brands had brought little returns. American brewers concentrated almost exclusively on their domestic market, with little serious attempt at even a significant export trade before the 1990s. Even in the case of the biggest brewer in the world, Anheuser–Busch, 12 of its 14 breweries were located in the United States and only 6% of sales came from outside the United States in 1999.

But none could deny that internationalization of brewing was established by the end of the century. Guinness, Heineken, and Carlsberg had been the leaders in this process. Long known as exporters of beer, and secure in their home markets (the Danes helped by long-standing cartel agreements even before the merger with rivals Tuborg in 1970),[75] these companies sought to expand by establishing breweries and trading partners worldwide from the 1960s onwards. By 1974, Carlsberg's foreign production exceeded exports from Denmark and, by 1976, sales in the international market exceeded domestic sales.[62] Later, the antipodean brewers, now known as Fosters and Lion Nathan, and the Belgian group Interbrew followed the international brewing path. Heineken were particularly successful in this strategy. In 2000, their beer was produced in more than 110 breweries in over 50 countries. Similarly, Carlsberg beer was produced in 67 breweries in 42 countries. Over 90% of the sales of both companies were from abroad. Guinness, as part of the world's biggest spirits group, had distribution in over 180 countries. The march to globalization continued in the 21st century. Interbrew, having already mopped up many old established brewers, including Labatt of Canada, swallowed a large chunk of the UK industry in Whitbread and Bass, before being forced to regurgitate part of the latter by the UK government and sell it to Coors in 2002. South African Breweries (SAB), enjoying new postapartheid respectability, specialized in emerging markets and by 2001 controlled 108 breweries in 24 countries to emerge as the sixth largest brewer in the world. SAB was one of the few brewing companies to make some success of its investment in China where the locals remained largely unimpressed by the sacerdotal approach of other giant international brewers. The purchase of the ailing U.S. giant, Miller, in the following year moved the new company, SAB–Miller, even higher up the pecking order. Meanwhile, in 2000, Carlsberg changed its corporate structure in order to compete more effectively and merged its brewing activities with those of Orkla, a Norwegian conglomerate owning Pripps and Ringnes breweries, before buying further breweries in Turkey and Poland a year later. Carlsberg disposed of its troubled German subsidiary, Hannen, in July 2003, only to bid for Germany's second biggest brewer, Holsten, in January 2004. In October 2002, the sleeping giant, Anheuser–Busch, announced that over the next seven year it would increase its holding in China's biggest brewer, Tsingtao, to 27%.[76] Almost every month brought news of a new example of the big battalions flexing their corporate muscles.

Science Applied and Technology Transformed

Investment in scientific research in brewing has never been high, even in the context of the relatively low-spending food industry, and in the first half of the 20th century expenditure was vanishingly small in most countries. In the

United Kingdom, for example, the brewing industry was spending an estimated 0.003% of its turnover on research in the late 1930s — a figure lower by a factor of three than any other industry surveyed.[77] Raw materials were the primary targets for the limited research effort of this period, the brewing process itself receiving little attention. Breeding of improved hop and barley varieties began during the early years of the century, starting a trend that would lead to hops containing much higher levels of bittering power and barleys that combined high extract and good agronomic properties. New barley varieties became generally available during the 1920s, and studies of the chemistry of hops revealed the basic structure of alpha acids in the same decade. Unraveling isomerization during wort boiling had to wait until the 1950s.

During the 1930s, chemical and microbiological analytical techniques were extended and improved, but science only impinged on the periphery of the average brewer's vision. The brewers' chemist retained a lowly place in the hierarchy, ranked somewhere between the second and third brewer to judge from his remuneration. As one insider was to note[78] rather sourly: "... brewers employed a chemist in an obscure laboratory as a sort of scientific chaplain in an otherwise unscientific industry...." Certainly, excluding the special circumstances of those employed in the Carlsberg Laboratory, the days had gone when the brewers' chemist could make a contribution to mainstream science as had been the case with Cornelius O'Sullivan, Horace Brown, and others. Indeed, the small community of scientists in the brewing industry became increasingly inward looking and took no part in the wider world of their disciplines. This insularity was to persist. Even in their heyday in the third quarter of the 20th century, scientists employed in breweries were rarely to be found publishing in journals, or participating in meetings and conferences, other than those specifically related to brewing.

Although women were closely associated with domestic and publican brewing (hence the term brewster), they have been much less prominent in industrial brewing; the laboratory being one of the few areas where they reached parity with men during the 20th century. Women provided much of the labor in bottling stores from the 1870s, were employed in clerical roles and as "typewriters" from the 1880s, and worked as technicians in laboratories from the 1920s. After the Second World War, more responsible roles in laboratories, marketing, information science, finance, and eventually as brewers followed. But industrial brewing has remained essentially a male preserve at the highest levels. Few women have become directors of brewing companies.

Practicing brewers have always been more interested in technology than in science and innovations of obvious practical utility were eagerly adopted: for example, the Seck mill, introduced in 1902, which had three or even four pairs of rollers as opposed to the one or two pairs the majority of brewers used until that time. Richard Seligman's countercurrent plate heat exchanger (originally used for milk and then adapted for use with beer) also met with wide acceptance within a few years of being patented in 1923. Other developments were more consumer driven. The habit in the United States of putting beer in the "ice-box" prompted Leo Wallerstein to patent the use of proteolytic enzymes to prevent chill haze in 1911.[79] Even so, radical technological change was only attempted by the most adventurous brewers, who experimented with mash filters, new designs of fermenters, and metal beer containers.

From the late 1940s, investment in scientific research into the brewing process was increased to unprecedented, if still modest levels. In the United Kingdom, the Brewing Industry Research Foundation (BIRF), paid for by a barrelage-based levy of the British brewers, was officially opened at Nutfield in Surrey in 1951.[80] The first director, Sir Ian Heilbron FRS, a prominent organic chemist from Imperial College, had firm views on what the Foundation should be about. In an early paper[81] outlining his aims for the new venture, he noted that it would be "a scientific headquarters furthering the application of science to the solution of tactical problems and to the strategic development of the industry." He saw these activities as "complementary and in no way conflicting" and stated that "the engagement in fundamental research is a duty, not a luxury which the industry can permit itself." In the first 25 years of its existence, BIRF had at its peak, over 100 scientists and support staff, and they published over 700 original papers. The postwar enthusiasm for science touched breweries in most countries, with specialist laboratories and pilot plant facilities established or extended in universities and technical institutes and by major brewers during the 1950s and 1960s. Detailed understanding of the chemistry, biochemistry, and microbiology of malting and brewing followed. By the end of the 1970s, highlights included knowledge of the enzymology of barley germination and mash conversion, the chemical structure of hop components, and the mechanism of formation of major beer flavor compounds, including diacetyl, esters, higher alcohols, and the prime determinant of lager flavor, dimethyl sulfide.

During the first half of the 20th century, although individual breweries and maltings had grown larger and output had increased, there had been little increase in batch or vessel size. The use of stainless steel, greater chemical, physical, and biochemical knowledge, better analytical control, increased availability of process aids (plant growth regulators, enzymes, coagulants, etc.), and then automated computer-controlled plants, led to step changes. Malting and brewing technology was transformed by new developments from both inside and outside the industry, or in some cases by adoption of techniques that had their origin many years earlier but had been held back by prevailing attitudes and difficulties of construction and operation.[82]

Until the 1950s, floor malting was the most usual procedure in many countries, with grain, after steeping, spread on a solid floor, kept cool during germination by hand-turning with a shovel, and dried in a kiln directly fired by coal or coke. This labor-intensive system was largely replaced in the next 30 years. Maltings became mechanized, access to air was given during steeping, and drums and then perforated rotating floors were adopted for germination. Oil and then methane gas were used for kilning. During the 1980s, it was found that the latter promoted the formation of carcinogenic nitrosamines and indirect firing was introduced as standard. In a totally new departure, the use of a plant hormone (gibberellic acid [GA]) together with a growth restrictor (potassium bromate) became popular during the 1960s as a means of accelerating malting without increasing losses.[83] The use of bromate ceased during the 1980s with improved temperature control, and the popularity of GA also decreased. A limited degree of battering or "abrasion" of barley prior to steeping as a means of accelerating malting gained transient popularity, at least amongst brewer–maltsters, during the 1970s but soon faded from view.[84] By the end of the century, malting was carried out in large plants (annual capacity 50,000 to 100,000 tons) and total processing time was about half of what it had been 50 years earlier. The malting industry also experienced consolidation: 30 companies made 60% of the world's malt in 1998.

Prior to the First World War, fermenters were usually fabricated in wood or slate. Aluminum was used during the 1920s and limited depth stainless steel during the 1930s. Cylindroconical fermenters became the norm everywhere after the 1960s, and fermenters of up to 6000 hl capacity replaced smaller (200 to 300 hl) box-shaped vessels. Patented in 1910 and introduced on a small scale by the 1930s in continental Europe, Australia, and America, cylindroconical vessels lent themselves to rapid batch processing. The new shape encouraged the use of sedimentary strains of yeast for ale as well as lager, and distinctions between the processing of the two became increasingly blurred as both became predominantly chilled and filtered beers. Process times generally were shortened, most notably in the case of maturation, where understanding of the chemistry of diacetyl production and removal allowed adjustment of fermentation conditions to give swift attainment of low levels. Storage of beer in the presence of yeast to stabilize, carbonate, and modify the flavor was increasingly replaced by the more rapid process, long practised in the United States, of cold filtration and injection of carbon dioxide.[85]

Steam boiling, as opposed to directly fired coppers, had been in use since the 1880s, but did not gain wide acceptance until well into the 20th century. For malt extraction the two-vessel (mash and lauter tun) system of brewing, which facilitated wort separation, became the norm even for ale brewing during the 1960s. Lauter tuns in turn began to lose ground in some quarters during the 1980s as mash filters, first used a century earlier, were increasingly adopted in an improved form. The breeding of hops with much higher levels of alpha-acids — a more than fourfold increase over the century — led to much lower hop rates.[86] Whole cone hops, the only method of bittering at the start of the century, came to be replaced by milled and later solvent-extracted preparations. Commercial preparations of preisomerized alpha-acids became available during the 1950s and 1960s, but failed to find wide acceptance other than for final beer bitterness adjustment in most companies. The whirlpool separator for removing hop and other residues after wort boiling was introduced to replace the more cumbersome "hop-backs" at Molson's brewery in Canada in 1960[87] and was ubiquitous in breweries by the 1980s. Syrups produced using enzymes were being used by brewers during the 1950s, and the first beer brewed using 100% barley converted with exogenous enzymes as a complete replacement of malt was sold in 1963.[88] Barley brewing never found wide acceptance, but the use of enzymes as palliatives became popular among nervous brewers who wished to avoid or prevent problems, particularly in wort production. Adjustment of the mineral composition of water by addition of salts was commonplace by the 1960s, and from the 1980s, demineralization followed by construction of the appropriate water from scratch, depending upon the type of beer to be brewed, increased in popularity. High-gravity brewing, popularized during the 1970s, was the norm by the mid-1980s, allowing better plant utilization. Sophisticated postfermentation treatments delayed haze development and improved techniques for excluding oxygen during packaging meant that beer flavor shelf-life was also extended, often to a year or more. Wood, then glass, lost out to metal as the keg and can, pioneered in the 1930s, gained ground and filling-line speeds increased.

Not all scientific and technological changes worked out well. Leaving aside the ill-starred use of cobalt salts as foam improvers from the late 1950s, which led to around 100 deaths in North America and its hasty withdrawal in 1966,[89] the most conspicuous example of misplaced enthusiasm is continuous brewing. "Much researched but little utilized," as one review puts it,[90] continuous brewing was seen as the technology of the future during the 1950s, but failed to live up to expectations. The first entirely continuous fermentation brewery in the world was New Zealand Breweries Palmerston North plant, which started commercial production in 1957. Despite footholds by the 1960s in the United States and the United Kingdom (where by one account it was responsible for 4% of the beer produced in the mid-1960s), the continued operation of a plant in New Zealand, and the introduction of continuous maturation in Finland during the 1990s, 99.99% of beer was still produced entirely by batch processes at the end of the century.[91] If during the 1950s hopes for continuous brewing went unfulfilled, then the same is true of the belief that took root in some quarters during the 1980s of the potential for utilizing genetically modified (GM) yeast strains.[92] Again, much research effort was directed at what became a hot topic and this was met by a measure of scientific achievement. A GM yeast for use in low-carbohydrate beer fermentation gained regulatory approval in 1993 and numerous other strains designed to give a technological advantage were constructed, but no GM strains have to date been used in commercial brewing. Similarly, targets for GM barley have been identified by scientists and progress made toward achieving these goals, but the drinks industry remains unconvinced.[29] While public opinion on genetic modification, and GM food in particular, remains so negative, companies are unwilling to imperil the expensively generated image of their brands for marginal advantage.[93] Only in the United States has there been apparent wide public acceptance, or at least indifference, to genetic modification. Transgenic maize became ubiquitous in the United States and was necessarily used in brewing. Elsewhere in the world, brewers took pains to reassure the public that their beers were free from GM material. Carlsberg, leading European users of maize for brewing,[94] turned their back on the cereal for this very reason in the 21st century.

With gathering pace, instrumentation transformed laboratory practice in academic and industrial laboratories in the second half of the 20th century. Wet chemical methods and laborious microbiological techniques largely disappeared and were replaced by sophisticated, sometimes automated, procedures. Productivity increased by several orders of magnitude such that control laboratories that had bustled with people during the 1970s had instead filled up with instruments by the 1990s. At-line, on-line, and in-line analysis were taken up by increasing numbers of breweries since the 1980s and seemed likely to lead to the eventual effective disappearance of control laboratories altogether.[95] Breweries increasingly moved toward use of in-line sensors and reliance upon external specialist laboratories. This latter move, having echoes of the widespread use of the consulting brewer and chemist of a century earlier with all "his dangers and his uses,"[96] was consistent with an industry that increasingly embraced the attitudes and jargon of "outsourcing," "best value," "externalization," etc. throughout its activities.

The belief in the utility of scientific research, which developed in the brewing industry after the Second World War, proved short-lived. Activity in breweries, never widely or firmly based, stalled during the 1980s[44] and all but evaporated during the 1990s.[97] Research laboratories and pilot plants closed and budgets were cut as companies became more secretive. Original publications from major breweries in the United Kingdom and North America, once major features in journals and at conferences, dried up. Remaining funds were directed primarily toward "near market" product and packaging innovations. What is now called Brewing Research International (BRi), at Nutfield, lost central funding from the increasingly unstable British industry, reduced staffing levels to half of those of its heyday, and refocused activities on service analysis, training, and contract work, rather than research. An exception to this move to what its adherents called "the new realism" and what its opponents called "short-termism" was Japan, where contributions to brewing science had been increasingly evident since the 1980s. In a highly competitive industry, Japanese companies heeded the often given but seldom taken advice of economists on the particular need to continue to find the money to support both short- and long-term research in a depressed economy.

References

1. Anderson, R.G., Beer, in *Alcohol and Temperance in Modern History: An International Encyclopaedia*, Blocker, J., Tyrrell, I., and Fahey, D., Eds., Vol. 1, ABC-Clio, Santa Barbara, 2003, pp. 92–100.
2. Katz, S.H. and Voigt, M.M., Bread and beer: the early use of cereals in the human diet, *Expedition*, 28:23–34, 1986.
3. Wilson, R.G. and Gourvish, T., The production and consumption of alcoholic beverages in Western Europe, in *Alcoholic Beverages and European Society. Annex 1. The Historical, Cultural and Social Roles of Alcoholic Beverages*, The Amsterdam Group, Eds., Elsevier, Amsterdam, 1993, pp. 1–34.
4. Mayer, K.T., Germany's Reinheitsgebot and brewing in the 20th century, *Tech. Q. Master Brew. Assoc. Am.*, 20:138–145, 1983.
5. Anderson, R.G., Samuel Allsopp & Sons, in *Alcohol and Temperance in Modern History: An International Encyclopaedia*, Blocker, J. Jr., Fahey, D., and Tyrrell, I., Eds., Vol. 1, ABC-Clio, Santa Barbra, 2003, pp. 37–38.
6. Mathias, P., *The Transformation of England. Essays in the Economic and Social History of England in the Eighteenth Century*, Methuen, London, 1979, pp. 209–230.
7. Mathias, P., *The Brewing Industry in England 1700–1830*, Cambridge University Press, Cambridge, 1959, pp. 6, 542–545, 551–552.
8. Cornell, M., Porter myths and mysteries. Brewery history, *J. Brew. Hist. Soc.*, 112:30–39, 2003. A fuller discussion of the origins of porter is given in the Cornell, M., *Beer: The Story of the Pint*, Headline Book Publishing, London, 2003, p. 89–120.
9. "O Poundage," The history of the London brewery, *London Chronicle*, November 4, 1760. (Quoted at length in Ref. 10).
10. MacDonagh, O., The origins of porter, *Econ. Hist. Rev.*, 16:530–535, 1964.
11. Corran, H.S., *A History of Brewing*, David and Charles, Newton Abbot, 1975, pp. 112–115.
12. Van Oevelen, D., De L' Escaille, F., and Verachtert, H., Synthesis of aroma compounds during the spontaneous fermentation of lambic and gueuze, *J. Inst. Brew.*, 82:322–326, 1976.
13. Spaepen, M., Van Oevelen, D., and Verachtert, H., Fatty acids and esters produced during the spontaneous fermentation of lambic and gueuze, *J. Inst. Brew.*, 84:278–282, 1978.
14. Claussen, N.H., On a method for the application of Hansen's pure yeast system in the manufacturing of well-conditioned English stock beers, *J. Inst. Brew.*, 10:308–331, 1904.
15. Simpson, W.J., Studies on the sensitivity of lactic acid bacteria to hop bitter acids, *J. Inst. Brew.*, 99:405–411, 1993.
16. Bennison, B., Adoption of steam by North-Eastern Breweries during the nineteenth century, *J. Brew. Hist. Soc.*, 98:49–56, 1999.
17. Sumner, J., John Richardson Saccharometry and the pounds-per-barrel extract: the construction of a quantity. *Br. J. Hist. Sci.*, 34:255–273, 2001.
18. Richardson, J., *The Philosophical Principles of the Science of Brewing*, New York, 1788.
19. Morrice, A., *A Treatise on Brewing*, 3rd ed., London, 1802.
20. Baverstock, J., Ed., *Treatise on Brewing*, London, 1824.

21. Gutzke, D.W., The social status of landed brewers in Britain since 1840, *Histoire Sociale — Social History*, 17:93–113, 1984.
22. Glamann, K., Founders and successors. Managerial changes during the rise of the modern brewing industry, *Proc. 18th Eur. Brew. Conv. Congr.*, European Brewery Convention, Copenhagen, 1981, pp. 1–10.
23. Barclay, C., Evidence to the report on the retail sales of beer, Parliamentary Papers, 1830, X, p. 16. (Quoted in Ref. 6, p. 210).
24. Wilson, R.G., The changing taste for beer in Victorian Britain, in *The Dynamics of the International Brewing Industry since 1800*, Wilson, R.G. and Gourvish, T.R., Eds., Routledge, London, 1998, pp. 93–104.
25. Owen, C.C., *The Development of Industry in Burton upon Trent*, Phillimore, Chichester, 1978, pp. 87–88.
26. Donnachie, I., *A History of the Brewing Industry in Scotland*, John Donald, Edinburgh, 1979, pp. 206–230.
27. Lynch, P. and Vaizey, J., *Guinness Brewery in the Irish Economy 1759–1876*, Cambridge University Press, Cambridge, 1960.
28. Anderson, R.G., Highlights in the history of international brewing science, *Ferment*, 6:191–198, 1993.
29. Anderson, R.G., Current practice in malting, brewing and distilling, in *Cereal Biotechnology*, Morris, P.C. and Bryce, J.H., Eds., Woodhead, Cambridge, 2000, pp. 183–214.
30. Teich, M., Bier, Wissenschaft und Wirtshaft in Deutschland 1800–1914. Ein *Beitrag zurdeutschen Industrialisierungsgeschichte*, Bohlau Verlag, Wein, 2000.
31. Halcrow, R.M., A look at the brewing industry in America 100 years ago, *Tech. Q. Master Brew. Assoc. Am.*, 24:121–128, 1987.
32. Protz, R., *The Taste of Beer*, Weidenfield & Nicholson, London, 1998, p. 127.
33. Anderson, W., *Beer, USA*, Morgan & Morgan, New York, 1986.
34. Ohnishi, T., Yoshida, H., and Kono, T., The history of automation in Japanese breweries and their future development, *Tech. Q. Master Brew. Assoc. Am.*, 31:26–31, 1994.
35. Jackson, M., *The World Guide to Beer*, Michael Beazley, London, 1979, p. 236.
36. Anon., Focus on Carlton and United Breweries, *The Brewer*, 75:827–830, 1989.
37. Bud, R., *The Uses of Life. A History of Biotechnology*, Cambridge University Press, Cambridge, 1993, pp. 17–21.
38. Kruger, R., The history of Weihenstephan, *Brauwelt Int.*, 4:296–297, 1990.
39. Manners, D.J., *Brewing and Biological Sciences at Heriot-Watt University 1904–1989*, Heriot-Watt University, Edinburgh, 2001.
40. Anderson, R.G., Louis Pasteur (1822–1895): an assessment of his impact on the brewing industry, *Proc. 25th Eur. Brew. Conv. Congr.*, Brussels, 1995, pp. 13–23.
41. Puddick, A.J., Changes in British bottling techniques, in *One hundred Years of British Brewing*. The Centenary Issue of the *Brewers' Guardian*, 1971, pp. 117–119.
42. Anderson, R.G., Yeast and the Victorian brewers: incidents and personalities in the search for the true ferment, *J. Inst. Brew.*, 95:337–345, 1989.
43. Anon., *The History of Anheuser-Busch Companies. A Fact Sheet*, Anheuser-Busch Inc., St. Louis, 1995, p. 12.
44. Anderson, R.G., The pattern of brewing research: a personal view of the history of brewing chemistry in the British Isles, *J. Inst. Brew.*, 98:85–109, 1992.
45. Gourvish, T.R., Wilson, R.G., *The British Brewing Industry 1830–1980*, Cambridge University Press, Cambridge, 1994, p. 295.

46. *The Brewers and Licensed Retailers Association Statistical Handbook*, Brewing Publications Ltd., London, 2000.
47. Gourvish, T.R., Economics of brewing, theory and practice: concentration and technological change in the USA, UK and West Germany since 1945, *Bus. Econ. Hist.*, 23:253–261, 1994.
48. Levison, J., Malting–brewing: a changing sector, *Bios*, 33:12–15, 2002.
49. Kerr, K.A., The American brewing industry 1865–1920, in *The Dynamics of the International Brewing Industry since 1800*, Wilson, R.G. and Gourvish, T.R., Eds., Routledge, London, 1998, pp. 176–192.
50. Ronnenberg, H.W., The American brewing industry since 1920, in *The Dynamics of the International Brewing Industry since 1800*, Wilson, R.G. and Gourvish, T.R., Eds., Routledge, London, 1998, pp. 193–212.
51. Barr, A.W., *Drink: An Informal Social History*, Bantam Press, London, 1995, pp. 307–308.
52. Katz, P.C., The state of the brewing industry 1933–1934, *Mod. Brew. Age*, 34:MS30–MS34, 1983.
53. *The Brewers' Almanac*, Review Press, London, 1919, Tables 1 and 11, and 1936, Tables 1 and 7.
54. Jones, S.R.H., The New Zealand brewing industry 1840–1955, in *The Dynamics of the International Brewing Industry since 1800*, Wilson, R.G. and Gourvish, T.R., Eds., Routledge, London, 1998, pp. 247–265.
55. Gourvish, T.R. and Wilson, R.G., *The British Brewing Industry 1830–1980*, Cambridge University Press, Cambridge, 1994, pp. 418–437.
56. Abbot, H., Bottled beer versus draught (Editorial), *Bottling*, No. 35, Jan. 1, 1933, pp. 1–2.
57. Mass Observation, *The Pub and the People*, Victor Gollancz, London, 1943, pp. 183–185.
58. Dunn, M., *The Penguin Guide to Real Draught Beer*, Penguin Books, Harmondsworth, 1979.
59. Cusack, J.C., The development of the brewing industry in the Far East, in *Brewing Room Book 1989–1991*, Lloyd, J., Ed., Pauls Malt, Ipswich, 1988, pp. 83–84.
60. Koide, K. and Inoue, T., The present situation of brewing technology in Asia, *Proc. 23rd Conv. Inst. Brew. Asia Pacific Section*, 1994, pp. 9–12.
61. Newman, P., Brewers expand frozen assets, *The Canner*, June 1994, 36–38.
62. Anderson, R.G., Carlsberg, in *Alcohol and Temperance in Modern History: An International Encyclopaedia*, Blocker, J., Tyrrell, I., and Fahey, D., Eds., Vol. 1, ABC-Clio, Santa Barbara, 2003, pp. 139–140.
63. Barr, A.W., *Drink: An Informal Social History*, Bantam Press, London, 1995, pp. 23–24.
64. Viner, B., Pundits in a league of their own, *Night & Day*, Jun. 21, 1998, p. 38.
65. Baxter, D., Beer is good for you — discuss. *The Brewer*, 82:63–66, 1996.
66. Editorial, The Butterfly and Cabbage Group, *Brew Int*, 3:3, 2003.
67. Anon., Bootleggers bounce back, *What's Brewing*, Jan. 2003, p. 1.
68. Gourvish, T.R. and Wilson, R.G., *The British Brewing Industry 1830–1980*, Cambridge University Press, Cambridge, 1994, pp. 447–497.
69. Merrett, D.T., Stability and change in the Australian brewing industry 1920–1994, in *The Dynamics of the International Brewing Industry since 1800*, Wilson, R.G. and Gourvish, T.R., Eds., Routledge, London, 1998, pp. 229–246.
70. Lewis, G.L., Morkel, A., and Hubbard, G., *Cases in Australian Strategic Management*, Prentice Hall, Sydney, 1991, pp. 38–99.

71. *The Economist*, Sep. 19, 1964.
72. Crompton, G., "Well-intentioned meddling." The beer orders and the British brewing industry, in *The Dynamics of the International Brewing Industry since 1800*, Wilson, R.G. and Gourvish, T.R., Eds., Routledge, London, 1998, pp. 160–175.
73. Verstal, I., "Brauereisterben" continues to shrink the industry, *Brew. Distill. Int.*, 26:20–21, 1995.
74. Baker, K., The battle for share of throat — perspectives on the global beverage consumption patterns, *Proc. 7th Conv. Inst. Brew. Africa Section*, Nairobi, 1999, pp. 1–3.
75. Johansen, H.C., Marketing and competition in Danish brewing, in *Adding Value. Brands and Marketing if Food and Drink*, Jones, G. and Morgan, N.J., Eds., Routledge, London, 1994, pp. 126–138.
76. Anonymous, *Brew. Int.*, 2:5, 2002.
77. Seligman, R., The research scheme of the institute — its past and its future, *J. Inst. Brew.*, 54:133–140, 1948.
78. Pyke, M., A biologist in industry, *Inst. Biol. J.*, 10:102–108, 1963.
79. Wallerstein, L., Chill proofing and stabilisation of beer, *Wallerstein Lab. Commun.*, 19:95–110, 1956.
80. Anderson, R.G., Brewing Research International: a survivor in changing times, *J. Brew. Hist. Soc.*, 108:56–58, 2002.
81. Heilbron, I., The Research Foundation and its place in the industry, *Brew. Guard.*, 80:12–16, 87, 1950.
82. Anderson, R.G., Past milestones in malting, brewing and distilling, in *Proc. Fourth Aviemore Conf. on Malting, Brewing and Distilling*, Campbell, I., Priest, F.G., Eds., 1994, pp. 5–12.
83. Ray, G.F., The application of gibberellic acid in malting, in *The Diffusion of New Industrial Processes: An International Study*, Nabseth, L. and Ray, G.F., Eds., Cambridge University Press, Cambridge, 1974, pp. 215–231.
84. Hudson, J.R., Impact of research in brewing practice: past, present and future, *Proc. 19th Eur. Brew. Conv. Congr.*, European Brewery Convention, London, 1983, pp. 13–32.
85. Munroe, J.H., Fermentation, in *Handbook of Brewing*, 1st ed., Hardwick, W.A., Ed., Marcel Dekker, New York, 1995, pp. 323–353.
86. Darby, P., A century of hop breeding, in *Brewing Room Book 1998–2000*, Beach, R.H., Ed., Pauls Malt, Ipswich, 1997, pp. 57–61.
87. Hudston, H.R., The story of the whirlpool, *Tech. Q. Master Brew. Assoc. Am.*, 6:164–167, 1969.
88. Crisp, J.W.M. and East, E., The use of added enzymes in the production of wort, *Proc. 13th Eur. Brew. Conv. Congr.*, European Brewery Convention, Estoril, 1971, pp. 185–190.
89. Lenihan, J., *The Crumbs of Creation*, Adam Hilger, Bristol, 1988, pp. 49–50.
90. Anderson, R.G., Brites Sanchez, A., Devreux, A., Due, J., Hammond, J.R.M., Martin, P.A., Oliver-Daumen, B., and Smith, I., *European Brewing Convention Manual of Good Practice: Fermentation and Maturation*, Fachverlag Hans Carl, Nuremberg, 2000, pp. 124–129.
91. Masschelein, C.A., A realistic view of the role of research in the brewing industry today, *J. Inst. Brew.*, 103:103–113, 1997.
92. Molzahn, S., Genetic engineering: yeast, *Proc. 21st Eur. Brew. Conv. Congr.*, European Brewery Convention, Madrid, 1987, pp. 197–208.

93. Verstl, I., Gone with the wind: will soft issues be relevant for international brewing in the future, *Proc. 26th Conv. Inst. Brew. Asia Pacific Section*, 2000, pp. 22–28.
94. Munck, L. and Lorenzen, K., *Proc. 16th Eur. Brew. Conv. Congr.*, European Brewery Convention, Amsterdam, 1977, pp. 369–376.
95. Pollock, J., The world's brewing industry in the twentieth century, in *Brewing Room Book 1998–2000*, Beach, R.H., Ed., Pauls Malt, Ipswich, 1997, pp. 28–30.
96. Bowly, J.E., The consulting brewer: his dangers and his uses, with some practical brewing notes, *J. Inst. Brew.*, 2:86–98, 1896.
97. Briggs, D.E., The Horace Tabberer Brown Memorial Lecture: Malt modification — a century of evolving views, *Proc. 26th Conv. Inst. Brew. Asia Pacific Section*, 2000, pp. 13–21.

2

Beer Styles: Their Origins and Classification

Charles Papazian

CONTENTS

Introduction	41
How Different Beer Styles are Created	42
Ingredients	42
Water	42
Fermentable Carbohydrates	42
Hops	43
Yeast	45
Processing	45
Equipment Configuration	46
Milling	46
Mashing	46
Lautering	46
Boiling Time	46
Fermentation Temperature	47
Maturation Time	47
Filtration	47
Packaging	47
Marketing	47
Cultural	48
Origins of Style	48
Analytical and Sensor Variables	50
Methods Used to Define Brewers Association's Beer Style Guidelines	52
Analysis	53
Tasting and Recording	53
Brewing Beer	53
Regular Suggestions from Brewing Industry Professionals	54
Brewers Association 2005 Beer Style Guidelines	54
The Beer Styles	55

Ales 56
  British Origin 56
    Classic English-Style Pale Ale 56
    English-Style India Pale Ale 56
    Ordinary Bitter 57
    Special Bitter or Best Bitter 57
    Scottish-Style Light Ale 57
    Scottish-Style Heavy Ale 58
    Scottish-Style Export Ale 59
    English-Style Dark Mild Ale 59
    English-Style Brown Ale 59
    Imperial Stout 60
    Other Strong Ales or Lagers 60
    English-Style Barley Wine Ale 60
  Irish Origin 61
    Irish-Style Red Ale 61
    Classic Irish-Style Dry Stout 61
  North American Origin 62
    American-Style Pale Ale 62
    American-Style India Pale Ale 62
    American-Style Amber/Red Ale 63
    Golden or Blonde Ale 63
    American-Style Stout 63
  German Origin 64
    German-Style Brown Ale/Düsseldorf-Style Altbier 64
    South German-Style Hefeweizen/Hefeweissbier 64
    South German-Style Kristal Weizen/Kristal Weissbier 65
    South German-Style Dunkel Weizen/Dunkel Weissbier 65
  Belgian and French Origin 65
    Belgian-Style Flanders/Oud Bruin or Oud Red Ales 65
    Belgian-Style Dubbel 66
    Belgian-Style Tripel 66
    Belgian-Style Pale Ale 67
    Belgian-Style White (or Wit)/Belgian-Style Wheat 67
    Belgian-Style Lambic 67
Lager Beers 68
  European-Germanic Origin 68
    German-Style Pilsener 68
    Bohemian-Style Pilsener 68
    European-Style Pilsener 69
    Traditional German-Style Bock 69
  North American Origin 69
    American Lager 69
    American-Style Light Lager 70
    American-Style Low-Carbohydrate Light Lager 70

    American-Style Premium Lager 70
    American Ice Lager 71
  Other Origin 71
    Australasian, Latin American, or Tropical-Style Light Lager 71
    Flavored Malt-Fermented Beverages 72
    Vegetable Beers 72
    Herb and Spice Beers 72
Acknowledgments 73
Bibliography of Resources 73
  Other Resources 74

Introduction

Beer is an expression of the human spirit. We use technical sciences as a tool to create it, psychology to market and help sell it, but its essence is and always will be a form of art. Beer style is the art of combining hundreds of factors to create a consistent combination of beer characters. Beer's complexity and all of the diversity it offers express the variety of the world's lifestyles.

There are an estimated 5000 commercial breweries in the world today. We can arguably estimate that each brewery may on an average produce eight different kinds of beer. That gives us 40,000 different beers available for sale around the world. While many beers may be similar in style, their individual creation and the *culture* surrounding their enjoyment and celebration help define each beer's individual uniqueness.

This chapter explores beer's stylistic diversity, and how our knowledge of the factors influencing variety can be useful in beer formulation, brewhouse management, beer evaluation, tempering government regulation, beer competitions, and improving the image of beer in the marketplace.

The word *style* is defined by the dictionary as "a particular manner or technique by which something is done, created or performed resulting in a distinctive quality, form of or type of something." This is applicable to all art forms, beer being one of them. I have identified and detailed almost 100 different beer styles of American, Belgian, British, German, Irish, and Japanese origin, most of which are available in the contemporary American beer market. The American beer market is perhaps the beer market that offers the most diversity in the world. There are indeed dozens of other styles popular in various regions of the world that remain to be "cataloged" in this ongoing project. With the expansion of international trade and the increase in intercultural experiences, there will be more opportunities to access, enjoy, and introduce new varieties of beer to the world's marketplaces.

How Different Beer Styles are Created

All beer types evolve from the combination of and relationships among:

1. Ingredients
2. Processing
3. Packaging
4. Marketing
5. Culture

When we vary these basics, we create variety and distinct styles. Where does one begin in order to understand how these basics create a product? Where does a brewing company begin? It begins with the training of key production, marketing (including the distribution network), and management people, helping them understand and appreciate these basics, and the role each contributes to the final qualities of the product. Only when these basics are appreciated can a beer be successfully created, introduced to the marketplace, sold, and fully appreciated by the consumer.

Ingredients

Most beer is made from four essential ingredients:
 (1) Water, (2) fermentable carbohydrates such as barley malt, starch, and sugar adjuncts, (3) hops, and (4) yeast. Not only can these ingredients differ, but they can also be used in numerous different combinations. Ingredient variables help to differentiate one beer from another and to define beer styles.

Water

The balance of minerals in brewing water (Table 2.1) will affect the flavor character and flavor perception of malt, hops, and by-products of fermentation. It may also influence the performance of yeast, which in turn influences the flavor, aroma, and mouthfeel of beer.

Fermentable Carbohydrates

The barley malt types listed in Table 2.2 are a sampling of the variety of malts often used in the American craft-brewing industry. Each malt type has its own unique specifications, resulting in unique contributions to the qualities of beer. The quality of malt is often unique to a given region. Its availability has often influenced the origin of a particular beer style. Color, flavor, aroma, alcohol, and mouthfeel are a few parameters influenced by malt, other fermentable

Beer Styles: Their Origins and Classification

TABLE 2.1

Approximate Ionic Concentrations (in ppm) of Classic Brewing Waters (Variations can be Expected to be ±25% and Sometimes More)

Brewing Area (Style of Beer Typifying Heritage)	Water Composition						
	Ca^{+2}	Mg^{+2}	Na^{+1}	Cl^{-1}	SO_4^{+2}	HCO_3^{-1}	Hardness
Pilzen (Light-colored lagers, zealously hopped)	7	2	2	5	5	15	30
Burton-on-Trent (Amber pale ales with distinctive sulfate-influenced hop character)	295	45	55	25	725	300	850
Dublin (Dark malty ales with medium bitterness)	115	4	12	19	55	200	300
Edinburgh (Dark malty strong ales with low bitterness)	120	25	55	20	140	225	350
Dortmund (Strong well-hopped amber lagers with full malty palate contributed by sodium and chloride, and sulfate-influenced hop character)	250	25	70	100	280	550	750
Munich (Dark malty lagers)	75	20	10	2	10	200	250
Vienna (Amber to brown malt-accented lagers with sulfate-influenced hop character)	200	60	8	12	125	120	750

carbohydrates, and sugar adjuncts. The choice, amount, and combination of malt types will create an extraordinary variety of characters in beer.

Table 2.3 indicates some of the variety of sugars and grain adjuncts used in the ingredient formulation of beer. The use of different grains is often influenced by its cost relative to barley malt and its availability in a given region. Availability has influenced the origin of some beer styles.

Hops

Hop types play an important and traditional role in recipe formulation. The choice, timing, amount, and combination of hop types create a variety of characters in beer. When identical hop varieties are grown in different areas of the

TABLE 2.2

Barley Malt Types

Pale malts	American Munich (dark)
	Canadian Munich (light)
American two-row	Canadian Munich (dark)
American six-row	British Munich
Belgian two-row	Caramelized Malts
British two-row	Dextrin malt
Canadian two-row	Belgian caramel Pils
British Pilsener	British CaraMalt
German Pilsener	American crystal 10°L
Belgian Pilsener	American crystal 20°L
British Lager (two-row)	American crystal 30°L
Lager malt (two-row)	American crystal 40°L
Lager malt (six-row)	American crystal 60°L
	American crystal 80°L
Other malts	American crystal 90°L
	American crystal 120°L
American wheat	British Light Carastan
American soft white wheat	British Carastan
Belgian wheat	British crystal 50–60°L
German wheat	British crystal 70–80°L
German dark wheat	British crystal 95–115°L
American rye	British crystal 135–165°L
	German Carahell
Specialty malts	German light caramel
	German dark caramel
Light	German wheat caramel
Sauer (acid)	Belgian CaraVienne
American Vienna	Belgian CaraMunich
German Vienna	Belgian Special "B"
British mild	Roasted Malts
German smoked (Bamberg)	American chocolate
British peated	Belgian chocolate
Scottish peated	British chocolate
	German black Caraffe
Dark	German Caraffe special
American victory	American black patent
British brown	British black patent
Belgian biscuit	Belgian black
Belgian aromatic	
British amber	*Unmalted*
Canadian honey	
American two-row (toasted)	American roasted barley
American special roast	American black barley
Melanoidin	Belgian roasted barley
Belgian Munich	British roasted barley
German Munich	German roasted wheat
American Munich (light)	Roasted rye
	German Caraffe chocolate

TABLE 2.3

Fermentable Grain and Sugar Adjuncts

Grains
Barley, raw or flaked
Corn/maize grits
Corn/maize flaked
Wheat, raw or flaked
Rice, raw or flaked
Millet/sorghum raw
Rye, raw or flaked
Oats, raw or flaked
Sugars
Cane or beet sugar
Belgian candy sugar
Corn sugar, dextrose
Invert sugars
Honey
Maple syrup
Corn syrups
Molasses
Rice extract

world, the resulting hop varies in character. Certain styles of beer derive their unique qualities from certain varieties of hops grown in specific areas of the world. Hop extracts, while contributing to the perception of bitterness and balance between sweetness and bitterness, do not contribute to flavor and aroma characters that are so important in the profile of many beer styles.

Yeast

Most beer is made from one of two different yeast types: lager yeast or ale yeast (originally referred to as *Saccharomyces uvarum* and *Saccharomyces cerevisiae*, respectively; see Chapter 8). There are hundreds of strains of these two types of yeast. When used in a traditional manner, a particular strain of yeast will behave somewhat predictably, producing distinctive characters in beer. For example, particular strains of yeast may produce desired levels of estery/fruity character, higher alcohols, sulfur compounds, diacetyl compounds (butterscotch or butter-like flavor and aroma character), and phenolic compounds (often described as clove-like and smoky).

Processing

There are hundreds of processing variables that can influence the perceived flavor, aroma, appearance, mouthfeel, balance, and overall character of beer. A few basic examples of process variables are:

1. Equipment configuration and design

2. Grain milling
3. Mashing
4. Lautering
5. Boiling — type and length
6. Temperature of fermentation
7. Time of maturation
8. Filtration

Equipment Configuration

The kettle design will influence the wort boiling characteristics and affect various aspects of the final beer, such as color (degree of caramelization), product stability (clarity and oxidation-related flavor and aroma compounds), and hop utilization (bitterness, flavor, and aroma). Similarly, fermenter design will influence the overall profile of a beer. For example, square fermenters versus cylindroconical shapes will affect yeast behavior, which in turn will affect many easily perceived beer characters. The size of fermenters will also influence yeast behavior and thus the production of by-products influencing flavor profile.

Milling

Type and extent of grain milling will affect beer character.

Mashing

The style of mashing (infusion, step infusion, and decoction methods; see Chapter 10) contributes to the final character of beer, such as alcohol levels, malt aroma, malt flavor character, mouthfeel due to residual dextrins, and flavor balance, especially between sweetness and bitterness. It will also affect the quality of the foam and head retention.

Lautering

Lautering methods will influence the final character of beer. Among several other factors, grain-bed depth, degree of raking, agitation, and temperature of sparge (rinsing) water will influence the ester (fruity character) level in the final product.

Boiling Time

The boiling time of hops will influence hop bitterness, flavor, and aroma (see Chapter 7). The vigor of the boil will affect various qualities (including stability) of the final product.

Fermentation Temperature

The temperature of fermentation will dramatically affect the behavior of yeast and influence many flavor and aroma qualities through the production or suppression of various compounds.

Maturation Time

Periods of cold lagering and "cellaring" of ales will influence the final quality and balance of beer.

Filtration

The final product is often filtered for clarity and to contribute to the microbiological stability of some beer types. However, filtration removes yeast and other flavor compounds critically important to the mouthfeel, flavor, and aroma profile of certain beer types.

Packaging

There are several packaging variables that can dramatically alter the overall character and stability of the end product. A few examples of these variables are associated with the following:

- Packaging lines/systems
- Bottles, cans, and polyethylene terephthalate (PET) containers
- Kegs/draft
- Bulk delivery
- Pasteurization

Packaging lines will introduce varying levels of oxygen into the final product, thus influencing stability of flavor. Packaging in clear or green glass bottles will influence access of light and consequently, the stability and character of flavor and aroma. Pasteurization will affect the quality of the product. Some beer styles are more sensitive than others. Packaging in a keg will influence the overall quality of the product, through the introduction of less oxygen per volume. Keg beer is usually not pasteurized, demanding increased attention and care through required refrigeration in the marketplace.

Marketing

Marketing creates a perception affecting not only how the beer is consumed but also helps create the mood and environment, which in turn encourages where and how beer is consumed. Marketing has a huge role in presenting a beer style and its subsequent culture in the consumer's mind.

For example, beer may be marketed in a manner encouraging consumption directly out of the bottle or can. The character of the product consumed in this manner is far different from consumption in a glass or served as a draft from the tap. The shape of the glass from which beer is consumed greatly influences the perception of the product and can dramatically define and accent the character of the style.

Cultural

Political, social, and religious factors have significant influence on beer character. Here are a few examples:

- The German Purity Law of 1516, often referred to as the Reinheitsgebot, had a major influence in developing styles of all-malt beer. The Nigerian restrictions on barley importation have resulted in the development of light lagers brewed from malted sorghum and the tax regime in Japan led to the development of soshu.
- The colonizing period of British culture that began centuries ago created unique styles of beer, which would be more stable during their exportation to other parts of the world.
- In cultures where alcohol has been stigmatized as a societal problem and taxed at extreme rates, brewers have developed types of beer to stay within the limits of the laws.
- The monasteries and other religious cultures of Western Europe developed types of beer that harmonized with their religious observances.

Origins of Style

So far, we have briefly examined several elements influencing the character of a beer. How does a particular character and resulting beer type emerge? What are its origins? From where have beer styles evolved? Every beer style we choose to recognize has emerged in its own unique time. All of the classic beer styles we know today were at one time in history a unique concept — an anomaly. People considered today's classic and popular beers crazy and strange at the time of their introduction. This is not unlike the origins of any stylistic art form.

For example, today's style of German bock beer is said to have originated as a style of dark, strong beer unique to the town of Einbeck. At the same time, a duke of an area south of Einbeck enjoyed this beer so much he had it exported to his hometown and eventually had it brewed locally. Its success as an "exported" beer could be attributed to its rich malty character and alcoholic strength, which helped preserve the character of the beer even when shipped relatively great distances in those times.

India pale ale is a style of ale brewed for a special need. During the colonizing period of the British Empire, it was necessary to ship beer great distances. In this particular instance, the style was pale ale and the destination India. The qualities of pale ale at normal strength and hopping could not survive the long sea voyage to India. Brewers developed a style of stronger pale ale with a high degree of hopping. The higher alcoholic strength and hop rate (resulting in an unusually bitter ale) helped preserve the beer during its journey. The original intention was to dilute this strong hoppy ale upon reaching its destination, but consumers expressed their preference otherwise.

Stout is dark ale whose origins lie with British porter, perhaps the first "national" style of beer in Britain (see Chapter 1). The current day style of stout emerged from the descriptive use of the word stout to describe a fuller flavored and stronger version of porter — stout porter. Over the years, the style of stout evolved into dark ale brewed with the addition of the distinctive character of unmalted roasted barley in the classic manner developed and still used by Guinness, and now by other brewers worldwide.

Pilsener lager was first brewed in the town of Plzen, now in the Czech Republic. It was original and unique because it was a cold-fermented, very pale, lager beer brewed with very soft water. Most other beers of the time were top-fermented, darker ales. Pilsener lager became popular as an original style, and several other breweries throughout Europe copied it.

These are four examples of classic beer styles that have survived through the years. There are several other types of beers that have been introduced which did not survive the test of time and popularity. One interesting example is a recipe for Cock Ale I discovered in the book *The Flowing Bowl*, published in Britain in 1899. Here is the recipe:

> In order to make this take 10 gallons of ale and a large cock, the older the better. Parboil the cock, flay him and stamp him in a stone mortar till his bones are broken (you must craw and gut him when you flay him), then put the cock into 2 quarts of sac [sixteenth-century dry Spanish white wine], and put to it 3 pounds of raisins of the sun stoned, some blades of mace, a few cloves; put all these into a canvas bag, and a little before you find the ale has done working, put the ale and a bag together into a vessel; in a week or 9 days bottle it up, fill the bottles but just above the neck, and give it the same time to ripen as other ale.

Who can we thank for sparing us the indulgences of this ill-fated tradition?

The principal point in these illustrations is that Bock beer, India pale ale, Stout, and Pilsener were once unique concepts in their time. Now, these styles of beer have a traditional value and are an asset in today's world of beverages.

But what is the value in being aware of the origins of beer traditions and styles? Beer traditions serve as a focus for developing and maintaining pride and history within the beer industry. They may also serve to establish consumer pride. With the development of stylistic awareness in the beer market, beer moves away from being viewed as simply a commodity or just another fad, trendy alcoholic beverage. Many brewmasters and beer enthusiasts already appreciate this, but consumers in most beer markets have little knowledge about the richness of beer's culture.

Beer tradition and stylistic awareness develop passion for the product. Subsequently, beer drinkers and beer enthusiasts become more actively involved in promoting a healthy beer culture. With the appreciation of beer styles and traditions, opportunities for developing unique products emerge and niche markets can be more easily addressed or developed. More opportunities can be created in an increasingly competitive market.

Progress in today's marketplace, however, cannot be addressed by simply modernizing existing facilities. Traditions and recognition of cultural uniqueness must be maintained. One can often observe brewing companies seeking to modernize their operations. In doing so, they have been able to produce, package, and distribute their products more efficiently, but the unique beers of their culture are often sacrificed and lost at the expense of improving the production standards of operation. Therein lies the potential for failure, particularly for a smaller brewing company. With modernization, there is a trend to produce lager beers similar to the competition in neighboring markets. At the same time, as trade barriers are reduced, markets that were traditionally viewed as foreign markets are not so foreign any longer. The strength of any brewing company becomes measured by how it is able to progress with modernization while maintaining an awareness of their unique brewing traditions. Brewing traditions are in danger of withering away in small, neglected, countryside breweries. At what ultimate cost will this be to the beer industry? The larger question becomes: What passion is there left for the consumer if all beers develop a similarity of style?

Analytical and Sensor Variables

There exist several quantitative variables that differentiate one beer from another and that help define beer styles. A list of variables measured analytically by Professor Anton Peindl of the Institut für Bräuerei-Technologie und Mickrobiologie der Technischen Universität München at Weihenstephan is given in Table 2.4. These measurements are a foundation from which we

TABLE 2.4

Beer Composition Parameters Used for Classification into Styles

Original gravity (balling)	Organic acids
Alcohol	Pyruvate
Real extract	Citrate
Water content	Malate
Caloric content	L-Lactate
Protein	D-Lactate
Raw protein	Acetate
Free amino nitrogen	Gluconate
Proline	Total polyphenols
Minerals	Anthocyanogens
Potassium	Bitterness
Sodium	Dissolved carbon dioxide
Calcium	Fermentation by-products
Magnesium	Glycerol
Total phosphorus	n-Propyl alcohol
Sulfate	i-Butyl alcohol
Chloride	i-Amyl alcohols
Silicate	2-Phenylethanol
Nitrate	Ethyl acetate
Copper	i-Amyl acetate
Iron	Acetaldehyde
Manganese	Diacetyl
Zinc	2,3-Pentanedione
Vitamins	Total sulfur dioxide
Thiamin	Hydrogen ion concentration (pH)
Riboflavin	Viscosity
Pyridoxine	Apparent degree of fermentation
Pantothenic acid	Attenuation limit, apparent
Niacin	Color
Biotin	

can begin to define beer types. Professor Peindl's work was published over a 13-year period during the 1980s and 1990s in the German brewing journal *Brauindustrie*.

In the United States, several serious beer competitions emerged during the 1980s, creating a need to outline a method to evaluate beer qualities with human senses. Here is an outline of those variables that helped evaluators in beer competitions differentiate one beer from another and helped define beer styles.

Appearance
  Clarity
  Foam quality
  Head retention
  Color

Aroma
    Malt related
    Hop related
    Fermentation related
    Packaging and handling related
    Age related

Flavor
    Malt related
    Hop related
    Fermentation related
    Packaging and handling related
    Age related
    Balance
    Conditioning (level of carbonation)
    Aftertaste

Mouthfeel/body (perceived viscosity)

Drinkability and overall impression

Methods Used to Define Brewers Association's Beer Style Guidelines

The origins of the Brewers Association Beer Style Guidelines developed from the need to develop standards of evaluation for judging a variety of beers in the Association's Great American Beer Festival, World Beer Cup, and other beer competitions throughout the world. Ever since the mid-1980s, these guidelines have evolved, with increased knowledge of traditions and the change in consumer trends. These guidelines emerged with the consideration of numerous brewing factors, especially technical, psychological, marketplace realities, analytical, cultural, historical, and practical.

The emergence and the continuing evolution of these guidelines involved the integration of four important foundations:

1. Analysis: a review of commercially produced examples of classic beer styles
2. Tasting and recording: years of sensory experiences of all beer types
3. Brewing experience
4. Annually submitted suggestions from international brewing industry professionals

Analysis

An initial comprehensive review of the data in *Brauindustrie* resulted in formulating an abbreviated list of factors considered to be the most important bases for determining perceived and analytical variation among beers. Fermentation by-products were mostly recognized through sensory perception of the beer and often confirmed with a closer observation of the data. From the extensive list given earlier, the chosen factors were:

- Color
- Original gravity (balling)
- Real extract
- Apparent degree of fermentation
- Attenuation limit, apparent
- Alcohol
- Bitterness
- Dissolved carbon dioxide
- Fermentation by-products
  - Diacetyl
  - Various esters
  - Various alcohol types
- Viscosity

With these parameters in mind, the data from a group of beers considered representative of their style were carefully examined and statistically considered to represent the general limits of the style in each field of analysis. The beers chosen were generally regarded to be classic examples of a style, popular styles in the current market or representative of current trends.

Tasting and Recording

There was an initial core committee involved in developing a structure for style descriptors. This involved professional brewers, active beer judges, authors, and experts in the field of beer evaluation and history. Michael Jackson and I, provided years of tasting, evaluating, and judging experience in the early years, and continue to do so. This combination of industry professionals and recognized experts provided invaluable experience for developing a language and format for the presentation of style guidelines.

Brewing Beer

As "project director," my years of actively brewing (since 1970) on a small, pilot brewing scale, 20–21 batches per year, and traveling the world, observing commercial brewing on large and small scales helped provide the foundation for my understanding the basis of beer style formulation and the effects of process, equipment, and packaging variables on the character of beer. My degree in engineering also fostered a respect for the contributions of science and engineering to beer quality.

Regular Suggestions from Brewing Industry Professionals

At least once a year, brewing industry professionals continue to submit their comments on the accuracy of style guidelines. Recognizing that some style definitions continue to evolve or may not be totally accurate, each style is subject to review and revision based on suggestions, commentary, and substantiation provided by beer judges involved in the professional brewing industry.

In summary, Style Guidelines have been developed and maintained on an annual basis using the laboratory, human senses, an awareness of the complexities of brewing, and including a system of annual review by industry professionals. The results of this research have evolved over the years to the current state and are presented here with the caveat that they will continue to evolve and develop with the consideration of information provided by brewers and beer enthusiasts from around the world. This list is currently a basis for many competitions' guidelines and is presented annually as the "Brewers Association's Beer Style Guidelines." The most current version can be accessed through the Brewers Association's website http://www.beertown.org.

This presentation of beer style guidelines is a unique assimilation of accurate scientific data and experiential information presented in clear language and accessible to all.

Brewers Association 2005 Beer Style Guidelines

Since 1979, the Brewers Association has provided beer style descriptions as a reference for brewers and beer competition organizers. Much of the early work was based on the assistance and contributions of beer journalist Michael Jackson. The task of creating a realistic set of guidelines is always complex. The beer style guidelines developed by the Brewers Association use sources from the commercial brewing industry, beer analyses, and consultations with beer industry experts and knowledgeable beer enthusiasts as resources for information.

The Brewers Association's beer style guidelines reflect, as much as possible, historical significance, authenticity or a high profile in the current commercial beer market. Often, the historical significance is not clear, or a new beer in a current market may be only a passing fad, and thus, quickly forgotten. For these reasons, the addition of a style or the modification of an existing one is not undertaken lightly and is the product of research, consultation, and consideration of market actualities, and may take place over a period of time. Another factor considered is that current commercial examples do not always fit well into the historical record, and instead represent a modern version of the style. Our decision to include a particular historical beer style takes into consideration the style's brewing traditions and the need to preserve those traditions in today's market. The more a beer style has withstood the test of time, marketplace, and consumer acceptance, the more likely it is to be included in the Brewers Association's style guidelines.

The availability of commercial examples plays a large role in whether a beer style "makes the list." It is important to consider that not every historical or commercial beer style can be included, nor is every commercial beer representative of the historical tradition (i.e., a brewery labeling a brand as a particular style does not always indicate a fair representation of that style).

Please note that almost all of the classic and traditional beer style guidelines have been cross-referenced with data from commercially available beers representative of the style. The data referenced for this purpose have been from Professor Anton Piendl's comprehensive work published in the German *Brauindustrie* magazine through the years 1982 to 1994, from the series "Biere Aus Aller Welt."

Each style description is purposefully written independently of any reference to another beer style. Furthermore, as much as it is possible, beer character is not described in terms of ingredients or process. These guidelines attempt to emphasize the final evaluation of the product and try not to judge or regulate the formulation or manner in which it was brewed, except in special circumstances that clearly define a style.

If you have suggestions for adding or changing a style guideline, write to us, making sure to include reasons and documentation for why you think the style should be included.

The bitterness specifications (IBUs) given in these guidelines are based on standard measurements for bitterness derived from kettle isomerization of naturally occurring alpha acids. As reduced isomerized hop extracts may produce substantially different perceived bitterness levels when measured by this technique, brewers who use such extracts should consider the perceived bitterness present in the finished product.

The Beer Styles

The full list of beer styles, updated annually is provided at http://www.beertown.org/education/styles.html. Here, we provide only some of the major styles to give the reader an idea of the comprehensive nature of this valuable resource.

Ales

British Origin

Classic English-Style Pale Ale

Classic English pale ales are golden to copper colored and display English-variety hop character. Medium to high hop bitterness, flavor, and aroma should be evident. This medium-bodied pale ale has low to medium malt flavor and aroma. Low caramel character is allowable. Fruity-ester flavors and aromas are moderate to strong. Chill haze is allowable at cold temperatures. The absence of diacetyl is desirable, though diacetyl (butterscotch character) is acceptable and characteristic when at very low levels.

Original Gravity (°Plato): 1.044–1.056 (11–14°Plato)

Apparent extract/final gravity (°Plato): 1.008–1.016 (2–4°Plato)

Alcohol by weight (volume): 3.5–4.2% (4.5–5.5%)

Bitterness (IBU): 20–40

Color SRM (EBC): 5–14 (10–28 EBC)7

English-Style India Pale Ale

India pale ales are characterized by medium–high hop bitterness with a medium to high alcohol content. Hops from English origins are used to contribute to a high hopping rate. The use of water with high mineral content results in a crisp, dry beer. This pale gold to deep copper-colored ale has a medium to high, flowery hop aroma and may have a medium to strong hop flavor (in addition to the hop bitterness). India pale ales possess medium maltiness and body. Fruity-ester flavors and aromas are moderate to very strong. Diacetyl can be absent or may be perceived at very low levels. Chill haze is allowable at cold temperatures. (English and citrus-like American hops are considered enough for justifying separate American-style IPA and English-style IPA categories or subcategories. Hops of other origins may be used for bitterness or approximating traditional American or English character. See American-style India pale ale.

Original gravity (°Plato): 1.050–1.064 (12.5–16°Plato)

Apparent extract/final gravity (°Plato): 1.012–1.018 (3–4.5°Plato)

Alcohol by weight (volume): 4–5.6% (5–7%)

Bitterness (IBU): 35–55

Color SRM (EBC): 6–14 (12–28 EBC)

Ordinary Bitter

Ordinary bitter is gold to copper colored with medium bitterness, light to medium body, and low to medium residual malt sweetness. Hop flavor and aroma character may be evident at the brewer's discretion. Mild carbonation traditionally characterizes draft-cask versions, but in bottled versions, a slight increase in carbon dioxide content is acceptable. Fruity-ester character and very low diacetyl (butterscotch) character are acceptable in aroma and flavor, but should be minimized in this form of bitter. Chill haze is allowable at cold temperatures. (English and American hop may be specified in subcategories).

Original gravity (°Plato): 1.033–1.038 (8–9.5°Plato)

Apparent extract/final gravity (°Plato): 1.006–1.012 (1.5–3°Plato)

Alcohol by weight (volume): 2.4–3.0% (3–3.7%)

Bitterness (IBU): 20–35

Color SRM (EBC): 8–12 (16–24 EBC)

Special Bitter or Best Bitter

Special bitter is more robust than ordinary bitter. It has medium body and medium residual malt sweetness. It is gold to copper colored with medium bitterness. Hop flavor and aroma character may be evident at the brewer's discretion. Mild carbonation traditionally characterizes draft-cask versions, but in bottled versions, a slight increase in carbon dioxide content is acceptable. Fruity-ester character is acceptable in aroma and flavor. Diacetyl (butterscotch character) is acceptable and characteristic when at very low levels. The absence of diacetyl is also acceptable. Chill haze is allowable at cold temperatures. (English and American hop may be specified in subcategories).

Original gravity (°Plato): 1.038–1.045 (9.5–11°Plato)

Apparent extract/final gravity (°Plato): 1.006–1.012 (1.5–3°Plato)

Alcohol by weight (volume): 3.3–3.8% (4.1–4.8%)

Bitterness (IBU): 28–46

Color SRM (EBC): 8–14 (16–28 EBC)

Scottish-Style Light Ale

Scottish light ales are light bodied. Little bitterness is perceived and hop flavor or aroma should not be perceived. Despite its lightness, Scottish light ale will have a degree of malty, caramel-like, soft, and chewy character. Yeast characters such as diacetyl (butterscotch) and sulfuriness are acceptable at very low levels. The color will range from golden amber to deep brown. Bottled versions of this traditional draft beer may contain higher amounts of carbon dioxide than is typical for mildly carbonated draft versions. Chill haze is acceptable at low temperatures. Though there is little evidence suggesting that traditionally made Scottish-style light ales exhibited peat smoke character, the current marketplace offers many Scottish-style light ales with peat or smoke character present at low to medium levels. Thus, a peaty/smoky character may be evident at low levels (ales with medium or higher smoke character would be considered a smoke flavored beer and considered in another category). Scottish-style light ales may be split into two subcategories: traditional (no smoke character) and peated (low level of peat smoke character).

> *Original gravity (°Plato)*: 1.030–1.035 (7.5–9°Plato)
> *Apparent extract/final gravity (°Plato)*: 1.006–1.012 (1.5–3°Plato)
> *Alcohol by weight (volume)*: 2.2–2.8% (2.8–3.5%)
> *Bitterness (IBU)*: 9–20
> *Color SRM (EBC)*: 8–17 (16–34 EBC)

Scottish-Style Heavy Ale

Scottish heavy ale is moderate in strength and dominated by a smooth, sweet maltiness balanced with low, but perceptible, hop bitterness. Hop flavor or aroma should not be perceived. Scottish heavy ale will have a medium degree of malty, caramel-like, soft, and chewy character in flavor and mouthfeel. It has medium body, and fruity esters are very low, if evident. Yeast characters such as diacetyl (butterscotch) and sulfuriness are acceptable at very low levels. The color will range from golden amber to deep brown. Bottled versions of this traditional draft beer may contain higher amounts of carbon dioxide than is typical for mildly carbonated draft versions. Chill haze is acceptable at low temperatures. Though there is little evidence suggesting that traditionally made Scottish-style heavy ales exhibited peat smoke character, the current marketplace offers many Scottish-style heavy ales with peat or smoke character present at low to medium levels. Thus, a peaty/smoky character may be evident at low levels (ales with medium or higher smoke character would be considered a smoke flavored beer and considered in another category). Scottish-style heavy ales may be split into two subcategories: traditional (no smoke character) and peated (low level of peat smoke character).

> *Original gravity (°Plato)*: 1.035–1.040 (9–10°Plato)
> *Apparent extract/final gravity (°Plato)*: 1.0010–1.014 (2.5–3.5°Plato)
> *Alcohol by weight (volume)*: 2.8–3.2% (3.5–4%)
> *Bitterness (IBU)*: 12–20
> *Color SRM (EBC)*: 10–19 (20–38 EBC)

Scottish-Style Export Ale

The overriding character of Scottish export ale is sweet, caramel-like, and malty. Its bitterness is perceived as low to medium. Hop flavor or aroma should not be perceived. It has medium body. Fruity-ester character may be apparent. Yeast characters such as diacetyl (butterscotch) and sulfuriness are acceptable at very low levels. The color will range from golden amber to deep brown. Bottled versions of this traditional draft beer may contain higher amounts of carbon dioxide than is typical for mildly carbonated draft versions. Chill haze is acceptable at low temperatures. Though there is little evidence suggesting that traditionally made Scottish-style export ales exhibited peat smoke character, the current marketplace offers many Scottish-style export ales with peat or smoke character present at low to medium levels. Thus, a peaty/smoky character may be evident at low levels (ales with medium or higher smoke character would be considered a smoke flavored beer and considered in another category). Scottish-style export ales may be split into two subcategories: traditional (no smoke character) and peated (low level of peat smoke character).

Original gravity (°Plato): 1.040–1.050 (10–12.5°Plato)
Apparent extract/final gravity (°Plato): 1.010–1.018 (2.5–4.5°Plato)
Alcohol by weight (volume): 3.2–4.2% (4.0–5.3%)
Bitterness (IBU): 15–25
Color SRM (EBC): 10–19 (20–38 EBC)

English-Style Dark Mild Ale

English dark mild ales range from deep copper to dark brown (often with a red tint) in color. Malt flavor and caramel are part of the flavor and aroma profile, while licorice and roast malt tones may sometimes contribute to the flavor and aroma profile. These beers have very little hop flavor or aroma. Very low diacetyl flavors may be appropriate in this low-alcohol beer. Fruity-ester level is very low.

Original gravity (°Plato): 1.030–1.036 (7.5–9°Plato)
Apparent extract/final gravity (°Plato): 1.004–1.008 (1–2°Plato)
Alcohol by weight (volume): 2.7–3.2% (3.2–4.0%)
Bitterness (IBU): 10–24
Color SRM (EBC): 17–34 (34–68 EBC)

English-Style Brown Ale

English brown ales range from deep copper to brown in color. They have a medium body and a dry to sweet maltiness with very little hop flavor or aroma. Roast malt tones may sometimes contribute to the flavor and aroma profile. Fruity-ester flavors are appropriate. Diacetyl should be very low, if evident. Chill haze is allowable at cold temperatures.

Original gravity (°Plato): 1.040–1.050 (10–12.5°Plato)
Apparent extract/final gravity (°Plato): 1.008–1.014 (2–3.5°Plato)
Alcohol by weight (volume): 3.3–4.7% (4–5.5%)
Bitterness (IBU): 15–25
Color SRM (EBC): 15–22 (30–44 EBC)

Imperial Stout

Dark copper to very black, imperial stouts typically have a high alcohol content. The extremely rich malty flavor and aroma are balanced with assertive hopping and fruity-ester characteristics. Bitterness can be moderate and balanced with the malt character or very high in the darker versions. Roasted malt astringency and bitterness can be moderately perceived but should not overwhelm the overall character. Hop aroma can be subtle to overwhelmingly hop-floral, -citrus, or -herbal. Diacetyl (butterscotch) levels should be very low. This style may be subcategorized into black and quite robust "American" versions and dark copper colored and caramel accented "European" versions.

Original gravity (°Plato): 1.080–1.100 (19.5–23°Plato)
Apparent extract/final gravity (°Plato): 1.020–1.030 (4–7.5°Plato)
Alcohol by weight (volume): 5.5–9.5% (7–12%)
Bitterness (IBU): 50–80
Color SRM (EBC): 20+ (40+ EBC)

Other Strong Ales or Lagers

Any style of beer can be made stronger than the classic style guidelines. The goal should be to reach a balance between the style's character and the additional alcohol. Refer to this guide when making styles stronger and appropriately identify the style created (e.g., double alt, triple fest, or quadruple Pilsener).

English-Style Barley Wine Ale

English-style barley wines range from tawny copper to dark brown in color and have a full body and high residual malty sweetness. Complexity of alcohols and fruity-ester characters are often high and counterbalanced by the perception of low to medium bitterness and extraordinary alcohol content. Hop aroma and flavor may be minimal to medium. Low levels of diacetyl may be acceptable. A caramel and vinous (sometimes sherrylike) aroma and flavor are part of the character. Chill haze is allowable at cold temperatures.

Original gravity (°Plato): 1.085–1.120 (21.5–28°Plato)
Apparent extract/final gravity (°Plato): 1.024–1.032 (6–8°Plato)
Alcohol by weight (volume): 6.7–9.6% (8.4–12%)
Bitterness (IBU): 40–60
Color SRM (EBC): 14–22 (28–44 EBC)

Irish Origin

Irish-Style Red Ale

Irish-style red ales range from light red–amber–copper to light brown in color. These ales have a medium hop bitterness and flavor. They often do not have hop aroma. Irish-style red ales have low to medium candy-like caramel sweetness and a medium body. The style may have low levels of fruity-ester flavor and aroma. Diacetyl should be absent. Chill haze is allowable at cold temperatures. Slight yeast haze is acceptable for bottle-conditioned products.

Original gravity (°Plato): 1.040–1.048 (10–12°Plato)
Apparent extract/final gravity (°Plato): 1.010–1.014 (2.5–3.5°Plato)
Alcohol by weight (volume): 3.2–3.6% (4–4.5%)
Bitterness (IBU): 20–28
Color SRM (EBC): 11–18 (22–36 EBC)

Classic Irish-Style Dry Stout

Dry stouts have an initial malt and light caramel flavor profile with a distinctive dry-roasted bitterness in the finish. Dry stouts achieve a dry-roasted character through the use of roasted barley. The emphasis of coffee-like roasted barley and a moderate degree of roasted malt aromas define much of the character. Some slight acidity may be perceived but is not necessary. Hop aroma and flavor should not be perceived. Dry stouts have medium-light to medium body. Fruity esters are minimal and overshadowed by malt, high hop bitterness, and roasted barley character. Diacetyl (butterscotch) should be very low or not perceived. Head retention and rich character should be part of its visual character.

Original gravity (°Plato): 1.038–1.048 (9.5–12°Plato)
Apparent extract/final gravity (°Plato): 1.008–1.012 (2–3°Plato)

Alcohol by weight (volume): 3.2–4.2% (3.8–5%)
Bitterness (IBU): 30–40
Color SRM (EBC): 40+ (80+ EBC)

North American Origin

American-Style Pale Ale

American pale ales range from deep golden to copper in color. The style is characterized by American-variety hops used to produce high hop bitterness, flavor, and aroma. American pale ales have medium body and low to medium maltiness. Low caramel character is allowable. Fruity-ester flavor and aroma should be moderate to strong. Diacetyl should be absent or present at very low levels. Chill haze is allowable at cold temperatures.

Original gravity (°Plato): 1.044–1.050 (11–12.5°Plato)
Apparent extract/final gravity (°Plato): 1.008–1.014 (2–3.5°Plato)
Alcohol by weight (volume): 3.5–4.3% (4.5–5.5%)
Bitterness (IBU): 28–40
Color SRM (EBC): 6–14 (12–28 EBC)

American-Style India Pale Ale

American-style India pale ales are characterized by intense hop bitterness with a high alcohol content. Hops from American origins are used to contribute to a high hopping rate. The use of water with high mineral content results in a crisp, dry beer. This pale gold to deep copper-colored ale has a full, flowery hop aroma and may have a strong hop flavor (in addition to the hop bitterness). India pale ales possess medium maltiness and body. Fruity-ester flavors and aromas are moderate to very strong. Diacetyl can be absent or may be perceived at very low levels. Chill haze is allowable at cold temperatures. (English and citrus-like American hops are considered enough of a distinction justifying separate American- and English-style India pale ale categories or subcategories. Hops of other origins may be used for bitterness or approximating traditional American or English character. See English-style India pale ale.)

Original gravity (°Plato): 1.050–1.070 (12.5–17°Plato)
Apparent extract/final gravity (°Plato): 1.012–1.018 (3–4.5°Plato)
Alcohol by weight (volume): 4–6% (5–7.5%)
Bitterness (IBU): 40–65
Color SRM (EBC): 6–14 (12–28 EBC)

American-Style Amber/Red Ale

American amber/red ales range from light copper to light brown in color. They are characterized by American-variety hops used to produce high hop bitterness, flavor, and medium to high aroma. Amber ales have medium–high to high maltiness with medium to low caramel character. They should have medium to medium–high body. The style may have low levels of fruity-ester flavor and aroma. Diacetyl can be either absent or barely perceived at very low levels. Chill haze is allowable at cold temperatures. Slight yeast haze is acceptable for bottle-conditioned products.

Original gravity (°Plato): 1.048–1.058 (12–14.5°Plato)
Apparent extract/final gravity (°Plato): 1.012–1.018 (3–4.5°Plato)
Alcohol by weight (volume): 3.5–4.8% (4.5–6%)
Bitterness (IBU): 30–40
Color SRM (EBC): 11–18 (22–36 EBC)

Golden or Blonde Ale

Golden or Blonde ales are straw to golden blonde in color. They have a crisp, dry palate, low hop floral aroma, light to medium body, and light malt sweetness. Bitterness is low to medium. Fruity esters may be perceived but do not predominate. Diacetyl should not be perceived. Chill haze should be absent.

Original gravity (°Plato): 1.045–1.056 (11–14°Plato)
Apparent extract/final gravity (°Plato): 1.008–1.016 (2–4°Plato)
Alcohol by weight (volume): 3.2–4% (4–5%)
Bitterness (IBU): 15–25
Color SRM (EBC): 3–7 (6–14 EBC)

American-Style Stout

Initial low to medium malt sweetness with a degree of caramel, chocolate, and roasted coffee flavor with a distinctive dry-roasted bitterness in the finish. Coffee-like roasted barley and roasted malt aromas are prominent. Some slight roasted malt acidity is permissible and a medium- to full-bodied mouthfeel is appropriate. Hop bitterness may be moderate to high. Hop aroma and flavor is moderate to high, often with American citrus-type and resiny hop character. The perception of fruity esters is low. Roasted malt/barley astringency may be low but not excessive. Diacetyl (butterscotch) should be negligible or not perceived. Head retention is excellent.

Original gravity (°Plato): 1.050–1.075 (13–17.5°Plato)
Apparent extract/final gravity (°Plato): 1.010–1.022 (2.5–5.5°Plato)

Alcohol by weight (volume): 4.5–7% (5.7–8.8%)
Bitterness (IBU): 35–60
Color SRM (EBC): 40+ (80+ EBC)

German Origin

German-Style Brown Ale/Düsseldorf-Style Altbier

Copper to brown in color, this German ale may be highly hopped and intensely bitter (although the 25–35 IBU range is more normal for the majority of Altbiers from Düsseldorf) and has a medium body and malty flavor. A variety of malts, including wheat, may be used. Hop character may be medium to high in the flavor and aroma. The overall impression is clean, crisp, and flavorful, often with a dry finish. Fruity esters can be low to medium–low. No diacetyl or chill haze should be perceived.

Original gravity (°Plato): 1.044–1.048 (11–12°Plato)
Apparent extract/final gravity (°Plato): 1.008–1.014 (2–3.5°Plato)
Alcohol by weight (volume): 3.6–4% (4.3–5%)
Bitterness (IBU): 25–48
Color SRM (EBC): 11–19 (22–38 EBC)

South German-Style Hefeweizen/Hefeweissbier

The aroma and flavor of a Weissbier with yeast is decidedly fruity and phenolic. The phenolic characteristics are often described as clove- or nutmeg-like and can be smoky or even vanilla-like. Banana-like esters are often present. These beers are made with at least 50% malted wheat, and hop rates are quite low. Hop flavor and aroma are absent. Weissbier is well attenuated and very highly carbonated, yet its relatively high starting gravity and alcohol content make it a medium- to full-bodied beer. The color is very pale to pale amber. Because yeast is present, the beer will have yeast flavor and a characteristically fuller mouthfeel and may be appropriately very cloudy. No diacetyl should be perceived. (The brewer may indicate a desire that the yeast be either poured or not poured when the beer is served.)

Original gravity (°Plato): 1.047–1.056 (11.8–14°Plato)
Apparent extract/final gravity (°Plato): 1.008–1.016 (2–4°Plato)
Alcohol by weight (volume): 3.9–4.4% (4.9–5.5%)
Bitterness (IBU): 10–15
Color SRM (EBC): 3–9 (6–18 EBC)

Beer Styles: Their Origins and Classification

South German-Style Kristal Weizen/Kristal Weissbier

The aroma and flavor of a Weissbier without yeast is very similar to Weissbier with yeast (Hefeweizen/Hefeweissbier) with the caveat that fruity and phenolic characters are not combined with the yeasty flavor and fuller-bodied mouthfeel of yeast. The phenolic characteristics are often described as clove- or nutmeg-like and can be smoky or even vanilla like. Banana-like esters are often present. These beers are made with at least 50% malted wheat, and hop rates are quite low. Hop flavor and aroma are absent. Weissbier is well attenuated and very highly carbonated, yet its relatively high starting gravity and alcohol content make it a medium- to full-bodied beer. The color is very pale to deep golden. Because the beer has been filtered, yeast is not present. The beer will have no flavor of yeast and a cleaner, drier mouthfeel. The beer should be clear with no chill haze present. No diacetyl should be perceived.

Original gravity (°Plato): 1.047–1.056 (11.8–14°Plato)
Apparent extract/final gravity (°Plato): 1.008–1.016 (2–4°Plato)
Alcohol by weight (volume): 3.9–4.4% (4.9–5.5%)
Bitterness (IBU): 10–15
Color SRM (EBC): 3–9 (6–18 EBC)

South German-Style Dunkel Weizen/Dunkel Weissbier

This beer style is characterized by a distinct sweet maltiness and a chocolate like character from roasted malt. Estery and phenolic elements of this Weissbier should be evident but subdued. Color can range from copper-brown to dark brown. Dunkel Weissbier is well attenuated and very highly carbonated, and hop bitterness is low. Hop flavor and aroma are absent. Usually dark barley malts are used in conjunction with dark cara or color malts, and the percentage of wheat malt is at least 50%. If this is served with yeast, the beer may be appropriately very cloudy. No diacetyl should be perceived. (The brewer may indicate a desire that the yeast be either poured or not poured when the beer is served.)

Original gravity (°Plato): 1.048–1.056 (12–14°Plato)
Apparent extract/final gravity (°Plato): 1.008–1.016 (2–4°Plato)
Alcohol by weight (volume): 3.8–4.3% (4.8–5.4%)
Bitterness (IBU): 10–15
Color SRM (EBC): 10–19 (20–38 EBC)

Belgian and French Origin

Belgian-Style Flanders/Oud Bruin or Oud Red Ales

This light- to medium-bodied deep copper to brown ale is characterized by a slight to strong lactic sourness and spiciness. A fruity-ester character is apparent with no hop flavor or aroma. Flanders brown ales have low to medium bitterness. Very small quantities of diacetyl are acceptable. Roasted malt character in aroma and flavor is acceptable at low levels. Oak-like or woody characters may be pleasantly integrated into overall palate. Chill haze is acceptable at low serving temperatures. Some versions may be more highly carbonated and, when bottle conditioned, may appear cloudy (yeast) when served.

Original gravity (°Plato): 1.044–1.056 (11–14°Plato)
Apparent extract/final gravity (°Plato): 1.008–1.016 (2–4°Plato)
Alcohol by weight (volume): 3.8–4.4% (4.8–5.2%)
Bitterness (IBU): 15–25
Color SRM (EBC): 12–20 (24–40 EBC)

Belgian-Style Dubbel

This medium- to full-bodied, dark amber to brown-colored ale has a malty sweetness and nutty, chocolate-like, and roast malt aroma. A faint hop aroma is acceptable. Dubbels are also characterized by low bitterness and no hop flavor. Very small quantities of diacetyl are acceptable. Yeast-generated fruity esters (especially banana) are appropriate at low levels. Head retention is dense and mousse like. Chill haze is acceptable at low serving temperatures.

Original gravity (°Plato): 1.050–1.070 (12.5–17°Plato)
Apparent extract/final gravity (°Plato): 1.012–1.016 (3–4°Plato)
Alcohol by weight (volume): 4.8–6.0% (6.0–7.5%)
Bitterness (IBU): 18–25
Color SRM (EBC): 14–18 (28–36 EBC)

Belgian-Style Tripel

Tripels are often characterized by a complex, spicy, phenolic flavor. Yeast-generated fruity banana esters are also common, but not necessary. These pale/light-colored ales may finish sweet, though any sweet finish can be light on the palate. The beer is characteristically medium to full bodied with a neutral hop/malt balance. Brewing sugar may be used to lighten the perception of body. Its sweetness will come from very pale malts. There should not be character from any roasted or dark malts. Very low hop flavor is okay. Alcohol strength and flavor should be perceived as evident. Head retention is dense and mousse like. Chill haze is acceptable at low serving temperatures.

Original gravity (°Plato): 1.060–1.096 (15–24°Plato)
Apparent extract/final gravity (°Plato): 1.008–1.020 (2–5°Plato)

Alcohol by weight (volume): 5.6–8.0% (7.0–10.0%)
Bitterness (IBU): 20–25
Color SRM (EBC): 3.5–7 (7–14 EBC)

Belgian-Style Pale Ale
Belgian-style pale ales are characterized by low, but noticeable, hop bitterness, flavor, and aroma. Light to medium body and low malt aroma are typical. They are golden to deep amber in color. Noble-type hops are commonly used. Low to medium fruity esters are evident in aroma and flavor. Low levels of phenolic spiciness from yeast byproducts may be perceived. Low caramel or toasted malt flavor is okay. Diacetyl should not be perceived. Chill haze is allowable at cold temperatures.

Original gravity (°Plato): 1.044–1.054 (11–13.5°Plato)
Apparent extract/final gravity (°Plato): 1.008–1.014 (2–3.5°Plato)
Alcohol by weight (volume): 3.2–5.0% (4.0–6.0%)
Bitterness (IBU): 20–30
Color SRM (EBC): 3.5–12 (7–24 EBC)

Belgian-Style White (or Wit)/Belgian-Style Wheat
Belgian white ales are very pale in color and are brewed using unmalted wheat and malted barley and are spiced with coriander and orange peel. Coriander and light orange peel aroma should be perceived. These beers are traditionally bottle conditioned and served cloudy. An unfiltered nearly opaque haze should be part of the appearance. The style is further characterized by the use of noble-type hops to achieve a low hop bitterness and little to no apparent hop flavor. This beer has low to medium body, no diacetyl, and a low to medium fruity-ester level. Mild acidity is appropriate.

Original gravity (°Plato): 1.044–1.050 (11–12.5°Plato)
Apparent extract/final gravity (°Plato): 1.006–1.010 (1.5–2.5°Plato)
Alcohol by weight (volume): 3.8–4.4% (4.8–5.2%)
Bitterness (IBU): 10–17
Color SRM (EBC): 2–4 (4–8 EBC)

Belgian-Style Lambic
Unblended, naturally and spontaneously fermented lambic is intensely estery, sour, and often, but not necessarily, acetic flavored. Low in carbon dioxide, these hazy beers are brewed with unmalted wheat and malted barley. Sweet malt characters are not perceived. They are very low in hop

bitterness. Cloudiness is acceptable. These beers are quite dry and light bodied. Characteristic horsiness (similar to wet horse blanket) from Brettanomyces yeast is often present at moderate levels. Versions of this beer made outside of the Brussels area of Belgium cannot be true lambics. These versions are said to be "lambic-style" and may be made to resemble many of the beers of true origin.

Original gravity (°Plato): 1.044–1.056 (11–14°Plato)
Apparent extract/final gravity (°Plato): 1.000–1.010 (0–2.5°Plato)
Alcohol by weight (volume): 4–5% (5–6%)
Bitterness (IBU): 11–23
Color SRM (EBC): 6–13 (12–26 EBC)

Lager Beers

European-Germanic Origin

German-Style Pilsener

A classic German Pilsener is very light straw or golden in color and well hopped. Hop bitterness is high. Noble-type hop aroma and flavor are moderate and quite obvious. It is a well-attenuated, medium-bodied beer, but a malty residual sweetness can be perceived in aroma and flavor. Low levels of sweet corn-like dimethylsulfide (DMS) character, if perceived, are characteristic of this style. Fruity esters and diacetyl should not be perceived. There should be no chill haze. Its head should be dense and rich.

Original gravity (°Plato): 1.044–1.050 (11–12.5°Plato)
Apparent extract/final gravity (°Plato): 1.006–1.012 (1.5–3°Plato)
Alcohol by weight (volume): 3.6–4.2% (4–5%)
Bitterness (IBU): 30–40
Color SRM (EBC): 3–4 (6–8 EBC)

Bohemian-Style Pilsener

Bohemian Pilseners are medium bodied, and they can be as dark as a light amber color. This style balances moderate bitterness and noble-type hop aroma and flavor with a malty, slightly sweet, medium body. Diacetyl may be perceived in extremely low amounts. A toasted-, biscuit-like malt character may be evident at low levels. There should be no chill haze. Its head should be dense and rich.

Original gravity (°Plato): 1.044–1.056 (11–14°Plato)
Apparent extract/final gravity (°Plato): 1.014–1.020 (3.5–5°Plato)

Alcohol by weight (volume): 3.2–4% (4–5%)
Bitterness (IBU): 30–45
Color SRM (EBC): 3–7 (6–14 EBC)

European-Style Pilsener

European Pilseners are straw/golden in color and are well attenuated. This medium-bodied beer is often brewed with rice, corn, wheat, or other grain or sugar adjuncts making up part of the mash. Hop bitterness is low to medium. Hop flavor and aroma are low. Residual malt sweetness is low; it does not predominate but may be perceived. Fruity esters and diacetyl should not be perceived. There should be no chill haze.

Original gravity (°Plato): 1.044–1.050 (11–12.5°Plato)
Apparent extract/final gravity (°Plato): 1.008–1.010 (2–2.5°Plato)
Alcohol by weight (volume): 3.6–4.2% (4–5%)
Bitterness (IBU): 17–30
Color SRM (EBC): 3–4 (6–8 EBC)

Traditional German-Style Bock

Traditional bocks are made with all malt and are strong, malty, medium-to full-bodied, bottom-fermented beers with moderate hop bitterness that should increase proportionately with the starting gravity. Hop flavor should be low and hop aroma should be very low. Bocks can range in color from deep copper to dark brown. Fruity esters should be minimal.

Original gravity (°Plato): 1.066–1.074 (16.5–18°Plato)
Apparent extract/final gravity (°Plato): 1.018–1.024 (4.5–6°Plato)
Alcohol by weight (volume): 5–6% (6–7.5%)
Bitterness (IBU): 20–30
Color SRM (EBC): 20–30 (40–60 EBC)

North American Origin

American Lager

Light in body and color, American lagers are very clean and crisp, and aggressively carbonated. Flavor components should be subtle and complex, with no one ingredient dominating the others. Malt sweetness is light to mild. Corn, rice, or other grain or sugar adjuncts are often used. Hop bitterness, flavor, and aroma are negligible to very light. Light fruity esters are acceptable. Chill haze and diacetyl should be absent.

Original gravity (°Plato): 1.040–1.046 (10–11.5°Plato)
Apparent extract/final gravity (°Plato): 1.006–1.010 (1.5–2.5°Plato)
Alcohol by weight (volume): 3.2–4.0% (3.8–5%)
Bitterness (IBU): 5–14
Color SRM (EBC): 2–4 (4–8 EBC)

American-Style Light Lager

According to the U.S. Food and Drug Administration (FDA) regulations, when used in reference to caloric content, "light" beers must have at least 25% fewer calories than the "regular" version of that beer. Such beers must have certain analysis data printed on the package label. These beers are extremely light colored, light in body, and high in carbonation. Corn, rice, or other grain or sugar adjuncts are often used. Flavor is mild and hop bitterness and aroma is negligible to very low. Light fruity esters are acceptable. Chill haze and diacetyl should be absent.

Original gravity (°Plato): 1.024–1.040 (6–10°Plato)
Apparent extract/final gravity (°Plato): 1.002–1.008 (0.5–2°Plato)
Alcohol by weight (volume): 2.8–3.5% (3.5–4.4%)
Bitterness (IBU): 5–10
Color SRM (EBC): 1.5–4 (3–8 EBC)

American-Style Low-Carbohydrate Light Lager

These beers are extremely light straw to light amber in color, light in body, and high in carbonation. They should have a maximum carbohydrate level of 3.0 g per 12 oz (356 ml). These beers are characterized by extremely low attenuation (often final gravity is less than 1.000 (0°Plato), but with typical American-style light lager alcohol levels. Corn, rice, or other grain adjuncts are often used. Flavor is very light/mild and very dry. Hop flavor, aroma, and bitterness is negligible to very low. Very low yeasty flavors and fruity esters are acceptable in aroma and flavor. Chill haze and diacetyl should not be perceived.

Original gravity (°Plato): 1.024–1.036 (6–9°Plato)
Apparent extract/final gravity (°Plato): 0.992–1.004 (−2 to 1°Plato)
Alcohol by weight (volume): 2.8–3.5% (3.5–4.4%)
Bitterness (IBU): 3–10
Color SRM (EBC): 1.5–10 (3–20 EBC)

American-Style Premium Lager

This style has low malt (and adjunct) sweetness, is medium bodied, and should contain no or a low percentage (less than 25%) of adjuncts. Color

Beer Styles: Their Origins and Classification

may be light straw to golden. Alcohol content and bitterness may also be greater. Hop aroma and flavor is low or negligible. Light fruity esters are acceptable. Chill haze and diacetyl should be absent. *Note:* Some beers marketed as "premium" (based on price) may not fit this definition.

Original gravity (°Plato): 1.044–1.048 (11–12°Plato)

Apparent extract/final gravity (°Plato): 1.010–1.014 (2.5–3.5°Plato)

Alcohol by weight (volume): 3.6–4% (4.3–5%)

Bitterness (IBU): 6–15

Color SRM (EBC): 2–6 (4–12 EBC)

American Ice Lager

This style is slightly higher in alcohol than most other light-colored, American-style lagers. Its body is low to medium and has low residual malt sweetness. It has few or no adjuncts. Color is very pale to golden. Hop bitterness is low but certainly perceptible. Hop aroma and flavor are low. Chill haze, fruity esters, and diacetyl should not be perceived. Typically, these beers are chilled before filtration so that ice crystals (which may or may not be removed) are formed. This can contribute to a higher alcohol content (up to 0.5% more).

Original gravity (°Plato): 1.040–1.060 (10–15°Plato)

Apparent extract/final gravity (°Plato): 1.006–1.014 (1.5–3.5°Plato)

Alcohol by weight (volume): 3.8–5% (4.6–6%)

Bitterness (IBU): 7–20

Color SRM (EBC): 2–8 (4–16 EBC)

Other Origin

Australasian, Latin American, or Tropical-Style Light Lager

Australasian, Latin American, or Tropical light lagers are very light in color and light bodied. Hop bitterness, aroma, and flavor should be negligible to very low. Sugar adjuncts are often used to lighten the body and flavor, sometimes contributing to a slight apple-like fruity ester. Sugar, corn, rice, and other cereal grains are used as an adjunct. Very low sweet corn-like DMS is sometimes characteristic of style. Chill haze and diacetyl should be absent. Fruity esters should be very low. Generally, these are highly carbonated.

Original gravity (°Plato): 1.032–1.046 (8–11.5°Plato)

Apparent extract/final gravity (°Plato): 1.004–1.010 (1–2.5°Plato)

Alcohol by weight (volume): 2.3–4.5% (2.9–5.6%)

Bitterness (IBU): 9–25

Color SRM (EBC): 2–4 (4–8 EBC)

Flavored Malt-Fermented Beverages

These are malt-fermented beverages that are enhanced with a variety of flavors, either natural or artificial. A minimum 51% of the fermentable carbohydrates must be derived from malted grains. Furthermore, the alcohol in these products must be 100% derived from the natural fermentation of the product. No distilled spirits in any form whatsoever are added to these products. Many of these beverages do not have any perceived hop character (and are most often made without hops). They are evaluated and assessed by their drinkability and balance of flavors. They most often do not have any beer-type character whatsoever.

Original gravity (°Plato): 1.030–1.110 (7.5–26°Plato)

Apparent extract/final gravity (°Plato): 1.006–1.030 (1.5–7.5°Plato)

Alcohol by weight (volume): 2–9.5% (2.5–12%)

Bitterness (IBU): 10 or less

Color SRM (EBC): 0–50 (10–100 EBC)

Vegetable Beers

Vegetable beers are any beers using vegetables as an adjunct in either primary or secondary fermentation, providing obvious (ranging from subtle to intense), yet harmonious, qualities. Vegetable qualities should not be overpowered by hop character. If a vegetable (such as chili pepper) has a herbal or spice quality it should be classified as herb/spice beer category. A statement by the brewer explaining what vegetables are used is essential in order for fair assessment in competitions. If this beer is a classic style with vegetables, the brewer should specify the classic style.

Original gravity (°Plato): 1.030–1.110 (7.5–26°Plato)

Apparent extract/final gravity (°Plato): 1.006–1.030 (1.5–7.5°Plato)

Alcohol by weight (volume): 2–9.5% (2.5–12%)

Bitterness (IBU): 5–70

Color SRM (EBC): 5–50 (10–100 EBC)

Herb and Spice Beers

Herb beers use herbs or spices (derived from roots, seeds, fruits, vegetables, flowers, etc.) other than or in addition to hops to create a distinct (ranging from subtle to intense) character, though individual characters of herbs and spices used may not always be identifiable. Under hopping often, but not always, allows the spice or herb to contribute to the flavor profile. Positive evaluations are significantly based on perceived balance of flavors. A statement by the brewer explaining what herbs or spices are used is essential in order for fair assessment in competitions. Specifying a style upon which the beer is based may help evaluation. If this beer is a classic style with an herb or spice, the brewer should specify the classic style.

Original gravity (°Plato): 1.030–1.110 (7.5–26°Plato)
Apparent extract/final gravity (°Plato): 1.006–1.030 (1.5–7.5°Plato)
Alcohol by weight (volume): 2–9.5% (2.5–12%)
Bitterness (IBU): 5–70
Color SRM (EBC): 5–50 (10–100 EBC)

Acknowledgments

These guidelines have been compiled for the Brewers Association by Charlie Papazian, copyright: 1993–2005. With Style Guideline Committee assistance and review by Ray Daniels, Paul Gatza, Chris Swersey, and suggestions from Great American Beer Festival and World Beer Cup judges.

Bibliography of Resources

The following books, magazines, and consultants were used to compile these style guidelines, along with personal knowledge. The guidelines have continually evolved through annual revisions recommended by colleagues worldwide.

Allen, F. and Cantwell, D., *Barley Wine*, Brewers Publications, Boulder, CO, 1998.
Daniels, R., *Designing Great Beers*, Brewers Publications, Boulder, CO, 1997.
Daniels, R. and Parker, J., *Brown Ale*, Brewers Publications, Boulder, CO, 1998.
Daniels, R. and Larson, G., *Smoked Beers*, Brewers Publications, Boulder, CO, 2000.
Dornbusch, H., *Altbier*, Brewers Publications, Boulder, CO, 1998.
Dornbusch, H., *Bavarian Helles*, Brewers Publications, Boulder, CO, 2000.
Eckhardt, F., *Essentials of Beer Style*, Fred Eckhardt Associates, Portland, OR, 1989.
Fix, G., *Vienna, Märzen, Oktoberfest*, Brewers Publications, Boulder, CO, 1992.
Foster, T., *Pale Ale*, Brewers Publications, Boulder, CO, 1990, 1999.

Foster, T., *Porter*, Brewers Publications, Boulder, CO, 1991.
Guinard, J.X., *Lambic*, Brewers Publications, Boulder, CO, 1990.
Jackson, M., *Simon and Schuster's Pocket Guide to Beer*, Simon and Schuster, New York, 1991.
Jackson, M., *World Guide to Beer*, Running Press, Philadelphia, PA, 1989.
Jackson, M., *The Beer Companion*, Running Press, Philadelphia, PA, 1993.
Kieninger, H., The influence on beermaking, in *Best of Beer and Brewing*, Vols. 1–5, Brewers Publications, Boulder, CO, 1987.
Lewis, M.J., *Stout*, Brewers Publications, Boulder, CO, 1995.
Miller, D., *Continental Pilsener*, Brewers Publications, Boulder, CO, 1990.
Narziss, L., Types of beer, *Brauwelt Int.* II/1991.
New Brewer, The, Institute for Brewing Studies, Boulder, CO, 1983–2000.
Noonan, G., *Scotch Ale*, Brewers Publications, Boulder, CO, 1993.
Piendl, A., *Brauindustrie* magazine, 1982–1994. From the series "Biere Aus Aller Welt." Schloss Mindelburg, Germany. *Note*: All styles in this guideline have been cross referenced with technical beer data compiled by Professor Piendl.
Rajotte, P., *Belgian Ales*, Brewers Publications, Boulder, CO, 1992.
Richman, D., *Bock*, Brewers Publications, Boulder, CO, 1994.
Sutula, D., *Mild Ale*, Brewers Publications, Boulder, CO, 1999.
Warner, E., *German, Wheat Beer*, Brewers Publications, Boulder, CO, 1992.
Warner, E., *Kölsch*, Brewers Publications, Boulder, CO, 1998.
Zymurgy®, American Homebrewers Association®, Boulder, CO, 1979–2000.

Other Resources

Karen Barela, past President, American Homebrewers Association, Boulder, CO.
Peter Camps and Christine Celis, formerly of Celis Brewing Co., Austin, TX.
Jeanne Colon-Bonet and Glenn Colon-Bonet, past co-directors of Great American Beer Festival Professional.
Ray Daniels, author *Designing Great Beers*, past Editor in chief *New Brewer* and *Zymurgy* magazines, Director Brewers Publications, Director Craft Beer Marketing (Association of Brewers), Boulder, CO and Chicago, IL.
Mark Dorber, British Guild of Beer Writers, and White Horse on Parson's Green, London, UK.
Dr. George Fix, Arlington, TX.
Paul Gatza, Director American Homebrewers Association and Institute for Brewing Studies, past Great American Beer Festival and World Beer Cup judge director.
Michael Jackson, author/journalist, London, UK.
Finn B. Knudsen, President, Beverage Consult International Inc., Evergreen, CO.
Gary Luther, Senior Brewing Staff, Miller Brewing Co., Milwaukee, WI.
Dr. James Murray, Brewing Research Foundation International (BRFI), Nutfield, UK.

Jim Parker, Director, American Homebrewers Association, Boulder, CO.
Brian Rezac, Former Administrator, American Homebrewers Association, Boulder, CO.
Fred Scheer, Brewmaster, Boscos, Nashville, TN.
Poul Sigsgaard, Scandinavian School of Brewing, Copenhagen, Denmark.
Pete Slosberg, Pete's Brewing, Palo Alto, CA.
James Spence, former National Homebrew Competition Director, American Homebrewers Association, Boulder, CO.
Chris Swersey, Competition and Judging Director Great American Beer Festival and World Beer Cup.
Keith Thomas, Campaign for Real Ale (CAMRA), St. Albans, UK.
Amahl Turczyn, Former Project Coordinator, American Homebrewers Association, Boulder, CO.
Thom Tomlinson, past director of Great American Beer Festival Professional Blind Panel Judging.
Brewers, homebrewers, beer enthusiasts, Scott Bickham, Al Korzonas, Al Kinchen, Jim Liddil, Marc Hugentobler, George De Piro, Hubert Smith, David Houseman, Stephen Klump, Ray Daniels, Jim Homer, Virginia Wotring.
J.E. Siebel and Sons, Chicago, IL.
Siebel Institute of Technology, Chicago, IL.
American Homebrewers Association National Competition Committee 1993–1999.
Judges of Great American Beer Festival® Judging Panel 1993–2004.
Judges of the World Beer Cup® 1996, 1998, 2000, 2002, 2004.
Personal travels, tastings, and evaluations of beer and brewing experience, 1974–2004.

3

An Overview of Brewing

Brian Eaton

CONTENTS

Introduction	77
Outline of the Brewing Steps	78
Discussion of the Individual Brewing Steps	79
Malting	79
Milling and Adjunct Use	80
Mashing	81
Wort Separation	82
Wort Boiling	82
Trub Removal	82
Wort Cooling/Aeration	83
Yeast Handling	83
Yeast Pitching	84
Fermentation	84
Yeast Removal	85
Aging	86
Clarification	87
Packaging	87
Warehousing and Distribution	89

Introduction

This chapter provides an introduction to the brewing process and forms a basis for more details of the individual processes involved and described in subsequent chapters. "Brewing" is defined in dictionaries as the making of beer or related beverages by infusion, boiling, and fermentation. Brewing is not unique to beer production: it is also the essential first part of the process of whisky production. We are also familiar with the term applied to the making of tea and herbal infusions and as a term in "brewing up trouble or mischief."

In this chapter, we will deal with the brewing of beer from barley malt, since this is the main raw material employed worldwide. Chapter 2, in particular, describes other beer types and the use of other materials while Chapter 23 covers innovative novel beers and beer-related products. In Africa, for example, the use of sorghum has flourished, partly due to its availability, but also due to restrictions on the import of barley malt that necessitated finding an alternative material. Nevertheless, the brewing practices employed around the world have become very similar, due largely to their origins being from Europe, predominantly Britain, Germany, and the Netherlands.

There are some notable differences; for example, the traditional production method of cask-conditioned beer in the UK is essentially unique, as too is the use of a lactic acid fermentation to produce lambic beers in Belgium. Government and social pressures have also had their influence on the types of beer being made. The high taxation on alcoholic content or original gravity in some countries, notably Britain, encouraged the sale of lower-strength beers but this trend was also affected by the industries in which the consumers were employed. Arduous manual labor in steel making, mining, and other heavy industries created a demand for low-strength, thirst-quenching beers. On a wider basis, concern about alcohol and its effect on the operators of machinery and car driving has led to the introduction of low-alcohol beers (LABs) and no-alcohol beers (NABs).

Outline of the Brewing Steps

The following is a brief introductory description of the malting and brewing processes and their subprocesses. More detailed descriptions can be found in individual chapters:

1. Malting — converting barley to malt
    a. Barley drying and dressing — removing debris, dry to store
    b. Barley storage — housing the barley to maintain its vitality
    c. Steeping — thoroughly soaking the barley in water
    d. Germination — allowing the barley to germinate naturally
    e. Kilning — stopping germination by heating; also to develop color and drying for storage
    f. Malt storage — housing the malt until required
2. Milling — grinding the malt (often with other cereals) to grist
3. Mashing — mixing grist with water (liquor), enzymic conversion

An Overview of Brewing

4. Wort separation — separating the liquid (wort) from the solids (draff)
   a. Mash tun separation
   b. Lautering
   c. Mash filtration
5. Wort boiling — sterilization, coagulation, hop extraction, concentration
6. Trub removal — removing coagulated material and hop debris
   a. Hop back, hop strainer
   b. Centrifugation
   c. Sedimentation
   d. Filtration
   e. Whirlpool
7. Wort cooling/aeration — aerate and cool the wort
8. Yeast handling
   a. Yeast propagation — preparation of pure culture
   b. Yeast storage — maintain yeast in good condition for use/reuse
   c. Acid washing — treat the yeast with acids to reduce bacteria
   d. Surplus yeast — sell excess to flavorings industries, pig feed, health foods, etc.
9. Yeast pitching — add the culture yeast to the wort
10. Fermentation — yeast growth; alcohol and CO_2 production
11. Yeast removal — reduces yeast level in the immature beer
12. Aging — matures and stabilizes the beer at low temperature
13. Clarification — removes particles to produce bright beer
    a. Fining — uses a coagulant (isinglass) to remove yeast, etc.
    b. Centrifugation — particle removal by centrifugal force
    c. Filtration — particles trapped on a filter bed or on sheets
14. Packaging — beer filled (racked) into its final container
15. Warehousing and distribution — storing and transporting the beer to final customer in perfect condition

Discussion of the Individual Brewing Steps

Malting

The malting process converts the raw barley by controlled steeping, germination, and kilning into a product that is much more friable, with increased enzyme levels and with altered chemical and physical properties. The first part of the malting process mimics what would occur in nature if the barleycorn is left to germinate in the field. The grain is first steeped in cool water, and drained occasionally to ensure that the corns are not asphyxiated. Once thoroughly wet, they are laid out as a shallow bed and the grains start to grow, producing roots and shoots. The grain bed is kept moist and cool by passing chilled wet air through the bed. The grain embryo produces and releases a plant hormone, gibberellin, which activates the aleurone layer of the grain to produce various enzymes. These enzymes, together with those already present in the grain, start to break down the food reserves of the grain. If allowed to continue, a new barley plant would be formed, but the maltster stops the germination just before the shoot emerges from inside the grain. This is achieved by heating (kilning) the grain with hot dry air. As well as drying the malt to preserve it, kilning also develops color and flavor. The roots are removed mechanically and the malted corns are then stored ready for use.

Although the main malted cereal for brewing is barley, other cereals are malted for specific purposes, mainly for the production of specialist beers: oats, maize, rice, rye, sorghum, and wheat. Wheat malt is used in the production of *weiss* beer. Rye malt is little used in brewing due to slow wort separation and haze instability in the resultant beer, but it is used in rye whisky and distillers using high proportions of adjunct benefit from the high levels of enzymes in rye malt. Sorghum malt, as mentioned previously, is widely used in Africa both in the brewing of native beers and also as a barley malt replacement in normal beers for economic and trade reasons. Chapter 5 covers the topic of barley and malt in detail.

Milling and Adjunct Use

Barley malt can be supplemented with other cereals, either malted or raw, for specific purposes (provided local legislation permits their use). Malted cereals are used as described in the previous section but raw cereals (barley, oats, maize, rice, rye, sorghum, and wheat) are added as an adjunct for one or a number of the following reasons:

- To produce a more stable beer, as they contain less protein
- To produce a different flavor — maize, for example, is said to give a fuller flavor
- To produce a better beer foam due to lower fat (lipid) levels and different proteins
- To improve the ease of processing in the brewhouse
- To produce beer at lower cost

An Overview of Brewing

When used as raw cereals in conjunction with barley malt, most cereals (and in particular maize, rice, and sorghum) need to be precooked before they can be mashed with the malt. The malt itself needs to be milled first to produce a range of smaller particles called grist. This makes the malt easier to wet at the mashing stage and aids faster extraction of the soluble components from the malt during the enzymic conversion.

Roll mills produce coarse grist for use with mash tuns or lauter tuns, but mash filters can use much finer grist, as produced by the more severe crushing of a hammer mill. The fineness of grind is checked by analysis through a series of sieves.

Mashing

Mashing is the process of mixing the crushed malt, and cereal adjuncts if used, with hot water and letting the mixture stand while the enzymes degrade the proteins and starch to yield the soluble malt extract, wort.

There are three main mashing methods that have been developed to suit the equipment and materials available to the brewer:

- *Infusion mashing.* This is the classic British thick mash system using a mash tun at a single temperature and without any stirring. It requires high-quality, well-germinated malt and is still used by smaller breweries today. The wort separation from the solids also takes place in the mash tun.
- *Decoction mashing.* This is the typical European mashing system that uses a series of different temperatures, more complicated brewing equipment, and a less well germinated malt. The increases in temperature are initiated by taking out (decocting) part of the mash, heating it to boiling point, and returning it back to the mash vessel. This can be one or more times: single, double, or triple decoction. In this way, optimal temperatures for proteolysis (about 40 to 50°C) can be followed by optimal temperatures for starch hydrolysis (54 to 65°C) and finally a high temperature for wort separation (about 70°C). Mash separation is usually carried out using other equipment — a lauter tun or a mash filter.
- *Double mashing.* This is the American double mash system. This process is used when cereal adjuncts require precooking (gelatinization) before addition to the main malt mash. Two separate vessels are used — the cereal cooker and the mash mixer. In the cereal cooker, the adjunct (rice, maize, etc.) is heated with water to about 85°C and held at this temperature for 10 min while the very viscous mash thins. Some malt, say 5 to 10%, can be added to assist in this reduction in viscosity. The cereal mash can then be raised to boiling point and transferred into the malt mash that

is already in the mash mixer at a temperature of about 35°C to raise the temperature for starch hydrolysis. Mash separation is usually carried out in a separate vessel, a lauter tun.

Wort Separation

For the classic infusion mash, the processes of mashing and wort separation take place in the same vessel, the mash tun. For other mash systems, the mash is transferred to either a second vessel (lauter tun) or to a mash filter. Both speed up the process: a lauter tun, due to its large diameter and shallow grain depth, gives fast separation; the mash filter uses pressure and a very thin bed depth of grain to give a rapid extraction. For both systems, the transfer enables a second mash to be started in the mash mixer while the first mash is being separated.

Typically, a lauter tun takes 2 h and a mash filter 90 min to complete the separation.

The principal objectives are to produce bright wort and to collect the maximum amount of sugars (extract) from the residual solid materials. These solids are sold as animal feed, draff, or spent grains for milk and beef cattle herds.

Wort Boiling

Wort boiling is the process unique to beer production, as it is not required in the distilling or vinegar production processes. Wort boiling satisfies a number of important objectives:

- Sterilization of the wort to eliminate all bacteria, yeasts, and molds that could compete with the brewing yeast and possibly cause off-flavors
- Extraction of the bittering compounds from hops added early to the boil and oils and aroma compounds from late additions
- Coagulation of excess proteins and tannins to form solid particles (trub) that can be removed later. This is important for beer stability and foam
- Color and flavor formation
- Removal of undesirable volatiles, such as dimethyl sulfide, by evaporation
- Concentration of the sugars by evaporation of water

Trub Removal

While there are differences of opinion as to whether wort should be brilliantly clear, most agree that the coagulated material formed during

An Overview of Brewing

the boil does need to be removed if the beer stability is not to suffer. The degree of removal is the debate. Hot break is the name for the particles present in the hot boiled wort, and they are quite large, 20 to 80 μm.

A much finer particle (cold break, <2 μm) appears when wort is cooled below 60°C and some brewers choose to remove it due to the suspected effects on beer flavor and coating of the yeast to inhibit efficient fermentation. There are several methods for removal, depending on the equipment available and also whether whole hops or hop products have been used:

- *Hop back/hop strainer.* If whole hops are used, then a hop back or strainer is needed to sieve out the hop debris. The hop back will also catch the trub, but a strainer will require a further stage for trub removal.
- *Centrifuge.* Hot wort centrifugation is an effective method for removing hot break, but expensive in capital and running costs. The machines are also prone to damage from small stones that find their way through the process from the barley at harvest.
- *Sedimentation/flocculation.* The large particle size of hot break means that it sediments quickly if the wort is run into a shallow vessel. This was the principle of the coolship — a large shallow open vessel that held the wort while it cooled naturally and clarified. The later sedimentation vessel achieved the same result but more hygienically. Alternatively, the particles can be made to float by attachment to air bubbles and then skimmed off the wort surface to separate (flocculation).
- *Filtration.* The trub particles can be removed by filtering through a bed of kieselguhr or perlite powder, which traps the solid particles.
- *Whirlpool.* The simplest and most elegant separation technique is the whirlpool, which makes use of the centrifugal/centripetal force acting on the particles when the wort rotates after tangential inlet into the cylindrical tank. The trub and hop debris are deposited as a mound in the center of the vessel; the bright wort can be taken away from the periphery of the vessel.

Wort Cooling/Aeration

The wort is cooled from almost boiling point to fermentation temperature through a heat exchanger using water as the main cooling medium. The temperature for fermentation is typically 8 to 13°C for lager and 14 to 17°C for ale.

Yeast Handling

The handling of the yeast is key to the efficiency of brewery fermentations and to the quality of the final beer. Important steps in maintaining the vitality and viability of the culture yeast are the proper pitching and cropping of the yeast, the propagation of the yeast from laboratory to full-scale pitching, its storage for reuse, and the techniques for acid washing the yeast to reduce infection. Surplus yeast is a valuable coproduct and is sold to the food flavorings industry, to distilleries, and to health supplement manufacturers.

Yeast Pitching

Yeast is pitched into the wort, either directly into the cooled wort in the fermentation vessel, or in-line en route from the heat exchanger to the fermenter. Typically, 5 to 20 million yeast cells are pitched per milliliter of wort. Air or oxygen is also added to the wort as an essential factor for the production of yeast membranes and, hence, new cells. It is usual to inject it in-line en route to the fermenter, but direct injection is also used, often if the vessels are of open design, which has the added benefit of mixing the wort so that a representative sample is obtained to determine the collection gravity.

Fermentation

There are two main classifications of fermentations, ale and lager, but a wide variety of different fermentation systems and equipment have been used over the years. It would to true to say that the definitions of ale and lager have become blurred and it is now difficult to define precisely what constitutes lager and what constitutes ale:

- *Ale fermentation.* In brief, ale uses a *Saccharomyces cerevisiae*, top-cropping yeast at a temperature of 14 to 17°C. The fermentation is fast and exothermic, so cooling is applied to maintain a constant temperature. The fermentation is not always taken to its full extent, but may be deliberately cooled early to flocculate the yeast and to leave residual sugar for palate sweetness and secondary fermentation (as in cask-conditioned ale). The yeast is cropped early from the vessel to prevent off-flavors from yeast autolysis and also to provide healthy, vital yeast for subsequent fermentations.

    The "green" (immature) beer is cooled slightly to encourage further yeast separation and then either filled direct to cask or further cooled and filled into cold tank for brewery conditioning. Isinglass finings are added to the casks to aid in clarification and often also dry hops to impart a strong, hoppy character to the beer.

    Early ale systems can still be found working in breweries in the UK, and these include the Burton Union system, Yorkshire Squares, and Open Squares. These systems evolved as methods of obtaining a good, fast fermentation, a good crop of the best yeast for repitching, and convenience for cropping the top-fermenting

yeast. However, ales are now mostly fermented in cylindroconical vessels and, under these conditions, the top-cropping yeasts become bottom sedimenting, but the processing remains similar.
- *Lager fermentation.* Lager ferments at a lower temperature, typically 8 to 13°C, and uses a bottom-cropping *Saccharomyces uvarum* yeast. The traditional lagering process involved a primary fermentation using flocculent yeast, which was followed by a secondary fermentation using nonflocculent yeast at a lower temperature, say 8°C (warm storage), followed by cold storage at less than 0°C to stabilize the beer. It is now more usual to "age" the beer at below 0°C after a single, complete primary fermentation.

For both ale and lager, the basic process of fermentation is similar; yeast uses sugars and proteins to produce alcohol, CO_2, new yeast cells, and flavor compounds. At the start of fermentation, the yeast appears dormant with very little activity — this is the lag phase. However, much is happening within the yeast cell as it adapts to its new environment; the cells bud to produce new cells using its carbon reserve of glycogen. Oxygen is also essential for this role. As the oxygen is consumed and anaerobic conditions prevail, the yeast transports sugars into the cell to form pyruvate, which it then metabolizes to ethanol and CO_2. These are both transported out of the cell and excess CO_2 from the fermentation is vented off from the vessel, either to the atmosphere or collected for use later in the packaging process. A level of CO_2 remains in the beer to give it effervescence.

Toward the end of fermentation, as the sugars are depleted, the yeast begins to flocculate. This can also be initiated by providing cooling to the fermentation system. A good separation of the yeast by flocculation is important in obtaining a clean, good-tasting beer, and for ease of processing through the subsequent stages. The fermentation also produces a range of flavor compounds, esters, alcohols, etc., that give character to the beer. There are, however, some flavor compounds that are unpleasant and need to be reduced or removed during the fermentation or later in the lagering or cask-conditioning processes. Such compounds include diacetyl: a rancid butter/butterscotch flavor produced by the yeast from pyruvate, but if given time, broken down by the yeast to acetoin and butanediol.

Sulfur compounds such as H_2S, SO_2, and dimethyl sulfide are volatile, and the CO_2-purging action during fermentation can be sufficient to reduce them to acceptable levels if the vessel is allowed to vent freely. The formation of excessive H_2S and SO_2, although determined primarily by the yeast strain, is associated with conditions that restrict yeast growth and in this respect the availability of oxygen at pitching is a critical factor.

Yeast Removal

It is important to remove the bulk of the excess yeast before maturation, generally by removing the beer from the sedimented yeast.

Aging

The maturation of the green beer to produce a stable, quality product suitable for filtration and packaging is called aging or, alternatively, cold conditioning or cold storage.

The objectives of beer aging are:

- Chill haze formation
- Clarification
- Carbonation (to a limited extent)
- Flavor maturation (again to a limited extent)
- Stored capacity for demand smoothing

Taking these in turn, in order to produce a beer of good colloidal stability in the final package, it is vital to promote, by storage at low temperature ($<0°C$), the formation of chill haze comprising flocs of polypeptides and polyphenols. These can be removed by sedimentation (slow) or by filtration to give a bright, stable product.

Long cold storage will bring about the clarification of the beer by sedimentation of residual yeast, chill haze material, and other debris, but it is a slow process even in horizontal storage tanks. For better tank utilization, it is usual to have a short period of cold storage to allow the flocs to form, say 2 days, and then to filter the beer to clarity.

At low temperatures of cold storage there is minimal yeast activity, so the beer will not gain appreciably in CO_2 content, even though the low temperature favors solubility. Similarly, flavor change is minimal. Removal of polyphenol material as chill haze will remove harsh, bitter flavors from the beer but low yeast count and low temperature cannot be expected to rectify a diacetyl problem, for example. Therefore, if aging is to be used, it is important that the primary fermentation runs to completion and produces beer with the correct final flavor profile. This is in contrast to the traditional "lagering" or warm storage process where higher yeast counts and a more favorable temperature, say $8°C$, facilitates some beer maturation.

The use of cold storage as a buffer stock in order to smooth the fluctuations of sales back into fermentation and brewing is commonplace. The low temperature minimizes any possible microbiological spoilage and changes to the beer character. It is therefore a convenient point in the process to create a small stock-holding, subject, of course, to the cost of providing the storage capacity.

Rapid maturation using immobilized yeast has been employed in several countries and by several companies. The Cultor company in Finland in association with Sinebrychoff AB developed a process of heat treatment (65 to $90°C$ for 7 to 20 min) followed by passage through a packed bed column of yeast immobilized on DEAE cellulose particles to accelerate the

An Overview of Brewing

breakdown of α-acetolactate to diacetyl and then to acetoin and, ultimately, to the taste-neutral molecule, butanediol A, resulting in high-quality beer. Alfa Laval of Belgium and the German company Schott Engineering have also jointly developed a rapid maturation process, using a porous glass bead called Siran® as the carrier. Both these processes reduce the maturation time from days or weeks to hours.

Cold storage is a very effective means of chillproofing beer, but there are still polypeptides and polyphenols present, and if a very long shelf life is required, then additional chillproof treatments may be required. These treatments reduce still further the polypeptides and polyphenols in the beer, so that colloidal haze formation will not occur and become visible during the beer's shelf life. In brief, proteins can be removed by absorption onto silica gel, which is then removed by filtration, or by addition of proteolytic enzyme, which prevents the polymerization of proteins by breaking them down to smaller units. Polyphenols can be removed by absorption onto polyvinylpyroporolidone beads, which are then filtered out.

Clarification

Some clarification is required for most beers although there are exceptions such as wiess beer, which is often served cloudy. Filtration will produce a bright, sparkling beer that will remain clear throughout its shelf life, provided that the stabilization has been correctly applied.

A coarse depth filtration using diatomaceous earth or perlite can remove most particles, but it can be followed by filtration through a cellulose filter sheet to give a polished, almost sterile, product. Health concerns about dust from filter powders and the cost of their disposal as spent filter cake to landfill is encouraging alternative methods. Ultrahigh speed centrifugation, deep-bed sand filtration, cross-flow filtration, and fining are some of the methods being tried, usually backed up with a cellulose sheet filter to give the final polish.

The filter is usually the last opportunity to make corrections to the beer before final packaging. Typically, adjustments can be made to the carbonation level, by adding or removing (by membrane diffusion) CO_2, adding color as caramel or farbe bier, hop products to correct bitterness, enhance foam, or prevent light-strike, and de-aerated water to cut the beer to its correct final alcohol strength.

Packaging

The packaging of beer can be conveniently divided into two categories:

- Large pack that includes kegs, casks, and demountable bulk tanks
- Small pack that covers cans and bottles

There are differences between countries in the relative volumes of beer packaged into these two categories for a host of reasons: historical, political, and geographical. In the UK, for example, ownership of the retail outlet by the breweries and the short delivery distances favored large pack. Also, the packaging itself did not have to sell or promote the product. In recent years, however, the increase in beer drinking at home has seen a shift toward small pack, although large pack still dominates the market.

For casks and kegs, stainless steel is the most widely used material for its cost, durability, and hygiene, although there are some aluminum and wood casks still in use.

Bulk tanks of 8-hl capacity have been used in the UK and Scandinavia for direct delivery to the retail outlet as a unit, or with a transfer of the beer from a road tanker to 4- or 8-hl bulk tanks in the outlet cellar. Although popular with larger retail outlets in the UK during the 1970s and 1980s, it has now almost disappeared, but is still used in Scandinavia.

Glass still predominates the bottle market although polyethyleneterephthalate (PET) and polyethylenenaphthalate are increasingly being used for their weight and safety benefits, now that the major obstacles of pasteurization and barrier properties to O_2 pick-up and CO_2 loss appear to be resolved. The reuse and recycling of glass is well established, although there has been a major switch from returnable to nonreturnable bottles (NRBs) in many countries. This has been driven by initiatives in the light-weighting of NRBs, the inconvenience of returnable cases, and the better payload of NRBs. Reuse of PET is not widely practiced, but has been successful in Scandinavia.

Cans are manufactured from aluminum and tin-plate. In America, aluminum is the main material but tin-plate is a strong contender in Europe and Mexico. Both materials are extensively recycled. A recent development of a can with a bottle-shaped neck is an attempt to capture the visual appeal and decorative opportunities that a bottle offers, but with a very low weight and full recyclability.

Two types of equipment are in use for keg filling (racking): a linear machine comprising multiple lanes (up to 24) and the rotary machine with up to 24 stations, looking very much like a large version of a can or bottle filler. Speed of operation, diversity of keg size, and shape are a few of the factors that might decide which type of machine to use. Speeds of operation from just a few kegs per hour to over 1000 kegs per hour are possible. Both types of washer/racker take kegs returned from trade, wash and sterilize the inside surfaces, counterpressure and cool slightly with an inert gas, usually CO_2, and then fill with sterile beer under carefully controlled and monitored conditions. Air pick-up is avoided and a microbiologically sound product can be produced. The beer for filling is either pasteurized or sterile-filtered. In comparison, the filling speeds of bottles and cans are very fast, with throughputs of up to 2500 containers per minute possible.

Warehousing and Distribution

Finally, it is important that the beer is distributed to the consumer in top condition. Today, this is an increasingly complex operation and the supply chain is a key element of successful distribution. This is a topic that could cover a book in its own right and is outside the scope of this handbook. Nevertheless, we must remember that an excellent product in the brewery does not necessarily mean that it will still be excellent when it reaches the consumer and that large companies in particular have struggled because the beer is in the wrong place at the wrong time. So careful attention to logistics and supply chain is essential to both the large international and small craft brewer.

4

Water

David G. Taylor

CONTENTS

Introduction	93
Background	94
Water Usages in the Brewery	94
Brewery Water Functions	95
Brewery Water Consumption	95
Brewery Water Categories	96
Brewing Water	96
Process Water	97
General-Purpose Water	97
Service Water	98
Water Standards for Ingredient Use	98
Water Sources	98
Surface Water	98
Ground Water	99
Abstraction of Surface Water	99
Abstraction of Ground Water	99
Seasonal Influences on Water Quality	100
Chemical Characterization of Water Types	100
Hard Water	100
Soft Water	100
Brackish/Salt Water	101
Peaty Water	101
Inorganic Constituents of Water	101
Major Constituents	101
Minor Constituents	102
Trace Constituents	102
Water Hardness	102
Organic Constituents of Water	103
Microbiological Constituents of Water	103
Legislation and Regulations Governing Water Quality	104

The Influence of Inorganic Ions on Beer Quality 107
  Historical Background 107
  Sources of Ions in Beer 108
  Effects of the Ionic Contents of Wort and Beer 108
  Direct Effects of Ions on Beer Flavor 109
    Sodium 109
    Potassium 109
    Magnesium 110
    Calcium 110
    Hydrogen 110
    Iron 111
    Chloride 111
    Sulfate 111
    Chloride: Sulfate Balance 111
  Indirect Effects of Ions on Beer Flavor and Quality 112
    Yeast Requirements 112
    Effects on Malt Enzymes 112
    Effects on Stability of Colloidal Systems 113
    Effects on pH Control 113
  Control of pH 113
    Buffer Systems 114
    Reactions Controlling pH 115
    pH Control during Wort Production 115
    Influences of pH on Wort Composition 116
    pH during Wort Boiling 117
    pH Control during Fermentation 117
  Summary of the Influences of Various Ions during Beer Production 118
    Hydrogen 118
    Calcium 118
    Magnesium 118
    Sodium 118
    Potassium 119
    Iron 119
    Zinc 119
    Manganese 119
    Copper 119
    Other Cations 119
    Bicarbonate/Carbonate 120
    Phosphate 120
    Sulfate 120
    Chloride 120
    Nitrate/Nitrite 120
    Silicate 120
    Fluoride 121
Organic Contamination 121

Organohalides 121
　　Phenols 121
Microbiological Control 122
Water Treatment Technologies and Procedures 122
　　Removal of Suspended and Semidissolved Solids 123
　　　　Coagulation and Flocculation 123
　　　　Sand Filtration 124
　　　　Oxidation 124
　　Adjustment of Mineral Contents 125
　　　　Decarbonation by Heating 125
　　　　Decarbonation by Treatment with Lime 125
　　　　Decarbonation by Acid Addition 126
　　　　Ion Exchange Systems 126
　　　　Reverse Osmosis 128
　　　　Electrodialysis 129
　　　　Addition of Calcium Salts 130
　　Removal of Organic Compounds 130
　　　　Aeration Systems 130
　　　　Granular Activated Carbon Filtration 131
　　Water Sterilization Techniques 131
　　　　Chlorination 131
　　　　Chlorine Dioxide 132
　　　　Bromine 133
　　　　Ozonation 133
　　　　Sterilization by Silver Ions 133
　　　　Ultraviolet Irradiation 134
　　　　Sterile Microfiltration 134
　　Gas Removal Applications 134
References 135

Introduction

Without question, water is the principal ingredient of beer. In reality, the "water" supplied to a brewery is actually a dilute solution of various salts, in which small quantities of gases and organic compounds may also be dissolved.

In any consideration of water in relation to brewing, it is usually accepted that the composition of the water used in beer production (or "liquor," as brewers are apt to call it) will have significant influence on the quality of the finished product.[1-6] However, what may be overlooked is the impact that sources of water other than that used directly for wort production may have on beer properties and that this "adventitious" water arising elsewhere during processing may still contribute to the finished product and, therefore, warrant appropriate treatment.

It is the author's intention in this chapter to review why there is often a need for water treatment for beer production; in other words, to understand how water quality can influence beer quality and then to review the appropriate processes that are currently available, the technologies involved, and their applications in achieving the desired quality standards, not only for water as a primary raw material, but also for the other process uses.

Background

Ninety-seven percent of the world's total water is in the oceans, meaning that only 3% is "fresh" water, supplied by the natural evaporation and precipitation process known as the water cycle. Of the total water on land, more than 75% is frozen in ice sheets and glaciers, with most of the remainder (about 22%) being collected below the earth's surface as "ground water." Consequently, only a comparatively small quantity exists as "surface water" in lakes, reservoirs, and rivers.

The continued availability of a supply of "good-quality" drinking water is one of the most important features of modern life and is of fundamental importance to the worldwide food and drink industries, including brewing. To this end, it is very significant that in several countries desalination of sea water to produce water of potable standard is now common.

The importance of a sustained water supply for brewing is reflected in the consumption ratio required, which can vary from 3 to 20 unit volumes of water per unit volume of beer, with an average of 6:1. This ratio depends on a number of factors and individual brewery constraints, but usually varies as a consequence of the differing packaging operations involved, with returnable bottling usually being the most water-use intensive.[3] In addition, the malting process uses some 30 to 40 hl of water to produce 1 tonne of malt.[5]

As (other than some exceptionally strong products) beer is composed of over 90% water, it is very apparent how important the quality of the water supply (as well as the quantity) will be to beer quality. The content of dissolved solids in the water used in brewing can have a significant influence on many beer properties, including flavor.[4,5]

Water Usages in the Brewery

The various water uses in the brewery may require different compositions or treatments appropriate to the requisite functions.

Water

Brewery Water Functions

In this respect, different functions can be categorized[5] as:

- *Brewing water*: Water actually contributing directly as an ingredient to beer. This is liquor used in wort production and for standardization to target alcohol content (as in high-gravity brewing). Brewing water requires treatment/adjustment to achieve the correct composition relevant to the beers being brewed.
- *Process water*: Comprises water used for washing and sterilizing of vessels and pipework and for container cleaning and rinsing (i.e., all beer process contact surfaces). It should be of potable standard and ideally softened; it may also be used for pasteurization and refrigeration.[6,7]
- *General-Purpose water*: Covers the water for general washing down, site hygiene, and office use, and will generally require no further on-site treatment.
- *Service water*: Includes water for boiler feed (i.e., steam raising). It should not produce scaling and so must be softened or, ideally, be fully demineralized.[6,7]

Brewery Water Consumption

On the basis of a water usage ratio of 6 hl/hl of beer, in broad terms, the following typical consumption proportions can be identified[3]:

Brewing water: 2.7 hl
Process water: 2.1 hl
General-purpose water: 1.0 hl
Service water: 0.2 hl

Kunze[3] provides details of water flows and usages at various process stages in a brewery and by various operations. Of particular note is the fact that of the total water used, about 1.7 hl is required for wort production, with 0.8 hl required for fermentation and maturation, and 0.4 hl for filtration, all per hectoliter of beer. In addition, up to 2.0 hl could be used for returnable bottle washing and refilling.

Of the 6.0 hl of water consumed per hectoliter of beer, it can be estimated that approximately 0.92 hl/hl contributes to finished beer, with up to 0.2 hl lost by evaporation, about 0.15 hl with by-products (such as spent grains and waste yeast), and the remainder (nearly 5.0 hl) lost as effluent (including about 0.1 hl as "domestic" waste water); water treatment losses could amount to nearly 1.0 hl of the total effluent output, depending very much on the particular treatment plant involved.

At this stage, it should be noted that I shall not cover waste water treatment in this chapter, this topic is reviewed elsewhere (e.g., Refs. 8–12 and Chapter 18). Suffice it to say that effluent treatment can be a serious cost to a brewery operation and thus will merit some attention, not least of which from the viewpoint of conforming to any relevant environmental and pollution legislation. It is interesting to realize that in calculating beer production costs,[9,10] water is the one ingredient for which the brewer pays twice (both before and after consumption)!

Brewery Water Categories

Brewing Water

Details of the influences of the composition of brewing liquor on beer quality parameters are presented in the Section "The Influence of Inorganic Ions on Beer Quality." However, it is appropriate here to indicate that water from several brewery activities can contribute to the finished product, sometimes adventitiously.[5]

The primary sources of ingredient water are, obviously, *mashing* and *sparging liquors* and, if high-gravity brewing is employed, then water used for alcohol/gravity adjustment (*dilution liquor*) can be a significant contributor to the beer volume. Dilution water, especially, must be deoxygenated and, ideally, have low mineral content.[6] It should also be sterile, if added postfilter. There are other potential contribution sources, which, although they may not represent large volumes, can have a significant effect on flavor and, arguably, deserve appropriate attention to composition and quality. Several can be identified and include:

- *Wort "breakdown" liquor.* It is common practice to adjust wort gravity after collection in the fermenter; clearly this should be of similar composition to sparge liquor.
- *"Chase" water.* This is the water used to "push" the product stream through the process on transfers between vessels, in order to reduce process losses.
- *Yeast washing/dilution water.* Breweries employing the principle of yeast washing to reduce CO_2 content or diluting yeast to reduce the alcohol content during storage, which is especially important for high-gravity fermentations,[13] should pay attention to the water quality, not only with respect to the effect on beer quality, but also considering yeast keeping quality. In addition to all other considerations, this water should be sterilized before use. This also applies to water used to prepare acid for yeast acid washing.
- *Packaging rinse water.* When using nonreturnable bottles and cans, the water used for rinsing the packages prior to filling will almost certainly leave residues that can influence beer quality on filling.

By the same token, *final rinse water* after returnable bottle and keg/cask, plus vessel and pipework washing should ensure no carry over of soil or detergent residues.

- *"Jetting" water.* Used for air pick-up control during bottle filling will leave a residue in the finished package and, although only a minor component, its quality should not be ignored. If sterile filling is employed, then clearly this water source must also be sterilized.
- *Steam residues.* The quality of any steam that can directly impact on the product stream, even as contact surface residues after steam sterilization, can represent a major risk to beer quality. Breweries that use direct steam injection into brewhouse vessels must seek assurance of the potable quality of the condensate, so use of "pure water" steam generation should be considered. The same applies to steam used to sterilize plant and large containers (kegs/casks), unless the brewer plans to remove any condensate residues by flushing with sterilized water.

In all these cases, it is pertinent for the brewer to ensure, at the very least, that these water sources are potable and also to consider the relevance of the mineral composition (see Section "The Influence of Inorganic Ions on Beer Quality").

Process Water

The water used for plant washing and sterilizing (i.e., contact surfaces) should be of potable standard and free from off-flavors. In addition, water for sterilizing must be free of microorganisms.

Usually, water for detergent and sterilant make-up should also be softened (i.e., hardness reduced). Water used for tunnel pasteurizers should also be treated to reduce its mineral content (at least, softened) and disinfected to reduce the growth of algae or microorganisms. The control of pH is important and anticorrosion treatment may also be required. Similar considerations will apply to water used for refrigeration and air compressors, where appropriate chemical and microbiological treatments will be specified by the plant suppliers.[7,14]

Finally, it should be noted that water used in externally located cooling towers for air-conditioning systems, etc., will require very stringent sterilization procedures, plus frequent, routine, appropriate microbiological monitoring for control of *Legionella* sp.

General-Purpose Water

The only consideration for this category is that it is likely to include water for effectively "domestic" use (i.e., offices, etc.) and consequently conventional potability standards will apply.

Service Water

Water for boiler feed must at least be softened, but ideally demineralized.[7,8,14] However, of primary concern will be the need to conform to the boiler manufacturers' recommendations, although the points raised earlier must take precedence in relation to the steam quality.

Water Standards for Ingredient Use

Water Sources

Various international standards and legislative requirements (e.g., WHO and EC) apply to the potability of water, and these are discussed in the Section "Legislation and Regulations Governing Water Quality." Potable water is derived from several fresh water sources and the quality and character will be directly related to the geology of the catchment area. All fresh water is derived from rain (or snow), which percolates through the upper layers of soil and, consequently, accumulates and dissolves inorganic salts, plus naturally occurring organic matter and microorganisms. In addition, pollutants arising from human activity may be included.

Water flowing through limestone or chalk will have a high content of dissolved solids, alkalinity, and total hardness, whereas water from sedimentary rock structures, such as granite, will be low in dissolved solids, hardness, and alkalinity. Water from peat areas can have a high content of organic matter and is often discolored yellow.[6]

Drinking water supplies are derived either from surface sources or from underground water courses via wells or boreholes, although municipal suppliers can use both source types and blend into reservoirs. The water is then treated to potable standards by filtration and sterilization for supply to consumers and industry via a system of water mains.

The two main sources of water are *surface* and *ground* waters, and they can differ widely in quality because of their differing routes through the environment.[6]

Surface Water

Water abstracted from lakes (natural and man-made) and rivers is termed surface water and can be subject to major variations in quality and composition, due to seasonal variations (arising from climatic conditions) and also as a result of contamination from effluent discharges and agricultural and urban runoff (which usually drains into rivers).

In the atmosphere, the water will dissolve gases, both natural and those derived from industrial activities, plus many types of air-borne particles. Once collected on the surface, the water is exposed to many materials, including mineral particles, natural organic matter (like decomposing vegetation), and artificial organic compounds, such as pesticides, herbicides, and biodegradation products (which, if derived from sewage, may be associated with fecal bacteria, plus other pathogens and viruses) from both urban and agricultural sources. Industrial effluent can also add to the biological and chemical oxygen demand of the water.

Ground Water

Water obtained from boreholes is called ground water and is usually of much more consistent quality than surface water, but does require a correctly constructed well that is of sufficient depth to eliminate pollution from the surface. The materials found in ground waters are also derived from the atmosphere, but additionally, as the water permeates through various rock types, it dissolves (depending on the geology of the area) various anions (bicarbonate, sulfate, nitrate, and chloride) and cations (such as calcium, magnesium, sodium, potassium, and perhaps traces of iron and manganese). Although likely to be less contaminated than surface waters, some soluble organic and inorganic materials can leach through to the ground waters.

Abstraction of Surface Water

River water is invariably polluted by waste water inflows from towns and industrial activity, so that processing river water is difficult and expensive. Consequently, it tends to be taken as far from waste outlets as possible, from un- or less-populated regions, and from straight stretches of the river or the outer side of river bends.

For abstraction from reservoirs and natural lakes, ideally, the collection area is far from industrial activity and the water collected remains in the reservoir for several months to improve purification by settling; this can be especially important in peaty areas.

Abstraction of Ground Water

Percolation of water depends on the topography and properties of the ground. Different types of rock differ in their resistance to water percolation, so that water collects in aquifers or water-retaining ground strata, having percolated down to a water-damming layer (such as slate, marl, granite). The height of the ground-water level can vary markedly. It is raised by increased rainfall and is lowered by natural outflow (underground streams and springs) in addition to well abstractions. Drought conditions can have a major impact on the water level in a well.

Ground water can be abstracted by boreholes (tube wells) dug to considerable depths, or where the water surfaces at springs. Ground water contains fewer suspended solids and microorganisms due to the filtration effect of percolation through the various rock strata.

The performance of boreholes can be dependent on a number of factors, including pump failure, clogging by sand influx, and the action of various bacteria, such as iron-oxidizing bacteria and sulfate-reducing bacteria.

Seasonal Influences on Water Quality

The quality of water entering the water treatment plants, prior to supply to consumers, may not be constant because of many factors that can affect the incoming water, such as seasonal variations. For example, runoffs in spring (especially from melt water) are likely to be soft, but have a high loading of mud and probably a high bacterial count. Runoff after a period of drought would likely produce hard water with a high mineral content. Water sources tend to increase in mineral level through dry summer periods, while in autumn dead and decaying vegetation will add color, taste, organic breakdown products, and bacteria. Aquatic organisms and algae are seasonal, especially in surface waters. Algal blooms can be particularly troublesome and lead to coloration, taints, and even toxins. The annual agricultural cycles and decay of vegetation can lead to an increase in nitrate levels in the autumn, in addition to the use of fertilizers. The daily cycle can also lead to composition variations, such as sewage or industrial effluent variations during the day and night, at weekends, or holiday periods.

Chemical Characterization of Water Types

A common way to characterize water is to describe its *hardness*, which is determined by the content of calcium and magnesium salts (the full chemical nature of hardness is considered in more detail in the Section "Inorganic Constituents of Water"). Thus, water can be described as *hard* or *soft* depending on its geological source, whereas other water sources can be described as *brackish/salty* or *peaty* (*humic*).[15]

Hard Water

Hard water contains calcium and magnesium salts in solution, in the form of bicarbonates when the water is drawn from a chalky or limestone source, or in the form of sulfates from sandstone sources. Hard waters have a fuller flavor and are thought to be most palatable and beneficial to consume.

Soft Water

Soft water is frequently obtained from surface sources flowing through rocky terrain or can be abstracted from underground sources where the aquifer is gravel or laterite. The mineral content may be very little and is usually sodium and potassium salts, such as bicarbonates, sulfates, chlorides, fluorides, or nitrates. The taste of soft water tends to be slightly soapy.

Brackish/Salt Water

Brackish water contains a high proportion of sodium or potassium chlorides in solution; this occurs in the neighborhood of underground salt deposits or in close proximity to the sea coast. Brackish water invariably tastes salty.

It should be noted here that there has been an increasing use of brackish water and sea water for potable supplies, especially in the Middle East and around the Mediterranean coast and islands (e.g., Malta), after desalination, usually, by reverse osmosis.

Peaty Water

Soft waters that are derived from moorland and heathland peat areas are referred to as humic and are obtained from surface sources draining from peat bogs or from borehole water abstracted from the marshland plain areas of river valleys. The water is colored yellow by the humic and fulvic acids present (arising from vegetation decomposition) and may also have an unpleasant odor. This water usually tastes bitter and often phenolic.

Inorganic Constituents of Water

If water is subjected to a fine enough analysis, it is likely that just about every inorganic element will be identified as present. The influence of these inorganic components on water quality and, hence, on the use of water as an ingredient is a major issue for all food and drink manufacturers, and the impacts on beer and the brewing process are discussed in detail in the Section "The Influence of Inorganic Ions or Beer Quality." However, at this point, it is pertinent to note that absolutely pure water actually tastes very flat and uninteresting and that some mineral content can be desirable. In addition, ultrapure water is a very aggressive material and would lead to corrosion problems in production.

The levels of the inorganic constituents in water acceptable as potable can be described in terms of major, minor, and trace quantities.[15]

Major Constituents

A major constituent is defined as present at levels above 10 mg/l (ppm) and may be up to levels of several hundred milligrams per liter. Major constituents include calcium, magnesium, sodium, sulfate, chloride, bicarbonate, nitrate, and silica.

Minor Constituents

A minor constituent is defined as present in the range of 0.01 to 10 mg/l (i.e., 10 μg/l to 10 mg/l or 10 ppb to 10 ppm). Minor constituents include potassium, iron, manganese, copper, aluminum, zinc, boron, carbonate, and fluoride.

Trace Constituents

Trace components are present below 0.01 mg/l (10 μg/l or 10 ppb) and include cadmium, lead, mercury, rare earths, and bromide.

Water Hardness

The definitions and analysis for water hardness are well documented (e.g., Refs. 3, 4, 6, 7).

Total hardness can be divided into two categories: *temporary* hardness (due to bicarbonates of calcium, magnesium, and sodium) and *permanent* hardness (composed of calcium and magnesium sulfates, calcium and magnesium chlorides, and calcium and magnesium nitrates).

Total hardness represents the total of all the calcium and magnesium salts present and is the sum of carbonate and noncarbonate hardness.

Hardness of water is classically defined in *hardness units* or *degrees of hardness* (°H), but different countries have their own definitions. For example:

- Germany (°D), where 1°D = 10 mg/l CaO
- France (°F), where 1°F = 10 mg/l $CaCO_3$
- USA (°USA), where 1°USA = 1 mg/l $CaCO_3$
- Britain (°GB), where 1°GB = 14.3 mg/l $CaCO_3$ (from one grain/imp. gallon)

A more modern classification of water hardness[3] records the analysis of total hardness as: mmol/l of $CaCO_3$, by titration of water against ethylenediaminetetra acetic acid (EDTA) and defines:

- Hardness range 1 — "soft"
  Up to 1.3 mmol/l = up to 7.3°D; 13.0°F; 130.0°USA; 9.1°GB
- Hardness range 2 — "average"
  1.3 to 2.5 mmol/l = 7.3 to 14°D; 13 to 25°F; 130 to 250°USA; 9.1 to 17.5°GB
- Hardness range 3 — "hard"
  2.5 to 3.8 mmol/l = 14 to 21.3°D; 25 to 38°F; 250 to 380°USA; 17.5 to 26.6°GB
- Hardness range 4 — "very hard"
  Over 3.8 mmol/l = >21.3°D; >38°F; >380°USA; >26.6°GB

Total alkalinity (temporary hardness) is also referred to as carbonate hardness and represents the carbonate plus bicarbonate content of water. It is measured by titration against standardized acid, and it is the pH-increasing effect of carbonate hardness that brewers have recognized for many years as the major water treatment procedure required for most efficient beer production.

Permanent hardness (noncarbonate hardness) collectively includes principally the chloride and sulfate salts of calcium and magnesium. These ions have a pH-decreasing effect in solution which opposes the pH-increasing effect of carbonate. These important reactions are, to a large extent, responsible for pH control throughout the brewing process, and this topic is developed further in the Section "The Influence of Inorganic Ions on Beer Quality."

Residual alkalinity is a concept that many brewers (especially in Germany) find valuable in defining the extent of treatment of water for brewing.[16] It represents the analytical difference between the carbonate and the noncarbonate hardness and is usually expressed as

Total carbonate ion concentration − (the sum of the calcium ion concentration + half of the magnesium ion concentration) ÷ 3.5

The residual alkalinity is essentially the result of the competition between the pH-increasing (carbonate + bicarbonate) and pH-lowering properties (all Ca and Mg ions, except those whose effect is neutralized by pH-increasing ions). The net effect is such that the higher the residual alkalinity, the more effective the carbonate hardness and so the higher the pH to be expected.

Organic Constituents of Water

Modern analytical instrumentation (such as mass spectrometry) allows detection of organic compounds at very low concentrations (as low as parts per trillion or ppt) and, consequently, a wide variety of different organic molecules (both natural and man-made in origin) can be found in water supplies (e.g., Refs. 5, 6, 17–19). Regulations concerning potable standards aim to identify those chemicals that are known to have toxic or carcinogenic properties. These include: pesticides, polycyclic aromatic hydrocarbons, trihalomethanes — THMs (trichloromethane, bromodichloromethane, dibromochloromethane, and bromoform), and many other halogenated compounds (such as tetrachloromethane, dichloromethane, 1,1,1-trichloroethane, trichloroethene, and tetrachloroethene).

Microbiological Constituents of Water

As water is very much part of the natural world, it will obviously be home to a wide variety of living organisms. Of principal interest is the content of pathogenic organisms, which is why so much emphasis is given to disinfection of water before use. Water is always contaminated when it first comes into contact with the ground. As water penetrates into the ground layers, there may be a gradual filtration to achieve a general improvement in the biological quality. However, the quality of surface waters may actually deteriorate with storage. There are potentially a very large number of contaminating microorganisms, so it is usual to select a few key indicators of the presence of both harmful and harmless species. Hence, water is examined for *Escherichia coli* as an indicator of fecal contamination; if it is present, it can be concluded that there is a possibility of other pathogenic organisms being present. Other microorganisms frequently monitored are sulfate-reducing bacteria, *Pseudomonas* sp., thermotolerant bacteria in general, and protozoa, such as *Cryptosporidium* sp. and *Giardia lamblia*. Other troublesome bacteria include *Thiobacillus* sp., which are able to oxidize sulfide and free sulfur to sulfate (*T. thiooxydans*) and oxidize ferrous to ferric iron (*T. ferrooxydans*); iron bacteria (such as *Gallionella ferruginea*) can precipitate iron to block pipes and bore holes.[6]

Legislation and Regulations Governing Water Quality

In recent years, there have been many laws passed relating specifically to the quality of water, but other legislations are also relevant. For example, in the UK, the Food Safety Act (1990) requires that any item of food or drink that is supplied for sale must be "safe," meaning that only potable water can be used in manufacturing foodstuffs, and that there is emphasis on maintaining its quality and not adding to it any undesirable materials.[20]

The base quality standards for water are derived from World Health Organization (WHO) findings, with *Guidelines for Drinking Water Quality* first published in 1984, but with revised advice issued as a second edition in 1993.[20] Individual countries have used these guidelines to set their own standards and a list of the latest UK Water Supply (Water Quality) Regulations 2000, transposed from the 1998 EU Directive (98/83/EC), is given in Table 4.1 (from personal communication from Dr DE Long, Technical Director of British Beer and Pubs Association). It is proposed that these UK standards came into force on January 1, 2004, and the inclusion of this table here is to indicate the scale of legislation applying to water quality, not to give a totally comprehensive account of all national regulations worldwide; individual countries have specific limits for a range of compounds, depending on different concerns.

These regulations all apply not only to drinking water, but also to water used in the production of food, with a basic requirement of being wholesome and clean. They set limits for a range of physical and chemical components and microbiological contaminants (requiring water "to be free from any pathogenic microorganisms and parasites and any substances, which constitute a

TABLE 4.1

Water Quality Legislation — Parametric Values

Parameter (1988)	Units	UK (2000)	EU
"Aesthetic" Qualities			
Turbidity	NTU (nephelometric turbidity units)	4[a]	1[a]
Taste/odor	—	Acceptable	—
Color	TCU (total color units)	20	Acceptable
Conductivity	μS/cm at 20°C	—	2500[a]
pH	—	6.5–10.0	—
Inorganic Constituents (Maximum Concentrations)			
Aluminum	mg/l	0.2	0.2
Ammonium	mg/l	0.5[a]	—
Antimony	μg/l	5	5
Arsenic	mg/l	0.01	0.01
Boron	mg/l	1.0	1.0
Bromate	mg/l	0.01	0.01
Cadmium	μg/l	5	5
Chloride	mg/l	250[a]	250[a]
Chromium	mg/l	0.05	0.05
Copper	mg/l	2.0	2.0
Cyanide	mg/l	0.05	0.05
Fluoride	mg/l	1.5	1.5
Iron	mg/l	0.2	—
Lead (to be reduced from December 2013)	mg/l	0.025 (0.01)	0.025 (0.01)
Manganese	mg/l	0.05	0.05
Mercury	μg/l	1	1
Nickel	mg/l	0.02	0.02
Nitrate	mg/l	50	50
Nitrite	mg/l	0.1	0.5
Selenium	mg/l	0.01	0.01
Sodium	mg/l	200	—
Sulfate	mg/l	250[a]	250[a]
Organic Contamination[b] (Maximum Concentrations)			
Acrylamide	μg/l	0.1	0.1
Aldrin and dieldrin	μg/l	0.03	0.03
Benzene	μg/l	1.0	1.0
Benzo(α)pyrene	μg/l	0.01	0.01
1,2-Dichloroethane	μg/l	3	3
Epichlorohydrin	μg/l	0.1	0.1
Heptachlor epoxide	μg/l	0.03	0.03
PAH[c]	μg/l	0.1	0.1

(*Table continued*)

TABLE 4.1 *Continued*

Parameter (1988)	Units	UK (2000)	EU
Pesticides			
Total	μg/l	0.5	0.5
Other	μg/l	0.1	0.1
Tetrachloroethene and trichloroethene	μg/l	10	10
1,1,2-Tetrachloromethane	μg/l	3	—
Total organic carbon	μg/l	No abnormal change[a]	No abnormal change[a]
Trihalomethanes (total)	μg/l	100	100
Vinyl chloride	μg/l	0.5	0.5
Radioactive Constituents			
Tritium	Bq/l	100	100
Microbiological Quality Parameters[d]			
Escherichia coli	per 100 ml	0	0
Enterococci	per 100 ml	0	0
Total coliforms	per 100 ml	0	0
Clostridium perfringens	per 100 ml	0	0
Total colonies		No abnormal change	No abnormal change

[a]Indicator parameter.
[b]Other countries (e.g., Australia, Canada, U.S.A.) set specification limits for a wider range of organic compounds than EU.
[c]PAH, polycyclic aromatic hydrocarbons.
[d]Other countries (e.g., Australia, Canada, U.S.A.) set limits for other organisms including bacteria (*Salmonella, Pseudomonas, Legionella*), viruses, and protozoa (*Giardia lambilia, Cryptosporidium*).

Source: All data provided by Dr. D.E. Long (BBPA). With permission.

potential danger to human health"). Most of the responsibility for meeting these directives is undertaken by municipal water suppliers, but the same regulations apply to food and drink producers who use their own water sources, and as a result are obliged to ensure that the ingredient water used in their processing also conforms to the parametric values set by these regulations.

In addition to these potability standards, there are environmental and pollution regulations that relate to the disposal of waste water generated during processing. All waste producers have a legal duty of care with respect to the disposal of all waste, including water. Since water is a prime raw material, it is imperative to keep all sources of water (ground, surface, and even tidal) as free from pollution as possible. As stated previously, there are several relevant publications on the topic of waste water management (e.g., Ref. 8).

The Influence of Inorganic Ions on Beer Quality

The literature contains many references on this topic (e.g., Refs. 1–6, 19). It is apparent that brewers have been aware of the importance of inorganic ions on beer flavor and quality for many years. In the following sections, many of the potential influences of inorganic ions on beer production will be described in detail.

Historical Background

Historically, breweries originated on sites having their own particular water supply and composition (such as wells and springs) and became grouped into regions producing beers whose characteristics came to be regarded as typical of those regions associated with the prevalent water supply. The data in Table 4.2 indicate how certain brewing centers became renowned for particular beer types as a consequence of the prevailing water composition.[2]

Burton water is very high in permanent hardness, due to the high content of calcium sulfate (gypsum) and, as a brewing center, Burton-upon-Trent was famous for strong, pale, bitter ales. London and Munich have alkaline water supplies and are associated with dark, soft-flavored beers — mild and brown ales in the case of London, but dark, relatively lightly hopped lagers from Munich. Pilsen has a very soft water, with very little mineral content, and is renowned for classical, highly hopped pale lagers.

The tendency for modern brewing operations is to be more dependent on municipal water supplies (which, although potable, may be less than ideal for brewing). In addition, the need to produce a wider range of products from one brewery requires a greater flexibility of liquor composition, so that one particular water supply may not be appropriate for the full product range. Further, one of the key decisions these days in locating a new brewery will be distribution access, as it will be assumed that an adequate water supply will be available and that appropriate treatment will be readily organized.

TABLE 4.2

Chemical Compositions (mg/l) of Various Brewing Waters

	Burton	Munich	London	Pilsen
Ca^{2+}	268	80	90	7
Mg^{2+}	62	19	4	1
Na^+	30	1	24	3
HCO_3^-	141	164	123	9
SO_4^{2-}	638	5	58	6
Cl^-	36	1	18	5
NO_3^-	31	3	3	0

Sources of Ions in Beer

In a consideration of the effects of inorganic ions on beer, one must take into account all the potential sources of ions, not just the mineral composition of the liquor, especially if the brewer is committed to maintaining consistent levels of certain ions in the finished product. To this end, it is necessary to carry out a full mass balance of all the major ions throughout the brewing process and to identify all likely sources not only from the liquor, but also all raw materials and processing aids and additives.[19] For example, Table 4.3 shows the ion contents analyzed in an all-malt wort (at 10° Plato or 1040° gravity) prepared using distilled water and in the beer fermented from this wort.

Several points are apparent from the data in Table 4.3, notably (1) malt will contribute significant proportions of various ions, especially phosphate, K^+, Mg^{2+}, and Cl^-; (2) during fermentation, several ions are carried through to beer unchanged (e.g., Cl^-), but yeast metabolism will lead to absorption of high levels of phosphate and K^+ and all available Zn^{2+}, with some Mg^{2+}.

Other raw materials may contribute to the total ionic content of beer,[19] but suffice to say here, it is important to note that, in addition to brewing water and any salts added by way of liquor treatment or ionic adjustment, the contribution from malt, adjuncts, and hops should be taken into account.

TABLE 4.3

Ion Contents in Wort (10°Plato) and Beer using Demineralized Water

	Wort (mg/l)	Beer (mg/l)
Na^+	10	12
K^+	380	355
Ca^{2+}	35	33
Mg^{2+}	70	65
Zn^{2+}	0.17	0
Cu^{2+}	0.15	0.12
Fe^{3+}	0.11	0.07
Cl^-	125	130
SO_4^{2-}	5	15
PO_4^{3-}		
Free	550	389
Total	830	604

Effects of the Ionic Contents of Wort and Beer

The ions present in water are from fully dissociated salts and there are significant interactions between all ions during brewing, from all sources. Some studies (e.g., Ref. 3) distinguish between chemically inactive and chemically reactive ions to indicate that some ions will pass unchanged into beer (but may influence beer flavor, either beneficially or adversely), because of no particular interaction with ions derived from other raw materials (e.g., the malt). Reactive ions in water (especially Ca^{2+} and Mg^{2+}) will react significantly with malt components, especially with regard to influencing pH.

It may be more appropriate, however, to distinguish between *direct* and *indirect* effects of ions on beer quality,[5,19] as the source of the ions involved (from water or added salts or malt, etc.) is likely to be irrelevant to the reactions involved. In subsequent sections, we will address these direct flavor effects and the indirect reactions in general, with a detailed analysis of pH control, followed by specific details of all ions in turn.

Direct Effects of Ions on Beer Flavor

By direct effects, we are referring to the taste sensations arising as a direct stimulation by ions of taste receptors (or "taste buds") in the buccal cavity and, especially, on the upper surface of the tongue. Taste buds are sensitive to four types of taste: sweet, salt, sour, and bitter. The four types of taste buds show distinct distribution patterns[21]; *sweet* is detected on the front of the tongue, with *salt* on the sides at the front, *sour* on the sides further back, and *bitter* on the top surface at the back of the tongue. It is interesting to note that inorganic ions, at least in humans, appear not to be detected readily by olfactory organs — the receptors (primarily in the nose) responsible for the sense of smell.

Knowledge of the direct effects of ions on flavor was gained over many years and is based, to some degree, but somewhat empirically, on the historic associations of various beer types with particular water supplies (as described in earlier in the Section "Historical Background"). If one attempts to quantify these direct effects, it is imperative to ensure that all other analytical parameters are identical, as flavor effects due to ions tend to be relatively subtle and easily masked.[19]

Sodium

Na^+ ions can contribute a *salty* taste at a concentration of 150 to 200 mg/l[19] and may be *harsh* and *sour* in excess, viz., greater than 250 mg/l.[18,22,23] However, at lower levels (up to 100 mg/l), sodium ions can produce a *palate-sweetening* effect, especially in association with chloride ions.

Potassium

K^+ ions, like Na^+, can taste salty, but only at concentrations greater than 500 mg/l.[19,22] In the main, the relatively high natural concentration of potassium in beer, some 300 to 500 mg/l,[4,19] principally extracted from malt, is essentially flavor neutral and, as such, additions of KCl (rather than NaCl) may be preferred as a means of increasing the chloride content of beers in order to influence perceived palate fullness.

Magnesium

Mg^{2+} ions can contribute a bitter and sour flavor,[19,24] which increases above 70 mg/l. These flavor effects appear to depend on a balance between Mg^{2+} and Ca^{2+} ions.[19]

Calcium

Ca^{2+} ions are essentially flavor neutral at the levels found in beer, other than for a slight moderating influence on the sour flavor produced by high levels of Mg^{2+}.[19]

Hydrogen

The influences of the concentration of H^+ ions on beer flavor are usually considered as the impact of pH on the production of flavor components, which can, of course, be considerable,[25,26] and the effect of various ions on pH control is discussed later in the Section "Control of pH." However, H^+ ions can also exert direct flavor effects[25]:

- At pH values below 4.0 (or 0.1 mg/l H^+ ions), beers tend to taste more sharp and acidic, with increased drying after-palate and a tendency for perceived bitterness to be enhanced
- At even lower pH (3.7 and below, or 0.2 mg/l H^+ and above), these effects increase in intensity rapidly, with markedly enhanced metallic after-palate
- Above pH 4.0, the palate effects relate to increased mouth coating, with enhanced scores for biscuit and toasted characters
- At pH 4.4 and above (or 0.04 mg/l or 40 µg/l H^+ ions and below), the mouth-coating effects become increasingly more accentuated, with soapy and even caustic characters developing

It should be noted that in the case of organic acids, their flavor contributions are not restricted merely to acting as suppliers of H^+ ions in order to produce the characteristic sourness, but the structural features of these molecules also determine their flavor threshold.[26] For example, organic acids such as acetic, lactic, and citric have very different flavor characteristics, but all contribute to perceived *acidic* and *sour* flavors. These effects can be very important in classical acidic beers, such as "Gueuze" and "Lambic" beers.

Iron

Fe^{3+} ions are notorious for contributing negative flavor characters, such as metallic and astringent,[19] even at very low concentrations, say 0.5 mg/l in most beers. Fe^{3+} can be detectable at less than 0.1 mg/l in more delicately flavored beers.

Chloride

It is well accepted that Cl^- ions give fullness and sweetness to beer flavor,[1,19,22,23] these effects being enhanced by increasing concentrations from 200 to 400 mg/l.

Sulfate

SO_4^{2-} ions impart dryness/astringency to beer and also increase bitterness, palate and after-palate, even at constant iso-humulone levels.[1,19,22,23] Again, these effects become more pronounced as the concentration increases from 200 to 400 mg/l.

Chloride: Sulfate Balance

Many authors (e.g., Refs. 1, 19, 22, 23) refer to the importance of the chloride to sulfate balance. From the previous discussion about chloride and sulfate, it can be seen that the relative flavor effects of these ions are somewhat antagonistic. In an attempt to quantify this point, it has been shown[19] that increasing the $Cl^-:SO_4^{2-}$ ratio from 1:1 to 2:1 (on a mg/l basis) achieved increased taste panel scores for body and sweetness, with commensurate reduction in drying, bitter, and metallic flavors. In contrast, when the $Cl^-:SO_4^{2-}$ ratio was changed from 1:1 to 1:2, the increased sulfate content achieved reduced body and sweetness, but increased bitterness and drying flavors.

These effects are repeatable at different absolute concentrations of chloride and sulfate. It appears that, in many cases, it is the relative ratio of the two ions that has the major flavor influence, often irrespective of the accompanying cations.

The examples listed previously serve to indicate the value of assessing the direct flavor contributions of inorganic ions. As stated previously, these flavor effects tend to be quite subtle. For instance, in a particularly sweet beer with a high perceived body (with a high content of added priming sugar or with a relatively high final gravity achieved from a wort of low fermentability), the influence of increasing the sodium or chloride content will be much less noticeable than in a slightly bitter, lower gravity, fully attenuated product.[19] However, taking cognizance of the inorganic ion contents can be particularly important in product development projects and, especially so, in beer flavor matching exercises involving different breweries in two or more locations, with different plant designs and different liquor supplies.

Indirect Effects of Ions on Beer Flavor and Quality

Some authors identify the effects of so-called "reactive ions" (e.g., Ref. 3), referring to those ions in water that react with malt constituents during mashing and have a major influence on pH control. These influence pH not only during wort production, but will also have a major effect on the progress of fermentation, with significant concomitant effects on beer flavor. In addition, there are other reactions that will influence other beer quality parameters. However, there is some merit in regarding these reactions as indirect effects of ions on beer flavor, as their influence is via chemical and enzymic reactions and interactions with other wort and beer components, all contributing in some way to beer quality. Further, it is also more appropriate to consider all the ions present, irrespective of their source (viz., directly dissolved in the liquor, added as deliberate liquor treatment/ionic adjustment, or derived from brewing materials), as contributing to the total effect.

In this way, one can identify four distinct categories of effects.[19]

Yeast Requirements

Yeast requires many inorganic ions for optimum growth and fermentation.[27] Appropriate concentrations of many elements allow for accelerated growth and increased biomass yield, enhanced ethanol production, or both. An imbalance in inorganic nutrition causes complex, and yet subtle, alterations of metabolic patterns and growth characteristics.

Inorganic ions are required in enzymic and structural roles. Enzymic functions include:

- As the catalytic center of an enzyme (e.g., Zn^{2+}, Mn^{2+}, Cu^{2+}, Co^{2+})
- As activators of enzyme activity (e.g., Mg^{2+})
- As metal coenzymes (e.g., K^+)
- As cofactors in redox pigments (e.g., Fe^{3+}, Cu^{2+})

Structural roles involve neutralization of electrostatic forces present in various cellular anionic molecules. These include:

- K^+ and Mg^{2+} ions bound to DNA, RNA, proteins, and polyphosphates
- Ca^{2+} and Mg^{2+} combined with the negatively charged structural membrane phospholipids
- Ca^{2+} complexed with cell wall phosphate ions

Effects on Malt Enzymes

In addition to influencing mash pH (as detailed later), calcium ions can directly stimulate amylolytic and proteolytic enzyme activities during wort production. Ca^{2+} ions protect malt α-amylase activity against inhibition by heat, leading to increases in extract[19,22] and can lead to an increase of endopeptidase activity at lower mash temperatures.[28]

Effects on Stability of Colloidal Systems

Several examples can be given on the effects or stability of colloidal systems:

1. Yeast flocculation is improved by Ca^{2+} [27,29,30]; most yeast strains require at least 50 mg/l Ca^{2+} ions for good flocculation. Calcium ions almost certainly act by binding to mannoproteins on the yeast cell walls and so crosslink cells in a lectin-like manner.[27]
2. The interactions between proteins, polyphenols, and hop iso-α-acids are influenced by several ions, including Ca^{2+}, Mg^{2+}, Fe^{3+}, and PO_4^{3-}. Formation of complexes such as these can lead to improved wort clarification during boiling and improved beer clarification during maturation, leading to enhanced haze stability.[4]
3. Protein precipitation during wort boiling (trub formation) occurs not only because of thermal denaturation, but also because of the neutralizing effect of cations (especially Ca^{2+}) on the negatively charged polypeptides. It has been estimated that a minimum level of 100 mg/l Ca^{2+} ions is required for good-quality protein break formation.[19]
4. Oxalate derived from malt is precipitated as calcium oxalate.[31] Ideally, this should occur during wort production, since subsequent formation of calcium oxalate crystals in beer can lead to gushing and haze formation.[18,32,33] It is recommended that 70 to 80 mg/l Ca^{2+} ions should be present during mashing to eliminate excess of oxalate during beer storage.

Effects on pH Control

Several ions can influence pH (i.e., the H^+ ion concentration) during brewing,[1-3,25,26] especially during wort production; the ionic interrelationships with the buffering systems involved are discussed in detail in the next section. Arguably, control of wort and beer pH is the single most important feature of the influence of inorganic ions on beer quality and flavor.

Control of pH

It is somewhat perverse that when assessing the influence of water on beer quality, it is the effects of materials (such as inorganic ions) dissolved in water that are principally identified, with little or no consideration given to the molecules and ions that actually compose water itself (viz., H_2O

and H^+ and OH^-). According to Bamforth,[26] assuming the molecular weight of water to be 18 and the density (at 25°C) to be 1000 gm/l, then *pure* water is actually 55.5 M (i.e., contains 55.5 mol/l). Despite this relatively high concentration, water is, however, only sparingly dissociated into ions, with both H^+ and OH^- ions being present at only 1.0×10^{-7} mol/l of each, but the influence of these dissociated ions can have far-reaching effects on all chemical and biochemical reactions.

With regard to brewing chemistry, the concentration of hydrogen ions, or more precisely, hydroxonium ions (H_3O^+ or $H_2O \cdot H^+$, as the active species is actually a hydrated proton) is particularly relevant in the range 200 μmol/l (pH 3.7) to 1.0 μmol/l (pH 6.0)[25,26]; but, of course, water is not the only source of dissociated hydrogen ions in solution. Consequently, it is most appropriate that full consideration is given to factors influencing pH as a key feature of the effects of water on beer, since, in many ways, control of pH throughout the brewing process (from mashing-in to final packaging) is fundamental to achieving end-product consistency. The maintenance of control of pH during wort production, fermentation, and conditioning, by ensuring reproducible conditions for the numerous enzymic and chemical reactions occurring during these key beer production stages, is of great importance to beer quality. In addition, finished beer pH impacts on beer flavor, physical stability, and microbiological stability.[25,26]

Clearly, the pH of wort and beer will be determined by the dissociated ions present, both inorganic and organic.[1-3,25,26] The actual pH values will be dependent on the concentration and nature of buffering components present, by the absolute concentration of H^+ and OH^- ions, by the concentration of the reactive ions, and by temperature.

Buffer Systems

There are three key buffering systems likely to be influencing wort and beer pH[19,25,26]:

- Carbonate/bicarbonate
- Phosphate; both inorganic and organic (especially phytates)
- Carboxylic acids (especially aspartate and glutamate side chains in proteins/polypeptides/peptides/free amino acids)

The influence on pH of these buffers is manifested through the relevant pK_a values of the acidic components involved[25,26]; viz.

- Carbonic acid (pK_1)   6.3
- Phosphoric acid (pK_2)   6.7
- Aspartic acid (β-carboxyl) (pK_2)   3.7
- Glutamic acid (γ-carboxyl) (pK_2)   4.3

Reactions Controlling pH

The reactive ions present in wort either from the liquor composition or added as a result of water treatment or derived from extract materials (malted barley, other cereals, adjuncts, etc.) and hops will interact with these buffering species to effect the pH value achieved. These effects will either increase the concentration of free H^+ ions (i.e., lower pH) or decrease H^+ concentration (i.e., increase pH). Calcium ions are most reactive in this respect, with Mg^{2+} ions also contributing, but to a lesser extent.

The key reactions are as follows:

$$H_2O + CO_2\uparrow \underset{}{\longleftrightarrow} H_2CO_3 \underset{H^+}{\longleftrightarrow} HCO_3^- \underset{H^+}{\longleftrightarrow} CO_3^{2-} \overset{Ca^{2+}}{\longleftrightarrow} CaCO_3\downarrow \quad (4.1)$$

$$3Ca^{2+} + 2HPO_4^{2-} \longrightarrow 2H^+ + Ca_3(PO4)_2\downarrow \quad (4.2)$$

$$Ca^{2+} + \text{polypeptide-}H_2 \longrightarrow 2H^+ + \text{polypeptide-}Ca\downarrow \quad (4.3)$$

pH Control during Wort Production

The key point for control of pH throughout the brewing process is during mashing. This is due to the major influence that can be exerted at this stage on the content and format of these buffer systems, which will operate subsequently in wort and beer. The grist composition selected for the beer type to be produced will be the key controlling parameter on the constituents present; the primary proportions of malt to adjunct, the malt protein content, the degree of malt modification, kilning characteristics, etc., will all be major determining factors of wort composition.

However, the liquor composition (possibly after treatment) used for mashing and sparging can have significant impact on the various buffer systems and so affect pH during mashing and wort runoff, which can, subsequently, have a modifying influence on brewhouse performance and wort composition.[25] For instance, the antagonistic reactions involving residual alkalinity (bicarbonate) and Ca^{2+} ions are clearly apparent from Equation (4.1),[19,25,26] with bicarbonate tending to increase pH and calcium (by precipitating not only carbonate, but phosphates and protein complexes also), as in Equation (4.2) and Equation (4.3), releasing H^+ ions and thus reducing pH. The phosphates involved are both free inorganic residues and organic phosphates, such as phytates (of varying degrees of phosphorylation). The protein complexes will range considerably in molecular size, from small peptides upwards.

The relative importance of these reactions in contributing to the level of free H^+ ions in solution is difficult to quantify,[19,25] but most authors agree that there is considerable merit in maintaining mash pH in the range 5.0 to 5.5,[1–3,19,25,26] especially so with regard to the pH optima of the malt enzymes involved (viz., 5.7 for α-amylase, 4.7 for β-amylase, and 4.2 to 5.0 for proteolytic enzymes). In addition, many authors suggest that this pH control is best achieved by maintaining low alkalinity (bicarbonate content) in brewing liquor (less than 50 mg/l), plus sufficient Ca^{2+} to achieve the desired pH level (not less than 100 mg/l).

Because of the precipitation of Ca^{2+} in these pH control reactions, there is a considerable reduction in the calcium ion concentration during wort production; about 50 to 60% of the Ca^{2+} ions present during mashing (either present in mashing liquor, or as added salts, or derived from grist materials) will be lost with spent grains and trub.

The pH value of the collected wort is a reflection of pH control during mashing, but it is worth noting that pH at actual mash temperatures is considerably lower (approximately 0.3 units) than the pH value determined at 20°C due to thermal encouragement of H^+ ion dissociation. Consequently, selecting the ideal pH for optimal activity of amylolytic and proteolytic enzyme systems is rather difficult, especially since conditions applying at mashing bear little resemblance to conditions usually employed for enzymological investigations; that is, assessment of initial reaction rates at relatively low temperatures and at high concentrations of pure substrates. However, if absolutely optimal pH is hard to define, reproducible pH conditions can be set, and judicious control of calcium content allows some measure of control on mash pH. This can be used to exert influence over wort composition and other properties.[19,25]

For example, wort runoff rate can be significantly increased as a consequence of improved mash bed permeability as a direct consequence of reduced pH or increased Ca^{2+} concentration during mashing.[19,25,34] This influence on mash bed permeability probably relates to effects of pH or other calcium reactions on the formation of gel proteins.[25,34]

Influences of pH on Wort Composition

Controlling mash pH, particularly by increasing the Ca^{2+} level, can significantly influence wort composition.[2,19,25,26,34] For example, it has been reported[25] that decreasing mash pH from 5.5 to 5.2 by increasing the Ca^{2+} content by 200 mg/l not only increased runoff rate, but also led to increased extract and increased levels of total soluble nitrogen (TSN) and free amino nitrogen (FAN). In addition, pH control during sparging can be of importance in limiting the excessive extraction of polyphenols and silica compounds (principally derived from malt husk), both of which increase as pH increases.[19,22] As extract gravity reduces during wort runoff, the pH of the wort tends to increase (thus favoring increased extraction of tannins and silica), unless the sparging liquor contains a relatively high level of Ca^{2+} ions (up to 200 mg/l), in order to ensure a consistent wort pH value throughout runoff.[25,34] Lloyd Hind[1] recommended that sparging water should contain sufficient Ca^{2+} ions to achieve a wort pH of 5.2 after boiling.

pH during Wort Boiling

The pH of wort drops by about 0.3 units during boiling,[35] due to precipitation of phosphates and proteins/polypeptides complexed with calcium.[19] Moreover, wort gravity will have a significant influence on wort pH, with lower values achieved as gravity increases.[26]

The recovery of hop bittering compounds is increased at higher pH values, since α-acids are more soluble at higher pH. Some brewers adjust wort pH by acidification, but only toward the end of boiling to ensure minimum impact on hop utilization.[26] Lower pH values during boiling will restrict the solution of tannins (this time from hops) and so reduce the risk of beer astringency.[25] In addition, color formation may also be reduced by lower pH or increased Ca^{2+} concentration.[2,25]

pH Control during Fermentation

pH falls during fermentation as a result of the consumption of buffering materials, principally FAN, the release of organic acids, and possibly direct excretion of H^+ ions by yeast.[25,26,36] As the nature and content of the buffering materials present in wort collected in the fermenter will be a direct consequence of pH control during wort production, the fundamental importance of the interrelationships between the reactive ions during mashing is again endorsed.

The rate and extent of the pH drop during fermentation are a balance between buffering capacity and factors stimulating yeast growth.[25,26] Studies have confirmed that a measure of wort FAN levels will correlate with the observed pH change during fermentation, but not in a simple fashion, since the FAN analysis includes factors responsible for both enhancing buffering capacity (e.g., aspartate and glutamate) and encouraging yeast growth (all amino acids), leading to increased proton (H^+ ion) excretion.[25] Consequently, low FAN levels may lead to reduced buffering potential (risking low beer pH), but may also limit the extent of yeast growth; conversely, high FAN may tend to drive yeast vigor (suggesting low beer pH again), but this may be offset by the enhanced buffering capacity. Of course, other factors must be taken into account, such as wort-dissolved oxygen content and levels of zinc ions, increases in both of which will stimulate yeast growth (leading to lower pH). Wort gravity is another factor to be considered.[26]

The important point here is that pH during fermentation can make a sizeable impact on the production of flavor components by yeast.[26] A shift in wort pH from 5.75 to 5.46 led to a 50% reduction in dimethyl sulfide production during fermentation.[37] Further, the conversion of acetolactate to diacetyl is favored by lower pH.[38]

Summary of the Influences of Various Ions during Beer Production

The following summarizes the influences of various inorganic ions on beer quality.

Hydrogen

Maintenance of an "ideal" level of H^+ ions (i.e., correct pH control) at various key stages throughout brewing, but especially during mashing, is fundamental to achieving consistent product quality.[25,26] Control of H^+ ion concentration is very dependent on interactions with several inorganic ions and organic molecules acting as pH buffering systems at various points during brewing, their concentrations and influences varying as wort production and fermentation/maturation (due to yeast action) progress.

Calcium

Ca^{2+} plays a key role in pH control, especially in mashing. It increases TSN and FAN levels in wort, improves wort runoff, limits extraction of polyphenols and silica,[19,25] and protects malt α-amylase from heat inhibition.[22] Ca^{2+} also improves wort clarification and protein coagulation,[23] accentuates yeast flocculation[27,29,30] and precipitates oxalate, prevents haze formation and gushing,[32,33] and stimulates yeast growth.[27]

Magnesium

Mg^{2+} has some influence on pH control, but since Mg^{2+} salts are more soluble than Ca^{2+}, they are less effective.[19] Mg^{2+} is the most abundant intracellular divalent cation in yeast cells. It has a central role in governing yeast growth and metabolism,[39] and acts primarily as an enzyme cofactor for enzymes such as pyruvate decarboxylase. Mg^{2+} neutralizes the anionic charges on nucleic acids and proteins.[27] It has bitter/sour direct flavor effects on beer.[5]

Sodium

Yeast cells do not accumulate Na^+ and continuously excrete it to ensure low intracellular levels.[27] Na^+ has sweet/mouthcoating/salty direct flavor effects.[5]

Potassium

K^+ is derived principally from malt.[19] Potassium ions are actively transported into fermenting yeast cells, where they neutralize charges on nucleic acids and proteins and contribute to osmoregulation.[27] K^+ has a salty direct flavor effect, but at higher concentration than Na^+.[5]

Iron

Fe^{3+} may be present in humic water supplies complexed with organic matter, and can produce slime deposits in wells and pipes.[2] Iron ions can oxidize polyphenols and produce haze.[2] While they may improve foam stability at 0.5 mg/l, they can have negative direct flavor effects (metallic/astringent) as low as 0.1 mg/l.[5] Fe^{3+} is an essential nutrient for yeast, acting as a cofactor in redox pigments in actively respiring cells.[27]

Zinc

Most yeast strains require 0.1 to 0.2 mg/l of Zn^{2+} ions for effective fermentation.[5,40] It is essential for the structure and function of many enzymes, where it can be involved in the active site (zinc-metalloenzymes); alcohol dehydrogenase is an example.[27] However, Zn^{2+} can inhibit yeast growth and fermentation at higher concentrations (greater than 0.6 mg/l) if wort contains less than 0.1 mg/l Mn^{2+} ions, but this inhibition is relieved by addition of 0.6 mg/l of Mn^{2+} ions.[41] Zn^{2+} may stimulate H_2S production at levels around 1.0 mg/l.[42]

Manganese

Mn^{2+} is essential in trace levels for yeast growth and metabolism. It acts as an intracellular regulator of key enzyme activities and acts as the catalytic center for several enzymes.[27] Like Mg^{2+}, Mn^{2+} ions accumulate in the yeast cell vacuole.[27] They may have a synergistic effect with Zn^{2+} ions.[41]

Copper

Copper is an essential micronutrient for yeast at low concentrations acting as a cofactor in redox pigments.[27] However, it is toxic to yeast above 10 mg/l,[2,22] and disrupts yeast cell plasma membrane integrity.[27] It can contribute to haze formation,[2] and may reduce the concentration and flavor effects of sulfur compounds (H_2S) in beer.[5]

Other Cations

Heavy metals, such as lead (Pb^{2+}) and tin (Sn^{2+}), can be inhibitory to certain yeast enzymes and can induce haze formation.[2]

Bicarbonate/Carbonate

These anions dramatically increase pH during mashing. The residual alkalinity should be reduced to 20 mg/l.[19]

Phosphate

High levels of phosphate (both inorganic and organic) are derived from malt.[19] Phosphate is essential to yeast cells and has many roles: for incorporation into structural molecules, such as phospholipids and phosphomannans and for formation of nucleic acids (DNA, RNA) and phosphorylated metabolites, such as ATP and glucose-6-phosphate.[27]

Sulfate

The key influence of sulfate on beer flavor is the production of a bitter/drying character.[5,19] While it is essential for synthesis by yeast of sulfur-containing amino acids, it is also a source of SO_2 and H_2S during fermentation.[2,27]

Chloride

The key influence of chloride on beer flavor is somewhat antagonistic to sulfate, producing smoothness and body effects.[5,19] It may be involved in regulation of yeast cell water content.[27]

Nitrate/Nitrite

The concentration of nitrate in water is restricted to less than 50 mg/l; this can still constitute a risk due to potential formation of nonvolatile nitrosamines (suspected carcinogens). The mechanism involves reduction of nitrate to nitrite (by bacterial nitrate reductase activity) and chemical reaction of nitrite with any wort and beer nitrogen compounds, such as amines.[43] Water represents the major source of nitrate in beer; the brewing process can only add to the nitrate content of the water used for brewing. Whole hops can contain up to 1.0% w/w nitrate.[43]

Reduction of nitrate content to less than 10 mg/l may be desirable and maintenance of high hygiene standards will reduce risk of bacterial contamination with nitrate reductase capability, especially during yeast handling procedures.[43]

The presence of nitrite in water indicates contamination by waste water. The concentration is restricted to less than 0.5 mg/l.[2]

Silicate

Silicate can be extracted from malt by sparging at high pH.[34] It is associated with Ca^{2+} and Mg^{2+} and may cause haze in beer and scaling of vessels and mains.[2]

Fluoride

Fluoride has no adverse effects on fermentation at concentrations below 10 mg/l.[2]

Organic Contamination

One aspect of water quality for brewing that is increasingly becoming important is the level of various organic compounds.[43] Of most concern are organohalides, such as trihalomethanes (THMs), various chlorinated hydrocarbons, and phenolic compounds, and it is interesting to note that chlorination is intimately involved in this area.

Organohalides

Because of the risks associated with products from reactions of chlorine with organic molecules, it is very important to ensure that residual active chlorine is eliminated from water prior to use in the brewery. However, it is quite possible that these harmful compounds may have already been generated in municipal water supplies prior to receipt by the brewery, and removal of these preformed organochlorine compounds will also be required.

There are several references in the literature to unusual beer taints having been generated by contaminated water supplies (e.g., Ref. 44). A common event relates to reservoirs exposed to drought conditions becoming more heavily loaded with accumulated organic matter (humic acids). Increased chlorination to overcome this heavy organic loading can generate high levels of THMs, leading to unpleasant flavor taints. In addition, THMs may represent a health hazard, being suspected carcinogens (possibly related to bladder cancer). Other chlorinated hydrocarbons, such as tetrachloromethane and trichloroethene, which are known to be toxic and are also possible carcinogens, can arise from industrial pollution.

Phenols

A major risk associated with chlorination is the production of those very undesirable flavors and aromas characteristic of chlorophenols. These classical flavor taints are medicinal, harsh, and astringent.[43] Some phenols are intensely flavor active. For example, trichlorophenol (TCP) can be detected at concentrations as low as 1 µg/l (1 ppb), while some cresols are tastable at a few parts per trillion (ppt or ng/l). It is interesting to note that there can be a very wide range of sensitivity to phenolic flavors; some people are extremely sensitive in their response to chlorophenols, while others are virtually "taste-blind" to these compounds.

A somewhat forgotten (or overlooked) area of considerable importance with respect to chlorophenolic taints is steam generation.[43] Contaminated steam condensate residues can contribute to chlorophenolic taints in beer. These can arise from any of the following two processes:

1. Inadvertent steam leaks (e.g., through heat exchangers) directly into product stream or into hot liquor supplies
2. From condensate residues left after steam sterilization of mains and vessels and beer containers (kegs and casks)

Chlorophenolic compounds can arise by continued concentration in the boiler feed water from initially low levels in the incoming water supply, or can be generated *in situ* from active chlorine residues, possibly even with boiler feed additive chemicals, such as tannin for oxygen scavenging; it may be safer to use sulfite systems, if possible. The continued build up in the boiler can lead to disastrous consequences, since the phenolic taints are so potent that no amount of blending of finished beer can achieve the necessary degree of dilution.

Microbiological Control

All liquors contacting the product stream should be sterile[3]; this is especially important with regard to water-borne organisms with nitrate reductase capability and the obvious potential for nitrosamine formation at any stage of the brewing process.

Bacterial contamination of water can occur at any time during storage, even at high temperatures. For example, contaminations with thermophilic bacteria (such as *Bacillus stearothermophilus*) have been identified in hot-brewing liquor tanks held at 80°C.[43] A small leak of wort across the paraflow heat exchanger plates into the liquor stream can provide the bacteria with sufficient nutrient to grow to considerable numbers within the hot water and can actually cause wort and beer clarity problems when the contaminated water is used for wort production.

Water Treatment Technologies and Procedures

In the previous sections, it has been established that water quality can influence the quality of beer, and so confirms the necessity for water treatment processes in order to obtain control over all aspects of liquor composition, for all the process requirements in the brewery. In order to achieve ultimate control on all quality aspects of water use, it may be necessary to design treatments that involve several key elements, such as removal of suspended solids, adjustment of mineral contents, removal of organic compounds, sterilization, and removal of gases.[3,5,6,43,45,46]

The overall design and content of the appropriate treatment will depend on many factors, including brewery location and beer types to be produced, but, fundamentally, the quality and consistency of the incoming water supply may well dictate the treatments required.[5,6,43,45] Indeed, the design may have to be robust enough to cope not only with seasonal changes in composition, but also with significant variations in incoming water at short notice. For example, if the supply is from a municipal supplier drawing on two or more discrete sources with differing characteristics, but usually providing a consistent blend, it is feasible that the blend proportions could vary, even on a daily basis. Usually, the water supplier is legally obliged only to supply water of a potable standard, but often, advance notice of a change may be provided, as well as full records of analytical data.

It may also be expedient for the treatment plant to incorporate sufficient storage capacity of both raw incoming water and treated water as contingency against potential interruption of supply, for whatever reason, including gross contamination. Economic factors will also have to be considered, including not only the costs of the incoming water and waste disposal costs, but also capital costs of various plants and their ongoing running expenses.[43]

Essentially, there is no simple formula for deciding on the optimum treatment (design criteria will be based on "horses for courses"), but if the brewer desires total control over all features of water quality, the necessary treatment methods are available, as detailed next.

Removal of Suspended and Semidissolved Solids

These procedures form the basic preliminary treatment of water supplies. Potable water purchased from a water company will have already been treated to remove suspended solids, but breweries drawing on owned borehole or spring supplies may have to apply such procedures.

Coagulation and Flocculation

This widely used treatment involves passing water through a reaction vessel and adding coagulating chemicals.[3,6] The most commonly used coagulants are ferrous sulfate or aluminum sulfate. These form flocs that trap insoluble impurities present in the water (such as undissolved earth and plant materials). The reaction products and impurities are precipitated and form a gelatinous sludge blanket, through which the treated water flows to a clear zone at the top of the settling tank. The sludge blanket absorbs many of the semidissolved or colloidal compounds responsible for off-flavors and color (such as humic acids). Sludge is periodically drawn off from the vessel in order to maintain a constant blanket volume. The settling tanks are sized such that the water flow velocity is decreased as low as possible to ensure optimal precipitation. The choice of coagulating chemical will depend on the nature of the water. Generally, iron (as ferrous sulfate) is the most suitable as plants using it tend to be simpler and easier to control, and residual levels of iron found in the treated water pose less of a health risk than residual aluminum. The pH of the water in the coagulation vessel is controlled between 9 and 10 in order that ferric hydroxide remains insoluble in the floc. Iron is usually dosed at approximately 20 mg/l. Alternatively, polyelectrolytic polymers, such as alginates or synthetic polymers of acrylamide or acrylic acid, may be used as flocculating agents to produce precipitates to trap impurities to be removed by filtration. This is the reason for the regulated maximum permissible levels set for acrylamide (possible residual monomer) in potable water standards (Table 4.1).

Not all suspended particles can be removed in the settling tanks, so it is often necessary to filter the water afterwards.

Sand Filtration

This is either used as a postcoagulation plant or as a stand-alone operation. Various types of material can be used, although sand and gravel filters predominate.[3,6,45]

In sand filtration, the water is fed through a layer of pure, calcined quartz sand of uniform grain size. Suspended particles (such as residual coagulated flocs) are retained in the pores of the sand as the water flows through.

A sand filter is typically constructed of a pressure vessel that contains a bed (up to 2 m deep) of fine sand, with a particle size of 0.8 to 1.2 mm.[3] The sand bed is supported on layers of gravel of increasing coarseness (up to 5 mm diameter at the bottom). Water is pumped in at a rate of 10 to 20 m^3/m^2 of filter surface per hour and is uniformly distributed over the entire filter surface. Retained material is removed from the filter by backwashing and thus loosening the filter bed.

Oxidation

Water containing excessive amounts of iron (ferrous) and manganese salts, either dissolved or in very fine particulate form (possibly from borehole supplies), can be treated by aeration techniques, such as direct injection of air or ozone, or spraying into a vessel against an air countercurrent.[2,3,6] The salts are oxidized to insoluble oxides and hydroxides, which precipitate out and are removed by filtration, as described previously. Manganese dioxide (because it is a strong oxidizing agent) can be incorporated into sand filters to remove ferrous and manganese salts by similar oxidation reactions.

Adjustment of Mineral Contents

The necessity to control the inorganic content of brewing liquor is clearly a well-established principle[5,43] and, following the pretreatments described earlier, most brewers will employ some form of treatment to reduce the content of certain ions (e.g., bicarbonate) and enhance the levels of other ions (such as calcium, chloride, and sulfate). Classically, there are several procedures available for reducing alkalinity (i.e., decarbonation), and addition of gypsum to replicate Burton-upon-Trent water or "Burtonization" is similarly well recognized. More modern approaches utilize ion exchange systems and, increasingly, reverse osmosis is established as an alternative procedure (and is especially attractive for desalination of sea water).

Decarbonation by Heating

This physical treatment (which can be expensive in terms of energy input) is one of the oldest, but is now rarely used.[3,43] When alkaline water is heated to 70 to 80°C (or even boiled), bicarbonate is converted into insoluble calcium carbonate and CO_2 is evolved:

$$Ca(HCO_3)_2 \xrightarrow{heat} CaCO_3\downarrow + CO_2\uparrow + H_2O$$

This process is, of course, prone to scale formation, which is why boiler feed water must be decarbonated (softened) before use.

Decarbonation by Treatment with Lime

This method effectively involves titrating the bicarbonate-containing water with slaked lime (calcium hydroxide) to form insoluble calcium carbonate.[3,43] In this case, CO_2 is not evolved, since reaction with calcium hydroxide again forms calcium carbonate:

$$Ca(HCO_3)_2 + Ca(OH)_2 \longrightarrow 2CaCO_3\downarrow + 2H_2O$$

A saturated lime solution is dosed into the water to be decarbonated in a reaction vessel, and the lime scale (calcium carbonate) slurry formed settles into the conical base of the vessel to be removed as a precipitate, but the treated water has to be clarified further in a sand/gravel filter to remove all suspended solids. Decarbonation by lime addition is still a widely used system, particularly in Germany (where acid addition is not permitted; see next section). Modern multistage plants are also available.[3]

The advantages of lime addition relate to the low costs of chemicals involved and the ability to remove iron, manganese, and other heavy metals as coprecipitates. Disadvantages relate to disposal of the lime scale slurry generated and the need to adapt an amount of lime dosing consistent with any variation in incoming water composition.

Decarbonation by Acid Addition

This represents the simplest method to reduce alkalinity and is widely practiced.[43] The reaction involved releases CO_2 from bicarbonate and usually requires a degassing system (such as a spray tower, against a forced air countercurrent) to ensure full removal of CO_2, so that the gas cannot redissolve:

$$Ca(HCO_3)_2 + 2H^+ \longrightarrow Ca^{2+} + 2H_2O + 2CO_2\uparrow$$

The acid used can be mineral, such as sulfuric or hydrochloric, or may be organic, such as lactic. Some brewers extend acid addition beyond water treatment and acidify during mashing (to achieve a desired pH value) or during wort boiling. If lactic acid is used, it may increase buffering capacity, and there is a risk that beer pH may increase.

In Germany, acid addition is not permitted according to the Reinheitsgebot and some brewers use "biological acidification" to achieve control of wort pH; this involves producing a lactic acid solution by fermenting malt wort with *Lactobacillus delbruckii* or *L. amylolyticus* and blending this "lactic" wort in precise proportions into the mash or wort to obtain the desired pH value.[3,5]

Ion Exchange Systems

The principle involved in these systems concerns the exchange of undesirable ions in the water for more acceptable ions.[6,43,45,47] This is achieved by the use of ion exchange resins, composed of a number of materials, some of which are naturally occurring, such as zeolites (hydrated aluminosilicate), or synthetic, such as synthetic zeolites, polystyrene, or polyacrylic acid. The polymeric resins have a crosslinked, three-dimensional structure to which ionic groups are attached. A cation exchange resin has negative ions built into its structure, ionized acidic side groups such as $-COO^-$ or $-SO_2O^-$, and will exchange positive ions (cations), whereas an anion exchange resin has ionized basic side groups, such as $-NH_3^+$, and will exchange negative ions (anions).

For water treatment, the resins are available in cationic form with associated H^+ or Na^+ ions and in anionic form with hydroxyl (OH^-) or Cl^- ions. The choice of resin is determined by the ion exchange process desired, viz., dealkalization (reduction of alkalinity by bicarbonate removal), full demineralization (to achieve virtually deionized water), softening (where sodium ions replace calcium and magnesium ions and so remove hardness without reducing alkalinity), or other special applications (such as nitrate removal).

Dealkalization involves the use of weakly acid cation exchange resins (WAC). These are able to remove all Ca^{2+} and Mg^{2+} associated with bicarbonate by exchange with the bound H^+ ions and so release CO_2:

$$Ca(HCO_3)_2 + 2\text{Resin-H} \longrightarrow (\text{Resin})_2\text{-Ca} + 2H_2O + 2CO_2\uparrow$$

This process is essentially similar to acid addition for decarbonation and, likewise, will require a degassing tower to remove evolved CO_2.

WAC exchangers have a high exchange capacity and are readily regenerated with either hydrochloric or sulfuric acid (although the use of H_2SO_4 has the risk of the formation and precipitation of gypsum, which is much less soluble than $CaCl_2$):

$$(\text{Resin})_2\text{-Ca} + 2HCl \longrightarrow 2\text{Resin-H} + CaCl_2$$

These ion exchange systems are readily controlled by monitoring pH, which increases as the resin becomes exhausted (pH 5.5 being approximately equivalent to a bicarbonate content of 50 mg/l), so that regeneration can be automatically initiated. (*Note*: as with all decarbonation procedures, there is some merit in maintaining a low level of residual alkalinity [at 20 to 40 mg/l] in order to avoid potential corrosion concerns).

Demineralization implies the full removal of all cations and anions to produce effectively deionized water. The first stage involves a strongly acidic cation exchanger (SAC), which will replace all cations (Ca^{2+}, Mg^{2+}, and Na^+) with H^+ ions, forming mineral acids (HCl, H_2SO_4). The pH value drops and bicarbonate generates free CO_2. Ca^{2+} and Mg^{2+} are removed in the same manner as WAC acts, but all cations are removed as follows:

$$\text{Resin-H} + NaCl \longrightarrow \text{Resin-Na} + HCl$$

Frequently, a typical treatment for brewing water may comprise a combination of weakly and strongly acidic resins (WAC/SAC) in a layer bed cation exchanger. For full demineralization (if the anionic content is too high), a weakly basic anion exchanger (WBA) could be employed as a second-stage treatment, which will ensure replacement of all anions (except silica and carbonate) by OH^- ions:

$$\text{Resin-OH} + NaCl \longrightarrow \text{Resin-Cl} + NaOH$$

Resin regeneration is achieved with caustic soda. If removal of silica and residual carbonate is also required then strongly basic anion (SBA) exchangers are available.

Many brewers may decide that full demineralization is unnecessary, but in this way, one can achieve total control on the inorganic content and so add back those ions (as appropriate salts), in whatever varying amounts required for a wide range of product types. Further, this water is ideal for many other uses on site (such as dilution water and boiler feed). One key deciding factor will be plant costs, both capital and revenue.

Softening involves exchanging Ca^{2+} and Mg^{2+} ions for Na^+. The resin is a SAC, but used in the sodium form (rather than H^+) and is regenerated with sodium chloride solution (brine):

$$2\text{Resin-Na} + Ca^{2+} \text{ (or } Mg^{2+}) \longrightarrow (\text{Resin})_2\text{-Ca (or Mg)} + 2Na^+$$

In this case, bicarbonate is not removed, but since the cations responsible for hardness have been removed, "soft" water (effectively a dilute solution of sodium bicarbonate) is provided and can be used in a number of applications as process water (e.g., pasteurizer and boiler feed water, bottle and plant washing, etc.). Softening systems are comparatively low-cost options.

Nitrate levels in water can be reduced by selective anion exchange resins. This may be the most appropriate treatment for water that contains only low levels of residual alkalinity, requiring little ionic adjustment other than nitrate removal.[6,43]

The operation of this system is similar to softening in that the resin is regenerated with brine, but in this case the resin is a modified WBA exchanger operating in the chloride form. Bicarbonate, chloride, and sulfate ions are initially exchanged, but during the operation equilibrium is established so that only nitrate is removed, with the treated water being essentially similar to the incoming water, other than the nitrate content.

Reverse Osmosis

Reverse osmosis represents the finest level of filtration available for water.[3,6,43,45,47] Modern systems use cellulose acetate or polyamide membranes in hollow fiber, spiral wound, or tubular format. These membranes act as barriers to all dissolved salts, as well as organic compounds with molecular weight greater than 100 Da (plus all microorganisms). Water molecules pass freely through the membrane to create a desalinated permeate. Removal of 95 to 99% of all dissolved salts is achievable, meaning that the permeate is essentially fully deionized. Indeed, reverse osmosis is proving particularly efficient for seawater desalination.[48]

Modern reverse osmosis systems can operate at relatively low energy inputs, requiring positive pressures of 8 to 12 bar, and it is usual to assemble the membrane units on a modular basis (in banks of two or more) in order to increase recovery rate and minimize waste water production.

One disadvantage of reverse osmosis plants is that the membranes are very susceptible to fouling (both organic and scaling) so that, in general, some form of protective pretreatment is necessary, involving sand filtration, or finer filtration, such as ultrafiltration at 0.2 or 0.45 µ, carbon filtration (to remove chlorine), or UV sterilization.

In comparison with ion exchange systems, it is usually accepted that a reverse osmosis plant will be the preferred water treatment system if the raw water has a dissolved solids content of greater than 800 mg/l, based on operating costs.[47] Basically, both technologies are capable of producing high-quality brewing water, but both have inherent advantages and disadvantages:

1. An ion exchange plant is capable of operating with a flexible output, whereas a reverse osmosis plant has a fixed capacity (that can be modularized), but usually requires sufficient treated water storage capacity to avoid frequent start up and shut down.
2. If the only ion exchange treatment required is cation exchange (i.e., dealkalization), then this plant will be less expensive to install than a reverse osmosis plant, but if full demineralization is required (i.e., both cation and anion exchangers), then the latter will tend to be less expensive.
3. Running costs of both systems are comparable in terms of routine maintenance and replacement costs of resins and membranes; for an ion exchange plant, the main running cost is regeneration of chemicals, whereas for a reverse osmosis plant the primary cost is electrical energy for pumping. For comparable throughputs, both resins and membranes can be expected to have 5 to 7 years of working life.
4. Reverse osmosis plants require significant raw water pretreatment compared to ion exchange, but are capable of removing virtually all organic contamination, whereas ion exchange systems will have to be coupled to carbon filtration systems for total organic removal.
5. Reverse osmosis plant effluent is less demanding (being concentrated "brine"), whereas ion exchange plant effluent may need pH adjustment or blending prior to disposal. However, the reverse osmosis process tends to produce more waste water and may require higher fresh water input than an ion exchange plant.

Electrodialysis

Like reverse osmosis, this is a membrane process, but here the driving force is electrical energy, rather than pressure.[6] The membranes used are ion selective, being impervious to water, but will allow ions to pass (anions or cations) when an appropriate electric current is applied. All ionic impurities can be removed in this way, but the process will not remove nonionic materials. In cost terms, it can be less expensive than a reverse osmosis plant for desalination and, although not common, has found application in some breweries.

Addition of Calcium Salts

As a consequence of all the procedures mentioned previously to reduce the inorganic content, the resultant water may require addition of certain ions to provide the ideal ionic composition for brewing, the precise requirements differing for specific beer types or brands.[5,19,43] Moreover, completely deionized water is corrosive, so that a low-level residual alkalinity (up to 20 mg/l) should be retained, possibly by blending back a small proportion of untreated water.

It is clearly established that adequate Ca^{2+} ions are essential for many reasons; also, other ions (such as Cl^- and SO_4^{2-}) have some flavor impact. Consequently, addition of salts is an old and widely used procedure for water treatment,[5,19,43] culminating in the process known as "Burtonization" where brewers attempt to replicate the calcium sulfate content of Burton-upon-Trent water by massive additions of gypsum.[49]

The practical addition of salts can be extended beyond addition to water for brewing as direct additions during mashing and boiling; as shown earlier, there is much merit in ensuring adequate Ca^{2+} during mashing for many reasons. German brewers tend to compensate low Ca^{2+} contents by additions to water postdealkalization treatment (ion exchange or reverse osmosis) of lime water (under pH control), with possible additions of mineral acids (HCl and H_2SO_4) to augment chloride/sulfate contents, in a specifically designed "calcium blending" plant.[47]

Removal of Organic Compounds

As mentioned in the Section "Organic Contamination," all water used for brewing should be free of chlorine and a number of contaminating organic compounds, many of which are chlorinated (or brominated), either already in the incoming water supply or may have been generated by the brewery's own activity.

Aeration Systems

Removal of volatile organic compounds can be achieved by aeration systems,[6,43,50] for instance, based on gas-stripping columns, where water flows down through a column packed with polypropylene packing or beads, against a countercurrent of forced air. About 80% removal of THMs can be achieved. Interestingly, it has been observed[43] that in ion exchange treatment plants (e.g., dealkalization) involving degasser towers for CO_2 removal, again, about 80% of THMs can be expelled from the treated water. However, the only truly completely effective method of organic removal is carbon filtration.

Granular Activated Carbon Filtration

Granular activated carbon (GAC) treatment is the preferred system for removal of chlorine and most organic contaminants, including halogenated compounds.[6,43,45,51,52] The plant is constructed of a pressure vessel, which houses the activated carbon, supported on a bed of graded gravel, and water flows down through the carbon bed. Activated carbon consists of porous particles (1 to 3 mm) with a high surface area to volume ratio. The plant will be sized and water flow rate adjusted to ensure adequate contact time for optimal absorption of contaminants (but especially chlorine).

Some of the key factors that affect GAC performance include the type and grade of carbon,[52] for example, coconut shell, bituminous coal, lignite, or bone. For brewing water, coal and coconut shell are preferred as they are less likely to be naturally tainted (like wood or peat charcoal). The properties of the carbon used for specification purposes include surface area–volume ratio, pore volume, and pore size distribution.[51] Carbon from coconut shells tends to have mainly small pores (less than 2 nm), whereas coal carbon has larger pores (over 5 nm). This means that coconut carbon is more effective for removal of small organic molecules (e.g., THMs), with coal carbon more effective for larger molecules (pesticides). Removal of chlorine may be catalyzed by chemical reduction, rather than just absorption. For brewery installations, a mixture of carbons may be the most appropriate.

Carbon filters can be cleaned by both forward flow and reverse flow hot water to regenerate the carbon and are usually equipped for direct steam injection at the base to allow for steam sterilization to eliminate any microbiological contamination actually growing in the carbon bed. Because the carbon will eventually be totally spent and unregenerable (with particle breakdown also occurring), capacity and performance for removal of chlorine are usually routinely monitored, to ensure acceptable water quality and to determine carbon replacement rates.

Water Sterilization Techniques

There are many sterilization techniques available for treating water for all uses in a brewery. These include chemical processes (e.g., addition of chlorine, chlorine dioxide, bromine, etc.) and physical methods (such as ultraviolet irradiation, microfiltration).

Chlorination

The use of chlorine is still by far the most widely used system for disinfection of potable water supplies.[4,6,7] In terms of the sequence of water treatments, chlorination is best carried out after pretreatment, such as flocculation and sand filtration. Addition to raw water risks the production of potentially unpleasant organochlorides (such as THMs) if organic matter is present, especially humic acids.[18] Treatment with chlorine is acceptable as a means of disinfection of water during distribution to the brewery and storage prior to any onsite water treatment, but cognizance must be taken of the flavor taint (and potential health) risks, and activated carbon filtration should be available for dechlorination prior to water usage.

Chlorine is an effective sterilant by virtue of its strong oxidizing capability.[53] It rapidly oxidizes protein constituents of bacteria (including spores), yeast, and viruses and probably acts by impairment of membrane function, preventing uptake of nutrients and disruption of protein synthesis.

Active chlorine is available to the brewer for water sterilization, most commonly in solution as sodium hypochlorite (NaOCl), or as a gas. When liquid chlorine or hypochlorite are mixed with water, they hydrolyze to form hypochlorous acid (HOCl):

$$Cl_2 + H_2O \longrightarrow HCl + HOCl; \quad HOCl \longleftrightarrow H^+ + OCl^-$$

HOCl in the nonionized form is 80 times more effective as a sanitizer than the hypochlorite ion.[53] Consequently, optimum activity occurs at lower pH values (e.g., pH 5.0). For effective treatment, the residual active chlorine concentration should be 0.5 to 1.0 mg/l, after a contact time of greater than 2 min. Chlorine may have to be dosed at higher levels (up to 7 mg/l) in order to achieve this residual active level, due to other chlorine reactions (with organic compounds and some ions, ammonia, sulphites, etc.), which will occur before any chlorine is available for antimicrobial action.

Chlorine Dioxide

Chlorine dioxide in solution in water is increasingly being used for disinfection of process water and is finding many applications in the brewing industry.[6,54] It is best used for water sterilization (up to 1.5 mg/l), but can be used (at 2 to 5 mg/l) for hard surface sanitizing (vessels and mains) as a detergent flush. It is particularly useful in packaging operations, such as rinsing cans, can ends, and nonreturnable bottles (glass and PET), and crowns and caps.

Chlorine dioxide is a powerful oxidizing agent (it has some 2.5 times the oxidizing power of chlorine), but is nontainting (because its action does not include the chlorine atom), noncorrosive, and nontoxic (at the normal use level in solution; the gas, however, is some 50 times more toxic than carbon monoxide). It is active as a biocide in a wide pH range (3.5 to 11.5) and is effective against a wide variety of beer spoilage organisms, including bacteria, yeast, and molds.

Chlorine dioxide must be generated *in situ*, commonly from sodium chlorite and hydrochloric acid:

$$5NaClO_2 + 4HCl \longrightarrow 4ClO_2 + 5NaCl + 2H_2O$$

Bromine

Bromine in solution produces hypobromous acid (as chlorine produces hypochlorous acid) and bromine biocides have been found to be more effective than chlorine for certain applications within breweries, such as *Legionella* control in cooling towers and against biofilm production in tunnel pasteurizers.[55] Several forms of bromine biocides are now available, but bromine is rarely supplied in gaseous form. It is more usually supplied in powder or tablet form, which dissolve in water to produce hypobromous acid or as solid-stabilized bromine/chlorine-based oxidizing agent (such as bromo-chloro-dimethylhydantoin) or as stabilized liquid bromine-based biocides.[56] Bromine biocides tend to be less aggressive to metal surfaces and other treatment additives than chlorine, but do risk the production of bromates (which are classified as potentially carcinogenic).

Ozonation

Ozone (O_3) possesses some advantages over other disinfectants, due to its high electronegative oxidation potential, and its rapid rate of reaction toward bacteria and organic compounds (actually removing taints and odors). It is a more expensive process than chlorination, but is unaffected by pH and does not risk formation of byproducts (such as THMs). Its bactericidal action is due to oxidation of cell membranes.

Ozone is generated from oxygen by high-voltage discharge in a current of clean air, which is injected into the water to achieve a dissolved ozone concentration of 0.5 mg/l.[6]

There are reports (e.g., Ref. 56) of ozone being used to achieve total sterility and removal of objectionable tastes in water used for brewing, blending, and product purging, as well as for washing purposes.

Sterilization by Silver Ions

Silver ions are bactericidal, at a level of only a few micrograms per liter.[3] The water stream is led to flow between silver electrodes across which an electrical potential is generated. Such systems are virtually maintenance free and, because only a low level of Ag^+ ions is required, will last for many years and can be used in many applications, such as cooling towers and even for brewing.[3]

Ultraviolet Irradiation

The use of ultraviolet (UV) irradiation to disinfect water for brewing avoids the costs of addition, removal, and disposal of chemicals involved in dosing systems, plus risk of overdosing and by-product generation associated with many of the treatments mentioned earlier.[6,43,57]

The system requires the passage of UV light (between 200 and 280 nm) through a relatively small depth of water, which should have good clarity and low turbidity. The principle of the action is destruction of cellular nucleic acids (DNA and RNA) by absorption of UV light at 256 nm. High-intensity UV lamps (constructed of quartz tubes) are used in standard compact units, sized so that water flow rates can be adjusted to give a minimum contact time of 20 to 30 sec.

The major advantages of UV irradiation are the nonchemical and noncorrosive aspects, with little or no risk of residues or taints being produced and the units are relatively inexpensive (both capital and maintenance) and are self-monitoring (the systems can be alarmed to indicate lamp failure). The one disadvantage is that there is no residual "kill-potential," so that multiple units may be required if several applications are needed, since the UV system should be located as close as possible to the point of use.

UV installations can be effective for all applications[43] requiring sterilized water (e.g., dilution and rinse liquors, water for yeast washing, in cooling towers, protection of activated carbon filters, etc.).

Sterile Microfiltration

Some applications may benefit from filtering water through very fine absolute filters at 0.45 μm (or less), so that there is no risk of flavor taint.[6,46] One such use would be for dilution water (to adjust beer alcohol content or gravity) if the blending procedure is carried out after beer filtration (i.e., "bright beer").

Gas Removal Applications

For effective decarbonation by acid addition or ion exchange with H^+ ions (see Section "Adjustment of Mineral Contents"), often removal of released CO_2 is required. This is readily achieved by a gas-stripping procedure, by purging treated water as a falling film in a packed column or as a spray against a countercurrent of air. Some volatile organic compounds (e.g., THMs) can also be removed in this way. However, the major application for degassing water is for dissolved oxygen removal or deaeration.

There are several applications in brewing for deoxygenated water,[6,19,43] the most obvious being dilution water (i.e., water used for alcohol or gravity adjustment, especially if high-gravity brewing is employed). Other uses[6] include pre- and postruns during filtration, filter aid make-up water, filtration counterpressure, and beer recovery from yeast and tank bottoms. Indeed, many brewers (e.g., Ref. 3) are now advocating the use of deoxygenated water in the brewhouse for wort production as a means of achieving "anaerobic" conditions designed to reduce wort and beer oxidation (to improve flavor stability).

Several technologies[3,6] can be employed for deoxygenation, the most common being gas stripping in which the countercurrent gas can be either nitrogen or carbon dioxide, usually in columns filled with polypropylene packing or beads. The operating principle involves reducing the partial pressure of oxygen by continual flushing with the alternative gas. These processes can readily reduce dissolved oxygen levels in water to less than 50 $\mu g/l$ (50 ppb).

Vacuum degassing includes CO_2 injection, followed by degassing in a vacuum vessel and may also include gas stripping. Dissolved oxygen can be reduced to less than 20 $\mu g/l$.

Finally, catalytic removal systems involve intimately mixing hydrogen gas (in carefully controlled proportions relative to the dissolved oxygen content) in water and passing over palladium (in granular form) in a reaction vessel, where oxygen and hydrogen react to form water molecules. In this way, virtually all oxygen is removed (to less than 5 $\mu g/l$).

References

1. Lloyd Hind, H., *Brewing Science and Practice*, Vol. 1, Chapman and Hall, London, 1938, pp. 411–494.
2. Briggs, D.E., Hough, J.S., Stevens, R., and Young, T.W., *Malting and Brewing Science*, 2nd ed., Vol. 1, Chapman and Hall, London, 1981, pp. 194–221.
3. Kunze, W., *Technology Brewing and Malting*, 2nd ed., VLB, Berlin, 1999, pp. 60–75.
4. Moll, M., Water in malting and brewing, in *Brewing Science*, Pollack, J.R.A., Ed., Vol. 1, Academic Press, London, 1979, pp. 539–577.
5. Taylor, D.G., The impact of water quality on beer quality, *The Brewer* 74:532–536, 1988.
6. Bak, S.N., Ekengreu, Ö., Ekstam, K., Härnulv, G., Pajunen, E., Prucha, P., and Rasi, J., *Water in Brewing — European Brewery Convention Manual of Good Practice*, EBC, 2001, pp. 1–128.
7. Degremont, *Water Treatment Handbook*, 5th ed., John Wiley, Chichester, 1979.
8. Benson, J.T., Coleman, A.R., Due, J.E.B., Henham, A.W., Twaalfhoven, J.G.P., and Vinckx, W., *Brewery Utilities — European Brewery Convertion Manual of Good Practice*, Brewing Research Foundation, 1997, pp. 157–174.
9. Askew, M., What price water? *Brew. Guard.*, 128:19–22, 1999.
10. Reed, R. and Henderson, G., Water and waste water management — its increasing importance to brewery and maltings production costs, *Ferment*, 12:13–17, 1999/2000.

11. Ward, J.A., Brewery wastewater treatment using membrane technology, *Proc. 26th Convention of Asia Pacific Section of Institute of Brewing*, Singapore, 2000, pp. 122–127.
12. Candy, E., Water — unlimited? *Brew. Int.* 3:12–18, 2003.
13. Stewart, G.G. and Russell, I., *Brewer's Yeast: An Introduction to Brewing Science and Technology*, Series III, The Institute of Brewing, London, 1998, p. 66.
14. Kemmer, F.N., and McCallion, J., Eds., *Water Handbook*, McGraw-Hill, New York, 1983.
15. Griffiths, A.R., *Water Quality in the Food and Drink Industries*, Chandos Publishing (Oxford) Ltd, Oxford, 1998, pp. 8–13.
16. Kolbach, P., Über brauwasser, *Monats. Brau.*, 13:77–86, 1960.
17. Richards, M., Quality hazards to brewery water supplies, *Brew. Dig.*, 51(8):34–42, 1976.
18. van Gheluwe, G., Bendiak, D.S., and Morrison, N.M., Considerations for future water systems, *Tech. Q. Master Brew. Assoc. Am.*, 21:33–38, 1984.
19. Taylor, D.G., How water composition affects the taste of beer, *Brew. Distill. Int.*, 11:35–37, 1981.
20. Griffiths, A.R., *Water Quality in the Food and Drink Industries*, Chandos Publishing (Oxford) Ltd, Oxford, 1998, pp. 19–23.
21. Briggs, D.E., Hough, J.S., Stevens, R., and Young, T.W., *Malting and Brewing Science*, 2nd ed., Vol. 2, Chapman and Hall, London, 1981, pp. 840–845.
22. Comrie, A.A.D., Brewing liquor — a review, *J. Inst. Brew.*, 73:335–341, 1967.
23. Harrison, J.G., Laufer, S., Stewart, E.D., Seibenberg, J., and Brenner, M.W., Brewery liquor composition — present day views, *J. Inst. Brew.*, 69:323–331, 1963.
24. Krauss, G., Waller, H., and Schmid, R., Brauwasserfragen der einfluss einiger brauwassersalze auf den geschmack des biers, *Brauwelt*, 95:617–624, 1955.
25. Taylor, D.G., The importance of pH control during brewing, *Tech. Q. Master Brew. Assoc. Am.*, 27:131–136, 1990.
26. Bamforth, C.W., pH in brewing: an overview, *Tech. Q. Master Brew. Assoc. Am.*, 38:1–9, 2001.
27. Stewart, G.G. and Russell, I., *Brewer's Yeast: An Introduction to Brewing Science and Technology*, Series III, The Institute of Brewing, London, 1998, pp. 41–46.
28. Narziss, L. and Lintz, B., Über das Verhalten eiweissabbaunder Enzyme bein Maischen, *Brauwiss* 28:253–260, 1975.
29. Rudin, A.D., Brewing liquor and beer quality, *Brew Guard.*, 105:30–33, 1976.
30. Amri, M.A., Bonaly, R., Duteurtre, B., and Moll, M., Interrelation between calcium and potassium ions in the flocculation of two brewers yeast strains, *Eur. J. Appl. Microbiol. Biotechnol.*, 7:235–240, 1979.
31. Comrie, A.A.D., Brewing water, *Brew Dig.*, 42:86–92, 1967.
32. Burger, M. and Becker, K., Oxalate studies in beer, *Annual Meeting of American Society of Brewing Chemists*, 1949, pp. 102–115.
33. Briggs, D.E., Hough, J.S., Stevens, R., and Young, T.W., *Malting and Brewing Science*, 2nd ed., Vol. 2, Chapman and Hall, London, 1981, pp. 821–823.
34. Laing, H. and Taylor, D.G., Factors affecting mash bed permeability, *Proc. 18th Convention of Australia & New Zealand Section of Institute of Brewing*, Adelaide, 1984, pp. 109–114.
35. MacWilliam, I.C., pH in malting and brewing, *J. Inst. Brew.*, 81:65–70, 1975.
36. Coote, N. and Kirsop, B.H., Factors responsible for the decrease in pH during beer fermentation, *J. Inst. Brew.*, 82:147–153, 1976.

37. Anness, B.J. and Bamforth, C.W., Dimethyl sulphide — a review, *J. Inst. Brew.*, 88:244–252, 1982.
38. Haukeli, A.D. and Lie, S., Conversion of α-acetolactate and removal of diacetyl: a kinetic study, *J. Inst. Brew.*, 84:85–89, 1978.
39. Stewart, G.G. and Rees, E.M.R., The effect of the divalent ions magnesium and calcium on yeast metabolism, *Tech. Q. Master Brew. Assoc. Am.*, 38:171–174, 1999.
40. Mändl, B., Mineral matter, trace elements, organic and inorganic acids in hopped wort, *Eur. Brew. Conv. Wort Symp.*, European Brewing Congress, Zeist, Monograph I, 1974, pp. 226–237.
41. Helin, T.R.M. and Slaughter, J.C., Minimum requirements for zinc and manganese in brewer's wort, *J. Inst. Brew.*, 83:17–19, 1977.
42. Hysert, D.W. and Morrison, N.M., Sulfate metabolism during fermentation, *J. Am. Soc. Brew. Chem.*, 34:25–29, 1976.
43. Taylor, D.G., The treatment of water at the brewery, *Ferment*, 2:76–79, 1989.
44. Jackson, A., Hodgson, B., de Kock, A., van der Linde, L., and Stewart, M., Beer taints associated with unusual water supply conditions, *Tech. Q. Master Brew. Assoc. Am.*, 31:117–120, 1994.
45. Mailer, A., Peel, R.G., Theaker, P.D., and Ravindran, R., Water treatment for brewing process purposes, *Tech. Q. Master Brew. Assoc. Am.*, 26:35–40, 1989.
46. Pratt, J. and Walker, G., Brewery process water, *Tech. Q. Master Brew. Assoc. Am.*, 39:71–73, 2002.
47. Eumann, M., Ion exchange or reverse osmosis? *Proc. 26th Convention of Asia Pacific Section of Institute of Brewing*, Singapore, 2000, pp. 186–187.
48. van der Berkmortel, H.A., Water desalination by reverse osmosis, *Tech. Q. Master Brew. Assoc. Am.*, 25:85–90, 1988.
49. Putman, R., The story of Burton water, *Brew. Int.*, 2(8):8–11, 2002.
50. Weaver, F.B., Controlling organics in brewing water, *Tech. Q. Master Brew. Assoc. Am.*, 21:140–142, 1984.
51. Gough, A.J.E., Factors affecting the choice of granular activated carbon for brewery filters, *Tech. Q. Master Brew. Assoc. Am.*, 32:195–200, 1995.
52. Sanchez, G.W., Granular activated carbon purification of brewing water supplies: selection, usage, monitoring and replacement, *Tech. Q. Master Brew. Assoc. Am.*, 38:99–104, 2001.
53. West, C., Chlorine, *Brew. Int.*, 1(11):34–35, 2001.
54. Dirksen, J., Chlorine dioxide for the brewing industry, *Tech. Q. Master Brew. Assoc. Am.*, 40:111–113, 2003.
55. Ballmier, A.W., King, M.R., Brundage, E.R., Gaviria Acuna, L.F., Herrera Andrade, F.J., and Gil, A., Bromine biocides for pasteurizer biofouling control, *Tech. Q. Master Brew. Assoc. Am.*, 36:93–100, 1999.
56. Egan, L.H., Taylor, K.R., and Hahn, C.W., Ozonation of brewing water, *Tech. Q. Master Brew. Assoc. Am.*, 16:164–166, 1979.
57. Whitby, E., Ultraviolet light and brewing, *Tech. Q. Master Brew. Assoc. Am.*, 24:28–32, 1987.

5

Barley and Malt

Geoffrey H. Palmer

CONTENTS

Introduction 139
Barley 142
   Structure and Function 142
   The Husk 142
   The Pericarp 143
   Testa 143
   Aleurone Layer 144
   Starchy Endosperm 146
   The Embryo 149
Malt Production 150
   Drying, Storage, and Handling 150
   Steeping, Germination, Kilning, and Malt Quality 151
   Malt Varieties 157
Acknowledgment 158
References 159

Introduction

Barley and other cereal grains are fruits because they contain a pericarp. The pericarp is the fruit wall (Figure 5.1). All cereals are members of the family Graminea and are monocotyledons. Plants such as beans and peas are dicotyledons because their seeds contain two cotyledons. In barley and other cereals the term monocotyledon is derived from the single cotyledon, which is called the scutellum.

Barley is grown in many parts of the world but it grows best in temperate climates. Malting barley is divided into two main species *Hordeum vulgare*

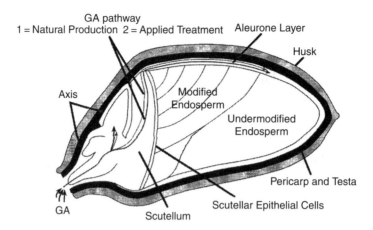

FIGURE 5.1
Relationship between transport of gibberellic acid (GA), aleurone activity, and enzymic modification of the endosperm.

and *Hordeum distichon*.[1,2] *H. distichon* is two-rowed barley and *H. vulgare* is six-rowed barley. The ear or inflorescence of barley contains six rows of grains. In six-rowed barleys, all six rows of grains develop. In two-rowed barleys, only two of the six rows of grains develop: the developed grains face each other and are subtended on each side by two rows of undeveloped grains. Although there are more grains per ear in six-rowed grains, most of these grains are too small for malting purposes. In two-rowed barleys, most of the grains are of a size suitable for malting. In this regard, two-rowed barleys produce more brewers' (starch) extract per hectare than six-rowed barleys.

Six-rowed and two-rowed barleys may be winter barleys or spring barleys.[1,2] Winter barleys are usually planted in about September and harvested in about July in the Northern Hemisphere whereas spring barleys are usually planted in about March and harvested in about August. In general, winter barleys are usually harvested before spring barleys, which is of economic benefit to the maltster in terms of availability of new grain. Although modern winter barleys are of equivalent quality to spring barleys, there is the perception in some quarters that winter barleys are of poorer quality. This is clearly not the case. For example, Maris Otter is a winter barley and its malting quality is equivalent to the best spring barleys.

Barley is a very old crop. Images of the ear (inflorescence) of barley are depicted on ancient Egyptian relics. Recent evidence[3] suggests that barley was malted in ancient Egypt and was likely to have been used to make beer and bread. Although all cereal grains, for example, barley, wheat, sorghum, rye, millet, oats, rice, and maize, can be changed into malt through the processes of steeping and germination,[1] barley malt is the preferred cereal used to make lagers, ales, and stouts worldwide. Wheat malt is mainly used to make weiss beer (see Chapter 2). Sorghum malt can be used to make European clear beer, but it is the primary raw material of traditional (African) opaque beer.[2,4] Millet malt is also an important raw material of some traditional African beers.

Barley malt is used in different quantities to make Scotch, Irish, Japanese, American, and Canadian whiskies. It is also used to produce vinegar and various kinds of foods such as bread, biscuits, confectionary, and malt drinks. Malt extract not only provides different colors and flavors to food products and beverages, it is also used as a growth medium for microorganisms.

Barley has been described as feed barley and malting barley. Malting barleys are usually "recommended." In the United Kingdom, malting barleys usually have grades that extend from a low grade (1) to a high grade (9). Grades are given for yield, disease resistance, and malting quality. These properties of the grain are under genetic control and are distinct in different barley varieties. It takes about 10 years to produce a new barley variety by traditional breeding methods. This time period can be reduced to 5 years, if two crops are achieved each year, by planting and harvesting in two countries each year. New varieties are tested for malting and brewing quality and must be distinct and uniform before they can be recommended by official bodies such as The National Institute of Agricultural Botany (NIAB), Cambridge, United Kingdom.

A significant amount of research interest is being directed at using new gene, transformation techniques to breed improved barley varieties. Transformation may involve substitution of genetic (chromosome) fragments or the insertion of specific genes into the genome of cells of present barley varieties.[5,6] The insertion of new genes is followed by cell division and the development of new plants containing the new genes. To date, no transformed barley plant has been recommended for malting. Indeed, many transformations have failed because transformed genes have failed to be inherited over succeeding generations. The use of genetically modified barley in the industry is still a controversial matter. However, acceptance will be based on product image, customer considerations, and the benefits derived from using transformed barley. Aspects of genetic transformations relate to the replacement of the genes of heat-labile enzymes such as β-amylase and endo-β-glucanase with genes from microorganisms that produce heat-stable versions of these enzymes. Potential improvements in malting barley quality also relate to the possibility of increasing disease resistance and crop yield, as well as improving malting quality.

Crop yield of malting barley is a very important factor in the economics of malt production worldwide. World production of barley is about 135 million tons but only about 21 million tons are considered suitable for producing the 17 million tons of malt required by the industry worldwide. Regarding the latter tonnage of malt, about 94% is required to produce 1.6 billion US hectoliters of beer: about 4% is required for distilling and about 2% for food and vinegar production.

The large amount of barley that is regarded as unsuitable for malting makes it necessary that available barley is malted effectively. Efficient malting depends on optimal operation of the physiological functions of the barley being malted. A better understanding of these functions is an essential feature of modern malting science and technology.

Barley

Structure and Function

Barley grains (Figure 5.1), except for naked (huskless) barley, contain: husk, pericarp, testa, aleurone layer, starchy endosperm, and embryo.[1,2,7] In terms of total dry weight of the barley grain: the husk is 10 to 12%, the pericarp and testa 2 to 3%, the aleurone layer 4 to 5%, the starchy endosperm 77 to 82%, and the embryo 2 to 3%.

The Husk

The husk is composed of two leaf-like structures. The dorsal half is called the lemma, the ventral half is the palea. The husk protects the underlying structures of the grain, especially the embryo. Husk damage is regarded as unacceptable and barley samples are rejected if husk damage is beyond specification requirements. Husk damage implies embryo damage, uncontrollable embryo growth, and loss of filtration potential in conventional mash tun operations. The husk carries background levels of microorganisms such as fungi and bacteria. These microorganisms may have invaded the grains in the field or during storage. For example, *Alternaria*, *Cladosporium* and *Fusarium* tend to invade the grain in the field, whereas *Aspergillus, Penicillium*, and *Rhizopus* can develop significantly during grain storage. Although microbial activity can be high during malting, grain drying and kilning reduce microbial levels on barley and malt, respectively.

Fungi (e.g., *Fusarium*) can release mycotoxins such as trichothecenes (e.g., deoxynivalenol, DON) and zearalenone. DON is associated with *Fusarium* infection and its presence in excessive quantities (>2 to 4 ppb) is associated with "beer gushing." *Aspergillus* produces aflotoxins and ochratoxins. Mycotoxins can be dangerous to the health and life of animals; aflotoxins from *Aspergillus* are potentially hazardous. In animals, the mycotoxin zearalenone is an estrogenic and tumor-producing toxin. Although zearalenone has been found in fermenting worts and would have been extracted from mashed cereals, it is changed to α-zearalenol during fermentation. The latter compound is more of a threat to human health than zearalenone. The levels of harmful fungi on malting barley are normally very low and should not be harmful to humans who drink beer. In contrast, spent grains from the mash tun can have concentrated levels of mycotoxins, especially if stored wet. However, spent grains can be detoxified using formaldehyde or by ammoniation before being fed to farm animals. Toxins from ergot infections are dangerous to human health and can cause nervous diseases. In general, sprouted, infected grain should not be used as animal feed. The control of microbial infection of cereal grains is of vital importance to the industry.

Excessive levels of microorganisms in the husk can produce uneven germination. This type of germination failure is called water sensitivity. Air-resting during steeping can reduce water sensitivity significantly. Barley husk contains significant quantities of lignin and cellulose, which are constituents of cell walls. Lignin is a complex noncarbohydrate polymer and is produced from phenolic compounds such as coniferyl alcohol. Residual phenolic compounds range from simple phenolic acids (e.g., ferulic acid and coumaric acid) to more complex phenols (polyphenols) such as the diamer procyanidin B_3. Phenols such as procyanidin B_3 have oxidative properties and can react with proteins to form hazes in beer.[1,2] Phenolic substances are leached from the husk during steeping. Insufficient leaching during steeping causes phenols to be extracted later during mashing and the resultant beer may have an astringent taste. Steeping with alkaline (lime) steep water reduces the phenol levels of malt.

Color pigments (anthocyanins) in the husk may assist barley variety identification. Structural features of the husk, such as shape of attachment points of the grain to the flower (ear) stalk, can also be used to identify barley varieties. The husk releases a significant quantity of dust during handling, storage, malting, and kilning. Dust can carry microorganisms such as *Aspergillus*, which together can produce lung diseases, if health and safety regulations are not followed.

The Pericarp

The pericarp is the fruit wall of the grain. Therefore, cereal grains are fruits and, strictly speaking, should not be referred to as seeds. Like the husk, it contains a waxy cuticle. Below this waxy layer is a compressed structure of cells. The pericarp is semipermeable. Certain chemicals will pass through it, others will not, such as gibberellic acid. Water can pass through the pericarp. Damage to the pericarp during the abrasion process allows gibberellic acid to enter the aleurone layer directly, rather than via the germinated embryo, thereby improving modification of the starchy endosperm by improving the efficiency of the aleurone to produce endosperm-degrading enzymes.[2,8]

Testa

Dehusking of barley grains using cold 50% sulfuric acid will remove the husk and pericarp but not the testa. The testa comprises two lipid layers that enclose cellular material. In contrast, dehusking by hand leaves both the pericarp and testa on the grain. Submersion of the distal (nonembryo) end of hand-dehusked barley into water causes hydration of the starchy endosperm, suggesting that water can pass directly through the pericarp and testa during processes such as steeping.[2]

The testa is permeable to gibberellic acid. Phenolic compounds such as anthocyanogens (proanthocyanidins) are associated with the aleurone and testa and can be seen clearly in some varieties of barley and sorghum. The small area of the pericarp–testa that lies over the coleorhiza (chit) is called the micropyle. The latter may facilitate the uptake of water and salts into the embryo during germination.

Aleurone Layer

The aleurone layer is about two- to three-cells deep over the starchy endosperm (Figure 5.2). It extends over the embryo as a single cell layer. Therefore, excised embryos contain aleurone cells that can produce endosperm-degrading enzymes that could be assessed wrongly as originating from the epithelial cells of the scutellum.[2] During malting, gibberellic acid from germinated embryos can induce aleurone cells to produce endosperm-degrading enzymes, such as α-amylase, endo-β-1,3:1,4-glucanases, limit dextrinases, endoproteases, and xylanases (pentosanases).[2] Biochemical evidence has established that α-amylase, endoprotease, and limit dextrinase are produced *de novo* in aleurone layers stimulated by gibberellic acid.

Aleurone layers have thick cell walls (3 μm) that contain mainly pentosans (ca. 60%) and β-glucans (ca. 30%), which are degraded in localized areas during malting. This degradation process may facilitate the release of endosperm-degrading enzymes into the starchy endosperm during malting.[7]

Gibberellic acid is produced in the germinated embryo but must be transported to the distal end of the grain (Figure 5.1) to produce an endosperm-wide distribution of enzymes that will convert the hard starchy endosperm of barley into friable malt. Transport of gibberellic acid through aleurone cells may occur through plasmodesmata that are present in aleurone cells (Figure 5.2).[7] The aleurone layer is therefore a large interconnected tissue of living cells. The plasmodesmata facilitate the transport of gibberellic acid from proximal aleurone cells to distal aleurone cells. Research has shown that factors that limit the rate of transport of gibberellic acid in the aleurone layer during malting can reduce the evenness of enzymic modification of the starchy endosperm.[8] For example, reduced transport of gibberellic acid can retard modification of the distal end of the malting grain (Figure 5.1). Aleurone layers of different barleys can produce different levels of endosperm-degrading enzymes to similar doses of gibberellic acid. This is usually a varietal characteristic. In addition, it is worth noting that endosperm-degrading enzymes, such as endo-β-glucanase require

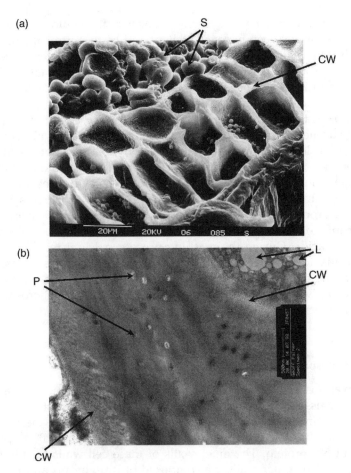

FIGURE 5.2
Cell walls of the aleurone layer (a) at low magnification and (b) at high magnification. S, starch granules of barley endosperm; CW, cell wall surface; P, plasmodesmata; L, lipid body.

higher concentrations of gibberellic acid than α-amylase to develop optimally.[2]

The levels of aleurone enzymes such as α-amylase, endo-β-1,3:1,4-glucanase, and protease, which are produced during malting, are not always related to the nitrogen (protein) content of the grain because the protein content of the aleurone layer can be different from the total protein content of the grain. The nitrogen (protein) content of the aleurone layers of steely grains tend to be higher than the nitrogen content of the aleurone layers of mealy grains. However, the amount of enzyme produced by the aleurone layer is not always linked to the nitrogen content of the aleurone layer (Koliatsou and Palmer, unpublished). The nitrogen levels of different barley varieties can be the same, but they will produce different levels of endosperm-degrading enzymes.[9] However, it should be noted that the relationship between nitrogen content and endosperm enzymes can be altered by the malting process. For example, a low nitrogen barley (e.g., 1.5%) can produce more β-amylase than a barley of higher nitrogen content (e.g., 1.7%) if endosperm modification was better in the low nitrogen barley.[4] The action of proteolytic enzymes during endosperm modification appears to activate enzymes, such as β-amylase, and limit dextrinase.[2]

The aleurone layer is a major constituent of bran. The aleurone layer contains high levels of lipid, proteins, and phytic acid. Phytic acid acts as a chelating agent and reduces the levels of available metal ions in the diets of humans and animals. The hydrolysis of phytic acid (inositol hexaphosphate) by phytase in the aleurone layers of malting grains produces phosphate products, which range from initial levels of phytic acid (inositol hexaphosphate), through intermediates, to final levels of inositol and phosphoric acids. During the mashing process, calcium reacts with phosphoric acids to produce calcium phosphate and hydrogen ions, which help to create the acid conditions of the mash from which a wort of about pH 5.2 is derived. Vitamin B and the nonreducing sugar sucrose are also present in the aleurone (bran) layer.

Starchy Endosperm

The term endosperm comprises the pericarp, testa, aleurone, and the starchy endosperm tissue. The starchy endosperm is the largest structure of the grain and is made up of thousands of cells. The walls of these cells are about 2-μm thick and contain about 70% β-D-glucans, 20% pentosans, and about 5% protein. The inner walls of these cell walls mainly contain β-D-glucans. The outer walls are composed of β-D-glucans and pentosans. β-D-Glucans have about 70% β-1,4 links and 30% β-1,3 links. About 50% of the cell wall is soluble in hot water (65°C). The cell wall is degraded by proteases (e.g., solubilase), endo-β-1,3:1,4-glucanases, endo-β-1,3-glucanases, exo-β-glucanases, arabinosidases, and xylanases to produce mainly glucose, cellulose, laminaribiose, and small amounts of arabinose, xylose, and small fragments of soluble arabinoxylans.[2] Although there is a view that the cell walls of the endosperm do not contain consecutive β-1,3 links,[10] it has been reported that this feature of β-glucans is associated with high molecular weight β-glucans and is varietal — the β-glucans of poor-quality barleys are more susceptible to the action of endo-β-1,3-glucanase than the corresponding high and low molecular size β-glucans of better-quality barleys such as Maris Otter and Proctor.[11]

The next major storage compound of starchy endosperm cells is protein. Hordein and glutelin are the major protein compounds (about 40 and 30%, respectively) of the protein matrix of the starchy endosperm. Albumins and globulins are also present (about 20 and 10%, respectively). Enzymes such as β-amylase and carboxypeptidases are part of the salt (sodium chloride)-soluble albumin and globulin fractions of the total protein of the grain. Hordein and glutelin are similar in structure and contain high levels of proline and glutamine. Hordein is soluble in hot (70%) ethanol or hot (60%) isopropanol. The main enzymes causing degradation of protein in the starchy endosperm are endoproteases (endopeptidases) — some are thiol-dependent, others are metallo-dependent. These enzymes convert hordein proteins into soluble proteins during malting. Carboxypeptidases release amino acids from solubilized proteins. The pH of the endosperm of malting barley varies between pH 5 and 6 and facilitates the activities of the protease enzymes. Proteases such as carboxypeptidases, in combination with enzymes such as endo-β-1,3-glucanase, may initiate the release of β-D-glucans from the cell walls of the endosperm during malting. These enzymes are probably part of the enzyme complex referred to solubilize.[1] Although the total soluble nitrogen/total nitrogen ratio of the malted grain is used as an index of malt (protein) modification, proline ratios and hordein ratios may also give useful indications of malt (protein) modification.[12]

Glutelin, like all the other protein fractions, is soluble in 4% sodium hydroxide. During malting, the hordein fraction is hydrolyzed extensively to produce the soluble proteins of brewers' hot water extract, which contain polypeptides, peptides, and amino acids. Structually, the starchy endosperm may be mealy, steely, or mealy/steely. These features of the grain are controlled to a great degree by field-growth conditions, such as temperature, moisture, and fertilizer application. However, some varieties tend to retain a high potential for mealiness or steeliness in different environmental conditions, suggesting that this structural feature of the grain is under heritable genetic control. Steely grains contain more proteins than mealy grains of the same sample.[2,9,13] Steely grains also contain higher percentages of hordein than corresponding mealy grains. The starchy endosperms of steely grains are compact (hard) whereas the starchy endosperms of mealy grains are looser (softer) in structure. Mealy grains have higher starch content than steely grains and will yield higher starch extracts.[2]

In barley, the starchy endosperm contains two types of starch granules, large (10 to 25 μm) and small (1 to 5 μm). Large starch granules are referred to as A-type and small starch granules are referred to as B-type. Both large and small starch granules are associated with lipoprotein that may limit their digestibility. About 10% of the starch content of the starchy endosperm is degraded during malting. Because the small starch granules comprise 10% of the weight of the starch, and as about half of the small starch granules are degraded during malting, equal weights of starch are degraded from the large and small starch granules during malting.

β-Amylase of the starchy endosperm is associated with endosperm protein. During malting, this enzyme is activated by reducing conditions and by the action of proteolytic enzymes in the endosperm of the malting grain. Although, β-amylase activity is supposed to be linked to the nitrogen (protein) levels of the grain, recent research suggests that β-amylase activity may, inter alia, be associated with the hordein content of the grain.[9] β-Amylase is a very important enzyme because it attacks gelatinized starch to produce maltose, and maltose is the main sugar (45%) of brewers' worts. About 90% of the diastatic power (DP) of malted barley is β-amylase activity. However, although the malted grain has a high β-amylase activity (DP), the 10% of raw starch degraded during malting is degraded by α-amylase. Limit dextrinase is another starch-degrading enzyme. It hydrolyzes the 1,6-links of gelatinized amylopectin. Its action is increased by proteolytic activity, reducing conditions that increase proteolytic activity, and by acid pH conditions.[14] Nevertheless, its action during malting is uncertain, and its action during mashing requires low temperatures that are below the minimum gelatinization temperature (63 to 64°C) of malt starch. This suggests that its role during malting and mashing is limited. Also, it is more heat labile than α-amylase and β-amylase and significant levels of the enzyme will be lost during kilning. However, unlike the beer fermentation, distillers' fermentations are not derived from a boiled wort, and limit dextrinase may have some activity during the fermentation stage of the distilling process where soluble starch dextrins are present, conditions are more acid, and the temperature is between 20 and 30°C.

Hot-water extract is derived mainly from hydrolyzed starch and protein. The weight of the large starch granules is an important index of malt quality. An increase in the protein content of the grain reduces the sizes and numbers of large starch granules.[2] Although the actions of cell wall-degrading and protein-degrading enzymes are important for extract development during malting and mashing, the sizes and numbers of large starch granules determine sugar extract yield during mashing and alcohol production during fermentation. Brewers' wort is about 75% fermentable and has about 10% glucose, 45% maltose, 15% maltotriose, 10% maltotetraose, 15% dextrins, and 5% sucrose. The soluble proteins that are released during malting constitute the other major component of brewers' worts. Soluble proteins include proteins, polypeptides, peptides, and amino acids. Of the protein fractions, lipid transfer proteins of 9000 Da and hordein-type proteins of 40 kDa are regarded as playing important roles in foam development and stability.[10] These proteins have hydrophobic properties that are enhanced after the proteins have been denatured by the wort-boiling process. However, although the biochemistry of foam is important, it is worth bearing in mind that malts will produce foam if their hordein proteins, some of which have molecular weights of 40 kDa, are not degraded excessivley during malting.[6,15] Excessive degradation of lipid transfer proteins would also limit foam development. Some proteins may combine with polyphenols to form haze. Although large quantities of soluble proteins are lost during beer production, functional qualities remain as part of the physicochemical properties of the beer.[1,2] Amino acids are about one fourth or one third of the soluble proteins of the wort and help to promote yeast growth and flavor development during the fermentation process.

The Embryo

The embryo contains about 30 to 35% protein in contrast to the endosperm, which contains about 10 to 12%. However, the protein level of the embryo is very small because the embryo is about 3% of the weight of the grain and the endosperm is about 87%. The lipid content of the embryo is about 15% and that of the entire grain is about 3%. The embryo absorbs large quantities of water during steeping and, for a grain with 45% moisture, the moisture content of the embryo is about 60%.

Structurally (Figure 5.1), the embryo comprises two major tissues: the axis (shoot, node, and roots) and the scutellum (a single cotyledon). During malting, gibberellic acid is synthesized at the nodal region of the axis and transported mainly through dorsally oriented vascular stands to the dorsally placed aleurone cells. This lopsided distribution of gibberellic acid initiates a lopsided (symmetric) pattern of enzyme production and endosperm modification.

Although the aleurone layer is the dominant enzyme-producing tissue of the grain, it has been proposed that the scutellar epithelial cell, in the absence of aleurone cells, can produce about 10% of endosperm enzymes, such as α-amylase. However, it has been shown that barleys such as Galant, which had dysfunctional aleurone layers, failed to modify its endosperm successfully.[2]

As the production of gibberellic acid is linked to the germination process, germination potential is a very important index of malting quality. Nevertheless, if germination is efficient but gibberellin production and transport to the aleurone layer are limited, malting performance will be suboptimal.[7] In such circumstances, small quantities of gibberellic acid (0.2 to 0.25 ppm) are sometimes added to the chitted (germinated) grain during malting to optimize the levels of gibberellic acid in the grain (Figure 5.1). Optimal levels of gibberellic acid are required at the beginning of the germination process (Figure 5.3), because a delay in the production and secretion of endosperm-degrading enzymes by the aleurone layer will delay endosperm degradation.

During germination, sucrose and raffinose are degraded in the embryo but are resynthesized when the sugar products of endosperm modification pass from the starchy endosperm to the embryo.[2] The embryo contains significant quantities of lipid.[1,2] Hydrolysis by lipases produces substrates that the embryo can use for energy or for starch synthesis in the scutellum. Of the 3% of lipid that is present in malted barley, only about 0.01% of lipid materials (mainly fatty aids) are extracted from malts into worts. Lipases such as lipoxygenases (LOX-1 or LOX-2) are found in malt. LOX-2 is present in barley. LOX-1 develops in the embryo during malting and can

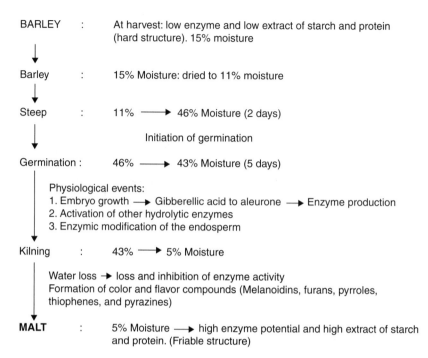

FIGURE 5.3
Basic steps of malt production.

cause lipid oxidation, producing beer-staling compounds, such as trans-2-nonenal. Lipoxygenases survive kilning and act quickly during mashing because of the high temperature.

Malt Production

Drying, Storage, and Handling

The drying, storage, and handling of barley must be managed by good practice in order to avoid damage to the living tissues of the grain,[16] that is, the embryo and aleurone layer. Barley should not be dried above an air-on temperature of 50 to 60°C. The lower end of this temperature range is recommended for high-moisture barleys (20 to 25°C). It is important that grains above 15 to 16% moisture are not stored for long periods. Irrespective of air-on temperature, the internal temperature of the grain driers should be significantly lower than the air-on and air-off temperatures. Only driers that are officially recognized as malting barley driers should be used to dry malting barley. Grains at 15 to 16% moisture should be malted or dried, soon after harvest. Airflow conditions in the drier should be optimal for the grain load being dried. Suboptimal airflow can cause stewing and damage germination potential. The embryos of stewed grains may give positive tetrazolium (viability) results, even though such damaged grains are unsuitable for malt production because embryo function is impaired. Long-term storage of high-moisture grain can also damage germination potential, especially in warm to hot climatic conditions.

Dried barley (<12% moisture) can be warm stored to remove dormancy. Warm storage can be carried out in carefully controlled storage conditions at 35 to 40°C for 2 to 4 weeks. Germination potential should be monitored regularly.

Stored barley should be aerated and extremes of heat or cold should be avoided during storage. Pests and extraneous materials should be eliminated and the development of fungi and bacteria avoided. Stored grains should be checked regularly for "hot spots" or condensation. There should be a scheme for tracing barleys as regards variety, total nitrogen, and germination potential. Grain damage (husk loss and broken grains) should be avoided. On the management of malting barley, good hygiene is preferable to chemical treatment.[16]

Steeping, Germination, Kilning, and Malt Quality

Steeping is designed to increase the moisture (water) level of the grain from the moisture of the stored grain to out-of-steep moisture (Figure 5.3). Air-resting procedures break up the submersion periods in different steeping regimes. The durations of submersion and air rest periods reflect sample requirements, water availability, and the production cycle. Each submersion steep can require about 900 l (200 gallons) of water per ton of barley. The disposal of such large quantities of water of high biological oxygen demand (BOD) can be expensive. However, short washing or limited steeping, supplemented by spray steeping, may reduce water usage and solve effluent problems, but may limit the development and action of cell wall-degrading enzymes such as endo-β-1,3:1,4-glucanase. Understeeping can limit endosperm modification and reduce malt quality. Spray steeping, after submersion steeping, should be a "top-up" exercise, where about 5% of water is added to obtain optimal moisture levels prior to the commencement of the germination process.

Steeping washes out a wide range of materials such as phenols, amino acids, sugars, mineral, and microorganisms.[1] Poststeep water is therefore active biologically. The reuse of steep water without prior treatment to reduce contaminating material is unwise because used steep water tends to inhibit germination. Various additives can be used during steeping, depending on acceptability. For example, calcium hydroxide (0.05 to 1.0%) and sodium hydroxide (0.05 to 0.1%), followed by alkali removal, may be used to assist phenol extraction. Formaldehyde (0.05 to 0.1%) can reduce microbial infection and hydrogen peroxide (0.1 to 1.0%) may assist oxygenation. Improved oxygenation can improve germination. Hypochlorites can reduce microbial infection but impart a disinfectant taint to the malt and the derived beer. Concentrated levels of detergent residues can inhibit germination. Germination can be slower in the conical regions of deep steeps than at the middle or top of such steeps. High-pressure pumping of slurries of fully steeped, high-respiration barleys can reduce germination potential. Understeeping may be a contributing factor as regards unexpected yeast precipitation that can occur during fermentation.

The loss of oxygen from the steep is very rapid and, therefore, aeration of the steep can improve germination. Air-resting reduces water sensitivity that inhibits germination. Usual extraction (displacement) of carbon dioxide may also improve germination. Steeping temperature varies and 16°C seems to be a good average temperature. In general, steeping at higher temperatures (e.g., 20°C) tends to reduce proteolysis but increases friability of the endosperm indicating inhomogeneity of malt modification. About 70 to 75% of the soluble protein in wort is produced during malting; the proteolysis, being more effective at 15 to 16°C. In general, modification is initially faster and more uniform at 16°C than at 20°C. Optimal starting moisture of the germination process is about 45 to 46% (Figure 5.3), and moistures of about 40% will retard the modification process, even in high-quality barleys. Optimal moisture levels over the first 2 days of the germination process are required for optimal initiation of the endosperm-modification process. This is the period when endosperm-degrading enzymes are produced in, and released from, the aleurone layer rapidly. At this early stage of malting, these enzymes are not only required to release extract, they are also required to reduce the high viscosity of solubilized β-glucan materials.

Recent studies have confirmed that steely (compact, high protein) endosperms absorb water more slowly than mealy (loose structure, low protein) endosperms. As a consequence of differences in hydration, mealy grains tend to malt faster than steely grains.[14] Because of the importance of mealiness and steeliness as quality parameters, methods have been developed to detect the degrees of mealiness and steeliness of barley samples.[13,17] However, the endosperm structure is not the only quality factor that must be considered in the complex process of malting.

Enzyme development in aleurone and enzyme distributions in the starchy endosperm layer are also important features of the malting process. Recent studies suggest that variety and out-of-steep moisture influence the extent to which the aleurone layer is activated during malting.[3,7] These studies showed that limited steeping retarded aleurone activation and the usually slow-modifying distal (nonembryo) end of the grain remained undermodified. This limiting effect on aleurone function was worse in low-grade barley varieties. Understeeping appears to delay the transport of gibberellic acid through the aleurone layer. Therefore, three of the most important aspects of steeping are: (1) to initiate germination and the development of gibberellic acid in the germinated embryo; (2) to facilitate the transport and action of gibberellic acid in the aleurone layer; and (3) to hydrate the starchy endosperm to a level that facilitates enzymic modification of the endosperm.

Limitations to endosperm modification caused by suboptimal moisture conditions during the first 1 to 2 days of germination are difficult to correct later in the germination (malting) process (Figure 5.3). Spraying during the later stages of germination can cause localized overmodification of the embryo end of the endosperm. Such malts can have high colors after kilning, but are nevertheless undermodified. For normal lager malts, high malt colors are usually associated with general overmodification of the malted grain. A germination temperature of 16°C produces better proteolysis than a higher temperature of 20°C. The latter temperature tends to give better friability but greater inhomogeneity of single grain modification.[7] In contrast, 16°C malts had better amylase development and better fermentabilities than malts produced at 20°C.

An essential feature of the germination process is that the relative humidity (RH) of the airflow through the grain bed should be as close to 100% as possible. This avoids water loss that reduces the rate of modification of the endosperm. Commercially, gibberellic acid is produced by the fungus *Gibberella fujikuroi* in fermentation systems, and processed, packaged, and sold to the industry. Maltsters have been using gibberellic acid since 1959 — long before definitive physiological work was carried out on the role that this hormone had on enzyme synthesis.[2] Gibberellic acid (0.2 to 0.25 ppm) is usually applied early during the first day of the malting process. The grain should have chitted (germinated) before gibberellic acid is applied — this facilitates the uptake of the hormone. When potassium bromate (50 to 100 ppm) and gibberellic acid were applied together, the increased proteolysis caused by the gibberellic acid was reduced by the potassium bromate.[1,2] Potassium bromate also reduced extract loss by the reducing root growth of the malting grain. The use of potassium bromate in the industry is now limited.

The levels of gibberellic acid produced naturally by some grains are insufficient to produce the malting rates required by many maltsters. However, although added gibberellic acid can increase the malting rate, the asymmetric pattern of enzymic modification of the endosperm remains the same (Figure 5.1 and Figure 5.4). Progression starts at the dorsal embryo–aleurone–endosperm junction and moves daily, downward and outward (Figure 5.1), toward the distal end of the malting grain over a period of about 5 to 6 days. The abrasion process was developed to accelerate the modification process.[2] Abrading machines selectively damage the pericarp layer, especially at the distal ends (regions) of the grain. Added gibberellic acid can now enter the embryo as well as the distal end of the malting grain simultaneously. Abraded grains, therefore, malt from both ends of the grain and, consequently, enzyme development and extract yield are

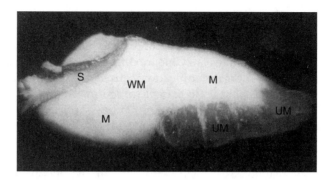

FIGURE 5.4
Asymmetric pattern of modification of the starchy endosperm of malting barley. Shoot and roots removed. Grain germinated (grown) for 4 days. WM, well-modified endosperm; M, modified endosperm; UM, under-modified (barley-like) endosperm; S, scutellum.

increased. Moreover, malting time is reduced. Physiologically, the abrasion process showed that the rate of *distribution* of endosperm-degrading enzymes in the starchy endosperm is a primary factor controlling the malting rates of barley grains. In this regard, gibberellic acid/aleurone layer efficiency is a key factor in malting efficiency.

In a well-modified malt, about 90% of the β-glucan is broken down. Although there is a positive correlation between the level of endo-β-glucanase and β-glucan breakdown during malting, the endo-β-glucanase levels of the individual grains of a barley sample do not correlate with β-glucan breakdown, suggesting that other enzymes may complement the action of endo-β-glucanases during malting.[18] Evidence for these complementing enzymes has not yet been found. However, the actions of proteases may be important. Studies of factors that influence the homogeneity of malt modification suggest that the undermodified grains of a malt sample always contain higher levels of β-glucans and nitrogen (protein) than well-modified grains.[7] This suggests that slow β-glucan breakdown seems to be associated with the high protein grains or grains that break down their proteins slowly.[19] The blending of high-protein barleys with low-protein barleys, or natural variations in nitrogen contents of the grains of a barley sample, is likely to produce malts whose modification is not homogeneous.[19] The inhomogeneity of modification in the grain population of a sample of malts is not usually exposed by standard malt analyses. However, malt processing in the brewhouse can cause unexpected problems such as slow wort separation and slow beer filtration.[7] Haze development may also be a feature of uneven malt modification. The addition of extraneous commercial enzymes to the mash tun can remove problems caused by the inhomogeneity (unevenness) of malt modification.[17]

Kilning inactivates many microorganisms and reduces the moisture content of the undried green malt from about 43 to about 5%. This reduction in moisture stabilizes the grain and permits long-term storage. During the early stages of kilning, rapid water loss reduces malt temperatures and malt modification continues to a limited degree, until excessive water loss stops enzymic action. Malt types are determined, in the main, by kilning or roasting procedures (Table 5.1 and Table 5.2). During kilning, there is a development of color, a reduction in enzyme activity, and an increase in acceptable flavors. Color development results from reactions between sugars and amino acids of the malt to form melanoidins (Maillard reaction).[2] Some of these products of roasting treatment are described as reductones and have antioxidant properties, which are important with regard to improving beer stability and as a source of antioxidants in the diets of beer drinkers. During kilning, the activities of endo-β-1,3:1,4-glucanase, β-amylase (DP), limit dextrinase, and endoproteases are reduced to greater degrees than the activities of enzymes such as α-amylase and carboxypeptidase.

Malts kilned for long periods at high moisture tend to have lower levels of heat-labile enzymes. The dimethylsulfide (DMS) levels in the beer (e.g., 50 to 100 ppb) are related to the levels of its precursor, S-methylmethionine (SMM), in the malt. DMS is an important flavor compound of lagers but is not usually found in ales. Malts kilned below 65°C can develop high levels of DMS in hot worts. Malts kilned between 80 and 82°C develop DMS, from dimethylsulfoxide (DMSO), during the fermentation process.[15]

Nitrosodimethylamine (NDMA) is a carcinogenic compound that can develop during malting, when hordenine of the embryo reacts with nitrogen oxides in kilns fired directly by sulfur-free natural gas.[2] Sulfur dioxide from burnt sulfur reduces nitrosamine development. Indirect kilning can do the same. Commercial malts tend to have less than 20 ppb of NDMA. Ethyl carbamate is another carcinogenic compound that originates from malted barley but develops during the distillation process.[2] Its control can be

TABLE 5.1

Average Analysis of Colored and Roasted Malts and Barley

	Extract (l°/kg)	Moisture (%)	Color (°EBC)	Final Kilning Temperature (°C)
Ale	305	4.0	5.0	100
Lager	300	4.5	2.0	80
Light Crystal[a]	265	7.0	25–35	75
Crystal[a] malt	268	4.0	100–300	75
Amber/Brown[a] malt	280	2.0	100–140	150
Chocolate[a] malt	268	1.5	900–1100	220
Roasted[a] malt	265	1.5	1100–1400	230
Roasted[a] barley	270	1.5	1000–1550	230

[a]These malts and barleys do not contain enzymes.

TABLE 5.2
Average Malt Analysis of Different Kinds of Malts

Analyses	Analytica EBC-Analyses[a]			Analyses[a] Wheat	Analyses	IOB Analyses[b]		
	Pilsner	Lager	Munich			Lager	Ale	Distillers'
Moisture (%)	4.0	4.0	4.0	5.0	Moisture (%)	4.5	4.0	4.0
Extract (%)	79.0	80.0	79.0	84.0	Extract (l°/kg)	300	305	81.0
					pH	5.9	5.6	5.9
Fine/coarse Difference (%)	2.0	1.5	1.0	1.0	Fine/coarse Difference (l°/kg)	5.0	2.5	1.0
Color (°EBC)	2.0	2.0	15.0	3.0	Color °EBC	2.0	5.0	2.0
α-Amylase (DU)	35.0	35.0	28.0	45.0	α-Amylase (DU)	35.0	30.0	38.0
Diastatic power Windish–Kolbach	250.0	250.0	100.0	300.0	Diastatic power	70.0	65.0	75.0
Diastatic power (°L)	76.0	76.0	33.0	90.0	β-Glucanase (IRV units)	700	500	700
Total protein (%) (N% ×6.25)	11.0	10.5	11.0	13.0	Total nitrogen % (N% × 6.25)	1.7	1.6	1.6
Kolbach index (TSN/TN)	40.0	42.0	45.0	42.0	Index of modification (TSN/TN)	38.0	40.0	39.0
Friability (%)	87.0	87.0	88.0	—	Friability (%)	88.0	92.0	90.0
Homogeneity (%)	98.0	98.0	98.0	—	Anthocyanogens (ppm)	55.0	50.0	60.0
Whole grain (%)	2.0	2.0	2.0	—	Polyphenols (ppm)	150.0	140.0	150.0

Notes: TSN, total soluble nitrogen %; TN, total nitrogen %.
[a]Programmed mashing (European and North American malts).
[b]65°C mashing.

effected through the use of barley varieties which have low levels of natural cyanide precursors. Levels of ethyl carbamate in whisky must be below 150 ppb.

Malt Varieties

Different kinds of malts are available to the industry. Distilling malts fall into two main categories: malts used for malt whisky production and malts used for grain whisky production. Both malts are lightly kilned and have high DP (β-amylase). Another type of distilling malt is unkilned and is known as green malt. Green malt has a higher enzyme content than kilned malt and is used to produce grain whisky. The amylase potential of grain-distilling malts tends to be higher than that of malt whisky malts because grain-distilling malts are required to convert about 90% of cooked, but unmalted, cereals such as wheat or maize. Distilling malts can be peated by passing peat smoke through them during kilning.[1,2] Peat levels reflect the quantities of phenols present on the malt.

Lager and Pilsener malts, like distilling malts, have low colors of about 2°EBC units. Ale malts can have colors of about 5°EBC units. Mild ale malts have colors of about 7°EBC units. The diastatic power (DP^0L) of lager and distillers' malts are usually higher than those of ale malts. Malt worts (10% solid materials, 1040 specific gravity) contain about 91% carbohydrate and about 6% protein. Distillers' worts are usually about 86% fermentable while brewers' worts are about 75% fermentable.[15] The high levels of maltose found in worts (45 to 50%) reflect the dominant actions of diastases (α-amylase and mainly β-amylase) during mashing.

Munich malts have colors of about 15 to 20°EBC units and enzyme levels are much lower than those of lager and ale malts. Vinegar or food malts, like lager malts, also have low colors and high enzyme potential. Special malts are made from kilned (white) ale-type malts (color 5°EBC). These are heated to different temperatures to produce the required types of malts. For example[1,2]: amber/brown malts (nutty, biscuit flavors) have colors of about 100 to 140°EBC units, chocolate malts' (treacle, chocolate flavors) colors range from 900 to 1100°EBC units; roasted (black) malts (smoky, coffee flavors) have colors that range from 1100 to 1400°EBC. Malts such as light crystal (sweet, nutty, toffee flavors), crystal (sweet, malty, caramel flavors), and dark crystal (burnt, caramel flavors) have colors that range from 20 to 60, 100 to 250, and 300 to 350°EBC units, respectively. These are produced from green (unkilned) malts. Roasted barley (burnt, smoky flavors) is an important ingredient of stouts and some ales, and is produced from barley and have colors of about 1100 to 1400°EBC. The color produced in these malts relate to the kilning temperatures used (Table 5.1). Because of the higher kilning temperatures used to produce them, special malts and roasted barley contain no enzymes.

All malts are used to provide extracts that are used by yeast to produce alcohol and flavor compounds. Some special malts provide mainly color and flavor compounds. In general, hot water extract is one of the most important parameters in malt specifications. To achieve the potential extract of a malt in the brewhouse is a vital part of the economics of the industrial processing of malt. However, recent studies show that although extract yield may be optimal, inhomogeneity of endosperm modification may conceal serious processing problems. For example, a malt sample with 0.3% β-glucans can have 0.2% β-glucans in 80% of its grains and 0.7% β-glucans in the remaining 20% of its grains. Irrespective of the details of the malt specification set by the brewer, the distiller, or a food company, such malts can cause processing problems. Therefore, one of the major problems facing the industry is that traditional analyses of malt do not reflect potential performance.[7] For example, the soluble nitrogen ratio (total soluble nitrogen divided by total nitrogen, TSN/TN) does not reflect overall protein modification in the malted grain. Also, it gives no indication of the overall modification of the grain. The friability test is very useful, but a *fragmented* endosperm could give similar friability results to that of a uniformly modified endosperm. In addition, the friable flours of different malt samples can have different levels of β-glucans. This should not be the case since friable flours are assumed to come from modified parts of the malted grain.[7,14]

Recent studies suggest that if a quality product is one that meets the expectations of the customer and if traditional malt analyses do not define the processing quality of purchased malt, then a radical improvement in the precision of the analyses that are used to define malt quality is required.[7] Accuracy relates to the repeatability of analyses of a malt sample; precision reflects the true state of modification of grains of a malt sample; and the overall potential of these gains to process well or badly in the brewhouse. This kind of analytical precision is essential if brewers or distillers are to produce high-quality products efficiently. The development of precision analyses requires scientific information that can only be derived from research work on the structure, function, and processing of cereal grains and malts derived from them. A recent study of the history[20] of the malting industry, in the United Kingdom, suggests that developments in processing were linked to innovative research. Future developments in the production and processing of malt in the industry can only be achieved on the basis of innovative malting research.

Acknowledgment

The author thanks C. Brown and Annie Hill for help with preparation of this manuscript.

References

1. Briggs, D.E., *Malts and Malting*, Blackie Academic & Professional, London, 1998.
2. Palmer, G.H., *Cereal Science and Technology*, Aberdeen University Press, Aberdeen, U.K., 1989.
3. Palmer, G.H., Structure of ancient cereal grains, *J. Inst. Brew.*, 101:103–112, 1995.
4. Agu, R.C. and Palmer, G.H., Some relationships between the protein nitrogen of barley and the production of amylolytic enzymes during malting, *J. Inst. Brew.*, 104:273–276, 1998.
5. Jacks, P.L., Application of DNA probes (RFLPs) in barley breeding, *European Brewing Convention, EBC Symposium, Plant Biotechnology*, Monograph XV, Helsinki, Finland, 1989, pp. 130–136.
6. Ritala, A., Aspegen, K., Kurtin, U., Salmentallio-Nattila, M., Mannonen, L., Hannus, R., Dauppinen, V., Teeri, T.H., and Enari, T.-N., Fertile transgenic barley by particle size bombardment of immature embryos, *Plant Mol. Biol.*, 24:317–325, 1994.
7. Palmer, G.H., Achieving homogeneity in malting, *European Brewery Congress*, Nice, 1999, pp. 323–363.
8. Palmer, G.H. and Sattler, R., Different ratios of development of α-amylase in the distal endosperm ends of germinated/malted Chariot and Tipper barley varieties, *J. Inst. Brew.*, 102:11–17, 1996.
9. Broadbent, R.E. and Palmer, G.H., Relationship between β-amylase activity, steeliness, mealiness, nitrogen content and nitrogen fraction of barley grains, *J. Inst. Brew.*, 107:349–354, 2001.
10. Ryder, D.S. and Vogelsang, F., Adopting malt quality to modern brewing techniques, *European Brewery Convention Symposium*, Malting Technology, Andernach, Germany, 1994, pp. 180–193.
11. Palmer, G.H., The influence of endosperm structure on extract development, *J. Am. Soc. Brew. Chem.*, 33:174–180, 1975.
12. Palmer, G.H., Maintaining progress in malting technology, *European Brewery Congress Proceedings*, Dublin, pp. 133–148, 2003.
13. Koliatsou, M. and Palmer, G.H., A new method to assess mealiness and steeliness of barley varieties and relationship of mealiness with malting parameters, *J. Am. Soc. Brew. Chem.*, 61:114–118, 2003.
14. Bryce, J.H., McCafferty, C.A., Cooper, C.S., and Brosman, J.M., Optimizing the fermentability of wort in a distillery — the role of limit dextrinase, Distilled Spirits: Tradition and Innovation, J. H. Bryce and G. G. Stewart eds. Nottingham University Press, pp. 69–78, 2004.
15. Chandra, G.S., Proudlove, N.J., and Baxter, E.D., The structure of barley endosperm — an important determinant of malt modification, *J. Food Sci. Agric.*, 79:37–46, 1999.
16. Brissart, R., Brauminger, U., Haydon, S., Morand, R., Palmer, G., Sanvage, R., and Seward, B., *European Brewing Convention Manual of Good Practice, Malting Technology*, Fachverlang Hans Carl, Germany, 2000.
17. Lalor, E., *Creating the Future in Brewing, Version One*, Quest International, Cork, Ireland, 2002, pp. 2–53.

18. De Sa Marins, R. and Palmer, G.H., Assessment of malt modification by single grain analysis, *J. Inst. Brew.*, 110:43–50, 2004.
19. Palmer, G.H., Malt performance is more related to inhomogeneity of protein and β-glucan breakdown than to standard analyses. *J. Inst. Brew.*, 106:189–192, 2000.
20. Clark, C., *The British Malting Industry since 1830*, The Hambledon Press, London, 1998.

6

Adjuncts

Graham G. Stewart

CONTENTS

Introduction 161
Corn Grits 163
Rice 163
Barley 164
Sorghum 166
Refined Corn Starch 167
Wheat Starch 168
Torrified Cereals 169
Liquid Adjuncts 169
Malt from Cereals Other than Barley 172
    Wheat Malt 172
    Oats and Rye Malt 172
    Sorghum 173
Conclusions 174
References 175

Introduction

The Germany purity law[1] defines an adjunct (or secondary brewing agent) as "anything that is not malt, yeast, hops, or water" (in fact, the original 1516 document does not mention yeast). However, for the purposes of this chapter, a much narrower definition of adjuncts will be employed. In the United Kingdom, the Foods Standards Committee[2] defines a brewing adjunct as "any carbohydrate source other than malted barley which contributes sugars to the wort." A wide range of materials fall within this definition

TABLE 6.1

Brewing Adjuncts and Their Preparation Processes, in Increasing Complexity

---
Basic raw cereal: barley, wheat
Raw grits: corn (maize), rice, sorghum
Flaked: corn, rice, barley, oats
Torrified/micronized: corn, barley, wheat
Flour/starch: corn, wheat, rice, potato, cassava, soya, sorghum
Syrup: corn, wheat, barley, potato, sucrose
Malted cereals other than barley: wheat, oats, rye, sorghum

Source: Adapted from Marchbanks, C. 1987. *Brew. Distill. Int.*, 17:16–18. With permission.

and, in this chapter, attention will be directed to three areas: (a) solid unmalted raw materials usually processed within the brewhouse, (b) liquid adjuncts usually added to the kettle and some specialty products used for priming, and (c) malted cereals other than barley, such as wheat and sorghum (Table 6.1). Adjuncts are usually considered nonmalt sources of fermentable sugars. They typically contribute no enzyme activity and little or no soluble nitrogen and are less expensive than malt. It is also sometimes stated that adjuncts do not contribute flavor to the finished product[4]; however, it will be discussed later that, in many beers, this is not really the case.

A great deal of effort has been expended to improve the performance of various adjuncts and to examine their contribution to the characteristics of the finished beer. In general, corn tends to give a fuller flavor to beers than wheat, which imparts a certain dryness. Barley will give a stronger harsher flavor. Both wheat and barley adjuncts can considerably improve head retention (foam). Rice will also give a very characteristic flavor to beer.

The overall brewing value of an adjunct may be expressed by the following equation[5]:

Brewing value = Extract + Contribution to beer quality − Brewing costs

The major benefit is extract.

In the United States, current use of nonmalt adjuncts averages about 38% of total brewing materials employed, excluding hops.[3] The most commonly used adjunct materials are corn (maize) (46% of total adjunct), rice (31%), barley (1%), and sugars and syrups (22%). Canales and Sierra[6] listed the following materials that are used as unmalted brewer's adjuncts: yellow corn grits, refined corn starch, rice, sorghum, barley, wheat, wheat starch, cane and beet sugar (sucrose), rye, oats, potatoes, tapioca (cassava), and triticale. In addition, processed adjuncts include corn, wheat and barley syrups, torrified cereals (see Section "Torrified Cereals"), cereal flakes, and micronized cereals.

Corn Grits

Corn grits are the most widely used adjunct in the United States and Canada. They are produced by dry milling yellow corn. The milling process removes the hull and outer layers of the endosperm along with the oil-rich germ, leaving behind almost pure endosperm fragments. These fragments are further milled and classified according to brewers' specifications. Corn grits produce a slightly lower extract than other unprocessed adjuncts and contain higher levels of protein and fat. The gelatinization temperature range for corn grits (62 to 74°C) is slightly lower than that of rice grits (64 to 78°C).

Flaked grits, flake micronized grits, and corn grits have been compared in a pilot brewhouse.[7] Extract yields were highest for the flaked and micronized flaked grits and they showed slightly higher fermentabilities. Untreated corn grits compared very favorably with flaked and micronized barley and wheat adjuncts with respect to extract yield, fermentability, alpha-amino nitrogen, and wort viscosity.

Meilgaard[8] has reviewed the composition of worts and beers prepared with a variety of adjuncts. He has shown that a carbohydrate profile similar to an all-malt wort can be attained with either 20% rice or 20% corn grits, although levels of sucrose and fructose decline as the adjunct level increases. Corn grits at the 30% level produce a volatile aroma compound similar to that of an all-malt beer. Wort protein, peptides, free amino acids, and nucleic acid derivatives decline in proportion to adjunct level. The amino acid profile of wort is not affected by a particular adjunct but by its level in the mash. High adjunct ratios lead to higher levels of diacetyl and related compounds at the end of fermentation. However, with the appropriate postfermentation processing, these levels return to normal after aging.

Rice

Rice is currently the second most-widely used adjunct material in the United States.[9] On an extract basis, it is approximately 25% more expensive than corn grits. Brewer's rice is a by-product of the edible rice-milling industry. Hulls are removed from paddy rice and this brown rice is then dry milled to remove the bran, aleurone layers, and germ. The objective of rice milling is to completely remove these fractions with a minimal amount of damage to the starchy endosperm, resulting in whole kernels for domestic consumption. However, up to 30% of the kernels are fractured in the milling process. The broken pieces ("brokens") are considered esthetically undesirable for domestic use and are sold to brewers at a price considerably less than the whole kernel or mill-run rice price. Rice is preferred by some brewers because of its lower oil content compared to corn grits. Rice has a very neutral aroma and flavor, and when converted properly in the brewhouse, yields a light, clean-tasting beer.

The quality of brewer's rice can be judged by several factors, including cleanliness, gelatinization temperature, mash viscosity, mash aroma, moisture, oil, and ash and protein content. Rice should be free of seeds and extraneous matter. Insect or mold damage should not be tolerated, as these indicate improper storage or handling conditions. It has been reported that rancidity in rice oil can be a problem, but with modern storage techniques this is a negligible factor. Laboratory mashes of rice samples should be conducted regularly and they should gelatinize and liquefy in a standard manner and should be clean and free from undesirable odors and tastes.

Not all varieties of rice are acceptable brewing varieties. Rice has a relatively high gelatinization temperature and is extremely viscous prior to liquefaction in the cereal cooker. Many rice varieties, such as Nato, will not liquefy properly and are impossible to pump from the cooker to the mash mixer. Other varieties, such as short-grain California Pearl, Mochi Gomi, and Cahose,[10] liquefy well in the cooker during a 15-min boil. Both amylose and liquid content of rice varies with the variety and the cultivation conditions, thus selection of suitable grades is important. Rice liquefies more easily the finer the particle grind, and particles less than 2 mm are considered adequate. Handling of rice is relatively easy, as the brokens contain little dust and flow easily through standard hopper bottoms and conveyoring equipment. Rice is milled in fixed roller mills. There is no difficulty in making the rice mash slurry at 64 to 76°C, although it is a common practice to mash and hold at 36 to 42°C as a protein rest. As with all cereal cooker operations, whatever the starch source, 5 to 10% of the malt grist is added to the cooker because the malt enzymes (amylases and proteinases) are essential for the partial liquefaction necessary to render the cooker mash fluid enough for pumping. Atmospheric boiling is required for gelatinization. Some brewers pressure cook at 112°C.

If properly converted, rice adjunct usage does not create runoff problems. As previously discussed, the extract is slightly lower in soluble nitrogen than corn grits.

Barley

Unmalted barley is an obvious adjunct for use in brewing. However, the raw grain is abrasive and difficult to mill, scattering to yield too high a percentage of fine material which gives problems during lautering. These difficulties disappear if the grain is conditioned to 18 to 20% moisture prior to milling although this process has not been widely employed in brewing.

In the past, barley has normally been partially gelatinized before use. The barley is gelatinized either by mild pressure cooking or by steaming at atmospheric pressure followed by passage of the hot grits through rollers held at approximately 85°C. Finally, the moisture content of the flakes is reduced to 8 to 10%. This process of pregelatinization can also be applied to corn. Pregelatinization of barley affects the ease of extraction of β-glucan during mashing and, hence, the β-glucan content of the wort. Prolonged steaming prior to rolling the barley produces a product which produces higher viscosity sweet worts. This can be controlled by measuring the viscosity of a cold-water extract of the flaked barley, which is a good indication of the extent of the steaming process. Barley starch is more readily hydrolyzed than corn or rice starch. Barley may be dehusked before use to increase extract yields, but this may lead to runoff difficulties because the husk provides material for filter bed formation. In the same way, fine grinding improves extraction efficiency but also leads to slow runoff. If significant proportions of barley are used in the mash, a malt with sufficient enzyme activity is required. Use of barley leads to a reduction in wort nitrogen content and decreased wort and beer color. No difficulties have been reported in fermentation. Foam stability is usually improved because of lower levels of proteolysis. However, a major difficulty associated with brewing with high levels of unmalted barley can be the increase in wort viscosity and runoff times caused by the incomplete degradation of β-glucans. Mashing at 65°C quickly destroys the malt β-glucanase activity. Suggestions to alleviate these problems have included pretreatment of the barley with β-glucanase and the use of a temperature-stable β-glucanase in the mash.

Raw (feed) barley can also be employed as an adjunct, and as high as 50% barley in the grist has been employed in some breweries in Australia. Use of raw barley requires significant modification to the brewing process. For reasons already discussed, conventional roller milling cannot be employed; consequently, hammer mills are necessary. This high level of malt replacement usually results in insufficient malt enzymes for the necessary hydrolysis of the starch, protein, and β-glucans. Consequently, a malt-replacement enzyme system is employed to compensate for the reduced level of malt enzymes.[11] A number of such enzyme systems have been developed and are usually a mixture of β-amylase, protease, and β-glucanase, which are obtained from microbial sources such as *Bacillus subtilis*.

In barley brewing, it is possible to approximate the starch hydrolysis profile and the degree of fermentability of 100% malt worts. This is possible by substituting malt with barley at levels of 50% (extract basis) and by controlling the main mash schedule (enzyme concentration, time, and temperature). Barley worts have been found to contain less fructose, sucrose, glucose, and maltotriose but more maltose than malt worts. No anomalies or difficulties in fermentation and aging have been noted. Most breweries can employ their normal fermentation and aging technology for barley brewing. In general, no significant difference in organoleptic properties between barley beers and 100% malt beers have been observed. A harshness of barley beers can be avoided by lowering the pH of the wort to 4.9 prior to boiling.

It would appear that, with the aid of microbial enzymes, today's brewer can increase the level of unmalted raw barley. It is to their economic advantage to do so and, at the same time, obtain the desired beer quality.

Sorghum

Although the potential for employing sorghum (millet) as a brewing adjunct goes back over 50 years, it is only in the past two decades that real interest has been shown in this cereal. Sorghum is the fifth most-widely grown cereal crop in the world; only wheat, corn, rice, and barley are produced in greater quantities.[12] Africa is a major source of sorghum as is Central America. However, the absence of a controlled "seed industry" has limited research into the wide range of sorghum genotypes that may be suitable in brewing. As well as its use as an adjunct, sorghum has recently been employed as a malt. This development will be discussed in a later section of this chapter.

Sorghum is the traditional raw material in Africa for the production of local top-fermenting beers that are known by various names.[12] Examples are "Bantu beer" in South Africa, "dolo" in Burkina Faso, and "billi billi" in Chad. These beers are produced without hops; they are slightly sour in taste, and they are drunk unfiltered, mainly in rural regions.

The U.S. brewing industry employed sorghum as an adjunct in 1943 when brewing materials were scarce. Unfortunately, sorghum was cracked and only partially dehulled and degerminated; consequently, brewers obtained poor yields and bitter-tasting beers plus a number of other quality problems. Modern milling techniques and better purification methods have changed the situation. Today, sorghum brewer's grits are considered by many to be of comparable quality to the best corn and rice grits.[6]

Probably because of the bad experience in 1943, sorghum brewer's grits are almost never used by brewers in the United States, but the interest is still alive, mainly because of economic factors. In Africa and Mexico, brewers are using an appreciable and a continuously increasing percentage of brewer's grits and, in most cases, producing beer of acceptable quality.

The advantage of sorghum in agronomic terms is its ability to survive under extreme water-stress conditions and, hence, the cereal is ideally suited for cultivation in tropical areas, including Africa and Central America. The current yield without fertilization is between 2 and 3.8 tons per acre, but this can easily be increased by adopting suitable measures. The Food and Agriculture Organization (FAO) recommends using sorghum to manufacture beer.

In the brewing process described by Canales and Sierra,[6] the dried sorghum is screened to remove extraneous material. The whole grain is then fed into a series of dehullers that produce two product streams. In one of these streams, the husks and embryonic material constituting some 48% of the original sorghum are removed and this fraction, together with the initial screenings, is sold as a by-product. The second stream, consisting of peeled sorghum together with a small amount of husk material is then passed through an aspirator in which the husk is removed. The purified pearled sorghum, now representing 47% of the original cereal processed, is milled to give 12% of the original material as flour and 35% as sorghum grits. Both of these components are used as brewing adjuncts contributing up to 45% of the total wort extract.

The chemical composition of sorghum grain is very similar to corn. Both grains contain starch consisting of 75% amylopectin and 25% amylose. Starch granules are similar in range, shape, and size. On an average, sorghum starch granules are slightly larger — 15 μm compared to 10 μm for corn. Sorghum starch has a higher gelatinization temperature (68 to 76°C) than corn starch (62 to 68°C). In the brewhouse, sorghum brewer's grits perform within acceptable limits. No special handling or cooking techniques are required. Five percent malt in the cooking mash is sufficient. Conversion of starch occurs within the mashing time allowed. The beers produced are fully equivalent in chemical analysis, flavor, and stability to beers produced with other adjuncts. Finally, in many areas (e.g., Africa and Central America) sorghum offers the lowest-cost source of available fermentable sugar.[13]

Refined Corn Starch

Refined corn starch is by far the purest starch available to the brewer.[9] It is a product of the wet-milling industry. It has not found widespread use because its price is higher relative to corn grits and brewer's rice and it is difficult to handle. An obvious drawback for refilled starch usage is handling. The starch powder is extremely fine, with 96% passing through a 200-mesh screen. It must be contained in well-grounded lines and tanks to prevent explosions resulting from static electricity sparks produced during conveying. The starch bridges easily and is nearly impossible to flow from tanks unless they have special fluidizing bottoms.

Refilled starch can be received by the brewer in 50 kg (100-lb) bags or in bulk air-slide hopper cars. For anything but experimental usage, bulk systems are essential to reduce the costs of starch and manpower to reasonable levels. The starch is removed from the rail car by vacuum and then is blown through an airveyor to a drop-out cyclone above a bulk storage bin. Bag house dust collection is essential for the airveyor.

Refilled corn starch can be utilized as the total adjunct or can be mixed with rice or corn grits at the option of the brewer. The gelatinization and liquefaction of the starch proceeds at lower temperatures than rice or grits, but is not sufficiently different to preclude its use as a blend with either. As a total adjunct, care must be taken to prevent sticking on the cooker bottom.

Refilled starch can easily be liquefied by the same process utilized for rice. The resultant extract cannot be organoleptically or chemically differentiated from an all-rice extract. Brewhouse yield can be increased 1 to 2% by the use of refilled starch in place of rice. There are no runoff problems. Fermentations tend to attenuate better, while colloidal stability is unaffected. Beer flavor is not affected, except that the beer is considered slightly thinner, because of higher attenuation limits.

The outlook for refilled starch as a brewing adjunct will depend to a large extent on relative pricing. It is unlikely that refined starch will ever capture any market from dry-milled corn grits, although a wet-milled starch plant in close proximity could produce a competitive slurry starch. The effect of a growing, high-fructose market could result in shortages with resultant high prices for starch until plant capacity catches up. A boom in the paper industry could also adversely affect starch availability. On the positive side, the by-products of a wet-milled low-protein meal and gluten could grow in demand and value, which would help pay for the starch fraction of the wet-milling operations.

Wheat Starch

Refined wheat starch is not presently attractive in the United States because of its high price compared with the more readily available adjuncts. It has been employed, in the past, in Canada, where it allowed the usage of surplus wheat grown in Canada. Chemically, wheat starch is very similar to refilled corn starch. An advantage is that its gelatinization temperature is similar to malt gelatinization and could be added directly to the malt mash; however, 10% higher brewhouse yields can be obtained by cooking in a conventional adjunct cooker. Lautering times are reported to run up to 10% longer than with corn grits.

Wheat starch has the same conveying and handling problems as refined corn starch. Slurrying should take place below 52°C to prevent lumping. The cooker temperature should not exceed 98°C, as the starch foams badly upon boiling.

Wheat starch is somewhat higher in β-glucans and it is suggested that the cooker mash, with 10% of the malt added, stand at 48°C for 30 min prior to the 66°C rise to give the β-glucanase time to break down the β-glucans at its optimal temperature. This procedure results in little or no runoff problems.

The beer is quite comparable to the beer brewed with corn grits in analysis and flavor. Should wheat starch be available at prices competitive to other adjuncts, it would be a perfectly suitable adjunct.

Torrified Cereals

Torrification is a process by which cereal grains are subjected to heat at 260°C and rapidly expanded or popped. This process renders the starch pregelatinized and thereby eliminates the cooking step in the brewhouse. It also denatures a major portion of the protein in the kernel such that the wort-soluble protein is only 10% of the total.

Both barley and wheat are potential candidates for torrification and for use as torrified adjuncts. The chemical analyses are quite similar for both. The torrified products have about 1.4% wort-soluble protein and could allow the use of lower-protein malts or higher adjunct levels, while maintaining soluble protein similar to worts produced with lower soluble protein adjuncts. Fat content is slightly higher than for other adjuncts, but again this would be negated in the final wort by using higher adjunct levels.

There are no handling or dust problems associated with the use of torrified cereals. It is possible to blend the torrified products with malt. They can then be ground simultaneously and mashed-in together. However, higher yields are found by cooking the torrified product separately at 71 to 77°C, prior to addition to the malt mash.

The use of torrified cereals leads to increased lauter grain bed depth and to slight runoff penalties. Torrified barley seems to be more refractory than torrified wheat in this respect. Particle size and mill setting are critical; large particle size leads to poor yield and too fine a grind leads to runoff problems. Because of its expanded nature, the torrified cereals absorb more water than other adjuncts and, especially in the case of torrified barley, higher ratios of water to cereal must be used. The flavor of beer produced with torrified adjuncts is reported to be unchanged, and one could easily conclude that if torrified cereals become economically competitive with other adjuncts they would be employed as an alternate adjunct source.

Liquid Adjuncts

The major liquid adjuncts used in brewing are glucose syrups, cane sugar syrups, and invert sugar syrups. Although differing in detail, the essential similarity is that they are all solutions of carbohydrates. The term glucose in this context can be misleading. Glucose is the commonly used name for dextrose, but the glucose syrups used in brewing are in fact solutions of a large range of sugars and will contain, in varying proportions depending upon the method of manufacture, dextrose, maltose, maltotriose, maltotetraose, and larger dextrins.

Cane sugar syrups contain sucrose derived from sugar cane and sometimes, depending upon the grade, small quantities of invert sugar. Invert syrups, as the name suggests, are solutions of invert sugar — a mixture of glucose and fructose. Invert sugar is produced, in nature and commercially, by the hydrolysis of sucrose which, together with glucose and fructose, occurs abundantly in nature. Commercially, sucrose is extracted from sugar cane or beet, and glucose syrups are usually manufactured from starch derived from corn or wheat grains.

Glucose syrups have been available since the mid-1950s. They were originally produced by straight acid conversion of starch to a 64 to 68 DE range. (The degree of starch conversion is usually expressed as "dextrose equivalent" or DE. This is a measure of the reducing power of the solution, expressed as dextrose in dry solids. For example, starch would have a DE of 0 and pure dextrose a DE of 100). During the mid-1960s, new developments in enzyme technology, in addition to the poor quality of straight acid-converted syrup, led to the switch to acid conversion to 42 DE followed by enzyme conversion to 64 DE. All were refiled by activated carbon filtration. Table 6.2 shows the carbohydrate profile of these syrups in comparison to a typical malt wort. It can be seen that the only similarity was the content of the higher saccharides or nonfermentables. At the time of this development, the brewer's main concern was that the apparent extract of the finished beer did not change with the addition of the liquid adjunct, that is, the syrup had to be approximately 20% nonfermentable.

As use of liquid adjuncts continued during the 1980s, it became apparent that they had shortfalls. The high level of glucose became a concern.[14]

TABLE 6.2

Sugar Spectrum (%) of First- and Second-Generation Liquid Adjunct (Acid and Acid/Enzyme Converted, Carbon Refined), Third-Generation Liquid Adjunct (Enzyme/Enzyme Converted, Ion Exchanged) Compared to All-Malt Wort

	Liquid Adjunct (First-Generation)	Liquid Adjunct (Second-Generation)	Liquid Adjunct (Third-Generation)	All-Malt Wort
Glucose	65	40	5	8
Maltose	10	28	55	54
Maltotriose	5	12	20	15
Dextrin	20	20	20	23

Source: Adapted from Chantler, J. 1990. *Tech. Q. Master Brewers' Assoc. Am.*, 27:78–82. With permission.

Conversion of starch with the aid of an acid produces predominantly glucose as the hydrolysis product. When brewer's yeast is exposed to high concentrations of glucose, a phenomenon referred to as the "glucose effect" may be experienced, which can result in sluggish and "hung" fermentations. This effect is discussed in greater detail in Chapter 8.

The brewer demands consistency. Acid and acid/enzyme syrups depend on the termination of a reaction by chemical or mechanical means when the desired endpoint is reached. It is difficult to attain the proper production consistency when operator judgment is required to predict this endpoint with such variables as temperature, time, pH, and concentration. In addition, there has been criticism regarding the physical properties of acid and acid/enzyme syrups. Large volume users and breweries located at long distances from the supplier experience inconsistent syrup color when the syrup is stored for lengthy periods at elevated temperatures. This is the "browning reaction of sugars," and with particularly high levels of glucose, the syrup will darken, catalyzed by the presence of metal ions and protein. To inhibit this reaction, the wet miller will add sulfites. However, sulfites have been shown to cause allergic reactions with some people and, consequently, food and drug agencies look upon them with some disfavor. In addition, these syrups can have a bitter aftertaste, stringent overtones, and a characteristic "corn" flavor. Finally, there has been concern regarding the chemical composition of acid/enzyme carbon-refined syrups. The production involves exposure to low pH conditions followed by neutralizing with sodium base. Consequently, residual compounds will include high levels of sodium and hydroxymethylfurfural (HMF). Although acid/enzyme carbon-refilled syrups possess disadvantages, they were acceptable and popular until the late 1980s when there was a need for a change. Why? Many changes have occurred in the brewing and wet-milling industries that have led to the development of a new liquid adjunct. The two most important have been the increase in adjunct levels in North American beer (up to 50% of the wort content) and high-gravity brewing — the latter being employed to ease capacity constraints and improve brewhouse efficiencies. The need for capital investment has been alleviated by increasing output as much as 50% through the practice of fermenting wort at 16°Plato or higher. The greatest need for change has been the need for brewing companies to compete on a world-market scale. This need is forcing the producer to become more flexible, more competitive, and more adaptive to the wide differences when marketing on an international scale.

With the advent of new technology in enzyme liquefaction and downline multistage enzyme hydrolysis, production of corn syrups of virtually any carbohydrate profile is possible. Table 6.2 shows the carbohydrate profile of a "new-generation" high-maltose corn syrup (enzyme/enzyme) and its comparison to a typical malt wort. The profiles are almost identical. This syrup now permits the brewer to introduce liquid adjuncts at any level without changing the carbohydrate profile of the wort. Brewing with these syrups is now routine, and no difficulties in either the brewhouse or fermentation cellar have been reported. Sensory evaluations of beer produced from third-generation syrups have revealed no significant differences from beers produced with acid/enzyme syrups. The sodium levels dropped by as much as 60% in comparison to beers produced from earlier-generation syrups. It would appear that these newer-generation syrups are meeting the current needs of the brewer. Already demands are being made for syrups with higher fermentables, lower glucose, and higher maltose. In the future, we will see the commercial ability to separate individual sugars according to molecular weight. The syrup manufacturer will store separately up to five or six sugars starting with the simple sugars, such as glucose, fructose, and maltose, and increasing the length of the polymer in each storage facility. By using these master grades, a blend of any carbohydrate profile will be immediately available, depending on the user's requirements.

Malt from Cereals Other than Barley

Although the principal cereal employed as the raw material for malt is barley, a number of other cereals are used including wheat, oats, rye, and sorghum.

Wheat Malt

Wheat malt is used in the production of some special types of beer, such as Berlin Weiss beer, in which it may constitute 75% of the grist, but only to a limited extent in ordinary beers. The limited use of wheat malt is mainly due to the difficulty experienced in malting the naked grain without damage to the exposed acrospire. As a result, much of the wheat malt made has been undermodified. However, the absence of the husk tends to result in a high extract. Wheat malt gives beer outstandingly good head retention.

Oats and Rye Malt

Malted oats are used to a limited extent and in some stouts are blended with barley malt. Malted rye does not seem to be used today although some 50 years ago it was used in specialty beers. Unmalted rye is sometimes used for vinegar brewing and also for certain distilled beverages (e.g., Canadian rye whiskey).

Sorghum

The use of unmalted sorghum has already been discussed. A relatively recent development in the use of sorghum in brewing is as a malt. For many years in southern Africa, malted sorghum has been used to brew a traditional alcoholic beverage of the region, known as sorghum or opaque beer.[12] Sorghum beer is characterized by its sour taste — it is flavored by lactic acid produced by bacterial fermentation and is not hopped. Sorghum beer is of moderate alcohol content (approximately 3% [w/w]) and the opaque appearance is due, in part, to incomplete hydrolysis of starch during mashing. The presence of high levels of complex carbohydrates in sorghum beer makes it a nutritious beverage as well as an alcoholic drink.

Malted sorghum differs in many aspects from barley malt, particularly in terms of the properties of its starch and diastatic enzymes (Table 6.3). Sorghum malt starch has a gelatinizing temperature in the range of 64 to 68°C, some 10°C higher than that of barley malt starch. The total diastatic activity of sorghum malt is less than half that of barley. This is probably because of the low β-amylase activity of sorghum malt. However, the α-amylase activity of sorghum malt is slightly higher than barley malt.

The development of sorghum for use as malt in conventional beer (ales, lagers, and stouts) production is in the process of rapid development. This development has been accelerated by the large foreign debt crisis in developing tropical countries, which has made it increasingly difficult for them to import either barley or barley malt for their existing breweries. For example, the government of Nigeria prohibited the import of barley malt since 1988. In order to reduce this economic difficulty, considerable research into local raw materials has been conducted (especially sorghum), not only for their use as adjuncts, but also as a complete replacement for barley malt. As a result of these research efforts, as much as 30% of the sorghum harvest in Africa is being used for malting and brewing.

There is general agreement that the "white" sorghum types are more suitable than the "red" types because of their lower content of polyphenols.[13] During the malting of sorghum, the pattern of endosperm enzyme breakdown is different from that of malting barley. For example, although

TABLE 6.3

Comparison between Sorghum and Barley Malt

	Sorghum Malt	Barley Malt
Starch gelatinization (°C)	64–68	55–59
Diastatic power	19	53
β-Amylase activity (%)	18	100
α-Amylase activity (%)	110	100

malting sorghum grains develop significant levels of endo-β-1,3-glucanase and pentosan enzymes, production of the cell wall-degrading endo-β-1,3:1,4-glucanase enzyme is significantly lower than the levels found in malting barley. Also, as previously discussed, sorghum has a lower enzyme complement, especially β-amylase and endo-β-glucanases, and even with the addition of these enzymes exogenously, starch extracts comparable with barley malt are rarely obtained.[15,16] Whereas mashing of barley malt at 65°C allows both starch gelatinization and enzyme solubilization of starch to occur simultaneously, mashing of sorghum malt at this temperature fails to gelatinize its starch, and despite high β-amylase activity, starch solubilization and extract development are still inadequate. Finally, the use of sorghum for lager and stout beer brewing will alter its taste and consequently, in many parts of Africa, the consumer has become acclimatized to a different type of beer.

Conclusions

The use of unmalted carbohydrates or adjuncts in brewing is widespread, except in those countries that adhere to the German purity law. In most countries, there are one or two dominant adjuncts and these are usually the cheapest suitable carbon source. Thus, in the United States and Canada, corn grits (and corn syrup) and rice grits are predominantly used; in France, Belgium, and Italy, corn grits are extensively used, while in the big cane sugar-producing countries such as Brazil, Australia, and the West Indies, cane sugar syrups are widely used together with corn grits. In Africa, corn remains the most popular adjunct; however, sorghum use (as an adjunct and a malt) is increasing. The brewing industry in the United Kingdom employs a wide range of cereals and sugars that have been processed by a number of methods. Although developments in the use of brewing adjuncts have been relatively stable for a number of years, the advent of "new-generation" syrups (produced principally, but not exclusively, from corn) is currently having a great impact on some parts of the brewing industry. Biotechnological advances such as wet milling, immobilized thermotolerant enzyme systems, and ion exchange downstream-processing techniques permit the production of syrup with virtually any carbohydrate profile. At the present time, syrups are available that allow the brewer to introduce them at any level without changing the carbohydrate profile of the wort. The future will see the commercial ability to separate and isolate individual sugars according to their molecular weight and, subsequently, produce a blended syrup of any sugar profile.

References

1. Narziss, L., The German beer law, *J. Inst. Brew.*, 90:351–358, 1984.
2. Collier, J., Trends in UK usage of brewing adjuncts, *Brew. Distill. Int.*, 16:15–17, 1986.
3. Marchbanks, C., A review of carbohydrate sources for brewing, *Brew. Distill. Int.*, 17:16–18, 1987.
4. Wilson, J., Brewing sugars — the versatile brewing adjuncts, *Brewer*, 76:139–143, 1990.
5. Lloyd, W.J.W., Brewers' sold adjuncts, *Brew. Guard.*, 117:23–25, 1988.
6. Canales, A.M. and Sierra, J.A., Use of sorghum, *Tech. Q. Master Brew. Assoc. Am.*, 13:114–116, 1976.
7. Pierce, J.S., Adjuncts and their effect on beer quality, *Proc. 21st Eur. Brew. Conv. Congr.*, Madrid, IRL Press, Oxford, 1987, pp. 49–60.
8. Meilgaard, M.C., Wort composition: with special reference to the use of adjuncts, *Tech. Q. Master Brew. Assoc. Am.*, 13:78–90, 1976.
9. Coors, J., Practical experience with different adjuncts, *Tech. Q. Master Brew. Assoc. Am.*, 13:117–123, 1976.
10. Teng, J., Stubits, M., and Lin, E., The importance of rice variety selection for optimum brewhouse operation, *Proc. 19th Eur. Brew. Conv. Congr.*, London, IRL Press, Oxford, 1983, pp. 47–54.
11. Allen, W.J., Barley and high adjunct brewing with enzymes, *Brew. Dig.* 62:18–26, 1987.
12. Taylor, J.R.N., Mashing with malted grain sorghum, *J. Am. Brew. Soc.*, 50:13–18, 1992.
13. Palmer, G.H., *Cereal Science and Technology*, Aberdeen University Press, Aberdeen, Scotland, 1989.
14. Chantler, J., Third generation brewers adjunct and beyond, *Tech. Q. Master Brew. Assoc. Am.*, 27:78–82, 1990.
15. Etim, M.U. and EtokAkpan, O.U., Sorghum brewing using sweet potato enzymic flour to increase saccharification, *World J. Microbiol. Biotechnol.*, 8:509–511, 1992.
16. Seidl, P., African sorghum millet — an alternative raw material for the malt and brewing industry, *Brauwelt. Int.*, 247–258, 1992.

7
Hops

Trevor R. Roberts and Richard J.H. Wilson

CONTENTS

Hop Growing	181
History	181
Hop Growing Today	181
The Hop Plant	182
Hop Classification	182
The Plant	182
Hop Cultivation	184
Conditions for Growing	184
Growing Infrastructure	185
Growing	185
Irrigation	185
Plant Protection	186
Harvesting	186
Drying and Packing	187
Hop Varieties	188
Hop Breeding	189
Objectives	189
New Variety Development	190
Super High-Alpha Hops	190
Low Trellis and Dwarf Hops	191
Gene Mapping	191
Hop Chemistry	191
Whole Hops	191
Hop Resins	192
Soft Resins	192
Hard Resins	197
Hop Oils	197
General	197
Terpenes	197

Oxygenated Compounds	198
Sulfur Compounds	200
Glycosides	200
Polyphenols	200
General	200
Proanthocyanidins	201
Flavonoids	201
Xanthohumol	201
Isoxanthohumol	202
8-Prenylnaringenin and Desmethylxanthohumol	204
Xanthogalenol	204
Pectins	204
Hop Waxes and Fats	205
Hop Storage	205
General	205
Hop Variety	205
Kilning and Moisture	206
Hop Oils	206
Sulfur Compounds	207
Hop Resin Acids	207
Bittering Value	209
Isomerization	209
The Isomerization Reaction	209
Significance of the Cohumulone Ratio	211
Significance of the *Cis:Trans* Ratio	211
Late Hopping	212
Bitter Flavor and Foam	212
Role of Iso-α-Acids	212
Reduced Iso-α-Acids	213
The "Lightstruck" Reaction	213
Inhibition by Reduction	214
Manufacture of Reduced Iso-α-Acids	216
Hop Products	217
Development of Hop Products	217
Benefits of Hop Products	218
Volume Reduction	218
Increased Stability	218
Reduction in Chemical and Heavy Metal Residues	219
Homogeneity	219
Reduced Extract Losses	219
Use of Automated Dosing Systems	219
Improved Efficiency	220
Quality Benefits	220
Classification of Hop Products	221
Nonisomerized Hop Products	221

Pellets	221
Type 90 Pellets	222
Type 45 Pellets	223
Kettle Extracts	224
Ethanol Extract	224
CO_2 Extracts	225
Comparison of Ethanol and CO_2 Extracts	227
Isomerized Hop Products	228
Stabilized and Isomerized Pellets	228
Isomerized Kettle Extracts	230
Magnesium-Salt Isomerized Kettle Extract (IKE)	230
Potassium-Form Isomerized Kettle Extract (PIKE)	231
Comparison of IKE and PIKE	231
Iso-Extract (Postfermentation Bittering)	231
Reduced Isomerized Hop Products	233
Rho-iso-α-acid (*Rho*)	233
Production of *Rho*	233
Quality Characteristics of *Rho*	233
Using *Rho*	234
Tetrahydroiso-*a*-Acid ("Tetra")	234
Production of α-Acids-based Tetra	234
Production of β-Acids-based Tetra	235
Quality Characteristics of Tetra	235
Using Tetra	236
Hexahydroiso-*a*-acid ("Hexa")	237
Production of Hexa	237
Quality Characteristics of Hexa	237
Using Hexa	237
Comparison of Reduced Products	237
Relative Foam Stabilization — Importance of Cohumulone Ratio	238
Other Foam Stabilizers	239
Other Products	239
Type 100 Pellets	239
Beta Extract	239
Oil-Enriched Extracts	240
Pure and Fractionated Hop Oils and Hop Essences	240
Storage Stability	241
Other Uses — New Products	241
Hop Usage	242
Choice of Hop Product	242
Hop Utilization	243
Calculation of Hop Additions	243
Example 1 — Single Product Added into the Kettle	243
Example 2 — Two Products, Both Added into the Kettle	244

Example 3 — Single Kettle Addition with Tetra Added
    Postfermentation 245
Cost of Bittering 246
Hop Analysis 248
  General 248
  Visual, Physical, and Olfactory Examination of Hops 248
    Hand Evaluation 248
      General 248
      Preliminary Physical and Visual Examination of
        Brewer's Cut 249
      Secondary Inspection for Quality Defects 250
      Assessing Aromatic Quality 250
    Hop Leaf and Stem 251
    Aphids 252
  Hop Resin Analysis 252
    General 252
    Hard, Soft, and Total Resins 253
    Polarographic Methods 253
    Lead Conductometric Methods 254
    Spectrophotometric Methods 256
    HPLC Methods 257
      Background 257
      Separation Principles 257
      Detection 258
      Obtaining Good Results 258
      Calibration 259
      Sample Preparation 260
    Capillary Electrophoresis 260
    NIR Reflectance Mode Analysis 260
      General 260
      Principles 261
      Preparation of Samples 261
      Pros and Cons 262
  Hop Oils 262
    General 262
    Steam Distillation Method (for Total Essential Oils) 262
    Presence of Sulfur Compounds 263
    Gas Chromatography Analysis (for Hop Oil Components) 263
      History and Principles 263
      Calibration and Peak Identification 264
      Peak Identification 264
      "Heart Cut" Analysis 265
    Adsorbent Fiber Technique (Solid Phase Micro-Extraction) 265
  Polyphenols 265
  Flavonoids 266

Analysis of Worts and Beers	266
BU Analysis	266
Hop Resin Acids by HPLC	266
Flavonoids	268
Hop Oils	268
References	269
Further Reading	278

Hop Growing

History

Although beer is first thought to have been brewed in Babylon as long ago as 7000 B.C., hops were almost certainly not used in brewing until very much later.[1,2] There is some evidence to suggest that hops were grown in central Europe before 1000 A.D., but it is unclear whether these were used in beer or merely for inclusion in early medicines and herbal remedies. Hops were probably first grown for brewing in Germany and the Czech Republic some time between 1000 and 1200 A.D. Their horticulture and use then gradually spread throughout Europe, eventually being imported into England during the 14th century. The famous *Reinheitsgebot*, or *Purity Law*, in which it was decreed that beer might be brewed only using (malted) barley, hops, and water, was issued by the Duchy of Bavaria in 1516. From Europe, hop-growing spread fairly rapidly with the early European settlers to the United States and South Africa (17th century), Australia, and New Zealand (early 19th century) and also during the 1800s and 1900s into several other countries, many of which no longer grow significant quantities of this perennial crop.

Apart from their obvious flavor benefits, the attraction of hops to early brewers appears to have been related to their preservative qualities, which were particularly relevant before the introduction of refrigeration.

Hop Growing Today

Today, hops are grown successfully in a number of countries in both the Northern and Southern Hemispheres. Location very much depends on achieving the combination of the right growing and climatic conditions, but is also influenced by the need for "local" production in countries situated a long way from the traditional growing areas. As is shown in Table 7.1, world hop production is dominated by Germany and the United States which between them account for around 60% of the total output.

TABLE 7.1

Principal Hop Producing Countries — Average Growing Area and Baled Hops Production 1999–2002

			% of Total	
Country	Hectares	Hops (mt)	Area	Hops
Germany	18,569	30,107	32.0	31.4
United States	13,715	29,015	23.7	30.3
China	4,689	11,935	8.1	12.5
Czech Republic	6,107	5,870	10.5	6.1
England	1,998	2,740	3.4	2.9
Poland	2,238	2,506	3.9	2.6
Slovenia	1,784	2,246	3.1	2.3
Australia	818	2,230	1.4	2.3
Spain	780	1,439	1.3	1.5
France	811	1,356	1.4	1.4
South Africa	491	891	0.8	0.9
New Zealand	385	790	0.7	0.8
The Rest	5,602	4,657	9.7	4.9
Total	57,987	95,782	100.0	100.0

Source: Taken from *Hopsteiner 2002 Guidelines for Hop Buying*. Simon H. Steiner GmbH, Mainburg, Germany. With permission.

The Hop Plant

Hop Classification

Thought to have originated in Asia (probably China), the hop plant is indigenous to the Northern Hemisphere, but is also now grown successfully in parts of the Southern Hemisphere. The classification of the hop plant (*Humulus lupulus* L.) used in brewing is shown in Figure 7.1.

The Plant

Humulus lupulus is a perennial, climbing plant with three- or five-lobed leaves. It is described as dioecious, meaning that it has separate male and female plants, but it is only the female plants that form the hop cones within which the all-important, yellow "lupulin glands" develop.

Each spring, the hop rootstock produces numerous shoots, which initially grow straight upwards. After a short time, they begin to twist in a clockwise direction around any available support (normally a string or wire, but in nature, typically, a bush or small tree) whilst at the same time the stems themselves twist to form a gently spiraling helix. The stems, or bines, are hexagonal in cross section, with six rows of hairs growing on the six-ridged angles of the bines, and it is these hairs that help the hop plant cling to the supporting string. As the hop grows, lateral shoots appear

Hops

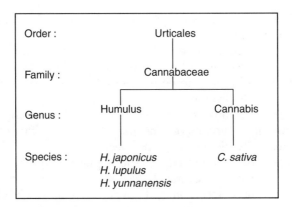

FIGURE 7.1
The hop plant — classification.

from buds in the axils of the leaves on the main stem and, eventually, flowers develop on these laterals. It is from these clusters of flowers that the hop cones subsequently form.

The hop cones consist of a central stalk or strig on which develop petal-like structures called "bracts" and "bracteoles" (Figure 7.2). Whereas the bracts appear to have only a protective role, it is at the base of the bracteoles that the sticky, yellow lupulin glands develop. It is within these lupulin glands that the most important constituents in brewing are found, namely the resins and the essential oils. If allowed to grow, any male plants present will flower and pollinate the female flowers, resulting in the eventual development of seeds in the folds at the base of the bracteoles. The male flowers

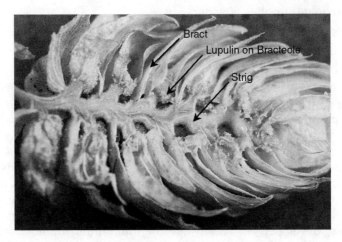

FIGURE 7.2
Cross section of a typical hop cone showing the general structure.

wither and drop off the bine soon after their appearance and are therefore of no use as such in brewing.

In most hop-growing areas, any male hops are physically removed from hop fields or hedgerows in order to avoid the production of seeds that are considered by some brewers to be undesirable, owing to the possibility of oxidation of seed fats producing off-flavors in beer.[4] However, in England and in some other regions, it is still common practice for male hops to be planted within hop gardens to fertilize the female flowers, which is known to increase the yield of hops per hectare.

Humulus japonicus is an annual, dioecious plant with lobed leaves originally found in Japan and China. It forms cones with very few lupulin glands and is not used in brewing. However, it is now widely cultivated by horticulturists and, as a strong climber, is often used in gardens as a decorative, leafy screen.

Humulus yunnanensis is a little-studied species thought to be native to the Yunnan province of southern China. As far as is known, it is not routinely cultivated and has no use in brewing.

Hop Cultivation

Conditions for Growing

As previously noted, hops are grown fairly widely throughout the world. However, in order to grow hops commercially the following conditions are extremely important:

- Deep and fertile soil
- Warm summer sun and cold winters
- Adequate supplies of water
- Absence of pests and diseases or else a means of controlling them
- Abundance of cheap labor or well-developed agricultural technology
- Adequately changing day length

The last point is of particular importance and is the reason why hops are normally only grown between latitudes 35 and 55°, where the changing day-length is most suitable. The hop plant is referred to as a "short-day plant" and during its growing cycle responds to different periods of daylight. Below about 13 h of daylight the plant will not grow and is dormant, but once daylight exceeds around 13 h, it begins to climb and eventually flowers and develops cones. However, flowering will not occur until the plant has developed a variety-specific, minimum number of nodes (distance between the axils on the main bine). By the time that this is achieved, day length in the temperate latitudes has increased beyond

15–16 h, under which conditions the plant will continue its vegetative growth but will not flower. Not until the day length drops back to below the critical point do conditions become suitable for flowering and the hop cones then develop and ripen.

In countries (e.g., South Africa) that are just outside of the normally acceptable latitudes for hop growing, hops are however grown successfully with the aid of artificial lights to extend "daylight" at key periods of the year.

Growing Infrastructure

Hop bines are normally supported on strings (typically coir or polypropylene) or fine wires, which are fixed between ground-anchored pegs and strong overhead wirework supported on tall wooden or concrete poles. In most countries, the height of the wirework varies between 4 and 6 m and is laid out in rows between 1.6 and 3.0 m apart, with hop plants spaced at 1.5–2.0 m intervals along each row. (China is a notable exception: because the hops are typically picked by hand, closer plant spacing is possible and hence high yields per acre are obtained. The wirework is also relatively low). The choice of spacing between rows and plants depends on practical issues such as tractor access, type of irrigation, and access of light to the growing hops. Several different string pattern systems are in use, with up to 4, steeply angled strings per plant.

Growing

Following wintering below ground, in early spring, the rootstock produces numerous shoots, which may be initially removed along with any dead plant material from the previous year. By late spring stringing is completed and, once shoots grow to around 50–80 cm, up to 4 shoots are manually trained clockwise around individual strings whilst unwanted shoots are removed. As the bines grow up the strings, the lower foliage (up to 1 m) is often removed, usually by use of chemical defoliants (or in Australia and New Zealand by grazing sheep), in order to discourage the spread of any pests and diseases and later to aid the harvesting process.

Soil inputs such as nitrogen, phosphates, potash, magnesium, straw, and dung are often added in order to restore essential nutrients and promote vigorous growth. However, as some of these applications can adversely affect levels of the all-important α-acids (see section titled "Hop Chemistry") as well as promoting the development of disease, the grower must exercise great care when deciding on amounts and times of addition.

Irrigation

In many temperate or maritime climates, irrigation is unnecessary but, nevertheless, may still be used to improve consistency and maximize yield. In other areas, such as the Yakima valley in Washington State, U.S.A., or Xinjiang in China, irrigation is essential, owing to insufficient rainfall. Three types of irrigation systems are commonly used — overhead (spray gun), rill (ditch), or trickle (piped) — each with its own advantages and disadvantages. Trickle irrigation methods offer the opportunity to include chemical additions with the irrigation water (a practice known as "fertigation") as well as minimizing wastage of water — an important consideration in areas of water shortage or high water costs.

Plant Protection

Although expensive, the application of agrochemicals in order to suppress pests and diseases is necessary in most Northern Hemisphere growing areas. Chemical applications are made throughout the growing season either as a preemptive measure or in response to the first signs of problems. However, in order to meet statutory minimum residue levels (MRLs) on hop cones at harvest, agrochemical applications are ceased at predetermined periods before harvest. Agrochemicals are either surface active, requiring good foliar contact and applied by using sprays, or systemic and therefore taken up into the root system via soil drenches or trickle irrigation. When using agrochemicals, the hop grower has also to consider health and safety issues, effects of spray drift on neighboring areas and possible contamination of ground water due to run-off.

The most common pests and diseases are shown in Table 7.2. Others can occur but to a much lesser extent and generally in response to specifically favorable conditions.

In most Southern Hemisphere growing regions and in a few isolated, newer hop-growing areas in the Northern Hemisphere, certain pests and diseases are rare or even absent, resulting in very low chemical applications. This added benefit presents the fortunate growers with a valuable selling opportunity, particularly at times when "green" and organic products are much in demand.

Harvesting

In late summer, when the hop cones have ripened, the hops are harvested. With the notable exception of the hop-growing regions in China, where hand picking is still the norm, the bines are severed approximately 1 m from the ground, and then cut or pulled from the overhead wires, either manually or by machine. The bines are then transported to local, static picking machines. There they are hung upside down on a moving rail to better expose the cones hanging from the lateral growths and passed through picking machines where a series of revolving, metal wire "picking fingers" strip the hop cones and leaves from the bines. The cones are then

Hops

TABLE 7.2

Common Pests and Diseases of Hops

	Comments
Pests	
Damson hop aphid (*Phorodon humuli*)	Can cause defoliation and significant cone damage; treated by foliar and systemic chemicals as well as natural predator programs
Red spider mite (*Tetranychus urticae*)	Silvery discoloration of leaves and cones resulting in total loss of crop in severe cases; can be controlled by chemical spraying; thrives in hot weather
Diseases	
Powdery mildew (*Podosphaera macularis*)	Fungal disease causing white pustules on leaves and severe cone damage; spreads rapidly, but can be controlled by early spraying before disease establishes a hold
Downy mildew (*Pseudoperonospora humuli*)	Germinating fungal spores on leaves cause black discoloration and severely reduced growth resulting in poor yields; can be treated by both spray and systemic chemicals
Verticillium wilt (*Verticillium albo-atrum*)	Fungal disease which can quickly devastate a hop field; no known chemical control available; good hygiene essential together with removal and destruction of infected plants
Virus diseases (*especially Hop Mosaic and hop latent viruses*)	Can cause significant reduction in yields and α-acid levels; no known chemical control; heavy contamination requires replacement of infected plants with virus-free rootstock

separated from leaves, bine fragments, etc. on a series of vibrating belts, sieves, and air-classifying devices with the cleaned hop cones being transported to the drier whilst the unwanted material is cut into small pieces for disposal or spreading back onto the land if there is no fear of spreading disease.

Drying and Packing

The freshly picked hop cones have a moisture content of approximately 80% and are dried in "oast house" kilns to a moisture level of 7–12%. Drying is typically achieved by blowing heated air for 6–8 h through a bed of cones of up to approximately 1 m depth. Normally, the air is heated using either oil or gas burners and can be passed either directly through the cone bed or indirectly via a heat exchanger. Drying temperatures are normally in the range 60–75°C, but this will vary depending on bed depth and cone size as well as on air speed. Physical damage and degradation of resins and oils can occur if drying is not carefully controlled, and better results are generally obtained if the air-on temperature is kept relatively low.

In former times, sulfur candles were often burned underneath the kiln floors, producing sulfur dioxide that had the effect of reducing browning and maintaining a good appearance to the hop cones. However, for environmental reasons, this practice has now been banned in most countries (see also the sections titled "Sulfur Compounds" and "Hop Resin Acids").

After completion of drying, subsequent cooling, and discharge from the kiln, residual moisture may be unevenly spread within and between the hop cones. It is therefore necessary to "condition" the hops by holding them in heaps at ambient temperatures for a few hours, during which time moisture is redistributed more evenly throughout the entire mass of material. Following completion of this conditioning period, the hops can then be packed, either for immediate transport to brewers or, more commonly, for cold storage prior to sale and processing into hop products. Normally, hops are compressed into tall, polypropylene sacks known as "hop pockets" or, more commonly, into rectangular bales that are then wrapped in burlap or polypropylene cloths. Bale size and weight typically depend on the country and eventual destination. In Germany, hops are initially gently compressed into "farmers'" bales weighing 50–60 kg prior to processing, and are often packed into relatively small, high compression, cylindrical "ballots" weighing up to about 150 kg, whereas in the United States, a rectangular, approximately 200 lb (90 kg) bale is the standard package, either for sale or for transport to the processing plant.

Hop Varieties

The range of hop varieties available to the brewer is both varied and extensive. In more recent years, the choice has been constantly changing as "older" varieties are phased out whilst "newer" varieties with increased disease resistance, higher yield, and more diverse quality characteristics are introduced.

Hop varieties have in the past been grouped into three broad categories — "Aroma," "Dual-purpose," and "High-alpha." However, more recently, this categorization has become blurred as very good aroma varieties with higher levels of α-acids have been introduced, and new "Super High-alpha" varieties with acceptable aroma characteristics have also found favor with brewers. In Table 7.3, a few examples of varieties from around the world have been loosely grouped together in the three categories mentioned previously.

Traditionally, brewers selected their hop varieties by hand evaluation (see the section "Visual, Physical, and Olfactory Examination of Hops"), possibly supported by limited analyses such as α-acid and total oil contents. Nowadays, the introduction of gas chromatographic (GC) and high-performance liquid chromatographic (HPLC) analyses provides additional information, which can be included in selection criteria (see the section "Hop Chemistry"). A high proportion of the specific α-acid cohumulone is considered by some to give a more harsh bitterness compared to the other α-acid homologs (see also the section "Role of Iso-α-Acids") and,

TABLE 7.3

A Few Examples of Aroma, Dual-purpose, and Alpha Varieties

	Hop Type		
Country	Aroma Excellent Aroma; α-Acids typically 3–7%	Dual-Purpose Good Aroma; α-Acids typically 6–10%	Alpha (and Super-Alpha) Acceptable Aroma; α-Acids typically 9–16%
United States	Mt Hood Cascade Willamette US Fuggle	Cluster	Chinook Nugget Galena Zeus
Germany	Tettnanger Hersbrucker Mittelfrüh Tradition	Perle Brewers' Gold	Northern Brewer Magnum Taurus Merkur
England	Golding Fuggle Progress	Challenger Northdown First Gold[a]	Target Admiral Pilgrim
New Zealand	Pacific Hallertau	NZ Hallertau Aroma	Green Bullet Pacific Gem Southern Cross

[a]Dwarf variety (see section "New Variety Development").

consequently, brewers often specify maximum permissible levels of α-acids. Similarly, ratios of other components, particularly within the oil fraction, can be used to determine both the acceptability of a variety as well as its quality in comparison with other varieties. However, in reality, the final evaluation of any new variety, either as a replacement for an existing hop or for the production of a new beer, relies on trial brewing and tasting of the resultant beer.

Hop Breeding

Objectives

The objectives of the plant breeder in developing new varieties of hops can be summarized as follows:

- Increased harvested weight yield of hop cones
- Higher α-acids leading to increased alpha yields
- Pest and disease resistance
- Better storage stability (of α-acids and aroma components)

- Improved aroma character through higher concentrations of certain hop oil components such as linalool (see the section "Oxygenated Compounds")
- Lower production costs

and, possibly in the future:

- Increased yields of constituents previously considered unimportant, such as β-acids and polyphenols (e.g., xanthohumol) (see the sections titled "Hop Resins" and "Polyphenols")

New Variety Development

The introduction of a new variety into commercial production is a lengthy and expensive process, which, in a country where diseases are prominent, can take up to 10 year from selection of parents to plant registration and release. The key stages are shown in Table 7.4.

In countries lacking disease problems, it is possible to shorten this process considerably and, in some instances, availability can be further accelerated by release ahead of organized brewing trials. Some varieties have been developed using laboratory techniques that alter the normal diploid chromosome content of the plant cells, facilitating formation of tetraploid plants, which may then be crossed with diploids to produce triploid offspring that are frequently more vigorous and always much less prone to seed production.[5]

The following two developments, arising from recent hop-breeding programs, demonstrate the benefits that can be achieved for both grower and brewer.

Super High-Alpha Hops

In order to meet the demand for cheaper α-acid derived from hops of acceptable brewing quality, plant breeders have successfully introduced a range of varieties that consistently produce α-acid levels of 15–16% (often peaking at

TABLE 7.4

Typical Stages in the Development of a New Hop Variety

Year 1	Selection of parents; pollinate and collect seeds
Year 2	Grow seedlings (often under exposure to mildew spores)
Years 3–5	Selection on the basis of vigor, disease resistance, cone production, and chemical analysis
Year 6	Propagation (via root splitting or meristem culture) of selected plants
Years 7–8	Further selection on the basis of yield, disease resistance and chemical analysis
Year 9	Commercial farm and pilot brewing trials
Year 10	Further field trials and commercial brewing trials; registration and release to growers

18% in good growing seasons) and very high yields of α-acids per hectare. However, most of the early releases of U.S. super high-alpha hops were later found to be highly susceptible to downy or powdery mildews and to show poor storage stability — problems that have been addressed in subsequent breeding programs.

Low Trellis and Dwarf Hops

On several occasions in the past, attempts have been made to grow naturally tall hop varieties on "low trellis" systems allowing vertical growth only to a height of about 3 m, with the intention of reducing set-up and growing costs. However, although costs were indeed found to be lower, a concomitant reduction in yield outweighed any operational savings. Subsequently, specific "dwarf" varieties were developed at Wye College in England. These hops possess short internodes, though good lateral growth is achieved despite the lower overall height, as well as the ability to produce hops well down the bine. Dwarf hops are grown up semipermanent netting as a dense hedge to a height of 2–3 m. Set-up costs are reduced by approximately 40% of the costs of a traditional tall wirework. Furthermore, considerable operational savings can be achieved, claimed to be up to 50% of the costs associated with growing tall varieties.[6] Most of these savings are made in reduced chemical additions (better coverage) and, most significantly, in lower manpower costs associated with the use of mobile field harvesters that pick the hops directly from the hop hedge, thereby eliminating the need to cut and transport bines to the picking machines. This also leaves the plant with much of the leaf growth intact and, hence, facilitates late-season strengthening of the rootstock.

Gene Mapping

All of the recently released new varieties have been developed using classic hybridization techniques. For reasons of public acceptability, genetic manipulation in the form of gene insertion methodology is not currently employed in hop-breeding programs; however, the use of gene mapping and markers is being introduced. These highly sophisticated techniques should shorten the breeding cycle whilst, at the same time, significantly improving the chances of successfully achieving the plant breeder's objectives.

Hop Chemistry

Whole Hops

Because of the presence of the oil- and resin-rich lupulin glands, the overall composition of fresh, dried hop cones shows them to be unlike that of other plant material, though the leafy nature of the hop petals ensures the presence of such ubiquitous substances as proteins and carbohydrates. In an excellent and thoroughly comprehensive review of hop chemistry in 1967, Stevens[7] quotes a typical analysis as being:

Resins	15%
Proteins	15%
Monosaccharides	2%
Tannins (polyphenols)	4%
Pectins	2%
Steam volatile oils	0.5%
Ash	8%
Moisture	10%
Cellulose, etc.	43%

No doubt this analysis was for a hop of quite modest resin content by today's standards, because some of the modern hop varieties contain nearly twice the amount quoted above. (For a more recent review of hop chemistry, see Verzele).[8] Of the aforementioned components, the lupulin glands contain virtually all of the hop resins (in the form of α-acids, β-acids, desoxy-α-acids, and uncharacterized "soft resins" — see the section "Soft Resins"), the essential oils and most of the hop cone fats and waxes (comprising perhaps 5% of the total and included previously under cellulose, etc). Seeded hops will contain a relatively large proportion of fats and nonvolatile "fixed" oils.

High-alpha hops (e.g., Galena or Magnum) will tend to have proportionally more of the resin and oil components, whereas a low-alpha hop such as Hersbrucker will have relatively more of the mainly petal-derived components such as polyphenols and nitrate. Hence, a brewer who for economic reasons switches from brewing with, say, a 7% α-acids Cluster hop to a 13% α-acids Galena will substantially reduce the amount of polyphenols added to his or her wort, though the balance and nature of the resin and oil components may not be much changed. Such changes can have consequences — for example, to beer haze stability — that are not easily predictable.

Hop Resins

Soft Resins

The resinous fraction of fresh hops contains mostly the α-acids and β-acids, each of which consists of analogous series of closely related homologs. Together with the desoxy-α-acids, they constitute the major portion of the so-called "soft resin" fraction of hops (see also the section "Hard, Soft, and

Total Resins"). The close similarities between these compounds arise as a result of the proposed common pathway to their formation (Figure 7.3)[9].

In their native state, the hop resin acids exist within the lupulin glands as fully protonated, nonionized species. In this form, they are virtualy insoluble in water, but they dissolve readily into methanol or less polar solvents such as dichloromethane (methylene chloride), hexane, or diethyl ether. At room temperature, a preparation of α-acids will be a pasty, yellow solid, while β-acids, also yellow unless highly purified, will be substantially harder and may become semicrystalline. When slowly heated, both become

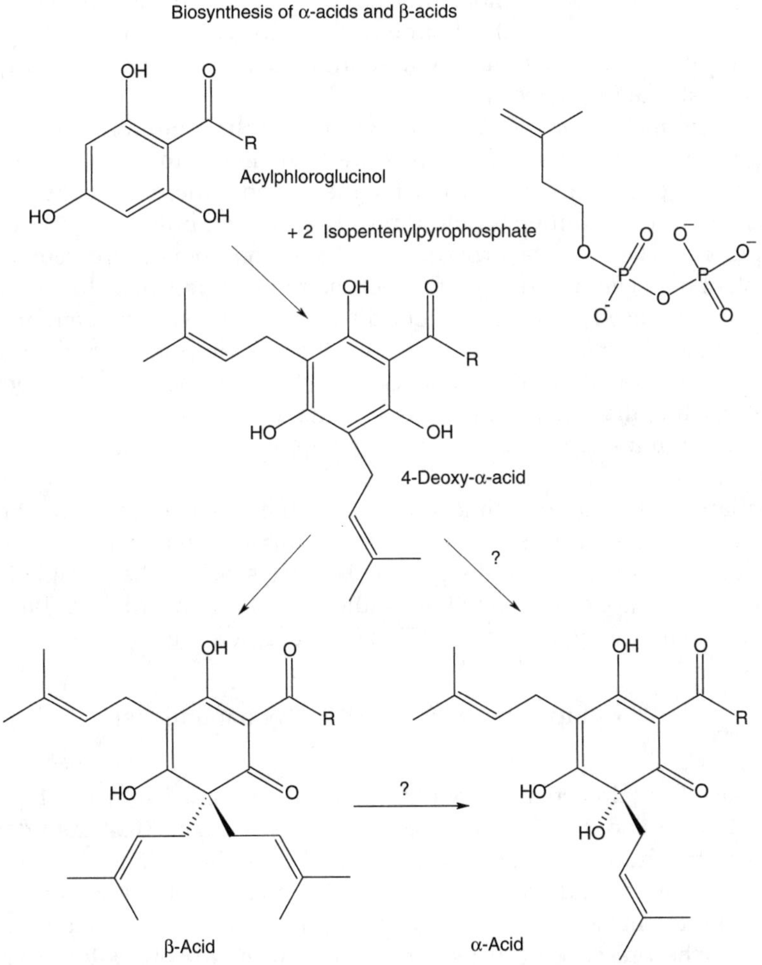

FIGURE 7.3
Biosynthetic pathway for hop resin acids. 4-Deoxy-α-acids are precursors to both the α- and β-acids, but it is not clear whether the α-acids are formed directly or via the β-acids.

progressively more mobile, though the temperature has to be raised to more than about 60°C before the β-acids will flow easily. When suspended in warm, demineralized water and titrated with strong alkali (NaOH or KOH), both will begin to dissolve, as their alkali metal salts are highly soluble. However, the pK_a values for the two types of resin acid are rather different, the α-acids having pK_a values close to pH 4.8, whereas the β-acids do not reach 50% dissociation until the pH value is around 5.7.[10] The consequence for practical brewing is that the solubility of α-acids in beer (pH usually in range 3.8–4.5) is significant, and this allows a meaningful proportion of any unisomerized α-acid in the wort to pass into the beer, whereas the amount of β-acid that can possibly remain is very small. Beers brewed with whole hops, normal pellets, or extracts, and especially those that were late- or dry-hopped, normally contain a few parts per million (ppm) of α-acids, but the amount of β-acids never exceeds 1 ppm and is most often undetectable.

The same major homologs are found in the resin from all hop varieties, though the proportions differ. There are three major forms of α-acid and three analogous forms of β-acids: for the α-acids, these are cohumulone, humulone, and adhumulone, and for the β-acids, colupulone, lupulone, and adlupulone. The relative proportions of the compounds are remarkably consistent between different growths of the same variety and their determination can be used as a tool to help distinguish one variety from another. Analysis of different hop varieties shows that the proportion of the α-acids that is found to be adhumulone is almost always less (and somewhat more variable) than that of the other two major homologs, of which humulone is predominant in most varieties. Table 7.5 illustrates the wide range of values found.[11]

Similarly, adlupulone is almost invariably the lesser component amongst the three analogous forms of β-acids, but, in this case, the relative content of colupulone as a fraction of the β-acids is always substantially higher than that of cohumulone as a fraction of the α-acids. Howard and Tatchell[12] derived the following relationship, held to be true for all varieties:

$$\% \text{Colupulone} = 20.2 + 0.943\, [\% \text{Cohumulone}]$$

Hops also contain small amounts of so-called minor α-acids, typically amounting to about 2–3% of the total α-acids content. Most abundant of these are posthumulone, prehumulone, and adprehumulone,[13] the former having the shortest side chain structure of the homologous series and, therefore, being the most polar and, in all probability, the best utilized. Evidence exists for the presence of at least four more forms of presently unknown structure.[14]

The ratio of total α-acids to total β-acids is also a varietal characteristic that is quite consistent for mature hops. Hop varieties can therefore be classified

TABLE 7.5

Relative Content of Major α-Acids in Different Hop Varieties (Ranking by % Cohumulone)

Hop Variety	% Cohumulone	% Humulone	% Adhumulone
Brewers' Gold	46	42	13
Bullion	46	43	11
Eroica	46	41	12
Galena	42	44	14
Cluster	40	50	10
Cascade	39	54	7
Perle	33	56	11
Nugget	31	58	11
Willamette	30	56	15
Goldings	28	54	9
Fuggles	27	60	12
Tettnanger	21	70	9
Mittelfrüh	17	72	11

Note: 1982 crop hops, grown in Oregon, United States.
Source: Data from Nickerson, G.B. and Williams, P.A., *J. Am. Soc. Brew. Chem.* 44: 91–94, 1986. With permission.

by various parameters connected to their resin content, of which the most important are the "Cohumulone Ratio" and the "α:β Ratio":

$$\text{Cohumulone Ratio (or CoH Ratio)} = \frac{\% \text{ Cohumulone}}{\% \text{ Total } \alpha\text{-acids}}$$

$$\alpha{:}\beta \text{ Ratio} = \frac{\% \text{ Total } \alpha\text{-acids}}{\% \text{ Total } \beta\text{-acids}}$$

Some typical values are shown in Table 7.6.

Not surprisingly, these differences have implications for the quality of the beer produced from different hops, as will later be described.

Whilst it is the α-acids that are the prime source of bittering (via their isomerization to iso-α-acids; see the section "The Isomerization Reaction"), the contribution of the β-acids is not necessarily insignificant. These latter do not undergo any kind of isomerization reaction and are mostly eliminated during the brewing process by precipitation, but a small proportion may be converted into hulupones, which are a much more soluble and substantially bitter derivative.[15] Generally of more significance, hulupones may already be present in the hops or kettle hop product used, and the amount of these compounds that ends up in the beer will therefore vary according to circumstance (see also the section "Hop Resin Acids"). A long-established rule of thumb that has been used by some brewers to assess the bittering

TABLE 7.6

Typical Key Values for Hop Resin Acids Content of Some Major Hop Varieties (Listing by α-Acids Content)

Hop Variety and Primary Growing Region	Total % α-Acids	Total % β-Acids	CoH Ratio	α:β Ratio
Hallertau Hersbrucker (Germany)	3.0	4.5	0.19	0.7
Saaz (Czech Republic)	3.0	2.5	0.24	1.2
Styrian Goldings (Slovenia)	5.5	3.0	0.29	1.8
Willamette (United States)	6.0	3.5	0.32	1.7
Perle (Germany)	7.5	3.0	0.29	2.5
Pride of Ringwood (Australia)	9.0	6.0	0.33	1.5
Target (United Kingdom)	12.0	5.0	0.36	2.4
Galena (United States)	12.5	7.5	0.36	1.7
Hallertauer Magnum (Germany)	13.0	5.0	0.25	2.6
Pacific Gem (New Zealand)	13.5	7.5	0.39	1.8
Nugget (United States)	14.0	4.0	0.26	3.5
Zeus (United States)	16.0	4.5	0.32	3.6

potential of hops states:

$$\text{Bittering potential} = \alpha\text{-acids} + \frac{\beta\text{-acids}}{9}$$

Clearly this is an oversimplification, but it is worth bearing in mind that (purely for experimental purposes) a drinkable, bitter beer was once prepared by a brewer with the addition in the way of kettle hops of nothing but a crude β-acids extract. It is also known by one of the authors of this chapter that a brewer once managed (albeit inadvertently) to sell beer that was bittered (on a BU basis) with hop pellets of such venerable age that they contained no measurable α-acids (by HPLC) whatsoever!

Although the properties of the different homologs within each hop resin acid series are similar, there are, nonetheless, some noticeable differences. Most importantly, in each case, the co-series homologs is found to be a little more soluble and chemically slightly more reactive. The practical consequence of this subtle distinction is that the utilization of cohumulone is usually significantly and often substantially higher than that of the humulone and adhumulone, which are in all respects extremely similar in their properties (see also section "Significance of the Cohumulone Ratio").

Desoxy-α-acids represent a significant proportion of the soft resin fraction, typically at a level of about 5% of the total. However, they do not seem to play any role in the brewing process, being substantially eliminated in the trub and any remainder almost completely removed by absorption onto the yeast during fermentation.

Hard Resins

In a fresh hop, the amount of the "hard resin" fraction (see the section "Hard, Soft, and Total Resins") is quite small and is mostly made up of the yellow/orange-colored prenylflavonoid, xanthohumol (see the section "Xanthohumol"). However, during storage, the hard resin fraction increases as the α-acids and β-acids from which they are mostly derived decline. With the exception of xanthohumol, the precise chemical identity of the hard resins is not well understood (and may include dimers and trimers), but generally they are more polar and therefore more water-soluble compounds and will tend to pass more readily into the beer than their precursors.

Hop Oils

General

Dependent on variety, hops contain from about 0.4 to 2.5 ml/100 g of steam volatile, essential oils. As a general rule, high-alpha, bittering hops contain more oil on a dry weight basis than do lower-alpha aroma hops, but this is really no more than a reflection of the greater amount of lupulin present in the former. Of much more significance is the spectrum of components, which is primarily genetically determined and varies substantially between varieties.[16] Gas chromatographic analysis of hydrodistilled oils from freshly harvested and dried hops reveals a vast number of compounds, but the chromatographic patterns are distinctive for each variety and generally enable reliable identifications to be made.[17,18]

Terpenes

The major components are hydrocarbon terpenes, of which the most abundant are the monoterpene myrcene, and the sesquiterpenes α-humulene and β-caryophyllene (Figure 7.4). Together, these three components alone may account for up to about 80% of the oil.

Other terpenes that may be of importance on a quantitative basis are the α- and β-selinenes and β-farnesene, some varieties being almost completely devoid of one or more of these compounds, others having substantial content. Table 7.7 gives an idea of the wide diversity of composition of the essential oil from different hop varieties.

Although the terpenes may together comprise well over 90% of the total oil of a fresh hop, their importance as such to the flavor of beer is generally inconsequential, as they are all virtually water-insoluble and have relatively high flavor thresholds. For the most part, they are driven off during the wort boil or later flushed out during fermentation (occasional exceptions may be the use of hops for dry-hopping and in the case of the brewing of "fresh hop" ales where freshly picked, undried hops are pitched into the kettle, often late

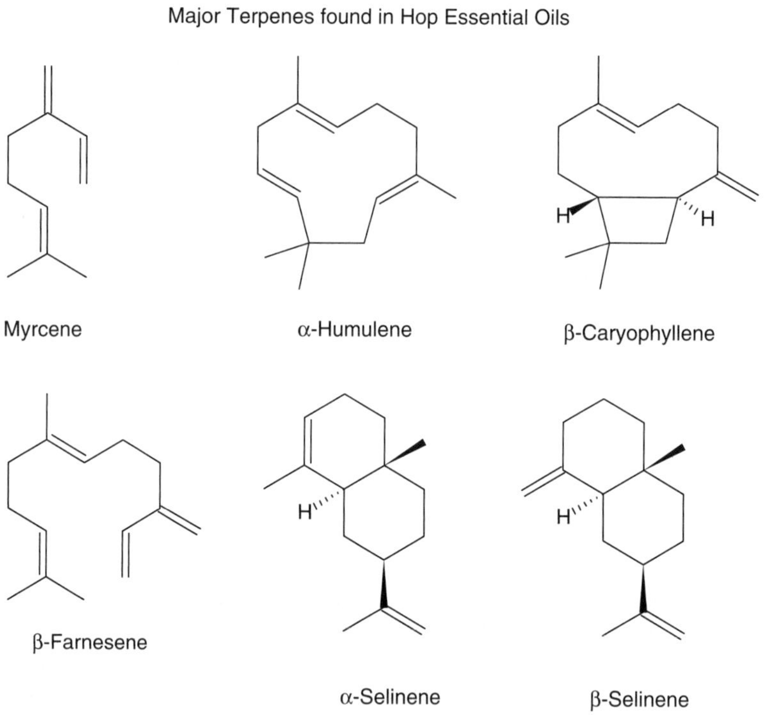

Major Terpenes found in Hop Essential Oils

FIGURE 7.4
Structures of some terpenes found in hop essential oil. The monoterpene mycrene is invariably the most abundant substance in the oil from fresh hops, but its proportion reduces substantially during storage, mostly due to volatilization and polymerization. Amongst the sequiterpenes, humulene in particular is a precursor to some oxygenated compounds that may positively influence beer flavor. Farnesene and the selinenes are mainly significant as markers for identification of different hop varieties.

in the boil. In these cases, flavor-active amounts of myrcene and perhaps also other terpenes can certainly be transferred to the beer). Notwithstanding the general unimportance of the terpenes as direct contributors to beer flavor, a relatively high humulene to caryophyllene ratio (the "H/C Ratio") is nevertheless considered by many to be a necessary hallmark of a good aroma hop, a value in excess of 3.0:1 being said to be necessary for a hop having so-called spicy/herbal/floral "noble" character.[20,21] The presence of β-farnesene is also generally held to correlate with good hop aroma in beer.

Oxygenated Compounds

Dry-hopping apart, a hoppy flavor in beer is usually associated with the late addition of hops to the kettle and is generally considered to be due to the presence in the hop oil — albeit in relatively small amounts — of oxygenated

TABLE 7.7

Typical % Content in Hop Essential Oil Fraction of Some Terpenoid Compounds as Related to Hop Variety (Ranking by Humulene/Caryophyllene Ratio)

Hop Variety	Myrcene	α-Humulene	β-Caryophyllene	β-Farnesene	α- + β-Selinenes	H/C Ratio	Variety Type
Hersbrucker	45	29	5.6	0.3	0.6	5.3	Aroma
Saaz	34	18	5.0	19	0.4	3.6	Aroma
Liberty	41	32	9.7	0.1	0.7	3.3	Aroma
Challenger	55	17	5.4	1.0	6.0	3.2	Dual-purpose
Cluster	61	13	4.7	0.0	0.7	2.8	Dual-purpose
Willamette	55	19	6.9	7.1	0.5	2.7	Aroma
Fuggles	54	20	7.6	4.9	0.4	2.6	Aroma
Cascade	68	10	4.1	5.3	1.0	2.5	Aroma
Perle	41	31	12.5	0.2	0.6	2.5	Dual-purpose
Nugget	61	15	6.3	0.0	1.2	2.4	High alpha, good aroma
Chinook	48	14	6.4	0.0	1.9	2.3	High alpha, good aroma
Galena	55	11	4.6	0.0	1.0	2.3	High alpha
Target	58	10	4.9	0.0	0.7	2.1	High alpha
Columbus	42	17	10.2	0.0	2.5	1.7	High alpha
Olympic	61	14	8.7	0.0	1.5	1.6	High alpha
Pride of Ringwood	47	2	8.7	0.0	19.5	0.2	Medium alpha
Eroica	62	<1	12.8	0.0	4.8	0.03	High alpha

Source: Data from Lewis, G.K., Zimmermann, C.E., and Hazenberg, H., U.S. Patent No. PP 10,956, 1999.

derivatives of the terpenes.[22–27] These components fall into several classes, including alcohols, ketones, and esters, and much research has been devoted to determining which of the various compounds has significance to beer flavor. Very little hop oil survives into beer and therefore to be of significance to beer flavor, a compound must have a flavor threshold measured in parts per billion (ppb) rather than ppm, though there is evidence to suggest that synergistic effects play an important role.[27] Undoubtedly, one component that often plays a major role is linalool, which is found in two stereoisomeric forms. These enantiomers have different organoleptic properties, though, owing to its much lower flavor threshold, only R-linalool (D (−)-linalool) is of significance to beer flavor.[28] The amount of linalool in beer seems generally to correlate well with hoppy character.[29] Some other hop oil (or oil-derived) components that have been suggested to be of possible importance are geraniol, humulenol II, α-terpineol, undecan-2-one, methyl-4-deca-4-enoate, humulene diepoxides, and citronellol.

Sulfur Compounds

Sulfur-containing compounds have also been found in hop oil and as many of these have very low flavor thresholds they may on occasion influence the beer flavor, generally in a negative manner.[30–32] Such compounds include dimethyl disulfide (DMDS), dimethyl trisulfide (DMTS), and methanethiol. The formation of many of the sulfur compounds has been shown to be linked to the spray application of elemental sulfur as an antifungal agent during the growing season and leads to the appearance of a range of polysulfides, particularly via a purely chemical reaction.[31] The former practice of burning sulfur on the kiln to discourage browning and ensure an attractive, pale green coloration to the dried hops also had implications for the formation of flavor-active sulfur compounds.[31,33]

Glycosides

In common with many other plants, hops contain a variety of glycosides,[34–36] and recent evidence[29,34] suggests that hop flavor may partly derive from hydrolysis of these covalently linked combinations of a wide range of organic substances with a sugar moiety, releasing flavor-active chemicals such as linalool that are better known as components of the hop oil. Relatively little is known about the occurrence of these compounds in hops, but their existence may account for some of the hop character and could explain the observation that adding aroma hops to wort before boiling (so-called "first wort hopping") can lead to unexpected, pleasant hoppy notes in the beer.[37]

Polyphenols

General

All hops contain a complex and to some extent varietal and even geographically specific[38] mixture of polyphenolic substances, a proportion of which is found to be in the form of glycosides.[36] This polyphenolic content is typically in quantity sufficient to be of significance to physical beer stability,[39,40] and sometimes also to a modest extent the beer flavor too, by contributing a degree of astringency, bitterness, and body to the beer.[41–44] Indeed, despite the very small quantity of hops used in the brewing process as compared to that of the major ingredient, malt, hops may be responsible for up to about 50% of the polyphenolic content of some heavily hopped beers. The hop polyphenols are to be found mostly in the hop cone petals and strig and, with the exception of the prenylflavonoids, not in the lupulin. Thus the brewery that uses low-alpha, aroma hops added to the kettle will tend to add more polyphenolic material to its wort than the brewery that uses the newer, high-alpha varieties. Obviously, it follows that purchase of aroma hops in the form of concentrated ("Type 45") hop pellets (see corresponding section) will also reduce the added amount of such material. Total elimination may readily be achieved by switching to the use of CO_2 extracts or postfermentation hop bittering.

Proanthocyanidins

A portion of the hop polyphenolic material is composed of such water-soluble substances as catechin and epicatechin (Figure 7.5), flavan-3-ols that are the monomeric building blocks of dimers, trimers, and higher polymeric structures, having as many as 20 or so basic units.[45] These polymers, known as proanthocyanidins or "condensed tannins," may also contain gallocatechin and epigallocatechin subunits, though these seemingly never appear as terminal units.[45] Much of the polyphenolic material added to the kettle associates and is precipitated with proteins in the hot and cold breaks but some survives, especially the structures of lesser molecular size. These residual compounds may then play a substantive role in determining the shelf life of the beer, often being responsible through interaction with beer polypeptides for the formation of haze. They are also recognized as antioxidants and are considered by many to have positive effects on beer flavor stability.[42-44,46,47] Furthermore, there is a growing body of evidence to suggest that their ability to scavenge free radicals may impart health benefits,[48] especially in conjunction with the moderate consumption of alcohol. They are, therefore, of considerable importance to brewers, there being both positive and negative aspects to their presence in worts and beers, though their chemistry is rather less well understood than that of the bitter hop resins.

Flavonoids

Xanthohumol

The yellow color of the lupulin is in part due to the presence of the prenylflavonoid xanthohumol (XN), a chalcone compound that is almost unique to hops. Although present at levels up to about 1.5% of the total weight of the dried hops,[49-51] until recently it was considered no more than a curiosity and of no particular value to brewers or hop growers. Indeed, if anything, its presence was regarded as decidedly negative on the presumption that it was implicated as a primary cause of beer haze and therefore best removed.[52] More recently, it has been the focus of great interest on account of the finding that it exhibits wide-ranging, positive activity in a variety of *in vitro* tests that might indicate various useful anticancer and antibacterial properties.[53,54] It remains to be seen whether such activity translates into real value in medicine, but sufficient interest has been aroused in the

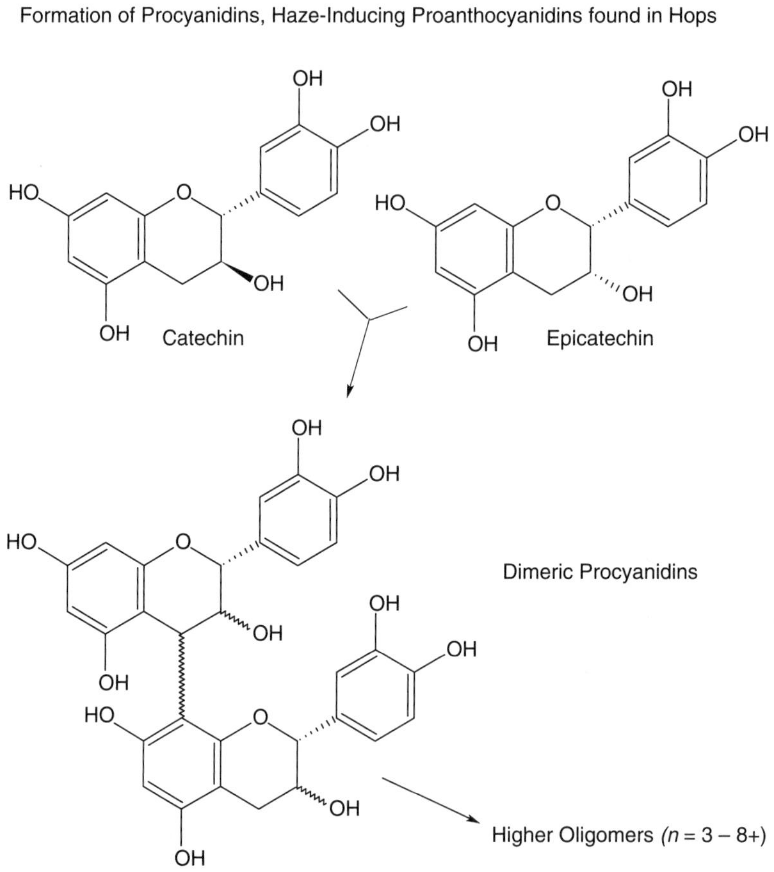

Formation of Procyanidins, Haze-Inducing Proanthocyanidins found in Hops

FIGURE 7.5
Structures of catechin and epicatechin, precursors of important haze-inducing, polyphenolic polymers. Catechin and epicatechin are relatively abundant hop polyphenol flavanols. As monomers, their antioxidant properties may be of value in retarding oxidative beer staling reactions, but in polymeric form their tendency to attach to proteins is a prime cause of beer haze. Other flavanols that may be found copolymerized with the catechins into haze-inducing proanthocyanidins include gallocatechin and epigallocatechin.

brewing industry to the extent that hop preparations containing enhanced levels of XN are now marketed to brewers who may wish to brew products having a healthy image[55,56] (see also the section "Other Uses, New Products").

Isoxanthohumol

Xanthohumol has poor aqueous solubility, and in traditional brewing very little survives into the beer, partly because it will readily be precipitated, but more particularly because it is largely isomerized during wort boiling to a more soluble, flavanone derivative, isoxanthohumol (IX) (Figure 7.6).

Isoxanthohumol has also been demonstrated to have positive *in vitro* properties in anti-cancer tests, though mostly at a lower level than for XN.[53,57] Whether or not a commercial beer contains XN or IX depends upon the brewer's choice of hopping method. Conventional brewing with kettle hops or hop pellets will introduce XN and also IX to the wort, a proportion of which compounds will survive into the beer, though losses during fermentation are generally substantial.[51,58–60] Late hopping will tend to increase the proportion of XN. Commercial beers have been found to contain up to about 3 ppm of IX and generally much lesser amounts of XN, though sometimes approaching 1 ppm.[48,51] However, many beers are completely devoid of either substance. This is simply because XN is not extracted by either liquid or supercritical CO_2, meaning that such beers

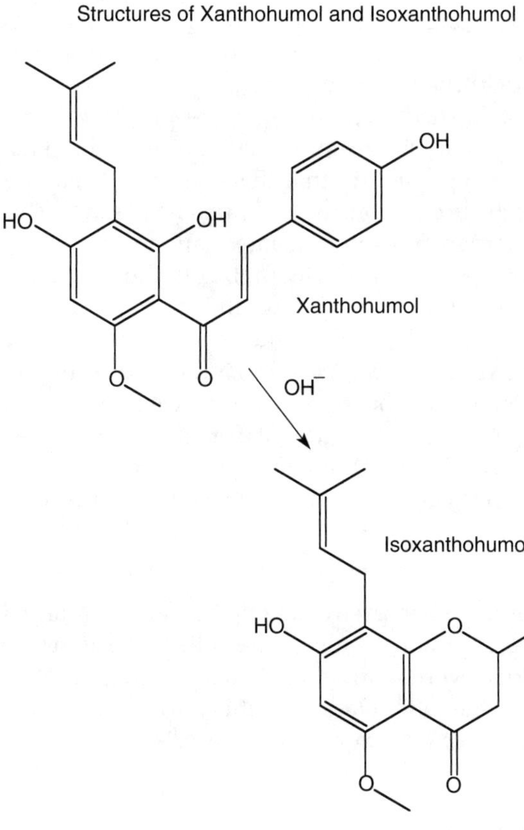

FIGURE 7.6
Formation of isoxanthohumol from xanthohumol during wort boiling. Xanthohumol, a prenylated chalcone, is less soluble in wort and beer than isoxanthohumol, which is an example of a prenylated flavanone compound. Both substances fall into the wide class of naturally occurring, polyphenolic and antioxidant substances collectively known as flavonoids.

will have been brewed only with CO_2 extract or postfermentation bittering products. On the other hand, use of ethanol extract for kettle hopping does introduce XN and IX to the beer, since XN is readily soluble in ethanol and more than 90% of the original XN content of the hops is typically found in this type of extract.[55]

8-Prenylnaringenin and Desmethylxanthohumol

Another compound of great interest to researchers is 8-prenylnaringenin (8-PN), which has also been given the trivial name "hopein."[54] Fresh hops contain very small amounts of this substance, as it appears not to be synthesized enzymically but forms via slow isomerization of another flavonoid chalcone, desmethylxanthohumol. The amounts of desmethylxanthohumol found in hops are much lower than those of XN and, hence, the levels of 8-PN are generally very low; nevertheless, the compound has significance. Hops have long been suspected to contain an estrogenic agent (or "phytoestrogen"), it having been noted long ago that picking hops by hand often disturbed the menstrual cycles of female pickers. Recent research has established that the agent responsible is 8-PN, and that its estrogenic activity is much greater than that of any other known phytoestrogen,[61] leading to the possibility that this compound may inhibit the formation of cancers of the breast, uterus, and prostate gland.[54] The appearance of 8-PN in beer is largely a result of isomerization during wort boiling of its weakly estrogenic precursor, desmethylxanthohumol. Hence, even though its concentration in hops is very low, and its concentration in beer also low (generally <0.1 ppm[62,63]), the amounts present have potential significance to the brewing industry, as moderate exposure to dietary phytoestrogens is thought to have generally desirable physiological effects.[64] Nevertheless, it might be prudent for any brewer considering taking steps to increase the XN or IX content of his beer to ensure that in so doing he does not also excessively elevate the content of 8-PN.

Xanthogalenol

Minor amounts of several other prenylflavonoids can be found in hops, some of which may be common to all, while others are variety specific. Xanthogalenol, first discovered in the American high-alpha variety Galena, is an example of the latter class, and this particular compound has been cited as a useful marker for taxonomic studies.[65]

Pectins

Although hops contain a considerable amount of pectins, these substances do not play a substantial role in normal brewing, normally being lost into the trub. However, their potential use as a nonbitter, natural foam stabilizer has been proposed and demonstrated.[66] About 30 ppm of hop pectins are claimed to be sufficient to improve the stability of beer foam.

Hop Waxes and Fats

Mostly concentrated in the lupulin are a substantial range of little-studied hop fats and waxes, including sterols and long chain, nonvolatile hydrocarbons and fatty acids.[67] In addition, hop seeds contain "fixed" oils that are widely held to be a potential source of rancid flavors — hence the preference of most brewers to purchase seedless hops.

Hop Storage

General

During storage of whole, baled hops, changes occur due to oxidation by atmospheric oxygen, enzymic reaction and, possibly in some cases, also as a result of microbiological activity. Dried hops may often be contaminated by an astonishing range of bacteria and molds, a fact that brewers of dry-hopped ales should always be aware of, though it appears that the risk of actually introducing beer spoilage organisms through dry-hopping is rather low.[68] The nature and rate of the changes that take place are dependent on a variety of factors, especially:

- Hop variety
- Kilning conditions
- Storage temperature
- Moisture content

These factors are discussed in the following sections.

Hop Variety

The importance of the hop variety can hardly be overstated. For reasons that even now are not clearly understood, some varieties are remarkably stable and deteriorate only slowly even at ambient temperatures, whereas at the other extreme there are varieties that are significantly unstable even in subzero storage conditions. The instability is most obviously manifested by loss of both α- and β-acids and a decline in the total content of the essential oils,[24,69,70] much of the latter due to oxidation and polymerization of myrcene, a substance that is also relatively volatile. The changes that take place are complex and not necessarily the same in all respects for different varieties. Particularly, the way in which the composition of the hop oil fraction changes through oxidation of the terpenes, possibly sometimes via enzyme mediation, is a varietal characteristic[69,71] and can have important consequences for the flavor of late- or dry-hopped beers.[24]

Kilning and Moisture

Unusual instability in hops can often be related to excessive drying temperatures or to poor air flow leading to stewing of the hops in the kiln. The first likely symptom will be a higher than normal value for the "Hop Storage Index" ("HSI") (see the section "Spectrophotometric Methods"), indicating that degradative chemical changes have already begun. Such changes can also be seen on HPLC chromatograms of the hop resin acids, where a relatively high proportion of so-called "S-fraction" compounds will be found.[72] Relatively low levels of myrcene in the oil fraction of a freshly dried hop may also be indicative of unsuitable kilning conditions, as will be the appearance of humulene epoxides (see following text). Low moisture in itself is most probably positive for hop stability, but, because it often indicates that the hops have been subjected to excessive heating, hops that have been dried to abnormally low moisture content (below 7%) are most likely to be unstable. High moisture levels (above 12%) promote instability of the hop resins[73] and are to be avoided also because there is a greater likelihood of microbiological activity occurring during storage. Both excessively high and excessively low moisture hops, particularly those of the super high-alpha varieties, may also present a serious fire hazard, being liable to spontaneous and often ferocious combustion.[74]

Hop Oils

It should be recognized that many of the chemical changes that take place are not necessarily undesirable. Indeed, for several varieties of aroma hops it is probably advantageous to allow a certain amount of oxidation in order to increase the level of the relatively polar (and therefore more soluble), flavor-active, oxygenated derivatives of terpenes such as α-humulene, which is the source of humulene epoxides and humulenol II,[24,69] though there is evidence suggesting that the most flavor-active derivatives have yet to be identified.[23] Because fresh hops contain relatively low levels of oxygenates, some brewers deliberately allow their hops to mature before use and may keep them for a certain number of months under ambient conditions before moving them into cold storage. In the absence of routine monitoring of the changes taking place to the chemical composition of the hop oils, this approach seems to carry a significant element of risk, since simultaneous oxidation of hop resin acids eventually leads to the development of rancid and cheesy flavors. Particularly, isobutyric and isovaleric acids are the compounds responsible.

Sulfur Compounds

The concentrations of some sulfur compounds may also increase over time. For example, the level of DMTS may rise considerably, especially if the hops were exposed to sulfur dioxide during kilning.[33] DMTS has an unpleasant, onion-like flavor, and an extremely low flavor threshold.[30]

Hop Resin Acids

The rate at which the hop α- and β-acids are lost via oxidative reactions during storage can be frighteningly fast, especially for an inherently unstable variety that has been subjected to excessive temperature during kilning. Not uncommonly, as much as 40% of the α-acids may be lost during only 6 months of storage at winter ambient temperature. Hop processors are acutely aware of this problem and endeavor to restrict the loss through cold storage of baled hops, sometimes at temperatures as low as $-12°C$ (10°F). More recently, cold stores have been built that also keep the hops under an oxygen-reduced atmosphere that is formed and maintained by the introduction of a proportion of pure nitrogen (which also helps to reduce the risk of fire). Pellets are normally much more stable than hops, though only because they are protected from direct aerial oxidation by virtue of packaging under vacuum or nitrogen gas.[73,75] However, experience has taught that the stability of pellets may be compromised if the hops are not in good condition at the time of pelleting. For reasons that are not properly understood, it seems that for each lot of hops there is a temperature-related safe period for which they may be kept before being processed into stable pellets. If pelleting is left too late, then the resultant pellets may continue to deteriorate, albeit in most cases at a slower rate than would have been the case if left as baled hops. This safe period, which roughly equates with a time during which the α-acid content remains relatively unchanged, may be quite short, measured in weeks rather than months for unstable varieties.[70,76]

The reasons for oxidative loss of the α- and β-acids is probably due to a combination of physical, chemical, and enzymic reaction factors. There is evidence to suggest that the presence of an α-acids oxidase may be responsible for much of the loss of α-acids,[77] though for the β-acids it is more likely that aerial oxidation is the prime or, possibly, only cause.[72,75] This oxidation of β-acids is responsible for the formation of the usefully bitter hulupones (Figure 7.7).[78]

Some hop oil components, including myrcene and myrtenol, can have a negative effect on the stability of α-acids.[79,80] The condition of the lupulin gland membrane is also important, as it acts as a natural barrier to oxygen from the surrounding air. Studies have shown that if excessive pressure is applied to the hops during baling, then a proportion of the lupulin glands may be ruptured and, although the rate at which oxygen penetrates into the bale itself may be reduced by the consequently higher degree of

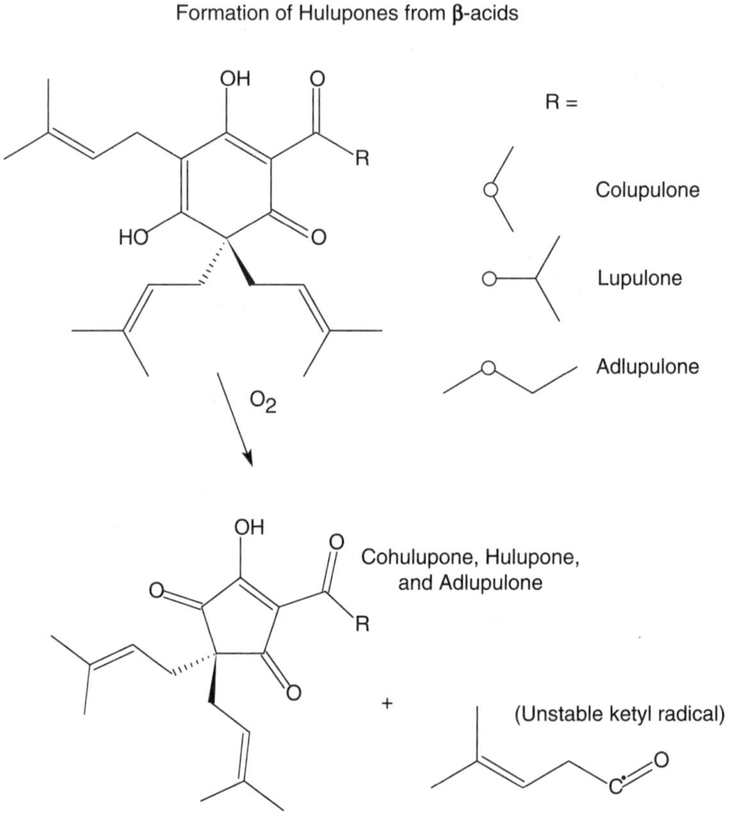

FIGURE 7.7
Formation of hulupones from β-acids. The hulupones are formed by auto-oxidation during storage of the hops and to a limited extent during the wort boil. They may contribute significantly to the bitterness of beers brewed with old hops.

compaction, nevertheless, the α-acids will probably deteriorate at a faster rate.[81,82] The former custom of burning sulfur on the kiln was known to improve retention of α-acids during storage, presumably due to the antioxidative effects of absorbed sulfur dioxide, but the general outlawing of this practice now prevents hop growers from improving the quality of their product in this important respect.

While most of the loss of α-acids is oxidative in nature, a small proportion of these compounds is transformed during storage into iso-α-acids, though as these are also unstable in air, the measurable amount is always low. To a small extent, a curious disproportionation reaction between two molecules of α-acids also occurs, leading to natural formation of dihydro-α-acids and dehydro-α-acids, the former then being subject to partial isomerization to dihydroiso-α-acids.[83] Interestingly, these latter compounds are also the

Hops

intermediates formed during the catalytic conversion of iso-α-acids by gaseous hydrogen to tetrahydroiso-α-acids (see section "Inhibition by Reduction").

Bittering Value

Although the loss of α-acids would seem to be a serious matter, brewing and tasting trials have clearly demonstrated that loss of α-acids during storage does not mean that the bittering value of the hops is proportionally diminished.[15,84,85] This is because certain of the degradation products of α-acids, especially tricyclodehydro-iso-α-acids,[86] are themselves bitter, as are some of those of the β-acids (most notably the hulupones).[15,78,87] Hence, for practical reasons, some brewers prefer to purchase and use hops and hop pellets on the basis of measuring the "alpha" content by a lead conductometric method (see the section "Lead Conductometric Methods"), since this value is found to better correlate with bittering value than does the "actual" or "true" alpha value given by HPLC.

Isomerization

The Isomerization Reaction

For a typical beer, the bitter character is largely derived from the α-acids via their conversion to their much more bitter, isomeric, forms, the iso-α-acids. In the traditional brewing process, this isomerization takes place in the kettle and is a key function of the wort boil (Figure 7.8). The isomerization reaction results in the formation of an analogous series of iso-α-acids: primarily isocohumulone, isohumulone, and isoadhumulone.

Because the reaction is facilitated by hydroxyl ions, the low pH of wort means that even at boiling point it is slow and inefficient. Side reactions take place, and especially in the presence of oxygen (which can enter the boiling wort if the kettle door is opened) there may be significant destruction of the formed iso-α-acids. Hence, it is typical that to achieve maximal formation of the α-acids requires around 90-min boiling at 100°C, though substantially less in a pressurized kettle. Indeed, breweries at high altitudes invariably achieve poor results unless the boiling temperature is raised by pressurization. Reaction efficiency is also related to the content of divalent metal cations, especially magnesium and calcium, which promote the isomerization,[88] though this positive effect may be partly counterbalanced by a greater tendency to precipitation in the trub, as iso-α-acids salts containing these metals are highly insoluble. Fundamentally, one may expect that lager worts — because they generally have higher pH values and lower salt content than ale worts — will allow better utilization of the α-acids. Wort gravity is also important: high gravity worts perform poorly because the higher content of trub encourages precipitation. Recovery of

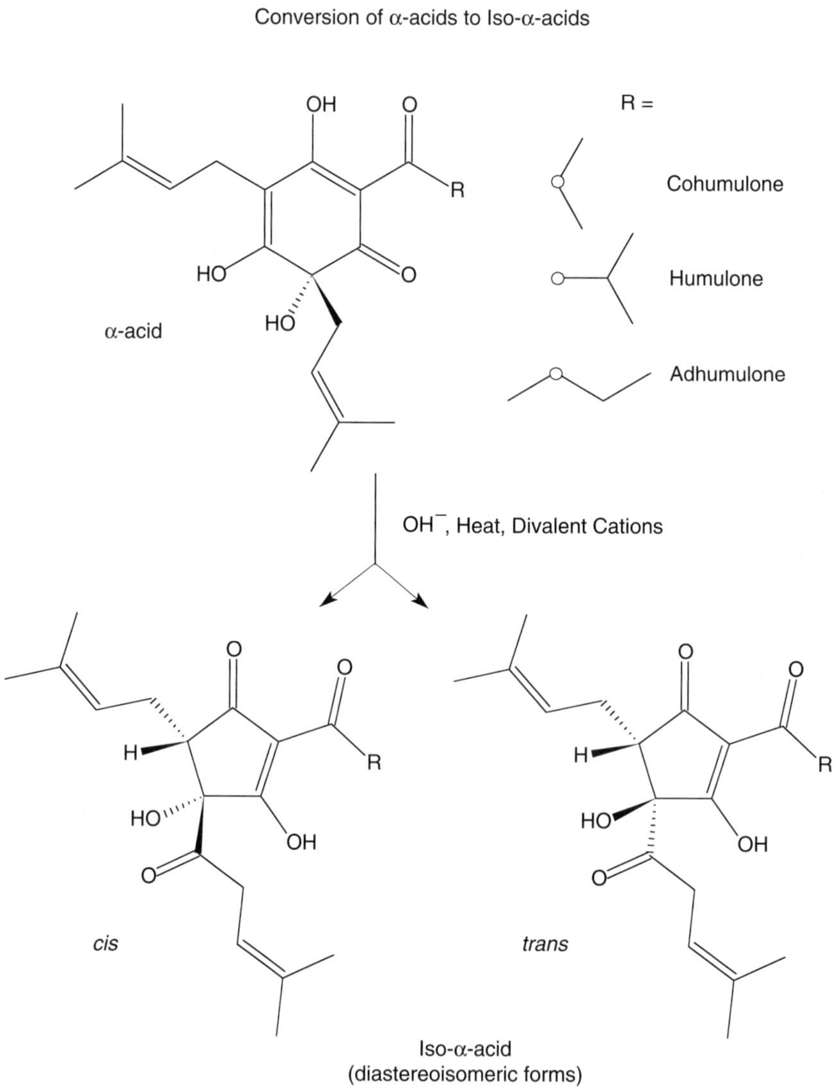

Conversion of α-acids to Iso-α-acids

FIGURE 7.8
Isomerization of α-acids to iso-α-acids during wort boiling. There are three main homologs of α-acids, of which humulone is generally the most abundant. Several minor homologs, usually representing about 1–3% of the total, are also present in the lupulin. Adhumulone normally accounts for a further 7–15%, while the proportion present in the form of cohumulone is considered a key varietal characteristic.

iso-α-acids into the pitching wort is also a function of the amount of α-acid added, lower bittering levels giving higher utilization values. Thus, the practice of high-gravity brewing inevitably means that utilization of kettle hops will be poor.

- Factors encouraging faster isomerization:
  - Higher wort pH
  - Higher boil temperature
  - Harder brewing liquor
- Factors encouraging better retention of iso-α-acids, once formed:
  - Lower wort gravity
  - Higher wort pH
  - Lower hopping rate
  - Softer brewing liquor
  - Higher adjunct ratio

Significance of the Cohumulone Ratio

Importantly, because they have much lower pK_a values than their precursors, the iso-α-acids are much more soluble in wort and beer, though still prone to precipitation and substantive loss through electrostatic association with the insoluble proteins of trub and by adsorption onto the surface of yeast cells during fermentation. As already mentioned, the α-acid cohumulone is both more soluble and more reactive than its humulone and adhumulone homologs, hence the formation of isocohumulone proceeds at a slightly but significantly greater rate, and the compound itself is less liable to precipitation with the trub.[89–91] Analysis through the stages of the brewing process will therefore generally show that the cohumulone ratio increases as between that of the hops or hop product added to the kettle and the beer ultimately produced. Hence, when hop varieties of differing cohumulone ratios are put through the same brewing process, those having a relatively high cohumulone ratio will return the best utilizations into beer.

Significance of the Cis:Trans Ratio

Because of the introduction of a new chiral center to the molecular structure, there are in fact two possible "diastereoisomers" of each iso-α-acid that may theoretically be formed from the nonracemic, natural α-acids. These are the *cis* and *trans* forms, both of which are formed in the wort and consequently found in the beer. For a traditionally brewed beer, the *cis:trans* ratio is normally in the range from 1.5:1 to 2:1, but where preisomerized products have been used as the source of bittering the ratio may sometimes be found to be as high as 4:1. This is a consequence of such products having been formed with the aid of Mg^{2+} catalysis, which favors the formation of the *cis* isomer. Brewers are generally aware that the bitter acids content of beer often reduces significantly over time, and this is now known to be largely due to the loss of *trans*-iso-α-acids, which have been determined to be substantially less stable than are their *cis* form counterparts.[92,93] This instability of the *trans*-iso-α-acids may also have adverse implications for beer staling.

Late Hopping

The addition of a portion of the kettle hops late in the boil for the purpose of imparting "late-hop" character to the beer results in very poor utilization of the α-acids for that part of the hop grist.[90,94] To some extent, this may be mediated by allowing the wort to remain for a substantial time at high temperature in the whirlpool. Typically though, the final utilization in the cooled wort of the late-added portion of hops may not be more than about 25%, compared to perhaps 45% for that part added at the start of the boil. Further reduction of the dissolved iso-α-acids will then take place during fermentation due to absorption onto the yeast and, to some extent, by precipitation onto the upper walls of the fermentation vessel (FV) in the collapsed foam, especially when the fermentation has been vigorous.

Bitter Flavor and Foam

Role of Iso-α-Acids

Apart from their role as the major bittering components of beers brewed from hops, hop pellets, or extracts, iso-α-acids are also key components of beer foam. Studies have shown that both the *cis* and *trans* forms of the iso-α-acids participate in foam formation and stabilization, which also requires the presence of particular, positively charged polypeptides derived from the malt and di- or trivalent metal ions such as manganese and aluminum.[95] Increasing the amount of iso-α-acids (or their hydrogenated derivatives) will normally increase the foam stability of a beer, though a point may be reached where little or no further improvement occurs because one or another of the other vital components is in short supply.

The increase in cohumulone ratio during brewing has significance for both the bitter flavor and foam stability of the beer.

- Isocohumulone differs from isohumulone in two respects:
  - It is less bitter
  - It is less foam-positive

The *cis*- and *trans*-forms of the iso-α-acids have fundamental differences, too.

- The *cis* forms of iso-α-acids are:
  - More bitter
  - More stable

Hops

The differences in bitterness between the major forms of the iso-α-acids are actually quite substantial. Referenced against *trans*-isohumulone, the relative bitterness in a model buffer system at pH 4.2 was measured by Hughes and Simpson.[96,97] They found that for the isohumulone (isoH) and isocohumulone (isoCoH) variants:

$$cis\text{-isoH}\,(1.82\times) > trans\text{-isoH} \sim cis\text{-isoCoH} > trans\text{-isoCoH}\,(0.74\times)$$

Hence, the practical effect of changing the source of bittering from one hop variety or hop product to another may not be entirely insignificant, and may therefore require an adjustment to the target BU or HPLC iso-alpha content of the beer.[97] Though Hughes and Simpson found no evidence to support the view, it is nevertheless widely held that isocohumulone bitterness is also harsher and less desirable than that derived from isohumulone, supposedly a consequence of the relatively greater degree of dissociation of isocohumulone at beer pH.[98] Indeed, it is worth noting that there is remarkably little scientific evidence to support the belief that isohumulone produces the finer quality of bitterness, and a contrary view based on brewing trials and indicating a preference for isocohumulone bitterness has even been expressed.[99]

Changes to the foam stability and lacing characteristics of a beer may also be expected if a change to the hopping regime results in the beer having a substantially different content of the six isomers. The comparative foam stabilizing abilities of isocohumulone and isohumulone have been studied, and it is clear that isocohumulone has a relatively inferior effect on foam.[100,101] Perhaps surprisingly in view of the considerable differences in the three-dimensional (3D) structures of the *cis* and *trans* isomers, experiments have not demonstrated associated practical differences in the stability or structure of beer foam, though there is some evidence to suggest that the *trans*-isomer is more readily transported into the foam.[96]

Reduced Iso-α-Acids

The "Lightstruck" Reaction

Bottled beer has traditionally been sold in brown, sometimes in green, bottles. Brewers, especially of pale-colored beers, doubtless noted long ago that beer exposed to direct sunlight was prone to develop unpleasant "sunstruck" or, more aptly for those unfortunate enough to have encountered the animal on a bad day, "skunky" flavors (see also Chapter 19). Use of brown glass very effectively suppressed such formation, green glass rather less so. Determination of the underlying chemistry of skunky flavor formation revealed that the offending compound was a small mercaptan, 2-methyl-3-butene thiol (MBT) (Figure 7.9). This rather volatile substance has an exceptionally low flavor threshold in beer, variously estimated as being in the range of about 2–20 parts per trillion (ppt).[102,103] MBT arises as a

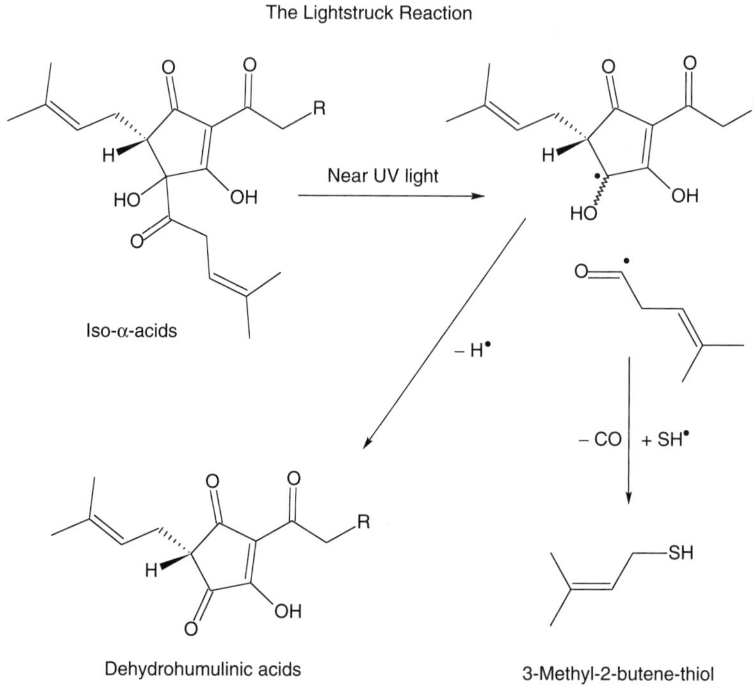

The Lightstruck Reaction

Iso-α-acids → (Near UV light) → [radical intermediate] → Dehydrohumulinic acids (−H•)

[radical intermediate] → (−CO, +SH•) → 3-Methyl-2-butene-thiol

FIGURE 7.9
The lightstruck reaction. Green glass bottles provide only a limited degree of resistance to the lightstruck reaction. Full protection requires the use of brown glass or substitution of the iso-α-acids by their reduced analogs — "Rho," "Tetra," or "Hexa".

result of reaction between the isohexenoyl side chain of iso-α-acids and small mercaptans naturally present in beer, probably including hydrogen sulfide, H_2S. The reaction requires activation by near UV, blue light and is also believed to require an electron-like transferring cofactor such as riboflavin. Interestingly, the reaction is blocked by molecular oxygen, which perhaps explains why outdoor drinkers of beer on a sunny day do not often complain of skunky taints forming in their beer.

Inhibition by Reduction

Driven no doubt by the desire of some beer marketers to present their brewery's product to better effect by bottling in clear glass, researchers turned their attention to ways by which the "skunk reaction" could be prevented. The problem was solved in the 1970s by the introduction of chemically reduced preparations of iso-α-acids.[104,105] By altering the structure of the reactive side chain on the iso-α-acids molecule, specifically either by reducing the carbonyl group or the C=C double bond, it was found possible to completely prevent the reaction. Today, three different types of chemically reduced, "light resistant" iso-α-acids, with two, four, or six hydrogen atoms, are commercially available. These modified iso-α-acid preparations are generally sold as slightly alkaline, potassium salt solutions and can be dosed directly into beer (see the section "Reduced Isomerized Hop Products"). In the course of investigations into the properties of these new compounds, it was also determined that their ability to improve foam stability was somewhat different to that of iso-α-acids. Particularly, reduction (by catalytic hydrogenation) of the side chain double bonds has the effect of increasing the hydrophobicity of the molecule, and, hence, its tendency to leave the beer and to concentrate in the foam was enhanced, thereby increasing foam stability.

With reference to Figure 7.10, reduction at points A and B gives tetrahydroiso-α-acids ("Tetra"). This is achieved by catalytic reduction of iso-α-acids using gaseous hydrogen and a palladium on carbon catalyst. The result is a product that has excellent foam-stabilizing characteristics and enhanced bittering power (Figure 7.11).

Reduction at point C is achieved by reaction of iso-α-acids with sodium borohydride (NaBH$_4$) under alkaline conditions, which reduces the carbonyl group to a hydroxyl one. The homologous series of compounds so formed are the *rho*-iso-α-acids (ρ-iso-α-acids), also sometimes known as dihydroiso-α-acids (a term to be avoided as it causes confusion with the chemically distinct dihydroiso-α-acid intermediates in the production of Tetra), or, more colloquially, simply as *Rho* (Figure 7.12).

Hexahydroiso-α-acids ("Hexa") may be formed either by catalytic reduction of *rho*-iso-α-acids or by borohydride reduction of Tetra. This

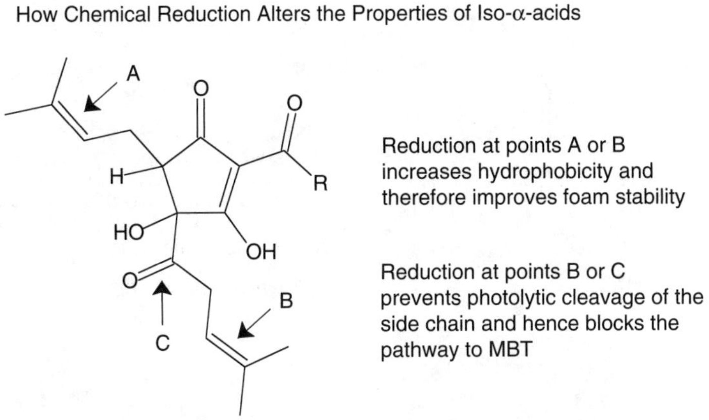

How Chemical Reduction Alters the Properties of Iso-α-acids

Reduction at points A or B increases hydrophobicity and therefore improves foam stability

Reduction at points B or C prevents photolytic cleavage of the side chain and hence blocks the pathway to MBT

FIGURE 7.10
Points at which reduction of iso-α-acids confers resistance to the lightstruck reaction.

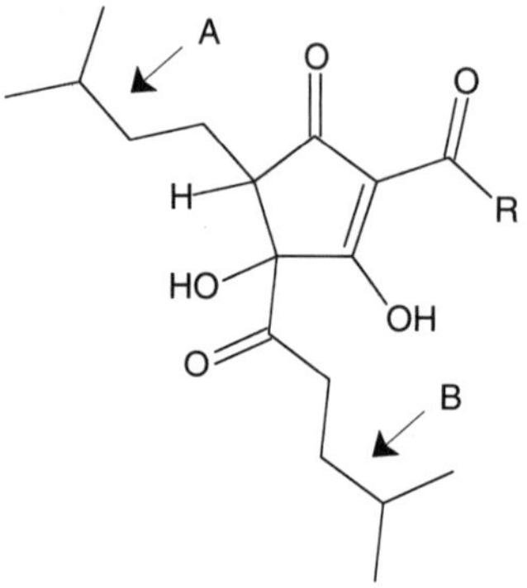

FIGURE 7.11
Structure and properties of tetrahydroiso-α-acids.

Tetrahydroiso-α-acids are:

More bitter than iso-α-acids (~1.1–1.7×)

At equivalent bitterness they give:

Enhanced foam stability

inevitably more expensive product has its own combination of properties (Figure 7.13).

Manufacture of Reduced Iso-α-Acids

Several commercial processes have been developed over the years for the production of these modified iso-α-acids, though in all cases the reductive stage is always either a borohydride reduction of the carbonyl or a catalytic reduction of $-CH=CH-$ to $-CH_2-CH_2-$, as already mentioned. However, it is worth noting that commercial processes for the production of Tetra have been developed that allow it to be made either from α-acids or from β-acids.[106] (Further details of the manufacture of the reduced isomerized iso-α-acids are given in the section titled "Reduced Isomerized Hop Products.")

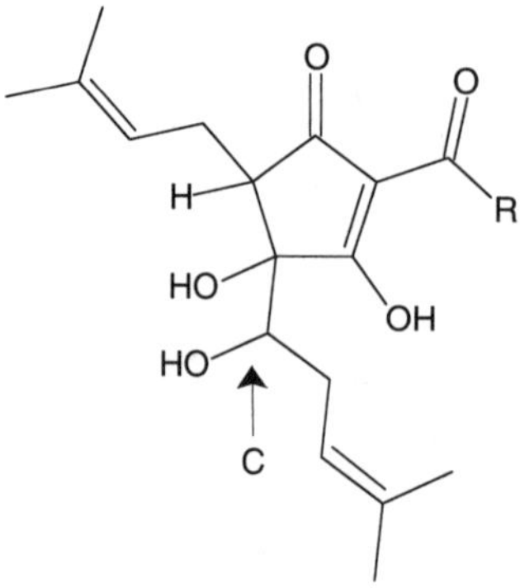

Rho-iso-α-acids are:

Less bitter than iso-α-acids (~0.7×)

At equivalent bitterness they give:

Slightly improved foam stability

FIGURE 7.12
Structure and properties of *rho*-iso-α-acids.

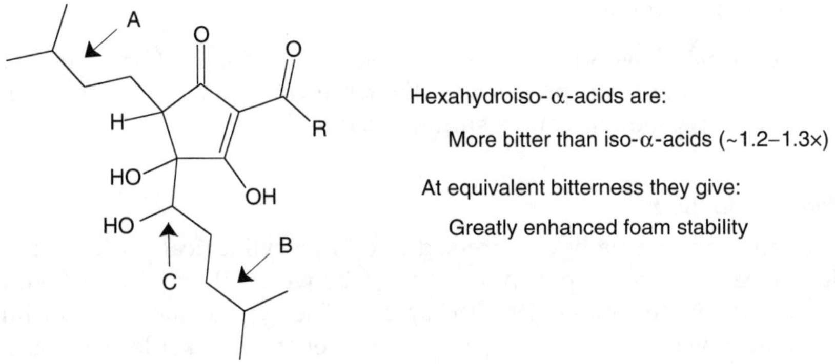

FIGURE 7.13
Structure and properties of hexahydroiso-α-acids.

Hexahydroiso-α-acids are:

More bitter than iso-α-acids (~1.2–1.3×)

At equivalent bitterness they give:

Greatly enhanced foam stability

Hop Products

Development of Hop Products

Hop products have been in existence for many years, with the earliest attempts at hop extraction dating back to the mid-19th century. Table 7.8 outlines some of the key steps in the chronological development of these products.

The development of hop products appears to parallel the increased understanding of hop chemistry and the introduction of new, better analytical techniques to define and quantify more accurately the changes taking place during the manufacture of hop products.

TABLE 7.8

The Development of Hop Products

1800s	First commercial, aqueous, hop extracts produced in Germany in the 1850s followed by petroleum ether (United States, 1870s) and alcohol extracts (1890s)
1910–1930	First hop oil emulsions produced
1960s	First iso-extracts produced (in the United Kingdom); hop powders and pellets introduced into Europe
1970s	Widespread use of Type 90 and Type 45 pellets; stabilized pellet patent and early patents on reduced products registered (United States)
1980s	Liquid and supercritical CO_2 extraction processes developed; increasing use of iso-extracts; iso-pellets patented and fractionated hop essences available
1990s	Further reduced product patents registered and the more general use of reduced products for foam enhancement and light stable beers
Early 2000s	Publication of patents for nonbrewing uses of hops

Benefits of Hop Products

Although some brewers still use only whole cone hops, most brewers prefer to use one or more hop products in their brewing processes. The general benefits of hop products can be summarized as follows:

Volume Reduction

Whole hops are normally compressed into large cylindrical pockets or rectangular bales, which typically can weigh between 50 and 100 kg, with a bulk density in the range 100–150 kg/m^3. The typical increases in bulk density achieved by processing hops into pellets and kettle extracts are shown in Table 7.9. In the case of extracts, the benefit associated with this greater density is then further enhanced by the typically fivefold or so increase in the α-acid content.

As a result of processing, storage and freight costs are reduced with manual handling normally made safer and easier. Further volume-related cost reductions may be achieved through the use of preisomerized products (see the sections titled "Stabilized and Isomerized Pellets" and "Isomerized Kettle Extracts").

Increased Stability

As already mentioned in the section "Hop Storage," the hop resins and oils present in whole hops are susceptible to oxidative degradation, though losses of essential components can be reduced by cold storage. However, this practice tends to be expensive owing to the relatively large volume occupied by the bales.

Processing of leaf hops into hop products encloses the product in airtight packs that are normally evacuated after filling or flushed with inert gas prior to sealing. It has been shown that some varieties can lose the majority (up to about 75%) of their initial α-acid content when stored at ambient temperatures for 12 months in bale form. By comparison, pellets and extracts made from unstable hop varieties may lose only 10–20% and 2–4%, respectively, when stored at the same temperature over the same period. These losses can be further reduced to about one third of these values by cold storage.

TABLE 7.9

Comparison of Typical Bulk Densities of Hop Products

Product	Typical Bulk Density (kg/m^3)
Baled hops	100–150
Pellets	450–550
Kettle extracts	960–1020

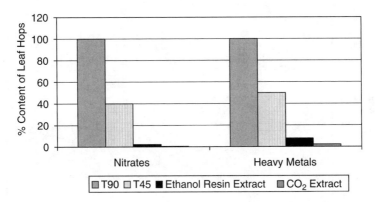

FIGURE 7.14
Typical reduction of nitrates and heavy metals in hop products.

Reduction in Chemical and Heavy Metal Residues

Much of the undesirable residues present in hops are found in the plant structural material rather than in the lupulin glands. If, during processing, this material is physically reduced (Type 45 pellets) or its components selectively left behind after extraction (extracts), chemical and heavy metal residues may be significantly reduced. With hop extracts, the degree of the reduction very much depends on the polarity of the solvent used. Figure 7.14 illustrates the degree of difference that may be found.

Homogeneity

Heterogeneity within batches of hops and even within any given bale can lead to considerable variation in bittering when using cone hops. The mixing and concentration processes involved in hop product preparation result in a much more homogeneous material for use in the brewery.

Reduced Extract Losses

Substantial retention of wort in spent, cone hops in the traditional hop back inevitably results in higher loss of extract compared to that achieved in whirlpool-based systems when using a hop product in which solid matter is reduced or eliminated altogether.

Use of Automated Dosing Systems

Both pellets and extracts (kettle and postfermentation additions) are capable of being dosed into the process stream using automated (or semiautomated) systems, thereby saving manpower costs and potentially achieving more consistent additions.

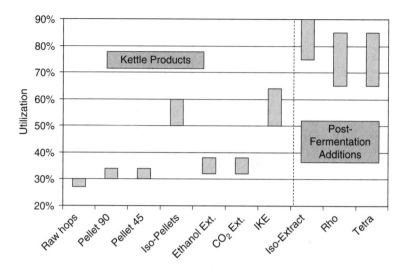

FIGURE 7.15
Comparison of typical per cent utilization of iso-α-acids in final beer using different hop products.

Improved Efficiency

Utilization of α-acids added in the brewhouse through to the finished beer is generally poor and usually in the range 25–30% for hops added at the start of wort boiling. Late addition of aroma hops will further reduce this efficiency, resulting in higher bittering costs. Poor utilization in the kettle is due to the relative inaccessibility of α-acids within the lupulin glands, their inherent low solubility and inefficient conversion to iso-α-acids at the low pH of wort, and the absorption of both α- and iso-α-acids onto the surface of precipitated trub and yeast cells.

Hop products generally improve the accessibility of the α-acids to the wort and, in the case of preisomerized products, require only that the iso-α-acids be dissolved into the wort or beer. Very significant improvements in utilization, often exceeding 100%, can therefore be achieved with preisomerized products added into rough beer (Figure 7.15).

Quality Benefits

The most recently developed products provide the brewer with important additional quality benefits, as well as offering the improvements in utilization outlined previously. Use of such materials, described later, can significantly enhance flavor and foam stability in the resultant beers.

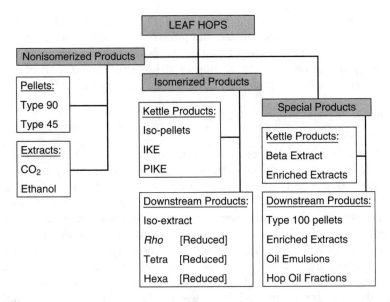

FIGURE 7.16
The family of hop products.

Classification of Hop Products

There are many ways of describing the range and interrelationship of hop products. The diagram in Figure 7.16 gives a broad outline of the key products, classifying them into three distinct groups:

- Nonisomerized products
- Isomerized (and in some cases also reduced) products
- Special products

With the continuing development of the range of options available to the brewer, some products have been formulated for use in both kettle and post-fermentation applications. For reasons of clarity these have not always been shown in the diagram.

Nonisomerized Hop Products

Pellets

The introduction of hop pellets superseded the earlier hop powders, which often proved difficult to package and handle. Introduced in the 1960s, hop pellets have become the most commonly used, kettle-added, hop products owing to their simplicity, ease of use, and low cost. Around 50% of the world's hops are used in this form.[3]

Pellets are normally added either during kettle filling or at the start of the boil. However, it is not uncommon that pellets are added in two or three separate charges at various stages of the boil, and sometimes even into the whirlpool. In general, the later the addition the lower the utilization of α-acids achieved and the better the retention of hop oil components in the wort.

Type 90 Pellets

Type 90 pellets are so called because originally the weight yield after the elimination of excess moisture and extraneous matter was normally around 90%. The process (Figure 7.17) involves a sequence of physical steps in which, after blending, redrying to 8–10% moisture (if necessary), and the removal of extraneous material, the cone hops are ground in a hammer mill. The resultant powder is mixed and then pelletized, by forcing with the aid of metal rollers through a multiplicity of holes in a metal die (normally bronze or steel), to form short pellets of approximately 6 mm diameter. Pelletizing machines may use flat, circular dies or else have drum-shaped dies in which the powder is pressed from the inside of the sideways-mounted, spinning die. The frictionally heated pellets thus produced are then cooled as quickly as possible on belt or vertical coolers, using cold air to minimize any degradation of the hop resins. The pellets may then be blended before being packed in aluminum foil (or metallized) plastic laminates under a vacuum (hard packs) or under an inert gas such as nitrogen (soft packs).

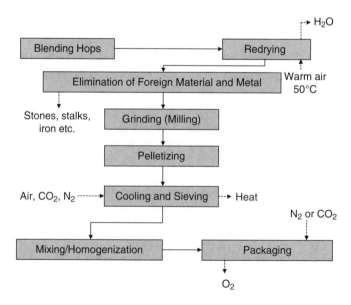

FIGURE 7.17
Type 90 pellet production.

Hops

The harder the pellet is pressed through the die the higher is the bulk density. However, as a result of the friction involved, the temperature of the pellet will inevitably rise and may cause loss of α-acids. Consequently, it is desirable to keep the bulk density of the pellets below 550 kg/m^3 and to maintain the exit temperature of the pellets below 55°C. This is often achieved by "flooding" the exit area of the die with liquid CO_2 or N_2. Hop resins and, particularly, the essential oils, normally act as lubricants during the passage through the die and, therefore, pelleting temperatures are generally lower when processing high-alpha hops as compared to low-alpha hops. Though generally difficult to measure (see the section "General" under "Hop Resin Analysis"), in most cases the yield of α-acids on a dry weight basis from cone hops to pellets is normally >98%.

Type 45 Pellets

In the production of "Type 45" pellets (also known as "enriched" or "concentrated" pellets) some of the unwanted "leaf" material is removed (Figure 7.18). This increases the relative proportion of lupulin glands and, consequently, the concentration of hop resins and oils in the resultant pellet. The processing also removes some of the undesirable chemical residues that generally reside in the removed plant material.

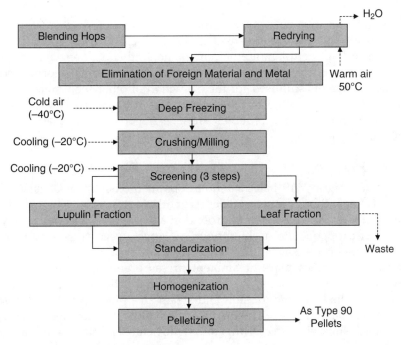

FIGURE 7.18
Type 45 pellet production.

Cone hops are blended, redried (if required), and extraneous matter removed, just as in Type 90 pellet production. However, in order to facilitate the separation of the sticky lupulin glands (the fine fraction) from the cellulosic material (the coarse fraction), the milling process is carried out at very low temperatures (ca. $-35°C$) so that the normally sticky lupulin glands become hard and easily separated. The two fractions are then separated by passing the powder through a series of sieves (minimum 0.5 mm mesh), the ambient temperature being maintained at about $-20°C$. Once the two fractions are separated, a calculated quantity of coarse fraction can then be added back to the lupulin fraction in order to achieve a desired α-acid content in the pellet. Following homogenization, the powder is pelletized, cooled, and packed in a similar way to Type 90 pellets.

Production of Type 45 pellets not only reduces the volume and weight of material while (largely) retaining basic brewing value, but substantially reduces undesirable residues of nitrates, heavy metals, and pesticides.[107] The hop polyphenols content is also reduced, which may have implications for beer quality and stability (as earlier mentioned in the section "General" under "Polyphenols"). The yield of α-acids on a dry-weight basis from cone hops to pellets is normally >95% while the weight yield is around 45–50%.

Kettle Extracts

Efforts at extracting hops on a commercial basis date back to the mid-19th century and, during the ensuing years, several organic solvents were tried, including petroleum ether, methanol, ethanol, benzene, hexane, dichloromethane, and finally CO_2 in liquid or supercritical form. Today, although hexane is still used for the extraction of other food materials, the extraction of hops is almost exclusively confined to the use of CO_2 and, to a lesser extent, ethanol. Together, these extracts nowadays account for about 30% of hop usage.[3]

Ethanol Extract

As shown in Figure 7.19, bales of whole hops are broken up on a bale breaker and, after separation of any metal debris or other extraneous matter, the hops are mixed with 90% ethanol and ground in a wet mill.[108] The hop slurry so formed is dropped into a continuous, two-deck, segmented extractor in which ethanol is percolated through the hops in a countercurrent direction. The extraction takes place at ambient pressure and at a temperature of 40–60°C. The spent hops are discharged from the extractor and the ethanol expelled by compression and steam stripping prior to being sent for rectification and reuse. The mixture of ethanol and extracted, polar hop material (the miscella) is centrifuged to remove any entrained solid material and immediately sent to a multistage vacuum evaporation system to remove the ethanol, which is then sent on to the rectification plant for concentration back to 90%. The remaining mixture is then separated into two

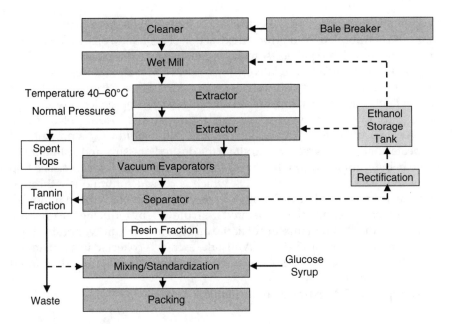

FIGURE 7.19
Ethanol extract production.

fractions — the resin and aqueous (or "tannin") fractions by centrifugation. Formerly, the aqueous fraction was used to standardize the α-acid concentration in the resin extract, but, as this fraction contains high concentrations of undesirable substances such as nitrates, ethanol extract is normally packaged as a pure resin extract. The residual ethanol content in this extract is typically <0.3%. If a standardized product is required, then glucose syrup may be used as the standardization medium.

Ethanol extraction is a high-capacity, continuous process with lower capital and operating costs than CO_2 extraction. The yield of α-acids from cone hops to extract is normally in the range 94–95%. During ethanol extraction, some changes in hop components take place:

- Isomerization of small amounts of α-acids into iso-α-acids (<2.0%)
- Formation of small quantities of polar degradation products of α-acids
- Removal of some volatile hop oil components (particularly the normally undesirable myrcene) and chemical modification of a small amount of other oil compounds

CO_2 Extracts

The extraction of materials with compressed carbon dioxide is today a well-developed technology, applied to such diverse applications as the decaffeination of coffee and the extraction of essential oils from plants such as mint.[109] When applied to hops for the extraction of resins and oils, the cone hops are first processed into pellets (sometimes smaller than normal Type 90 pellets at ca. 3–4 mm diameter). If relatively high extraction temperatures and pressures are to be used, then resins and oils are extracted directly from the pellets as such, whereas at lower temperatures and pressures, the pellets are initially remilled to a coarse powder immediately prior to the loading of one of a series of pressure vessels (extractors) (Figure 7.20). Liquid CO_2 is pressurized and heated or cooled as required to attain the desired pressure and temperature before being pumped through the bed of hops within the extractors. The hop resins and oils dissolve into the CO_2, which is passed continuously through the extractor train until virtually all of these components have been removed. The dissolved material is then separated from the CO_2 by reducing the pressure and passing the resultant mixture of liquid and gas through an evaporator, typically at a temperature of 45–50°C. Whilst the extract is collected in a separation vessel, the gaseous CO_2 is passed through a condenser, liquefied, and returned to a storage tank for reuse.

Two types of CO_2 extracts are available:

- Liquid CO_2 extract
- Supercritical CO_2 extract

The use of CO_2 at lower temperatures (down to about 10°C) and pressures is generally referred to as liquid CO_2 extraction and under these conditions the

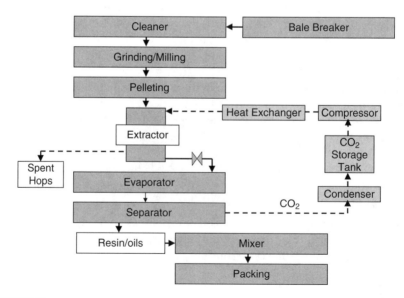

FIGURE 7.20
CO_2 extraction of hops.

CO_2 acts as a nonpolar solvent. However, its polarity can be increased by raising the extraction pressure and temperature above the critical point (74 bar and 31°C), at which point the vapor and liquid phases coincide. Above the critical point, CO_2 becomes a supercritical fluid and its dissolving power is progressively increased with increase of pressure and temperature, resulting in the extraction of some polar degradation products and hard resins. As the total yield of hop components is normally greater when using supercritical CO_2, this is the more commonly used option by commercial hop extraction plants, though because of the wide range of possible operating conditions, a distinction is often drawn between so-called "mild" and "hard" supercritical conditions. Table 7.10 shows the typical operating temperature and pressure ranges for each type of extraction, together with an indication of the yields of α-acids commonly achieved.

As for ethanol extract, CO_2 extracts can be standardized with glucose syrup but this is not recommended because these materials are difficult to mix and readily separate during and after packing. A more satisfactory alternative is to request that the extract be packed with a specified amount of α-acid filled into each container.

Comparison of Ethanol and CO_2 Extracts

The compositional differences between liquid and supercritical extracts are fairly small (Table 7.11), though the "cleaner" profile of liquid CO_2 extracts is certainly preferred by some brewers. "Mild supercritical" extracts are hardly distinguishable from liquid CO_2 extracts, but if more extreme temperatures and pressures are used (the "hard supercritical" conditions) then color differences can be apparent, and analysis can show more significant differences in the amounts of hard resin and oxidation products present. Hard supercritical extracts are also substantially more viscous, hence liquid or mild supercritical CO_2 extracts are preferable for use in automated dosing systems.

Both ethanol and CO_2 extracts can be added at any time during the wort boil; however, the later the addition the poorer the utilization of α-acids. Although there are obvious analytical differences between the two types of extract, trials have usually shown that no significant flavor differences can be

TABLE 7.10

Typical Operating Temperatures and Pressures for CO_2 Extraction, together with Indication of Effect on Alpha Yields

Parameter	Liquid CO_2 Extraction	Mild Supercritical CO_2 Extraction	High Supercritical CO_2 Extraction
Pressure (bar)	60–65	140–170	250–320
Temperature (°C)	10–15	35–45	50–80
Yield of α-acids (%)	90–94	92–96	94–98

TABLE 7.11

General Comparison of the Compositions of Ethanol and CO_2 Extracts

Parameter	Ethanol Extract	CO_2 Extracts
Iso-α-acids (%)	0.5–2.0	0.0–0.2
Depending on Variety and Extraction Technology		
α-acids (%)	20–55	35–55
β-acids (%)	15–40	15–40
Hop oils (%)	3–9	4–12
Fats, waxes, other soft resins (%)	15–25	10–20
Hard resins (%)	5–11	0–4
Solvent residue (%)	<0.3	Nil
Density (g/ml)	ca. 1.0	ca. 0.9–1.0
Viscosity at 40°C (Pa. sec)	1.3–3.5	0.4–0.9
Heavy metals (ppb)	50–150	10–30
Nitrates (mg/100 g)	30–70	0–10
Xanthohumol (%)	0.4–2.5	0.0
Color	Dark green (chlorophyll extracted)	Yellow to greenish-brown

Note: Liquid CO_2 extracts have highest α-, β-acids content and lowest viscosity, heavy metals, and nitrates content.

found in comparative beers brewed with each extract.[110] From a practical viewpoint, there is evidence to show that ethanol extract gives slightly faster conversion of α-acids in the kettle owing to its better dispersion into the wort,[111] though this is not necessarily of significance to the end result.[110]

Isomerized Hop Products

The conversion of α-acids to iso-α-acids during wort boiling is not only slow but also incomplete, leading to poor utilization of α-acids and possibly a need for extended boiling times, both of which result in money wasted. The problem is even more acute when adding hop products late in the boil in order to achieve hop aroma in the final beer. Preisomerization of the α-acids by the hop processor, using more advantageous conditions for conversion, can greatly improve efficiency and significantly reduce the impact of this situation (Figure 7.21). Furthermore, if the iso-α-acids are purified by removing the other resins and oils, bittering products capable of being added directly into beer after fermentation can be produced.

Since the late 1970s, a range of preisomerized hop products suitable for adding to the kettle and dosing postfermentation into beer have been successfully developed and are now widely used by brewers.

Stabilized and Isomerized Pellets

During the 1970s, a patented process was developed to produce what are called "stabilized pellets."[112] During the production of these pellets small

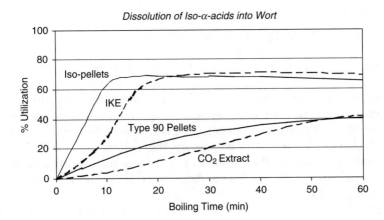

FIGURE 7.21
Comparison of the typical rate of dissolution of α-acids during boiling using isomerized and nonisomerized hop products.

amounts of a magnesium salt (normally food-grade magnesium oxide) are added to the hop powder prior to pelleting and packing. The presence of magnesium ensures that, when subsequently exposed to heat either during storage or wort boiling, the α-acids are preferentially and more rapidly converted to iso-α-acids than into other nonbitter, degradation products. During the 1980s, this idea was taken a step further by encouraging this isomerization process under controlled conditions so as to produce pellets in which the α-acids were almost completely preisomerized.[113] Although stabilized pellets are still commercially available, they have now largely been superseded by these preisomerized pellets, known most commonly as "isomerized pellets" or simply as "iso-pellets."

The production of iso-pellets is very similar to the Type 90 pelleting process. However, in order to achieve isomerization of the α-acids within the pellet during processing, two additional steps are required:

- Addition of magnesium oxide (MgO, 1–3% by weight) to the hop powder after milling, followed by adequate mixing to ensure a good dispersion of the magnesium salt throughout the powder
- Conditioning of the vacuum-packed and palletized pellets in a hot room at a temperature of 45–55°C for a period of 8–14 days

The amount of MgO powder added is varied according to:

- The α-acid content of the hops
- The particle size of the powder
- The design of the pellet press and dimensions of the die

- The grade and type of magnesium oxide
- The hop variety (and seasonal variation)

During the conditioning period within the hot room, it is essential that good air circulation be maintained around the pallet stacks in order to achieve even isomerization of the complete batch. Normally, the progress of isomerization is monitored by analyzing small, vacuum-packed, test samples placed amongst the pallets. A minimum of 95% conversion of α-acids to iso-α-acids is desirable before the pellets are allowed to cool down naturally. The yield of iso-α-acids plus residual α-acids compared to the original α-acid before isomerization is typically around 90–95% (depending on the methods of analysis used).

There is little difference in physical appearance between standard Type 90 pellets and isomerized pellets, and both products can be handled in the brewhouse in similar ways. Although during the heating process there are small changes to the oil components of the pellets, practical brewing experience has shown that normally no flavor differences can be detected in comparable beers brewed with either material.[114]

Because the α-acids in the isomerized pellets are already isomerized and present as the magnesium salt of iso-α-acids, solution into wort is both rapid and efficient. Increases in utilization of approximately 50% are typically achieved in commercial brewing plants when compared to standard hop products. Consequently, boil times can often be reduced (subject to other quality criteria being met) and late addition of isomerized pellets does not result in poor utilization.[114–116] Thus, late hop aroma can be achieved from late-added hops without expensive loss of bittering. Increases in the magnesium content of beers brewed with isomerized pellets can be shown to be typically less than 5 mg/l.

Isomerized Kettle Extracts

Similar benefits to those achieved by using isomerized pellets can be obtained by the preisomerization of the α-acids in kettle extracts. Two forms of isomerized kettle extracts are currently available to the brewer.

Magnesium-Salt Isomerized Kettle Extract (IKE)

Similar to the production of isomerized pellets, a magnesium salt may be used to catalyze the isomerization of the α-acids in CO_2 extract. In one of the several processes in use, magnesium sulfate solution is intimately mixed with warmed CO_2 extract and deionized water in a jacketed, enclosed vessel fitted with a stirrer.[117] The mixture is adjusted to pH 8.5–9.5 and the vessel contents held at 75–85°C for several hours (under a blanket of nitrogen), during which time the α-acids are isomerized to iso-α-acids. At the completion of this isomerization stage, the mixture is reduced to about pH 2 by the addition of mineral acid. This reduction in pH causes the hop resins and oils to precipitate and separate from the aqueous phase. After a period of settling, the resin and aqueous phases are run off separately and the IKE, containing free-acid form iso-α-acids, β-acids, hop oils, and some uncharacterized soft resins, is packed ready for use in the brewery.

IKE is more mobile than normal CO_2 extract and can readily be poured or pumped into the kettle. However, owing to its low pH, IKE is mildly corrosive and care must be taken when selecting materials for packing and when handling in the brewhouse.

Potassium-Form Isomerized Kettle Extract (PIKE)

Potassium-form isomerized kettle extract (PIKE), is produced from IKE by the addition of a near stoichiometric quantity of potassium hydroxide (KOH), which reacts with the free-acid form iso-α-acids in the IKE to produce their neutral salts. PIKE is slightly more viscous than IKE and has a tendency to separate out into resin and waxy phases if left to stand. Hence, when used in the brewery in a bulk-dosing system, it is necessary to ensure that the material is preheated and well mixed. However, PIKE does have the significant housekeeping benefit of being more miscible with water than either normal CO_2 extract or IKE and can be washed away easily with warm water. It also has a near-neutral pH and is therefore less aggressive and safer to handle in the brewery.

Comparison of IKE and PIKE

Both IKE and PIKE can be added into the kettle at any stage during the boil, with improvements in utilization similar to that achieved by iso-pellets (approximately 50% increase on standard hop products). Because of its better dispersion in water, trials have shown that PIKE can even be added into the whirlpool without any reduction in utilization.[94] Such late addition could be of significant help in achieving cost-effective, late-hop character in beers brewed with extracts.

Owing to the dilution effect of adding the KOH, the hop resin and oil content of PIKE will inevitably be lower than the IKE from which it is produced (Table 7.12). The amount of hop oil remaining in the isomerized kettle extracts very much depends on the process conditions employed. As with iso-pellets, some oils are lost or modified during processing; however, trial brews have shown that beers comparable with those brewed with normal CO_2 extracts can be produced with both forms of isomerized extract.[94]

Iso-Extract (Postfermentation Bittering)

Iso-α-acids are more soluble than α-acids and, hence, in a purified form, can be added successfully into beer with good utilization (often up to 90%).

TABLE 7.12

Compositions of CO_2 Extract and IKE and PIKE Produced from that Extract

Parameter	CO_2 Extract	IKE	PIKE
α-Acids (%)	55.9	0.1	0.2
Iso-α-acids (%)	Nil	56.8	47.4
β-Acids (%)	17.6	16.4	13.7
Total oil (ml/100 g)	9.0	8.6	8.0
pH	3.0–3.5	2.0–2.5	7.0–8.0
Density (g/ml)	1.0	1.0	0.9

Source: From Wilson, R.J.H., Roberts, T., Smith, R.J., and Biendl, M., *Tech. Q. Master Brew. Assoc. Am.*, 38:11–21, 2001.

"Iso-extract," also known as Postfermentation Bittering or PFB, is normally dosed into rough beer before filtration.

Several ways of producing iso-extract have been described over the years, the more recent achieving efficient isomerization of α-acids by using added magnesium salts under alkaline pH conditions and at elevated temperatures. The starting material is invariably CO_2 extract, and the separation of β-acids and hop oils can either take place before or after isomerization. Partition is achieved by pH adjustment and relies on the fact that the iso-α-acids are more acidic than the other hop components. Typically, the separation involves at least two steps such that the amount of residual β-acids is very low (Table 7.13). The end product is a solution of the potassium salts of iso-α-acids, adjusted to a concentration of either 20% or 30% (of actual iso-α-acids), with deionized water prior to sale. The β-acids and hop oils collected during the processing can be used to produce a nonbitter, aroma extract that can be added to the kettle (see the section "Beta Extract").

Iso-extract only provides bitterness and is rarely used as the sole source of bittering. However, it can be used to provide a fixed percentage (say

TABLE 7.13

Typical Analysis of Iso-extract

Typical Isomerized Extract	
Parameter	Value
Iso-α-acids	30.0%
α-Acids	0.2%
β-Acids	<0.1%
Hop oils	<0.1%
pH (as is)	9.0
Density (at 20°C)	1.065 g/ml

30%) of the total bitterness or, most commonly, for the adjustment of bitterness to within specification prior to packing. With the aid of rigorous injection, iso-extract can be added at sales strength directly into beer, but more normally, it is diluted to 1–5% with deionized, preferably deaerated, and slightly alkaline water before addition. In order to achieve optimum utilization, iso-extract must be accurately metered in over at least 70% of the beer flow.

Reduced Isomerized Hop Products

Whereas isomerized products present significant cost-saving opportunities, the range of reduced isomerized extracts variously offers two important quality benefits — light stability and foam enhancement. As indicated in the section "Hop Chemistry", the term "reduced" isomerized extract describes commercially available products containing (sometimes as mixtures): rho-iso-α-acids, tetrahydroiso-α-acids, or hexahydroiso-α-acids.

Rho-*iso*-α-*acid* (Rho)

Production of Rho

Rho-iso-α-acids or Rho can be produced in either of two ways:

- The reduction of purified iso-α-acids in the presence of sodium borohydride
- The simultaneous isomerization and reduction of α-acids, also using sodium borohydride

Both processes are carried out at elevated temperatures and at alkaline pH value. In each case, after the reduction stage is complete, the residual boron is removed by reducing the pH with mineral acid, resulting in a separation into two distinct phases — a resinous, "free acid" rho phase and an aqueous phase containing the boron residues. Once separated, the resin phase is washed with deionized water to remove virtually all traces of boron. Potassium hydroxide is then added to the rho-iso-α-acids, producing a soluble, potassium salt form product that can be standardized to the desired concentration (typically 35% by spectrophotometric analysis) using deionized water.

Quality Characteristics of Rho

The key quality characteristics of Rho products are as follows:

- Beers brewed with Rho products can be used to avoid the development of "skunky" flavors in beers filled into clear or green glass bottles.

- Organoleptic comparisons show that *rho*-iso-α-acids in beer give a less bitter taste than unreduced iso-α-acids. When using *rho*-iso-α-acids, it is usual to assume that perceived (tasted) bitterness is only 0.7–0.8 times that of the perceived bitterness of normal iso-α-acids. Rho is therefore often described as giving a "softer" bitter flavor than that achieved by normal hopping.
- Although *rho*-iso-α-acids themselves are not normally considered to have any significant impact on foam stability (over and above that of normal iso-α-acids), there is some practical evidence to suggest that small improvements in foam are achieved when using Rho products, presumably due to the need to add a greater quantity to the beer in order to achieve the same level of perceived bitterness.

Using Rho

Rho, at its sales strength of 35%, can be either added to the kettle or injected into a flow of beer after fermentation. Dilution of *Rho*-35% is not recommended, as it typically produces a milky suspension that can cause problems with dosing equipment owing to fouling with sticky deposits. Also, the product itself can be physically unstable and may need warming and agitating to get any precipitate back into solution prior to use. Alternatively, a more highly purified, essentially nonprecipitating 10% product is available, which can be dosed directly into a beer stream. In the case of kettle addition, more concentrated, resinous *Rho* products are now available (some containing purified, *rho*-iso-α-acids in their potassium salt form for better stability and dispersion into the wort). These convenient products are physically similar to CO_2 extracts and may be handled in the same way.

A further alternative, specifically for use in the kettle, is "light stable kettle extract" (LSKE) or "beta aroma *Rho* extract" (*BARho*), which contains β-acids, hop oils, and iso-α-acids in the *Rho* form.[94] This type of product is particularly suitable for use in cases where full hop flavor as well as light stability are required, or where an existing beer (possibly currently produced using normal CO_2 extract) needs to be flavor-matched but made light stable.

Tetrahydroiso-α-Acid ("Tetra")

Tetrahydroiso-α-acids can be produced by a number of differing patented processes using either α- or β-acids as the starting material.

Production of α-*Acids-based Tetra*

Typically, pure iso-α-acids in aqueous or solvent solution are hydrogenated using gaseous hydrogen in the presence of a palladium/carbon catalyst to produce tetrahydroiso-α-acids (though α-acids can also be used, in which case, the tetrahydro-α-acids produced are subsequently isomerized). In each case, the Tetra is separated from the catalyst by filtration before further processing. The end product is an aqueous, slightly alkaline solution of the tetrahydroiso-α-acids in potassium salt form, usually adjusted to a sales strength of 10%.

Production of β-Acids-based Tetra

The rationale for using β-acids as the starting material is simply an economic one: β-acids are a by-product of the manufacture of iso-α-acids and have in the past had little commercial value. Therefore, it makes sense for a manufacturer of hop products to take advantage of the availability of cheap β-acids, though this has to be weighed against the greater processing cost and lower yield of the process. The production of tetrahydroiso-α-acids from β-acids is a relatively complex process involving three steps — hydrogenation using a palladium/carbon catalyst, oxidation of the hydrogenated product (tetrahydrodeoxy-α-acids) by peracetic acid or air and, finally, isomerization (of tetrahydro-α-acids) to produce Tetra (Figure 7.22).

Although the α- and β-acids based Tetra products have broadly similar characteristics in beer, as explained elsewhere, there are, nevertheless, subtle differences between them, notably in respect of foam stabilizing activity (see the section "Relative Foam Stabilization — Importance of Cohumulone Ratio") and probably also in bittering power (see the section "Role of Iso-α-Acids").

Quality Characteristics of Tetra

The key quality characteristics of Tetra products are as follows:

- They may be used to produce light stable beers when used on their own or in combination with other reduced hop products.
- They may be added to beer in small quantities to improve foam appearance and stability. Very noticeable improvements in foam performance (as measured analytically by NIBEM meter and by visual comparison) can be achieved by as little as 3–4 mg/l present in final beer.[118] However, beers that have relatively high levels of Tetra are often considered to have unnatural, "creamy" foam.
- When Tetra was first introduced, it was considered to have a much more bitter taste than normal iso-α-acids.[105] However, since then, many brewers have concluded that the difference is not as great as first thought and that it depends on a number of factors, such as beer type and bitterness levels. There does seem to be fairly strong evidence to suggest that in water tetrahydroiso-α-acids have a perceived bitterness of around 1.7 × more than normal iso-α-acid, but that in beer this figure can range between 1.1 and 1.7 ×.[119, 120] Brewers should arrive at a figure to suit their own

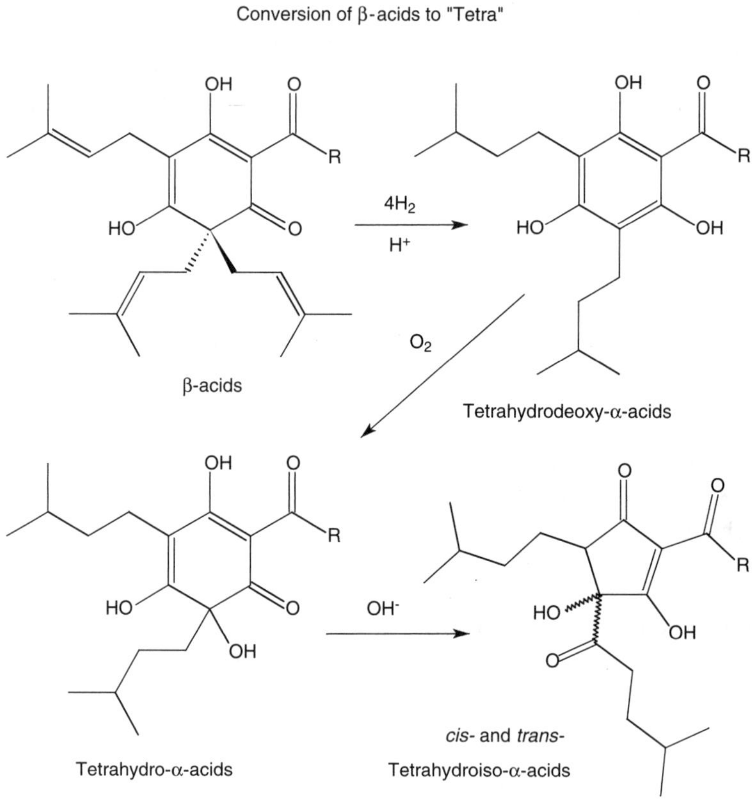

FIGURE 7.22
Conversion of β-acids to tetrahydroiso-α-acids. Tetrahydroiso-α-acids may be prepared from β-acids by a process of catalytic hydrogenation and hydrogenolysis, followed by oxidation and alkaline isomerization. The end product will have a higher cohumulone ratio than "Tetra" prepared by isomerization and hydrogenation of α-acids obtained from the same hop variety.

particular application by comparative flavor assessment. Perhaps because of this potentially more intense bitterness, Tetra is often described as giving a "harsher" flavor than normal iso-α-acids.

Using Tetra

Tetra-10% is normally injected into a beer main during beer transfer. It can be diluted using alkaline, pH-adjusted, deionized water to a concentration of 2–5% prior to addition or, if the dosing equipment is capable of inducing a vigorous injection, it can be added directly into beer at sales strength. Occasionally, Tetra is added directly into the kettle. Whereas this avoids the need for expensive dosing equipment, it is highly inefficient in terms of utilization and is therefore best avoided. However, in the absence of post-fermentation dosing equipment, a "tetra concentrate" product in which the

Tetra is in the potassium salt form is available for kettle addition. Having a Tetra concentration of between 60% and 70%, this product can be used in similar ways to other kettle extracts, and its use avoids the need to ship, store, and handle large quantities of water. The same product is also proposed for postfermentation use, though this may require the installation of special dosing equipment.[121]

Hexahydroiso-α-acid ("Hexa")

Production of Hexa

During the production of Hexa, hydrogen atoms are introduced into the iso-α-acids molecules by a combination of reduction using sodium borohydride and catalytic hydrogenation using a palladium on carbon catalyst. The starting point for the commercial production of Hexa may be either *rho*-iso-α-acids, subsequently hydrogenated by catalytic hydrogenation or, alternatively, tetrahydroiso-α-acids that are then subjected to borohydride reduction.

Quality Characteristics of Hexa

The key quality characteristics of Hexa products are as follows:

- Hexa provides both light stability and excellent foam enhancement. However, beers produced using significant quantities of Hexa demonstrate very "stiff" foam, which can be described as unnatural or atypical when compared to normal beer foam. For this reason, Hexa is not suitable as the sole source of bitterness (with the possible exception of very low BU beers), but rather should be used in combination with other isomerized and reduced hop products.
- Hexa is considered to provide a slightly more bitter flavor relative to unreduced iso-α-acids ($1.0-1.2\times$).[105] However, its "softer" palate makes it an excellent material to use in combination with the so-called "harsher" flavored Tetra.

Using Hexa

Because of its relatively poor solubility, Hexa is normally sold as a 10% aqueous solution of the potassium salts form or as a 20% solution in aqueous propylene glycol or ethanol. As with other reduced iso-products, Hexa is normally diluted with deionized water prior to addition into beer. However, it can be added at sales strength provided that suitable dosing systems are used.

Comparison of Reduced Products

Table 7.14 shows a comparison of the key characteristics of all three reduced iso-α-acid products.

TABLE 7.14
Comparison of the Characteristics of Reduced Iso-α-acid Products

Parameter	Rho	Tetra	Hexa
Normal sales strength	10%, 35% or Concentrate	10% or Concentrate	5%, 10% or 20%
Relative solubility in beer	Reasonable	Poor	Very Poor
Perceived bitterness (Relative to iso-α)	0.6–0.8×	1.1–1.7×	1.0–1.3×
Light stability	Good	Very good	Very good
Foam enhancement (Relative to iso-α at equivalent perceived bitterness)	Slight	Good	Excellent

Relative Foam Stabilization — Importance of Cohumulone Ratio

The question arises as to whether or not Tetra prepared from β-acids is fundamentally different from that made by the simpler route from α-acids. Insofar as the product is highly unlikely to contain any measurable amount of iso-α-acids, it may be said to be absolutely safe to use for the purpose of brewing a light-stable beer. However, there is a possibility that the foam-stabilizing ability of the β-acids-based product might be noticeably inferior to that prepared from α-acids obtained from the same hops. The reason for this lies with the cohumulone ratio. As explained earlier, the cohumulone form of any hop resin acid series is more polar and, hence, less inclined to migrate into the foam than are the corresponding humulone and adhumulone homologs. Also, it was noted (in the section "Soft Resins") that for any particular hop the proportion of colupulone as a fraction of the total β-acids is always greater than the proportion of cohumulone as a fraction of the α-acids. Hence, Tetra prepared from β-acids will always be found to have a high proportion of tetrahydroisocohumulone, often exceeding 60%, whereas α-acids-based Tetra normally has no more than 45% of the total tetrahydroiso-α-acids in the form of this isomer and may be as low as 30%. All else being equal, the potential effect of this difference can be inferred from a study by Wilson and coworkers of the relative foam-enhancing ability of pure compounds, in which five mainstream American commercial lager beers, each having inherently rather low content of iso-α-acids by world standards, were spiked with 5 ppm of the purified hop compounds.[101] The improvement in foam stability (shown here as the average percent increase) was measured and a clear relationship established:

$$\text{TetrahydroisoH (42\%)} > \text{TetrahydroisoCoH (26\%)} > \text{IsoH (10\%)} > \text{IsoCoH (3\%)}$$

In practice, the relative foam-stabilizing abilities of two Tetra products having markedly differing cohumulone ratios are unlikely to be more than fractionally different; especially perhaps because the recovery of added tetrahydroisocohumulone is anyway likely to be higher than that of the tetrahydroisohumulone and tetrahydroisoadhumulone.

Other Foam Stabilizers

As already noted, the quality of the foam produced by tetrahydro- and hexahydroiso-α-acids is sometimes considered as having an unnatural appearance. A possible alternative would be a product based on the dihydroiso-α-acids that are the intermediates in the formation of Tetra from iso-α-acids. Such a product would be expected to have properties that are intermediate between commercial preparations of iso-α-acids and tetrahydroiso-α-acids, but would not offer protection against the lightstruck reaction. That dihydroiso-α-acids are good foam promoters has indeed been demonstrated and experiments suggest that it would produce foam having particularly good lacing properties.[13,83,101]

The production of a hop-derived foam stabilizer that is nonbitter — and therefore can be added independently of the normal bittering — has also been the subject of investigation by researchers. Hop pectins (see the section "Pectins") have been demonstrated to have possibilities in this respect, but no pectin-based commercial products are currently available.

Other Products

Type 100 Pellets

In order to achieve a characteristic "hoppy" aroma in cask-conditioned beer (that is mainly produced in the United Kingdom), compressed leaf hops are introduced into the casks immediately before filling. These compressed hops are referred to as "hop plugs," "whole hop" pellets, or "Type 100" pellets and are produced by breaking up baled hops and compressing weighed amounts into a single, very large pellet. The hop plugs (either 0.5 or 1 oz — 14.2 and 28.3 g, respectively) are then vacuum-packed in laminate bags in order to reduce the loss of essential oils.

Beta Extract

"Beta extract" (also referred to as "beta aroma extract" or "Base Extract") is a mixture of β-acids and hop oils, together with uncharacterized soft resins. It is a coproduct of the manufacture of purified iso-α-acids (iso-extract). Beta extract may be used as a means of introducing largely nonbitter hop material into the kettle in order to increase hop character and flavor. A light-stable form, in which nearly all of the residual α-acids and iso-α-acids have been removed, is often used as a kettle addition in conjunction with reduced hop products added postfermentation.

Oil-Enriched Extracts

Produced from normal CO_2 extract, these enriched extracts can be added to unfiltered beer to enhance hop character and were initially developed to replace dry hops in processed beers.[122] They are less viscous than normal extracts and, generally, contain hop oil in the range of 10–30% (usually according to customer specification). Oil-enriched extracts can be produced by any of the following three methods:

- Addition of (typically) molecularly distilled hop oil to pure resin extract
- Fractionation of pure resin extract with supercritical CO_2
- Partial extraction of hops with liquid CO_2

In order to achieve solution of hop oil into beer, dispersion of the extract is normally achieved with the help of a carrier or solvent. Options available include redissolving the extract into liquid CO_2 before injection into beer,[123,124] dispersion of the extract into a food grade emulsifier,[122] or dissolving in alcohol (in which case the solution should be used within a few hours to avoid possible development of undesirable flavors[122]).

Pure and Fractionated Hop Oils and Hop Essences

Pure hop oil products are normally produced from pure resin extracts, either by hydrodistillation (commonly referred to as steam distillation), low-temperature vacuum distillation, or molecular distillation. These products can be further fractionated into separate, highly flavor-active elements, which are described as producing "spicy," "estery," "herbal," "floral," or "citrussy" aromas when added to beer.[125,126] Such products are normally sold as "hop essences," being dilute (e.g., 1%), ethanolic solutions that may be added directly to bright beer. Pure hop oil products themselves are usually dissolved in ethanol or premixed (generally by the hop processor) with a food grade emulsifier before making the addition — which is preferably done before final filtration to remove any haze that may be formed by partial precipitation of the terpene content.

Attempts to duplicate the subtlety of late-hop character by preparation of fractionated hop oils have had limited success, at least in part due to the influence of yeast, which can modify certain compounds by enzymic means, to some extent by creating ethyl esters via transesterification reactions and also by enzyme-mediated hydrolysis.[24,127] Clearly, it follows that selection of yeast strain and fermentation conditions may often play important roles in determining hop character in beer. That said, hop-derived aroma in beer is, however, notoriously unstable, especially in bottles, where some components may be slowly absorbed by the plastic liner of the crown cork.[128]

Hops

TABLE 7.15

Typical "Best Before Dates" and Recommended Storage Temperatures of Hop Products

Hop Product	Best Before Date (Years from Production Date)	Storage Temperature (°C)
Raw hops	N/A	<5
Pellets — Type 90, Type 45, Stabilized	2	<5
Isomerized pellets	3	<10
CO_2 and ethanol extracts	4	<10
IKE, PIKE	1	<5
Beta aroma extract	2	<10
Iso-extract 30%	2	5–15
Light stable kettle extract	2	<10
Rho — 10%	1	10–25
Rho — 35%	2	10–25
Tetra — 10%	1	10–15
Tetra and *Rho* concentrates	2	<10
Steam distilled hop oils	1	<10

Storage Stability

Although processing of raw hops into hop products helps to preserve their important constituents, it is desirable that products are:

- Stored according to supplier's recommendations
- Used quickly after opening the package
- Used before the recommended "best before date"

However, it is also worth pointing out that hop products that are past their "best before date" may still be suitable for use, albeit they might provide a slightly altered brewing value. Recommended "best before dates" and temperature of storage may vary from supplier to supplier; however, typical values are shown in Table 7.15.

Other Uses — New Products

In recent years, there has been considerable focus on the potentially beneficial properties of hop resins and hop polyphenols (particularly the flavonoids) when used as antibacterial agents and as antioxidants, respectively, for applications within the food and beverage industries. This work has resulted in the introduction, both on the commercial and pilot scale, of a range of new hop products aimed both for the brewing and nonbrewing markets, such as standardized β-acid products, a "xanthohumol-enriched extract," and purified xanthohumol. The xanthohumol-enriched product is commercially available and has been used to produce a beer of acceptable flavor, containing up to 6–7 ppm of isoxanthohumol (W. Mitter, personal communication).

Much research and development effort continues to be expended on the exploration of these opportunities and several products and processes have been patented. The pharmaceutical value of certain hop compounds is also beginning to be manifested, and it is to be hoped that this interest too will continue, be turned into much-needed commercial reality, and become significant to the future viability of the world hop industry.

Hop Usage

Choice of Hop Product

The range of hop products is both varied and extensive, presenting the brewer with the opportunity to produce a diverse selection of beer types and flavors. The selection of hop variety and product depends on the following criteria:

- Bitterness levels
- Hop flavor and aroma
- Flavor stability
- Foam stability

Other practical considerations include:

- Design and specification of brewing plant
- Availability and quality of labor
- Geographical location (ambient temperatures, supply chain costs, etc.)
- Cost of bittering

Hops can be added as a single product and single addition, or as multiple products added at different times within the process. In the latter case, the calculation of quantities of each product required can be complicated and requires a detailed knowledge of products, plant, and addition efficiencies. In practice, such variable factors as utilization and perceived bitterness of the reduced hop products will need to be initially established by trial and error within individual plants.

Hops

Hop Utilization

The utilization of α-acids (normally referred to as hop utilization) is calculated as follows:

$$\% \text{ Utilization} = \frac{\text{Weight of iso-}\alpha\text{-acids in the wort or final beer}}{\text{Weight of }\alpha\text{-acids added to the wort or beer}} \times 100$$

Care must be taken to ensure that the methods of measuring α-acids and iso-α-acids are clearly specified, as the result can be significantly affected by the choice of analysis methods. In principle, the most accurate calculation of the utilization will be achieved by using HPLC analysis to measure both α- and iso-α-acids in both product and beer (see also the section "General" under "Hop Resin Analysis").

Calculation of Hop Additions

The following three examples show very simply how to calculate the quantities of hop products to be added in order to achieve the desired level of final bitterness. It must be remembered though that, as shown, an adjustment has to be made for the variable "perceived" bitterness of the reduced, isomerized downstream hop products, but that this adjustment may vary according to beer type and flavor.

Example 1 — Single Product Added into the Kettle

Assumptions:

Brew volume	100 hl
Target bitterness	20 mg/l iso-α-acids in final beer
Hop product	Type 90 aroma pellets at 5% α-acids
Time of addition	At start of the kettle boil
Utilization of α-acids	30%.

Calculation:

(a) Total quantity of iso-α-acids required in 100 hl:

$$20 \, (\text{mg/l}) \times 100 = 2000 \, \text{mg/hl}$$
$$2000 \, (\text{mg/hl}) \times 100 \, (\text{hl}) = 200{,}000 \, \text{mg} = 0.20 \, \text{kg iso-}\alpha\text{-acids}$$

(b) Adjustment for utilization:

$$0.20 \times 100/30 = 0.67 \, \text{kg } \alpha\text{-acids}$$

(c) Adjustment for α-acids content of aroma pellets (5%):

$$0.67 \times 100/5 = 13.4 \text{ kg of aroma pellets.}$$

Example 2 — Two Products, Both Added into the Kettle

Assumptions:

Brew volume	100 hl
Target bitterness	20 mg/l iso-α-acids in final beer
Hop products	Type 90 alpha pellets at 10% α-acids
	Type 90 aroma pellets at 5% α-acids
Alpha proportions	Alpha pellets — 70% of total bitterness
	Aroma pellets — 30% of total bitterness
Time of addition	Alpha pellets at start of the kettle boil
	Aroma pellets 15 min before end of boil
Utilization of α-acids	Pellets at start of boil — 30%
	Pellets added late — 15%

Calculation:

1. Alpha Hop Addition:
    (a) Total quantity of iso-α-acids required in 100 hl:

    $$20 \text{ (mg/l)} \times 100 = 2000 \text{ mg/hl}$$
    $$2000 \times 100 \text{ (hl)} = 200{,}000 \text{ mg} = 0.20 \text{ kg iso-α-acids}$$

    (b) Total iso-α-acids from alpha pellets:

    $$\frac{0.20 \times 70}{100} = 0.14 \text{ kg iso-α-acids}$$

    (c) Adjustment for utilization:

    $$\frac{0.14 \times 100}{30} = 0.47 \text{ kg α-acids}$$

    (d) Adjustment for α-acid content of alpha pellets (10%):

    $$\frac{0.47 \times 100}{10} = 4.7 \text{ kg of alpha hop pellets}$$

2. Aroma Hop Addition:
    (a) Total quantity of iso-α-acids required in 100 hl:

    $$20 \text{ (mg/l)} \times 100 = 2000 \text{ mg/hl}$$
    $$2000 \times 100 \text{ (hl)} = 200{,}000 \text{ mg} = 0.20 \text{ kg iso-α-acids}$$

(b) Total iso-α-acids from aroma pellets:

$$\frac{0.20\,(\text{kg}) \times 30}{100} = 0.06\,\text{kg iso-}\alpha\text{-acids}$$

(c) Adjustment for utilization:

$$\frac{0.06 \times 100}{15} = 0.40\,\text{kg }\alpha\text{-acids}$$

(d) Adjustment for α-acids content of aroma pellets (5%):

$$\frac{0.40 \times 100}{5} = 8.0\,\text{kg of aroma hop pellets}$$

Example 3 — Single Kettle Addition with Tetra Added Postfermentation

Assumptions:

Brew volume	100 hl
Target bitterness	Equivalent perceived bitterness — 20 mg/l iso-α-acids in final beer
Hop products	CO_2 (kettle) extract — 30% α-acids Tetra-iso-α extract — 10% tetrahydroiso-α-acids
Alpha proportions	CO_2 extract — 85% of total perceived bitterness Tetra — 15% of total perceived bitterness
Time of addition	CO_2 extract — at start of the kettle boil Tetra — added in-line before filter
Utilization of α-acids	CO_2 extract at start of boil — 35% Tetra — postfermentation addition — 80%
Perceived bitterness	Tetra — 1.6 × normal iso-α-acids.

Calculation:

1. Alpha Hop Addition:

    (a) Total quantity of iso-α-acids required in 100 hl:

    $$20\,(\text{mg/l}) \times 100 = 2000\,\text{mg/hl}$$
    $$2000 \times 100\,(\text{hl}) = 200{,}000\,\text{mg} = 0.20\,\text{kg iso-}\alpha\text{-acids}$$

    (b) Total iso-α-acids from CO_2 extract:

    $$\frac{0.20\,(\text{kg}) \times 85}{100} = 0.17\,\text{kg iso-}\alpha\text{-acids}$$

(c) Adjustment for utilization:
$$\frac{0.17\,(\text{kg}) \times 100}{35} = 0.49\,\text{kg}\,\alpha\text{-acids}$$

(d) Adjustment for alpha content of CO_2 extract (30%):
$$\frac{0.49 \times 100}{30} = 1.6\,\text{kg of kettle extract}$$

2. Tetra Addition:
   (a) Total quantity of iso-α-acids required in 100 hl:
   $$20\,(\text{mg/l}) \times 100 = 2000\,\text{mg/hl}$$
   $$2000 \times 100\,(\text{hl}) = 200{,}000\,\text{mg} = 0.20\,\text{kg iso-}\alpha\text{-acids}$$

   (b) Total iso-α-acids from Tetra:
   $$\frac{0.20\,(\text{kg}) \times 15}{100} = 0.03\,\text{kg tetrahydroiso-}\alpha\text{-acids}$$

   (c) Adjustment for perceived (tasted) bitterness:
   $$\frac{0.03}{1.6} = 0.019\,\text{kg tetrahydroiso-}\alpha\text{-acids}$$

   (d) Adjustment for utilization:
   $$\frac{0.019 \times 100}{80} = 0.024\,\text{kg tetrahydroiso-}\alpha\text{-acids}$$

   (e) Adjustment for tetrahydroiso-α-acid content of Tetra product (10%):
   $$\frac{0.024 \times 100}{10} = 0.24\,\text{kg of Tetra-10\%}$$

Cost of Bittering

The cost of bittering using different hop products depends on the following:
- Cost of α-acids of source hops — mainly affected by variety and demand
- Processing charge — usually quoted either on a weight of hops or weight of α-acids processed basis
- Processing yield — usually quoted as a guaranteed minimum based on either weight or quantity of α-acids in finished product
- Utilization within brewery — variable depending on the type of product and the point of addition

- Methods of analyses used for product and beer
- The perceived relative bitterness of reduced isomerized products

It is difficult to make precise comparisons of the relative bittering costs of hop products because of the variation in the factors shown above. However, by making some fairly broad assumptions, a rough comparison of the cost of hop products in terms of the cost per kilogram iso-α-acid in the final beer is shown in Figure 7.23. All costs are compared to leaf (baled) hops based at 100%.

These figures are intended only as a guide, and brewers should make their own assessments based on actual data relevant to the type and price of hops used, together with the operating conditions within their plants, especially insofar as these determine the utilization of each product. Some other factors that can have significant impact on both the absolute and relative costs of bittering using hop products are:

- Reduced labor costs
- Savings in freight, storage, and handling
- Reduced losses and effluent charges
- Switching to high gravity brewing
- Selection of point of addition
- The stability of the product
- Capital costs of handling and dosing equipment

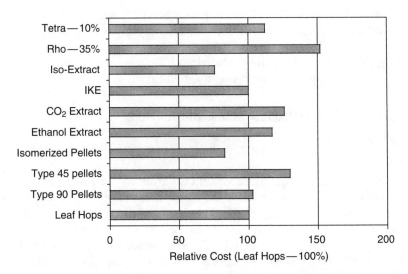

FIGURE 7.23
Comparison of the relative costs per kilogram of iso-α-acids in beer when using different hop products.

Incorporating these factors into the calculation can change the overall picture considerably.

Hop Analysis

General

Over the course of the last century, a variety of methods have been developed for the analysis of hops and hop products, primarily, of course, with the objective of obtaining a useful measure of the bittering power of hops. Many of these methods have been subjected to collaborative testing and subsequent approval by brewing organizations, and are published as recommended methods. These methods, many of which have been adopted as international methods, should generally be preferred for use by the brewer who is not thoroughly familiar with the subject, though it is true to say that improved methods are often to be found in the literature, and also that there are specific areas that are poorly or not at all covered. The most comprehensive and widely used analytical compilations are:

- ASBC *Methods of Analysis*
- *Analytica-EBC*
- *Methods of Analysis* of the Institute of Brewing (now the Institute of Brewing & Distilling)
- *MEBAK Methods* (*Brautechnische Analysenmethoden* compiled by the Mitteleuropäischen Brautechnischen Analysenkommission)

These guides are periodically updated and the user should always ensure that he or she is equipped with the latest edition.

Visual, Physical, and Olfactory Examination of Hops

Hand Evaluation

General

For many brewers, comparative visual and aromatic examination of different lots of dried and baled hops is the single most important factor governing their choice of lots to purchase. This is especially true for purchasers of aroma hops intended for late kettle or dry-hopping use, whether as loose hops or in the form of hop pellets. An experienced buyer will be able to make a qualitative assessment of many features that may impact on hop quality and ultimately beer quality, such as the existence of infection by downy or powdery mildew, or overdrying, though it is also important to recognize that hops having visible or even aroma defects will not necessarily produce tainted beer. Much depends on the cause of the apparent damage; for example, browning of the hop bracts may cause a particular lot to have a poor appearance, yet if this is simply the result of windburn, then the brewing quality may well be perfectly satisfactory. It is also important to note that every variety has its own particular characteristics in terms of such features as typical cone size and shape, color of lupulin, liability to browning and shatter, and so forth, so that judgment of relative quality must take some account of varietal and even geographical features. Clearly, assessing hops is a task for which there is no substitute for experience, though certain principles can be set down.

Proper examination, first of all, requires the right environment. Hop dealers will normally provide brewers with the opportunity to examine samples under diffuse, bright fluorescent lighting of a type that gives near-daylight spectral emission. The room will be quiet, gently ventilated with fresh, odor-free air, and equipped with tables that have been painted matt black. "Brewer's cut," rectangular samples, typically measuring about $15 \times 10 \times 10$ cm and wrapped in a square of stiff, light-proof paper held together by a pin, will be carefully opened and set out for examination. These samples are cut from the sides of representative bales or pockets using a sharp knife and a special pair of hinged tongs, which grips the sample on the pressed sides of the cones and enables the sample to be withdrawn without breaking apart. For aroma hops, perhaps one in ten bales will be sampled in this way, whereas for alpha hops, it is more typical to sample one in 30 or so. A particular lot of hops may number 200 or more bales, so it is important for the buyer to establish that the samples from a single lot are consistent, as it is possible for there to be substantial variations if the farmer does not do a good job of blending hops that are picked from different yards and dried in different kilns.

Preliminary Physical and Visual Examination of Brewer's Cut

The first examination that the buyer should make is to inspect the surfaces of the sample. The block should be gently pressed on the uncut side, where the cones have been pressed by the baling machine. A correctly dried and pressed sample should exhibit a noticeable degree of springiness. A hard sample may have been overpressed, causing rupture of some of the lupulin glands, or it may have a relatively high moisture content. Next, an examination should be made of a side where the cones have been cut across by the knife, exposing the lupulin. The color and general appearance of the lupulin are particularly important. Fresh lupulin is pale yellow to light orange in color, though somewhat dependent on variety (and to some extent, also on agronomic or geographical factors), hence varietal differences in the color must be taken into account. But a relatively dark, orange-colored lupulin likely indicates drying at too high a temperature, in which case the hops may be relatively unstable in storage and have a lower than normal oil content. Use of a hand lens will help to determine whether the lupulin glands have suffered damage through overpressing, directly exposing the hop oils and resins to the air and microorganisms, and thereby accelerating oxidative deterioration.[81,82,129] Of course, the cut will also expose any seeds, so that excessive seed content will be readily apparent.

Further information may now be obtained by breaking open the sample and breaking off some of the pressed hops. If the cones shatter easily, then the hops may have been harvested rather late, although if the sample already contains a lot of broken cones, then the cause may be more related to overdrying or rough handling, especially after leaving the kiln. Although distinctly varietal dependent, as a class, low-alpha hops are most likely to show this phenomenon, as higher levels of the sticky lupulin helps to hold the bracts and bracteoles in place. However, though generally considered a defect, shattering in itself may not actually indicate poor brewing quality, as oil and alpha contents usually peak late in the season. The break may also reveal evidence of excessive "leaf and stem," material that contributes nothing to the brew and indicates a poorly designed or operated picking machine, suggesting the possible presence of other extraneous material, such as mud and stones.

Secondary Inspection for Quality Defects

The best quality hops will exhibit only green, varietal-dependent coloration, though, for some varieties, it would be normal that the petals are not all of uniform color. In times past, the color was often improved and to some extent stabilized by the deliberate burning of elemental sulfur in the kiln, thereby exposing the hops to the preservative effects of sulfur dioxide. However, this practice has now been discontinued or actually banned in most countries. Hop samples will commonly show a certain degree of browning, and it is important to determine the likely cause. Browning that is superficial is most likely windburn or symptomatic of stress due to use of chemical sprays and probably does not influence the brewing quality. However, browning may also be an indicator of damage caused by mildew or red spider mites, the latter causing whole cones to turn brown or reddish-brown. Since the mites are very small, their presence is more easily inferred from the damage done than by finding their physical remains.

By breaking open the cones, it is possible to obtain further information. For example, if there has been serious infestation by aphids, then it is likely that there will be black areas inside the cones, which are due to fungal growth on the sugar-rich honeydew exuded by the aphids. The experienced buyer will even be able to gain an approximation of the moisture content by twisting the strig, as the degree of brittleness is related to the moisture content.

Assessing Aromatic Quality

Buyers normally place great emphasis on the aromatic qualities of the hops, especially those destined for use as the source of hoppy aroma in their beer. The assessment of aroma is something that is even more dependent upon experience than that of the visual examination, though it has to be pointed out that the aroma of the hops is not necessarily directly related to that of the beer itself. The primary objective of the buyer is, therefore, to establish whether the hops are true to type, have no aromatic defects, and are of good intensity, as only through brewing trials will it be reliably established whether or not a particular variety will produce the desired flavor in the beer.

Descriptors for hop aroma are generally considered as falling into two, simple categories: desirable and undesirable. However, what constitutes desirable is necessarily modified by the buyer's perception of the ideal hop of the variety he is seeking to purchase. A typical set of primary descriptors used in the assessment would be:

- Desirable: Citrus, Spicy, Piney, Grapefruit, Floral, Herbal, Estery, Resinous
- Undesirable: Musty, Cheesy, Smoky, Hay-like, Earthy, Solvent-like, Rubbery

Before starting the aromatic examination, it is obviously important to ensure that the hands are free from interfering odors that may arise, for example, as a consequence of cleaning with a perfumed or antibacterial soap or hand cleanser. The desirable aromatic compounds of hops reside in the lupulin glands, so it is often possible to detect some off-notes by smelling the whole hops first, before releasing the aroma by rubbing the sample between the palms of the hands. This action breaks open the lupulin, exposing the oils and resins. The aroma may then be assessed by inhaling with cupped hands held close to the nose and a note made of the descriptors that are detected. Trueness to type may also be noted. Some assessors prefer to split the hand rub into two parts, initially assessing the aroma released from a light rub, then making a more vigorous rub. This may make it possible to first detect subtle aromas that are otherwise overwhelmed.[130]

According to individual preference, the experienced buyer may make assessments of offered lots of hops purely by visual and aromatic comparisons, and without the aid of pencil and paper, sorting out the good from the bad, and narrowing down the choices by going back to the samples more than once. Others may prefer to make at least their primary assessments on the basis of scoring various qualities. Since the assessment is necessarily somewhat subjective, there is no correct approach, but the wise buyer would certainly not rely wholly upon hand evaluation, as much valuable information may also be obtained through chemical analysis.

Hop Leaf and Stem

Hops that have been badly picked will contain a significant amount of dried leaves and pieces of hop bine. Quantification of the amount may be accomplished by following the methods described in the ASBC Methods of Analysis, Analytica-EBC or the Methods of the Institute of Brewing & Distilling. Excessive leaf and stem normally leads to a downgrading of quality and price, even though the quality of the hop cones themselves may be entirely satisfactory.

Aphids

ASBC Methods of Analysis describes a technique for assessing the amount of aphids in hops. However, aphids added into the brewing process are in themselves entirely inconsequential to product quality, unless the possibility exists that they may appear in unfiltered, dry-hopped beer. Hence, of more importance than the physical presence of the remains of these pests is the associated damage that is likely due to the growth of molds.

Hop Resin Analysis

General

Methods for the analysis of hop resins in hops and hop products fall mostly into one of the following categories:

- Classification into "hard," "soft," and other resin fractions via solvent dissolution
- Polarographic methods (for α-acids)
- Lead conductometric value (LCV or CLV) methods (for α-acids)
- Spectrophotometric methods (for α-acids, β-acids, and hop storage index)
- HPLC methods (for α-acids, β-acids, and iso-α-acids)
- Capillary electrophoresis (for α-acids and β-acids)
- Near infrared (NIR) methods (for moisture, α-acids, β-acids, and hop oil)

With the exception of NIR-based techniques, all these methods can be considered to have two prime components:

- Extraction into solvent
- Analysis of concentration in the solvent

It is important to note that many methods do not fully extract the hop resin acids, hence, even though the analysis of the concentration in the solvent may be absolutely correct, the value obtained may be somewhat erroneous.[131] From the scientific viewpoint, this is particularly important for HPLC analysis. However, for the practical brewer, perhaps what is more important is that the method used is: (a) repeatable and (b) meaningful for his or her purpose. In respect of repeatability, it is worth mentioning here the extreme difficulty of quantitatively evaluating the composition of a run of baled hops — or even of a single bale. This is simply due to the heterogeneous nature of the product, which makes it a challenging exercise to obtain even a moderately representative sample for analysis, leading to a necessity to take large numbers of samples. Such a situation led Verzele et al. to conclude: "We believe now that precise α-acids analysis in hops is impossible."[132] These authors further declared: "It is most imperative that all parties involved in hop transactions become aware of this fact," a sentiment with which all hop dealers will heartily concur. Of course, batches of processed hop products, especially extracts, should not be so difficult to analyze — but it always remains problematic to truly assess the yield of any production process that starts with baled hops.

Hard, Soft, and Total Resins

The so-called "hard" and "soft" resins are primary subclasses of the "total" resin content of hops. These classes of resins are defined by their propensity to dissolve in various solvents, the hard resins (mostly xanthohumol and oxidation products of α- and β-acids) being more polar than the soft resins (mainly α-acids, β-acids, and desoxy-α-acids):

- Total Resins – *soluble* in diethyl ether and methanol and consisting of the:
  - Hard resins — *insoluble* in hexane
  - Soft resins — *soluble* in hexane

The classical method for analysis of hop resins depends upon subjecting hops (or hop extracts) to sequential dissolution in different solvents and is tedious to perform. It fractionates and quantifies the hop resin into the above classes and, in addition, provides values for:

- % Hop waxes
- % "Beta-fraction"
- % Lead conductance value

The method is of limited value today because, with the exception of the % LCV (which can anyway be separately obtained, see following text), the various values produced can hardly be translated into very meaningful indications of bittering power.

Polarographic Methods

Because the naturally formed α-acids are all of the same enantiomeric structure, they exhibit the ability to rotate polarized light. This property may be put to use as a means to estimate the content of α-acids. A weighed amount of ground hops is extracted by ether in a Soxhlet extractor and the ether solution then removed by vacuum evaporation. The residue is then dissolved in light petroleum in a volumetric flask. Finally, a portion of the petroleum solution is clarified by centrifugation, filled into a 10-cm long polarimeter tube, and the degree of rotation of polarized light is read. The α-acids percentage is then calculated. Useful in its day, this method has now fallen out of general use.

Lead Conductometric Methods

Many years ago, it was noted that solutions of hop resin acids in organic solvents readily formed insoluble, yellow precipitates when treated with methanolic lead acetate. The precipitate was found to consist of nearly pure "humulones," the "lupulones" remaining in solution. This discovery was soon put to good use as the basis of various related means to determine the α-acids content of hop samples. In essence, the methods first require the hops to be extracted into a nonpolar organic solvent or solvent mixture, followed by partial dilution into methanol. This latter solution is then titrated with methanolic lead acetate and the conductivity monitored. Because the lead salt is so insoluble, any free α-acids are immediately precipitated out of solution, and the conductivity hardly changes. However, as soon as the α-acids are consumed, further addition of lead acetate markedly increases the conductivity, this increase quickly rising to the point of being more or less proportional to the excess quantity of the relatively fast-moving lead ions added to the solution. A chart may therefore be obtained that consists of two approximately straight lines connected by a short, curved section (Figure 7.24). By subsequently drawing two straight lines, the analyst may then obtain an interception point (the "end point"), one locus of which is the quantity of lead acetate solution added. This amount is then entered into a simple equation from which is obtained the percent lead conductance value or % LCV for the hop sample.

For an absolutely fresh hop, and assuming 100% extraction efficiency, the % LCV comes very close to being a true measure of its percentage content of α-acids. Automatic titration machines are available that take the drudgery out of the method and the more sophisticated are preprogrammed to automatically determine the end point of the titration and make the calculations. Such machines may be coupled to automatic samplers and allow up to about 20 samples to be analyzed per hour. However, the preparation of the samples still requires substantial manual input.

Several variations to the aforementioned principle have been described and adopted as official methods. At one time, the most popular basic extraction solvent was benzene, but this has long since been abandoned on safety grounds. The simplest variants of the basic method now involve extraction

Hops

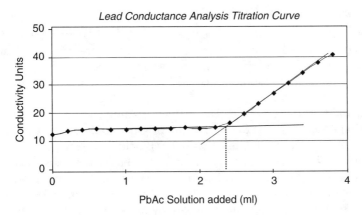

FIGURE 7.24
Titration curve obtained during determination of percent lead conductance value (%LCV).

into low-sulfur-containing toluene and dilution into methanol. However, studies have shown that the extraction of the α-acids into toluene alone is incomplete,[131] probably in part because a small proportion of these homologues is bound up as relatively insoluble metal salts. Hence other, more effective, solvent mixtures have been recommended, especially methanol/diethyl ether. The addition of mineral acid (normally HCl) then ensures that the α-acids are fully converted into their more soluble, free acid forms and better extracted. Consequently, these variant methods give higher values, typically by about 5–10%.[131] However, they are neither simpler, nor inherently more repeatable — an important practical point, as it is the comparison *between* samples that is of most practical value for brewing purposes.

As already noted, the % LCV is a good measure of the per cent α-acids content of fresh hops. But as the hops deteriorate, progressively more of the original α- and β-acids content is converted to compounds that may or may not also react with lead acetate to form insoluble salts. Hence, the rate at which the % LCV declines is considerably less that of the α-acids themselves. This is an important and valuable feature of the method, because brewing tests have clearly established that the bittering power of hops, pellets or extract, when used in the kettle, does not decline in exact proportion to their loss of α-acids, indicating that some of the degradation compounds are themselves bitter.[15,85,133] Thus, although % LCV is often equated with per cent α-acids content for commercial purposes, in reality, it is better considered simply as a measure of bittering power. Many brewers, therefore, continue to purchase hops or nonisomerized hop products on this basis, eschewing the more modern, compound specific, HPLC analysis.

Spectrophotometric Methods

The spectral characteristics of dissolved α- and β-acids are dependent upon their degree of dissociation and are substantially different from one another, especially in alkaline solution. This fact is used as the basis for an analytical method in which the absorbance of a dilute, alkaline solution in methanol of organic, solvent-extracted hops is first measured at three different wavelengths in the UV region.[134] The values obtained are then entered into regression equations that determine the original content in the hop sample of not only the α-acids, but also that of the β-acids. The method is especially popular in the U.S., where it forms the basis for most commercial transactions. In the official ASBC method (Hops 6, known colloquially as "ASBC Spectro"), the primary extraction is done with toluene and the residual hop material separated simply by allowing the extraction mixture to stand for about 30 min. The toluene solution is next diluted into alkaline methanol prior to reading in a 1-cm path length cell in a variable wavelength spectrophotometer. Readings are then taken at 275, 325, and 355 nm and inserted into two regression equations:

$$\% \, \alpha\text{-Acids} = D \times [-19.07 A_{275} + 73.79 A_{325} - 51.56 A_{355}]$$
$$\% \, \beta\text{-Acids} = D \times [5.10 A_{275} - 47.59 A_{325} + 55.57 A_{355}]$$

where D is the dilution factor.

The method lends itself to automation at the analytical stage, where it is much quicker to perform than the conductometric methods. Importantly, it also avoids the use of poisonous lead salt and, hence, presents less of a waste disposal problem. A further benefit is the fact that it allows the analyst to also determine the approximate degree to which the hops have deteriorated. This is achievable because the majority of the degradation compounds absorb more strongly at 275 nm — as also do the iso-α-acids that may be present to a limited extent — than do either the α- or the β-acids. The frequently quoted "Hop Storage Index" (HSI) is simply obtained:

$$\text{HSI} = \frac{A_{275}}{A_{325}}$$

In an absolutely fresh hop that has not been damaged in any way during the kilning process, the HSI value may be around 0.220, but a more typical value for a sample taken from a bale of recently dried hops would be 0.240. Values above about 0.260 generally indicate that a small, but significant amount of deterioration has taken place, though the base (i.e., completely fresh) value is to some extent variety-dependent. Via a simple equation, the HSI may be used to determine the "% transformation" of the α-acids content during kilning and storage, and hence, in conjunction with the determined α-acids value, enables a reasonably accurate back-calculation to the original α-acids content of the fresh hops[135]:

$$\% \text{ Transformation} = 102 \log [\text{HSI}] + 61.8$$

Example: For a hop sample measuring 12.2% α-acids by ASBC Spectro and having an HSI value of 0.350:

$$\% \text{ Transformation} = 102 \times \log [0.350] + 61.8 = 15.3\%$$

Therefore, original α-acid content = 12.2 × 100/(100 − 15.3) = 14.4%

HSI values are frequently used by hop buyers as a tool to help establish the quality of the hops they purchase.

Spectrophotometry may also be used to determine the content of isomerized or reduced isomerized α-acids in the purified solutions of Iso, Rho, Tetra, or Hexa sold for postfermentation bittering,[136] though such analysis is nonspecific and, hence, will give no indication of purity. HPLC methods are therefore to be preferred.

HPLC Methods

Background

The development of high pressure liquid chromatography (also referred to as high performance liquid chromatography or HPLC) during the late 1970s and 1980s has revolutionized hop resin acid analysis. Although prior chromatographic methods did exist for the analysis of extracts of hops, they were rather unreliable, slow, and the separation of the various compounds was relatively crude. The quantification of the constituents was also hampered by the absence of recognized standards against which the recorded peaks could be calibrated. Advances in computing have now also allowed the development of sophisticated software for peak recognition and integration, completely displacing former methods that relied on literally cutting out and weighing the drawn peaks on the recorder paper, or on mechanical disc integration.

Separation Principles

Separation in chromatography depends upon exploiting differences in the relative affinity of solutes for absorption onto the solid phase packing of the column. For solutes of similar chemical structure, these differences may be extremely small, and the high efficiency of HPLC is achieved by using column packing that has an extremely small and even particle diameter (usually around 5 μm). This ensures that the solutes passing down the column are exposed to a very large number of "theoretical plates," each equating to single-stage equilibration between solution in the solvent and absorption onto the solid surface of the packing. HPLC columns used in hop analysis have silica-based packings that are treated to have long chains of hydrocarbon molecules extending out from the silica base. Commonly used packing for hop resin acid analysis has octadecyl hydrocarbon chains and is termed "C18."

The chromatography usually takes place in mixtures of organic solvent and water (the "mobile phase") and is referred to as "reverse phase" chromatography. A typical column, normally made of high quality stainless steel, is 4.6 mm in internal diameter and has a length of 250 mm. Samples for analysis are prepared in conventional ways, such as for spectrophotometric or lead conductance analysis, and then greatly diluted, often simply with methanol. A very small volume (typically 10 µl) of this diluted solution is then injected onto the top of the column while the mobile phase is pumped through at a steady rate. In order to improve the separations, some methods demand that the solvent mixture be changed during the course of the analysis, either as a stepwise change, or by gradual change from one to the other (gradient elution).

Detection

In hop resin analysis, the acids are normally detected as they leave the column by passage through a very small, in-line quartz cell through which a beam of UV light at a particular wavelength is passed. As in a normal spectrophotometer, the degree of attenuation of the light is determined by a photoelectric detector cell, the amount of light absorption being dependent upon the spectral characteristics of the solute in the mobile phase and its concentration. The information from the detector is passed electronically to an integrator, which typically draws the peak on a monitor screen and integrates its area. Usually, by virtue of prior "external" calibration of the system by use of pure standards run through the chromatograph, coupled with knowledge of the amount and dilution of the sample injected, the concentration of the components in the test sample can be calculated.

Obtaining Good Results

In essence, the HPLC methods are straightforward, but the practice is often found to be more difficult than the principles suggest.[137] Pressures of up to about 400 bar may be generated across the column, so it is essential to watch closely for leaks in the fittings and also important to ensure that the solutions are finely filtered to avoid clogging of the packing. If the packing is not tight and even in the column, then voids or channels may occur that badly distort the emerging peaks. Silica is very slightly soluble in acidic, aqueous solutions and analytical columns will last longer if the mobile phase is first passed through a short column packed with particulate silica. This column then acts to presaturate the mobile phase with dissolved silica. A similarly short, easily replaced "guard" column, containing packing similar to that of the analytical column, may be inserted immediately before the latter to protect it from highly absorbent substances that may be present in the sample. Impurities in the column packing, especially residual iron in the silica or minute amounts of nickel ions leached from the metal components of the system, may cause chemical degradation of some hop acids, particularly iso-α-acids.[137] For similar reasons, only high-purity solvents should be used,[138] though the addition of ethylenediamine tetraacetic acid (EDTA) to the mobile phase can often improve the chromatography by chelating these metal ions.[137,139]

The analyst also needs to know how to set the integration parameters that determine the start and end points of the peaks on the chromatogram, otherwise the machine may draw peak baselines that are unrealistic. This can be a substantial source of error, particularly when one peak tails on the top of another, much larger one. Choice of detection wavelength is also important. Normally, it is best to set the detector to the wavelength at which the compound to be analysed exhibits a peak in its absorbance spectrum in the mobile phase, though sometimes it is better to choose a wavelength at which interfering compounds that coelute have relatively poor absorbance. For the analysis of iso-α-acids, selecting a wavelength of 270 nm in combination with use of an acidic mobile phase is recommended, as, under these conditions, the extinction coefficients of the homologs and the *cis* and *trans* diastereoisomers are virtually identical.[140,141]

By judicious choice of packing type and size, column dimensions and mobile phase, it is possible in a single analysis to make clear separations of several different hop resin acid analogs and homologs. Enantiomeric separations of, for example, *cis*- and *trans*-iso-α-acids are also possible,[137,142,143] as is the simultaneous separation of all of the four types of commercially used iso-α-acids[144]; however, as a general rule, the better the separation, the slower the analysis. Therefore, the analyst must decide on priorities and may elect to sacrifice accuracy for speed.

Calibration

Of crucial importance to any HPLC method is the means by which the method is calibrated. There are two standard approaches to calibration: (a) internal calibration using a pure compound of known extinction coefficient to which the samples of unknown concentration can be related and (b) external calibration, in which a standard solution containing known amounts of the target compound is first put through the chromatographic procedure. Both approaches have their merits, but for hop resin acid analysis it is now generally accepted as better to use the external calibration method. Internationally agreed standards for all the major hop resin acids are now available. For α-acids and β-acids, an International Calibration Extract (ICE) has been prepared from a de-oiled CO_2 extract of hops. For iso-α-acids and reduced isomerized α-acids, there are four International Calibration Standards (ICS) that are prepared as semicrystalline mixtures of the major homologs of purified iso-α-acids and reduced isomerized α-acids.[145] For technical reasons, three of these are prepared in the form of their dicyclohexylamine (DCHA) salts and one as a mixture of the pure compounds themselves[146]:

- ICS-I (DCHA-iso-α-acids)
- ICS-R (DCHA-*rho*-iso-α-acids)
- ICS-T (tetrahydroiso-α-acids)
- ICS-H (DCHA-hexahydroiso-α-acids)

All of these standards are further designated by a suffix number, since the supply of each one is limited and they will necessarily be replaced occasionally by new preparations. HPLC-based commercial transactions will therefore typically indicate the standard to be used (e.g., ICE-2 or ICS-R1).

Sample Preparation

The precision of any HPLC method is, of course, in part dependent upon the method of sample preparation. For the measurement of hop resin acids in hops, hop pellets, or hop extracts, it is often convenient to use the same primary methods that are well established for lead conductance or spectrophotometric tests. The extraction solution can then be diluted as necessary into methanol, acidic methanol, or mobile phase before injection. Calibration solutions can also be prepared in one or other of these solvents, split into vials and held in the freezer for use as required, though stability is best if the standards are diluted into acidic methanol (unpublished data, International Subcommittee for Isomerized Hop α-Acids Standards).

Capillary Electrophoresis

The possibility of separating the hop resin acids by capillary electrophoresis has been described and the suggestion made that it offers advantages over HPLC.[147] However, methods based on this principle are not in widespread use and have not been adopted by any of the brewing organizations responsible for the publication of recommended analytical methods.

NIR Reflectance Mode Analysis

General

Near Infrared (NIR) analysis is a powerful technique that has the ability to analyze samples very quickly and simply, entirely without the aid of solvents. It is therefore attractive for a laboratory that needs to examine large numbers of samples. The most important thing to note regarding NIR analysis is that the results produced are essentially predictions of what would be achieved had an unknown sample been subjected to a more conventional test. Hence, in the case of hops, the NIR analyzer can, for example, be programmed to produce predicted results for % LCV, % α-acids by HPLC, and % α-acids by ASBC Spectro all at the same time.

Principles

All organic substances exhibit unique, complex spectra in the near infrared region, as indeed does water. The spectrum of a mixture of substances essentially comprises an addition of the individual spectra of the components and will naturally be dependent upon the relative amounts of each one. Hence, differing mixtures will each have their unique spectrum and, akin to the principles of the spectrophotometric determination of α- and β-acids, it is, therefore, possible to determine the amounts of the components in simple mixtures by measuring the absorption of the radiation at suitable wavelengths and entering the values into previously determined regression equations. The power of the NIR technique arises because the spectra of pure compounds are much more complex in the NIR region than they are in the UV region, enabling the analyst to obtain much more information out of mixtures by examining the absorption at a great many different wavelengths.[148] The mathematics is necessarily handled by sophisticated computer software, and programs have been developed to enable appropriate algorithms to be obtained. In practice, these algorithms must first be developed by feeding the computer with data obtained from conventional analysis of samples that are also run in the NIR analyzer. Once sufficient data have been supplied, the program can then predict the result of a conventional analysis simply by examination of the NIR spectrum of the unknown sample. The more wavelengths from which data points are taken, and the more information that is fed in from samples of known conventional analysis, the more reliable will be the predictions that the analyzer will make. The power of the technique may be enhanced still more by use of first- and second-order differentials of the absorption spectrum at each data point wavelength.

Preparation of Samples

The preparation of samples for analysis could hardly be more straightforward. Whole hops or pellets are usually ground to a powder and placed into a special cell, wherein the powder is pressed beneath a quartz disk, though, surprisingly, it is actually possible to omit the grinding stage.[149] The cell is then placed in the analyzer and subjected to IR radiation of successively changing wavelength. The analyzer then records the amount of radiation that is reflected from the surface of the sample at each selected wavelength. With the aid of more than one sample cell, the whole process can be accomplished within a minute or two. Hop extracts can also be analyzed in the same way.

Pros and Cons

While NIR analysis clearly has enormous advantages, and certainly works well for moisture analysis, its predictive nature always leaves open the possibility of substantially erroneous results for, say, % α-acids, in a material as complex as hops. Indeed, it has been found that algorithms that work for one hop variety may not work well for another, implying a need for prior (and potentially incorrect) determination by the analyzer of the likely variety so that the most appropriate predictive algorithms can be applied. Seasonal factors have also been noted, which then necessitate running a substantial number of conventional analyses so that the algorithms can be corrected. Hence, the NIR technique has thus far failed to gain acceptance for commercial transactions involving hops or hop products, being generally used instead for rapid moisture checks and fast screening for the % α-acids content of samples at harvest or during processing in the larger hop-processing facilities.

Hop Oils

General

Analysis of the hop oils consists of two elements:

- Determination of total essential oils
- Compositional analysis of the oil fraction

The oil fraction of hops is usually defined as that portion of the hop that is steam volatile, and its amount in any sample of hops, hop pellets, or extracts is readily (but slowly) determined by simple hydrodistillation, condensation of the vapors, and collection of the entire oil fraction. Compositional analysis is then commonly made by gas chromatography of the collected oil, though other methods are now appearing that offer some advantages.

Steam Distillation Method (for Total Essential Oils)

In the standard methods, a weighed amount of hops, pellets, or extract is placed in a round-bottomed flask and boiled for 3 h. The vapors are continuously condensed and the water returned to the flask by means of a cohobation head. The condensed oil separates in the cohobation head, forming a floating layer, the volume of which can be read directly in a calibrated section of the head at the end of the boiling period. Results are expressed as milliliters of oil per 100 g of sample. The method is reasonably reproducible within a laboratory, but because the less volatile substances are not necessarily fully extracted during the 3-h boil, interlaboratory results may sometimes deviate significantly. It is also important to ensure that vapors are condensed efficiently to prevent any of the more volatile components (especially myrcene) from escaping the apparatus. A good flow of cold water to the condensing section of the cohobation head is, therefore, an essential requirement.

Because of the fairly high temperature involved, some of the compounds found in the distillate may be artefacts formed in consequence of the method itself.[150] Geraniol, for example, may be derived in part from hydrolysis of linalyl acetate or from geranyl esters such as geranyl acetate and geranyl isobutyrate. However, in conventional brewing at least, as the hop oils will necessarily be exposed to boiling conditions in the wort kettle, this may be no bad thing.

Presence of Sulfur Compounds

A crude, but simple and effective test for the presence of excessive amounts of sulfur compounds in hops is to incubate hop oil with a yeast slurry and test for the generation of hydrogen sulfide (H_2S) by use of filter paper containing lead acetate.[151] If the paper turns black, then the hops may have been heavily sprayed with elemental sulfur and could cause unpleasant taints in the beer. An alternative, quantitative test utilizes the property of copper gauze to react with sulfur compounds to form copper sulfide.[152] A quantitative method for estimating sulfur dioxide in hops has also been described.[152]

Gas Chromatography Analysis (for Hop Oil Components)

History and Principles

In comparison with HPLC, GC is a far more powerful technique — fortunately so, in view of the myriad of compounds to be found in hop oils. The principles of the method are similar though, separations depending upon repeated absorption and desorption of compounds out of the gaseous phase and onto a solid, stationary phase. In the earlier GC systems, the column would typically consist of a 5 m long, 1/8 in bore, coiled glass tube packed with an inert particulate material and held inside an oven at perhaps 150°C. The sample to be analyzed would be dissolved in a carrier solvent (e.g., methanol or hexane) and injected into a stream of hydrogen or helium gas, which was then passed through the heated column. The volatile compounds would be separated according to their relative propensity to adsorb onto the hot packing. The presence of organic compounds in the exit gas was then detected by burning in a hydrogen flame, the light emissions created being converted into an electrical signal. By today's standards, the separations were crude, but nevertheless sufficient to make reasonable estimates of the major hydrocarbon components of, say, a hydrodistilled hop oil sample.

Modern GC machines have greatly improved the separations, and detection of hundreds of compounds is now possible in a single run of perhaps 40 min. A major advance was the introduction of capillary glass tubing for the separating coil. The principle of the capillary tube is to do away with the particulate packing, replacing it with a suitable coating on the inside wall that acts as the stationary phase. This change also enables the length of the coil to be extended to 50 m or more, both modifications serving to considerably increase the separating capacity of the coil. Coupled with the development of much more sensitive detection and integration devices, the necessary amount of sample injected was also greatly reduced, also assisting to improve the separation. Together with improvements in the method of sample injection came development of very accurately controlled ovens that have enabled reproducible analysis to be run using thermal gradients, helping to shorten the time for a single analysis whilst maintaining a high degree of separation. Finally, the development of alternative detection methods has allowed certain groups of compounds to be specifically detected. Commonly used detectors are:

- Flame ionization detector (FID) — for detecting hydrocarbons
- Flame photometric detector (FPD) — for detecting sulfur compounds
- Nitrogen/phosphorous detector (NPD) — particularly for detection of N-heterocyclics
- Electron capture detector (ECD) — for detecting organochlorine compounds

Sulfur compounds may also be detected by use of a chemiluminescence (Sievers) detector.[153]

Calibration and Peak Identification

Calibration of GC analysis is often related to an internal standard, but because of the vast number of compounds detected, quantification is often made on the basis of assuming that the detector response is the same for each compound. Fortunately, this is approximately true for FID detection of hydrocarbons, so that the content of the various components of interest is normally reported as a percentage of the total sample, on the presumption that the entire sample is composed of hydrocarbon substances. This means that the use of an internal standard may be unnecessary, as the integrator can be set to discount the solvent peak and then calculate the percentage of each component by comparing it against the total area of the nonsolvent peaks. Dispensing with the internal standard does, however, entail the risk that a significant portion of the unknown sample may not be eluted from the column before the run is terminated, or that some compounds may be hidden within the (normally fast-running) solvent peak.

Peak Identification

Clearly, identification of the compounds eluting from the GC is dependent upon knowing the retention time under the run conditions; hence, the ability of the equipment to exactly duplicate gas flow-rates and thermal gradients is crucial. Hop oil analysts will usually have no problem determining the position of the major peaks, such as myrcene, α-humulene, and β-caryophyllene, and these peaks can act as useful markers. The identification of minor constituents, such as undecan-2-one, can then be established by comparison of their relative retention times against, say, α-humulene. In the absence of coupled GC–mass spectrometry (GC–MS), positive identification will depend upon use of pure compounds to establish the correct retention times and, usually, cross-comparison of samples and data with analysts who already know the relative positioning of the compounds of interest. Obviously, misidentification is all too easy, and the analyst needs to be vigilant, not relying entirely upon preprogramming of the integrator to report results. Enantiomers may also be difficult to separate, and this can potentially lead to false assessment of the significance of a flavor-active compound such as linalool, where its two forms evoke differing organoleptic responses and have different flavor thresholds,[28] as earlier mentioned (in the section "Oxygenated Compounds").

"Heart Cut" Analysis

It should also be borne in mind that the peaks on GC analysis do not necessarily represent single compounds. Recent work by Dufour and coworkers[154] shows that expansion of short sections of a GC chromatogram — by collecting the vapors exiting over a short period from an analytical column and subjecting them to rechromatography on a second column — usually reveals a vast number of components that are normally hidden from view.

Adsorbent Fiber Technique (Solid Phase Micro-Extraction)

This more recent technique involves suspending an absorbent fiber in the headspace above a sample of hops in a closed vessel. The absorbed volatiles are then desorbed by heating the fiber in a special injector coupled to a GC machine.[155]

Polyphenols

Although methods exist for the determination of tannins as a class, their value to the brewer may be somewhat limited, as detailed knowledge of the composition of this fraction is necessary before conclusions as to likely brewing consequences can be drawn. HPLC methods for separation of hop polyphenols have been described,[38,41,42,156,157] but there are no internationally recognized methods and their routine application is not something to be undertaken lightly.

Flavonoids

Recent interest in xanthohumol (XN) has begun to cause brewers to seek information regarding its content in hops and hop products and also its utilization (mostly in the form of isoxanthohumol, IX) in beer. Measurement of XN in hops, pellets, and extracts is quite easily facilitated by use of HPLC. The same extraction and chromatographic separation conditions may be used as are used for hop resins. Dependent upon the column packing used, and under the recommended conditions for analysis of α-acids and β-acids, XN normally runs within or close to the area that would be occupied by iso-α-acids, when present. However, even in the presence of significant, apparently interfering amounts of the latter compounds, it can be integrated successfully by setting the detector wavelength to 370 nm, at which the XN exhibits a strong peak of absorbance, but where the iso-α-acids are hardly detected. For the accurate calibration of XN analysis it will be necessary to obtain a sample of purified XN from a research laboratory or commercial source.

Analysis of Worts and Beers

BU Analysis

Bittering unit (BU) values have been the brewer's primary tool for monitoring and controlling bitterness for almost 50 years and, although nonspecific, the method has stood the test of time. The principle is simple: beer or wort (optionally clarified by filtration or centrifugation) is acidified to convert hop resin acids into their nonionic, water-insoluble forms and then extracted gently against iso-octane. The hop acids migrate into the organic layer, the UV light absorption of which may then be read in a spectrophotometer at 275 nm. The BU value is obtained by multiplying the absorbance by a factor that was originally intended to give an approximation of the iso-α-acid content of the original sample. Somewhat in the manner of the % LCV for hops and hop products, the value is rather dependent on the presence of other hop-derived substances,[158] some of which (e.g., the hulupones) are bitter. Therefore, whilst it certainly does not provide a direct or totally reliable measure of taste bitterness, BU analysis as applied to beer will normally be found to correlate well with panel scores for bitterness of a particular product where the process is otherwise under good control and the hops used are not badly deteriorated.[85] Where preisomerized or reduced isomerized products have been used, for quality control purposes it is often helpful to apply a correction factor to the standard BU figure, in order to obtain results that better correlate with taste perceptions based upon brewing with unmodified (and reasonably fresh) hops, pellets, or extracts.[114,159,160]

Hop Resin Acids by HPLC

The analysis of wort or beer samples for (primarily) iso-α-acids can often be done very simply and satisfactorily by direct injection of the sample after clarification (in the case of wort) or degassing (in the case of beer). In the case of wort, the analyst first needs to decide whether or not to include or exclude that portion of the hop acids that are coprecipitated with the hot or cold breaks. If inclusion is desired, then prior, low-rate dilution into methanol may be indicated. Success with direct injection depends upon sensitive equipment, a good analytical column, and the ability to integrate the iso-α-acids as relatively small peaks tailing on a very much larger peak of malt-derived substances (Figure 7.25a). Avoidance of unsuitable filter media is necessary, as some (such as nylon) will absorb hop acids in the absence of organic solvent.[137] Degassing of the sample must be done very carefully to avoid losing the surface-active hop compounds. Addition

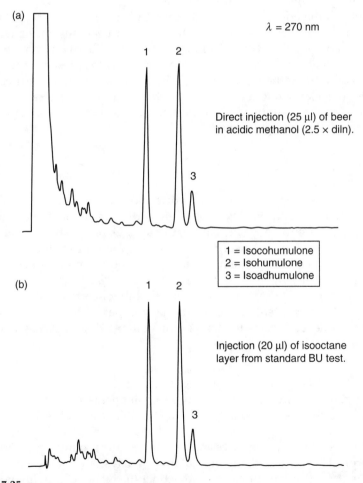

FIGURE 7.25
HPLC charts of a degassed (stout) beer sample by: (a) direct injection and (b) use of the isooctane layer obtained during routine BU analysis.

of no more than a drop or two of *n*-octanol per bottle of beer is recommended to break the foam, though any more risks a significant loss of the target compounds into the undissolved octanol.

Some analysts prefer to concentrate the sample first, by adsorbing the iso-α-acids (and any residual α-acids and traces of β-acids) from a suitable sized aliquot onto a silica-based medium that is prepacked into convenient cartridges, then eluting the target compounds with a smaller amount of solvent (e.g., methanol). This procedure enables a larger quantity of the hop resin acids to be injected onto the column and also mostly eliminates interfering, malt-derived, water-soluble compounds. The downside is the extra cost in materials and time, and a likely reduction in reproducibility. Another approach is to prepare samples as if for a BU analysis, then to inject the octanol layer into the chromatograph (Figure 7.25b). This method has the added attraction that the BU value can be obtained at the same time, acting as a check that there has not been any major failure in the calibration of the HPLC. Examination of the chromatograms will also illustrate rather well how BU values are affected by the presence of hop compounds other than the iso-α-acids.[158] The use of a photodiode array (PDA) detector (also known as a diode array detector or "DAD") is recommended for checking the identity of minor peaks, such as suspected iso-α-acids in a beer intended to be light stable.[158,161]

Flavonoids

Analysis of the prenylflavonoids in worts and beers is naturally made difficult by their low concentration, but is nevertheless probably best done by direct injection into an HPLC chromatograph. Quantification of xanthohumol may depend upon ensuring its clear separation from interference by any of the iso-α-acids, though advantage can be taken of its much higher extinction coefficient at its peak maximum (in acidic mobile phase) of 368 nm (Figure 7.26). Analysis of isoxanthohumol certainly requires good separation from the iso-α-acids, as its spectral peak (287 nm)[57] is too close to those of the iso-α-acids to allow of integration without clear separation.

Hop Oils

Detection and quantification of hop oil components is particularly tricky because of their exceptionally low levels. No internationally accepted methods exist, though several authors have described techniques for oil analyses — albeit the objective may have been to produce qualitative GC "fingerprints" rather than quantitative analysis of individual components.[21,162–166] One approach is to extract the oils into a highly nonpolar solvent such as pentane and, if necessary, concentrate the solution before injection into a GC or coupled GC–MS. Another is to use solid phase

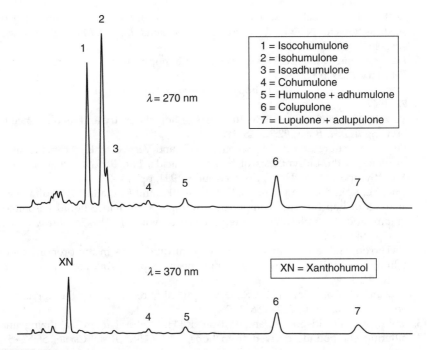

FIGURE 7.26
HPLC analysis of an organic solvent extract of isomerized hop pellets as recorded at two detection wavelengths and illustrating the possibility to detect and quantify different types of compounds according to their varying spectral characteristics.

micro-extraction to collect the volatiles from the headspace of a beer sample, followed by desorption and GC analysis.[167] The specific analysis of linalool (the substance considered by some to be a good indicator of "hoppy" aroma) has been described,[29] in this case the linalool being first collected from the wort or beer via solid phase extraction and subsequently analyzed by GC–MS.

References

1. Canales Gajá, A.M., Beer: art and science, *Tech. Q. Master Brew. Assoc. Am.*, 20: 53–67, 1983.
2. Moir, M., Hops — a millennium review, *J. Am. Soc. Brew. Chem.*, 58:131–146, 2000.
3. Anon., *Hopsteiner, 2002 Guidelines for Hop Buying*, Simon H Steiner GmbH, Mainburg, Germany, 2002.
4. Harrison, J., Effect of hop seeds on beer quality, *J. Inst. Brew.*, 77:350–352, 1971.

5. Beatson, R.A., Ansell, K.A., and Graham, L.T., Development and performance of seedless hops for New Zealand growing conditions, *Tech. Q. Master Brew. Assoc. Am.*, 40:7–10, 2003.
6. Darby, P. and Walker, C., Where next for the hop industry? *Brewers' Guardian*, February 2003, pp. 22–25.
7. Stevens, R., The chemistry of hop constituents, *Chem. Rev. Am. Chem. Soc.*, 67:19–71, 1967.
8. Verzele, M., Centenary review. 100 Years of hop chemistry and its relevance to brewing, *J. Inst. Brew.*, 92:32–48, 1986.
9. Fung, S.Y., Zuurbier, K.W.M., Scheffer, J.J.C., and Verpoorte, R., Aromatic intermediates in the biosynthesis of hop bitter acids, *Eur. Brew. Conv. Monograph XXII*, Symposium on Hops, Zouterwoude, 1994, pp. 14–23.
10. Ono, M., Kakudo, Y., Yamamoto, Y., Nagami, K., and Kumada, J., Simultaneous analysis of hop bittering components by high-performance liquid chromatography and its application to the practical brewing, *J. Am. Soc. Brew. Chem.*, 43:136–144, 1985.
11. Nickerson, G.B. and Williams, P.A., Varietal differences in the proportions of cohumulone, adhumulone, and humulone in hops, *J. Am. Soc. Brew. Chem.*, 44:91–94, 1986.
12. Howard, G.A. and Tatchell, A.R., Development of resins during the ripening of hops, *J. Inst. Brew.*, 62:251–256, 1956.
13. Smith, R.J., Davidson, D., and Wilson, R.J.H., Natural foam stabilizing and bittering compounds derived from hops, *J. Am. Soc. Brew. Chem.*, 56:52–57, 1998.
14. Forster, A., Beck, B., Koberlein, B., and Schmidt, R., Hochaufloesende Gradienten-HPLC mit Dioden-Array-Detektor zur verbesserten Auftrennung von Hopfenbitterstoffen, (Highly dissolving gradient HPLC with diode array detector for improved separation of hop bitter substances), *Proc. Eur. Brew. Conv.*, Maastricht, 1997, pp. 223–230.
15. Aitken, R.A., Bruce, A., Harris, J.O., and Seaton, J.C., The bitterness of hop-derived materials in beer, *J. Inst. Brew.*, 76:29–36, 1970.
16. Kovačevič, M. and Kač, M., Determination and verification of hop varieties by analysis of essential oils, *Food Chem.*, 77:489–494, 2002.
17. Peacock, V.E. and McCarty, P., Varietal identification of hops and hop pellets, *Tech. Q. Master Brew. Assoc. Am.*, 29:81–85, 1992.
18. Walsh, A., The mark of nobility. An investigation into the purity of noble hop lineage, *Brewing Techniques Mar/Apr*, 1998, pp. 60–69.
19. Lewis, G.K., Zimmermann, C.E., and Hazenberg, H., Hop variety named 'Columbus.' U.S. Patent No. PP10,956, 1999.
20. Haunold, A. and Nickerson, G., Development of a hop with European aroma characteristics, *J. Am. Soc. Brew. Chem.*, 45:146–151, 1987.
21. Peacock, V.E. and Deinzer, M.L., Chemistry of hop aroma in beer, *J. Am. Soc. Brew. Chem.*, 39:136–141, 1981.
22. Deinzer, M. and Yang, X., Hop aroma: character impact compounds found in beer, Methods of formation of individual compounds. *Eur. Brew. Conv. Monograph XXII*, Symposium on Hops, Zouterwoude, 1994, pp. 181–197.
23. Goiris, K., De Ridder, M., De Rouck, G., Boeykens, A., Van Opstaele, F., Aerts, G., De Cooman, L., and De Keukeleire, D., The oxygenated sesquiterpenoid fraction of hops in relation to the spicy hop character of beer, *J. Inst. Brew.*, 108:86–93, 2002.

24. Lam, K.C., Foster II, R.T., and Deinzer, M.L., Aging of hops and their contribution to beer flavor, *J. Agric. Food Chem.*, 34:763–770, 1986.
25. Moir, M., Hop aromatic compounds. *Eur. Brew. Conv. Monograph XXII*, Symposium on Hops, Zouterwoude, 1994, pp. 165–180.
26. Murray, J.P., Westwood, K., and Daoud, I., Late-hop flavour, *Proc., 21st Cong. Eur. Brew. Conv.*, Madrid, 1987, pp. 321–328.
27. Siebert, K.J., Sensory analysis of hop oil-derived compounds in beer; flavor effects of individual compounds. Quality control, *Eur. Brew. Conv. Monograph XXII*, Symposium on Hops, Zouterwoude, 1994, pp. 198–215.
28. Kaltner, D., Untersuchungen zur Ausbildung des Hopfenaromas und technologische Massnahmen zur Erzeugung hopfenaromatischer Biere. (Investigations into the formation of hoppy aromas and technological procedures for the production of hop aromatic beers), Dissertation, Technische Universität München, 2001.
29. Kaltner, D., Thum, B., Forster, C., and Back, W., Investigations into hoppy aroma in pilsner beers under variation in technological parameters, *Monatsschr. Brauwiss.*, 54:199–205, 2001.
30. Peppard, T.L., Dimethyl trisulphide, its mechanism of formation in hop oil and effect on beer flavour, *J. Inst. Brew.*, 84:337–340, 1978.
31. Peppard, T.L., Volatile sulfur compounds in hops and hop oils: a review, *J. Inst. Brew.*, 87:376–385, 1981.
32. Peppard, T.L. and Laws, D.R.J., Hop derived sulfur compounds and their effect on beer flavour, *Proc. 17th Cong. Eur. Brew. Conv.*, Berlin, 1979, pp. 91–104.
33. Pickett, J.A., Peppard, T.L., and Sharpe, F.R., Effect of sulfuring on hop oil composition, *J. Inst. Brew.*, 82:288–289, 1976.
34. Goldstein, H., Ting, P., Navarro, A., and Ryder, D., Water-soluble hop flavor precursors and their role in beer flavor, *Proc. 27th Cong. Eur. Brew. Conv.*, Cannes, 1999, pp. 53–62.
35. Sägesser, M. and Deinzer, M., HPLC-ion spray-tandem mass spectrometry of flavonol glycosides in hops, *J. Am. Soc. Brew. Chem.*, 54:129–134, 1996.
36. Vancraenenbroeck, R., Callewaert, R., and Lontie, R., Identification of flavonol glycosides in hops, *Cerevisiae*, 5:69–73, 1980.
37. Preis, F. and Mitter, W., The re-discovery of first wort hopping, *Brauwelt Int.*, 13:308, 310–311, 313–315, 1995.
38. Forster, A., Beck, B., Schmidt, R., Jansen, C., and Mellenthin, A., Über die Zusammensetzung von nieder-molekularen Polyphenolen in verschiedenen Hopfensorten und zwei Anbaugebieten (On the composition of low molecular polyphenols in different varieties of hops and from two growing areas), *Monatsschr. Brauwiss.*, 55:98–108, 2002.
39. McMurrough, I., Chromatographic procedures for measuring polyphenol haze precursors, *Proc. 17th Cong. Eur. Brew. Conv.*, Berlin, 1979, pp. 321–334.
40. Vancraenenbroeck, R. and Devreux, A., Effect of malt and hop polyphenols on wort and beer properties, *Cerevisiae*, 6:81–88, 1981.
41. Bellmer, H.G., Galensa, R., and Gromus, J., Bedeutung der Polyphenole fuer die Bierherstellung. Analytik, Ergebnisse und Diskussion (Teil 1). (Significance of polyphenols for beer production. Analysis, results and discussion (Part 1)), *Brauwelt*, 135:1372–1379, 1995.
42. Forster, A., Beck, B., and Schmidt, R., Untersuchungen zu Hopfenpolyphenolen (Investigations on hop polyphenols), *Proc. 25th Cong. Eur. Brew. Conv.*, Brussels, 1995, pp. 143–150.

43. Forster, A., Beck, B., and Schmidt, R., Hop polyphenols — do more than just cause turbidity in beer, *Hopfenrundschau Int.*, 68–74, 1999.
44. Mikyška, A., Harabák, M., Hašková, D., and Šrogl, J., The role of malt and hop polyphenols in beer quality, flavour and haze stability, *J. Inst. Brew.*, 108:78–85, 2002.
45. Taylor, A.W., Barofsky, E., Kennedy, J.A., and Deinzer, M.L., Hop (*Humulus lupulus* L) proanthocyanidins characterized by mass spectrometry, acid-catalysis, and gel permeation chromatography, *J. Agric. Food. Chem.*, 51:4101–4110, 2003.
46. Buggey, L., A review of polyphenolic antioxidants in hops, brewing and beer, *Brew. Int.*, 1:21–25, 2001.
47. Lermusieau, G., Liegeois, C., and Collin, S., Reducing power of hop cultivars and beer ageing, *Food Chem.*, 72:413–418, 2001.
48. Piendl, A. and Biendl, M. Physiological significance of polyphenols and hop bitters in beer, *Brauwelt Int.*, 310, 312–313, 316–317, 2000.
49. Biendl, M. Research on the xanthohumol content in hops, *Hopfenrundschau Int.*, 72–75, 2002/2003.
50. Kammhuber, K., Ziedler, C., Seigner, E., and Engelhard, B., Stand der Erkenntnisse zum Hopfeninhaltsstoff Xanthohumol. (The situation regarding knowledge of the xanthohumol content of hops), *Brauwelt*, 36:1633–1636, 1998.
51. Stevens, J.F., Taylor, A.W., and Deinzer, M.L., Quantitative analysis of xanthohumol and related prenylflavonoids in hops and beer by liquid chromatography-tandem mass spectrometry, *J. Chromatogr. A*, 832:97–107, 1998.
52. Gardner, D.S.J. Purification of hop extracts., U.S. Patent No. 3,794,744, 1974.
53. Gerhauser, C., Alt, A., Heiss, E., Gamal-Eldeen, A., Klimo, K., Knauft, J., Neumann, I., Scherf, H-R., Frank, N., Bartsch, H., and Becker, H., Cancer chemopreventive activity of xanthohumol, a natural product derived from hop, *Mol. Cancer Thera.*, 1:959–969, 2002.
54. De Keukeleire, D., Beer, hops, and health benefits, *Scand. Brew. Rev.*, 60(2):10–17, 2003.
55. Biendl, M., Eggers, R., Czerwonatis, N., and Mitter, W., Investigations into the production of a xanthohumol-enriched hop product, *Cerveza Malta.*, 38:25–27, 29, 2001.
56. Biendl, M., Mitter, W., Peters, U., and Methner, F-J., Use of a xanthohumol-rich hop product in beer production, *Brauwelt Int.*, 39–42, 2002.
57. Stevens, J.F., Miranda, C.L., Buhler, D.R., and Deinzer, M.L., Chemistry and biology of hop flavonoids, *J. Am. Soc. Brew. Chem.*, 56:136–145, 1998.
58. Forster, A., Gahr, A., Ketterer, M., Beck, B., and Massinger, S., Xanthohumol in Bier — Möglichkeiten und Grenzen einer Anreicherung. (Xanthohumol in beer — feasibility and limits for enrichment), *Monatsschr. Brauwiss.*, 9/10: 184–194, 2002.
59. Forster, A. and Köberlein, A. What happens to xanthohumol from hops during beer preparation, *Brauwelt Int.*, 35–37, 2000.
60. Stevens, J.F., Taylor, A.W., Clawson, J.E., and Deinzer, M.L., Fate of xanthohumol and related prenylflavonoids from hops to beer, *J. Agric. Food Chem.*, 47: 2421–2428, 1999.
61. Milligan, S.R., Kalita, J.C., Heyerick, A., Rong, H., De Cooman, L., and De Keukeleire, D., Identification of a potent phytoestrogen in hops (*Humulus lupulus* L.) and beer, *J. Clin. Endocrinol. Metab.*, 83:2249–2252, 1999.

62. Rong, H., Zhao, Y., Lazou, K., De Keukeleire, D., Milligan, S.R., and Sandra, P., Quantitation of, 8-prenylnaringenin, a novel phytoestrogen in hops (*Humulus lupulus* L), hop products, and beers, by benchtop HPLC-MS using electrospray ionization., *Chromatographia*, 51:545–552, 2000.
63. Tekel, J., De Keukeleire, D., Rong, H., Daeseleire, E., and Van Peteghem, C., Determination of the hop-derived phytoestrogen, 8-prenylnaringenin, in beer by gas chromatography/mass spectrometry, *J. Agric. Food. Chem.*, 47:5059–5063, 1999.
64. Walker, C.J., Phytoestrogens in beer — good news or bad news? *Brauwelt Int.*, 38–39, 2000.
65. Stevens, J.F., Taylor, A.W., Nickerson, G.B., Ivancic, M., Henning, J., Haunold, A., and Deinzer, M.L., Prenylflavonoid variation in *Humulus lupulus*: distribution and taxonomic significance of xanthogalenol and 4'-O-methylxanthohumol, *Phytochemistry*, 53:759–775, 2000.
66. Wubben, M.A. and Doderer, A., Hop pectins as foam stabilizers for beverages having a foam head, European Patent Specification No. EP 0 772 675 B1, 2001.
67. Delanghe, L., Strubbe, H., and Verzele, M., On the composition of hop wax, *J. Inst. Brew.*, 75:445–449, 1969.
68. Guinard, J-X., Woodmansee, R.D., Billovits, M.J., Hanson, L.G., Gutiérrez, M-J., Snider, M.L., Miranda, M.G., and Lewis, M.J., The microbiology of dry-hopping, *Tech. Q. Master Brew. Assoc. Am.*, 27:83–89, 1990.
69. Foster II, R.T. and Nickerson, G.B., Changes in hop oil content and hoppiness potential (sigma) during hop aging, *J. Am. Soc. Brew. Chem.*, 43:127–135, 1985.
70. Green, C.P., Kinetics of hop storage, *J. Inst. Brew.*, 84:312–314, 1978.
71. Green, C.P., Effect of storage changes on the identification of hops by essential oil analysis, *Eur. Brew. Conv. Monograph XXII*, Symposium on Hops, Zouterwoude, 1994, pp. 239–246.
72. Ono, M., Kakudo, Y., Yamamoto, R., Nagami, K., and Kumada, J., Simultaneous analysis of hop bittering components by high-performance liquid chromatography. II. Evaluation of hop deterioration, *J. Am. Soc. Brew. Chem.*, 45:61–69, 1987.
73. Forster, A., Hops and hop pellets. The influence of the water content on the storage behaviour, *Brauwelt Int.*, 151–154, 1985.
74. Hysert, D.W., White Jr, J.A., Cuzzillo, B.R., and Garden, S.W., Fire loss prevention, self-heating and spontaneous combustion of hops, Abstracts of the, 67th Meeting, Am. Soc. Brew. Chem., Tucson, 2002, in *ASBC Newsletter* 62(2), O-3, pp. PB11.
75. Likens, S.T. and Nickerson, G.B., Factors controlling the storage stability of hops. *Proc. Am. Soc. Brew. Chem.*, 1973, pp. 62–66.
76. Wain, J., Bath, N.A., and Laws, D.R.J., Effects of storage on hops and hop pellets, *Proc, 16th Cong. Eur. Brew. Conv.*, Amsterdam, 1977, pp. 167–177.
77. Menary, R.C., Williams, E.A., and Doe, P.E., Enzymic degradation of alpha acids in hops, *J. Inst. Brew.*, 89:200–203, 1983.
78. Laws, D.R.J., Hop resins and beer flavour. V. The significance of oxidised hop resins in brewing, *J. Inst. Brew.*, 74:178–182, 1968.
79. Menary, R.C., Williams, E.A., and Nickerson, G.B., Effect of myrtenol on the rate of oxidation of alpha- and beta-acids in hops, *Proc, 5th Int. Symp. on Medicinal, Aromatic and Spice Plants*, Mungpoo/Darjeeling, 1985, in *Acta Horticult.* 188(Sect V):149–156, 1986.

80. Pickett, J.A. and Sharpe, F.R., Effect of reduction in hop oil content on rate of deterioration of alpha-acid in hops, *J. Inst. Brew.*, 82:333, 1976.
81. Peacock, V., Coleman, M., Buholzer, W., and Smith, D., Optimal drying, conditioning and baling of hops to be stored and used as whole cones, *Proc. Tech. Comm. Int. Hop. Grow. Conv.*, Int. Hop. Cong., Munich, 1996, pp. 18–28.
82. Weber, K.A., Jangaard, N.O., and Foster II., R.T., Effects of postharvest handling on quality and storage stability of Cascade hops, *J. Am. Soc. Brew. Chem.*, 37:58–60, 1979.
83. Moir, M. and Smith, R.J., Foam-enhancing and bitter, reduced iso-α-acids derived from hops, *Proc. 25th Cong. Eur. Brew. Conv.*, Brussels, 1995, pp. 125–134.
84. Hums, N., The bitter flavour of beer, *Brauwissenschaft*, 26:159–164, 1973.
85. Rehberger, A.J. and Bradee, L.H., Hop oxidative transformations and control of beer bitterness, *Tech. Q. Master Brew. Assoc. Am.*, 12:1–8, 1975.
86. Laws, D.R.J. and McGuinness, J.D., Analytical and brewing significance of a tricyclic oxidation product of humulone (tricyclodehydroisohumulone), *J. Inst. Brew.*, 80:174–180, 1974.
87. Palamand, S.R. and Aldenhoff, J.M., Bitter tasting compounds of beer. Chemistry and taste properties of some hop resin compounds, *J. Agric. Food. Chem.*, 21:535–543, 1973.
88. Koller, H., Magnesium ion catalysed isomerisation of humulone; a new route to pure isohumulones, *J. Inst. Brew.*, 75:175–179, 1969.
89. Hughes, P. and Marinova, G., Variations in iso-α-acid distribution and behaviour of different hop products. *Brewers' Guardian*, March 1997, pp. 26–29.
90. Irwin, A.J., Murray, C.R., and Thompson, D.J., An investigation of the relationship between hopping rate, time of boil, and individual alpha-acid utilization, *J. Am. Soc. Brew. Chem.*, 43:145–152, 1985.
91. Wackerbauer, K. and Balzer, U., Hop bitter compounds in beer. Part I. Changes in the composition of bitter substances during brewing, *Brauwelt Int.*, 144–146, 148, 1992.
92. De Cooman, L., Aerts, G., Witters, A., De Ridder, M., Boeykens, A., Goiris, K., and De Keukeleire, D., Comparative study of the stability of iso-α-acids, dihydroiso-α-acids, and tetrahydroiso-α-acids during beer ageing, *Proc. 28th Cong. Eur. Brew. Conv.*, Budapest, 2001, pp. 566–575.
93. De Cooman, L., Aerts, G., Overmeire, H., and De Keukeleire, D., Alterations of the profiles of iso-α-acids during beer ageing, marked instability of *trans*-iso-α-acids and implications for beer bitterness consistency in relation to tetrahydroiso-α-acids, *J. Inst. Brew.*, 106:169–178, 2000.
94. Wilson, R.J.H., Roberts, T., Smith, R.J., and Biendl, M., Improving hop utilization and flavor control through the use of pre-isomerized products in the brewery kettle, *Tech. Q. Master Brew. Assoc. Am.*, 38:11–21, 2001.
95. Hughes, P.S. and Simpson, W.J., Interactions between hop bitter acids and metal cations assessed by ultra-violet spectrophotometry, *Cerevisia Belg. J. Brew. Biotechnol.*, 20:35–39, 1995.
96. Hughes, P., The significance of iso-α-acids for beer quality, *J. Inst. Brew.*, 106:271–276, 2000.
97. Hughes, P. and Simpson, W.J., Bitterness of congeners and stereoisomers of hop-derived bitter acids found in beer, *J. Am. Soc. Brew. Chem.*, 54:234–237, 1996.
98. Rigby, F.L., A theory on the hop flavor of beer, *Proc. Am. Soc. Brew. Chem.*, 1972, pp. 46–50.

99. Wackerbauer, K. and Balzer, U., Hop bitter compounds in beer. Part II. The influence of cohumulone on beer quality, *Brauwelt Int.*, 116–118, 1993.
100. Diffor, D.W., Likens, S.T., Rehberger, A.J., and Burkhardt, R.J., The effect of isohumulone/isocohumulone ratio on beer head retention, *J. Am. Soc. Brew. Chem.*, 36:63–65, 1978.
101. Wilson, R.J.H., Roberts, T.R., Smith, R.J., Bradley, L.L., and Moir, M., The inherent foam stabilising and lacing properties of some minor hop-derived constituents of beer. *Eur. Brew. Conv. Monograph XXVII*, Symposium on Beer Foam Quality, Amsterdam, 1998, pp. 188–207.
102. Irwin, A.J., Bordeleau, L., and Barker, R.L., Model studies and flavor threshold determination of, 3-methyl-2-butene-1-thiol in beer, *J. Am. Soc. Brew. Chem.*, 51:1–3, 1993.
103. Templar, J., Arrigan, K., and Simpson, W.J., Formation, measurement and significance of lightstruck flavor in beer: a review, *Brew. Dig.*, May: 18–25, 1995.
104. Todd, P.H., Held, R.W., and Guzinski, J.G., The development and use of modified hop extracts in the art of brewing, *Tech. Q. Master Brew. Assoc. Am.*, 33:91–95, 1996.
105. Todd, P.H., Johnson, P.A., and Worden, L.R., Evaluation of the relative bitterness and light stability of reduced iso-alpha acids, *Tech. Q. Master Brew. Assoc. Am.*, 9:31–35, 1972.
106. *EBC Manual of Good Practice*, Hops and Hop Products, Gertränke Fachverlag Hans Carl, Nürnberg, Germany, 1997, pp. 99–100.
107. Forster, A., Beck, B., and Schmidt, R., Problemstoffe des Hopfens — Gedanken und Untersuchungen. (Problem materials in hops — thoughts and investigations), *Brauwelt* 130:930–932, 934–935, 938–940, 1990.
108. Biendl, M., Optimisation of ethanol hop extraction, *Brauwelt Int.*, 12:311–315, 1994.
109. Rajaraman, K., Narayanan, C.S., and Mathew, A.G., Extraction of natural Products with liquid and super critical carbon dioxide, *Indian Food Ind.*, 3(April/June):48–51, 1984.
110. Forster, A., Balzer, U., and Mitter, W., Commercial brewing tests with different hop extracts, *Brauwelt Int.*, 14:324–326, 1996.
111. Biendl, M., Hug, H., and Anderegg, P., Halbtechnische brauversuche mit ethanol- und CO_2-Reinharzextrakten (Semiscale brewing trials with ethanol and carbon dioxide pure resin extracts), *Brau Rundsch.*, 104:7–9, 1993.
112. Grant, H.L., Stabilized hop pellets, *Tech. Q. Master Brew. Assoc. Am.*, 16:79–81, 1979.
113. Burkhardt, R.J. and Wilson, R.J.H., Process for the preparation of isomerized hop pellets., U.S. Patent No. 4,946,691, 1990.
114. Taylor, D., Humphrey, P.M., Yorston, B., Wilson, R.J.H., Roberts, T.R., and Biendl, M., A guide to the use of pre-isomerized hop pellets, including aroma varieties, *Tech. Q. Master Brew. Assoc. Am.*, 37:225–231, 2000.
115. Wilson, R.J.H., Properties and value of isomerised hop products, *Grist Int.*, March/April: 23–25, 1988.
116. Wilson, R.J.H., The future for pellets and related products, *Eur. Brew. Conv. Monograph XIII*, Symposium on Hops, Freising, 1987, pp. 216–229.
117. Smith, R.J. and Wilson, R.J.H., Production of isomerized hop extract, U.S. Patent No. 5,370,897, 1994.

118. Weiss, A., Schönberger, Ch., Mitter, W., Biendl, M., Back, W., and Krottenhaler, M., Sensory and analytical characterisation of reduced, isomerised hop extracts and their influence and use in beer, *J. Inst Brew.*, 108:236–242, 2002.
119. Goldstein, H. and Ting, P., Post kettle bittering compounds: analysis, taste, foam and light stability, *Eur. Brew. Conv. Monograph XXII*, Symposium on Hops, Zouterwoude, 1994, pp. 141–164.
120. Seldeslachts, D., De Winter, D., Mélotte, L., De Bock, A., and Haerinck, R., The use of high tech hopping in practice, *Proc. 27th Cong. Eur. Brew. Conv.*, Cannes 1999, pp. 291–297.
121. Roberts, T.R., Smith, R.J., and Wilson, R.J.H., Expanding the options for achieving bitterness in beer through the use of potassium salt concentrates of isomerised and reduced isomerised hop extracts, *Proc. 27th Conv. Inst. & Guild of Brew. (Asia Pacific Sect.)*, Adelaide, 2002, Poster No. 12.
122. Scott, R.W., Theaker, P.D., and Marsh, A.S., Use of extracts prepared with liquid carbon dioxide as a substitute for dry hops, *J. Inst. Brew.*, 87:252–258, 1981.
123. Bett, G., Grimmett, C.M., and Laws, D.R.J., Addition of hop extracts to beer using liquid carbon dioxide, *J. Inst Brew.*, 86:175–176, 1980.
124. Grimmett, C.M. and Reed, R.J., Production-scale apparatus for the injection of hop extracts into beer using liquid carbon dioxide, *J. Inst. Brew.*, 91:166–168, 1984.
125. Marriott, R., Hop products — opportunities for product development and diversification, *Proc. Conv. Inst. Brew. (Africa Sect)*, Nairobi, 1999, pp. 140–146.
126. Wilson, R.J.H. and Fincher, J.M., Unlocking the potential of hops. *Proc. 5th Central & South African Sect. Inst. Brew.*, 1995, pp. 255–263.
127. Seaton, J.C., Moir, M., and Suggett, A., The refinement of hop flavour by yeast action, *Curr. Dev. Malt Brew Distill.* (Proc. 1st Aviemore Conf.), 1983, pp. 111–128.
128. Peacock, V.E. and Deinzer, M.L., Fate of hop oil components in beer, *J. Am. Soc. Brew. Chem.*, 46:104–107, 1988.
129. Forster, A. and Kaltner, D. Der Einfluss der Hopfenpressung auf die Lupulindrüsen. (The effects of hop-pressing on the lupulin glands), *Hopfenrundschau Int.*, 52–55, 2001/2002.
130. Harris, J. Hop evaluation and selection, *Tech. Q. Master Brew. Assoc. Am.*, 37: 97–103, 2000.
131. Anderegg, P. and Pfenninger, H., The influence of sample size and preparation on the conductometric value and alpha acid content of hop products, *Brew. Dig.*, October: 28–33, 1986.
132. Verzele, M., Van Dyck, J., and Claus, H., On the analysis of hop bitter acid, *J. Inst. Brew.*, 86:9–14, 1980.
133. Laufer, S., Hop bitter flavour of beer and its control, *Am. Brewer*, June: 21–29, 1965.
134. Alderton, G., Bailey, G.F., Lewis, J.C., and Stitt, F., Spectrophotometric determination of humulone complex and lupulone in hops, *Anal. Chem.*, 26:983–992, 1954.
135. Likens, S.T., Nickerson, G.B., and Zimmermann, C.E., An index of deterioration in hops (*Humulus lupulus*), *Proc. Am. Soc. Brew. Chem.*, 68–74, 1970.
136. Maye, J.P., Mulqueen, S., Xu, J., and Weis, S. Spectrophotometric analysis of isomerized α-acids, *J. Am. Soc. Brew. Chem.*, 60:98–100, 2002.
137. Wilson, R.J.H., HPLC analysis of hops and hop products: some problems and solutions, *Ferment*, 2:241–248, 1989.

138. Green, C.P. and Osborne, P., Effects of solvent quality on the analysis of hops, *J. Inst. Brew.*, 99:223–225, 1993.
139. Verzele, M. and De Keukeleire, D., *Chemistry and Analysis of Hop and Beer Bitter Acids*, Elsevier Science, Amsterdam, 1991, pp. 358–360.
140. Biendl, M. and Hartl, A., Dicyclohexylamine-*trans*-Iso-alpha-Saeuren-Komplex als externer Eichstandard zur HPLC-Bestimmung von Iso-alpha-saeuren. (Dicyclohexylamine/*trans*-iso-alpha-acid complex as an external calibration standard for the HPLC determination of iso-alpha-acids), *Monatsschr. Brauwiss.*, 48: 102–107, 1995.
141. Verzele, M. and De Keukeleire, D., *Chemistry and Analysis of Hop and Beer Bitter Acids*, Elsevier Science, Amsterdam, 1991, pp. 352–353.
142. Harms, D. and Nitzsche, F., High-performance separation of unmodified and reduced hop and beer bitter compounds by a single high-performance liquid chromatographic method, *J. Am. Soc. Brew. Chem.*, 59:28–31, 2001.
143. Verzele, M., Steenbeke, G., Dewaele, C., Verhagen, L.C., and Strating, J., Advances in the liquid chromatography of hop bitter compounds, *J. High. Res. Chromatogr.*, 13:737–741, 1990.
144. Burroughs, L.J. and Williams, P.D., A single HPLC method for complete separation of unmodified and reduced iso-alpha acids, *Proc. 27th Cong. Eur. Brew. Conv.*, Cannes, 1999, pp. 283–290.
145. Wilson, R.J.H., The new international calibration standards (ICS), *Proc, 28th Cong. Eur. Brew. Conv.*, Budapest, 2001, pp. 1018–1023.
146. Maye, J.P., Mulqueen, S., Weis, S., Xu, J., and Priest, M., Preparation of isomerized α-acid standards for HPLC analysis of iso-α-acids, *rho*-iso-α-acids, tetrahydro-iso-α-acids, and hexahydro-iso-α-acids, *J. Am. Soc. Brew. Chem.*, 57:55–59, 1999.
147. Szücs, R., Vindevogel, J., Everaert, E., De Cooman, L., Sandra, P., and De Keukeleire, D., Separation and quantification of all main hop acids in different hop cultivars by microemulsion electrokinetic chromatography, *J. Inst. Brew.*, 100:293–296, 1994.
148. Garden, S.W., Pruneda, T., Irby, S., and Hysert, D.W., Development of near-infrared calibrations for hop analysis, *J. Am. Soc. Brew. Chem.*, 58:73–82, 2000.
149. Halsey, S.A., Near infrared reflectance analysis of whole hop cones, *J. Inst. Brew.*, 93:399–404, 1987.
150. Pickett, J.A., Coates, J., and Sharpe, F.R., Improvement of hop aroma in beer, *Proc. 15th Cong. Eur. Brew. Conv.*, Nice, 1975, pp. 123–140.
151. *Analytica — EBC*, 3rd ed., 1975, Method 6.4: the hydrogen sulphide test for hops. (Method taken from Brenner, M.W., Owades, J.L., and Golyzniak, R., *Proc. Am. Soc. Brew. Chem.*, 1953, p. 53).
152. Buckee, G.K. Detection of the sulphuring of hops during kilning and cultivation, *J. Inst. Brew.*, 87:360–364, 1981.
153. Owades, J.L. and Plam, M., The influence of brewing processes on volatile sulfur compounds in beer, *Tech. Q. Master Brew. Assoc. Am.*, 25:134–136, 1988.
154. Dufour, J-P., Marriott, P., Reboul, E., Leus, M., Beatson, R., and Silcock, P., Application of comprehensive multidimensional gas chromatography for high resolution analysis of hop essential oil, *Proc. Conv. Inst. Guild Brew. (Asia Pacific Section)*, Adelaide, 2002, Presentation No. 9 (5 pp.).

155. Field, J.A., Nickerson, G., James, D.D., and Heider, C., Determination of essential oils in hops by headspace solid-phase microextraction *J. Agric. Food Chem.*, 44:1768–1772, 1996.
156. Jerumanis, J., Separation and identification of flavanoids by high-performance liquid chromatography (HPLC), *Proc. 17th Cong. Eur. Brew. Conv.*, Berlin, 1979, pp. 309–319.
157. Jerumanis, J., Quantitative analysis of flavanoids in barley, hops and beer by high-performance liquid chromatography (HPLC), *J. Inst. Brew.*, 91:250–252, 1985.
158. Moir, M., Raw materials – prediction of their brewing behaviour hops, *Ferment*, 3:97–102, 1990.
159. Jorge, K. and Trugo, L.C., Quality control of beer hopped with reduced isomerized products, *Tech. Q. Master Brew. Assoc. Am.*, 37:219–224, 2000.
160. Taylor, D.G., Cecil, M.J., and Humphrey, P.M., A practical guide to the use of modern hop products, *Inst Brew. Proc. 5th C & SA Sect.*, Victoria Falls, 1995, pp. 248–254.
161. Clark, J.G., Burroughs, L.J., and Guzinski, J.A., High-performance liquid chromatography analysis of tetrahydro-iso-α-acids in beer, *J. Am. Soc. Brew. Chem.*, 56:76–79, 1998.
162. Kaltner, D., Forster, C., Thum, B., and Back, W., Untersuchungen zur Ausbildung des Hopfenaromas Während des Brauprozesses. (Investigations into the development of hop flavour during the brewing process), *Proc. 27th Cong. Eur. Brew. Conv.*, Cannes, 1999, pp. 63–70.
163. Irwin, A.J., Varietal dependence of hop flavour volatiles in lager, *J. Inst. Brew.*, 95:185–194, 1989.
164. Irwin, A.J., A comparative analysis of hop flavour constituents in lagers brewed with single hop varieties, *Proc. 21st Cong. Eur. Brew. Conv.*, Madrid, 1987, pp. 329–336.
165. Murakami, A.A., Rader, S., Chicoye, E., and Goldstein, H., Effect of hopping on the headspace volatile composition of beer, *J. Am. Soc. Brew. Chem.*, 47:35–42, 1989.
166. Sanchez, N.B., Lederer, C.L., Nickerson, G.B., Libbey, L.M., and McDaniel, M.R., Sensory and analytical evaluation of beers brewed with three varieties of hops and an unhopped beer in *Food Science and Human Nutrition* (Dev Food Sci 29), Charalambous, G. Ed., 1992, pp. 403–426.
167. Newton, J. and Cairns, L. The development of a solid phase micro-extraction method for the analysis of hop aroma compounds in beer, *Proc. Conv. Inst. Brew. (Asia Pacific Section)*, Singapore, 2000, pp. 155–157.

Further Reading

1. Burgess, A.H., *Hops*, Interscience Publishers, New York, 1964.
2. *European Brewery Convention Manual of Good Practice — Hops and Hop Products*. Getränke-Fachverlag Hans Carl, Nürnberg, 1997.
3. The Institute of Brewing. *An Introduction to Brewing Science & Technology*, Series II, Vol. 1, Hops.

4. Neve, R.A., *Hops*, Chapman and Hall, London, 1991.
5. *European Brewery Convention Monograph XIII*, Symposium on Hops, Friesing, 1987.
6. *European Brewery Convention Monograph XXII*, Symposium on Hops, Zouterwoude, 1994.
7. Barth, H.J., Klinke, C., and Schmidt, C., *The Hop Atlas*, Joh. Barth & Sohn, 1994.
8. Verzele, M. and De Keukeleire, D., *Chemistry and Analysis of Hop and Beer Bitter Acids*, Elsevier Science, Amsterdam, 1991.

8
Yeast

Inge Russell

CONTENTS

Taxonomy of Yeast	282
Structure of Yeast	283
Cell Wall	284
Plasma Membrane	286
The Periplasmic Space	287
Nucleus	287
Mitochondria	287
Other Cytoplasmic Structures	287
Vacuoles	287
Peroxisomes	288
Endoplasmic Reticulum	288
Golgi Complex	288
Cytosol	288
Cytoskeleton	289
Life Cycle and Genetics	289
Vegetative Reproduction	289
Genome	290
Strain Improvement	290
Mutation and Selection	290
Hybridization	292
Rare Mating	294
Cytoduction	294
Killer Yeast	295
Genetic Manipulation	296
Spheroplast Fusion	296
Recombinant DNA	297
Nutritional Requirements	298
Oxygen Requirements	298
Lipid Metabolism	299
Uptake and Metabolism of Wort Carbohydrates	301

    Maltose Uptake 302
    Uptake and Metabolism of Wort Nitrogen 304
    Yeast Excretion Products 306
        Alcohols 306
        Esters 307
        Sulfur Compounds 307
        Carbonyl Compounds 308
        Diacetyl and Pentane-2,3-dione 309
Flocculation 310
Glycogen and Trehalose — The Yeast's Carbohydrate
  Storage Polymers 312
Pure Yeast Cultures 313
    Introduction 313
    Strain Selection 314
    Storage of Cultures 314
    Propagation and Scale-Up 316
    Contamination of Cultures 317
    Yeast Washing 318
    Yeast Pitching and Cell Viability 319
    Yeast Collection 320
    Yeast Storage 321
    Shipping of Yeast 321
Glossary 322
Acknowledgments 323
References 324

Taxonomy of Yeast

Interest in brewing yeast centers around its different strains and there are thousands of unique strains of *Saccharomyces cerevisiae*. These strains encompass brewing, baking, wine-distilling, and laboratory cultures. There is a problem classifying such strains in the brewing context; the minor differences between strains that the taxonomist dismisses can be of great technical importance to the brewer. *Saccharomyces*, Latin for sugar fungus, is the name first used for yeast in 1838 by Meyen, but it was the work of Hansen at the Carlsberg laboratory in Denmark during the 1880s that gave us the species names of *S. cerevisiae* for head-forming yeast used in ale fermentations and *S. carlsbergensis* for non-head-forming yeast associated with the lower temperature range of lager fermentations. Historically, lager yeast and ale yeast have been taxonomically distinguished on the basis of their ability to ferment the disaccharide melibiose. Strains of lager yeast possess the *MEL* genes, produce the extracellular enzyme α-galactosidase (melibiase), and are able to utilize melibiose, whereas ale strains do not produce α-galactosidase and, therefore, are unable to utilize melibiose. Traditionally, lager is produced by bottom-fermenting yeasts at fermentation temperatures between 7 and 15°C and, at the end of fermentation, these yeasts flocculate and collect at the bottom of the fermenter. Top-fermenting yeasts, used for the production of ale at fermentation temperatures between 18 and 22°C, tend to be somewhat less flocculent, and loose clumps of cells adsorbed to carbon dioxide bubbles are carried to the surface of the fermenting wort. Consequently, top yeasts are collected by skimming from the surface of the fermenting wort, whereas bottom yeasts are collected, or cropped, from the fermenter bottom. The differentiation of lagers and ales on the basis of bottom and top cropping has become less distinct with the advent of vertical bottom fermenters and centrifuges. There is much greater diversity between ale strains than between lager strains.

The taxonomy surrounding the yeast *Saccharomyces* is confusing and still changing. *Saccharomyces sensu stricto* is a species complex that includes most of the yeast strains relevant in the fermentation industry as well as in basic science (i.e., *S. bayanus*, *S. cerevisiae*, *S. paradoxus*, and *S. pastorianus*). Names of species and single isolates have, and are still undergoing, changes that cause confusion for yeast scientists and fermentation technologists.

There are three main reasons that taxonomists change yeast names. These are: (i) the yeast was incorrectly described or named; (ii) new observations warrant such change; or (iii) a prior published legitimate name is uncovered.[1] Unfortunately, in the past, changes in nomenclature varied with changes in the criteria taxonomists believed should be used. These varied from phenotypic characteristics (such as microscopic appearance and ability to use certain substrates), nutritional characteristics (which could mutate), or the criteria of interfertility within strains of the same species. Newer techniques of molecular taxonomy and DNA relatedness are now being used for yeast classification.[2]

In 1970, taxonomists repositioned the lager yeast *S. carlsbergensis* as *S. uvarum*. In 1990, taxonomists repositioned *S. uvarum* as part of *S. cerevisiae*. Then *S. cerevisiae* var. *carlsbergensis* was classified as *S. pastorianus* and often written as *S. pastorianus/carlsbergensis* for clarity. There was an argument for keeping the name *S. carlsbergensis* for lager-brewing yeast rather than using the name *S. pastorianus*.[3] Recent findings have demonstrated that *S. bayanus* and *S. pastorianus* are not homogeneous and do not seem to be natural groups.[4] For simplicity, brewing yeast will be referred to as *S. cerevisiae* throughout this chapter unless distinct reference is made to lager yeast.

Structure of Yeast

The group of microorganisms known as "yeast" is by traditional agreement limited to fungi in which the unicellular form is predominant (Figure 8.1).[5]

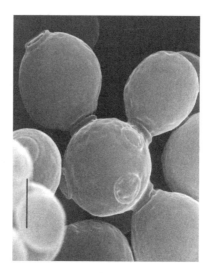

FIGURE 8.1
Electron micrograph of budding yeast cell. (With thanks to Alastair Pringle Anheuser Busch, Inc.) — The bar represents 5 μm.

Figure 8.2 illustrates the main features of a typical yeast cell. Yeasts vary in size, from roughly 5 to 10 μm length to a breadth of 5–7 μm. The mean cell size varies with the growth cycle stage, fermentation conditions, and cell age (older cells are larger in size).

Cell Wall

The yeast cell wall is a multifunctional organelle of protection, shape, cell interaction, reception, attachment, and specialized enzymic activity. The cell wall, which is 100–200 nm thick, constitutes 15–25% of the dry weight of the cell and consists primarily of equal amounts of phosphomannan (31%) and glucans (29%). There are three glucans present in the wall (Figure 8.3). The major component is an alkali-insoluble, acid-insoluble β-1,3-linked polymer, which helps the wall maintain its rigidity. There is also an alkali-soluble branched glucan, with predominantly β-1,3-linkages, but with some β-1,6-linkages as well. Finally, the cell wall also contains a small portion of predominantly β-1,6-linked glucan. Chitin, a polymer of N-acetylglucosamine, present in small quantities (2–4%) is almost always restricted to the bud scar. Mannans are present as an α-1,6-linked inner core with α-1,2- and α-1,3-side chains. Lipid is present at about 8.5% and protein at about 13%. The carbohydrate portion of the mannoprotein on the yeast cell surface determines the immunochemical properties of the cell. The exact composition of the cell wall is dependent on growth conditions, age of the culture, and the specific yeast strain.

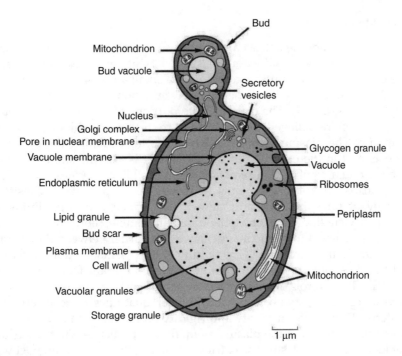

FIGURE 8.2
Main features of a typical yeast cell.

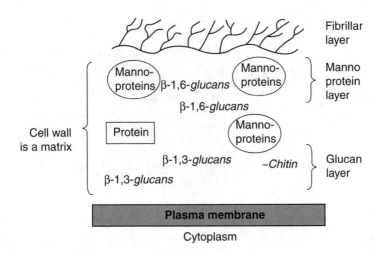

FIGURE 8.3
Simplified structure of the yeast cell wall.

Plasma Membrane

The plasma membrane acts as a barrier to separate the aqueous interior of the cell from its aqueous exterior. It consists of lipids and proteins, in more or less equal amounts, together with a small amount of carbohydrate. It is 8–10 nm thick, with occasional invaginations protruding into the cytoplasm (Figure 8.4). The carbohydrate portions of the membrane-bound glycoproteins are believed to extend only from the external surface of the membrane. The plasma membrane has a role in regulating the uptake of nutrients and in the excretion of metabolites. It is also the site of cell wall synthesis and assembly, and secretion of extracellular enzymes.

The intrinsic membrane proteins are inserted in a lipid bilayer consisting primarily of phospholipids and sterols. The principal phospholipids are phosphatidylcholine and phosphatidylethanolamine, and their primary function is believed to be the maintenance of a barrier between the external and internal cell environment. The fluidity of the plasma membrane is regulated by phospholipid unsaturation and is an important factor in ethanol tolerance. The sterols in the plasma membrane are ergosterol, 24(28)-dehydro-ergosterol, zymosterol, and smaller quantities of fecosterol and lanosterol. They confer integrity and rigidity on the membrane.

The main function of the plasma membrane is to dictate what enters and what leaves the cytoplasm. The extrinsic proteins, those which interact with membrane lipids and proteins by polar binding, cover part of the bilayer

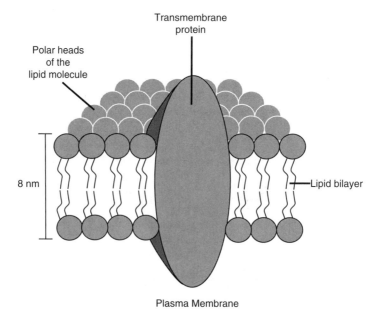

FIGURE 8.4
Yeast plasma membrane.

surface and a large group of these proteins are involved in solute transport.[6] Brewer's yeast requires initial oxygenation, prior to fermentation, to ensure correct plasma membrane synthesis, as oxygen is absolutely required for the synthesis of the unsaturated fatty acids and sterols.

The Periplasmic Space

This is the thin area between the outer surface of the plasma membrane and the inner surface of the cell wall. Secreted proteins, which are unable to permeate the cell wall, are located here. This includes enzymes such as invertase, acid phosphatase, and melibiase. For example, sucrose is broken down in the periplasmic space by the enzyme invertase to fructose and glucose.

Nucleus

The cell nucleus is roughly spherical, about 2 μm in diameter, and is visible with phase contrast microscopy. In resting cells, it is usually situated next to a prominent vacuole. The nucleus consists primarily of DNA and protein and is surrounded by the nuclear membrane. Sixteen individual, linear chromosomal DNA molecules have been identified using pulsed-field gel electrophoresis. The nuclear membrane, which is perforated at intervals with pores, remains intact throughout the cell cycle.

Mitochondria

Electron micrographs of yeast cells reveal round or elongated mitochondrial structures composed of two distinct membranes, the outer and the inner, within the cytoplasm. The cristae within the mitochondria are formed by the folding of the inner membrane. Specific enzymes are associated within four distinct locations: the outer membrane, the intermembrane space, the inner membrane, and the matrix. Most of the enzymes of the tricarboxylic acid cycle are present in the matrix of the mitochondrion. The enzymes involved in electron transport and oxidative phosphorylation are associated with the inner membrane. When yeast is grown aerobically on a nonfermentable carbon source, the mitochondria of the fully respiring cells are rich in the cristae structures. Under aerobic conditions, yeast mitochondria are primarily involved in ATP synthesis during respiration. In glucose-repressed cells, only a few mitochondria with poorly developed cristae can be seen.

Other Cytoplasmic Structures

Vacuoles
Vacuoles are a part of an intramembranous system that includes the endoplasmic reticulum, and are easily seen under the light microscope.

They serve as dynamic stores of nutrients and they provide a site for the breakdown of macromolecules. They are also key in intracellular protein trafficking in yeasts. The form and size of the vacuoles change during the cell cycle. Mature cells contain large vacuoles, which fragment into small vesicles when bud formation is initiated. Later in the cell cycle, the small vacuoles fuse again to produce a single vacuole in the mother and daughter cell. The vacuoles contain proteases and hydrolases, and they also store metabolites such as amino acids. Vacuoles are bounded by a single membrane called the tonoplast. The secretory pathway in yeast is believed to work in the following fashion: endoplasmic reticulum → Golgi complex → vesicle → cell surface or vacuolar compartment. This sequence allows the transport of soluble and membrane-bound proteins, both for extracellular secretion and for assembly of the vacuole.[7]

Peroxisomes

Peroxisomes are sealed vesicles surrounded by a single membrane. Yeast peroxisomes perform a variety of functions and are also the sites of fatty acid degradation. The number and volume of the peroxisomes change depending on the external conditions. Dedicated membrane proteins are required to allow communication across the peroxisomal membrane.

Endoplasmic Reticulum

The endoplasmic reticulum (ER) is a large organelle formed as a network of tubules and sacs that extends from the nucleus. The whole ER is surrounded by a continuous membrane. The ER is divided into two parts: the rough ER on which the ribosomes are attached and the smooth ER. Secreted proteins are synthesized by the membrane-bound ribosomes and transported into the ER, where they are processed, and then either retained or transported further to the Golgi apparatus.

Golgi Complex

The Golgi complex is an organelle built of several flattened, membrane-enclosed sacs. It packages large molecules for secretion in secretory vesicles.

Cytosol

The cytosol is the prime site for protein synthesis and degradation. The cytosol is contained within the plasma membrane and surrounds the nucleus. It makes up more than half of the cell volume and consists of, among other things, all the free ribosomes and proteasomes. Proteasomes are responsible for the digestion of proteins that may be detrimental to the cell. The cytosol together with the rest of the cell content, except the nucleus, constitutes the cytoplasm.

Cytoskeleton

The cytoskeleton gives the cell its mobility and support. The shape and movement of the cell is controlled by three main components. The microtubules and actin filaments give the cell support and movement. The intermediate filaments are built of several different proteins and play a supportive role.

Life Cycle and Genetics

Vegetative Reproduction

Most brewing strains are diploid, polyploid, or aneuploid, whereas most laboratory strains are haploid. The yeast cell cycle refers to the repeated pattern of events that occur between the formation of a cell by division of its mother cell and the time when the cell itself divides. Individual yeast cells are mortal. Yeast aging is a function of the number of divisions undertaken by the individual cell and is not a function of the cell's chronological age. All yeast cells have a set lifespan determined both by genetics and environment. The maximum division capacity of a cell is called the "Hayflick limit." Once a cell reaches this limit, it cannot replicate further and enters a stage of senescence and then death. A yeast cell can divide to produce 10–33 daughter cells. Research with industrial ale strains showed a high of 21.7 ± 7.5 divisions to a low of 10.3 ± 4.7 divisions.[8] When yeast is examined under the microscope, the following are signs of aging: an increase in the number of bud scars, an increase in the cell size, surface wrinkles, granularity in the cytoplasm, and the retention of daughter cells. There is a progressive impairment of cellular functions that results in the yeast having a reduced ability to adapt to stress.

As a cell reproduces by budding, a birth scar is formed on the mother cell but not on the daughter cell. As budding continues, an individual cell ages and birth scars accumulate (usually 10–40). Because at any specific time a particular cell gives rise to exactly one cell without a birth scar, while itself assuming a birth scar, at any generation, 50% of the cells have birth scars. By counting the number of birth scars on a cell it is possible to estimate a particular cell's age. The important point to remember is that although individual cells age and die, the total ensemble of cells does not age, and, in theory, a well cared for yeast culture could be used indefinitely. A haploid yeast cell in a rich medium such as wort and at its optimum temperature has a doubling time of approximately 90 min.

The cell cycle can be divided into a number of different phases, with the two major phases being the S phase (synthesis phase where DNA is duplicated) and the M phase (mitosis phase where cell division occurs). Mitosis and nuclear division occupy a relatively short time in the cell cycle. The remainder is divided into a G_1 period, (telophase to the beginning of DNA synthesis), an S period (time of DNA synthesis), and a G_2 period (end of DNA synthesis to prophase). The G_1, S, and G_2 phases together constitute the interphase, and this occupies more than 95% of the cell's time. The gap phases, G_1 and G_2, give the cell time for growth and duplication of organelles.

Genome

Brewing yeasts are polyploid and, in particular, triploid, tetraploid, or aneuploid. Polyploidy brings benefits to the strain in that extra copies of genes for sugar utilization can improve fermentation performance. Polyploid yeasts are also more genetically stable, as it takes multiple mutational events to change them.[9] The lager yeast *S. carlsbergensis*, now reclassified as *S. pastorianus*, is speculated to have arisen from a natural hybridization of two yeast strains, possibly *S. cerevisiae* and *S. bayanus*.[10]

The chromosomes are located in the cell nucleus, and the total haploid yeast genome consists of 12.1 million base pairs. A widely studied laboratory haploid yeast, strain S228C,[11] was chosen for the yeast genome project and its DNA sequence has now been fully characterized. The *Saccharomyces* genome database (SGD) provides a wealth of information at http://www.yeastgenome.org/. To date, over 6116 yeast genes and 96 intergenic regions have been identified. The chromosomes vary in size from 230 to 2000 kb. *Saccharomyces* also contains mobile genetic elements called *Ty*, which can cause rearrangement of the yeast genome. Extrachromosomal elements are also present: the most common being the 2-μm DNA present as a circular plasmid, mitochondrial DNA, a 75-kb circular molecule, and double-stranded RNA (dsRNA).

Strain Improvement

Mutation and Selection

Classical approaches to strain improvement include mutation and selection, screening and selection, and cross-breeding. Mutation is any change that alters the sequence of bases along the DNA molecule, thus modifying the genetic material. The average spontaneous mutation frequency in *S. cerevisiae* at any particular locus is approximately 10^{-6} per generation. Chemical mutagens and physical treatments such as ultraviolet light are used to induce mutation frequencies to detectable levels.

Problems are often encountered with the use of mutagens for brewing strains, as mutagenesis is a destructive process and can cause gross rearrangement of the genome. The mutagenized strains often no longer exhibit many of the desirable properties of the parent strain, and, in addition, may exhibit a slow growth rate and produce a number of undesirable taste and aroma compounds during fermentation. Moreover, mutagenesis is seldom employed with industrial strains: their polyploid nature obscures mutations because of the nonmutated genes. Casey[12] cautions on the use of mutagens, as these hidden undesirable mutations can become expressed after a time lag by such events as chromosome loss, mitotic recombination, or mutation of other wild type alleles. He suggests an analogy to computer viruses in computer software that can surface months later. In a similar way, hidden mutations can later spoil the efforts of a protracted strain improvement program.

Screening of cultures to obtain spontaneous mutants or variants has proved to be a more successful technique, as this avoids the use of destructive mutagens. Some early examples follow. To select for brewery yeast with improved maltose utilization rates, 2-deoxyglucose, a glucose analog, was employed and spontaneous mutants selected, which were resistant to 2-deoxyglucose. These isolates were also found to be derepressed for glucose repression of maltose uptake, thus allowing the cells to take up maltose without first requiring a 50–60% drop in wort glucose levels. This resulted in faster fermentation rates and no alternation in the final flavor of the product.[13,14] Galván et al.[15] reported the isolation of brewing yeast strains with a dominant mutation and resistance to the herbicide sulfometuron methyl (SM). Yeast mutants resistant to SM are dominant and showed a decreased level of acetohydroxyacid synthase. This enzyme is the first enzyme in the isoleucine–valine biosynthetic pathway and produces acetolactate, the precursor of diacetyl in brewing fermentations. Workers at the Carlsberg Research Institute have isolated low diacetyl-producing strains and conducted successful plant trials at the 4500 hl scale.[16]

There are three characteristics routinely encountered resulting from yeast mutation that can be harmful to a fermentation. These are: the tendency of yeast strains to mutate from flocculence to nonflocculence; the loss of ability to ferment maltotriose; and the presence of respiratory deficient mutants. This last group usually consists of cytoplasmic mutants.

The most frequently identified spontaneous mutant found in brewing yeast strains is the respiratory deficient (RD) or "petite" mutation. The RD mutant arises spontaneously through a rearrangement of the mitochondrial genome. It normally occurs at frequencies of between 0.5 and 5% of the yeast population but, in some strains, figures as high as 50% have been reported. The mutant is characterized by deficiencies in mitochondrial function, resulting in a diminished ability to function aerobically and, as a result, these yeasts are unable to metabolize nonfermentable carbon sources, such as lactate, glycerol, or ethanol. Respiratory deficient mutants can range from point mutations (mit^-) to deletion mutations (rho^-) to the complete elimination of mitochondrial DNA (rho^0). Many phenotypic effects occur due to this mutation, and these include alterations in sugar uptake, metabolic by-product formation, and tolerance to stress factors, such as ethanol and temperature. Flocculation, cell wall and plasma membrane structure, and cellular morphology are affected by this mutation.

It should be remembered that beer produced with an yeast RD or that produces a high number of RD mutants is more likely to have flavor defects and fermentation problems.[17,18] Ernandes et al.[19] have reported that beer produced from these mutants contained elevated levels of diacetyl and some higher alcohols. Wort fermentation rates were slower, higher dead cell counts were observed, and biomass production and flocculation ability were reduced.

To test for the presence of respiratory deficient mutants, 5-day-old colonies of yeast grown on peptone-yeast (PY) extract medium are overlaid with triphenyl tetrazolium agar, according to the method of American Society of Brewing Chemists.[20] Respiratory sufficient colonies stain red and RD mutants remain white. Respiratory deficient colonies growing on an agar plate are often smaller in size and hence the name "petites." Since RD mutants lack respiratory chain enzymes, but can still grow by employing glycolysis as a source of ATP, a confirmative test involves streaking a suspect colony onto a PY glucose plate and a PY plate containing a nonfermentable carbon source, such as lactate or glycerol. An RD colony typically grows on the glucose plate but not on the lactate plate.[21]

Hybridization

The study of yeast genetics was pioneered by Winge[22] and his coworkers at the Carlsberg laboratory in Denmark and in 1935 they established the haploid–diploid life cycle (Figure 8.5). *Saccharomyces* can alternate between the haploid (a single set of chromosomes) and diploid (two sets of chromosomes) states. Yeast can display two mating types, designated *a* and α, which are manifested by the extracellular production of an *a*- or an α-mating pheromone. When *a* haploids are mixed with α haploids, mating takes place and diploid zygotes are formed. Under conditions of nutritional deprivation, diploids undergo reduction division by meiosis and differentiate into asci, containing four uninucleate haploid ascospores, two of which are *a* mating type and two of which are α mating type. Ascus walls can be removed by the use of glucanase preparations such as snail gut enzyme. The four spores from each ascus can be isolated by use of a micromanipulator, induced to germinate, tested for their fermentation ability, and subsequently employed for further hybridization work. Both haploid and diploid organisms can exist stably and undergo cell division via mitosis and budding.

Although the technique of hybridization fell into disfavor for a number of years, when recombinant DNA was thought to be the solution to all future gene manipulation requirements, it has gradually come to be accepted again as a valuable technique. There are three prerequisites for the production of a hybrid yeast by this technique. First, it must be possible to

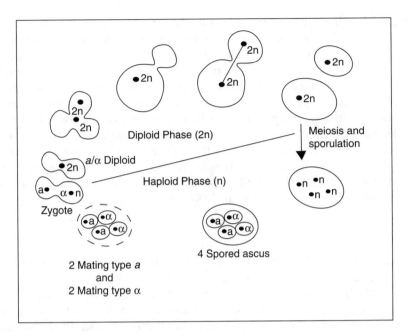

FIGURE 8.5
Haploid/diploid life cycle of *Saccharomyces* spp.

induce sporulation in the parent strains; second, single spore isolates must be viable; and, third, hybridization (mating) must take place when the spores of single-spore cultures are placed in contact with one another. The great stability of industrial strains has been attributed to its following characteristics: little or no mating ability, poor sporulation, and low spore viability. Nevertheless, it is possible to increase sporulation ability in many of these industrial strains, thus making them much more amenable to hybridization.[23]

Employing the classical techniques of spore dissection and cell mating, it has been possible to produce diploid strains with multiple genes for carbohydrate utilization: for example, a diploid that is homozygous for all three known starch-hydrolyzing genes DEX1/DEX1, DEX2/DEX2, STA3/STA3 or triploid strains containing multiple copies of the genes that code for maltose utilization. These strains can then be employed for specific purposes or further improved by fusing them to industrial polyploids with additional desirable characteristics.[24]

Gjermansen and Sigsgaard[25] carried out extensive cross-breeding studies with a *S. uvarum* (*carlsbergensis*) strain and produced hybrids, which were tested at the 575 hl scale and which produced beer of acceptable quality. Similarly, crosses between ale and lager meiotic segregants produced hybrids with faster attenuation rates and produced beers of good palate, which lacked the sulfury character of the lager but retained the estery aroma of the ale.[26] The aforementioned examples demonstrate that although hybridization is an old and "classical" technique, it is still very useful. One recent publication reported on reconfiguring the S. cerevisiae genome so that it was collinear with that of S. mikatae. Researchers demonstrated that this imposed genomic collinearity allowed the generation of inter-specific hybrids, which produced a large proportion of spores that were viable, but extensively aneuploid. They obtained similar results in crosses between wild-type S. cerevisiae and the naturally collinear species S. paradoxus.[27] One of the major advantages to cross-breeding is that this technique carries none of the burden of ethical questions and fears that can sometimes accompany the use of recombinant DNA technology.

Rare Mating

Rare mating, also called forced mating, is a technique that disregards ploidy and mating type and, thus, is ideal for the manipulation of polyploid/aneuploid strains where normal hybridization procedures cannot be utilized. When nonmating strains are mixed at a high cell density, a few hybrids that have fused nuclei form, and these can usually be isolated using appropriate selection markers. A possible disadvantage to this method is that while incorporating the nuclear genes from the brewing strain, the rare mating product can also inherit undesirable properties from the other partner, which is often a nonbrewing strain. A good example of this is the work of Tubb and coworkers[28] who constructed dextrin-fermenting brewing strains, but introduced the *POF* gene (*phenolic-off-flavor*), which imparted the ability to decarboxylate wort ferulic acid to 4-vinyl guaiacol, and gave the resulting beer a phenolic clove-like off-flavor. This made the hybrid product unsuitable for commercial use from a taste perspective but acceptable from a dextrin utilization standpoint.

Cytoduction

Cytoduction is a specialized form of rare mating in which only the cytoplasmic components of the donor strain are transferred into the brewing strain, that is, the cell receives the cytoplasm from both parents but retains the nucleus of only one parent. The process of cytoduction requires the presence of a specific nuclear gene mutation designated *Kar*, for karyogamy defective. This mutation impairs nuclear fusion.[29] Cytoduction can be used in three ways: substitution of the mitochondrial genome; introduction of DNA plasmids; or transfer of dsRNA species. When used in the substitution of the mitochondrial genome, it is possible to study the effects of these genetic elements on various cell functions. Mitochondrial substitution has been demonstrated to bring about variations in respiratory functions, cell surface activities, and various other strain characteristics.[30] In addition, rare mating, used to introduce DNA plasmids, has been successful in the introduction of specific genetic elements constructed by gene-cloning experiments.[31] Lastly, rare mating has also been used to transfer a "zymocidal" or "killer" factor from laboratory haploid strains to brewing yeast strains, without altering the primary fermentation characteristics of the brewing yeast strain.[31–33]

Killer Yeast

In 1963, Bevan and Makower discovered the killer phenomenon in a *S. cerevisiae* strain, and it was isolated as a brewery contaminant.[34] During the past four decades, intensive investigations of the various killer systems in yeast have resulted in substantial progress in many different fields of biology, providing important insights into basic and more general aspects of eukaryotic cell biology, virus–host cell interactions, and yeast virology.

In *Saccharomyces*, killer strains secrete a protein toxin that is lethal to sensitive strains of the same genus and, less frequently, strains of different genera. Among the yeasts, killer, sensitive, and neutral strains have been described. The killer toxin of *S. cerevisiae* kills sensitive cells of the same species by disturbing the ion gradient across the plasma membrane after binding to the receptor at cell wall β-1,6-glucan. The "killer" character of *Saccharomyces* spp. is determined by the presence of two species of cytoplasmically located dsRNA plasmids. The M-dsRNA (1.0–1.8 kb) "killer" plasmid is killer-strain specific and codes for killer toxin and also for the immunity factor, which is a protein or proteins that make the host immune to the toxin (i.e., prevents self-killing). The L-dsRNA, which is also present in many "nonkiller" yeast strains, codes for the production of a protein that encapsulates both forms of dsRNA, thereby yielding virus-like particles. These virus-like particles are not naturally transmitted from cell to cell by any infection process. The killer plasmid behaves as a true cytoplasmic element, showing dominant non-Mendelian segregation. It depends, however, on a number of chromosomal genes for its maintenance in the cell and for expression. Cells of killer strains normally contain about 12 copies of the M-dsRNA and 100 copies of the L-dsRNA. The yeast can be cured of the M-dsRNA by growth at elevated temperature or by treatment with cycloheximide.[35]

Based on the lack of cross-immunity, their molecular mode of action, and their killing profiles, toxin-producing *S. cerevisiae* killer strains have been classified into three major groups (K1, K2, and K28). Each secretes a unique killer toxin as well as a specific but as yet unidentified immunity components that render the killer cells immune to their own toxins. The production of killer toxin K1, K2, or K28 is associated with the presence of a cytoplasmically inherited M-dsRNA satellite virus (designated ScV-M1, ScV-M2, or ScV-M28 for *S. cerevisiae* virus). This virus depends on the coexistence of L-A helper virus to be stably maintained and replicated within the cytoplasm of the infected host cell. For a recent review see Schmitt and Breinig.[36]

Brewing strains can be modified such that they are both resistant to killing by a zymocidal yeast and so that they themselves have zymocidal activity, thereby eliminating contaminating yeasts. Rare mating has been successfully employed to produce brewing killer yeast by crossing a brewing lager yeast with a *Kar* killer strain.[32] Beer produced with this strain was acceptable but contained an ester note that was not present in the control. This suggested that the cytoplasm of the killer strain appeared to exert more influence on the brewing strain than originally predicted. A question often asked is whether the toxin is still active in the finished beer. The toxin is extremely heat sensitive, and a brewery pasteurization cycle of eight pasteurization units was shown to completely inactivate it.[33]

To determine the effect of the zymocidal lager strain on a typical brewery fermentation, "killer" lager yeast was mixed at a concentration of 10% with an ale brewing strain. The control was the ale strain mixed with 10% "nonkiller" lager. Within 10 h the killer lager strain had almost totally eliminated the ale strain. When the concentration of the killer yeast was reduced to 1%, within 24 h the ale yeast was again eliminated.[32] The speed at which this occurs may well make a brewer apprehensive about using such a yeast in the fermentation cellar, particularly where several yeasts are employed for the production of different beers. An error on an operator's part in keeping lines and yeast tanks separate could have serious consequences. In a brewery with only one yeast strain, this would not be a cause for concern.

An alternative to the killer strain would be to produce a yeast strain that does not kill but is "killer resistant." That is to say, it has received that genetic complement that makes it immune to zymocidal activity. The construction of such a yeast would perhaps be a good compromise, because it would not itself kill; it would allay the brewer's fear that this yeast might kill all other production strains in the plant; and, at the same time, it would not itself be killed by a contaminating yeast with "killer" ability.

Genetic Manipulation

Spheroplast Fusion

Spheroplast (protoplast) fusion, first described by van Solingen and van der Plaat[38] is a technique that can be employed in the genetic manipulation of industrial strains and circumvents the mating/sporulation barrier. The method does not depend on ploidy and mating type and, consequently, has great applicability to brewing yeast strains because of their polyploid nature and absence of mating type characteristic. Examples of fusions with commercial brewing strains include the construction of a brewing yeast with amylolytic activity by the fusion of *S. cerevisiae* with

S. diastaticus,[32] a polyploid capable of high ethanol production by fusion of a flocculent strain with saké yeasts,[39] and the construction of industrial strains with improved osmotolerance by fusion of S. diastaticus with S. rouxii.[40,41]

Two yeasts, one a S. diastaticus strain capable of dextrin fermentation (DEX) and the other a flocculent brewing lager strain (FLO), are converted to spheroplasts by enzymic removal of the cell wall. This results in osmotically fragile spheroplasts, which must be maintained in an osmotically stabilized medium such as 1 M sorbitol. The spheroplasting enzyme is removed by thorough washing, and the spheroplasts are then mixed and suspended in a fusing agent consisting of polyethylene glycol (PEG) and calcium ions in buffer. Subsequently, the fused spheroplasts must be induced to regenerate their cell walls and recommence division. This is achieved in solid media containing 3% agar and sorbitol. The action of PEG as a fusing agent is not fully understood, but it is believed to act as a polycation, inducing the formation of small aggregates of spheroplasts. The PEG can be replaced by an electrical current or by short electric field pulses of high intensity.[42,43]

Although spheroplast fusion is an extremely efficient technique, it relies mainly on trial and error and does not modify strains in a predictable manner. The fusion product is nearly always different from both original fusion partners because the genome of both donors becomes integrated. Consequently, it is difficult to selectively introduce a single trait such as flocculation into a strain using this technique. For example, hybrid strains created by fusing lager yeast with S. diastaticus had unsatisfactory beer flavor/taste profiles, but could survive higher osmotic pressure and temperature and produced higher ethanol yields, and thus were of use to the fuel alcohol industry.[44-46] Similarly, a hybrid from a saké yeast with a brewing strain fermented high-gravity wort effectively, but the beers contained more ethanol and esters than the brew produced with the parental brewing strain.[47]

Recombinant DNA

Although the techniques of hybridization, rare mating, and spheroplast fusion have met with success, they have their limitations, the principal one being the lack of specificity in genetic exchange. Since 1978, when a transformation system for yeast became available,[48] great strides have been made in yeast transformation. It is now possible to modify the genetic composition of a brewing yeast strain without disrupting the many other desirable traits of the strain, and it is possible to introduce genes from many other sources.

Hammond[49] reviews the applications of recombinant DNA methods to brewing yeast in great detail. Examples include the production of brewing strains with the ability to use a wider range of carbohydrates. The glucoamylase, α-amylase, and pullulanase genes from various sources have been cloned into brewing strains. Work has been carried out cloning genes coding for β-glucanase into brewing strains. Research has been conducted on improving fermentation efficiency by increasing the gene dosage of the *MAL* genes for maltose utilization. The flocculation properties of brewing strains have been modified using cloning techniques. There is much interest in the use of transformation in terms of producing strains that have a reduced capacity to produce compounds such as diacetyl, H_2S, SO_2, and dimethyl sulfide (DMS) as well as the increased ability to produce higher levels of esters such as isoamyl acetate. Genes for this cloning work come from a number of sources including *Aspergillus niger, Bacillus subtilis*, and *Trichoderma reesii*.

Strains have been constructed with the desirable traits expressed, but these strains are not used commercially at this time, as there is still concern over consumer acceptance of beer made using recombinant DNA. The availability of alternative, inexpensive, traditional solutions for many of the problems that it was hoped that a genetically modified (GM) yeast could solve, such as inexpensive sources of β-glucanase and gluco- and α-amylase, has also retarded the implementation of these new strains. It is speculated that as people become accustomed to pharmaceuticals produced by recombinant DNA, and more plants with improved characteristics for farming/food gain regulatory approval and consumer acceptance, the current reluctance to use this technology in the brewing industry will slowly disappear.

The first GM brewing yeast to receive official government blessing was reported in the United Kingdom. in 1994.[50] This organism has not been used commercially to date. A gene coding for a glucoamylase was transferred from a diastatic strain of *Saccharomyces* to a brewing strain of *Saccharomyces*. The construct was considered "self-cloning" and was thus not treated as a recombinant organism.[51,52]

The first commercially sold GM-labeled beer in the world was launched in Sweden in 2004. Rather than using a recombinant DNA yeast, however, the beer is promoting the GM corn used in its manufacture. The company claims to have launched the first beer to use genetic modification as a marketing tool. The beer, brewed in the southern Swedish town of Ystad, is made from corn — supplied by U.S. biotech giant Monsanto — genetically modified to resist attacks from pests.[53]

Nutritional Requirements

Oxygen Requirements

Molecular oxygen has a multifaceted role in yeast physiology. Wort fermentation in beer production is largely anaerobic, but this is not the case when the yeast is pitched into the wort, and at this time, some oxygen must be made available to the yeast. There is a need for oxygen because brewing yeasts require molecular oxygen to synthesize sterols and unsaturated fatty acids that are present in wort at suboptimal concentrations. Thus, yeast is capable of growth under strictly anaerobic conditions only when there is an exogenous supply of these compounds. Under aerobic conditions, the yeast can synthesize sterols and unsaturated fatty acids *de novo* from carbohydrates. Sterols and unsaturated fatty acids are abundant in malt, but normal manufacturing procedures prevent them from passing into the wort. When ergosterol and an unsaturated fatty acid, such as oleic acid, are added to wort, the requirement for oxygen disappears.

Cells prepared aerobically can grow to some extent anaerobically. Growth of yeast during the anaerobic phase of fermentation dilutes the preformed sterol pool between the mother cell and progeny. Cells can continue to divide until sterol depletion limits growth. Optimization of the dissolved oxygen (DO) supply for any brewing yeast strain is important to achieve good fermentation and a high-quality end product.

The quantity of oxygen required for fermentation is yeast-strain-dependent. Ale yeast have been classified into four groups based on the oxygen concentration required to produce satisfactory fermentation performance.[54] Similar groupings for lager yeast were reported by Jacobsen and Thorne.[55] The amount of oxygen required for sterol synthesis and satisfactory fermentation varies widely, not only with the particular yeast employed but also with time of addition, whether in increments, etc.

Lipid Metabolism

Lipids are sparingly soluble in water but readily soluble in organic solvents. They are an integral part of the plasma membrane where they are involved in the regulation of movement of compounds in and out of the cell, regulation of the activities of membrane-bound enzymes, and enhancement of the yeast's ability to resist high ethanol concentrations. *Saccharomyces* yeast are able to take up fatty acids at low concentrations via facilitated diffusion and at high concentrations via simple diffusion.[56]

Free fatty acids are powerful detergents and once taken up by the cell are quickly esterfied to coenzyme A to reduce their potential for nonspecific enzyme inactivation. The addition of lipids, especially ergosterol and unsaturated long chain fatty acids, has a pronounced effect on the growth and metabolism of yeast. The addition of the unsaturated fatty acids: oleic, linoleic, and linolenic, has been reported to be a mechanism for the regulation of the concentration of flavor-active compounds in beer.[57]

The oxygen content of the wort at pitching is important with regard to lipid metabolism, yeast performance, and beer flavor. Underaeration leads to suboptimal synthesis of essential membrane lipids and is reflected in limited yeast growth, a low fermentation rate, and concomitant flavor problems. Overaeration results in overextending nutrients for the production of unnecessary yeast biomass, thus lowering fermentation efficiency because of the excess biomass and lower ethanol production. Studies by Devuyst et al.[58] have shown that the key factor in the successful application of wort or yeast aeration techniques is the physiological state of the yeast harvested at the end of the wort fermentation cycle. They suggest that the current regimes of yeast management produce variations in the physiological condition of the yeast, and that fluctuations in the ratio of glycogen to sterol may result in fermentation inconsistency, unless appropriate pitching rate and wort oxygen concentration are selected. They describe a procedure where the cropped yeast is suspended in water and aerated until maximum oxygen uptake rate is reached. The authors suggest that the formation of yeast sterols during the oxygenation process is fueled by glycogen dissimilation and that when the yeast is subjected to vigorous oxygenation pre-pitching, it provides a pitching yeast of standardized physiology. They believe that the best approach to improve fermentation control is to eliminate variability and that the oxygenation process they recommend produces pitching yeast of consistent and stable physiology, with no requirement for subsequent wort oxygenation to achieve satisfactory fermentation performance.

High metabolic activity of the cropped yeast is critical for efficient mobilization of reserve materials.[59] Boulton et al.[60] describe experiments to produce pitching yeast, replete in sterol, and thereby ostensibly remove the requirement for subsequent wort aeration. Yeast suspended in beer, cropped from the fermenter was forcibly exposed to oxygen and was allowed to accumulate sterol. They found that it was necessary to increase pitching rates by 30% in order to achieve the same vessel residence times as control fermentations. Oxygenated yeast withstood the rigors of storage at elevated temperature less well than the untreated control.

Sterols are taken up by the yeast only while the yeast is growing under aerobic conditions. They are not taken up by stationary phase cells under anaerobic conditions. The poor physiological quality of cropped yeast is due to the depletion of sterols and unsaturated fatty acids. To restore the physiological activity of the plasma membrane, the yeast must take up its requirements from the wort, or synthesize the required lipids *de novo*, or convert them from the available pool of precursors. Callaerts et al.[61] oxygenated an anaerobic brewing yeast to improve its physiological condition and measured the glycogen, sterol, and trehalose content of the yeast cell over time. In addition to the degradation of glycogen, and the expected synthesis of sterols, they observed an unexpected accumulation of trehalose and found that the trehalose and sterol concentrations correlated. The authors speculated that perhaps excessive contact with oxygen provoked a certain stress for the yeast and that the trehalose acted as a general stress protectant.[62,63]

Fujiwara and Tamai[64] showed that adequate aeration accelerates lager yeast fermentation and inhibits acetate ester formation. They demonstrated that complete depletion of trehalose adversely influences fermentation properties and ester synthesis and warned that aeration or agitation should be avoided during storage, as high levels of dissolved oxygen enhanced the consumption of trehalose. Excess aeration resulted in trehalose exhaustion and did not stimulate anaerobic growth or decrease the synthesis of volatile esters. They suggest that intracellular trehalose concentration could be used as a marker that can reflect the appropriate level of aeration.

Uptake and Metabolism of Wort Carbohydrates

When yeast is pitched into wort, it is introduced into an extremely complex environment consisting of simple sugars, dextrins, amino acids, peptides, proteins, vitamins, ions, nucleic acids, and other constituents too numerous to mention. One of the major advances in brewing science during the past 25 years has been the elucidation of the mechanisms by which the yeast cell, under normal circumstances, utilizes, in a very orderly manner, the plethora of wort nutrients.

Wort contains the sugars sucrose, fructose, glucose, maltose, and maltotriose together with dextrin material. In the normal situation, brewing yeast strains [i.e., *S. cerevisiae* and *S. uvarum* (*carlsbergensis*)] are capable of utilizing sucrose, glucose, fructose, maltose, and maltotriose in this approximate sequence, although some degree of overlap does occur. The majority of brewing strains leave the maltotetraose and other dextrins unfermented, but *S. diastaticus* is able to utilize some dextrin material. The initial step in the utilization of any sugar by yeast is usually either its passage intact across the cell membrane or its hydrolysis outside the cell membrane, followed by entry into the cell by some or all of the hydrolysis products. Maltose and maltotriose are examples of sugars that pass intact across the cell membrane, whereas sucrose (and dextrin with *S. diastaticus*) is hydrolyzed by an extracellular enzyme and the hydrolysis products are taken up into the cell (Figure 8.6).

Brewing yeast has several mechanisms for sensing the nutritional status of the environment, allowing it to adapt its uptake and metabolism to the surrounding conditions. Glucose and sucrose are always consumed first. Sucrose is hydrolyzed into glucose and fructose and the monosaccharides are taken up by facilitated diffusion involving common membrane carriers.[65] The yeast prefers glucose over fructose. Glucose slows down the uptake of fructose, as the sugars have the same carriers and these have a greater affinity for glucose. The presence of glucose and fructose in a fermentation causes the repression of gluconeogenesis, the glyoxylate cycle, respiration, and the uptake of less preferred carbohydrates, such as maltose and maltotriose. In addition, these two sugars activate cellular growth, mobilize

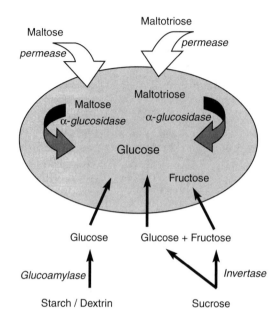

FIGURE 8.6
Carbohydrate uptake by *Saccharomyces* spp.

storage compounds, and lower cellular stress resistance (reviewed by Verstrepen et al.[66]). Since yeast consumes fructose at a different rate from glucose, when large amounts of sucrose are used as adjunct it can result in some of the very sweet fructose remaining in the beer and affecting the flavor profile.

Maltose and maltotriose are the major sugars in brewer's wort and, therefore, the brewing strain's ability to use these two sugars is vital and depends upon the correct genetic complement to transport the two sugars across the cell membrane into the cell. Once inside the cell, both sugars are hydrolyzed to glucose units by the α-glucosidase system.

Maltose Uptake

Maltose fermentation in *Saccharomyces* requires at least one of five unlinked *MAL* loci. These five, nearly identical regions are located at telomere-associated sites on different chromosomes: *MAL1*, chromosome VII; *MAL2*, III; *MAL3*, II; *MAL4*, XI; and *MAL6*, VIII.[67] Different maltose-fermenting strains carry at least one of these fully functional alleles, but often two or more loci are present in a strain.[68] A typical *MAL* locus is a cluster of three genes, all of which are required for maltose fermentation. Gene 1, a member of the 12-transmembrane domain family of sugar transporters, encodes maltose permease, gene 2 encodes maltase (α-glucosidase), and gene 3 encodes the regulator/activator of transcription of the other two genes.

The expression of the maltose permease and α-glucosidase are induced by maltose and repressed by glucose.[69] The constitutive expression of the maltose transporter gene is critical and the maltose-fermentation ability of brewing yeasts depends mostly on maltose permease activity.[70] The maltose uptake system is an active process requiring cellular energy. Uptake involves a proton symport system and potassium is exported to maintain electrochemical neutrality.

Brewing strains carry multiple copies of the maltose transporter gene, and it is speculated that the high number of *MAL* loci in brewing strains may have evolved as a mechanism to adapt to the high maltose environment of wort, giving the yeast strain a selective advantage.[71]

Maltose and maltotriose share the same transporter,[72-74] the gene for which is closely linked to the maltase gene, and it is believed that brewing strains probably contain several copies of each of the two genes scattered as pairs around several different chromosomes. Yeast strains constructed with multiple *MAL* genes show increased rates of maltose uptake up from wort.[75] The *AGT1* (*MAL1*) permease is capable of transporting maltotriose as well as maltose, but although it is 57% identical to the *MAL 6* permease gene, the *MAL 6* permease gene cannot transport maltotriose, only maltose.[76]

During the brewing process, the rate and extent of wort sugar uptake are controlled by numerous factors including: the yeast strains employed; the concentration and category of assimilable nitrogen; the concentration of ions; the fermentation temperature; the pitching rate; the tolerance of yeast cells to ethanol; the wort gravity; the wort oxygen level at yeast pitching; and the wort sugar spectrum. These factors influence yeast performance either individually or in combination with others. The kinetics of maltose transport by an ale and a lager strain was affected by the presence of other sugars, ethanol, high gravity, and temperature, but maltose uptake was the dominant factor controlling the rate of maltose utilization.[77]

The uptake and hydrolysis of maltose and maltotriose from the wort is also dependent on the glucose concentration. When the glucose concentration is high [greater than 1% (w/v)], the *MAL* genes are repressed, and only when 40–50% of the glucose has been taken up from the wort will the uptake of maltose and maltotriose commence. Thus, the presence of glucose in the fermenting wort exerts a major repressing influence on the wort fermentation rate. Mutants of brewing strains have been selected in which the maltose uptake was not repressed by glucose and, as a consequence, these strains have increased fermentation rates.[78]

The presence of residual maltotriose in beer is not due to the inability of yeast to utilize the sugar, but rather to the lower affinity for maltotriose uptake in conjunction with deteriorating conditions present at the later stages of fermentation.[79] Indeed, transport, rather than hydrolysis of maltose, is the rate-limiting step determining fermentation performance. Meneses et al.[80] used a model fermentation system to define the abilities of 25 industrial *S. cerevisiae* strains, to utilize maltose and sucrose in the presence of glucose and fructose. Their survey exposed a number of novel phenotypes that could be harnessed as a means of producing strains with rapid and efficient utilization of fermentable carbohydrates.

Uptake and Metabolism of Wort Nitrogen

The nitrogen content of a yeast cell varies between 6 and 9% (w/w), and active yeast growth requires nitrogen, mainly in the form of amino acids, for the synthesis of new cell proteins and other nitrogenous components. Nitrogen uptake slows or ceases later in the fermentation as yeast multiplication stops. Wort nitrogen levels affect yeast growth and at levels below 100 mg/l free amino nitrogen, growth is nitrogen-dependent. In wort, the main source of nitrogen for the synthesis of proteins, nucleic acids, and other nitrogenous cell components is the variety of amino acids formed from the proteolysis of barley protein. Lager wort nitrogen has been reported as ~30–40% amino acids, 30–40% polypeptides, 20% protein, and 10% nucleotides.[81] Assimilable yeast nitrogen is the aggregation of the individual wort amino acids and small peptides (di-, tripeptides). These compounds are essential for the formation of new amino acids, synthesis of new structural and enzymic proteins, cell viability and vitality, fermentation rate, ethanol tolerance, and carbohydrate uptake.

Wort contains 19 amino acids and, under brewery fermentation conditions, brewing yeast takes them up in an orderly manner, different amino acids being removed at various points in the fermentation cycle.[82] There are at least 16 different amino acid transport systems in yeast. In addition to permeases specific for individual amino acids, there is a general amino acid permease (GAP) with broad substrate specificity. Short chains of amino acids in the form of di- or tripeptides can also be taken up by the yeast cell. Uptake patterns are very complex, with a number of regulatory mechanisms interacting (Figure 8.7).[83–85] Amino acids, like a number of sugars, do not permeate freely into the cell by simple diffusion; instead, there is a regulated uptake facilitated by a number of transport enzymes. At the start of fermentation, arginine, aspartic acid, asparagine, glutamic acid, glutamine, lysine, serine, and threonine are absorbed rapidly. The other amino acids are absorbed only slowly, or not until later in the fermentation. Under strictly anaerobic conditions, such as those encountered late in a brewery fermentation, proline, the most plentiful amino acid in wort, has scarcely been assimilated by the end of the fermentation, whereas over 95% of the other amino acids have disappeared. Proline is still present in the finished product at 200–300 mg/ml; however, under aerobic laboratory conditions, proline is assimilated after exhaustion of the other amino acids.

The inability of *Saccharomyces* spp. to assimilate proline under brewery conditions is the result of several phenomena. When other amino acids or

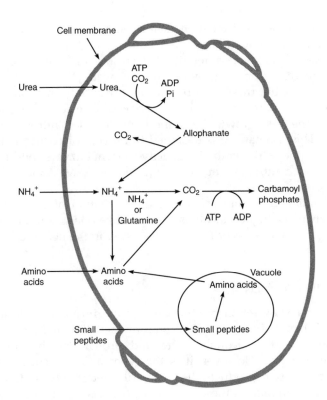

FIGURE 8.7
Nitrogen uptake by yeast.

ammonium ions are still present in the wort, the activity of proline permease, the enzyme that catalyzes the transport of proline across the cell membrane, is repressed. Once inside the cell, the first catabolic reaction of proline involves proline oxidase, which requires the participation of cytochrome c and molecular oxygen. By the time the other amino acids have been assimilated, thus removing the repression of the proline permease system, conditions are strongly anaerobic. As a result, the activity of proline oxidase is inhibited and proline uptake does not occur.

Different yeast strains exhibit different amino acid absorption rates and preferences. Amino acid uptake by yeast depends on a number of factors, including percentage of total assimilable nitrogen, individual amino acid concentrations, quality and absorption rate, amino acid competitive inhibition, yeast strain and generation, and yeast growth phase.

Free amino nitrogen or FAN refers to free α-amino nitrogen and is expressed as milligrams N per liter assimilable nitrogen. FAN includes all of the amino acids minus proline (proline is not an α-amino acid and is not utilizable by *Saccharomyces* under anaerobic conditions). FAN affects a great range of other fermentation factors, such as cell growth, biomass, viability, pH, and attenuation rate.[86]

S. cerevisiae is able to use different nitrogen sources for growth but not all nitrogen sources support growth equally well. Yeast prefers to use ammonium salts, but these are present in wort in only small amounts and yeast cannot use inorganic nitrogen. *S. cerevisiae* selects nitrogen sources that enable the best growth by a mechanism called nitrogen catabolite repression. This mechanism is designed to prevent or reduce unnecessary divergence of the cell's synthetic capacity to form enzymes and permeases for nonpreferred nitrogen sources (reviewed by Magasanik and Kaiser[87] and Dickinson).[88]

The utilization of amino acids and the formation of fermentation by-products such as higher alcohols, esters, diketones, and organic acids and their importance to beer flavor is discussed in the next section.

Yeast Excretion Products

One of the major excretion products produced during wort fermentation by yeast is ethanol. This primary alcohol impacts the final beer mainly by intensification of the alcoholic taste and aroma and by imparting a warming character.[89] However, it is the types and concentrations of the other yeast excretion products that primarily determine the flavor of the product. The formation of these excretion products depends on the overall metabolic balance of the yeast culture, and there are many factors that can alter this balance and consequently the flavor of the product. Yeast strain, incubation temperature, adjunct level, wort pH, buffering capacity, wort gravity, oxygen, pressure, etc. are all influencing factors.

Some volatiles are of great importance and contribute significantly to beer flavor, while others are of importance merely in building the background flavor of the product. The composition and concentration of beer volatiles depend upon the raw materials used, brewery procedures in mashing, fermentation parameters, and the yeast strain employed. The following groups of substances are to be found in beer: alcohols, esters, carbonyls, organic acids, sulfur compounds, amines, phenols, and a number of miscellaneous compounds.

Alcohols

In addition to ethanol, there several other alcohols found in beer, and these higher alcohols or fusel oils contribute significantly to flavor. Their formation is linked to yeast protein synthesis. Higher alcohols can be synthesized via two routes: *de novo* from wort carbohydrates (the anabolic route) or as by-products of amino acid assimilation (the catabolic route). The contribution of the two routes is influenced by a number of factors, but generally, when low levels of amino acids are available, the anabolic route predominates and when high concentrations of amino acids are present the catabolic pathway is favored. For n-propanol, there is only the anabolic route, as there is no corresponding amino acid for the catabolic pathway. The transport of branched-chain amino acids is important in brewing as the metabolites of these compounds are converted to higher alcohols.[90] The composition of the wort,[91] in particular the amino nitrogen content, influences the formation of these compounds. The yeast strain chosen for fermentation is of great significance and the levels of alcohols are dependent upon the fermentation temperature, with an increase in temperature resulting in increased concentrations of higher alcohols in the beer.

Esters

Of the flavor-active substances produced by yeast, esters represent the largest and most important group. Esters are responsible for the highly desired fruity/floral character of beer. Esters are formed intracellularly by an enzyme-catalyzed condensation reaction between two cosubstrates, a higher alcohol and an activated acyl-coenzyme A (acyl-CoA) molecule.[92] The regulation of ester synthesis is complex (see Verstrepen et al.[93]). Beer contains over 100 different esters with the key ones being ethyl acetate (fruity/solvent), isoamyl acetate (banana/apple/fruity), isobutyl acetate (banana/fruity), and 2-phenylethyl acetate (honey/rose) aroma. The C6–C10 medium-chain fatty acid ethyl esters, such as ethyl hexanoate (ethyl caproate) and ethyl octanoate (ethyl caprylate), have sour apple/aniseed aromas. Ethyl esters are present in the highest quantity presumably because ethanol is present in large amounts.

A number of factors have been found to influence the amount of esters formed during fermentation. The yeast strain is very important, as are the fermentation parameters of temperature, pitching rate, and top pressure. Wort composition affects ester production: assimilable nitrogen compounds, the concentration of carbon sources, dissolved oxygen, and fatty acids all have an effect. Wort components, which promote yeast growth, tend to decrease ester levels.[94] High-gravity worts can lead to disproportionate amounts of ethyl acetate and isoamyl acetate,[95] and the use of large-scale cylindroconical fermentation vessels causes a dramatic drop in ester production, with a resultant imbalance in the ester profile.[96] In order to obtain better control over ester synthesis, much research is being focused on the elucidation of the biochemical mechanisms of ester synthesis and on the factors influencing ester synthesis rates.

Sulfur Compounds

Sulfur is an area of importance in brewing because traces of volatile sulfur compounds such as hydrogen sulfide, dimethyl sulfide (DMS), sulfur dioxide, and thiols significantly contribute to the flavor of the beer.

Sulfur compounds are also one of the more difficult variables to control. Although small amounts of sulfur compounds can be acceptable or even desirable in beer, in excess they give rise to unpleasant off-flavors, and special measures such as purging with CO_2 or prolonged maturation times are necessary to remove them. Although volatile organic sulfur compounds are contributed to the wort and beer by hops, adjuncts, and malt, a significant proportion of those present in finished beer are formed during or after fermentation. During fermentation, yeasts usually excrete significant amounts of hydrogen sulfide and sulfur dioxide. Sulfur dioxide can give indirect flavor effects by binding to compounds associated with beer flavor staling. In addition, sulfite acts as a natural antioxidant. Sulfur dioxide is usually present at concentrations below taste threshold. Excessive concentrations of hydrogen sulfide arise from deficiencies in wort composition, poorly controlled fermentations, and stressed yeast. Normal beers contain low levels of free hydrogen sulfide.

DMS is an important beer flavor compound derived from the wort production process and via yeast metabolism. The flavor of DMS is described as cooked sweet corn or cooked vegetable. At low levels, it is considered an essential flavor component contributing to the distinctive flavor and aroma of lager beer, but at high concentrations it is objectionable. Precursors of DMS are S-methylmethionine (SMM) and dimethylsulfoxide (DMSO), both of which originate from malted barley. DMS in beer may be derived through thermal degradation of SMM to DMS during kilning and wort preparation. Further, yeast during fermentation can reduce DMSO to DMS.[94,97]

Carbonyl Compounds

Carbonyl compounds are important because they have a high flavor potential and a significant influence on the flavor stability of beer. Over 200 carbonyl compounds have been detected in beer. Excessive concentrations of carbonyl compounds are known to cause a stale flavor in beer. The carbonyl found in highest concentration in beer is acetaldehyde.

The effects of aldehydes on finished beer are a grassy aroma (propanol, 2-methyl butanol, pentanal) and a papery taste (*trans*-2-nonenal, furfural). Acetaldehyde is formed by yeast during fermentation in the final step of alcoholic fermentation. Acetaldehyde is reduced to ethanol by an enzymic reaction. The concentration of acetaldehyde varies during fermentation and aging/conditioning, reaching a maximum during the main fermentation, and then decreasing. Removal of acetaldehyde is favored by a vigorous secondary fermentation, warmer maturation, sufficient wort aeration, and increased yeast content during maturation. High levels of acetaldehyde can also be caused by high air levels during fermentation. Acetaldehyde levels in bottled beer have been observed to increase during pasteurization and storage, especially if there is a high air content in the bottle headspace.

Excessive quantities of acetaldehyde in beer can also be the result of bacterial spoilage, especially by strains of *Zymomonas*.

Diacetyl and Pentane-2,3-dione

The vicinal diketones, diacetyl and pentane-2,3-dione, are normal products of brewery fermentations that impart to beer a characteristic aroma and taste described as butterscotch and honey or toffee. The taste threshold concentration for diacetyl in lagers is 0.1–0.14 mg/l and the levels are somewhat higher in ales. Over the years, there has been a great deal of interest in the factors that influence the concentration of diacetyl in beer[98,99] and in the methodology for measuring diacetyl.[100]

Diacetyl and pentane-2,3-dione are formed outside the yeast cell by the oxidative decarboxylation of α-acetolactate and α-acetohydroxybutyrate, respectively. These α-acetohydroxy acids are intermediates in the biosynthesis of leucine and valine (acetolactate), and isoleucine (acetobutyrate), and are leaked into the wort by yeast during fermentation. In the wort, they are chemically decarboxylated into diacetyl and pentane-2,3-dione, respectively, and subsequently converted to acetoin or pentane-2,3-diol by the yeast. Thus, the final concentration of diacetyl in beer is the net result of three separate steps: (i) synthesis and excretion of α-acetohydroxy acids by yeast; (ii) oxidative decarboxylation of α-acetohydroxy acids to their respective diketones; and (iii) reduction of diacetyl and pentane-2,3-dione by yeast.

The presence of diacetyl in beer at above threshold levels occurs when α-acetolactate has decomposed to give diacetyl, at a time when the yeast cells are either absent or have lost their ability to reduce diacetyl to acetoin. Commonly, the fault arises because α-acetolactate breakdown has been curtailed by the use of temperatures conducive to yeast settling when the potential to produce diacetyl remains. When the beer becomes warm, which is usually when it is packaged and pasteurized, but may not be until the beer is disposed at the point of sale, diacetyl is produced and, in the absence of yeast, this diacetyl is not converted to acetoin and, therefore, accumulates. Diacetyl levels can thus be controlled by ensuring that there is sufficient active yeast in contact with the beer at the end of fermentation to reduce diacetyl to acetoin. Diacetyl formation from α-acetolactate has been shown to be dependent upon pH, the concentration of α-acetolactate, temperature, the presence of oxygen, the vigor of the fermentation, and certain metal ions. Vigorous fermentations produce more acetohydroxy acids, but the decomposition of acetohydroxy acids to vicinal diketones is also more rapid. In addition, since diacetyl is formed earlier in the fermentation, there is more time for diacetyl removal by the yeast. Excessive levels of diacetyl can also be the result of beer spoilage by certain strains of bacteria such as *Pediococcus* and *Lactobacillus*.

Flocculation

Flocculation, a property of the yeast cell wall, is also the characteristic of a brewing yeast that allows the separation of yeast from beer, and is strongly correlated to the physical surface properties of the cell. Flocculation can be defined as the phenomenon wherein yeast cells adhere in clumps and either sediment rapidly from the medium in which they are suspended or rise to the surface. This definition excludes other forms of aggregation, such as chain formation, where daughter cells do not separate from mother cells.

Individual strains of yeast differ considerably in their flocculating power. At one extreme are the very flocculent strains referred to as "gravelly" and at the other extreme are totally nonflocculent strains sometimes referred to as "powdery." The strongly flocculating yeasts can sediment out of the fermentation broth prematurely, giving rise to sweeter and less fermented beers, whereas the weakly flocculating strains can remain in the beer during aging and cause yeasty flavor and filtration difficulties.

To produce a high-quality beer, it is axiomatic that not only must the yeast culture be effective in removing the required nutrients from the wort, be able to tolerate the prevailing environmental conditions (e.g., high ethanol levels), impart the desired flavor to the beer, but the yeasts themselves must be effectively removed from the wort by flocculation, centrifugation, and filtration after they have fulfilled their metabolic role. Consequently, the flocculation properties, or conversely, lack of flocculation, of a particular brewing yeast culture is very important when considering important factors affecting wort fermentation.

The measurement of yeast flocculation is a controversial topic. There is little standardization of the wide variety of tests available (for reviews see Speers et al.).[101,102] Helm's test has been adopted as a recommended method by various brewing societies, and suggestions for improvements continue to be published by researchers.[103,104] A concern with flocculation tests is that laboratory test results often do not mimic what happens at the plant scale.

In addition to flocculation, there is the phenomenon of coflocculation. Coflocculation is defined as the phenomenon where two strains are nonflocculent alone but flocculent when mixed together. To date coflocculation has only been observed with ale strains and there are no reports of coflocculation between two lager strains of yeast. There is a third flocculation reaction that has been described, where the yeast strain has the ability to aggregate and cosediment with contaminating bacteria in the culture. Again, this phenomena is confined to ale yeast and cosedimentation of lager yeast with bacteria has not been observed.[105,106] In yeasts, flocculation may occur as a prelude to sexual reproduction or as a protective response under adverse environmental conditions.[107]

The exact mechanism of yeast flocculation onset and flocculent bond formation are far from understood.[108–112] Flocculation is determined by the genotype of the cell. Dominant and recessive flocculation genes and flocculation suppresser genes have been described, but much remains to be discovered regarding regulation of flocculation. It is generally agreed that there are at least two dominant flocculation genes present in brewing yeast. The existence of multiple gene copies and the possibility of there being more than one flocculation mechanism emphasize the difficulties researchers encounter when studying the genetics of brewing yeast flocculation.

The "lectin theory," that yeast flocculation is mediated by specific interactions between cell wall proteins and carbohydrates from neighboring cells, is generally accepted to explain flocculation.[113] It is believed that flocculation is due in part to the mannose-specific lectin-like adhesion, but that it is also modulated by electrostatic and hydrophobic interactions. The yeast cell wall consists of an outer layer of o-mannosylated proteins, which overlays a layer of glucan to which hyperglycosylated proteins are covalently attached. Stratford and Carter[114] describe a specific lectin–receptor interaction and van de Aar et al.[115] describe a correlation between cell hydrophobicity and flocculence. Kock et al.[116] observed the accumulation of hydrophobic carboxylic acids (3-hydroxy oxylipins) on the cell surfaces of a *S. cerevisiae* strain during the initiation of flocculation. Mitochondrial function also appears to be important for flocculation. Flocculation induction is repressed in the presence of uncouplers and glycolytic and respiratory inhibitors.[117]

Yeast flocculation types can be classified into two groups distinguished by sugar inhibition: the NewFlo phenotype, which is inhibited by mannose, glucose, maltose, and sucrose, but not by galactose, and the Flo1 type, which is inhibited by mannose, but not by glucose, maltose, sucrose, or galactose.[118] These two phenotypes are thought to be caused by two different lectin-like proteins. In some NewFlo-type yeasts, flocculation develops only towards the early stationary growth phase through the binding of zymolectins with the cell wall sugar receptors of neighboring cell surfaces. These receptors are available throughout the growth of the yeasts.[119,120] The presence of glucose inhibits the flocculation of the yeast by binding to specific lectins.[118] The onset of flocculation coincides with the termination of budding and glucose limitations with a concomitant increase in cell surface hydrophobicity.[115,121]

In other yeasts, flocculation of the *FLO 1* type occurs throughout the growth curve in all types of media.[122] The *FLO 1* gene comprises a 4.6-kb open reading frame, which includes repetitive sequences. The *FLO 1* gene product is believed to be a hydrophobic cell wall protein, located on fimbriae-like structures, which are absent in nonflocculent cells.[113,123] Straver et al.[124] suggest that cell surface hydrophobicity is a major determinant for yeast cells to become flocculent during growth in wort. Increased hydrophobicity of the cells may facilitate cell contact leading to calcium-dependent lectin–sugar binding. They also suggest that oxygen may be an indirect growth-limiting factor and that shortage of sterols and unsaturated fatty acids precedes flocculence under brewing conditions. Verstrepen et al.[125] have reviewed in detail the origins of variation in the fungal cell surface and discuss how in *S. cerevisiae*, the *FLO* genes confer adhesion to agar and plastic as well as to other yeast cells. The genome sequence shows that there are five unlinked *FLO* genes in the adhesin family (*FLO 1, FLO 5, FLO 9, FLO 10,* and *FLO 11*). This paper discusses the generation of diversity in strains in regard to the flocculation genes, the differential regulation of the *FLO* genes, and the complexity of the interactions and suggests that much more research will be required before the interaction of all the factors is understood.

Glycogen and Trehalose — The Yeast's Carbohydrate Storage Polymers

Yeast cells produce two storage polymers, glycogen and trehalose. Both are accumulated by the yeast towards the end of the fermentation. Glycogen, the yeast's major reserve carbohydrate, is a multiple-branched molecule consisting of numerous chains of α-1,4-linked glucose residues. Chains containing 10–14 glucose residues form a tree-like structure, the branching being formed by α-1,6-linkages. Typically, under appropriate conditions, 20–30% of the yeast dry weight can consist of glycogen. The glycogen serves as a store of biochemical energy for use during the lag phase of fermentation, when energy demand is high for the synthesis of compounds such as sterols and fatty acids. This intracellular source of glucose fuels lipid synthesis at the same time that oxygen is available to the cell. The dissimilation of glycogen and lipid are both rapid. Hydrolysis of glycogen from $\sim$25% to 5% and the corresponding production of lipid from $\sim$5% to 11.5% of cell dry weight occurs in the first few hours after pitching. Toward the later stages of fermentation, the yeast restores its reserve of glycogen.

Once wort constituents are taken up by the yeast cell, the glycogen level increases to a maximum and then it decreases slightly toward the end of the fermentation. Glycogen accumulation occurs where growth is limited by a nutrient other than carbon. Yeast glycogen can be determined by a number of methods, but a simple indication of glycogen concentration can be obtained by staining the yeast with tincture of iodine (Lugol's stain). A yeast cell rich in glycogen stains deep brown whereas a yeast cell depleted in glycogen stains a pale yellow.

Ideally, the glycogen content of the pitching yeast should be high. Since glycogen reserves are depleted rapidly in storage, yeast that is repitched 24–48 h after collection is preferred, but it should be remembered that the rate of glycogen depletion is dependent on a number of factors, including yeast strain and brewing conditions. Low glycogen levels in the pitching yeast result from unsatisfactory yeast handling practices (such as high storage temperature and extended storage time) and this correlates with low cell viability, extended fermentation times, and high end-of-fermentation diacetyl, acetaldehyde, and sulfur dioxide levels.[126]

Trehalose is a nonreducing disaccharide, consisting of two glucose residues linked by an α-1,1-glycosidic bond. It is present in high concentrations in resting and stressed yeast cells and serves as a stress protectant.[127] It has the ability to increase the thermotolerance of proteins and is capable of stabilizing cellular membranes. This results in increased tolerance of the cells to stresses such as desiccation, dehydration, and high temperatures. When cells of *Saccharomyces* spp. encounter starvation conditions, ATP is produced by the catabolism of trehalose, which involves the action of the enzyme trehalase to yield glucose.[128] Trehalose is utilized only after prolonged starvation conditions and when the glycogen pool is relatively depleted. Disappearance of trehalose correlates with a rapid loss in viability.

Under certain conditions, yeast cells can store up to 10–20% of their dry weight as trehalose, but in the standard brewing process, levels of less than 5% are more typical.[129] Very high gravity brewing leads to higher levels of trehalose. Trehalose accumulation occurs in late stationary phase or in response to environmental changes. Trehalose's stabilizing effect on the yeast membrane ensures yeast viability during germination, starvation, and dehydration by protecting the plasma membrane against autolysis. It plays a protective role in osmoregulation and nutrient depletion, against toxic compounds, and improves cell resistance to temperature changes.

Pure Yeast Cultures

Introduction

Yeasts belonging to the genus *Saccharomyces* are often referred to as the oldest plants cultivated by man. The practice of using a pure yeast culture for brewing was started by Emil C. Hansen in the Carlsberg laboratories over 100 years ago. Hansen, using dilution techniques, was able to isolate single cells of brewing yeast. This allowed one to test and then select the specific yeast strains that gave the desired brewing properties. The first pure yeast culture was introduced into a Carlsberg brewery on a production scale in 1883 and the benefits of pure culture quickly became clear. By 1892, 23 countries had installed Hansen's pure culture plant, and in North America, Pabst, Schlitz, Anheuser Busch, and 50 smaller breweries were using pure lager yeast cultures.[130] For a detailed history of research on yeasts, see articles by Barnett[131,132] and Barnett and Lichtenthaler.[133]

Strain Selection

In a laboratory or on a small pilot plant scale, it is not difficult to keep a culture pure and healthy. However, once the culture enters the brewery environment, large-scale management of the yeast is a more difficult task. Henson and Reid[134] identified three basic needs:

1. A regular pure yeast culture source
2. Maintenance of the brewery yeast supply
3. Microbiological control

In many breweries, fresh yeast is propagated every 8–10 generations (fermentation cycles), or earlier if contamination or a fermentation problem is identified. The systematic use of clean, pure, and highly viable cells ensures that bacteria, wild yeast, or yeast mutants, such as respiratory deficiency petites, do not lead to inconsistent fermentations and off-flavor development.

Lager yeast is normally a pure culture, whereas ale yeasts have often been a mix of strains. The pure culture practice is invaluable in ensuring that any wild yeast are quickly detected and not allowed to proliferate into a significant problem for the brewer. Moreover, undesirable mutations of the parent strain, which often occur over long usage, are kept to a minimum. Today, various procedures are used to isolate pure cultures from pitching yeast including culturing from a single cell by Linder's hanging drop technique, micromanipulation, or culturing from a single colony isolate from an agar or gelatin plate.

Storage of Cultures

The most important consideration in the maintenance of a culture collection of brewing yeasts is that the stored cultures and their subsequent progeny continue to accurately represent the strains originally deposited. The yeast preservation method should confer maximum survival and stability and be appropriate to the laboratory facilities available. There are many methods available to store yeast and bacteria, and a book entitled *Maintenance of Microorganisms and Cultured Cells — A Manual of Laboratory Methods*[135] outlines the various methodologies in detail and is a valuable resource book. The most common preservation methods currently in use are subculture, drying or desiccation, freeze drying, and freezing or cryopreservation.

Subculture, a traditional and popular method, involves the use of two vials — one for transfer and one for laboratory use, that is, for inoculation to scale up the culture for plant use. The cultures are maintained on a medium suitable for yeast growth, such as MYGP or PYN,[136] incubated between 20 and 30°C to stationary phase (~72 h), and then stored for up to 6 months at 1–4°C. At 6 months, the culture is transferred to two fresh slopes from the vial reserved exclusively for transfer. Few cultures are lost using this method, but the cultures do change over time. Studies have shown that in 600 yeast strains studied, after 10–25 years of storage, 46% of the ascosporogenous strains had lost their ability to sporulate and 50% of the strains that carried amino acid markers had lost some of their nutritional markers. In addition, of the 300 brewery strains studied, 25% of these strains lost their ability to utilize maltotriose, and 10% showed a change in flocculation ability.[137] In summary, this method is inexpensive and versatile, and the slopes are convenient for distribution purposes, but the method can lead to unacceptable levels of strain degeneration and is not recommended for long-term storage. Another concern is the danger of poor technique and cross-contamination, compromising the strain identity or purity.

There are a number of methods that use drying or desiccation. For example, silica gel can be used as a desiccant, but this method is generally reported to be more successful for genetically marked research strains rather than for industrial strains. The damaging effects appear to be very strain-specific, and substantial changes in fermentation patterns have been observed. Another popular drying method uses squares of filter paper and tinned milk as the suspending medium. Again, this method is favored for use by culture collection curators because of the ease of mailing cultures and is used primarily for genetically marked strains.

Freeze drying or lyophilization is also a popular technique. It differs from desiccation in that water is removed by sublimation from the frozen material using a centrifugal dryer. The yeast is sealed under vacuum in a glass ampoule. Survival levels tend to be low using this method and when 580 strains of *Saccharomyces* were examined, the mean percentage of survival was only 5%. There is also the question as to whether the surviving cells represent the original population. Studies have shown little change in morphological, physiological, or industrial characteristics, one exception being the increased level of RD mutants in some strains of *Saccharomyces*.[136,137] Long-term survival is generally satisfactory, and loss of viability is usually 1% per year.[137] The advantages of this method include longevity of the freeze-dried culture and easy storage and distribution of ampoules. The major disadvantage is the initial diminished activity of the culture. In addition, the technique is labor intensive and requires special equipment.

Cryopreservation is the method of choice, as little molecular activity takes place at the lower temperatures. For long-term storage, with maximum genetic stability, storage at −196°C in liquid nitrogen is ideal. Storage at −20 to −90°C is acceptable but only for shorter storage periods. At very low temperatures, there are few reports of genetic instability, phenotypic and industrial characteristics are reported to be unchanged, yeast plasmids are retained, and the petite mutation is not a problem. Of 75 *Saccharomyces* strains studied, the mean survival rate was 66%.[137] This method clearly yields the highest viability and superior stability, but this must be balanced against the disadvantages of using liquid nitrogen (cost, handling, delivery) and the inconvenience of culture distribution. Mechanical freezers that operate below $-130°C$ are now available and this eliminates many of the disadvantages associated with the use of liquid nitrogen. When this method is employed, it is wise, as a safeguard, to keep a duplicate set of the most critical cultures on solid medium at 4°C in case of mechanical failure or a prolonged interruption of the electrical supply.

Propagation and Scale-Up

The first yeast propagation plant was developed by Hansen and Kuhle and consisted of a steam-sterilizable wort receiver and propagation vessel equipped with a supply of sterile air and impeller. The basic principles of propagation devised by Hansen in 1890 have changed little.[138] The propagation can be batch or semicontinuous. There are usually three stainless steel vessels of increasing size equipped with attemperation control, sight glasses, and noncontaminating venting systems. They are equipped with a clean-in-place (CIP) system and often have in-place heat sterilizing and cooling systems for both the equipment and the wort. The yeast propagation system is ideally located in a separate room from the fermenting area with positive air pressure, as well as humidity control and air sterilizing systems, disinfectant mats in doorways and limited access by brewing staff.

During yeast propagation, the brewer wishes to obtain a maximum yield of yeast but also wishes to keep the flavor of the beer similar to a normal fermentation so that it can be blended into the production stream. As a result, the propagation is often carried out at only a slightly increased temperature and with intermittent aeration to stimulate yeast growth. The propagation of the master culture to the plant fermentation scale is a progression of fermentations of increasing size (typically 4–10×), until enough yeast is grown to pitch a half size or full commercial size brew.

Wort sterility is normally achieved by boiling for 30 min, or the wort can be pasteurized using a plate heat exchanger and passed into a sterile vessel and then cooled. Wort gravities range from 10°Plato to 16°Plato. Depending on the yeast, zinc or a commercial yeast food can be added. Aeration is important for yeast growth and the wort is aerated using oxygen or sterile air and antifoam is added if necessary. Agitation is not normally necessary as the aeration process and CO_2 evolved during active fermentation are sufficient to keep the yeast in suspension.

The exact details of the yeast propagation will vary whether it is a small brewery[139] or a larger brewery utilizing high-gravity fermentation[140] and depending on the propagation equipment available. Typically, the initial inoculum from the slope/plate of fresh yeast goes into 10 ml of sterile hopped wort for 24 h at 25°C. This is then scaled up to approximately 100 ml in a 200 ml shake flask, 1000 ml in a 2000 ml shake flask, and 5 l in a 10 l Van Laer flask or equivalent using 24–48 h increments. The steps can be larger and the temperature varied from 12°C to 25°C with resultant longer propagation times at the lower temperature. Scale-up steps are kept small at the early stages to ensure good growth. In the yeast propagation plant, use can be made of a three-vessel procedure (i.e., 10 hl at 16°C → 30 hl at 14°C → 300 hl at 12–14°C for 4–5 days), or two vessels of 10 and 100 hl are also commonly used with the yeast inoculum being transferred from an 18 l Cornelius Spartan vessel. Yields can vary from 8 to 25 g yeast/l depending on growth conditions. A recent paper by Kurz et al.[141] describes a model for yeast propagation in breweries and presents the basis for a control strategy aimed at the provision of optimal inoculum at the starting time of subsequent beer fermentations.

Contamination of Cultures

Various bacteria can contaminate the pure culture pitching yeast (see Chapter 16). These organisms originate from a number of sources: the wort, the yeast inoculum, or unclean equipment. Great care must be taken to ensure that there is no contamination during yeast propagation. For a detailed review of the bacteria encountered during propagation and beer fermentation, and the media required for their isolation, see Priest and Campbell.[142]

Wild yeasts can originate from very diverse sources and, in addition to various *Saccharomyces* strains, include species of the genera *Brettanomyces, Candida, Debaromyces, Hansenula, Kloeckera, Pichia, Rhodotorula, Torulaspora,* and *Zygosaccharomyces*[143] (see Chapter 16). The potential of the wild yeast to cause adverse effects varies with the specific contaminant. If the contaminant wild yeast is another culture yeast, the primary concern is with rate of fermentation, final attenuation, flocculation, and taste implications. If the contaminating yeast is a nonbrewing strain and can compete with the culture yeast for the wort constituents, inevitably problems will arise as these yeasts can produce a variety of off-flavors and aromas often similar to those produced by contaminating bacteria. Some wild yeasts can utilize wort dextrins, resulting in an overattenuated beer that lacks body. These yeasts are found as both contaminants of fermentation and as postfermentation contaminants. In addition, wild yeasts often produce a phenolic off-flavor due to the presence of the *POF* gene.[144] However, under controlled conditions, such as in the production of a German wheat beer or "weiss beer," this phenolic clove-like aroma, produced when the yeast decarboxylates wort ferulic acid to 4-vinylguaiacol, can be a positive attribute of the beer.

Yeast Washing

If there is evidence of bacterial contamination, the yeast can be washed to purify it. Some breweries incorporate a yeast wash into their process as a routine part of the operation, especially if there are concerns over eliminating bacteria responsible for the production of apparent total N-nitroso compounds (ATNC). There has been much controversy over the use of yeast washing and the effects on subsequent fermentations but these problems, that is, reduced cell viability, vitality, reduced rate of fermentation, changes in flocculation, fining, yeast crop size, and excretion of cell components are generally only a problem if yeast washing is carried out incorrectly.[145,146]

Historically, there are three commonly used procedures for washing yeast:

1. *Sterile water wash*: With the water wash, cold sterile water is mixed with the yeast slurry, the yeast is allowed to settle, and the supernatant water is discarded. Bacteria and broken cells are removed through this process. This can be repeated a number of times.

2. *Acid wash*: There are a number of acids that can be used. Most common are phosphoric, citric, tartaric, or sulfuric. The yeast slurry is acidified with diluted acid to a pH of 2.0 and it is important that agitation is continuous through the acid addition period. The yeast is usually allowed to stand for a maximum period of 2 h at a temperature of less than 4°C.

3. *Acid/Ammonium persulfate wash*: An acidified ammonium persulfate treatment has been found to be effective and can yield material cost savings. It is recommended that 0.75% (w/v) ammonium persulfate is added to a diluted yeast slurry (2 parts water:1 part yeast) and then the slurry acidified with phosphoric acid to pH 2.8.[145,147,148] This treatment is more effective than acid alone at a pH of 2.2. If a pH of 2.0 is employed, a 1-h contact time is the maximum.

Many brewers have a strong preference for a certain regime of yeast washing, and a number of factors must be taken into account when choosing the method, such as food grade quality of the acid, hazards involved in using the acid, and cost. Phosphoric and citric acid offer the advantage of being weak acids and yeast pH is more easily controlled, whereas with strong acids, such as sulfuric acid, there are special handling procedures required for the operators and a slight overdose will yield excessively low pH values.

Simpson and Hammond[146] have listed those criteria, which if followed, should alleviate many of the problems that are associated with the yeast washing process. They include:

1. Use a food grade acid — phosphoric or citric acid are good choices.
2. Wash the yeast as a beer or water slurry.
3. Chill both the yeast slurry and the acid to less than 4°C.
4. Stir constantly, and slowly while adding the acid to the yeast.
5. If possible, stir throughout the wash.
6. Never let the temperature exceed 4°C during the wash.
7. Check the pH of the yeast slurry.
8. Do not wash for more than 2 h.
9. Pitch yeast immediately after washing.
10. Do not wash unhealthy yeast or yeast from fermentations with greater than 8% ethanol present (if a wash is unavoidable, use a higher pH and/or a shorter contact time).

Yeast Pitching and Cell Viability

Microscopic examination of brewery pitching yeast is as important today as it was when first described by Pasteur. It is a rapid way to ensure that there is not a major contaminant or viability problem with the yeast. When a sample of pitching yeast in water, wort, or beer is examined under the microscope, it can be difficult to distinguish a small number of bacteria from the trub and other extraneous nonliving material. The trub material, however, is irregular in size and outline, and dissolves readily in dilute alkali.

A trained microbiologist becomes very familiar with the typical appearance of the production yeast: the appearance of the cytoplasm, the shape of the yeast cells, whether the cells are chain formers, etc. and thus, one can sometimes identify the presence of a wild yeast due to the presence of cells with unusual shapes or differences in budding or flocculating behavior.

The use of a viability stain such as methylene blue[149] gives a good indication of the health of the cells. Although there are a number of other good stains and techniques available, in experienced hands, methylene blue will still quickly identify a viability problem before the yeast is pitched. For a review of the various yeast viability and vitality methods see Heggart et al.[150,151]

To accurately determine the health of a culture yeast with a low viability, the slide culture technique is the method of choice.[152] A suitably diluted suspension of yeast is applied to a microscope slide covered with a thin layer of nutrient medium. A sterile cover slip is positioned over the yeast and the slide is incubated for no longer than 18 h at room temperature. The slide is examined at a magnification of 200×. Cells that give rise to microcolonies are viable. Single cells not giving rise to microcolonies are scored as dead.

Yeast pitching is governed by a number of factors, such as wort gravity, wort constituents, temperature, degree of wort aeration, and previous history of the yeast. Ideally, one wants a minimum lag in order to obtain a rapid start to fermentation, which then results in a fast pH drop, and ultimately assists in the suppression of bacterial growth. Pitching rates employed vary from 5 to 20 million cells per ml, but 10 million cells per ml is considered an optimum level by many and results in a lager yeast reproducing four to five times. Increasing the pitching rate results in fewer doublings, as yeast cells, under given conditions, multiply to a maximum number of cells/unit volume, regardless of the original pitching rate. The pitching rate can be determined by various methods, such as dry weight, turbidimetric sensors, hemocytometer, and electronic cell counting. More recently, commercially available in-line biomass sensors have been introduced that utilize the passive dielectric properties of microbial cells and can discriminate between viable and nonviable cells and trub.[153] The amount of yeast growth is limited by a number of factors including oxygen supply, nutrient exhaustion, and accumulation of inhibitory metabolic products.

Yeast Collection

Yeast collection techniques differ between traditional ale top-fermentation systems, traditional lager bottom-fermentation systems, and the cylindroconical fermentation system. With the traditional ale top fermentation, although there are many variations on this system, a single, dual, or multistrain yeast system can be employed, and the timing of the skimming can be critical to maintain the flocculation characteristics of the strains. Traditionally, the first skim or "dirt skim" with the trub present is discarded, as is the final skim in most cases. The middle skim is normally kept for repitching. With the traditional lager bottom fermentation, the yeast is deposited on the floor of the vessel at the end of fermentation. Yeast cropping is nonselective and the yeast contains entrained trub. With the cylindroconical fermentation vessel, now widely adopted for both ale and lager fermentations, the angle at the bottom of the tank allows for effective yeast plug removal.

Today, the use of centrifuges for the removal of yeast and the collection of pitching yeast is commonplace. There are a number of advantages, such as shorter process time, cost reduction, increased productivity, and reduced shrinkage. Care must be taken to ensure high temperatures (i.e., >20°C) are not generated during centrifugation and that the design ensures low dissolved oxygen pickup and a high throughput. This is usually accomplished by use of a self-desludging and low heat-induction unit. Timing control of the desludge cycle is important: it allows for a more frequent cycle for yeast for the pitching tank and resultant lower solids, or a longer frequency for yeast being sent to waste with higher solids and resulting reduced product shrink.

Yeast Storage

Ideally the yeast is stored in a room that is designed to be easily sanitized, contains a plentiful supply of sterile water, a separate filtered air supply with positive pressure to prevent the entry of contaminants, and a temperature of 0°C. Alternatively, insulated tanks in a dehumidified room are employed. When open vessels were commonly used, great care had to be taken to ensure that sources of contamination were eliminated. Reduction of moisture levels to retard mold growth and elimination of difficult to clean surfaces and unnecessary equipment and tools from the room are useful.

Yeast is most commonly stored as a slurry at 2–4°C under 6 in. of beer, or under a water or 2% potassium dihydrogen phosphate solution. With high-gravity brewing, it is important to remember that the ethanol levels are significantly higher and that this can affect the viability of the stored yeast. As more sophisticated systems have become available, storage tanks with external cooling and equipped with low shear stirring devices have become popular. Reduction of available oxygen is important during storage and minimal yeast surface areas exposed to air is desirable. Low dead cell counts and minimum storage time are desirable with the yeast being cropped "just-in-time" for repitching if possible.

Shipping of Yeast

The pure culture is often maintained and supplied from a commercial laboratory and sent through the mail growing on a slope or in a pressed form or the brewery can supply a slope from its own central laboratory to branch plants. More often, if a brewery has a number of branch plants, a pure yeast culture system is situated in only one plant and this plant supplies the other branch plants with the yeast needed to pitch a larger fermentation vessel. For breweries with multiple plants, interplant homogeneity of the culture yeast is always a concern when consistency of flavor between plants is critical. Sending yeast from a central culture plant in sufficient quantity ensures conformity between plants.

Maule[154] published general principles to be observed when shipping yeast from a central brewery or culture plant to other breweries. These include:

1. Selecting yeast with the highest viability
2. Absence of contaminants
3. Removal of fermentable matter by washing with chilled sterile liquor prior to pressing or centrifugation
4. Chilling the yeast and storing at 0°C prior to transport
5. Shipping of slurry for short distances (i.e., short time span) is acceptable, but for longer distances (time) a pressed cake of low moisture (around 70%) is more suitable
6. Smaller quantities can be kept cooler during transport
7. Monitoring of the entire transfer, from collection of the yeast to use in the recipient brewery, including quality assessments, such as temperature, viability, and bacterial contamination

Glossary

Anabolism	Metabolic synthesis of proteins fats and other constituents — requires energy
Aneuploid	Where the cell does not contain an exact multiple of a haploid set of chromosomes
Catabolism	Metabolic breakdown of molecules often with the liberation of energy
Cytoplasmic	Pertaining to cell material that is outside the nucleus but within the plasma membrane
Diploid	Contains $2n$ chromosomes — twice the haploid number which pair normally
Genotype	The sum total of the genetic information (genes) contained in the chromosomes
Haploid	Contains a single set of chromosomes
Lipid bilayer	Two layers of lipids arranged so their hydrophobic parts interpenetrate, whereas their hydrophilic parts form the two surfaces of the bilayer
MAL11 (AGT1)	Part of *MAL1* locus; gene encoding maltose permease
MAL12	Part of *MAL1* locus; gene encoding maltase (α-glucosidase)
MAL13	Part of *MAL1* locus; gene encoding transcriptional activator
MAL31	Part of *MAL3* locus; gene encoding maltose permease
MAL32	Part of *MAL3* locus; gene encoding maltase (α-glucosidase)

MAL33	Part of MAL3 locus; gene encoding transcriptional activator
Mating type	In *Saccharomyces* either *a* (MAT*a*) or α (MATα) alleles determine the ability of cells to hybridize with those of the opposite type
Meiosis	Reduction division where the diploid number of chromosomes is reduced to the haploid number
Mitosis	Duplication and division of chromosomes during vegetative growth, the initial chromosome number is maintained
Mutagen	A chemical or physical agent which alters the structure (base sequence) of DNA
Mutant	The "natural" or nonmutated form of a strain, differs from the wild type by one or more mutations
Mutation	A change in the structure (base sequence) of DNA
Permease	General term for a membrane protein, which increases the permeability of the plasma membrane to a particular molecule
"Petite"	Respiratory deficient yeast
Phenotype	All the characteristics of an organism that result from the interaction of its genetic make-up with the environment.
Polyploid	Having three (triploid), four (tetraploid), or more integral sets of the haploid number of chromosomes. A ploidy greater than two
Rare-mating	A low-frequency mating event that occurs without involvement of the mating type
Respiratory deficient	Yeast is unable to oxidize glucose to CO_2 and H_2O and lacks complete mitochondrial function. Small ("petite") colonies will grow on glucose media but not on glycerol, lactate, or ethanol
Respiratory sufficient	Yeast is able to oxidize glucose to CO_2 and H_2O and has complete mitochondrial function. Large ("grande") colonies will grow on glucose media as well as on glycerol, lactate, or ethanol
Wild-type	The "natural" or nonmutated form of a strain. Originally meant to denote the form in which the organism was usually found in nature (the "wild")

Acknowledgments

The author wishes to thank Ms Lesley Bell for the figure artwork.

References

1. Barnett, J.A., The taxonomy of the genus *Saccharomyces* Meyen *ex* Reess: a short review for non-taxonomists, *Yeast*, 8:1–23, 1992.
2. Pulvirenti, A., Nguyen, H.-V., Caggia, C., Giudici, P., Rainieri, S., and Zambonelli, C., *Saccharomyces uvarum*, a proper species within *Saccharomyces* sensu stricto, *FEMS Microbiol. Lett.*, 192(2):191–196, 2000.
3. Pederson, M.B., Recent views and methods for the classification of yeasts, *Cerevisia — Belg. J. Brew. Biotechnol.*, 20:28–33, 1995.
4. Rainier, S., Zambonelli, C., and Kaneko, Y., Review *Saccharomyces* sensu stricto: systematics, genetic diversity and evolution, *J. Biosci. Bioeng.*, 96:1–9, 2003.
5. Walker, G.M., Yeast cytology, in *Yeast Physiology and Biotechnology*, John Wiley & Sons Ltd, West Sussex, UK, 1998, pp. 11–42.
6. Schweizer, M., Lipids and membranes, in *The Metabolism and Molecular Physiology of Saccharomyces cerevisiae*, Dickinson, J.R. and Schweizer, M., Eds., CRC Press, London, 2004, pp. 140–223.
7. Schwencke, J., Vacuoles, internal membranous systems and vesicles, in *The Yeasts*, Rose, A.H. and Harrison, J.S., Eds., Vol. 4, 2nd ed., Academic Press, London, 1991, pp. 347–755.
8. Smart, K., The death of the yeast cell, in *Brewing Yeast Fermentation Performance*, Smart, K., Ed., 1st ed., London, Blackwell Science, 2000, pp. 105–113.
9. Bond, U., Neal, C., Donnelly, D., and James, T.C., Aneuploidy and copy number breakpoints in the genome of lager yeasts mapped by microarray hybridization, *Curr Genet.*, 45:360–370, 2004.
10. Tamai, Y., Momma, T., Yoshimoto, H., and Kaneko, Y., Co-existence of two types of chromosomes in the bottom fermenting yeast, *Saccharomyces pastorianus*, *Yeast*, 14:923–933, 1998.
11. Mortimer, R.K. and Johnson, J., Genealogy of principal strains of the yeast genetic stock center, *Genetics*, 113:35–43, 1986.
12. Casey, G.P., Yeast selection in brewing, in *Yeast Strain Selection*, Panchal C.J., Ed., Vol. 7, Marcel Dekker, New York, 1990, pp. 65–111.
13. Jones, R.M., Russell, I., and Stewart, G.G., The use of catabolite derepression as a means of improving the fermentation rate of brewing yeast strains, *J. Am. Soc. Brew. Chem.*, 44:161–166, 1986.
14. Novak, S., D'Amore, T., Russell, I., and Stewart, G.G., Sugar uptake in a 2-deoxy-D-glucose resistant mutant of *Saccharomyces cerevisiae*, *J. Ind. Microbiol.*, 7:35–40, 1991.
15. Galván, L., Pérez, A., Delgado, M., and Conde, J., Diacetyl production by sulfometron resistant mutants of brewing yeast, *Proc. 22nd Eur. Brew. Conv. Congr.*, Zurich, IRL Press, Oxford, 1987, pp. 385–392.
16. Kielland-Brandt, M.C., Gjermansen, G., Nilsson-Tillgren, T., Holmberg, S., and Petersen, J.G.L., Diacetyl and brewer's yeast, *Yeast*, 4:470, 1988 (special issue).
17. Slaughter, J.C., The biochemistry and physiology of yeast growth, in *Brewing Microbiology*, Priest, W.G. and Campbell, I., Eds., 3rd ed., Kluwer Academic, New York, 2003, pp. 19–66.

18. Rosenfeld, E. and Beauvoit, B., Review: role of the non-respiratory pathways in the utilization of molecular oxygen by *Saccharomyces cerevisiae.*, *Yeast*, 20:1115–1144, 2003.
19. Ernandes, J.R., Williams, J.W., Russell, I., and Stewart, G.G., Respiratory deficiency in brewing yeast strains — effects on fermentation, flocculation, and beer flavor components, *J. Am. Soc. Brew. Chem.*, 51:16–20, 1993.
20. American Society of Brewing Chemists, Yeast-3.C., Triphenyltetrazolium chloride for identification of respiratory deficient cells, in *Methods of Analyses*, 9th ed., American Society of Brewing Chemists, St. Paul, MN, 2004.
21. Ogur, M., St. John, R., and Nagai, S., Tetrazolium overlay technique for population studies of respiration deficiency in yeast, *Science*, 125:928, 1957.
22. Winge, O., On haplophase and diplophase in some *Saccharomycetes*, *C. R. Trav. Lab. Carlsberg, Ser. Physiol.*, 21:77–111, 1935.
23. Bilinski, C.A., Russell, I., and Stewart, G.G., Analysis of sporulation in brewer's yeast: induction of tetrad formation, *J. Inst. Brew.*, 92:594–598, 1986.
24. Jones, R.M., Russell, I., and Stewart, G.G., Classical genetic and protoplast fusion techniques in yeast, in *Yeast Biotechnology*, Berry, D.R., Russell, I., and Stewart, G.G., Eds., Allen and Unwin, London, 1987, pp. 55–79.
25. Gjermansen, C. and Sigsgaard, P., Construction of a hybrid brewing strain of *Saccharomyces carlsbergensis* by mating of meiotic segregants, *Carlsberg Res. Commun.*, 46:1–11, 1981.
26. Bilinski, C.A., Russell, I., and Stewart, G.G., Crossbreeding of *Saccharomyces cerevisiae* and *Saccharomyces uvarum (carlsbergensis)* by mating of meiotic segregants. Isolation and characterization of species hybrids, *Proc. 21st Eur. Brew. Conv. Congr.*, Madrid, IRL Press, Oxford, 1987, pp. 499–504.
27. Delneri, D., Colson, I., Grammenoudi, S., Roberts, I.N., Louis, E.J., and Oliver, S.G., Engineering evolution to study speciation in yeasts, *Nature*, 422:68–72, 2003.
28. Goodey, A.R. and Tubb, R.S., Genetic and biochemical analysis of the ability of *Saccharomyces cerevisiae* to decarboxylate cinnamic acids, *J. Gen. Microbiol.*, 128:2615–2620, 1982.
29. Conde, J. and Fink, G.R., A mutant of *Saccharomyces cerevisiae* defective for nuclear fusion, *Proc. Natl. Acad. Sci., USA*, 73:3651–3655, 1976.
30. Conde Zurita, J. and Mascort, S.J.L., Effect of the mitochondrial genome on the fermentation behavior of brewing yeast, *Proc. 18th Eur. Brew. Conv. Congr.*, Copenhagen, IRL Press, Oxford, 1981, pp. 177–186.
31. Young, T.W., Brewing yeast with anti-contaminant properties, *Proc. 19th Eur. Brew. Conv. Congr.*, London, IRL Press, Oxford, 1983, pp. 129–136.
32. Russell, I. and Stewart, G.G., Valuable techniques in the genetic manipulation of industrial yeast strains, *J. Am. Soc. Brew. Chem.*, 43:84–90, 1985.
33. Hammond, J.R.M. and Eckersley, K.W., Fermentation properties of brewing yeast with killer character, *J. Inst. Brew.*, 90:167–177, 1984.
34. Bevan, E.A. and Makower, M., The physiological basis of the killer character in yeast, in *Genetics Today*, XI International Congress on Genetics, Geerts, S.J., Ed., Vol. 1, Pergamon Press, Oxford, 1963, pp. 202–203.
35. Soares, G.A.M. and Sato, H.H., Killer toxin of *Saccharomyces cerevisiae* Y500-4L active against Fleischmann and Itaiquara commercial brands of yeast, *Rev. Microbiol.*, 30:253–257, 1999.

36. Schmitt, M.J. and Breinig, F., The viral killer system in yeast: from molecular biology to application, *FEMS Microbiol. Rev.*, 26:257–276, 2002.
37. American Society of Brewing Chemists, Yeast-8 killer yeast identification, in *Methods of Analyses*, 9th ed., American Society of Brewing Chemists, St. Paul, MN, 2004.
38. Van Solingen, D. and van der Plaat, J.B., Fusion of yeast spheroplasts, *J. Bacteriol.*, 130:946–947, 1977.
39. Seki, T.S., Myoga, S., Limtong, S., Uedono, S., Kamnuanta, J., and Taguchi, J., Genetic construction of yeast strains for high ethanol production, *Biotechnol. Lett.*, 5:351–356, 1983.
40. Spencer, J.F.T., Bizeau, C., Reynolds, N., and Spencer, D.M., The use of mitochondrial mutants in hybridization of industrial yeast strains. VI. Characterization of the hybrid, *Saccharomyces diastaticus* and *Saccharomyces rouxii*, obtained by protoplast fusion, and its behavior in simulated dough-raising tests, *Curr. Genet.*, 9:649–652, 1985.
41. Crumplen, R.M., D'Amore, T., Russell, I., and Stewart, G.G., The use of spheroplast fusion to improve yeast osmotolerance, *J. Am. Soc. Brew. Chem.*, 48:58–61, 1990.
42. Weber, H.W., Forster, W., Jacob, H.E., and Berg, H., Enhancement of yeast protoplast fusion by electric field effects, in *Current Developments in Yeast Research*, Stewart, G.G. and Russell, I., Eds., Pergamon Press, Toronto, 1981, pp. 219–224.
43. Halfmann, H.J., Emeis, C.C., and Zimmermann, U., Electro-fusion of haploid *Saccharomyces* yeast of identical mating type, *Arch. Microbiol.*, 134:1–4, 1983.
44. Stewart, G.G., Panchal, C.J., and Russell, I., Current developments in the genetic manipulation of brewing yeast strains — a review, *J. Inst. Brew.*, 89:170–188, 1983.
45. Whitney, K., Murray, C.R., Russell, I., and Stewart, G.G., Potential cost savings for fuel ethanol production by employing a novel hybrid yeast strain, *Biotechnol. Lett.*, 7:349–354, 1985.
46. Stewart, G.G., Russell, I., and Panchal, C.J., Genetically stable allopolyploid somatic fusion product useful for the production of fuel alcohols, Canadian Patent No. 1,199,593, 1986.
47. Mukai, N., Nishimori, C., Wilson-Fujishige, I., Mizuno, A., Takahashi, T., and Sato, K., Beer brewing using a fusant between a sake yeast and a brewer's yeast, *J. Biosci. Bioeng.*, 91:482–486, 2001.
48. Hinnen, A., Hicks, J.B., and Fink, G.R., Transformation of yeast, *Proc. Natl. Acad. Sci., USA*, 75:1929–1933, 1978.
49. Hammond, J.R.M., Yeast genetics, in *Brewing Microbiology*, 3rd ed., Priest, F.G. and Campbell, I., Eds., Kluwer Academic, New York, 2003, pp. 67–112.
50. Hammond, J.R.M. and Bamforth, C.W., Practical use of gene technology in food production, *Brewer*, 90:65–69, 1994.
51. Walgate, R., Genetic manipulation: Britain may exempt 'self cloning,' *Nature*, 277:589, 1979.
52. Hino, A., Safety assessment and public concerns for genetically modified food products: the Japanese experience, *Toxicol. Pathol.*, 30:128, 2002.
53. Anon., News in brief, *Nat. Biotechnol.*, 22:259, 2004.
54. Kirsop, B.H., Oxygen in brewery fermentations, *J. Inst. Brew.*, 80:252–259, 1974.

55. Jacobsen, M. and Thorne, R.S.W., Oxygen requirements of brewing strains of *Saccharomyces uvarum (carlsbergensis)* — bottom fermentation yeast, *J. Inst. Brew.*, 86:284–289, 1980.
56. Finnerty, W.R., Microbial lipid metabolism, in *Microbial Lipids*, Ratledge, C. and Wilkinson, S.G., Eds., Vol. 2, Academic Press, London, 1989, pp. 525–566.
57. Rosi, I. and Bertuccioli, M., Influences of lipid addition on fatty acid composition of *Saccharomyces cerevisiae* and aroma characteristics of experimental wines, *J. Inst. Brew.*, 98:305–314, 1992.
58. Devuyst, R., Dyon, D., Ramos-Jeunehomme, C., and Masschelein, C.A., Oxygen transfer efficiency with a view to the improvement of lager yeast performance and beer quality, *Proc. 23rd Eur. Brew. Conv. Congr.*, Lisbon, IRL Press, Oxford, 1991, pp. 377–384.
59. Boulton, C.A., Jones, A.R., and Hinchliffe, E., Yeast physiological condition and fermentation performance, *Proc. 23rd Eur. Brew. Conv. Congr.*, Lisbon, IRL Press, Oxford, 1991, pp. 385–392.
60. Boulton, C., Clutterbuck, V., and Durnin, S., Yeast oxygenation and storage, in *Brewing Yeast Fermentation Performance*, Smart, K., Ed., Blackwell Science, Oxford, 2000, pp. 140–151.
61. Callaerts, G., Iserentant, D., and Verachtert, H., Relationship between trehalose and sterol accumulation during oxygenation of cropped yeast, *J. Am. Soc. Brew. Chem.*, 51:75–77, 1993.
62. Wiemken, A., Trehalose in yeast, stress protectant rather than reserve carbohydrate, *Antonie van Leeuwenhoek*, 58:209–217, 1990.
63. van Laere, A., Trehalose, reserve and/or stress metabolite? *FEMS Microbiol. Rev.*, 63:201–210, 1989.
64. Fujiwara, D. and Tamai, Y., Aeration prior to pitching increases intracellular enzymatic and transcriptional responses under nonnutritional conditions, *J. Am. Soc. Brew. Chem.*, 61:99–104, 2003.
65. Kruckeberg, A.L. and Dickinson, J.R., Carbon metabolism, in *The Metabolism and Molecular Physiology of Saccharomyces cerevisiae*, Dickinson, J.R. and Schweizer, M., Eds., CRC Press, London, 2004, pp. 42–105.
66. Verstrepen, K.J., Iserentant, D., Malcorps, P., Derdelinckx, G., Van Dijck, P., Winderickx, J., Pretorius, I.S., Thevelein, J.M., and Delvaux, F.R., Glucose and sucrose: hazardous fast-food for industrial yeast? *Trends Biotechnol.*, 22:531–537, 2004.
67. Charron, M.J., Read, E., Haut, S.R., and Michels, C.A., Molecular evolution of the telomere-associated MAL loci of Saccharomyces, *Genetics*, 122:307–316, 1989.
68. Naumov, G.I., Naumova, E.S., and Michels, C.A., Genetic variation of the repeated MAL loci in natural populations of *Saccharomyces cerevisiae* and *Saccharomyces paradoxus*, *Genetics*, 136:803–812, 1994.
69. Wang, X., Bali, M., Medintz, I., and Michels, C.A., Intracellular maltose is sufficient to induce MAL gene expression in *Saccharomyces cerevisiae*, *Eukaryot. Cell*, 1:696–703, 2002.
70. Kodama, Y., Fukui, N., Ashikari, T., and Shibano, Y., Improvement of maltose fermentation efficiency: constitutive expression of *MAL* genes in brewing yeasts, *J. Am. Soc. Brew. Chem.*, 53:24–29, 1995.

71. Ernandes, J., Williams, J., Russell, I., and Stewart, G.G., Effect of yeast adaptation to maltose utilization on sugar uptake during the fermentation of brewer's wort, *J. Inst. Brew.*, 96:67–91, 1993.
72. Benito, R. and Lagunas, R., The low affinity component of *Saccharomyces cerevisiae* maltose transport is an artifact, *J. Bacteriol.*, 174:3065–3069, 1992.
73. Day, R.E., Rogers, P.J., Dawes, I.W., and Higgins, V.J., Molecular analysis of maltotriose transport and utilization by *Saccharomyces cerevisiae*, *Appl. Environ. Microbiol.*, 68:5326–5335, 2002.
74. Day, R.E., Higgins, V.J., Rogers, P.J., and Dawes, I.W., Characterization of the putative maltose transporters encoded by YDL247w and YJR160c, *Yeast*, 19:1015–1027, 2002.
75. Stewart, G.G., Panchal, C.J., and Russell, I., Current developments in the genetic manipulation of brewing yeast strains — a review, *J. Inst. Brew.*, 89:170–188, 1983.
76. Han, E.K., Cotty, F., Sottas, C., Jiang, H., and Michels, C.A., Characterization of AGT1 encoding a general alpha-glucoside transporter from *Saccharomyces*, *Mol. Microbiol.*, 17:1093–1097, 1995.
77. Rautio, J. and Londesborough, J., Maltose transport by brewer's yeasts in brewer's wort, *J. Inst. Brew.*, 109:251–261, 2003.
78. Russell, I., Jones, R.M., Berry, D.R., and Stewart, G.G., Isolation and characterization of derepressed mutants of *Saccharomyces cerevisiae* and *Saccharomyces diastaticus*, in *CRC Biological Research on Industrial Yeasts II*, Stewart, G.G., Russell, I., Klein, R.D., and Hiebsch, R.R., Eds., CRC Press, Boca Raton, FL, 1987, pp. 7–65.
79. Meneses, F.J. and Jiranek, V., Expression patterns of genes and enzymes involved in sugar catabolism in industrial *Saccharomyces cerevisiae* strains displaying novel fermentation characteristics, *J. Inst. Brew.*, 108:322–335, 2002.
80. Meneses, F.J., Henschke, P.A., and Jiranek, V., A survey of industrial strains of *Saccharomyces cerevisiae* reveals numerous altered patterns of maltose and sucrose utilization, *J. Inst. Brew.*, 108:310–321, 2002.
81. Ingledew, W.M., Utilization of wort carbohydrates and nitrogen by *Saccharomyces carlsbergensis*, *Tech. Q. Master Brew. Assoc. Am.*, 12:146–150, 1975.
82. Jones, M. and Pierce, J.S., The role of proline in the amino acid metabolism of germinating barley, *J. Inst. Brew.*, 73:577–583, 1967.
83. da Cruz, S.H., Cilli, E.M., and Ernandes, J.R., Structural complexity of the nitrogen source and influence on yeast growth and fermentation, *J. Inst. Brew.*, 108:54–61, 2002.
84. da Cruz, S.H., Batistote, M., and Ernandes, J.R., Effect of sugar catabolite repression in correlation with the structural complexity of the nitrogen source on yeast growth and fermentation, *J. Inst. Brew.*, 109:349–355, 2003.
85. Campbell, I., Yeast and fermentation, in *Whisky: Technology, Production and Marketing*, Russell, I., Ed., Elsevier, London, 2003, pp. 117–154.
86. Shimizu, H., Mizuno, S., Hiroshima, T., and Shioya, S., Effect of carbon and nitrogen additions on consumption activity of apparent extract of yeast cells in a brewing process, *J. Am. Soc. Brew. Chem.*, 60:163–169, 2002.
87. Magasanik, B. and Kaiser C.A., Review — nitrogen regulation in *Saccharomyces cerevisiae*, *Gene*, 290:1–18, 2002.
88. Dickinson, J.R., Nitrogen metabolism, in *The Metabolism and Molecular Physiology of Saccharomyces cerevisiae*, Dickinson, J.R. and Schweizer, M., Eds., CRC Press, London, 2004, pp. 104–116.

89. Bamforth, C.W., The flavor of beer — mouthfeel, *Brewers Guardian*, 130:18–19, 2001.
90. Kodama, Y., Omura, F., Miyajima, K., and Ashikari, T., Control of higher alcohol production by manipulation of the *BAP2* gene in brewing yeast, *J. Am. Soc. Brew. Chem.*, 59:157–162, 2001.
91. Younis, O.S. and Stewart, G.G., Effect of malt wort, very-high-gravity malt wort, and very-high-gravity adjunct wort on volatile production in *Saccharomyces cerevisiae*, *J. Am. Soc. Brew. Chem.*, 57:39–45, 1999.
92. Nordström, K., Formation of esters from alcohols by brewer's yeast, *J. Inst. Brew.*, 70:328–336, 1964.
93. Verstrepen, K.J., Van Laere, S.D.M., Vanderhaegen, B.M.P., Derdelinckx, G., Dufour, J.-P., Pretorius, I.S., Winderickx, J., Thevelein, J.M., and Delvaux, F.R., Expression levels of the yeast alcohol acetyltransferase genes ATF1, Lg-ATF1, and ATF2 control the formation of a broad range of volatile esters, *Appl. Environ. Microbiol.*, 69:5228–5237, 2003.
94. Boulton, C. and Quain, D., Formation of flavor compounds, in *Brewing Yeast and Fermentation*, Blackwell Science, Oxford, 2001, pp. 116–142.
95. Pfisterer, E. and Stewart, G.G., Some aspects on the fermentation of high gravity worts, *Proc. 15th Eur. Brew. Conv. Congr.*, Nice, Elsevier, Amsterdam, 1975, pp. 255–267.
96. Meilgaard, M.C., Effects on flavor of innovations in brewery equipment and processing: a review, *J. Inst. Brew.*, 107:271–286, 2001.
97. Duan, W., Roddick, F.A., Higgins, V.J., and Rogers, P.J., A parallel analysis of H_2S and SO_2 formation by brewing yeast in response to sulfur-containing amino acids and ammonium ions, *J. Am. Soc. Brew. Chem.*, 62:35–41, 2004.
98. Bamforth, C.W. and Kanauchi, M., Enzymology of vicinal diketone reduction in brewer's yeast, *J. Inst. Brew.*, 110:83–93, 2004.
99. Petersen, E., Margaritis, A., Stewart, R.J., Pilkington, P.H., and Mensour, N.A., The effects of wort valine concentration on the total diacetyl profile and levels late in batch fermentations with brewing yeast *Saccharomyces carlsbergensis*, *J. Am. Soc. Brew. Chem.*, 62:131–139, 2004.
100. Buckee, G.K. and Mundy, A.P., Determination of vicinal diketones in beer by gas chromatography (headspace technique) — collaborate trial, *J. Inst. Brew.*, 100:247–253, 1994.
101. Speers, R.A., Tung, M.A., Durance, T.D., and Stewart, G.G., Biochemical aspects of yeast flocculation and its measurement: a review, *J. Inst. Brew.*, 98:292–300, 1992.
102. Speers, R.A., Tung, M.A., Durance, T.D., and Stewart, G.G., Colloidal aspects of yeast flocculation: a review, *J. Inst. Brew.*, 98:525–531, 1992.
103. American Society of Brewing Chemists, Yeast-11 A yeast flocculation. Helm assay, in *Methods of Analyses*, 9th ed., American Society of Brewing Chemists, St. Paul, MN, 2004.
104. Hautcourt, D. and Smart, K.A., Measurement of brewing yeast flocculation, *J. Am. Soc. Brew. Chem.*, 57:123–128, 1999.
105. White, F.H. and Kidney, E., The influence of yeast strain on beer spoilage bacteria, *Proc. 17th Eur. Brew. Conv. Congr.*, Berlin, DSW, Dordrecht, 1979, pp. 801–815.
106. Zarattini, R. de A., Williams, J.W., Ernandes, J.R., and Stewart, G.G., Bacterial-induced flocculation in selected brewing strains of *Saccharomyces Cerevisia Biotechnol.*, 18:65–70, 1993.

107. Stratford, M., Lectin-mediated aggregation of yeasts — yeast flocculation, *Biotechnol. Genet. Eng. Rev.*, 10:283–341, 1992.
108. Jin, Y.-L., Ritcey, L.L., and Speers, R.A., Effect of cell surface hydrophobicity, charge, and zymolectin density on the flocculation of *Saccharomyces cerevisiae*. *J. Am. Soc. Brew. Chem.*, 59:1–9, 2001.
109. Nishihara, H., Miyake, K., and Kageyama, Y., Distinctly different characteristics of flocculation in yeast, *J. Inst. Brew.*, 108:187–192, 2002.
110. Gouveia, C. and Soares, E.V., Pb^{2+} inhibits competitively flocculation of *Saccharomyces cerevisiae*. *J. Inst. Brew.*, 110:141–145, 2004.
111. Strauss, C.J., Kock, J.L.F., van Wyk, P.W.J., Viljoen, B.C., Botes, P.J., Hulse, G., and Nigam, S., Inverse flocculation patterns in *Saccharomyces cerevisiae* UOFS Y-2330, *J. Inst. Brew.*, 109:3–7, 2003.
112. Strauss, C.J., Kock, J.L.F., Viljoen, B.C., Botes, P.J., Hulse, G., and Lodolo, E., Lipid turnover during inverse flocculation in *Saccharomyces cerevisiae* UOFS Y-2330, *J. Inst. Brew.*, 110:207–212, 2004.
113. Miki, B.L.A., Poon, N.H., James, A.P., and Seligy, V.L., Possible mechanism for flocculation interactions governed by the FLO 1 gene in *Saccharomyces cerevisiae*, *J. Bacteriol.*, 150:878–889, 1982.
114. Stratford, M. and Carter, A.T., Yeast flocculation: lectin synthesis and activity, *Yeast*, 9:371–378, 1993.
115. van der Aar, P.C., Straver, M.H., and Teunissen, A.W.R.H., Flocculation of brewers' lager yeast, *Proc. 24th Eur. Brew. Conv. Congr.*, Oslo, IRL Press, Oxford, 1993, pp. 259–266.
116. Kock, J.L.F., Venter, P., Smith, D.P., Van Wyk, P.W.J., Botes, P.J., Coetzee, D.J., Pohl, C.H., Botha, A., Riedel, K.-H., and Nigam, S. A, novel oxylipin-associated 'ghosting' phenomenon, *Antonie van Leeuwenhoek*, 77:401–406, 2000.
117. Soares, E.V., Teixeira, J.A., and Mota, M., Influence of aeration and glucose concentration in the flocculation of *Saccharomyces cerevisiae*, *Biotechnol. Lett.*, 13:207–212, 1991.
118. Stratford, M. and Assinder, S., Yeast flocculation: Flo1 and NewFlo phenotypes and receptor structure, *Yeast*, 7:559–574, 1991.
119. Jin, Y.-L. and Speers, R.A., Flocculation of *Saccharomyces cerevisiae*, *Food Res. Int.*, 31:421–440, 1998.
120. Stratford, M., Yeast flocculation: flocculation onset and receptor availability, *Yeast*, 9:85–94, 1993.
121. Straver, M.H., van der Aar, P.C., Smit, G., and Kijne, J.W., Determinants of flocculence of brewers' yeast during fermentation in wort, *Yeast*, 9:527–532, 1993.
122. Stratford, M., Induction of flocculation in brewing yeasts by change in pH value, *FEMS Microbiol. Lett.*, 136:13–18, 1996.
123. Day, A.W., Poon, N.H., and Stewart, G.G., Fungal fimbriae III. The effect on flocculation in *Saccharomyces cerevisiae*, *Can. J. Microbiol.*, 21:558–564, 1975.
124. Straver, M.H., Kijne, J.W., van der Aar, P.C., and Smit, G., Determinants of flocculence of brewer's yeast during fermentation in wort, *Yeast*, 9:527–532, 1993.
125. Verstrepen, K.J., Reynolds, T.B., and Fink, G.R., Origins of variation in the fungal cell surface, *Nat. Rev. Microbiol.*, 2:533–540, 2004.

126. Pickerell, A.T.W., Hwang, A., and Axcell, B.C., Impact of yeast-handling procedures on beer flavor development during fermentation, *J. Am. Soc. Brew. Chem.*, 49:87–92, 1991.
127. Dawes, I., Stress response, in *The Metabolism and Molecular Physiology of Saccharomyces cerevisiae*, Dickinson, J.R. and Schweizer, M., Eds., CRC Press, London, 2004, pp. 376–388.
128. Reinman, M. and Londesborough, J., Rapid mobilization of intracellular trehalose by fermentable sugars: a comparison of different strains, in *Brewing Yeast Fermentation Performance*, Smart, K., Ed., Blackwell Science, Oxford, 2000, pp. 20–26.
129. Boulton, C., Trehalose, glycogen and sterol, in *Brewing Yeast Fermentation Performance*, Smart, K., Ed., Blackwell Science, Oxford, 2000, pp. 10–19.
130. Von Wettstein, D., Emil Christian Hansen Centennial Lecture: From pure yeast culture to genetic engineering of brewer's yeast, *Proc. 19th Eur. Brew. Conv. Congr.*, London, IRL Press, Oxford, 1983, pp. 97–120.
131. Barnett, J.A., A history of research on yeasts 1: work by chemists and biologists 1789–1850, *Yeast*, 14:1439–1451, 1998.
132. Barnett, J.A., A history of research on yeasts 2: Louis Pasteur and his contemporaries, 1850–1880, *Yeast*, 16:755–771, 2000.
133. Barnett, J.A. and Lichtenthaler, F.W., A history of research on yeasts 3: Emil Fischer, Eduard Buchner and their contemporaries, 1880–1900, *Yeast*, 18:363–388, 2001.
134. Henson, M.G. and Reid, D.M., Practical management of lager yeast, *Brewer*, 74:3–8, 1988.
135. Kirsop, B.E., Maintenance of yeasts, in *Maintenance of Microorganisms and Cultured Cells — A Manual of Laboratory Methods*, 2nd ed., Kirsop, B.E. and Doyle, A., Eds., Academic Press, San Diego, 1991, pp. 161–182.
136. Russell, I. and Stewart, G.G., Liquid nitrogen storage of yeast cultures compared to more traditional storage methods, *J. Am. Soc. Brew. Chem.*, 39:19–24, 1981.
137. Kirsop, B., Yeast identification and maintenance, in *Yeast Biotechnology*, Berry, D.R., Russell, I., and Stewart, G.G., Eds., Unwin Hyman, London, 1987, pp. 1–32.
138. Andersen, A., Yeast propagation plants — development in relationship to history, process and construction, *Tech. Q. Master Brew. Assoc. Am.*, 31:54–57, 1994.
139. Edgerton, J., A primer on yeast propagation technique and procedures, *Tech. Q. Master Brew. Assoc. Am.*, 38:167–175, 2001.
140. Cholerton, M., Yeast management under high-gravity brewing conditions, *Tech. Q. Master Brew. Assoc. Am.*, 40:181–185, 2003.
141. Kurz, T., Mieleitner, J., Becker, T., and Delgado, A., A model based simulation of brewing yeast propagation, *J. Inst. Brew.*, 108:248–255, 2002.
142. Priest, F.G. and Campbell, I., Eds., *Brewing Microbiology*, 3rd ed., Elsevier Applied Science, London, 2003, 399 p.
143. Back, W., Detection and identification of wild yeasts in the brewery, *Brauwelt Int.*, II:174–177, 1987.
144. Russell, I., Hancock, I.F., and Stewart, G.G., Construction of dextrin fermentative yeast strains that do not produce phenolic off-flavors in beer, *J. Am. Soc. Brew. Chem.*, 41:45–51, 1983.

145. Simpson, W.J., Kinetic studies of the decontamination of yeast slurries with phosphoric acid and acidified ammonium persulphate and a method for the detection of surviving bacteria involving solid medium repair in the presence of catalase, *J. Inst. Brew.*, 93:313–318, 1987.
146. Simpson, W.J. and Hammond, J.R.M., The response of brewing yeasts to acid washing, *J. Inst. Brew.*, 95:347–354, 1989.
147. Bruch, C.W., Hoffman, A., Gosine, R.M., and Brenner, M.W., Disinfection of brewing yeast with acidified ammonium persulphate, *J. Inst. Brew.*, 70:242–248, 1964.
148. Brenner, M.W., Disinfection of brewing yeast with acidified ammonium persulfate, *J. Inst. Brew.*, 71:290, 1965.
149. American Society of Brewing Chemists, Yeast-3. Yeast stains, in *Methods of Analyses*, 9th ed., American Society of Brewing Chemists, St. Paul, MN, 2004.
150. Heggart, H.M., Margaritis, A., Pilkington, H., Stewart, R.J., Dowhanick, T.M., and Russell, I., Factors affecting yeast viability and vitality characteristics, *Tech. Q. Master Brew. Assoc. Am.*, 36:383–406, 1999.
151. Heggart, H., Margaritis, A., Stewart, R., Pilkington, H., Sobczak, J., and Russell, I., Measurement of yeast viability and vitality, *Tech. Q. Master Brew. Assoc. Am.*, 37:409–430, 2000.
152. American Society of Brewing Chemists, Yeast-6. Yeast viability by slide culture, in *Methods of Analyses*, 9th ed., American Society of Brewing Chemists, St. Paul, MN, 2004.
153. Carvell, J.P., Austin, G., Matthee, A., Van de Spiegle, K., Cunningham, S., and Harding, C., Developments in using off-line radio frequency impedance methods for measuring the viable cell concentration in the brewery, *J. Am. Soc. Brew. Chem.*, 58:57–62, 2000.
154. Maule, D.R., Propagation and handling of pitching yeast, *Brew. Dig.*, 55:36–40, 1980.

9

Miscellaneous Ingredients in Aid of the Process

David S. Ryder and Joseph Power

CONTENTS

Introduction	334
Ingredients Employed: By Function	336
Brewing Liquor (Wort) Treatment	336
Brewing-Water Treatment	336
Addition of Acids	337
Addition of Salts	339
Exogenous Brewing Enzymes	340
Enzyme Preparations	340
Enzymes Used for Cooking Starchy Adjuncts	341
Enzymes Used to Aid Wort Runoff	343
Enzymes that Supplement Malt in the Mash	344
Enzymes Used in Fermentation	345
Enzymes Used During Storage of Beer	347
Yeast Nutrients	349
Defoaming Agents (Control of Foam During Fermentation)	350
Foam Suppressants of Natural Origin	351
Foam Suppressants of Synthetic Origin	351
Clarifiers and Fining Agents	352
Carrageenans	354
Isinglass and Gelatin	355
Chips	358
Colloidal Stabilizers	358
Enzymic Chillproofing of Beer	360
Chillproofing with Precipitants	363
Gallotannins	363
Silica Sol (Colloidal Silica)	364
Insoluble Adsorbents for Chillproofing	364
Amorphous Silicas	365
Production of Silica Gels	365

　　　　Production of Silica Precipitates 366
　　　　Polyamides 367
　　　　Natural Clays 368
　　　Oxidation Reactions that Enhance Chill Stability 370
　　　Novel Future Possibilities 370
　　Flavor Stabilizers 371
　　　Effects of Reducing Agents 372
　　　Enzymic Deoxidation 373
　　Foam Stabilizers 374
　　　Propylene Glycol Alginate 376
　　　Metal Salts 376
　　　Quillaia and Yucca 376
　　Biological Stabilizers 377
　　　Heptylparaben 377
　　　Nisin 377
　　　Hop Resins 377
　　　Other Novel Possibilities 378
Acknowledgments 378
References 378

Introduction

Brewers have always had a basic curiosity concerning their products: to better understand them and improve their properties, to optimize physical and flavor components, to extend shelf life, to emulate different beer styles, and to innovate novel beer types. Strategically, the goal is most often to produce increasingly higher and more consistent quality products in an improved and more cost-effective framework. To help reach these goals, the brewer has traditionally used processing aids.

Processing aids were deemed in 1980 by the Codex Alimentarius Committee of the Food and Agricultural Organization of the United Nations and the World Health Organization as "a substance or material ... intentionally used in the processing of raw materials, foods or its ingredients, to fulfil a certain technological purpose during treatment or processing and which may result in the non-intentional but unavoidable presence of residues or derivatives in the final product." Of course, the use of processing aids must provide an advantage to the overall product or process economic equation but not compromise product safety or consumer acceptance.

Since 1958 in the United States, by the Food Additives Amendment, the Food and Drug Administration (FDA) has had legal control over the safety of all ingredients employed in manufacturing food products. For effectiveness, the Amendment focused on new or relatively unknown food ingredients and excluded from official FDA preclearance any common food ingredient that at that moment were "generally recognized ... to be safe under the conditions of its intended use." This standard is usually abbreviated as "GRAS."

Consequently, the basic ingredients used in brewing, such as malt, cereal grains, sugars, hops (including hop oil and hop extracts), starches, yeast, and soy bean products (used by some brewers particularly in earlier years), were excluded from regulation and could be employed without seeking FDA approval. In addition, the Amendment permitted continued use of any substance approved by the FDA prior to the Amendment's enactment; for example, certain enzymes were also considered to be GRAS under specified conditions of use. This led to an advisory FDA list of substances or "adjuncts" (used here in a different sense to the more common understanding of adjuncts in brewing) that were "generally recognized as safe for their intended use."

The term "adjuncts" does not include materials that are peripheral to beer and do not become part of the final malt beverage product, such as filter aids, insoluble colloidal stabilization agents, brewing water treatment materials, consumed yeast nutrients, yeast wash and care materials, and coagulant and flocculants that settle and are filtered out.[1] However, any "adjunct" or processing aid to be employed in brewing must also be approved for its intended use by the Alcohol, Tobacco, and Firearms Division (ATFD) of the Treasury Department. Information on such approved materials is listed in the *Adjunct Reference Manual* originally published by the United States Brewers Association in cooperation with the (then) Brewers' Association of America (now published by the Beer Institute). This three-part manual, which is continually updated, lists (a) adjuncts reported to be employed in brewing, their maximum level of reported use, and GRAS status; (b) materials on the FDA advisory GRAS list and materials affirmed as GRAS by the FDA; and (c) alphabetical list of adjuncts and other materials available for use in brewing. For clarification and information, refer to the first part of the *Adjunct Reference Manual*.

The distinction between processing aids and "adjuncts" (as used in the *Adjunct Reference Manual*) is a fine one. Processing aids are used in production and may become part of the final product only unintentionally as residues or derivatives. Adjuncts do become part of the final product. A processing aid, which is completely removed from, and is not detectable in, the final product is by strict definition not an adjunct. A processing aid, which leaves a slight residue, would be an adjunct. For the sake of completeness, the *Adjunct Reference Manual* also lists the processing aids used in brewing, which are not present in the final product and are not strictly adjuncts. For the sake of uniformity, here we shall refer to both adjuncts and processing aids under "processing aids."

Processing aids are used for one or more of the following functions:

- Improved yield or plant capacity
- Product type, quality, stability, or appearance
- Reduced process/product losses and process time

The earliest use of processing aids in fermented beverages is lost in antiquity. Pliny (about 60 A.D.) and Dioscorides (about 120 A.D.) describe the use of processing aids for producing barley wine; the ancient Egyptians similarly recorded their use, and we also know that ancient Mayans in Central America also applied processing aids in the production of beer. More recently, our brewing technology and brewing skills, exacerbated by market pressures and demands, has stimulated marked developments in the type and variety of processing aids now available. This chapter addresses some of the more topical developments in processing aids.

Ingredients Employed: By Function

Brewing Liquor (Wort) Treatment

Brewing-Water Treatment

A number of chemicals may be added to brewing water to make it more suitable for brewing purposes, particularly for mashing. The primary reason for treating water is to reduce or counter the effects of its alkalinity. A mash made with treated water will have a lower pH and, as a result, will have benefits traditionally associated with lowered pH: better enzyme action, better wort separation and clarity, and finer hop flavor.

Treatment chemicals added to the mash principally for their effect on pH may also directly affect reactions in the mash or in fermentation. Calcium stabilizes certain enzymes, such as α-amylase, encourages yeast flocculation, and enhances oxalate precipitation. This eliminates possible calcium oxalate formation in packaged beer, which is a common cause of gushing. Magnesium is an essential mineral nutrient for yeast, and added magnesium can sometimes stimulate fermentation. Sodium chloride (salt) is sometimes added to brewing water in low concentration exclusively because it enhances flavor.

Chemicals used for brewing-water treatment should be approved for use by the appropriate government agency, which is the Bureau of Alcohol, Tobacco, and Firearms in the United States. The chemicals should also meet appropriate purity specifications for use in food products. In the U.S., most of these specifications are published in the *Food Chemicals Codex* (FCC) published by the National Academy of Sciences Press. Chemicals meeting these specifications and produced by good manufacturing practices may be labeled FCC. The terms USP and reagent grade may also be encountered on chemicals. USP indicates that the material meets specifications published in the *United States Pharmacopeia* and reagent grade usually means meeting specifications published in *Reagent Chemicals* by *the American Chemical Society*. Food chemical specifications generally include limits on arsenic, lead, and heavy metals, and give minimum assay values for the specific chemical and possibly for other requirements.

Since food and pharmaceutical requirements may be similar, chemicals are sometimes certified as both FCC and USP. Reagent grade chemicals will also qualify as FCC, but are more expensive than those meeting FCC alone because reagent specifications are more stringent. Chemicals cheaper than FCC grade may also be found, but it is up to the user to establish that they meet the FCC requirements before they can be used in brewing. The cost of performing all of the required analyses and keeping strict records on all lots of the chemicals used is usually high enough to make purchase of FCC grade cheaper in the final analysis.

Water treatment begins with analysis of the water supply. Sufficient analysis should be performed to be certain that the water is not changing in composition or to compensate for changes that occur from time to time. Analysis of pH, alkalinity (which is usually due to the bicarbonate content), and hardness, preferably both calcium and magnesium hardness measured separately, supply the basic information on which water treatment is based. Calculation of a property called residual alkalinity is useful in predicting the effect of the untreated water on pH of the mash. Residual alkalinity is calculated as

$$\text{Residual alkalinity} = \frac{\text{Bicarbonate}}{3.5} - \frac{\text{Magnesium}}{7.5}$$

where all ion concentrations are measured in milliequivalents. A residual alkalinity of zero indicates that the water will not change the pH of the mash from the pH obtained when completely pure water is used. A positive value is most often obtained and indicates that the mash pH will be increased when the water being analyzed is used. A negative value indicates that the pH will be decreased when the water is used.

Addition of Acids

Addition of acid to water, or to the mash containing the water, is the most direct way to counteract alkalinity in a water supply. Lactic, phosphoric, and sulfuric acids are suitable and commonly used for this purpose. Lactic acid use is decreasing. The amount of acid to be used depends on the pH desired in the mash or in the treated water. Experimental data obtained by Schwarz Laboratories on a variety of natural waters allow an estimation of the amounts of phosphoric (Table 9.1) and sulfuric (Table 9.2) acids needed to acidify alkaline waters with differing degrees of hardness.

The pH of a mash after water treatment with acid will depend on the grain used, so the amount of acid needed to reach the desired pH must be determined by experimentation.

TABLE 9.1

Acidification of Alkaline Brewing Water with 85% Phosphoric Acid[a]

Hardness	Alkalinity (ppm as $CaCO_3$)	pH at Start	lb/100 bbl to Reach pH			
			6.5	6.0	5.5	5.0
0	100	8.4	2.2	4.1	5.2	5.6
	200	8.55	4.5	8.7	10.6	11.3
	300	8.6	7.4	13.1	15.9	16.9
	400	8.6	9.0	16.6	20.7	22.4
100	100	8.25	2.0	3.9	5.1	5.6
	200	8.4	4.3	8.3	10.5	11.3
	300	8.5	6.7	12.7	15.7	16.9
	400	8.5	8.6	16.3	20.9	22.5
300	100	8.2	1.9	3.8	5.1	5.6
	200	8.3	4.1	7.7	10.2	11.4
	300	8.35	6.2	12.9	15.6	16.8
	400	8.35	9.5	16.5	21.0	22.6

[a]Concentrations of phosphoric acid which were found necessary to reduce the pH of water of listed alkalinity and hardness to the given pH value.

Acidification with sodium bisulfate, the partially neutralized salt of sulfuric acid, is also possible. Extra added sodium from this salt addition is usually undesirable, and it is more costly. There may be occasions where the solid salt rather than a liquid acid shows some advantage. There may

TABLE 9.2

Acidification of Alkaline Brewing Water with Concentrated Sulfuric Acid[a]

Hardness	Alkalinity (ppm as $CaCO_3$)	pH at Start	lb/100 bbl to Reach pH			
			6.5	6.0	5.5	5.0
0	100	8.3	1.1	1.9	2.3	2.5
	200	8.45	2.4	3.8	4.7	5.1
	300	8.5	3.5	5.9	7.3	7.7
	400	8.55	4.9	7.7	9.6	10.1
100	100	8.35	1.2	1.9	2.3	2.5
	200	8.5	2.5	4.0	4.8	5.2
	300	8.5	3.6	6.2	7.3	7.8
	400	8.55	5.2	7.9	9.7	10.2
300	100	8.2	1.2	2.0	2.4	2.6
	200	8.35	2.5	4.0	4.8	5.2
	300	8.4	3.9	6.3	7.4	7.7
	400	8.45	5.5	8.1	9.7	10.2

[a]Concentrations of concentrated sulfuric acid which were found necessary to reduce the pH of water of listed alkalinity and hardness to the given pH value.

also be some occasions where an increase in alkalinity of the water is desired. Sodium bicarbonate, sodium carbonate, sodium hydroxide, and calcium hydroxide may all be used for this purpose. Sodium bicarbonate is the least alkaline of these and is the easiest to handle.

Addition of Salts

As the equation for calculation of residual alkalinity given in the previous section shows, calcium ions in the mash will counteract alkalinity found in water supplies. Gypsum or calcium sulfate dihydrate has most commonly been used as a source of calcium ions for this purpose. Calcium sulfate is naturally found in many water supplies. The common name, burtonizing salt, used for gypsum, stems from the fact that this compound is found in the famous English brewing waters of Burton-upon-Trent. Calcium sulfate has limited solubility in water, so it should be added with good mixing. It can be added to the mash as a concentrated slurry or directly but slowly to the mash as it is being stirred. To encourage quick dissolution, the gypsum should be purchased as a finely ground powder. Gypsum ground so that at least 90% of it passes through a U.S. No. 100 sieve is readily available. Calcium chloride is also used as a source of calcium for water treatment; 27% by weight of the more common dihydrate form is calcium. Anhydrous calcium chloride is also available; it is very soluble in water. It tends to pick up moisture from the air very quickly, so should be stored in tightly sealed containers and resealed promptly after use.

Magnesium has a similar effect to calcium in reducing the pH of a mash, but as the calculation of residual alkalinity showed, it is only about half as effective as calcium in this respect. However, magnesium is a mineral required by yeast for growth, and the addition of magnesium can sometimes speed up fermentation activity. For this reason, the addition of some magnesium, as magnesium sulfate, to brewing water may be desirable. The amount should not exceed 20 mg/l (parts per million), measured as free magnesium.

The major effect of water treatment chemicals is to control pH. The pH of the mash affects biological reactions, especially protease enzyme activity occurring in the mash. The pH of wort when it is hopped during boiling will affect isomerization and solubility of the hop bittering acids. Often, the most desirable pH for mashing is higher than the pH desired for wort boiling. Therefore, water treatment is often done in two stages, with part of the treatment being added to mash water and part added to the kettle so that pH can be optimized at each stage.

Salt (sodium chloride) can also be used to treat brewing water: it enhances the flavor of beer. Experience has shown that when salt is present in beer at 300 mg/l or higher, the beer tastes definitely salty. Treatment with salt normally should not exceed 150 mg/l.

Exogenous Brewing Enzymes

Enzyme Preparations

The major source of enzymes for brewing is the malt used in making beer; exogenous enzymes have been applied to supplement or replace the malt enzymes. These enzymes may be of plant, animal, or microbial origin. Enzymes are normally prepared according to applicable legal requirements. In the U.S., these requirements are found in the *Food Chemicals Codex*. Enzyme preparations meeting these specifications can be labeled FCC grade. Similar standards apply in other countries; some have even adopted FCC specifications. Cooperative efforts are under way to establish international standards.

All FCC enzymes must meet certain standards of microbiological purity because of the potential for contamination and subsequent growth of microorganisms. Nevertheless, where there is potential introduction of spoilage organisms into beer, such as with enzymes added during fermentation, it is wise to check or to specify that the number of potential beer spoilers be low. Modern enzyme preparations are, however, generally sterile when they leave the manufacturer.

Because enzymes are proteins produced by living material, it is very difficult to prepare a pure enzyme free of the other enzymes produced at the same time by the same biological entity. Commercial enzymes are not pure enzymes; in fact, the desired enzyme is usually a minor portion of the commercial preparation. Other enzymes, material added to increase total solids concentration (which help prevent growth of microorganisms), and preservatives are normally present. The presence of the desired enzyme is detected and quantified by measuring the rate at which a suitable dilution of the enzyme preparation catalyzes the desired chemical reaction. The amount of enzyme present per unit of weight or volume is declared as units of enzyme per gram or milliliter. An enzyme unit is the amount of enzyme that will catalyze its reaction at a certain rate under specified conditions. Unfortunately, these specified conditions are not always the same for different enzyme manufacturers and for the application conditions of the brewer. Therefore, the user must consider the type of units used, not just the magnitude of the numbers, when comparing enzymes from different suppliers. The user must try to establish a correlation factor if the units are not the same. Some methods for enzyme analysis have been established and published in the *Food Chemicals Codex*. Not all manufacturers use the FCC units to quantify their enzymes; they may have sound reasons for not doing so.

Difficulties involved in measuring some enzymes and the instability of some enzyme preparations over time have caused the *Food Chemicals Codex* to recommend that enzymes have between 85 and 115% of their stated activity when sold. Because activity decreases with time, the initial activity provided is usually slightly greater than the stated activity.

The presence of extraneous enzymes can sometimes be a problem. Enzymes catalyzing side reactions that lead to deterioration of beer foam, haze formation, and loss of floc formation by yeast are present in many commercial enzymes. Sometimes they can be removed during processing of the desired enzyme. It is important to remember that such enzymes may be present when a new enzyme is introduced and checked for possible side effects, particularly on beer foam and clarity.

All enzymes are unstable and subject to inactivation or denaturation by unfavorable conditions, such as heat, inappropriate pH, mixing with excess foaming, and sometimes oxygen or metal ions. A supplier's information sheet should give instructions for storage and handling. Often, the information sheet will give suggestions on conditions under which the enzyme works most effectively. The terms "optimum pH" and "optimum temperature" are often used. These optima should be used as rough guides only. They have no absolute meaning. They pertain only to a certain set of conditions; the conditions of actual use may be quite different. For example, the β-amylase enzyme of malt has an optimum temperature of 45 to 50°C and an "optimum" pH of just below 5 when tested with soluble starch, the standard substrate. Under actual brewing conditions with malted barley starch as substrate, this enzyme is almost totally inactive, and it is so under these "optimum" conditions because of the different physical state of the malt starch.

Enzymes Used for Cooking Starchy Adjuncts

When brewing adjuncts are prepared by boiling for use in a mash, α-amylase enzyme must be present to hydrolyze the starch as it gelatinizes, so that the adjunct mash does not get too thick. This ensures that viscosity of the adjunct mash does not get too high and prevents possible retrogradation of the starch as it is added to the colder malt mash.

The α-amylase in malt is most often used to thin the starch slurry as the adjuncts are cooked. Exogenous enzymes can be substituted for the malt enzyme during cooking and may show certain advantages. The most effective enzymes are α-amylase from several species of bacteria. Because they are highly concentrated, these microbial enzymes are used at about 1% of the amount of malt used. This allows more adjunct to be used in the cooking vessel. A flavor advantage is claimed for the microbial enzymes; they are free of certain harsh-tasting compounds that are extracted from the malt husk during boiling of the adjunct mash. The bacterial α-amylases are more heat stable than those from malt, which allows them to more completely degrade and thin down the adjunct starch as it gelatinizes during the cooking operation. This makes it possible to use a lower ratio of water to grain in the cooker. This can increase cooker capacity and permit more flexibility in controlling the heat balance when the cooker mash is added to the malt mash.

The α-amylases produced by the bacteria *Bacillus amyloliquefaciens* and *Bacillus licheniformis* have similar properties and are available from a number of suppliers. Unfortunately, not all suppliers use the same units for this enzyme. The commonly encountered assays all use soluble starch as a substrate and the same iodine comparator used in the ASBC method for malt α-amylase to determine the reaction endpoint, so comparisons between the various units are reliable once a correlation is established.

The *Food Chemicals Codex* describes bacterial α-amylase units (BAU) for measuring this enzyme. The "optimum" conditions for these enzymes with a soluble starch substrate are 70°C and a pH of 6 to 8. With adjunct grains, the enzymes perform best under approximately the same conditions. The enzyme requires calcium ions for activity; at least 200 ppm of $CaCl_2$ or 300 ppm of gypsum should be present in the cooker water before the enzyme is added. The amylases work best at about 70°C, but the enzyme from *B. licheniformis* can hydrolyze starch up to 90°C. They are very effective in liquefying the starch from any gelatinized adjunct material and they work very well on malt. They can be used to strengthen the action of a low malt mash enzyme.

Although various suppliers measure this enzyme in different units, most of the currently available concentrated liquid preparations are of approximately the same strength. About one part of enzyme per 1000 parts of grain is used for cooking adjunct, and one part per 10,000 is usually recommended to supplement malt in the mash. The brewer should check with suppliers to make certain of the concentrations recommended for their products. The enzyme has only α-amylase activity. Bacterial amylase reduces starch to dextrins very effectively, but does not produce much fermentable sugar (Table 9.3). It cannot be used as a complete replacement for malt amylase, but it will help with conversion and runoff of a mash where there is a deficiency in enzyme for some reason. A supply of this enzyme is sometimes kept available for problem situations.

The heat-stable amylase from *B. licheniformis* is not very effective at temperatures below 70°C. The enzyme is not recommended for use in a

TABLE 9.3

Comparison of the Products of Different Amylases, in Percent (Reaction on Soluble Starch at pH 5.5 at 50°C)

	Bacterial α-Amylase	Fungal α-Amylase	Malt Enzymes	Glucoamylase
Glucose	4	3	1	83
Maltose	10	50	60	7
Maltotriose	18	26	8	3
Dextrins	68	21	31	7

normal brewer's malt mash, where the temperature cannot be above 70°C because of the limited heat stability of barley β-amylase. When corn starch (refilled corn grits) is used as an adjunct, significant gelatinization of the starch occurs below 70°C. If heat-stable amylase is used to liquefy the starch, the temperature must be increased very slowly around the 70°C range to adequately liquefy starch during the early stages of gelatinization. As the temperature rises, the enzyme becomes much more active and eventually gives a very complete breakdown of the starch. Its great heat stability allows the enzyme to completely break down starch from all adjuncts as they are cooked. The adjunct mash generally shows a negative iodine reaction by the time it boils, and measurements have shown that corn starch is broken down to dextrins averaging ten glucose units in length. In spite of the very complete starch digestion, very little fermentable sugar is produced by this enzyme (Table 9.3).

Measurements of enzyme activity in mashes of corn grits made with thermostable α-amylase show that a small percentage of the added enzyme activity survives in the boiled adjunct. In boiling wort, the enzyme is 90% inactivated in 5 min. During normal wort boiling, all of this extremely stable enzyme will be completely inactivated, and there is no danger that enzyme activity will remain during fermentation of the cooled wort.

Enzymes Used to Aid Wort Runoff

Adequate α-amylase activity during mashing is absolutely critical for normal and free runoff of the wort during separation from the grains. Malt normally contains more than enough α-amylase to allow good runoff. If a problem with starch degradation does occur, the bacterial amylase of *B. amyloliquefaciens* can be used as described in the previous section to improve runoff.

More recently, enzymes digesting β-glucans have been used as aids to runoff, particularly in cases where the malt may not be properly modified. Commercial literature sometimes claims that poor runoff may be due to an increase in viscosity because of a high level of undigested β-glucans in the wort. This is not true. Viscosity of β-glucan solutions at the concentration and high temperature of wort runoff is not significantly higher than that of wort from a very well modified malt. Experimental investigation of the causes of poor wort filtration[2] indicate that the presence of small particles, such as undigested starch granules and pieces of barley cell wall, that have not been broken down are the main cause of poor wort separation. Addition to the mash of microbial enzymes capable of breaking down β-glucans and other hemicelluloses, such as xylans and pentosans, has often been shown to improve runoff,[3,4] especially with difficult malts or when barley is used as an adjunct. Sources of enzymes digesting barley hemicelluloses include the bacterium *Bacillus subtilis* and the fungi *Aspergillus niger*, *Penicillium emersonii*, and *Trichoderma reesei*.

The β-glucanase from *B. subtilis* was the first to be applied to brewer's mash. It shows moderate heat resistance, having an optimum of 50°C for activity on purified β-glucan at pH 6.5. This enzyme has a mode of action very similar to that of barley endoglucanase. It gives limited hydrolysis at many different points on the interior of the β-glucan molecule, resulting in a quick size reduction with very little production of fermentable sugar.

The various fungal enzymes were not added to the mash in any early attempts to use them in brewing, because it was thought that their greater activity at lower pH values made them more suitable for addition to fermented beer where the pH is lower.[5] Later experience showed that they were effective in a mash. These enzymes all show optimal activity at a pH between 4 and 5, but they still show good activity at higher mash pH values. They are, however, all significantly more heat stable than the bacterial enzyme. The *Aspergillus* and *Trichoderma* enzymes are stable at 60°C and still show good activity at 65°C. These enzymes will reduce the size of β-glucan molecules very quickly, hydrolyzing the glucan at several interior points. They do, however, tend to produce some glucose by cleaving the molecule repeatedly at the same location. Limited experience with the enzyme in mashing indicates that an increase in fermentable sugar of at most 0.1°Plato occurs.

Different suppliers of the various β-glucanase enzymes use many different units to measure enzyme activity, and they are difficult to compare. There is a method published in the *Food Chemicals Codex*, but it is cumbersome and has gained little acceptance in the trade. The use rates for various β-glucanase preparations vary, but as little as 0.01% of the weight of the grains can be an optimal dose in some cases, especially with the fungal enzymes, which tend to be higher in activity than those prepared from bacteria. A supplier can make suggestions on the amount to use, but some experimentation is usually necessary.

Enzymes that Supplement Malt in the Mash

There is a limit to the amount of adjunct that can be used to make a wort, which will provide enough nutrient for the yeast to produce an acceptable beer. The limit is around 50% adjunct. At this point, the amount of nitrogenous material, commonly measured as free amino nitrogen, becomes so low that yeast nitrogen is significantly altered,[6] leading to flavor changes in the beer. Beers can be made with higher adjunct levels if enzymes are added to the mash, which will increase the amount of available nitrogen in the wort. Various protease enzymes have been used to increase protein digestion in the mash and provide more soluble nitrogen. The plant proteases ficin and papain have sometimes been used for this purpose: these enzymes have good temperature stability. Papain will hydrolyze proteins at temperatures as high as 65°C.

Most often, when brewing systems with very low malt levels have been developed, barley is used as an adjunct. Barley contains β-amylase and does not require cooking before it is used. Methods for brewing with large amounts of barley and appropriate enzymes to replace most of the malt have been developed.[7,8] These systems depend on *Bacillus* proteases to provide extra soluble nitrogen derived from barley protein. These bacteria are capable of producing two different proteases. The neutral protease works best at neutral pH but will also work in the slightly acid pH of the mash up to a temperature of about 50°C to provide both soluble protein and smaller free amino compounds from both malt and barley.[9]

Alkaline protease may also be produced by the same bacteria. This enzyme is so classified because it works at a high pH, but it will also work at a neutral and slightly acid pH. It is sensitive to an inhibitor, however, and because of this inhibition it is almost inactive in a mash. If the *Bacillus* strain used to make enzyme for brewing produces both kinds of protease, it is important to recognize that the alkaline protease portion of the total protease activity is useless as far as brewing application is concerned. The two enzymes can be distinguished because serine protease inhibitors inactivate only the alkaline protease, while chelating agents, such as EDTA, inactivate only the neutral protease, which requires zinc ions for its activity.

When high levels of barley are used as an adjunct to replace malt, additional α-amylase may be required in the mash. In addition, β-glucanase will be required to prevent the high levels of β-glucan in the barley from interfering with wort and beer filtration. The same *Bacillus* species, which produce protease for the mash, may also produce the other two enzymes when needed. For brewing wih barley there are a number of commercial preparations available that contain all three enzymes: protease, amylase, and glucanase.

Enzymes Used in Fermentation

Enzymes can be added during fermentation to increase the amount of fermentable sugar available in the wort in order to produce special beers with more alcohol or lower calories if the alcohol level is then reduced by dilution. The most effective enzyme for this purpose is the glucoamylase from the fungus *A. niger*. Glucoamylase is the currently accepted common name for the enzyme previously called amyloglucosidase or abbreviated to AG. This enzyme is capable of digesting starch or dextrins derived from starch completely to glucose (Table 9.3). The enzyme hydrolyzes the α-1,4-bonds and also the branch point bonds of starch. Malt α- and β-amylases cannot break the α-1,6-branch point bonds of starch. Some glucoamylases have limited action on the α-1,6-bonds, and when it is used during fermentation up to 95% of the extract derived from starch can become fermentable sugar.

Glucoamylase is most active at acid pH values (optimum is usually between pH 4.0 and 4.5). The enzyme is therefore most active at pH values found in fermentation. The optimum temperature is given as 60°C. Various methods are used by different suppliers to measure the enzyme, usually involving digestion of a substrate such as soluble starch, dextrin, or maltose. The amount of glucose formed from all three substrates (and from beer dextrins when a regular beer is used as a substrate) is roughly the same; consequently, the units used by the different suppliers are often comparable. The method published in the *Food Chemicals Codex* uses dextrin as a substrate, and the activity of many commonly used preparations is presently around 200 FCC units/ml.

Under most circumstances, the addition of 5 to 10 FCC units of glucoamylase per liter of fermenting wort (600 to 1200 units per bbl) will give maximum conversion of nonfermentable dextrins to glucose during fermentation and the maximum degree of fermentation needed to make a low-calorie beer. Toward the end of fermentation, the yeast uses the glucose formed by the enzyme almost as soon as it is released.

For most complete utilization of the dextrins, the yeast should not settle from the beer too quickly. Yeast growth should be encouraged by ensuring good wort aeration and similar practices. The presence of a small amount of protease side activity in the glucoamylase will delay the onset of flocculation with a flocculant yeast strain, but this activity should not be high; it might have an adverse effect on foam quality. Other side activities present in the glucoamylase should be checked. Transglucosidase will produce nonfermentable sugars, and enzymes are routinely produced with low transglucosidase activity. In the past, other impurities have led to problems with haze and foam instability in treated beers. Improved purification procedures during enzyme production have led to much purer enzymes and problems are now rare. It is a good idea, however, to thoroughly test a new source of enzyme experimentally before using it on a large scale.

A pH of 4.0 to 4.5 and a temperature of 60°C are optimal for hydrolysis of beer dextrins to glucose. These are exactly the conditions that will occur when beer is pasteurized; sweetness will develop rapidly. For this reason, it is important never to blend beers containing glucoamylase with regular beers, which still contain considerable amounts of unfermented dextrin. As little as 5% of enzyme-treated beer will provide enough enzyme to make a regular beer noticeably sweet if the two are mixed prior to pasteurization.

Because potential problems can occur when active glucoamylase is present in beer, there are advantages to using this enzyme during wort production for low-calorie beer and having it inactivated during wort boil. The problem with doing this is that the enzyme does not function well at mash pH, which normally is above 5.0. If the enzyme is to be used during mashing, the best thing to make it work more effectively is to reduce the pH of the mash to the lowest level possible. A low mash pH favors malt protease activity and decreases malt α-amylase activity, so a compromise has to be made on the pH selected. A pH of 5.3 or 5.4 is suggested.

The initial saccharification temperature should be just above 60°C, the optimum for the enzyme. A temperature of 63°C is satisfactory. After initial saccharification for no more than 30 min, the temperature should be slowly raised to a final mash conversion temperature of 70°C or above. During the rise, more starch will become available for digestion by enzymes present in the mash, but the glucoamylase will be progressively inactivated. A slow rise in temperature results in a compromise between the two processes. The amount of enzyme that must be added is higher than the amount that would be added during fermentation, but the exact level must be determined by experimentation.

Other enzymes are available that will increase the level of fermentable sugar in wort, but these are less effective than the glucoamylase described. A glucoamylase produced by the fungus *Rhizopus* will increase fermentability to the same extent as the *Aspergillus* enzyme and has the advantage of being inactivated at pasteurization temperature. Unfortunately, the *Rhizopus* glucoamylase has several side activities that create problems with beer foam and clarity; this enzyme has not been adopted for brewing.

The α-amylase of *Aspergillus oryzae*, often called fungal α-amylase, produces considerable glucose and maltose (Table 9.1) even under fermentation conditions. This enzyme increases the attenuation of wort, but will neither digest the branch points of dextrins derived from starch nor increase attenuation to the same extent as glucoamylase. Commercial preparations vary considerably in strength and in units used to measure strength, so it is necessary to check with suppliers for recommendations on the amounts of enzyme required.

Fungal α-amylase is less heat stable than glucoamylase, so it is inactivated at pasteurization temperature. The β-amylase of barley is also available commercially and can be used to increase attenuation.[10] Barley β-amylase is relatively stable at acid pH, so it will partially digest dextrins under fermentation conditions.

Enzymes Used During Storage of Beer

There is a long history of the use of enzymes to improve clarity and haze stability of beer.[11] Enzymes are usually added to beer during storage. Addition at this point can also make the beer easier to filter; choice of the point of addition is usually a matter of convenience. Yeast cells are susceptible to digestion of proteins on their outer surface by proteases, and other enzymes potentially may also solubilize material from the yeast cell's surface. The materials made soluble are a potential source of haze. For this reason, it is a good practice to add enzymes after the yeast has been removed from the beer or, at least after most of it has settled out of the beer.

Clarification of beer via a chillproofing enzyme, particularly papain, is discussed in the section "Flavor Stabilizers." Papain is a protease, and measurement of protease activity has always presented difficulties. Today, proteases are generally assayed by measuring the rate at which a dyed protein is solubilized and releases the dye. Papain is subject to inactivation by oxidation or heavy metals. Inactive enzymes may be activated by reversing oxidation with so-called sulfhydryl reagents or by chelating the heavy metals with EDTA. *The Food Chemicals Codex* papain unit uses activating agents to detect total papain, both active and inactive. When papain is used for chillproofing, however, it is not activated, so only the active enzyme will be of any use. It is more accurate to base enzyme dosage for chillproofing on assay values determined without activation.

While papain has been known for many years to be extremely effective in preventing the development of protein haze in beer, haze less frequently can be caused by the presence of high-molecular-weight dextrins (called starchy haze because it gives a color reaction with iodine) or β-glucans. Hazes caused by dextrins are effectively prevented by fungal amylase, the α-amylase of *A. oryzae*. This enzyme will increase the level of fermentable sugars in fermentation when used at a high level. At a much lower level, there is just enough enzyme hydrolysis to bring about partial hydrolysis of the dextrins without significant increase in sugar. Preparations of this enzyme vary widely in concentration. The most concentrated ones require as little as 0.04 lb of enzyme per 100 bbl (1 bbl = 31 gal) of beer to reduce haze. The enzyme will reduce starchy hazes that have formed in beer during storage within a day, even at cold storage temperatures.

The β-glucans, which are in all beers, can cause filtration problems or formation of a haze, usually in packaged beer. Enzymes can effectively deal with the β-glucans, which may be troublesome even after fermentation. Haze formation because of β-glucan content is relatively rare and probably only occurs when their content is very high. The β-glucans of beer have been blamed for filtration difficulties for many years without significant substantiating evidence. Recent research has rigorously correlated filtration characteristics of beer with high-molecular-weight β-glucan content. Letters et al.[3] have shown that cold temperatures and shear, which develops as beer flows through a filter, can lead to formation of a gel by β-glucan molecules. Glucans accumulate on a filter surface as flow rates decrease.

The same β-glucanase enzymes, which can be added to the mash, can also be added to beer during storage. The relatively high pH optimum for bacterial β-glucanase makes it less effective in the beer than in the mash, but it will have some effect. The fungal enzymes from *A. niger* and *T. reesei* work better at the lower pH values found in beer. The *Trichoderma* enzyme is particularly effective in rendering beers easier to filter.[4] Only small amounts of fungal enzyme may be required, 0.1 lb/bbl or less. Treatment at cellar temperatures typically requires several days before the beer is filtered.

Yeast Nutrients

Flavor variation in beers made by a set formula and identical brewing process is a problem that most brewers inevitably experience. Longer yeast lag phases, slower specific fermentation rates, and variable times to reach the desired attenuation or incomplete fermentations often accompany changes in yeast by-product concentrations, that is, ethanol, higher alcohols, esters, organic acids, fatty acids, carbonyl compounds such as diacetyl or acetaldehyde, or sulfur compounds such as hydrogen sulfide or mercaptans. The rate and extent of fermentation, and thus the consistency of beer flavor, are directly related to the rate and extent of yeast growth; this provides some indication that flavor anomalies are often associated with yeast metabolic limitations.

When all aspects of yeast management and pitching are found to be in order, and the more obvious wort compositional variations, such as assimilable nitrogen, desired carbohydrate spectrum, and dissolved oxygen level, are found to be satisfactory, the investigation inevitably focuses on essential elements, particularly trace metals. Zinc limitation, for example, has been implicated in yeast growth and fermentation problems.

Progressive decline in fermentative activity or premature termination of exponential growth is invariably complicated by all the associated physiological and enzymic changes occurring in the yeast cell as a result of, or limited by, nutrient uptake. For example, a direct role of the free cation concentration in facilitating ethanol tolerance in high-gravity brewing, because of its importance in regulating overall cellular metabolism and cell division, can be significant. Of these cations, magnesium is necessary for many of the enzymes in general metabolic processes and is a significant cofactor and counterion in many glycolytic reactions. Magnesium deficiency has been found to be important for ethanol toxicity when cellular growth and conversion of carbohydrate to ethanol is limited.[12]

Proprietary blended "yeast foods" are often used in brewing as a processing aid, either to avoid fermentation problems or simply to ensure more consistent fermentations; particularly so as wort original gravities increase or brewers use larger fermentation vessels. Such nutrient blends are designed to provide a supplement that ensures the availability of essential nutrients throughout the fermentation cycle. A typical blend of yeast nutrients, as reported by Hsu et al.,[13] might contain some essential amino acids, proteolipids, various vitamins, such as inositol, niacin, pyridoxin, pantothenic acid, folic acid, riboflavin, thiamine, biotin, and trace metals including potassium, calcium, magnesium, zinc, and manganese. These investigators demonstrated that the addition of 40 ppm of their yeast food to wort before pitching provided greater final product flavor uniformity than when compared to controls where the yeast nutrients were not added.

A study of 15 commercially available yeast foods was conducted by Ingledew et al.[14] These investigators assayed protein, free amino nitrogen (FAN), amino acid profiles, and the trace minerals, sulfur, phosphorus, zinc, copper, magnesium, and calcium. Although in model systems some of these "yeast foods" stimulated fermentations to a greater degree than others, the authors concluded that only one of the studied "yeast foods" showed significant levels of constituents, other than trace minerals, which would aid yeast nutrition.

A further study of "yeast foods" was conducted by Axcell et al.[15] Their studies suggested that "yeast foods" do not function by providing key nutrients to the wort substrate. On the contrary, they concluded that these supplements supply particulate matter to the fermenting wort to provide nucleation sites for CO_2, thereby preventing supersaturation during fermentation. Supersaturation of the fermenting wort with CO_2 would otherwise have a negative impact on yeast growth. In agreement with the work of Rice et al.[16] and Siebert et al.,[17] these investigators found that by using "yeast food" during fermentation, an approximate 31% decrease in CO_2 levels could be achieved. They also showed that similar effects and similarly enhanced fermentation patterns (yeast growth and attenuation) could be achieved using soy flour, pea flour, or activated charcoal.

There is obviously some variation of opinion as to precisely how yeast foods exercise their effects. Some technologists, after having been involved in yeast food formulations for a number of years, subscribe to both nutrient and particulate effects.

Studies have also been conducted to determine whether yeast foods contribute to the formation of ethyl carbamate (urethane) in beer or wine. It has been suggested that ethyl carbamate occurs naturally because of a reaction of carbamyl phosphate with ethanol. Fermentation experiments conducted by Ingledew et al.,[18] however, showed that ethyl carbamate was not formed during fermentation, either in the presence of "yeast foods" or when the fermentations were directly supplemented with urea.

Defoaming Agents (Control of Foam During Fermentation)

It has been considered desirable in beer production to prevent excess foaming in the brewhouse or, more particularly, during fermentation. Excess foaming during production has a negative impact on the final foaming capability of the malt beverage being produced by (a) physically removing the foaming components from the product and (b) denaturing or destabilizing beer-foaming components. In addition, excessive foaming requires headspace, which limits vessel capacity (those of high aspect ratio where increased turbulence during fermentation takes place being a typical example) or may lead (particularly during fermentation) to unwanted beer loss. Furthermore, such (excessive) foaming may substantially remove hop-bittering components from the product.

The inhibition of foaming during processing would allow the maximization of (a) vessel capacity (large cylindroconical tanks can be filled to over 95% of their available capacity when efficient antifoams are employed), (b) product-foaming components, and (c) hop usage, and minimize product (extract) loss. To this end, a variety of foam suppressants can be used during processing. These agents may be conveniently divided into those of either natural or synthetic origin and are designed to be effectively removed from the process stream at some point before final packaging.

Antifoams act in two different ways; either to displace the stabilizing components from the bubble walls or to locally burst the bubble. Displacement of the stabilizing components occurs when a thin layer of the antifoam comes into contact with the bubbles. Differences in surface tensions (dispersion of antifoam across the bubble surface) can only be achieved if the surface tension of the antifoam is less than the foaming product, and interfacial tensions are critical for efficient action. Localized depression of surface tension is responsible for bubbles bursting and preventing the foam-stabilizing components in the bubble wall to pack together efficiently.

Foam Suppressants of Natural Origin

Roberts[19] reported that an antifoaming agent extracted from spent grain was very efficient in controlling the foam head during fermentation. Concentrated methanol/chloroform (1:1) extractions from fresh spent grains provided an emulsion, containing a relatively high percentage of unsaturated fatty acids. These suppressed foam during fermentation, provided more bitterness units, and improved chill stability of the final beer. However, the nonfood-grade status of the extractants used prevented any commercial use of this potential processing aid.

Dadic et al.[20] described a concentrated ethanolic extract obtained from ground malt that had five times the antifoam activity of that obtained from spent grains. This invention contained predominantly unsaturated fatty acids, with some lipoproteins and phenolic compounds.

Foam Suppressants of Synthetic Origin

For brewing, only two basic types of synthetic antifoams are generally in use: those based on silicone and those based on mineral oil (Table 9.4). Over the past 20 years, the most popular silicone-based antifoam used in brewing is one containing dimethylpolysiloxane.[21] This agent is usually commercially available as an emulsion containing an emulsifying agent such as a sugar alcohol long-chain ester and also some finely divided fume silica to aid dispersion. Typically, this processing aid is used at a few parts per million added directly to the kettle to prevent foaming during boil and later during fermentation. It may alternately be added to wort after wort boiling but before or during early fermentation.

The antifoam is removed from the beer stream by various physical mechanisms: (a) adsorption to the brewing yeast at the end of fermentation

TABLE 9.4

General Classes of Antifoaming Agents

Class	Examples	Comments
Alcohols	2-Ethylhexanol Polyalkylene glycols	Branched-chain alcohols and polyols of low solubility; low surface tension
Fatty acid esters	Sorbitan trioleate Diglycol stearate	Water insoluble
Amides	Distearoyl-ethylene diamine	Moderately high molecular weight
Phosphate esters	Trioctyl phosphate Tributyl phosphate Sodium octyl phosphate	Effective at low concentrations
Metallic soaps	Calcium stearate Magnesium palmitate	Water insoluble
Multiple polar group molecules	Di-*tert*-amylphenoxy ethanol Castor oil	Molecule possessed two or more polar functional groups
Silicone oils	Polysiloxanes	Very effective at low concentrations

(brewing yeast has a high adsorptive capacity for such substances), (b) layering on the sides of the vessel as the vessel is being emptied (as the antifoams are active at the beer surface, it can be appreciated that much is removed by this mechanism), (c) removal by adsorbents, such as silica gels or aluminosilicate clays, and (d) removal by filtration and clarification aids, such as isinglass, kieselguhr, and perlite. The synthetic particles present in popular synthetic foam suppressants localize the effects of hydrophobic components. Thus, removal of the particles ensures that foam depression no longer occurs in the beer.

Clarifiers and Fining Agents

Clarity has always been regarded as an essential quality for most beer types. Clarifiers and fining agents have been traditionally used in the brewhouse to clarify wort, in fermented beer to aid clarification and maturation, and in the final packaged product as typified during cask conditioning.

With reference to wort, a clean and clear fermentable substrate has always been appreciated by brewers to achieve a consistent quality product. A desirable characteristic of hot break formation during wort boiling is to have rapid flocculation with strong cohesion and compactness of the trub in the whirlpool or wort-sedimentation vessel, allowing clear wort production.

Early communications stressed the importance of calcium ions; in model systems with the absence of calcium ions, practically no protein coagulation occurred during boiling. Sulfites were also shown in early communications to aid coagulation during wort boiling. Consequently, calcium sulfite became an important kettle aid until it was appreciated that potassium metabisulfite provided excellent effects. Despite years of research, reducing agents remain the most effective aids to coagulation during wort boiling. It is interesting to note that formulated kettle aids such as a combination of potassium metabisulfite and tannic acid (gallotannins) improved not only wart clarity but also beer stability. Here, carrageenans are highlighted as not only kettle coagulants but also as an auxiliary fining material during beer aging.

More traditional beer-fining materials, such as isinglass and chips, are still significant processing aids used by the industry. Isinglass is particularly effective for cask beer fining—it is used extensively in the UK for this purpose—and speciality beer clarification. It is also effective as a general prefiltration clarification aid, and in many international breweries significantly contributes to savings in filtration costs. Table 9.5 lists the raw material sources for the fining substances, identifies the active components, lists some critical characteristics, and indicates the stage of brewing targeted.

TABLE 9.5

Characteristics of the Various Types of Finings

Type of Finings/ Characteristics	Kettle Finings	Auxiliary finings			Isinglass Finings
		Organic	Mineral		
Raw material used to prepare finings	Red and brown seaweeds (Irish moss) + tannins	Red and brown seaweeds (Irish moss)	Sodium silicate with mineral acid		Swim bladders of tropical fish plus cutting acid and preservative
Active component(s) of finings	Alginate, carrageenan tannic acid	Alginate, carrageenan	Silicate		Collagen (protein)
Charge of finings	Negative	Negative	Negative		Positive
Precipitated materials	Protein	Protein	Protein		Yeast, tannins, excess auxiliary finings
Charge of precipitated materials	Positive	Positive	Positive		Negative
Brewing stage employed	Wort boiling	Aging, at least 1 h and often several days before Isinglass finings added			Aging and finishing

Carrageenans

Carrageenan and furcellaran are traditional fining materials[22] obtained from extracts of marine algae. The carrageenan family of polymers consists of linear-sulfated polysaccharides found at levels of 2 to 7% in the red seaweeds. They have high molecular weights of 100 to 500 kDa and are made up of repeating galactose units and 3,6-anhydrogalactose, both sulfated and nonsulfated, joined by alternating -1,3–1,4-glycosidic linkages. The carageenan portion in these seaweeds comprises κ, ι, and λ fractions. The three main carrageenan-producing seaweeds are *Chondrus crispus*, *Eucheuma* spp., and *Gigartina* spp. *C. crispus* is commonly known as Irish moss and yields mainly κ and λ fractions with small amounts of ι fraction. *Eucheuma* spp. yields κ and ι fractions. *Gigartina* spp. yields more κ and λ fractions than any other species.

Furcellaran, or Danish agar, is a galactose polysaccharide similar to carrageenan but differs in its concentration of 3,6-anhydrogalactose. It is extracted from the red seaweed, *Furcellaria jastigiata*, and yields a fraction resembling κ carrageenan. Discriminating between κ, ι, and λ carrageenans and furcellaran is important when formulating any kind of fining material for wort or beer to provide the most desirable functional properties.

When κ carrageenan forms protein–carrageenan complexes, the carrageenan moiety appears to form brittle gels, resulting in the formation of rather compact but not tightly bound sediments. Iota carrageenan, in the presence of Ca, Mg, and K ions, forms elastic gels. Consequently, when protein molecules are bound to the ι fraction, the flocculant particles aggregate to form an elastic sponge-type sediment. The fraction of carrageenan does not form a gel and stays soluble independent of inorganic ion effects. Lambda carrageenan is an effective foam stabilizer, and advantage is taken, of these functional properties for applications other than brewing.

In practical brewing studies, κ carrageenan strongly removes from wort the hot and cold break proteins and other proteins that would eventually cause chill haze. It produces compact but not strongly cohesive hot trubs. The cohesive nature of a carrageenan product can be enhanced by the incorporation of ι fractions. Iota carrageenan can bind undesirable wort protein strongly, but does not substantially reduce chill-haze proteins from the wort. Iota carrageenan is a good kettle aid for producing a cohesive hot trub. Although influenced by wort pH, some investigators believe λ carrageenan shows little functionality for trub flocculation or for any significant reduction of the chill-haze proteins. Indeed, the presence of excess λ carrageenan fraction in a seaweed containing κ–λ carrageenans may reduce the functional capacity of the κ fraction. Consequently, the concentrations of λ and κ fractions in Irish moss or *Gigartina* spp. may reflect upon the performance of these seaweeds as processing aids for beer production.

Since carrageenan is a natural product, commercial extracts do not always conform to idealized structures; the nature of the raw material and the extraction conditions are obviously important. In addition, conformation of 6-sulfated residues introduces kinks into the chains that inhibit molecular alignment. The presence of one kink in 200 residues has been found to have a dramatic effect on gel strength as well as resistance toward gelation. Interestingly, dekinking will occur naturally to a certain extent by an enzyme known trivially as "dekinkase." However, the extent of chains straightening will, to a large degree, depend upon the conditions employed during alkaline extraction when the 6-sulfate is cleaved with a concomitant formation of the 3,6-anhydrogalactose ring structure, effectively removing kinks from the chain.

It is evident, therefore, that qualitative as well as economic benefits require optimized extraction and blending of carrageenans from carefully chosen raw materials. Optimized formulated kettle (copper) finings are able to provide:

- Reduced wort boiling time, depending on usual process practice
- Faster trub compaction; therefore, faster wort runoffs
- Tighter trub compaction, allowing reduced losses
- Better, more consistent fermentations, giving better beer flavor
- Improved colloidal stability
- Improved beer filtration throughput

Irish moss and formulated carrageenan blends are also effectively used as auxiliary finings in beer and have been used with or without isinglass (see the Section "Isinglass and Gelatin") for beer fining operations. The charge on carrageenans is opposite that of isinglass (positive) in beer and similar to that of yeast (negative); the use of auxiliary finings to precipitate unwanted colloidal material, such as chill-haze proteins, improves the economics of this function. Auxiliary finings may also assist in the deposition of isinglass finings.

Isinglass and Gelatin

The use of isinglass, a very pure form of collagen protein from fish swim bladders, as a clarifying agent is steeped in tradition. The ancient world was certainly acquainted with its clarifying properties because Herodotus and Pliny refer to it. During the first century after Christ, the Greek "doctor," Pedacius Dioscorides, mentions its use in his *Materia Medica*. Circumstantial evidence, however, for its first use probably points to the ancient Mayans in Central America who, some thousands of years before Christ, indicate the use of this clarifying agent during their various travels from the Yucatan across the Pacific and through the East Indies, Malaya, and India to the Euphrates valley. In those times, water and wine were generally stored in animal skins, and it is reasonable to suppose that when such skins were not available, fish swim bladders were used as an alternative, particularly when they were found to efficiently clear wine.

The use of isinglass in the UK for clarification, where it is still used in some considerable quantity for fining beer, particularly cask-conditioned ales, dates back to before 1765 (see Chapter 1).

Not all fish have swim bladders, and those from certain tropical or semitropical areas (roughly 20°N and 15°S) provide the best-quality isinglass for fining purposes. Of these fish, the threadfin family (Polynemoidia) provide excellent quality. In order to gain a desirable fining quality for a particular beer type, two or more kinds of isinglass may be blended. To illustrate this more precisely, *Polynemus* isinglass provides a fining that settles densely, whereas *Silurus* isinglass, from the great catfish family (Siluridae) in South America, gives a more flocculant fining, but which settles less densely and consequently is more easily disturbed.

Preparation of isinglass finings for use by the brewer is dependent on the physical form used as starting material. Isinglass has to be "cut" using suitable acids before use. To explain the cutting process, we must first consider the chemical structure of isinglass. Isinglass is a very pure form of collagen protein, which has a very high proportion of glycyl and prolyl residues. These residues contribute to the formation of a helical structure which, for collagen, involves three chains of amino acids. In collagen, the imino group of every third amino acid residue in each of the three chains is hydrogen bonded to a carbonyl in one of the other chains. The α-carbon atom of this same residue lies close to the axis of the triple helix where there is no room for a side chain, so that this is always a glycyl residue. The side chains and imino groups of the other two residues in the repeating unit of each chain point away from the helix axis where there is room for a side chain. A typical collagen structure is when at least one of these residues is proline and the molecule has the repeating sequence $(Gly-Pro-X)_n$ or $(Gly-Pro-Pro)_n$. This provides a very strong twisted rope-type structure.

Cutting, very simply, is the solubilization of the triple helix collagen molecule using acid to cause the individual helices to become "unraveled," so that the molecule is best able to perform a fining action. It is pertinent to the optimal fining action of isinglass that, although unraveling occurs, the triple helix *per se* is still intact. Various acids have been used for cutting; there is some evidence that organic acids provide a better product than inorganic acids, and although tartaric, citric, lactic, and acetic acids have all been used, the generally preferred acid is tartaric.

For an efficient cutting action, the pH is brought to 2.5 to 3.0 in the tanks used for this purpose, and the concentration of isinglass is typically between

0.5 and 1.0%; the temperature of cutting is typically in the range of 10 to 15°C. The parameters of cutting crucial to the correct preparation of an isinglass solution are pH, time, isinglass concentration, and the use of a shear mixer at appropriate stages.

Fining with isinglass probably depends on at least two distinct properties; the ability to attract oppositely charged particles, such as yeast cells (negatively charged, whereas isinglass is positively charged in beer) as well as the physical property of trapping particles in the collagen mesh during sedimentation of the isinglass.[23]

Experienced observations have also shown that different beers may fine more or less easily. For example, beers with higher calcium and hop rates have been found to fine less easily than beers where these values are less. The lower the beer pH, the more difficult it is to fine. This is probably due to the fact that lower pHs are further away from the isoelectric point of the finings. It is also important to use the correct amount of isinglass for a particular type of beer and its concentration of yeast in suspension. Too much or too little can cause a phenomenon called "fluffy bottoms,"[24] or loose sediment at the bottom of a tank after fining.

In addition to its fining properties, isinglass can improve the physical properties of some beers, particularly foam stability. In fact, Ballard[24] found a 15 to 25% enhancement of foam stability by isinglass treatment of beers with poor foaming properties (90 to 105 Sigma units). However, beers with good foaming properties (135 to 150 Sigma units) showed no significant improvement after treatment with isinglass finings. Investigations have shown that any improvement to beer foam is due primarily to the removal of foam-negative factors, such as lipids, by the isinglass fining action.[24,25] This invites speculation that lipid removal by isinglass might also contribute to a better flavor stability in some beers.

Freeze-dried powdered finings give ready-to-use finings in less than 30 min, controlled hydrolysis having been done prior to the freeze-drying process.[25] The finings are then metered into the beer stream at rates of typically 1.0 to 2.5 g/hl. This addition rate has been shown to effectively fine 3000 hl of beer in a vertical tank in less than 3 days.

Bovine collagen has been periodically proposed and debated[26] as an alternative to the piscine isinglass finings. Collagen from mammalian tendons, hides, and bone are readily available, and a number of preparation steps are involved to prepare finings of suitable quality for beer. These preparation steps yield an insoluble native collagen powder suitable for fining purposes but clearly reflect a relatively expensive manufacturing process when compared to that of isinglass. It has been suggested that the molecular weight profile is important for good fining potential[26] and mammalian finings have been produced where this parameter has been strictly controlled. In addition, bovine collagen has a lower isoelectric point than isinglass, which might help fining potential in specific products.

It should be noted, however, that "gelatin" was fairly extensively used as a fining agent in the Americas, particularly during the first half of the 20th century. Gelatin is defined in the *United States Pharmacopeia* as "a product obtained by the partial hydrolysis of collagen derived from the skin, white connective tissue, and bones of animals." It can be appreciated, therefore, that bovine collagen and gelatin are essentially the same product, albeit differing by degree of hydrolysis. Doubtlessly, different gelatins could vary widely in their degree of efficacy for fining beer.

Generally, isinglass is today's most popular fining agent and the most likely to be used in the world's larger and smaller breweries.

Chips

The use of chips to aid beer clarification is almost obsolete as a brewers' art. Nevertheless, this traditional practice is still used by a few brewers. Wood chips, principally well-seasoned beech (although maple and, to a certain extent, oak were also used) typically 15 to 30 cm in length with an average thickness of a few centimeters (although historically the length, width, and thickness have varied considerably), provide a large porous surface area on which suspended particles in the beer eventually adhere. In this way, wood chips act as an adjuvant for removing colloidal particles that might otherwise be associated with the eventual formation of haze. Probably, however, their most important action is to provide a substantial surface area for yeast cells to settle and thereby clarify the beer. Suffice to say that the enormous surface area of yeast provided to the beer by using chips would also aid beer maturation. Obviously, the greater the surface area, the greater the efficiency of the chips. In the past, therefore, in UK and Germany, many different forms of chips were tried, some with corrugated or grooved surfaces, to increase surface area.

It is perhaps of historical interest to note that both chips and isinglass (see the Section "Isinglass and Gelatin") were used in what was called the "chip cask." This was a procedure employed before the introduction of the beer filter; after its introduction, the chip cask was still used by some brewers as a final clarification.

Colloidal Stabilizers

It is a basic observation that beer that has been stored cold and filtered to clarity in the cold will become turbid if it is simply warmed to a moderate temperature and then returned to its original chilled condition. The haze formed during the chilling is commonly termed "chill haze." The cloudy appearance of beer with a chill haze is esthetically objectionable to the majority of consumers. Consequently, there has been, over the years, a continuous development of various methods for abolishing or reducing the chill haze that forms in beers. These procedures are usually termed "chillproofing" treatments. During the 20th century, the increasing popularity and sales of packaged beers has inevitably demanded a need for effective chillproofing treatments. Packaged beer may be stored at varying ambient temperatures for relatively long periods of time, so it has more opportunity to exhibit maximum development of chill haze than beer kept under refrigeration until it is served.

While it is outside the scope of this chapter to thoroughly review the general knowledge of chill haze, it is appropriate to review the properties of haze to help with understanding the methods that have been used to inhibit its formation.

If an untreated beer is repeatedly subjected to cycles of warming and chilling, the chill haze will form in the cold and dissolve when the beer is rewarmed. Gradually, as the cycles are repeated, the haze will become less and less soluble during warming. If beer is stored warm for longer periods of time, a haze that is insoluble at warm temperatures is also found. The haze or portion of haze that is insoluble at warm temperatures is often termed "permanent haze" to distinguish it from chill haze. However, there is a general consensus that both chill hazes and permanent hazes are formed by the same basic mechanism from the same type of components. Permanent-haze components may be more tightly bound or more oxidized than chill-haze components.

There is basic agreement that a polyphenol component and a proteinaceous component are both absolutely necessary to get production of a chill haze in beer. Other components may be included in the haze and may help or accelerate its formation. These include polysaccharides and various metallic ions.

The polyphenols involved in beer-haze formation are sometimes called tannins. Tannins are generally recognized by their ability to form precipitates with the protein gelatin. There are two main groups of polyphenols that possess good tanning ability: condensed tannins and hydrolyzable tannins; beer tannins belong to the condensed tannin group. They are brightly colored pigments: red cyanidins or related pigments when heated with acids, which is why they are called procyanidins. Only polymerized procyanidins,[27,28] usually with a degree of polymerization of 2 or 3 in beer, are capable of tanning, that is, reacting to precipitate proteins.

The basic haze-forming mechanism in beer appears to be an interaction of tannin-type polyphenols and specific proteins to form a complex. Haze formation is usually envisioned as a stepwise process because the final particles are large compared to individual protein or polyphenol molecules in solution. At first the complex would be a small, soluble one, gradually increasing in complexity until a relatively large microscopic size particle develops. The light-scattering properties of these particles leads to the haze typically seen and measured in beer. Many of the bonds holding these particles together are weak hydrogen bonds, which can be broken by warming the beer, allowing the chill haze to at least partially redissolve. Eventually, however, with the help of temperature fluctuations and oxidation, the bonds holding haze particles together become stronger, and some covalent bonds are formed. Strongly bonded particles do not dissociate at warmer temperatures and appear as permanent haze.

There is evidence that some haze particles contain carbohydrate groups. Carbohydrates from beer could incorporate into the haze particle by incidental inclusion or by direct linkage as it is being formed. Either mechanism would account for the carbohydrate measured in haze material. Metal ions can also incorporate into the growing haze particle as it is formed, stabilizing negative groups present in the protein components.

Despite the qualitative knowledge of the biochemical nature of colloidal particles contributing to beer haze, physical characterization is limited.[29] The size and behavior of such colloidal particles are not precisely known, so physicochemical relationships, which would ultimately affect in-process separation of haze-forming particles from the beer stream, need additional study in order to optimize clarification techniques.

Some years ago, extensive studies undertaken in the Siebel Laboratories by P. Glenister demonstrated that colloidal particles, approximately 1 μm in diameter, appear to play a principal role in haze formation. The term "point particles"[30] was assigned to these initial haze interactions. Interestingly, these particles appear to be very similar to the particles of "cold break," which form in wort when it is cooled below about 60°C. When viewed under the microscope, these point particles are roughly spherical and seem to be the result of a polyphenol–protein interaction. Point particles may represent an initial precipitation product of the least stable components in wort, similar to beer-haze particles that form later from slightly more stable components.

Point particles settle slowly because of their low density. Many of them probably remain in beer after fermentation to be removed by filtration. Similar particles are found in beer ready for filtration and even in beer filtered through diatomaceous earth.

Enzymic Chillproofing of Beer

Chillproofing materials are usually thought of as materials that remove from beer either protein or polyphenol, the two principal components of beer haze, to prevent or greatly retard its formation. The application of proteolytic enzymes to chillproof beer was patented by Leo Wallerstein in 1911.[31] His original patent described addition of proteolytic enzymes either to cooled wort or to beer at any stage of production in order to allow modification of the proteins in beer during pasteurization. The most effective enzyme for this purpose was soon established to be papain from the green fruit of the papaya plant (*Carica papaya*). Bromelain from pineapple stems, ficin from figs, and pepsin from animal sources have also had a history of use for chillproofing beer. These, however, have not been widely adopted. Today, it is fair to say that chillproofing beer with enzyme is almost synonymous with the use of papain in beer.

Commercial papains contain a mixture of at least ten different enzyme activities. The main proteolytic enzymes are papain and chymopapain; both are present in a number of forms representing partially hydrolyzed products of a higher-molecular-weight enzyme. Smaller amounts of other proteases, lysozyme, and peroxidase are also present. It is not precisely known exactly how each of these enzyme activities can contribute to the final chillproofing activity of the preparation. Strictly speaking, papain is a sulfhydryl protease with a relatively broad specificity, hydrolyzing proteins to peptides and amino acids, the idea being to degrade the haze-forming proteins to such small sizes that associations with polyphenols either will not occur or, if they do, will not lead to visible haze.

Many attempts have been made to find a microbial enzyme that will chillproof beer. Extracellular chillproofing proteases have been reported from various yeasts,[32,33] including a genetically altered strain of *Saccharomyces cerevisiae*.[34,35]

Papain for chillproofing beer is sold mainly as a liquid preparation. Usually protease activity is standardized at approximately the "double strength" level originally established by Wallerstein. Exact enzymic strength from different sources will vary somewhat because of differences in methods of assay and raw materials. Since papain is a sulfhydryl protease, it is sensitive to oxidation. Sulfur dioxide is added to the enzyme preparation to establish and maintain a reducing environment and to inhibit microbial growth. Water activity in the liquid is reduced by addition of polyols, most commonly sugar syrups. Small amounts of fungal amylase, which can hydrolyze dextrinous haze material occasionally found in beer, may also be sometimes included in the formulation.

The amount of "double strength" papain needed to successfully chillproof a beer can vary over approximately a fivefold range. A concentration of 10 to 50 mg/l of papain preparation is needed to stabilize most beers against chill haze. Papain is normally added to beer after fermentation and, at least, after a preliminary clarification of the beer is completed. Although Wallerstein's original patent specified addition of enzyme to either wort or beer, papain does have an effect on yeast. When flocculant yeast cells are exposed to papain, they become temporarily nonflocculant, presumably because of hydrolytic cleavage of yeast cell wall peptides involved in the flocculation process. Papain, as well as other proteases, has also been found to give a slight decrease in the rate of fermentation when it is present with yeast.

Wallerstein's patent on enzymic chillproofing claimed that enzymes "become active during pasteurization" and that proteins of the beer are "modified by proteolysis so that the resulting beer will remain clear and brilliant, being no longer sensitive to cold." This statement still remains an accurate summary of our knowledge of the mechanism of chillproofing by papain. Even though the enzyme is usually added to beer at some time prior to its final packaging, the action of papain seems to be very minimal at cold storage temperatures. Below 20°C it is not very active. Optimum activity of chillproofing papain occurs at a temperature of 50 to 60°C, depending upon the substrate, and papain is denatured at 60 to 70°C. Since most beer pasteurization is carried out at a temperature of 60°C, papain is most active in beer during the period in which beer is being heated during pasteurization.

Even though proteolysis by papain occurs most readily under pasteurization conditions, the relative amount of hydrolysis that actually occurs in the treated beer is relatively small. There is no detectable increase in free amino nitrogen, which would be generated during peptide bond hydrolysis in successfully papain-treated beer. Therefore, the number of bonds hydrolyzed by the enzyme is minimal. Herbert et al.[36,37] have proposed that as hydrolyzed peptides in beer become insoluble soon after they are formed, they be quickly removed from solution, especially during pasteurization — much as milk protein is clotted by the action of rennin-type enzymes on casein. Thus, the peptides would no longer be available to form a chill haze. As no other enzyme has been found that is as effective as papain in chillproofing beer, it must be assumed that its efficacy stems from specific hydrolysis of those proteins that ultimately become components of chill haze.

Treatment of beer, with levels of papain normally used for chillproofing, decreases peptide material with isoelectric points between 4 and 5. The fraction removed appears to be an acidic and possibly glycosylated peptide derived from barley that has been associated with beer haze. If amounts of enzyme are added beyond those needed for chillproofing, a decrease in the head retention value of the treated beer will result. Presumably this is due to hydrolysis of proteins important for foam formation.

Papain is partially inactivated by pasteurization but some residual activity remains. The presence of residual papain activity, detected, for example, by its milk-clotting reactivity, is an indication that beer has been treated for chill stability. The residual activity, however, precludes the use of papain in areas where laws prohibit or discourage the use of additives in beer. Residual activity in packaged beer, while possibly contributing to increased colloidal stability, might also be detrimental to foam stability in that beer. Consequently, attempts have been made to use an immobilized form of papain[38] to treat beer during storage so that no enzyme activity would remain in the packaged product. The basic problem encountered in the use of immobilized papain, apart from microbiological aspects, has been the low activity of the enzyme at the low temperatures ordinarily associated with beer storage temperatures.

Enzymic chillproofing of beer is relatively simple, reliable, and inexpensive. Over the years, it has been a commonly accepted method used worldwide for this purpose. It continues to be the standard by which many brewers judge the effectiveness of beer chillproofing.

Chillproofing with Precipitants

There are a number of soluble or nearly soluble compounds that have been used to chillproof beer. These compounds are normally added to beer as a solution at a convenient time after fermentation is over. They then become insoluble in beer and precipitate, removing precursors of haze production and also haze particles already present.

Gallotannins

The most traditional precipitating compound used to chillproof beer is tannic acid[39] or, more correctly, gallotannins. Gallotannins may be deemed as normal metabolic plant products that are water-soluble phenolic compounds having molecular weights between 750 and 3000 Da. Principal commercial gallotannins are Chinese gall nuts (*Rhus semialata*), sumac leaves (*Rhus coriaria, Rhus typhina*), Tara fruit pods (*Caesalpinia spinosa*), Turkish alleppo gall nuts (*Quercus infectoria*), and myrobolans fruit (*Terminalia chebula*-hydrolyzable ellagitannin). Gallotannins belong to the hydrolyzable group of tannins.[40] They have a long history of use in various food products and are considered safe. Their commonly accepted mode of action for chillproofing beer is precipitation with the proteins in beer, which would eventually react with beer tannins to form haze. These proteins can then be removed by settling out the precipitate formed. The degree of stability that can be achieved by this treatment is extremely good.

Tannic acid is usually purchased as a powder and is put into solution with (usually deaerated) water a short time before being added to beer, usually by proportioning during transfer of beer to special treatment tanks. Tannic acid (usual use rate 2 to 10 g/hl) reacts quickly with compounds in the beer and becomes insoluble probably within minutes of addition. It is perhaps interesting to note that with reference to our previous discussions on point particles, the addition of gallotannins to beer almost instantaneously promotes the formation of large numbers of these complexes. Within a couple of hours, large flocs of insoluble material are visible in the beer, which are composed of many smaller point particles tightly aggregated in a rather random fashion. Using traditional commercial products, the individual particles and even their larger flocced masses tend to settle rather slowly in the beer because of their low specific gravity. After approximately 24 h, a voluminous sediment is formed at the bottom of the settling tank. Clear beer can be removed from the top of the settled material, but because of the loose nature of the sediment, a considerable quantity of beer will be left in the tank bottoms. This beer can be partially reclaimed by collecting it from the tank bottoms and letting further settling take place over longer periods of storage.

A recent alternative treatment with a high-purity tannic acid involves an inline type of process. Tannic acid solution is proportioned into beer as it is being transferred to the diatomaceous earth filter. The efficiency of this high-purity form of gallotannins is dependent on temperature and contact time. Thus, a contact time of 15 to 20 min at a temperature below 0°C would be considered ideal for this treatment. Although the aggregated particles do not have time to form larger flocs, they are removed by filtration. It is, therefore, advisable that reappraisal of filter aid quality and quantity be made to maximize filter throughput and to remove the tannin–protein complexes. The exact filtration regime is determined by experimentation and compromise. Beer treated with this new generation of gallotannins shows excellent colloidal stability as well as highly acceptable initial beer clarity values, with no negative effects to any quality attributes. Indeed, some measurements have indicated[41] that this gallotannin treatment gives very low resistance to staling values (RSVs) measured with thiobarbituric acid and, by implication, longer flavor stability.

Silica Sol (Colloidal Silica)

Silica sol, otherwise known as colloidal silica or silicic acid, has also been used to chillproof beer. Its characteristics during use are somewhat similar to those of tannic acid. The silica sol is composed of subcolloidal size particles of silicon dioxide suspended in an aqueous base. The particles used may be from 3 to 20 nm in size. Silica sol can be formed by acidifying a sodium silicate solution to allow the silica particles to form. The suspension is then stabilized before a silica gel can form by adjusting the pH to a slightly alkaline level where, because of negative charges developed on the silica, a gel will not form. The pH should not be so high that the particles will begin to dissolve. Ion exchange to form silicic acid from sodium silicate is a more commonly used method to generate a more acid-stable preparation. When the silica sol is added to beer, it is unstable at the pH and ionic conditions of the beer and reacts with materials in beer, such as protein, and with itself to form aggregated particles similar to a silica gel. Again, these particles are about 1 mm in diameter and look very similar to normal haze particles found in beer. They give a proteinaceous reaction to microscopic staining techniques.

The amount of silica sol used is about 200 to 1000 mg/l of sol as received. Insoluble particles formed after addition to beer tend to aggregate slowly and sediment to give large amounts of a loose sediment. As is true with tannic acid, the precipitated material tends to block the filter, so the filter aid used must be a type that allows a coarser filter bed to be formed. In general, only fairly good chill stability is obtained with silica sol even when maintaining a tight filtration with such materials.

Insoluble Adsorbents for Chillproofing

Insoluble adsorbents have the advantage of retaining their physical integrity when added to beer. If the added material is of a size that can be completely removed by subsequent filtration, there is a certainty that the filtered beer will be free of such chillproofing material. Consequently, the material used is not considered to be an additive in the beer. Insoluble adsorbents can be used in areas where soluble materials cannot legally be used for beer treatment. For this reason, the insoluble adsorbents, especially silica gel[42-46] and crosslinked polyvinylpolypyrrolidone,[47] have enjoyed great acceptance for chillproofing beer in many areas of the world.

Amorphous Silicas

Amorphous silicas, particularly silica hydrogels and xerogels, have been popularly used by the brewing industry for over 25 years as a colloidal stabilization treatment in the adsorption of protein haze precursors (of protein–polyphenol hazes, that is, so-called nonbiological haze) from beer.

It is well known that the production of amorphous silicas basically involves the reaction of sodium silicate with mineral acid, and that there are two main ways of achieving this: the gel route and the precipitate route.

Production of Silica Gels: Sodium silicate and sulfuric acid are mixed under carefully controlled conditions. Complex molecules of silicic acids are formed, which condense to form polysilicic acid. The complex polysilicic acids present in the solution are then destabilized and liberated in a colloidal form, popularly referred to as "colloidal silica" or "silica sol." Size of the colloidal particles is influenced by silica concentration, electrolyte level, amount of excess acidity, and temperature.

The highly "active" colloidal particles (diameter 15 to 20 Å) continue to interact to form a solid, close-packed structure with the water phase immobilized within the matrix. Calculations reveal that in a hydrogel structure, containing about 18% SiO_2, approximately two layers of water are associated with each colloidal silica particle. In this condition, the "silica hydrogel" is mechanically broken into small pieces and washed for some hours under controlled conditions of temperature and pH to remove residual sodium silicate and sodium sulfate (if sulfuric acid is used). This washing step is critical to regularize the spaces (pore diameter and pore volume) and the surface area within the hydrogel structure.

This so-called regular density hydrogel (about 70% water and 30% SiO_2) is then milled, under proprietary conditions, to obtain the desired particle size and, as importantly, the desired particle size distribution before packaging. Particle size/distribution will have an important bearing on stabilization efficacy as well as optimizing filter bed permeability for beer filtration throughput.

Alternatively, the hydrogel can be dried to become a xerogel before milling and packaging. This dehydration process will cause the pore volume to contract until the colloidal particles achieve such a dense close-packed structure that further contraction is resisted. It will therefore be appreciated that xerogel and hydrogel can readily be distinguished by the fact that pore volume as well as surface area irreversibly change according to the hydrothermal treatment of the hydrogel. In the xerogel state, milling then becomes exceptionally critical; the xerogel is brittle and is more prone to shatter, creating a higher percentage of the particles that can reduce permeability of the Cuter bed.

Production of Silica Precipitates: In comparison with the 'gel route,' this route involves destabilization of the complex polysilicic acid anions present in the silicate solution by partial neutralization of the Na_2O present in the solution. In this situation, because of the combination of high pH and high electrolyte content, more complex "structuring" occurs. This makes absolute control of the mixing regime essential during the so-called precipitation stage. This is usually conducted in a specifically designed stirred reaction vessel, where manipulation of the product profile can be achieved.

Evidently, silicas having a wide range of surface areas and pore volumes can be produced, depending on the unit conditions employed. It is of interest that the principal patents concerning amorphous silicas for colloidal stabilization use these parameters.[48] Despite this wide range of patented values, effective colloidal stabilization of beer essentially resides in two basic principles:

1. Selective adsorption and removal of haze-forming proteins, precursors, or associated products with polyphenols because of the interaction with active adsorption sites, that is, isolated surface silanol (SiOH) groups on the amorphous silica particle
2. Selective removal of such components by permeation based on the pore structure and pore diameter distribution

It is considered that the binding of protein to silica gel is analogous to the binding of haze-forming proteins to polymerized polyphenols, that is, via hydrogen bonding of protein carbonyl groups to hydroxyl groups on either the polyphenol catechol ring or the silica gel surface. The conditions required for both reactions are similar and favor the same types of proteins in each.[49] Initially, therefore, the adsorption stability is due to the multiplicity of bonds formed between the silica gel surface and the haze-forming protein.

It is, of course, axiomatic that only the colloidal components associated with haze are removed, leaving behind those proteins that enhance foam, product flavor, or mouthfeel. Thus, the suitability of a silica gel for use as a beer stabilizer depends not only on its ability to adsorb haze proteins, but also on its inability to adsorb proteins associated with foam.

An important factor, therefore, is the exact pore size. Pores must be large enough to accept the haze-forming proteins but cannot be large enough to remove the foam-active proteins. Haze particle precursors are therefore removed by surface absorption onto and permeation into the silica gel and by physical entrapment during filtration. It is a frequent observation, therefore, that the tighter the filtration, the better will be the haze stability of the treated beer. Breweries in the U.S. place considerable emphasis on optimizing grades of diatomaceous earth, particularly for precoating, in order to enhance colloidal stability of their products when using silica gel as the chill-proofing agent.

Although it often appears that there are many similarities between different brands of beer, there are often just as many differences. It can be appreciated that differing qualities of raw materials and differing processing conditions will cause malt beverages to vary in their response to colloidal stabilization using silica hydrogels or xerogels. The development of silicas "customized" to specific products offers an attractive option.

To meet this need, calcined xerogels have been developed, where the total number of isolated silanol groups (to react with the haze proteins) are increased by the calcining process.[48] Alternatively, supplementation or incorporation of amorphous silicas with one or more other sorbents, such as polyvinylpolypyrrolidone (PVPP), synthetic magnesium, calcium or zinc silicates, or natural clays, can be done.

Polyamides

Insoluble polyamides were first used by Harris and Ricketts.[21] These investigators showed that Nylon 66 effectively removed anthocyanidins from beer. However, practical use of nylons as a colloidal stability treatment has essentially been eliminated; they cost too much and are difficult to recover and regenerate.

The use of PVPP has largely eliminated the economic and performance disadvantages previously experienced using nylons. When beer is chill-proofed with PVPP, it is usually the crosslinked, insoluble form, PVPP, which is used rather than the original soluble form polyvinylpyrrolidone or PVP. PVPP is a very effective chillproofing agent at low concentrations. It can be ground to different size particles and incorporated into filter sheets or other materials. The most common method of use is to add a slurry of PVPP particles to beer by proportioning and to remove the particles after a few minutes by filtration. PVPP is expensive, so it is often cleaned and regenerated with caustic soda after being used in beer, and reclaimed by filtration. Regenerated PVPP can be used many times for chillproofing beer. At normal use rates, PVPP has not usually been shown to have any deleterious effects to beer attributes such as flavor or foam, although many brewers using PVPP will monitor the polyphenolic content of the treated beer to be sure that overtreatment is not occurring. It should be noted, however, that McMurrough et al.[50] did demonstrate in ales treated at various rates (5 to 1000 g/hl) of PVPP that increases in chill-haze stability were accompanied with decreases in astringency. This correlated with decreases in the beer content of simple and polymeric flavanols.

PVPP is usually considered to be a relatively specific adsorbent for polyphenols in beer. Beer treated with PVPP shows lowered content of polyphenols, as measured in a number of different ways. For example, PVPP treatment has been shown to reduce anthocyanidin and total polyphenols in beer by 56 and 48%, respectively. As with silica gel, there may be some adsorption of protein–polyphenol interaction products that are in the early stages of haze formation. The mechanism by which PVPP adsorbs polyphenols is probably through hydrogen bonding between the phenolic hydroxyl and PVPP nitrogen.

Beer may also be treated simultaneously with both PVPP and silica gel. The effects of these two adsorbents are additive and (according to studies) even complementary. The chillproofing effect obtained is often better than that obtained with larger amounts of either material alone. Both adsorbents can be mixed into the same slurry and added to the beer simultaneously. If this is done, mixing may have to be more vigorous than with either product used separately. The two tend to aggregate to each other (we suggest by hydrogen bonding).

Natural Clays

Various types of clays, and especially bentonite in the montmorillonite family, have been used as adsorbents to impart chill stability to treated beers. The montmorillonite family refers to a mineral species of the smectite clay group. There are five mineral species within the smectite group that, in addition to montmorillonite, includes beidellite, nontronite, saponite, and hectorite. Montmorillonite is the most abundant of the smectite clay minerals and has many unique physiochemical properties, namely, a large chemically active surface area, a high cation exchange capacity, and interlamellar surfaces having unusual hydration characteristics. In order to understand the mechanism of chillproofing beer, it is important to have an appreciation of the structure of bentonite and how it impacts on the physiochemical properties.

Bentonite, in the dry form, is made up of many individual bundles of very small aluminosilicate platelets or structural units.[51] Each structural unit or platelet consists of three pieces — two tetrahedral sheets "sandwiching" one octahedral sheet. The tetrahedral sheet contains primarily tetravalent Al cations and usually some trivalent Al cations. The apexes of the tetrahedra point toward each other; negatively charged oxygen anions at these apexes also form part of the octahedral sheet, which is primarily composed of trivalent Al cations, with some divalent Fe and Mg and trivalent Fe present in various amounts.

Interlayer exchangeable Na cations are present on the surface of each 2:1 platelet and individual platelets are separated by a layer of water molecules. These individual platelets are dipolar: positive charges are on the outer edges and negative charges are on the flat surfaces or "faces." These "counter-balancing" Na cations are obviously attracted to the negative surfaces. The overall effect is a weak net negative charge on the 2:1 structure resulting from the interval chemical substitutions in the tetrahedral and octahedral sheets.

When the bentonite is mixed with water, the water molecules penetrate into interlayer spacing. As the sodium cations hydrate, they dissociate from the surface of the platelet allowing the negatively charged platelets within the bundle to repel each other. It follows that the individual platelets within the bundle "pop apart" to become separate discrete entities.

Subsequent to this separation, attractive forces between the surface oxygen atoms of the tetrahedra and the hydrogen ions of the water molecules develop. This allows a hydrogen-bonded, hexagonal water structure to build up and form a rigid network made up of many water layers and which further separates the platelets. A "house of cards" structure is then assumed through the matching of positively and negatively charged platelet surfaces (edge to face). This structure can be disrupted through mixing but reforms when mixing stops. The large surface area that is formed can be well appreciated.

Bentonite is considered to act mainly by attraction and adsorption of the positively charged chill-haze proteins in beer to the large surface area of the negatively charged bentonite. In essence, the basic amino groups of many organic cations are actually ammonium cations in which one or more of the hydrogens have been substituted by organic groups. These organic cations, therefore, will exchange with the sodium cations of bentonite and, assisted by van der Waal forces, "coat" the platelets. Each platelet is rendered hydrophobic and oleophilic, which causes flocculation and settling. In addition, the "house of cards" network has the additional ability to encase these proteins either by physical entrapment or by entanglement. Beer treated with bentonite is traditionally allowed to remain in storage until most of the clay has settled to the bottom of a tank. The beer can then be filtered free of residual particles. Subsequent filtration can be quite difficult because of incomplete settling of the clay particles, which have the ability to "blind" the filter and cause short filtration runs.

As appreciated from the foregoing, bentonite clays must be hydrated before use. Water enters slowly between the layers of the clay's crystalline structure to create the internal surfaces where the haze proteins can be adsorbed. The concentrations used are invariably rather high, usually 40 to 100 g/hl. It should also be appreciated that the iron content in the clay, which can be picked up by the treated beer (beer-soluble iron), can be an associated problem to blinding of the filter screens.

Hectorite, another member of the smectite clay family, was first discovered south of the town of Hector in California. It is distinguished from bentonite by its high magnesium, lithium, and fluorine content and low aluminum content, but otherwise has a similar 2:1 structure as previously described for bentonite. Hectorite has higher gel strength and viscosity when compared to bentonite and is more effective at chillproofing than the same beer treated with PVPP or sulfhydryl protease.

Such clays are not now in wide use for colloidal stabilization because of the previously discussed reasons. Nevertheless, it is possible that clays could be incorporated together with other stabilizers to take advantage of different properties. For example, the simultaneous addition to beer of silica xerogel and bentonite (25 g/hl of each) has been shown to have as good a stabilizing effect as using silica xerogel alone (at 50 g/hl) and at a more economic cost.[52] Other clays might show other enhancements with silica gels. Because of the magnesium silicate content of hectorite (minimum 80% to be called hectorite), this clay might be useful in actually scavenging Fe ions, by substitution, from beer, for its fining ability, and for improving colloidal stability.

Oxidation Reactions that Enhance Chill Stability

Polyphenols are susceptible to oxidation. Oxidized polyphenols seem to be more reactive in forming haze than the reduced components. This is integral to the tendency of beer to form more haze as it ages. A number of chillproofing treatments have been suggested that promote stabilization of beer by encouraging oxidation and precipitation of polyphenols in wort. The precipitated material is then removed by normal clarification procedures; the resulting beer is stabilized because haze precursors have been removed from the beer. None of these treatment methods have been successful commercially, but the basic concept appears sound.

Novel Future Possibilities

Internationally, there is still research being conducted to identify new processing aids for colloidal stabilization. For example, enomelanine, a condensed phenolic compound, has been shown to be useful in removing proteins, polyphenols, and lipids in wine. Whether such a compound could be useful in beer has not yet been demonstrated.

We subscribe to the view that value-added products such as those based on silica gels with multiple properties, for example, to use as a filtration medium[42,46,48] and to combine colloidal stability would be most useful. Such gels, where improvements could also be made to flavor stability, would also be useful. Alternatively, cogels, where silica gels might be manufactured with components that remove polyphenols as well as haze-forming proteins, and therefore provide greater colloidal stability to the product, would also prove useful.

Whether techniques based on genetic modification of brewing yeast strains to produce extracellular sulfhydryl proteases during fermentation[32,34,35,53] becomes a popular choice remains to be seen. This is now a proven technology but it is still associated with consumer concern.

As has been discussed at different times in the past,[27,54] proanthocyanidin-free barley varieties offer a further pathway to achieve a product with an extended shelf life against haze. Such varieties may well be used by brewers in the future, providing there is an agronomic and agricultural impetus to farmers to grow these varieties.

Flavor Stabilizers

Flavor perception may be defined as the synchronous sensation of taste and aroma, modified by the simultaneous perception of tactile properties in the mouth. Aroma, in itself, has a profound and complicated effect on the quality of an alcoholic beverage. With over 1300 volatile components having been identified in beer, wine, and whiskey, it can be appreciated that beer flavor is complex and dynamic, subject to a wide variety of influences that can promote sensory changes.

The technical brewer, therefore, has a formidable task to produce a consistent palatable product. Despite unavoidable variations in the quality and composition of brewing raw materials, the brewer's responsibility is to select processing conditions that will result consistently in the desired flavor profile. The irony for the brewer is that beer produced under the most stringently defined set of parameters will eventually alter in flavor on aging, particularly after the product has left the brewery. Beer flavor is never stable and, therefore, as increasing shelf life is demanded by economic and market considerations, flavor instability is still one of the most pressing problems in the brewing industry.

The pattern of flavor change during beer aging was comprehensively described by Dalgliesh.[55] The main feature of Dalgliesh's schematic was the relative intensity changes in the sweetness/bitterness continuum. Sensory bitterness decreased, shifting the sensory balance toward a sweeter flavor. This coincided with the development of burned sugar and caramel notes. Dalgliesh also described at length the intensification of ribes flavor before an eventual decrease, followed by the onset of a typical cardboard flavor.

Having regularly tasted beers from around the world over a number of years, Ryder et al.[56] add slight differences to the descriptors used by Dalgliesh, but the overall trend is similar. In addition to the sweetness/bitterness balance changing over time, they found that tactile properties also change significantly. This greatly depends upon the type of beer; for example, palate smoothness increases for well-attenuated adjunct lagers with original gravities in a typical range for "light" or "regular" American beers, whereas for higher-gravity ales and lagers, such as American malt liquors, these authors found that palate harshness increases. Yeasty sulfury notes tend to decrease rather abruptly at first, followed by a slower decrease. Concomitantly, fresh hop aroma decreases, to be gradually replaced with a dull hop presence. Following this trend is a gradual loss of "fresh beer character." After a certain lag period, cidery/aldehydic flavor notes appear and increase as do typical papery/cardboard flavors.

Effects of Reducing Agents

Flavor stability can be enhanced by judicious addition of a number of "antioxidants." Sulfites and ascorbates (or its analogs or salts such as isoascorbate or sodium erythorbate) are the most common. There are a number of problems associated with the use of sulfite. It is generally believed that free SO_2 directly and very quickly scavenges free oxygen. This is perhaps an oversimplification. Sulfite reacts quickly with oxygen in a water solution, but more slowly in an environment containing ethanol. In fact, and as a highlighted example, despite the relatively large quantities of sulfite (e.g., 45 ppm) added to wine, it would take in the order of 60 days to scavenge all of the oxygen from a wine bottle. In beer, SO_2 will promptly bind with carbonyls and, at a slower rate, phenols, thereby preventing their involvement in primary or secondary staling reactions; it offers a very limited security as an oxygen scavenger.

A second problem associated with sulfite is its addition when yeast is present in the product. This can easily be demonstrated by adding sulfite to storage beer, where, under these very reduced conditions, significant quantities of hydrogen sulfide can be formed.

Probably the most important current problem is that certain segments of the population are allergic to SO_2. Recent food safety legislation in the U.S. mandates the labeling of more than 10 ppm total SO_2 in package. Consequently, for those brewers who use SO_2 as the sole antioxidant, there is much current interest in controlling endogenous SO_2 production in fermentation in order to keep within legislated limits following sulfite addition in package. Despite these disadvantages, judicious use of SO_2 still remains a fairly popular means of controlling oxidation. One of its chief advantages is that it will react very quickly with hydrogen peroxide, which is formed as a by-product of phenol oxidation. In addition, sulfite has been shown to effectively hinder oxidation of higher alcohols to aldehydes. If endogenous total sulfite is controlled to low limits, and total carbonyl concentrations in finished beer also reflect low concentrations, the judicious use of sulfite continues to offer the brewer reasonable security.

Ascorbic acid (or the less-expensive isoascorbate, sodium erythorbate) was first proposed for use in beer as an "antioxidant" in 1939.[57,58] These compounds have been shown to effectively protect isohumulones from decomposition. In addition, ascorbates are known to slow down the degradation of amino acids and also the oxidation of fatty acids to aldehydes. Furthermore, ascorbates will react very quickly with molecular oxygen. In addition, the effectiveness of ascorbates in preventing the oxidation of polyphenols (tannins) points to not only an antioxidative advantage, but also a possible mode of action for beer-haze prevention.

Unfortunately, one of the oxidation products of ascorbic acid is dehydroascorbic acid; it acts as a potent oxygen carrier, which on degradation, produces hydrogen peroxide. In addition, Cu^{2+} ions are reduced by ascorbates to Cu^{2+} ions which, through reaction with oxygen, again promotes the formation of hydrogen peroxide. Therefore, advanced radical oxygen is formed, which will subsequently form the more highly reactive hydroxyl groups. In addition, less-soluble Fe^{3+} is converted to Fe^{2+}, which may lead to higher levels of iron in ascorbate-treated beers exposed to iron during processing. Iron also promotes oxidation.

From this evidence, it can be appreciated that ascorbates can only be used in limited quantities, depending on the redox state of the final beer, before the potential exists to negatively affect the end product. It is often useful, therefore, to use SO_2 whenever ascorbates are used. In combination, therefore, the dual mechanisms afforded by the use of ascorbates and sulfites are attractive and, indeed, in the U.S. this option is gaining in popularity. In this regard, Klimovitz and Kindraka[59] demonstrated that a split addition of potassium metabisulfite (KMS) and sodium erythorbate to beers was more flavor stable than the treatment of the same beers with sodium erythorbate alone. Their suggestion was to commence trials with a 40:60 blend of KMS to sodium erythorbate. Essentially, the only question that remains for using these combined processing aids is the real availability of sulfite. Sulfite will bind with dehydroascorbate as well as with carbonyls and phenols.

Sulfites can also create reducing conditions in the brewhouse. Some brewers, who do not have modern brewhouses where air ingress can be controlled, add sulfites to the mash tun and also at the beginning of kettle filling. The overall effect protects the mash and wort against oxidative effects, as well as aiding mash separation (by prevention of oxidation of the malt gel proteins). In addition to impeding the formation of stale aldehydes in beer, such practice helps in retarding the formation of light-struck flavor. This is particularly noticeable when beer is packaged in flint-colored or green bottles. The sulfite is eliminated during kettle boil. The only effect that might be considered negative is that color formation is reduced during wort boiling.

Other reducing agents have been used and continue to be used by some brewers. For example, sodium dithionite is very effective, particularly when used in combination with sodium erythorbate. Dithionite reacts more quickly than sulfites or ascorbates and reduces dissolved oxygen in beer very quickly, before it can react in other ways. Use rates of this combination at 20 to 30 ppm have been found to be very effective.

Enzymic Deoxidation

Yeast is, of course, an ideal oxygen scavenger for naturally conditioned ales and lagers, but this practice is limited for both technical and commercial reasons.

Another possibility for reducing both dissolved oxygen and headspace oxygen in beer is by using a combination of an acid-stable glucose oxidase and catalase. Glucose oxidase has its traditional application in foods and soft drinks. This enzyme performs its function by oxidizing glucose, in the presence of water and molecular oxygen, to gluconolactone and hydrogen peroxide. The gluconolactone is hydrolyzed nonenzymically to gluconic acid and, in the presence of catalase, the hydrogen peroxide is reduced to water and a one-half molecule of oxygen.

Glucose oxidase was first proposed for use in beer by Ohlmeyer. Ohlmeyer demonstrated in his experimental system that treatment of beer with glucose oxidase-bound oxygen decreased Fe ions in the beer, partially inhibited Maillard-type reactions, and inhibited microbiological spoilage. It was shown that the minute quantities of glucose that he found present in beer were sufficient to enable the glucose oxidase system to function and catalyze the reaction of free oxygen with the glucose to form gluconic acid.

The most widely known oxygen carrier found in nature is hemoglobin. Hemoglobin is a tetramer composed of two polypeptide chains and four heme groups, and incorporates a central iron atom. Recent developments have enabled a synthetic "mimic" of this structure and the synthesis of a number of oxygen carrier complexes based on hemoglobin's fundamental structure. This has led to the development of bottle crowns, containing seals in which these complexes are in an immobilized form. Model experiments have shown that the oxygen level in a 5% ethanol solution over a 24-h period using such immobilized metal/ligand complexes reduced from 2000 ppb to less than 50 ppb. These experiments are encouraging; they demonstrate that even when the oxygen scavenger is not in direct contact with ethanol solution, it is still effective in removing oxygen. As oxygen is removed from the headspace, further oxygen is removed from the fluid and is released into the headspace to re-establish equilibrium. This oxygen is then scavenged, and the process continues until only traces are left in the product.

A major issue in the development of this scavenger is its activation after packaging. This is currently achieved when the crown seal is subjected to the high humidity conditions typically found in the package headspace. In addition, pasteurization would be expected to increase the rate of scavenging.

Foam Stabilizers

Foam stability and adhesion have always been of fundamental interest to the technical brewer[60]—albeit differing in degree between international locales because of consumer preference.

There are two origins of beer foam when it is poured or dispensed from a typical package to a glass: the air entrapped by the falling liquid and the carbon dioxide (or in some beers, a mixture of carbon dioxide and nitrogen) released by the ensuing lowering of pressure. As the mass of tiny bubbles rises, it presents, a tremendous surface area to the liquid beer, and the many surface-active or surface-seeking components present in the beer tend to collect on the bubbles and are subsequently carried to form the head of the foam. In terms of optimizing beer foam, the technical brewer is cognizant of the favorable and unfavorable processing and raw material aspects, which can ultimately influence product foam formation and cling, and strives to maximize foam-positive components accordingly.

Beer foam is never stable and follows a dynamic and complex course of events from formation to collapse. Closely akin to the subject of emulsions, the two fields are similar in that their properties both depend on surface effects, changes in interfacial tension, electrolyte composition, and manner of preparation. However, despite the vast amount of recorded literature since the last century, the fundamental theory of foams is not as well formulated as the theory of emulsions.

Formation of foam requires the presence of surfactants in solution. Surfactants are compounds that are amphipathic, that is, that have both hydrophobic and hydrophilic properties in the same molecule. Such compounds can satisfy both these characteristics simultaneously by being situated with the proper orientation at the surface of a water-based solution.

The proteins and iso-alpha acids present in beer are surfactants and have long been recognized as being of prime importance in the formation of beer foam. Other components of beer, such as polysaccharides, melanoidens, and metal ions, are also foam positive. Foam-negative compounds, for example, lipids, are also important in determining the overall head-forming characteristics of a particular beer.

Proteins or polypeptides are well known as surfactants. The foam-forming properties of egg albumen in egg whites are an obvious and extreme example of the tendency of protein solutions to form foam. Proteins have a tendency to concentrate at the surface of their solutions. The strength of this tendency and their capability to interact with themselves or with other compounds at the surface in beer determines how effective the various types of proteins are in forming a stable foam.

Surface activity of proteins or protein complexes is important in the ability of a beer to foam. Once the foam is formed, the surface viscosity determines how quickly liquid can drain from the foam and thus the stability of the foam that is formed. The glycoproteins present in beer have been suggested to be particularly important for the stability of beer foam because of their ability to interact with each other, increasing the viscosity at the surface of the foam.

Glycoproteins have side chains of polysaccharide attached by a variety of linkages. The polysaccharides are hydrophilic and are not particularly surface active in themselves. When attached to a surface-active protein, they will be concentrated just below the surface when the protein portion of the molecule migrates to the liquid surface. The polysaccharide chains tend to interact with each other or with other proteins because of their ability to form many hydrogen bonds. The glycoproteins, therefore, tend to form a concentrated and interlinked network of macromolecules at the liquid surface, increasing the viscosity and tending to form a semisolid layer that resists liquid drainage.

Propylene Glycol Alginate

It is probably true to say that the only processing aid to gain international acceptance for increasing the quality of beer foam attributes is propylene glycol alginate (PGA). The charged nature of PGA is reported to provide greater foam stability than equal amounts of neutral polysaccharides. Electrostatic interaction between carbonyl groups on the glycol alginate molecules and amino groups on the peptides within the bubble wall was suggested to be responsible for the stabilizing effect of PGA against the harmful effects of foam-negative materials, such as lipids, to beer foam. Too much negative charge, however, leads to an overreaction with protein and may cause haze in beer. Addition of propylene glycol esters via reaction of alginic acid with propylene oxide neutralizes most of the carboxyl groups of the alginate. Many investigators believe that 80 to 85% esterification is optimal, leaving some negative charge, but not enough to cause haze. Propylene glycol also has the effect of making alginic acid more hydrophobic, giving it a surface-active characteristic.

Similar to PGA, melanoidins have also been reported to assist in forming ionic bonds with proteins in the bubble wall. It should be noted that the positive effects of propylene glycol alginate are minimized at higher pH or if the proteins are acetylated.

Metal Salts

Metal salts, particularly iron salts, have also been used as post-beer-storage processing aids to improve foam stability. Their use has now largely been discontinued because of the potential risk of metallic flavors to the final product, or because of the fact that certain metal ions are toxic.

Quillaia and Yucca

Quillaia extract is the concentrated purified extract of the outer cambium layer (bark) of the *Quillaia saponaria* molina tree, native to Chile. Yucca extract is the concentrated, purified extract of the Mohave yucca plant, *Yucca schidigera*, which is native to Baja, California, the southwestern deserts of North America, and also parts of Africa. Both have been used as foam stabilizers for beverages. In beer, however, although they are very effective, the color of the foam is off-white and, therefore, not as appealing as when using propylene glycol alginate.

Biological Stabilizers

The quest for effective broad-spectrum biological stabilizers for beer and other malt beverages remains a challenge in brewing research. This desire has been regularly re-emphasized by the increasing popularity of non-pasteurized products, the growth of nonalcoholic and alcohol-free beverages, and the trend toward lower bitterness in beers (and, hence, reduction in any antimicrobial effects from hops).

Heptylparaben

The only exogenous antimicrobial agent approved by the U.S. Food and Drug Administration for use in beer is *n*-heptyl-*p*-hydroxybenzoate, or heptylparaben.[61] The permitted use rate in the U.S. is a maximum of 12 ppm. Since its proposed use in 1966, however, it has not seen widespread use because of a number of deficiencies, such as low solubility at the normal pH of beer, negative surface-active effects to beer foam stability and cling, removal from the beer by can-lining compounds and crown seals, and haze formation.

Nisin

Nisin is one of several bacteriocins formed by lactic acid bacteria. This compound is a polypeptide of 34 amino acid residues and has a molecular weight of 3510 Da. Nisin is produced by certain strains of *Lactococcus lacis*.[62] Its antimicrobial activity is restricted and limited against other lactic acid bacteria and a few other gram-positive bacteria. This antibiotic has been proposed for use in beer[63,64] and wine,[62] though, to date, it has not found widespread use.

Hop Resins

Although hops have been traditionally known for their antimicrobial effects in brewing, it is only relatively recently that interest in precisely characterizing the antimicrobial components within hops has become popular to establish whether it might be possible to take further advantage of these "natural" antimicrobial agents. The hop resins, lupulone (β-acids) and humulone (α-acids), have been shown[65,66] to have remarkable potency against gram-positive bacteria. However, as iso-acids, this potency is 15 to 30 times less.

The antimicrobial action of the hop resins has been shown[67] to act at the level of membrane leakage and also membrane perturbations, which prevent amino acid uptake. It has been suggested that impairment of membrane permeability is the primary site of attack by these hop components. Inhibition of protein synthesis could be a secondary event. It appears that these antibiotic properties[68] are mainly dependent on the hydrophobic parts of the hop components. Thus, the acyl-lupuphenones [2-acyl-4,4′,6-tri(3-methyl-2-butenyl)cyclohexane-1,3,5-triones] having three prenyl and one acyl side chains are the most active substances.

Despite the restricted antimicrobial properties of hops, research in this area will continue — essentially to establish whether further advantage can be taken of these components in hop-breeding programs or in the production of new hop products for the future.

Other Novel Possibilities

One area that could be advantageous for future use in brewing is the "natural" antimicrobial properties of substances of vegetable origin. For example, plumbagine (2-methyl-5-hydroxy-1,4-naphthonquinone) isolated from various plants is one possibility. Another might be juglone (5-hydroxy-1,4-naphthoquinone), which can be isolated from the exocarp of walnuts. Both compounds have shown remarkable antimicrobial properties. Of course, another possibility is the cloning of wide-spectrum antimicrobial peptides from natural sources (e.g., plants) into brewing yeast; this field continues to receive active attention.

Acknowledgments

The authors wish to sincerely thank the following for their support in the preparation of this chapter: Ron Siebel, J. E. Siebel Sons' Co. Inc.; Robert Fernyhough and Tony Lovell, Joseph Crosfield Ltd.; Roger Mussche, Omnichem S.A.; J. J. Scott Cowper and Robert Taylor, James Vickers Ltd.; George Sheppard, Rhone-Poulenc Inc.; and Mark Steward, American Colloid Co. Thanks also to our colleagues at Miller Brewing Company and the Siebel Institute for their support.

References

1. Beer Institute, *Adjunct Reference Manual*, 1989.
2. Barrett, J., Clapperton, J.F., Divers, D.M., and Rennie, H., Factors affecting wort separation, *J. Inst. Brew.*, 79:401–413, 1973.
3. Letters, R., Byrne, H., and Doherty, M., The complexity of beer β-glucans, *Proc. 20th Eur. Brew. Conv.*, Helsinki, 1985, pp. 395–402.
4. Oksanen, J., Ahvenainen, J., and Home, S., Microbial cellulase for improving filterability of wort and beer, *Proc. 20th Eur. Brew. Conv.*, Helsinki, 1985, pp. 419–425.
5. Ducroo, P. and Delecourt, R., Enzymic hydrolysis of barley beta-glucans, *Wallerstien Lab. Commun.*, 35:219–226, 1972.
6. Button, A.H. and Palmer, J.R., Production scale brewing using high proportions of barley, *J. Inst. Brew.*, 80:206–213, 1974.

7. Nielsen, E.B., Brewing with barley and enzymes — a review, *Proc. 15th Eur. Brew. Conv.*, Estoril, 1971, pp. 149–170.
8. Weig, A.J., Brewing adjuncts and industrial enzymes, *Tech. Q. Master Brew. Assoc. Am.*, 10:79–86, 1973.
9. Denault, L.J., Glenister, P.R., and Chau, S., Enzymology of the mashing step during beer production, *J. Am. Soc. Brew. Chem.*, 39:46–52, 1981.
10. Norris, K. and Lewis, M.J., Application of a commercial barley β-amylase in brewing, *J. Am. Soc. Brew. Chem.*, 43:96–101, 1985.
11. Wallerstein, L., Chillproofing and stabilization of beer, *Wallerstein Lab. Commun.*, 24:158–167, 1961.
12. Dombek, K.M. and Ingram, I.O., Magnesium limitation and its role in apparent toxicity of ethanol during yeast fermentation, *Appl. Environ. Microbiol.*, 52:975, 1986.
13. Hsu, W.-P., Vogt, A., and Bernstein, L., Yeast nutrients and beer quality, *Tech. Q. Master Brew. Assoc. Am.*, 17:85, 1980.
14. Ingledew, W.M., Sosulski, F.W., and Magnus, C.A., An assessment of yeast foods and their utility in brewing and enology, *J. Am. Soc. Brew. Chem.*, 44:166, 1986.
15. Axcell, B., Kruger, L., and Allan, G., Some investigative studies with yeast foods, *Proc. Inst. Brew., Australia and New Zealand Section*, Brisbane, 1988, p. 201.
16. Rice, J.F., Helbert, J.R., and Garver, J.C., The quantitative influence of agitation on yeast growth during fermentation, *Proc. Am. Soc. Brew. Chem.*, 1974, p. 94.
17. Siebert, K.J., Blurn, P.H., Wisk, T.S., Stenroos, L.E., and Anklam, W.J., The effect of trub on fermentation, *Tech. Q. Master Brew. Assoc. Am.*, 23:37, 1986.
18. Ingledew, W.M., Magnus, C.A., and Patterson, J.R., Yeast foods and ethyl carbamate formation in wine, *Am. J. Enol. Vitic.*, 38:332, 1987.
19. Roberts, R.T., Use of an extract of spent grains as an antifoaming agent in fermenters, *J. Inst. Brew.*, 82:96, 1976.
20. Dadic, M., Knudsen, F.B., Van Gheluwe, G.E., and Weaver, R.L., Antifoaming agent from malt, V.S. Patent No. 4,339,466, 1982.
21. Hall, R.D., Foam reduction during preparation of beer, U.S. Patent No. 3,751,263, 1973.
22. Montgomery, G.W.G., Hough, J.S., Mathews, A.J.D., Morrisoll, K.B., and Morson, B.T., The effects of kettle additives on the biochemical composition of wort and beer, *Proc. Inst. Brew., Australia and New Zealand Section*, Hobart, 1986, p. 171.
23. Taylor, R., The fining of cask beer, *Brewer*, May:20, 1993.
24. Ballard, G.P.S., Isinglass types in relation to foam stability, *Proc. Inst. Brew., Australian and New Zealand Section*, Adelaide, 1987, p. 127.
25. Cowper, J.J.S. and Taylor, R., Making things clear, *Tech. Q. Master Brew. Assoc. Am.*, 25:90, 1988.
26. McCreath, A., New bovine finings of "predictable consistency," *Brew. Dist. Int.*, June:27, 1962.
27. Baxter, E.D., Morris, T.M., and Picksley, M.A., Low proanthocyanidin malts for production of chilled and filtered or fined beer, *J. Inst. Brew.*, 93:387, 1987.
28. Outtrup, H. and Erdal, K., Haze forming potential of proanthyocyanidins, *Proc. Eur. Brew. Conv.*, 1983, p. 307.
29. Morris, T.M., The relationship between haze and the size of particles in beer, *J. Inst. Brew.*, 93:13, 1987.
30. Power, J. and Ryder, D.S., Some principal points of colloidal stabilization, *Brew. Room Book 1989–1991, Malt*, St. Paul's, 1988, p. 55.

31. De Clerck, J., The use of proteolytic enzymes for the stabilization of beer, *Tech. Q. Master Brew. Assoc. Am.*, 6:136, 1969.
32. Bilinski, C.A., Russell, I., and Stewart, G.G., Applicability of yeast extracellular proteinases in brewing: physiological and biochemical aspects, *Appl. Environ. Microbiol.*, 53:495, 1987.
33. Rosi, I., Costamagna, L., and Bertuccioli, M., Screening for extracellular acid protease(s) production by wine yeasts, *J. Inst. Brew.*, 93:322, 1987.
34. Sturley, S.T. and Young, T.W., Extracellular protease activity in a strain of *Saccharomyces cerevisiae*, *J. Inst. Brew.*, 94:23, 1988.
35. Sturley, S.T. and Young, T.W., Chill proofing by a strain of *Saccharomyces cerevisiae* secreting proteolytic enzymes, *J. Inst. Brew.*, 94:133, 1988.
36. Hebert, J.P., Scriban, R., and Strobbel, B., Action de la papaine sur les protides de la biere au cours de la pasteurization, *Proc. Eur. Brew. Conv.*, 1975, p. 405.
37. Hebert, J.P., Scriban, R., and Strobbel, B., Analytical study of proteolytic enzymes for beer stabilization, *J. Am. Soc. Brew. Chem.*, 36:31, 1978.
38. Witt, P.R., Sair, R.A., Richardson, T., and Olson, N.F., Chillproofing beer with insoluble papain, *Brew. Dig.*, 45:70, 1970.
39. De Clerck, J., The stabilization of beer through tannin addition, *Tech. Q. Master Brew. Assoc. Am.*, 7:1, 1970.
40. Delcour, J.A., VanLoo, P., Moerman, E., and Vancraenenbroeck, R., Physicochemical stabilization of beer — use of hydrolysable tannins for production of chill-proof beers, *Tech. Q. Master Brew. Assoc. Am.*, 25:62, 1988.
41. Mussche, R., Physico-chemical stabilization of beer using a new generation of gallotannins, *Proc. Inst. Brew., Australia and New Zealand Section*, Auckland, 1990, p. 136.
42. Clark, B.E., Crabb, D., Lovell, A.L., and Smith, C.J., The use of silica hydrogels for combined filtration and stabilization, *Brewer*, 66:168, 1980.
43. Halcrow, R.M., Silica hydrogels — their properties and uses in chillproofing beer, *Brew. Dig.*, 51:44, 1976.
44. Hough, J.S. and Lovell, A.L., Recent developments in silica hydrogels for the treatment and processing of beers, *Tech. Q. Master Brew. Assoc. Am.*, 13:34, 1979.
45. Patterson, R.E. and Siebel, R.E., Stabilization of beer using silica gel, *Proc. 18th Conv. Inst. Brew., Australia and New Zealand Section*, 1984, p. 133.
46. Smith, C.J., The simultaneous filtration and stabilization of beer using silica hydrogels, *Brewer*, 68:60, 1982.
47. Dadic, M. and Lavallee, J.G., The use of polyclar (PVPP) in brewing, *J. Am. Soc. Brew. Chem.*, 41:14, 1983.
48. Fernyhough, R. and Ryder, D.S., Customized silicas — a science for the future, *Tech. Q. Master Brew. Assoc. Am.*, 27:94, 1990.
49. Fitchett, C.S., The adsorption of beer proteins by silica gels, Thesis, University of Birmingham, 1981.
50. McMurrough, I., Roche, G.P., and Hennigan, G.P., Effects of PVPP dosage on the flavanoid contents of beer and the consequences for beer quality, *Brew. Dig.*, 59:28, 1984.
51. Farmer, V.C. and Russell, J.D., Interlayer complexes in layer silicates, *Trans. Faraday Soc.*, 67:2737, 1971.
52. Raible, K.J., Method of treating beer with adsorbing agents, U.S. Patent No. 3,436,225, 1969.

53. Nelson, G. and Young, T.W., The addition of proteases to the fermenter to control chill haze formation, *J. Inst. Brew.*, 93:116, 1987.
54. Delcour, J.A., Vandenberghe, M.M., Dondeyne, P., Schrevens, E.I., Wijnhoven, J., and Moerman, E., Flavour and haze stability differences in unhopped and hopped all-malt pilsener beers brewed with proanthocyanidin-free and with regular malt, *J. Inst. Brew.*, 90:67, 1984.
55. Dalgliesh, C.E., Flavour stability, *Proc. Eur. Brew. Conv.*, Amsterdam, 1977, p. 623.
56. Ryder, D.S., Power, J., and Siebel, R.E., A question of flavour stability or instability, *Proc. Inst. Brew., Central and Southern African Section*, Johannesburg, 1989, p. 386.
57. Gray, P. and Stone, I., Oxidation in beers, *Wallerstein Lab. Commun.*, 2:5, 1939.
58. Gray, P. and Stone, I., Oxidation in beers, *Wallerstein Lab. Commun.*, 2:49, 1939.
59. Klimovitz, R.J. and Kindraka, J.A., The impact of various antioxidants on flavor stability, *Tech. Q. Master Brew. Assoc. Am.*, 26:70, 1989.
60. Ryder, D.S., Power, J., Foley, T., and Hsu, W.-P., Critical interactions-beer foam, *Proc. Inst. Brew., Central and Southern African Section*, Victoria Falls, 1991, p. 104.
61. Brenner, M.W. and Iffland, H., Economics of microbial stabilization of beer, *Tech. Q. Master Brew. Assoc. Am.*, 3(3):193, 1966.
62. Radler, F., Possible use of nisin in winemaking. I. Action of nisin against lactic acid bacteria and wine yeasts in solid and liquid media, *Am. J. Enol. Vitic.*, 41:1, 1990.
63. Ogden, K. and Waites, M.J., The action of nisin on beer spoilage lactic acid bacteria, *J. Inst. Brew.*, 92:463, 1986.
64. Ogden, K. and Tubb, R.S., Inhibition of beer-spoilage lactic acid bacteria by nisin, *J. Inst. Brew.*, 91:390, 1985.
65. Teuber, M., Low antibiotic potency of isohumulone, *Appl. Microbiol.*, 19:871, 1970.
66. Kokubo, E. and Kuroiwa, Y., Tetrahydrohumulones as a new source of bitter flavor in brewing, *Rep. Res. Lab. Kirin Brew.*, 11:33, 1968.
67. Teuber, M. and Schmalreck, A.F., Membrane leakage in *Bacillus subtilis* 168 induced by the hop constituents lupulone, humulone, isohumulone and humulinic acid, *Arch. Mikrobiol.*, 94:159, 1973.
68. Schmalreck, A.F., Teuber, M., Reininger, W., and Hartl, A., Structural features determining the antibiotic potencies of natural and synthetic hop bitter resins, their precursors and derivatives, *Can. J. Microbiol.*, 21:205, 1973.

10

Brewhouse Technology

Kenneth A. Leiper and Michaela Miedl

CONTENTS

Introduction	385
General Layout of a Brewhouse	386
Heat Transfer in the Brewhouse	388
Introduction	388
Heat Transfer	388
Materials	390
Raw Materials Intake	390
Storage	390
Removal of Foreign Objects	391
Milling	392
Reasons for Milling	392
Roll Mills	392
Roll Milling with Conditioning	393
Low-Pressure Steam Conditioning	393
Hot Water Conditioning	394
Wet Milling	394
Wet Milling with Steep Conditioning	395
Hammer Milling	397
Mash Conversion and Separation	398
Purpose of Mashing	398
Basic Principles of Mash Separation	398
Mash Tuns	399
Mashing-in Systems	401
The Mash Conversion Vessel	402
Adjunct or Cereal Cookers	403
Mash Kettle	404
Mash Acidification	404
Mash Separation Systems	405
Lauter Tuns	406

Strainmaster	408
Mash Filters	409
Membrane Mash Filters	411
The Nortek Mash Filter	412
Comparison of Separation Systems	413
Wort Boiling	414
Overview	414
Introduction	414
Principles of Boiling	414
Types of Boiling	416
Objectives and Events	417
Wort Preheating	419
Types of Wort Boiling System	419
Direct Fired Kettles	419
Kettles with Internal Steam Coils	420
Kettles with External Heating Jackets	421
Other Systems	421
Internal Heater with Thermosyphon	421
External Heaters	423
Dynamic Low-Pressure Boiling	425
Merlin System with Wort Stripping	425
Conventional Boiling with Stripping	427
Energy Recovery Systems in the Brewhouse	428
Hop Addition	429
Addition of Hop Products	429
Addition of Whole Hops	430
Addition of Hop Pellets and Powders	430
Addition of Hop Extract	430
Wort Clarification	431
Introduction	431
Separation Systems	432
Hop Back	432
Hop Strainer	432
Settling Tanks	433
Filtration	434
Centrifugation	434
Whirlpool Separators	434
Whirlpool Kettle	436
Wort Recovery from Trub	436
Wort Cooling and Aeration	436
Introduction	436
Plate and Frame Heat Exchangers	437
Removal of Cold Break	438
Cold Sedimentation Tank	438
Anstellbottich	438

Centrifugation 438
Filtration 439
Flotation 439
Aeration 439
Ceramic or Sintered Metal Candles 440
Venturi Pipes 440
Static Mixers 440
Centrifugal Mixers 441
Yeast Addition 441
Brewhouse Efficiency 441
Brewhouse Yield 441
Brewhouse Capacity 442
Brewhouse Cleaning 443
References 444

Introduction

This chapter will examine the range of equipment and processes that are currently in use in brewhouses throughout the world. This includes old-fashioned equipment that is still used for historical reasons, current best practice technology, and new systems that have recently come into use. The advantages and disadvantages of the various systems will be compared and contrasted. The brewer must be aware that whatever system is in use, the brewhouse process must be optimized in a way that ensures both energy efficiency and the production of good quality wort.

A brewery has a brewhouse in order to produce wort in a process starting with the raw materials and terminating with wort cooling. This series of complex and costly procedures takes place to convert the raw materials — water, malt, adjuncts, and hops — into a fermentable liquid, wort, which will become beer.

There are three linked requirements of a brewhouse:

- To produce good quality wort
- At a high throughput
- With optimal extract recovery from the raw materials

These processes must be carried out consistently and efficiently. This is difficult to achieve; if one of the three requirements is disrupted, the other two will also be affected.

As brewhouses normally operate on a batch system to allow for frequent cleaning, the production of wort with a consistent composition is a challenge

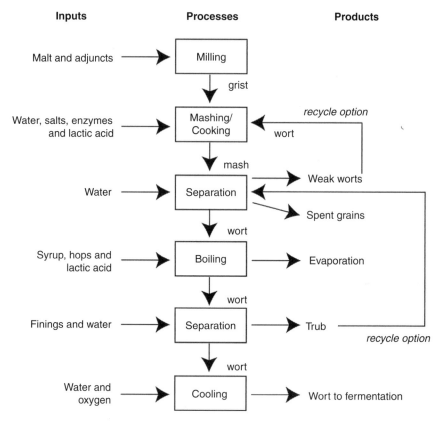

FIGURE 10.1
Inputs, processes, and products of a brewhouse.

to the brewer. The various processes, inputs, and products of a brewhouse are shown in Figure 10.1.

General Layout of a Brewhouse

In traditional brewing, mashing was simply the process of mixing warm water with ground malt, and after a period of standing, as much of the liquid as possible was recovered. Brewers soon realized that varying results depended on the temperature of the water used, but were unable to control this accurately, as thermometers were not available at that time. The brewer is said to have overcome this challenge by using brewing liquor whose temperature was such that his or her face was best reflected in the water surface. Below this temperature of 65–71°C, the ability of the water to reflect declines, whereas above this temperature water vapor fogs the air. In the absence of a thermometer, consistent temperatures can also be achieved by mashing with water from a well, boiling a defined portion of this mash, and then mixing the hot and cold mashes.

It was probably the need for consistent temperatures throughout the mashing process that inspired the decoction mashing procedures mainly employed in continental Europe. Another factor promoting decoction was the slightly less modified and less strongly kilned malts from central Europe compared to British malts, which give good results with the less costly infusion mash. Infusion mashing requires only one vessel, which is a combination mash and lauter tun.

In decoction mashing, the unit operations are extended by the additional heating, pumping, and mixing steps; thus, there is some variation from brewery to brewery in the number of vessels used. Figure 10.2a shows the simplest arrangement for decoction mashing, with one vessel for mixing and one for boiling used in conjunction with a vessel for solid–liquid separation. It is, however, more usual to find a conventional double brewhouse with a mash conversion tun or heated mash-conversion vessel, a mash copper, wort copper, and a lauter tun or mash filter, as depicted in

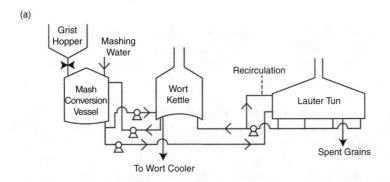

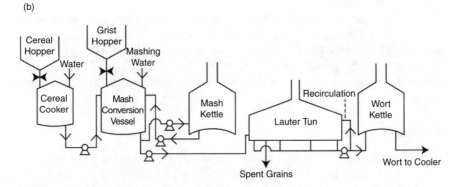

FIGURE 10.2
Layouts of brewhouses suitable for decoction mashing (a) Two vessel brewhouse, (b) Four vessel brewhouse.

Figure 10.2b. In order to utilize equipment to the greatest extent, there may be one lauter tun shared by two mash-mixing vessels and two coppers. In order to increase the number of brews in a given period, the brewhouse may also contain a wort buffer tank to hold the first wort until the wort kettle is available and a whirlpool if no whirlpool kettle is present. Typically, vessels are located on one level, but in smaller and older breweries at least the lauter tun is above and close to the wort kettle. The wort flows through the wort bed by gravitational force into the kettle.

The various brewhouse operations, events, and pieces of equipment will now be described in the order that they are encountered as the raw materials are transformed into wort. Where necessary, background information on the engineering principles is provided; this begins with a general overview of heat transfer, a concept crucial to the whole brewing process.

Heat Transfer in the Brewhouse

Introduction

Liquids are required to be heated and cooled at various points in the brewhouse, and heat exchange is therefore of great importance. Liquor needs to be heated for mashing and cooking; mash vessels and cereal cookers need to be heated; wort must be preheated, boiled, and then cooled. Vapors from the kettle require to be cooled for reuse.

Heating inside vessels such as mash-conversion vessels and kettles is carried out using steam jackets, coils, or tube heaters. When heating or cooling liquids during transfer between vessels such as during wort preheating or wort chilling, the method of choice is the plate and frame heat exchanger. Waste heat can be used for other purposes; thereby conserving energy and resources.

When the system is used to heat a liquid, a heating medium such as hot water or low-pressure steam is run against the liquid. Heat from the medium is transferred to the liquid, increasing its temperature and decreasing the temperature of the medium. When the system is used for cooling, cold water, glycol, or ammonia is used. This cools the liquid, and the medium leaves the cooler unit at a higher temperature.

Heat Transfer

Wort boiling and cooling require rapid changes in temperature and, therefore, efficient heat transfer is of major importance. Heat transfer (Q) is a function of surface area, driving force, and resistance, and is defined as:

$$Q = UA\Delta T \quad \text{(expressed as kJ or MW)}$$

where U is the overall heat transfer coefficient (kW/m² K), A is the surface area (m²), and ΔT is the overall temperature difference (K).

In wort boiling, the factor UA (kW/m²) gives a value that indicates the amount of heat exchange. This is important, as this will affect the type of boiling that will take place and the extent of fouling that can occur. ΔT takes into account the amount of surface film (h) and fouling (R) that occurs on both the steam (s) and water (w) sides of the heat exchange wall (thickness t with conductivity k). As the wort side of the heater will foul quicker than the steam side, this will control the heat transfer efficiency.[1] The origins of the various values are shown in Figure 10.3.

In plate and frame heat exchangers, the heat transfer will occur between two liquids with a temperature gradient, such as cold wort and hot water, with the heat flowing from the hot liquid to the cold liquid. The heat flow will be faster the higher the temperature difference. This difference is known as the driving temperature. The amount of heat (Q) to be transferred depends on three factors: mass (m), specific heat (c_p), and the temperature change of the liquid (Δt). Thus:

$$Q = m \times c_p \times \Delta t$$

Friction between the two fluids and the plate separating them induces the formation of layers of static fluid on both plate surfaces. These are referred to as "boundary layers" and heat transfer occurs here by conduction only. The flow of the rest of the liquid is more turbulent, and heat exchange occurs by both conduction and convection. Assuming an even

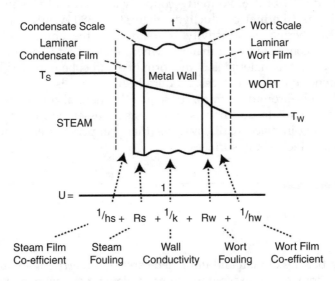

FIGURE 10.3
Heat transfer coefficients.

temperature throughout the wort flow, the opposite situation exists on the cold side of the plate, with the liquid away from the boundary layer being colder.

The area (A) required to transfer a given amount of heat is a function of three parameters: the amount of heat to be transferred (Q), the average driving temperature (taken as a log of the mean temperature difference, Δt_{LM}), and the overall heat transfer coefficient (U) as a measure of heat transfer efficiency. Thus:

$$Q = U \times A \times \Delta t_{LM}$$

This can be combined with the first equation to calculate the required heat transfer area. Thus:

$$A = \frac{m \times c_p \times \Delta_t}{U \times \Delta t_{LM}}$$

Wort contains solid material, some of which is important later when the wort is fermented. Shear stress has to be avoided to minimize damage to these particles. The measure used for sheer stress in plate heat exchangers is τ. A high τ value increases heat exchange with increased turbulence, but may damage solids and flocs.

Materials

The modern material of choice for heating surfaces is austenitic stainless steel of grade 304 or 316. Formerly, copper was used and was popular, as it has a higher thermal conductivity than steel: 380 W/mK compared to 167 W/mK. Thermal conductivity is defined as the amount of heat in watts that is conducted by a material where there is a temperature difference of 1 K and the material is in the form of a 1-m cube with a 1° temperature difference on opposite faces. Modern heating vessels, however, tend to have thin walls (around 1.6 mm); consequently, thermal conductivity is not a major problem.[1] In addition, stainless steel vessels are easier to clean using cleaning-in-place (CIP) systems than copper vessels, and unlike copper retain their shiny surfaces.

Raw Materials Intake

Storage

Generally, batches of malt and unmalted cereal grain arriving in bulk transporters are conveyed into separate stores by elevator or conveyor. Except for special materials, such as amber, crystal, and chocolate barley malts, the

Brewhouse Technology

grain is usually held in silos or storage bins of steel and concrete construction with hopper bottoms. Adequate facilities for receiving and storing malted and unmalted cereal grains are required to cover for unforeseen delays in delivery, but storage of malt and adjuncts at a brewery is expensive, as it requires capital expenditure and takes up valuable space. A diagram of a malt-handling system is shown in Figure 10.4. Syrups are delivered either in bulk or in drums. The bulk syrup tanks in the brewery have to be heated and insulated to keep the viscosity of the syrup low. The syrup is added directly to the wort kettle or added to a special, sugar-dissolving vessel prior to addition to the kettle.

Removal of Foreign Objects

Before storage, cereal grains are conveyed by pneumatic or mechanical means past magnetic separators to rotating, cylindrical, oscillating, or flat-bed screens. Cereal corns of abnormal size are rejected, and foreign matter such as straw, stones, metal, and string are removed. Grain dust can be a highly explosive substance, so ignition sources are avoided by removing stones and metal from the malt before milling. Careful attention should also be paid to mechanical and electrical installations wherever there is dust. Where possible, dust is removed by aspiration with air cyclones and trapping in cloth-screen filters. Before milling, malt is removed from the silo and transferred to a hopper above the mill, being weighed en route. Once the mill is ready, the hopper is opened and the malt enters the mill.

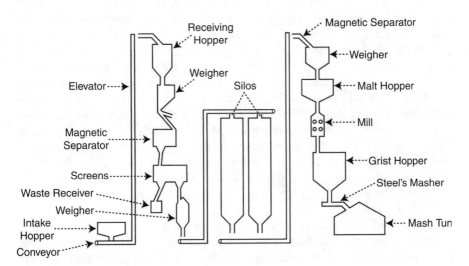

FIGURE 10.4
Typical malt handling system.

Milling

Reasons for Milling

Milling is carried out for two reasons: particle size reduction and particle size control. The type of milling carried out depends on the separation process that is to be used. If a mash or lauter tun is to be used, then a roll mill should be chosen, as this will create a coarse grist and preserve the husk material required to create the filter bed. If a mash filter is to be used then a hammer mill should be chosen as this will produce a fine grist.

Roll Mills

The traditional method used for preparing malt for mashing in a mash or lauter tun is by using a roll mill. It crushes the malt to produce the particle size distribution desirable for optimal extract recovery, but preserves the husk material that is required for filter bed formation and subsequent solid–liquid separation. Malt for dry milling should have a moisture content of 2.5–4%.[2]

Roll mills work by crushing the malt as it is drawn through the gap between the rollers exerting pressure and shear forces on the kernels. The rollers are commonly fluted to increase friction. The capacity and efficiency of the mill depends on the length, diameter, speed, and gap distance of the rollers. Crushing has two effects, compression and shear. Compression is related to the gap between the rollers, and shear depends on their speed.

Mills are available with two, four, five, or six rollers. Two-roll mills are not very flexible, as reducing the gap too much will cause damage to the husk and will not give a proper grist size distribution. Such mills are only useful for well-modified malt or for use in small breweries where the running costs are low. The distance between the rollers is normally 1.3–1.5 mm.[3] Mills with more rollers are fitted with vibrating screens to sort the various fractions that are produced from the malt — husks, coarse grits, fine grits, and flour. The husks and flour do not require to be milled a second or third time and are separated and channeled away. Revolving beaters can be installed to speed up the sorting process. The various arrangements of roll mills are shown in Figure 10.5. Four- and five-roll machines are suitable for breweries carrying out a large-scale process or a full production schedule. The top pair of rollers is generally 1.3–1.5 mm apart and the lower set 0.25–0.4 mm apart. Six-roll machines are suitable for those breweries requiring 6–12 brews per day. The top set of rollers is normally 0.75–1.5 mm apart, the middle set 0.7–0.9 mm apart, and the lower set 0.3–0.6 mm apart. An advantage of the dry milling system is that samples of the milled malt can be easily taken and checked for composition. This is not possible in wet milling.

Brewhouse Technology

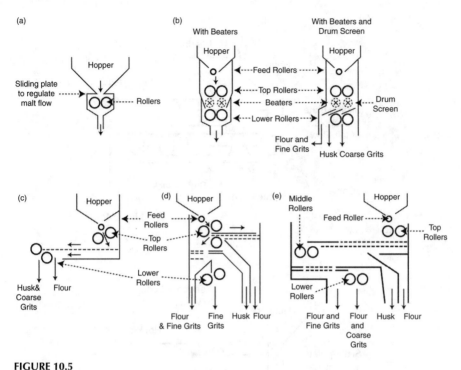

FIGURE 10.5
Types of roll mills. (a) two-roll, (b) four-roll, (c) four-roll with flat screen, (d) five-roll with flat screen, (e) six-roll with flat screens.

Roll Milling with Conditioning

In order to reduce husk damage during dry milling, the malt can be conditioned before milling by increasing the moisture content of the husk. This can be achieved by treating the malt with hot water or low-pressure steam. This process is only worthwhile on large-scale processes using six-roll mills.

Low-Pressure Steam Conditioning

The steam system consists of a screw conveyor fed by a hopper situated above the mill (Figure 10.6a). The screw is fitted with a steam jacket on the lower half and has a series of steam injection points on the top half. As the malt passes along the conveyor, steam at 0.5–1.0 bar (112–121°C) is injected into the malt. Residence time in the conveyor is 40–60 sec, which gives time for the malt moisture to be increased by 0.5–1.0%, the bulk of this being in the husk, which has its moisture increased by about 23%.[3]

The malt is milled straight after conditioning; a feedback loop is fitted so that if the mill becomes blocked, the steam supply will be shut off. The function of the heating jacket is to keep the conveyor dry. Due to the potential for

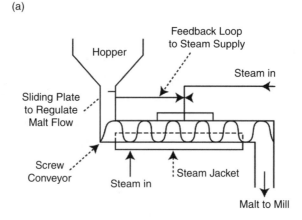

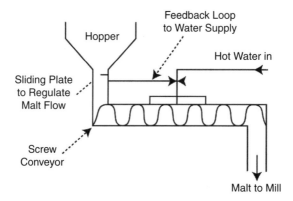

FIGURE 10.6
Conditioning systems for roll milling. (a) Low pressure steam conditioning, (b) hot water conditioning.

enzyme damage from the use of steam, this system has been superseded by hot water systems.

Hot Water Conditioning

In this process, water at 30–40°C is sprayed onto the malt as it passes along the conveyor. The rest of the system is similar to the steam system (Figure 10.6b). The lower temperatures avoid enzyme damage.[3]

Wet Milling

Wet milling is a further development of dry roll milling. In this process, the malt is wetted before milling, in order to minimize husk damage by making the husk more pliable, and so it can pass between the rollers without being broken. The rollers are closer together than in dry milling (0.35–0.45 mm apart). Wet milling systems normally have two rollers. Before milling, the malt is steeped in water in a hopper above the mill to raise the moisture to around 30%, — the water can range in temperature from cold to that used for mashing (Figure 10.7a). This water is normally recirculated for 15–30 min. Following steeping, the water can be removed or passed through the mill with the malt.[3]

The first stage of milling is to turn on the mashing water that flows from the mill to the mash mixer, where it covers the base of the vessel and the agitator. The mill is then started, and the malt is drawn between the rollers by a feed roller. The crushed malt is mixed with the mashing water and pumped to the mash mixer. Once the hopper is empty, it is rinsed out with water to clean the equipment and to wash all the remaining grist into the mash mixer.

Wet milling is advantageous because the combination of large husk and small endosperm particles leads to rapid run-off, and high extracts are obtained. However, it has several disadvantages. It is difficult to obtain good mixing and uniform wetting in the steeping hopper. Removing the steeping liquor can also remove some enzymes; mashing times of 35–45 min can give too long a proteolytic stand and thus will affect wort gravity and possibly foam stability. The mash will be thick during milling and due to the need to use large amounts of chase liquor later, it is difficult to alter the mill settings, and oxygen pick-up can be high.[3] Due to these problems, this type of wet milling technology is no longer manufactured, but is still in use.

Wet Milling with Steep Conditioning

This is a development of the wet milling process and is a combined milling and mashing system. In this system, the feed hopper contains dry malt; mashing liquor at 60–70°C is added to the system below the hopper and above the rollers (Figure 10.7b). Once the mash mixer has a suitable layer of water in it, the hopper is opened and milling begins. The malt is mixed with the hot water in the conditioning chamber for about 1 min, and the husk moisture content is increased to around 20%. The malt is then drawn between the rollers. More liquor is added after milling, and chase liquor is used to clean the system at the end of the process.[3]

This system has the advantage of a feedback loop from the rollers, which regulates the speed of the feed roller; thus, compensations can be made for undermodified malt, etc. This system shortens the process to about 20 min, but still has certain disadvantages related to flexibility. It is not easy to obtain the 20% husk wetting desired; the problems with thick mash milling and oxygen pick-up still exist, and due to the faster speed of the process, more powerful motors and pumps are required.

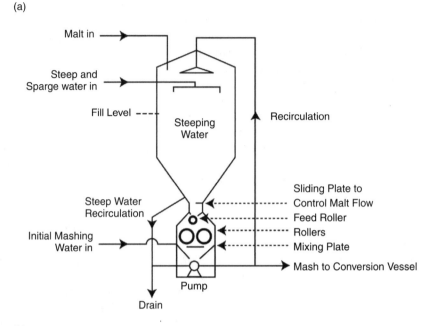

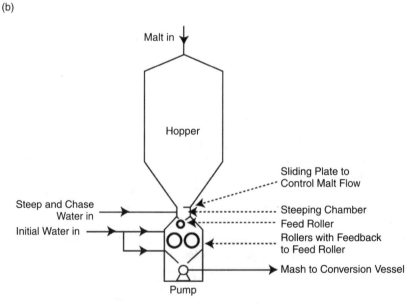

FIGURE 10.7
Wet milling systems. (a) Wet milling, (b) wet milling with steep conditioning.

Hammer Milling

This process is suitable for mash filter separation as it produces the flour required. Hammer milling is carried out on dry malt. Malt is placed in a feed hopper which is opened when milling begins, allowing the malt to fall into a chamber containing a rotor spinning at approximately 1500 rpm fitted with free swinging hammers. The malt grains are smashed against a perforated screen that surrounds the rotor — the perforations are between 2.0 and 4.0 mm diameter. The milled particles pass through the screen and fall into the collection hopper below (Figure 10.8). The particles are three times smaller than roll milled malt.[2]

Hammer milling combined with mash filtration produces more extract than the lauter tun method. However, there are disadvantages; this type of mill is very noisy and requires to be enclosed to protect brewery personnel. High-powered motors are required to drive the rotors and these must be fitted with emergency braking systems. The motors are set up to operate in both directions to even the wear on the hammers. Both hammers and screens wear out quickly and should be inspected frequently. Hammer milling is known to increase levels of β-glucan in wort due to the finer grind produced; this can lead to separation and possibly to filtration problems at a later stages of the brewing process.[3]

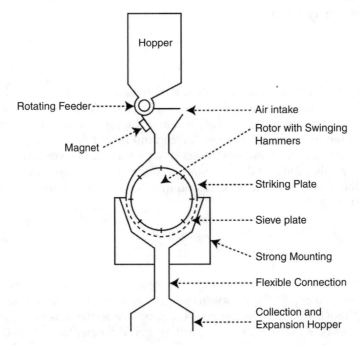

FIGURE 10.8
Hammer milling.

Mash Conversion and Separation

Purpose of Mashing

Mashing is the process in which malt grist, solid adjuncts, and water are mixed together at a suitable temperature for the malt enzymes to convert the various cereal components into fermentable sugars and other nutrients. The liquid containing the nutrients is referred to as wort or extract.

Different vessels are used for the various processes. Infusion mashing requires one vessel, whereas systems using lauter tuns require two — one for mixing and one for separation. Systems using mash filters require a vessel for mixing before the mash is transferred to the filter. Decoction systems require an extra mash kettle to heat part of the mash or a mash mixer vessel fitted with heaters. If solid adjuncts are used, an extra cereal, cooker will be needed.

Basic Principles of Mash Separation

The basic principles of filtration as shown in Equation (10.1) were established by Henry Darcy in 1856.

$$Q = \frac{KA(h_1 - h_2)}{L} \qquad (10.1)$$

where Q is the total volume of liquid percolating in unit time, A is the constant cross-sectional area, L is the height of the filter medium, K is the constant for the properties of the fluid and the porous filter medium, and $h_1 - h_2$ is the pressure drop across the column height L.

These principles were modified by Huite and Westermann for process applications, taking into account the factors of viscosity, permeability, and particle size distribution in a mash. This is shown in Equation (10.2).[4]

$$Q = \frac{KA\Delta P}{\mu L} \qquad (10.2)$$

where Q is the wort volume flowing, A is the mash bed cross-sectional area, L is the mash bed depth, K is the mash bed permeability (average volume) given as $Y^3 d_e^2/(180(1-Y)^2)$, ΔP is the pressure drop across the mash bed, μ is the wort viscosity, d_e is the effective particle size diameter, and Y is the bed porosity (wort volume/mash volume). The interaction of these variables determines the flow from the filter.

Given the influence of the effective particle size diameter (d_e) in Equation (10.2), and the wide range of grist particle size distributions for use in the lauter tun and membrane mash filter, it is clear that there would be a major difference in the wort flow rate performance, unless the other factors were correspondingly adjusted to compensate. In addition, wort viscosity changes dramatically across the wort collection phase and the mash bed is compressible by the applied ΔP; therefore, the only factor that remains constant in the modified equation is A, the filtration area. It is important to note that Darcy's equation describes the conditions for optimum flow, not optimum wort quality. However, the brewer is also interested in extracting sugars and producing bright wort.

As previously discussed, the particle size distribution plays an important role in extraction efficiency, but diffusion contact time and filtration speed also become increasingly important, as the law of diminishing returns has to be considered in the sparging phases. This topic is addressed through the total water/grist ratio applied. The larger the amount of sparging water passed through the spent grains, the higher the extract yield, but the larger the amount of water which must be evaporated during boiling. Thus, a compromise must be found between lautering times, yields, boiling times, and energy costs. Prolonged sparging and reuse of last runnings may improve the yield, but increase the amount of undesirable materials (polyphenols and bitter substances from the husks, etc.) passing into solution. For roller milled grist, a water/grist ratio of 7.5 l/kg, and for hammer milled grist a target ratio of 5.3 l/kg, is applied in order to reach the optimum extract yield.[5]

Mash Tuns

Infusion mashing differs from the other mashing and separation systems as it takes place in one vessel, which is used for both conversion and separation.

A mash tun is the traditional piece of equipment used for mash conversion and separation in breweries in United Kingdom and other countries. It has the form of a round insulated enclosed vessel of approximately 10 m diameter and 2 m depth, although this will vary according to the size of the brewery. The mash tun is fitted with a false bottom with slots of 1 mm width sitting just above the floor of the vessel. The floor is fitted with a series of pipes that are used to run off the wort during separation. A system for recirculating wort is normally provided, and the vessel is fitted with a sparge arm to introduce sparge liquor toward the end of separation (Figure 10.9).

Infusion mashing begins when the milled malt is mixed with the mashing liquor. This takes place in a piece of equipment known as Steel's masher. This is a tubular structure fitted with a screw conveyor and mixing blades as depicted in Figure 10.9. Temperature control is most important here, and the temperature of the mash entering the tun should be 62–65°C. This is influenced by the temperature of the grist, water, and the vessel, and there is no easy way of adjusting this temperature at a later stage as the vessel is not fitted with any form of temperature control. Mashing-in is a rapid process, taking around 20 min with a liquor/grist ratio of 1.7–2.5 l/kg.[6]

To reduce heat loss in the mash tun, the bottom part of the tun to just above the false floor is filled with mashing liquor through the sparging system just prior to mashing-in. This also prevents the slots from being clogged as the mash enters the vessel from Steel's masher. Once the total volume of the mash has been transferred into the vessel, it will be 1–1.5 m deep and will float on top of the layer of water. The mash sits for 20–60 min to allow enzymic action to take place before runoff can begin. At this point, there is the option of increasing the mash temperature by the process of underletting in which hot water is pumped under the false bottom. This can be used to speed up runoff by reducing mash viscosity.

Runoff is controlled by taps on the pipes described earlier. Wort can be runoff to the kettle or it can be recirculated to the top of the mash until the wort being run off from the bottom of the mash has reached sufficient clarity. Runoff is carried out slowly at first and is done over a control system fitted with a series of weirs, which is designed to reduce differential pressure and to avoid pulling down the mash bed. Filtration takes place within the grain bed, not at the false bottom, which acts only as a support.

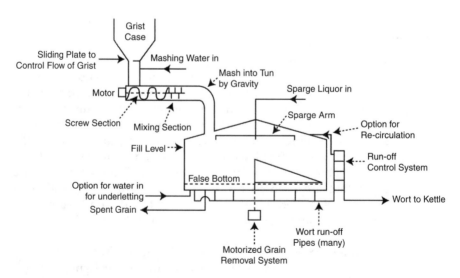

FIGURE 10.9
Layout of infusion mash tun with Steel's masher.

At the start of runoff, the gravity of the wort is high, but this decreases during the process due to sparging.

Sparge liquor is hot (around 75–78°C) and as it is sprayed on top of the mash, the weight of this liquor and gravity push the wort through the bed. Very little mixing takes place, so the gravity of the collected wort will remain high until the later stages of sparging. As the gravity falls, the runoff rate can be increased without damaging the bed. Gravity falls off slowly rather than abruptly, as might be expected, due to sugars being slowly leached from the grains by the sparge liquor. Runoff can continue until the gravity of the wort is too low to be of use, or enough wort has been collected. Once the wort has been collected, the bed is drained and the spent grains removed. In large operations, this is accomplished using a motorized removal system, but in smaller breweries removal is typically done manually.

The traditional mash tun is still widely used in medium-sized British breweries. It is also the method of choice in smaller breweries in many countries due to its simplicity. It is the most cost-effective system in terms of capital outlay and is the simplest to operate with little or no automation. The system, however, has some disadvantages. Mash tuns can only use a single temperature for mash conversion and, as a result, poor quality malts or malts requiring a protein or glucanase stand cannot be handled. Furthermore, it is unsuitable for mashing recipes that include more than small amounts of adjunct that must be cooked elsewhere. Control of the system is difficult due to the absence of temperature control and mixing equipment such as rakes and knives (as found in lauter tuns). Mash tuns are also less well suited to modern large batch production, where high brewhouse utilization and extract efficiency are expected.

Cycle times for infusion systems are around 4–6 h, which is slow by modern standards, only allowing a cycle of two to six brews per day. This is due to the slow runoff caused by the deepness of the bed and the small surface area; these factors affect Darcy's equation as explained earlier.

Extract recovery rates are comparatively low due to the coarseness of the grist, with rates of 96–97% recovery possible.[7] However, low extract recovery can be compensated for by the small amount of mashing liquor used; the consequent higher volumes of sparge liquor will recover more extract from the bed. Despite these drawbacks, such systems produce wort of excellent quality and will remain in use for many years.

Mashing-in Systems

Mashing-in for infusion mash tuns has been described previously. We now consider systems that are used in conjunction with lauter tuns and mash filters. Mashing-in is the process where the grist and mashing liquor are mixed prior to conversion. This is a crucial stage, as it is important to obtain an even mix without any lumps of dry material. Traditionally, the mash conversion vessel was filled with liquor and the grist dropped in during mixing with the agitator. However, this is a dusty process and results in high oxygen pick-up.

A variety of mashing-in systems is now available. If a wet or steep conditioned milling system is used, then the mash can be pumped to the conversion vessel, where it enters the vessel through the bottom to avoid turbulence and potential oxygen pick-up. If dry grist is being used, a premasher can be fitted on top of the vessel. This is similar to the Steel's masher described previously. Water at mashing temperature is injected into the grist stream as it passes into the conversion vessel (Figure 10.9). An alternative system is the inclined disk mashing vessel designed by BTE (Essen, Germany). Here, a motor rotates inclined disks on a shaft in the lowest third of a horizontal tank. The adjacent disks rapidly mix the grist and liquor with very low shear damage (Figure 10.10).

The Mash Conversion Vessel

This vessel is where mash conversion takes place. As this is a temperature-dependent process, such vessels are insulated and fitted with heaters. Mash conversion vessels are normally circular with a height/diameter ratio of 0.6 and should have no unnecessary obstructions inside (Figure 10.11).[3] Rectangular vessels were once in common use but the contents of these are difficult to mix properly.

To ensure good mixing of the vessel contents, a low shear agitator is fitted to provide gentle but intensive mixing. Normally, these have two blades

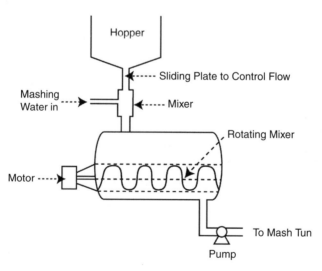

FIGURE 10.10
Layout of typical premashing vessel.

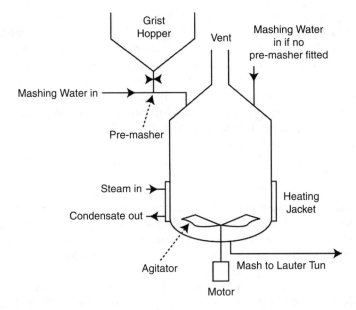

FIGURE 10.11
Layout of typical mash conversion vessel.

covering 85% of the vessel's diameter and are fitted at the bottom of the vessel.[6] Variable speed agitators are advantageous. The most common heating system is a heating jacket using steam at 3.0–4.0 bar.[8] Tubular heaters can also be used. Hot water heated vessels are also in use, but are no longer recommended for economic reasons.

The mash is added to the vessel as described earlier. If the malt has been milled using a roller mill, the liquor/grist ratio should be 2.7 l/kg and the vessel should be mixed at 3.8 m/sec. If a hammer mill was used, the values should be 2.3–2.5 l/kg and 3.0 m/sec, respectively. Mash pH should be in the range 5.2–5.4.[3] Once fully mixed, the temperature of the mash can be raised either by using the heating jacket or by adding cooked cereal from an adjunct cooker. An initial step of around 48°C can be provided to allow proteolysis, then the temperature can be increased to around 65°C for saccharification and finally to 75–78°C to reduce viscosity. The mash is then transferred to the separation system. A variable speed agitator is of use here as mixing can be fast during mashing-in, slow during proteolysis, fast during temperature increases or cereal additions, slow during saccharification (to reduce release of β-glucans), and during transfer to the separation process.[3]

Adjunct or Cereal Cookers

These vessels are similar to mash-conversion vessels and can be used interchangeably in a brewhouse if suitably arranged. In cereal cookers, solid adjuncts such as maize or rice are cooked at high temperature to solubilize the starch. Cereal, water, and some malt (5–10% of total weight) are added to the vessel at 35–50°C. The temperature is increased slowly at a rate of 1°C/min, and the malt enzymes solubilize the starch. The contents of the vessel are then boiled for 20–40 min to gelatinize the starch. The boil should not be too vigorous, to avoid foaming. Agitation must be constant and fast throughout the cooking process to prevent the contents from setting. Agitation can be slowed down during the transfer of the cooker contents to the mash mixer.

Mash Kettle

If a decoction system is in use, a separate vessel is required to boil part of the mash if a temperature controlled conversion vessel is not available. The mash kettle has a similar design to the mash-conversion vessel, but is smaller. Usually, the kettle holds about 66% of the volume of the conversion vessel.

In decoction processes, part of the mash is withdrawn and boiled. When it is transferred back to the main mash, it raises the temperature. A distinction is made between three, two, and single decoction mash processes depending on the number of boiled mashes. The type and amount of boiled mash is of major importance for the breakdown processes. The stirrer is switched off and the mash particles sink to the bottom of the vessel to form a thick mash, whereas the dissolved components form a thin mash in the upper part of the vessel. The thick mash is drawn off and boiled to break open the particles still contained within it. The boiled mash has to be pumped back under continuous stirring into the main mash to protect the enzymes in the unboiled mash. The boiling of the mash increases the formation of melanoidins, the extraction of husks, and the removal of dimethyl sulfide (DMS). Furthermore, gelatinization and saccharification of the starch are intensified, and higher brewhouse yields are possible compared to an infusion mash. Figure 10.12 shows the temperature profile for a three-decoction mash.

Mash Acidification

The pH value of the mash is essential for enzymic activity and thus for maximum extract recovery. Lowering the mash pH to 5.4–5.6 leads to higher attenuation limits, reduction of viscosity, rapid lautering, and less increase in wort color during boiling. However, the activity of phosphatases is enhanced at this lower pH, and the phosphatases increase the buffering capacity by releasing phosphate ions. Consequently, the pH drop during fermentation is lower and the effect of acidification is reduced. For these reasons, wort acidification to a pH of 5.1–5.2 is advisable. Combined mash and wort acidification leads to higher brewhouse yields, rapid lautering, softer beer taste, better foam stability, and less color formation in the wort

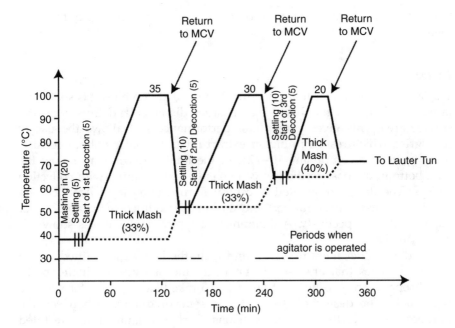

FIGURE 10.12
Temperature profile of a triple decoction mash.

boiling stage. A disadvantage of acidification is the lower bitterness yield due to slower isomerization of α-acids at low pH values.[9]

Except in countries which produce beer according to the Reinheitsgebot (the German purity law), acid can be added to the mash and wort. A variety of mineral acids can be used, but most commonly phosphoric acid is employed.

The addition of foreign substances is not allowed according to the Reinheitsgebot and, therefore, mineral acids cannot be used for acidification. Malt contains large populations of lactic acid bacteria on its surface, which can be used to produce "natural" lactic acid by acidification of unhopped wort. Lactic acid produced in this way can be used in accordance with the Reinheitsgebot for mash or wort acidification.

Mash Separation Systems

Once mash conversion is completed and the starch has been broken down to sugars, the aqueous extract solution has to be separated from the insoluble malt solids to produce clear sweet wort. The method and the equipment used are mainly a matter of choice on the part of the individual brewer, and sometimes of tradition. Wort separation may be carried out by a number of different methods — the mash tun (described earlier), the lauter tun, the Strainmaster — or by several types of mash filtration methods.

Lauter Tuns

The lauter tun is similar to a mash tun, but the bed depth used is shallower (around 0.5 m) and the vessel has a larger diameter and thus a greater surface area. This gives better filter performance and allows the use of finer grist, which results in higher extract rates. The false bottom consists of an open area comprising 10–22% of the total surface area.[7] Below the false bottom, the base of the tun can be either flat or fitted with collection valleys. The design originated in central Europe and North America and is suited to these areas due to the use of undermodified malts and higher adjunct rates, respectively. A diagram of a typical lauter tun is shown in Figure 10.13a.

In order to filter more finely ground malt, the lauter tun is fitted with a system of rakes that are used to break up the bed and facilitate solid–liquid separation. A concentration of 1–1.5 rake blades per meter squared is normal.[7] The design of the blades is a major difference between the European and North American versions. European lauter tuns are fitted with zig-zag-shaped blades with "shoes" at the lower ends. These shoes are used to lift the bottom of the bed. The North American design has straight blades with lifting properties and is intended for continuous operation.

Lauter tuns are normally loaded with mash at a liquor/grist ratio of 7.5 l/kg, which includes underlet liquor and sparge. High-gravity brewing requires this ratio to be reduced. Malt is normally milled using a six-roll mill with conditioning as described previously. Lauter tuns can filter recipes containing 100% malt to those containing 50% malt and 50% adjunct. Below 50% malt, there will be insufficient husk material to form an adequate filter bed.

Operation of a lauter tun begins by preheating with hot liquor at around 75°C; this is added to the tun until the plates are just covered. The mash from the mash mixer is then transferred to the tun either entering through the side wall tangentially or through the vessel bottom to avoid turbulence and oxidation. The rakes can be switched on at this stage to aid even distribution. Filling the tun takes around 10 min, bed loading can range from 153 to 339 kg/m^2 depending on the number of brews per day desired.[7]

Wort recirculation can commence as soon as the plates are covered, or can wait until the tun is full. Recirculation is operated until the desired wort clarity is achieved (less than 5 EBC) after which the wort is collected. The wort is transferred to the underback or kettle and the husk materials remain in the tun. Following initial collection, sparging will begin, operating continuously or intermittently as desired. At this point, the rakes will be operated to prevent channels forming in the bed, or the bed totally

(a)

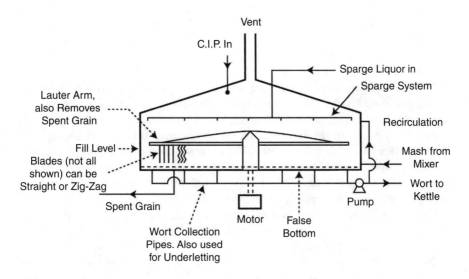

(b)

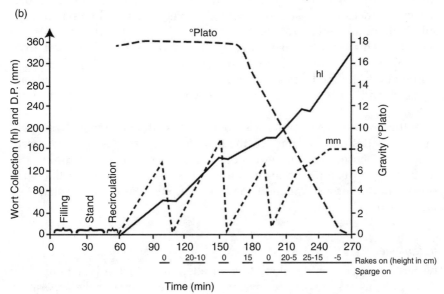

FIGURE 10.13
The lauter tun. (a) Cross section of a lauter tun, (b) profile of lauter tun operation.

collapsing. During runoff, differential pressure (DP) should be monitored to avoid this situation; failsafe systems are available to control this. Wort collection speed can be increased after sparging is in progress, and a complete breakup and remash of the bed can be carried out if desired. Once all the wort has been collected, the spent grains are removed by being pushed into a grain port by the raking arm. The spent grains are normally dried and sold as animal feed. The tun is then cleaned before the next mash. Most lauter tun operations are now fully automated, controlling mash loading, runoff rates, sparging cycles, and DP. A diagram of the runoff process from a lauter tun is shown in Figure 10.13b.

A lauter system can be operated to a cycle time of 2–4 h, allowing 6–12 brews per day. Wort collection normally occupies 65–67% of this time, the rest of the time being taken by filling (10–15 min), recirculation (5–10 min), draining (5 min), grain removal (8–18 min), and flushing (9 min).[5] The system is flexible as it can be operated at a range of bed depths and can accommodate various adjunct levels. It can be operated as a continuous process and can be considered the rate-limiting step in the brewhouse. Lauter tuns are currently the most commonly used system for wort separation.

There have been a number of significant developments in lauter tun design in recent years, no doubt aided by the competition from mash filters. Steinecker (Freising, Germany) launched its new Pegasus system in 2002.[10] The ring shape of this lauter tun resulted from the realization that the innermost zone of normal tuns makes little contribution to performance and can be dispensed with. The Pegasus tun thus has a large central pillar, and this space has been utilized to contain the mashing-in pipework. This allows mash entry to the vessel side rather than from below, allowing a gentle transfer with minimal oxygen pick-up. Steinecker claim that despite the loss of surface area, wort runoff is quicker, yields are increased, and spent grains are drier. Twelve brews a day are possible with this system.

Another recent development is concerned with the removal of spent grains. It is desirable to remove these as quickly as possible, and recent work has focused on large-diameter grain exit ports and trapezium plough bars to push the grains into the exit ports. It is important that such ports do not occupy so much of the false bottom area of the tun that they inhibit wort collection.[7]

Strainmaster

The Strainmaster was developed as an alternative to the lauter tun. The concept of this system was to increase the filtration area of the separation vessel by fitting a series of perforated straining tubes through which the wort is drawn leaving the grains inside the vessel.

The vessel consists of a rectangular hopper-bottomed tank with the separation pipes running longitudinally down its length (Figure 10.14). Mash from the conversion vessel is pumped into the top of the vessel. As soon as the top row of tubes is covered, recirculation is started using a pump. This creates a filter bed around the tubes in much the same manner that occurs in a lauter tun. Recirculation continues until the desired wort brightness is

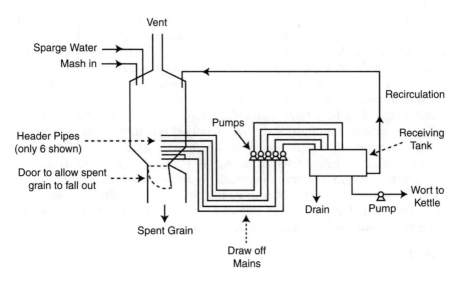

FIGURE 10.14
The Strainmaster.

achieved, the wort is then run to the kettle by gravity. As the level of the remaining first wort approaches the mash level, but while the mash is still covered with wort, the sparge is initiated over the top of the bed. Upon completion of the sparge, the grain bed is allowed to drain. The hopper doors are then opened, and the grains fall into the grain receiving tank.

Capacity of up to 15 brews per day has been claimed for the Strainmaster.[11] However, low extract yields, high water usage, and problems with cleaning meant that this system was never widely adopted.

Mash Filters

Mash filters provide an alternative separation system to the lauter tun, but are not yet as widely used. Mash filters are similar to plate and frame filters; they consist of a series of grid-type plates alternating with hollow frame plates suspended on side rails. Each grid plate of the filter is covered on both sides with a monofilament polypropylene cloth. The mash filter optimizes the filtration conditions defined in the Darcy equation and is therefore able to handle very fine grist, thus ensuring excellent extract recovery. Mash filter grist is produced using a hammer mill.

The closed chamber system is the modern successor to systems that were available as long ago as 1891.[7] The system has always used much finer grist than the lauter tun and also operates with a shorter cycle time. The mash filter consists of a series of vertical frame chambers laid out horizontally, with filter cloths interspaced between the frames. Mash is pumped into the chambers under pressure, and the wort leaves through the filter cloth to adjacent frames, followed by the sparge water. The chambers are closed with a hydraulic ram and opened for grain removal with a mechanical latching/tractor device. The plates used are illustrated in Figure 10.15a, and the separation process is shown in Figure 10.15b.

All of the chambers have to be filled for efficient wort separation, and specific chamber loading (kilograms/chamber) has to be used. This value is related to the solids replacement in the liquid solution compared specifically to malt, frequently referred to as the "malt equivalent." Only a small allowance of a solid tolerance of typically $\pm 10\%$ on the total loading is possible. However, this does not resolve the issues of inflexibility adequately, and poor wort clarity and high cleaning requirements are the reasons why the conventional mash filter never gained real favor.

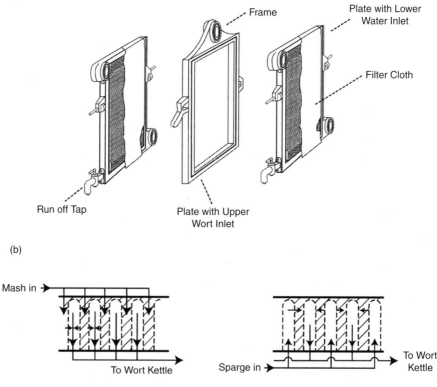

FIGURE 10.15
Conventional closed system mash filter. (a) Types of plates, (b) operation.

Membrane Mash Filters

The modern membrane mash filter is defined by the incorporation of an inflatable rubber membrane into the polypropylene chambers that can squeeze the mash against a cloth.[7]

To initiate the wort separation cycle, the mash filter is flushed, then preheated with hot water. The mash filter is charged through the top channel, completely filling the filter frames with mash at 75–78°C from the mash-conversion vessel at a controlled constant pressure of 0.5–1.0 bar. Underfilling of the filter will diminish extract recovery, because the sparge water will flow through the empty portion of the chamber. Overfilling, in contrast, results in high wort viscosities, adversely affecting filtration efficiency. When the filter is full, the wort collection system is opened, and the wort is drawn horizontally through the filter cloths (fine pore polypropylene filter sheets suitable for fine grist) without particles bleeding through the sheets. High-gravity first worts leave the filter initially at high flow, typically 900 hl/h, which rapidly reduces as the mash vessel is emptied and the filter fills with mash/grain. The fine filter sheets and grind result in a tight filter bed, which means that no recirculation is required before first worts are drawn off, which can run straight to the kettle. The large number of plates and shallow depth give a high filter flow rate, and the fine grind coupled with a thin filter bed result in high extract efficiency without reduction in wort quality. As the mash conversion vessel is emptied, the vessel and the transfer line to the filter are flushed. This takes about 30 min and defines the end of mash transfer and first wort running. Precompression is applied to the grains by inflating the filter membranes, which removes the residual amount of first worts and makes the mash bed more permeable.

Sparge water between 75 and 78°C is pumped into the filter at a steady flow rate of typically 300 hl/h as the membranes are continuously deflated with the start of second wort removal. When the total sparging volume has been applied, the membranes are inflated again to complete wort collection, followed by deflation of the membranes and grain removal. Typically, grains at a moisture level of 73% are discharged to a hopper. A typical wort collection profile is illustrated in Figure 10.16, demonstrating the various operating phases described. Total cycle time is around 2 h, allowing up to 12 brews per day.[5]

The membrane mash filter is produced by Meura (Tournai, Belgium) and is known as the Meura 2001.[7] This filter has a large surface area because of the number of filter plates. It uses a thin filter bed in the chamber (40–42 mm) and operates at up to 1.5 bar pressure, which provides a significant driving pressure to aid filtration. The principal advantages of the 2001 mash filter are the high brewhouse yield (equal to or even greater than typical laboratory yields), low sparge liquor consumption, low spent grains moisture content, and rapid throughput (12 brews/24 h). A variety of machines has been produced, ranging from a 7-kg pilot plant unit up

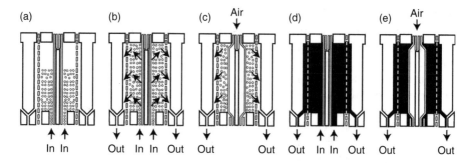

FIGURE 10.16
Operation of a meura 2001 mash filter system. (a) Filling, (b) filtration, (c) precompression, (d) sparging, (e) compression.

to a single-ended unit with 12,000 kg malt loading. Double-ended machines with increased capacity are avaiable. The "malt equivalent" chamber loading is currently 175 kg and will operate from 100% malt through to 100% adjunct. The low ratio of total water/grist of typically 5.2–5.3 l/kg is very attractive for high-gravity brewing without the necessity to recycle weak worts. Wort composition is generally similar to the lauter tun, except for lower concentrations of dextrins (because the finely milled grist facilitates enzymic activity during mashing) and fatty acids (resulting in a clearer wort and a higher ester content in the finished beer). The service life of the filter cloths is stated to be 1500 brews and that of the membranes to be over 2 years.

The membrane mash filter is an attractive alternative for producing high-gravity wort at high extract efficiency and a very short cycle time. The disadvantages are the limited recipe flexibility due to the requirement to fill all the chambers, the need to clean the cloths regularly, and the high levels of effluent produced.

The Nortek Mash Filter

Nordon (Nancy, France) have developed the Recessed Chamber Plate filter replacing the traditional elements of frame, plate, and membrane.[12] Figure 10.17 illustrates the differences between membrane mash filter sections and those used in Nortek filters. With this device, the bladder expansion can be better controlled during working conditions. Also, the mechanical resistance and the tightness have been improved compared to the traditional mash filter and the membrane mash filter. The material used for the production of these filtration elements is polypropylene instead of cast iron or stainless steel due to weight and sensitivity to corrosion. Polypropylene is resistant to CIP solutions, has low density, low conductivity, and higher shock resistance. The filter cloths in the Nortek mash filter are fitted inside the plates with an O-ring, ensuring better tightness, fewer torn filter cloths, reduction in maintenance, and high

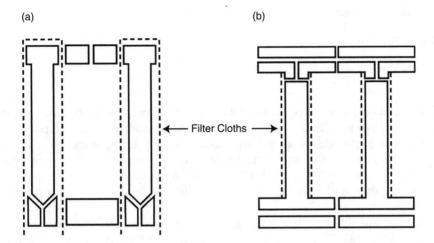

FIGURE 10.17
Segments of membrane and Nortek mash filters. (a) Membrane filter, (b) Nortek filter recessed chamber plate.

CIP efficiency as claimed by the supplier. This filter can be installed in any brewhouse without regard to the milling system used, because the cake thickness of the mash filter can be adapted according to the particle size distribution of the grist. Industrial yields of 100 ±0.5% and filtration times of below 120 min have been reported. Furthermore, fully automatic filtration, including the removal of the spent grains and low maintenance costs due to the absence of membranes have been claimed as advantages.

Comparison of Separation Systems

A comparison of five lauter tuns, three mash filters, and two Strainmasters with a range of 2.5–11.5 t malt load showed that the process costs for the lauter tun and the Strainmaster declined with increasing capacity, whereas the costs for the mash filter increased in proportion to its capacity.[13] Maintenance costs are considerably lower for the Strainmaster than for the mash filter and the lauter tun. The personnel costs are dependent on the grade of process automation and capacity.

The Strainmaster produced the most wastewater, and cleaning was most expensive for the mash filter but this was strongly dependent on the operating philosophies. The highest costs are the extract losses during mash separation where the Strainmaster showed the poorest performance, but this can be compensated for by recycling the last sparging water and the liquid contained in the spent grains. The total costs of mash separation with the Strainmaster and the mash filter are similar if the extract recovery in the Strainmaster is optimized. The lauter tuns are typically 10–15% more expensive due to their lower capacity.[13]

Wort Boiling

Overview

Introduction

Wort for boiling is collected after the separation process. If the brewhouse process takes less than 5–6 h, wort has to pass from the mash separation stage to a buffer vessel, called an underback or prerun, before transfer to the kettle because the wort kettle is still in use; if not, the wort can pass directly to the kettle. These buffer vessels have to be insulated, and the wort entrance and exit must be fitted with a baffle to minimize oxygen uptake.

Kettles are fitted with a heating system that heats the wort from mash temperature (65–78°C) to boiling temperature, which is just above 100°C (at sea level) due to dissolved solids. Boil length can range from 30 to 120 min. Liquid adjuncts and hops can be added at various points during boiling. Following boiling, the solid material precipitated is removed and the clear wort is cooled ready for fermentation. This process stabilizes the wort, removes unpleasant flavors, and extracts hop components that give beer its distinctive flavor.

Principles of Boiling

Any wort boiling process includes the following stages: in-fill, preheating, rise to boil, boil, and transfer out.[14] Some of these events may overlap in some systems. All systems should incorporate certain design features to prevent unnecessary fouling or damage to wort. Wort should be filled into the bottom of the kettle to reduce splashing and oxidation. Preheating should begin as soon as possible during filling and be applied slowly to reduce the risk of fouling. Some of the hop input should be added during preheating as the oils present help to reduce surface tension and foaming. Liquid adjuncts should be added later in the boil to avoid fouling due to poor mixing, but to ensure sterilization takes place.

As boiling is the most energy-consuming part of the brewhouse process, cost and energy recovery must be considered. The energy requirements for wort boiling can range from 24 to 54 mJ/hl, depending on the equipment size.[15] The energy requirement for producing finished beer ranges from 145 to 285 mJ/hl. If 81–128 mJ/hl is required in the brewhouse, then wort boiling consumes around 18% of the total energy requirements of a brewery. However, to gain maximum value for the user, the choice should really be based on a balance between optimum wort production, quality, energy usage efficiency, and total capacity costs.

It is important to be able to measure process efficiency in wort boiling systems to ensure economical use. The measurements normally made are boil time, percentage evaporation, and energy consumption.[16] Evaporation is usually measured as a percentage of the total volume at the start of boil. It is difficult to measure the volume of boiled wort accurately due to the turbulent nature of hot wort and the amounts of any additions made. The amount of evaporation desired must be converted to the quantity of steam required and the rate at which it is required during the boil.

This can be calculated by the method known as steam mass flow control outlined in the subsequent figures.[14] The initial volume of the wort at 75–78°C (temperature after mashing) must be determined along with contributions from any adjuncts. The figures include the steam mass flow and the totalizer formulae. It is important to have a totalizer value for the steam in order to ensure that each boil receives the same amount of steam regardless of fouling or steam supply. This can be used to increase the length of the boil until the target amount of steam has been applied. This will increase over time as the equipment becomes fouled and, eventually, the length of time will become constant as the steam valves cannot be opened any further. This event indicates that a clean is required.

$$\text{Steam flow required for control} = \frac{\text{All} - \text{in volume} \times 1.017 \times \text{total evaporation} \times 65}{\text{Total time of boil}}$$

$$\text{Totalizer accumulated steam target} = \text{Steam flow} \times \frac{\text{total boil time}}{60}$$

where the steam flow is expressed in kilograms per hour (kg/h), steam totalizer in kilograms (kg), all-in volume in hectoliters (hl), total evaporation in %, total type of boil in minutes (min), 1.071 is the factor for temperature expansion 75/100°C, and 65 is the factor for adjustment of units, based on 3.0 bar steam enthalpy combined with an assumed typical radiant heat loss allowance.

Evaporation is important but so is vigor, and this parameter is also difficult to measure. Normally, the recycle ratio is used to measure boil vigor. This is the total content of the kettle moved across the heaters between six and ten times an hour.

Heating surfaces will foul progressively during use, and this will lengthen the time required to achieve boiling. This can be compensated for by having a much larger surface area than is actually required. The heat transfer coefficient (U) will decrease based on $Q = UA\Delta T$, so the pressure and temperature of the steam must be increased.

Most of the energy used to heat wort to boiling point can be recovered by the use of heat exchangers. Heat recovery efficiencies up to 99% have been reported.[17] Energy loss during evaporation is more difficult to avoid as it can be lost up the brewhouse chimney if condensers are not fitted. The main way to reduce energy consumption is to reduce evaporation itself.

Types of Boiling

Three forms of heat transfer can take place inside the heating tubes: film boiling, forced convection, and nucleate boiling[1] (Figure 10.18). Film boiling is not desirable, as a vapor film forms on the heating surface and prevents heat transfer to the wort, and causing rapid fouling. This can occur at fairly low temperature differences, as stainless steel (unlike copper) is not a wettable surface. The use of steam at a pressure of 4 bar or above should be avoided.

Forced convection will occur at low temperature differences (between the wort and the heater) or when bubble formation is suppressed by the application of backpressure. To achieve the high rate of heat exchange required during wort boiling, the flow through the tubes must be turbulent and this requires multipass systems with pumping. This could cause damage to the wort.

Nucleate boiling makes use of bubble formation to cause the required turbulence. The presence of bubbles also helps with trub formation and removal of undesirable volatiles.

Most modern wort boiling systems use dry saturated steam as the heating medium. Contact with wort is achieved via vertical tube elements that can be situated inside or outside the kettle and be arranged as pumped or thermosyphon systems. Boiling takes place at the heater surface; nucleate boiling (phase change) can be either encouraged or surpressed depending on the pressure maintained. Both these types of boiling are catered for in modern systems and it is a matter of debate which is the more effective.[14]

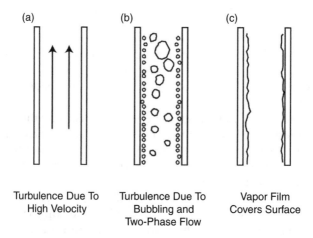

FIGURE 10.18
Types of boiling. (a) Forced convection, (b) nucleate boiling, (c) film boiling.

Objectives and Events

Wort boiling has a number of objectives, which are summarized below:

- Evaporation, concentration, and removal of volatiles
- Destruction of enzymes
- Sterilization, killing of spoilage organisms
- Extraction and conversion of hop material (flavors and preservatives)
- Coagulation of proteins (hot break)
- Promotion of reactions between proteins and hop constituents
- Completion of salt reactions
- Caramelization of sugars, especially in dark wort
- Color and flavor formation
- Cooking of nonfermentable extracts

Of these, evaporation and concentration are the most important and all wort boiling installations are designed with these in mind. Concentration is of less importance nowadays; formerly, this was used to create high gravity worts for strong beers, but there are now more cost-effective ways to achieve this such as high-gravity mashing. Evaporation rates of 12–20% were average 30 years ago compared to modern levels of 4–8%.[17] Evaporation is still of great importance, as this is the mechanism used to drive off undesirable volatiles, mainly DMS.

A vigorous boil must be created for evaporation to be effective. The precursor of DMS, S-methylmethionine, originates from lager malts, and DMS is formed during boiling by thermal decomposition of this precursor. DMS formed in this way is rapidly lost by evaporation. However, formation can continue after boiling and thus DMS can survive into finished beer. Further reduction of DMS can be achieved at the whirlpool stage (described later).

Enzyme activity after the normal mashing period can alter wort fermentability. Normally, a temperature rise is incorporated into a mash program to inhibit further enzymic action. Enzyme denaturation has normally taken place by the time the wort reaches boiling point.[17] Any other nonmalt enzymes added to mashing, such as β-glucanase, are also destroyed.

Wort can be contaminated with a wide variety of microorganisms originating from the raw materials. Survival of these would cause spoilage in the final product and thus must be eliminated. Wort boiling destroys practically all microbial contaminants within about 15 min. Only the spores of a few spore-forming bacteria are able to survive, and fortunately these are unable to germinate and grow in beer, although care should be taken with low-alcohol beers.

Hops are added to the wort at various points during boiling to provide flavor, aroma, and antimicrobial attributes to beer. Addition is made to the boil, because hops require to be heated in order to convert their α-acids to iso-α-acids in a process known as isomerization. This is a rapid process and 90% of the final wort bitterness will be produced during the first 30 min of boiling and the maximum level will be reached by 60–70 min. It is important to achieve good isomerization as some iso-α-acids will be lost later in the brewing process. Hop oils will be largely lost during boiling, which is of benefit as these will cause a bitter, vegetable flavor in the beer if present in too high levels.[18]

Wort contains high levels of nitrogen in the form of proteins and polypeptides. If allowed to persist into the later stages of brewing, some of these proteins will cause problems with pH, fining, filtration, and colloidal stability. Loss of proteinaceous material takes place throughout the brewing process with a significant amount being removed during boiling. Approximately 6% of wort nitrogen is lost during a 1-h boil.

The mechanism of removal involves the aggregation of proteins by intermolecular bonding. Providing these are not damaged later in the process, these large molecules can be removed from the wort during whirlpool separation. The likelihood of protein molecules colliding and bonding together is increased by a vigorous boil. It has been shown that given enough turbulence, the removal of high molecular weight protein is a function of time and vigor and is largely independent of evaporation.[17] High pressure and temperature promote protein coagulation, but should not be allowed to continue for too long as this can lead to the formation of excess color.

Wort proteins can also be removed by promoting interactions with polyphenols. Wort contains a high level of polyphenol material: 40% of this originates from hops and the rest from the malt. Polyphenols are readily oxidized during boiling, and polymerize and bind to proteins. Proteins that bind to oxidized polyphenols become insoluble and are removed with the hot break. Proteins that bind to unoxidized polyphenols remain soluble but will precipitate when cooled to form cold break.

Removal of protein can be increased by the addition of copper or kettle finings to the boiling wort. These are made from carrageenan (Irish moss) and work by causing the protein flocs to aggregate to form larger molecules. During subsequent clarification, these molecules can be removed more easily to give brighter wort.

Several other events take place during boiling. The most noticeable of these is the formation of color. This increases due to three reasons: Maillard reactions between carbonyl and amino groups, caramelization of sugars, and oxidation of tannins. The last reason is the most important, and high pH and the presence of air will aid this reaction. This has the added benefit of decreasing the reducing power of the wort, which will improve the stability of the final beer.

Wort pH falls during boiling by 0.2–0.3 to reach a final pH of between 5.2 and 5.3. This is mainly due to loss of Ca^{2+} that binds to phosphates and polypeptides to form insoluble compounds that release H^+ ions. The increased acidity is important, as it will improve protein coagulation, encourage diacetyl reduction, encourage yeast growth, inhibit microbial contaminants, and reduce formation of excess color.[17] Too low a pH, however, will reduce hop extraction and is thus undesirable. To aid pH reduction, calcium sulfate, calcium chloride, phosphoric acid, or sulfuric acid can be added to the kettle. An alternative is to add lactic acid or wort fermented with lactic acid bacteria if purity laws must be adhered to.

Reducing compounds are formed during boiling. These are mostly reductones and melanoides formed via the Maillard reaction. These compounds, combined with others originating from the raw materials can protect the final beer against oxidation, provided they are not oxidized themselves. Fortunately, the level of dissolved oxygen in the wort will be reduced, particularly if the system is open, mostly in preheating.

Heater surfaces will progressively foul with salts and protein, especially, with high-gravity worts. This can be compensated for by increasing the steam input or boil length.

Wort Preheating

The length of time taken to heat wort to its boiling point can be reduced if the wort is preheated during transfer from the separation stage to the kettle. Here, the method of choice is to use a plate and frame heat exchanger. This is arranged to heat the wort from around 72 to 100°C and uses a counterflow of steam or hot water. Due to the amount of solids present in the wort, the selection of a channel gap of 4–16 mm and a τ value of between 50 and 120 is advisable. These parameters provide heating that is vigorous enough to avoid excessive fouling, but gentle enough to avoid shear stress damage to wort particles.[19]

Types of Wort Boiling System

Direct Fired Kettles

Traditionally, wort was boiled in large open vessels manufactured from copper. The term "copper" is still in use today as an alternative to "kettle" although few genuine copper vessels are still in use. Heating is by direct firing by either coal or gas, restricting the area of heat exchange to the base of the vessel, thus limiting the size of such vessels to not more than 330 hl[20] (Figure 10.19a). Another disadvantage is that the heating area becomes very hot causing rapid fouling with burnt wort. Cleaning can be required every two to five brews. To obtain sufficient vigor, boiling times

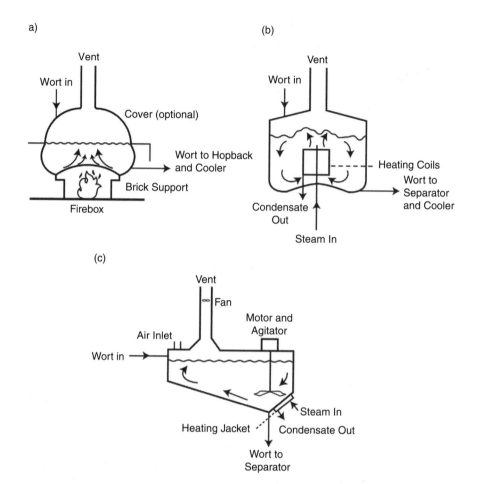

FIGURE 10.19
Older methods of wort boiling. (a) Direct fired copper, (b) kettle with internal steam coils, (c) kettle with external heating jacket.

are usually long, up to 90 min, thus evaporation rates are high especially from open vessels, normally 10%/h.[20]

More modern versions of this system are still in use. These take the form of cylindrical vessels with dished bottoms usually with a higher wort depth to diameter ratio of around 1 : 1.[1]

Kettles with Internal Steam Coils

The invention of steam coils allowed the construction of larger kettles with larger heating areas and more efficient heat exchange. In the most common designs, the heater is situated in the center of the vessel to give good mixing (Figure 10.19b). Some designs feature wort preheating.

Although offering advantages over direct fired coppers, steam coil heated vessels are difficult to clean, tend to suffer from corrosion, and give poor mixing resulting in wort caramelization and excess color formation.[20]

Kettles with External Heating Jackets

This design was intended to avoid problems with internal heaters by providing an external heating jacket. In order to promote turbulence, the jacket was placed on one side of the vessel only, resulting in a rather small heating area (Figure 10.19c). To compensate for this, a mechanical mixing system is provided to ensure good heat exchange. Such kettles require less cleaning (every 6–12 brews), but tend to have problems with excessive foaming. Extractor fans are often installed in an attempt to control this.[20]

Other Systems

Other methods have been tried over the years. The use of high pressure hot water (HPHW) systems with hot water as the heating medium was popular. However, the water at 171°C and 15 bar pressure was too hot and caused wort damage. In order to reduce energy costs, continuous boiling systems were installed in some brewhouses, but proved inflexible.

Internal Heater with Thermosyphon

This system of wort boiling operates by using the natural circulation that occurs when a thermosyphon is created to produce a well-mixed boil at a pressure of 3.0–3.5 bar. A thermosyphon is established by the difference between the product of the wort head and density on the inlet to the boiling tubes (98°C) and the product of the two-phase head and density within and on the outlet side of the heater (105°C). As the two-phase density is much lower, a significant differential pressure exists, which is sufficient to generate flow rates equivalent to six to ten times the kettle volume per hour. This is aided by concentrating devices and spreaders.[16]

The heater is in the form of a bundle of tubes contained in a drum. Wort flows through the tubes that are surrounded by steam, and nucleate boiling takes place. The tubes are normally 1–2 m long and between 25.4 and 63.5 mm in diameter. The heater is positioned in the center of the kettle, often in a well sunk into the bottom of the vessel so it can be quickly covered by wort during filling, because the heater should not be started until it is submerged. For flexibility, the heater is placed lower than the vessel's half-full capacity. To speed up the process, the wort can be preheated before entering the vessel. During boiling, the total wort volume will pass through the tubes approximately six to ten times an hour.[16] Evaporation is around 5 – 6%/h and 16–32 brews can be achieved between cleans.

A diagram of the equipment is shown in Figure 10.20a in which the central position of the heater tubes can be seen. A sheath is placed around the drum to increase wort turbulence and mixing. Above this is a conical feature that concentrates the wort leaving the heater onto the venturi spreaders, which helps draw in more wort. The spreaders control foaming at the surface and help remove volatiles. The wort is directed to fall back into the main body of liquid in a way that the radius of the circular wort flow is two thirds of the radius of the vessel, so that wort should not strike the kettle wall.

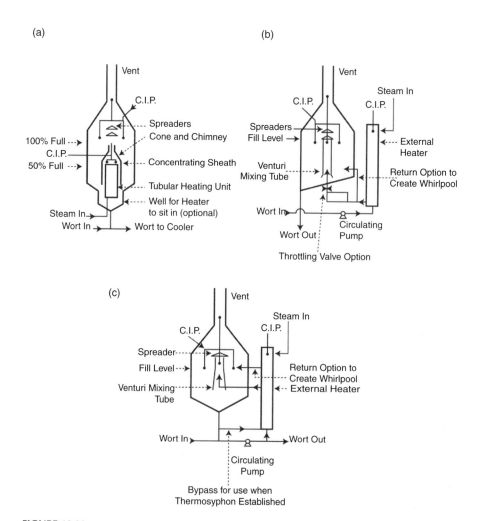

FIGURE 10.20
Modern methods of wort boiling. (a) Internal heater with thermosyphon, (b) external heater with forced circulation, (c) external heater with thermosyphon.

The advantage of this system is that it makes use of a thermosyphon to mix the wort and no pumping is required. Therefore, this system is energy efficient and less shear damage occurs. However, there are several disadvantages. Wort boiling at less than full capacity can be difficult and fouling can occur in the tubes of the heating unit. Care must be taken when using liquid adjuncts, because if they are not totally dissolved, they can caramelize and cause blocked tubes.

Internal heating systems with pumped circulation exist, but are almost identical to external heater systems (see following text); the only difference being that the heater unit is inside the vessel.[1]

There have been two recent developments in internal heaters made by Steinecker. The Ecotherm system uses a pump for circulation (the same one as used to empty the vessel after boiling and to add hops), and the speed of this can be altered to vary the circulation rate during the boil. Steam pressure can also be controlled during the boil. A double spreader is provided above the boiler to increase mixing. The manufacturers claim less energy usage, shorter boil times, and more even wort temperature changes.

The more recent (2003) Stromboli system has the novel feature of separate wort circulation loops allowing the removal of trub as it is formed and thus reducing fouling within the kettle. Like the Ecotherm, this system allows control of pressure and the rate of circulation.

External Heaters

Briggs of Burton (Burton-on-Trent, England) introduced these types of heater arrangements, often referred to as Kalandria systems, during the late 1960s. Initially, pumped circulation was used with short wide tubes to cater for the use of whole hops, which was common at the time. Since then the system has developed into forms promoting either forced convection or nucleate boiling.[1]

The external wort boiler with forced circulation is a development of the above system, in which the heater is placed outside the vessel. Wort is pumped out of the kettle through the boiling tubes, where it may pass several times up and down the unit before being returned to the vessel. Wort is returned to the kettle through a venturi mixing tube with spreaders fitted above. The system is usually constructed as a combined kettle and whirlpool.[16]

In contrast to the 6–10 circulations found in a thermosyphon system, in this system the wort will pass through the heater only three to four times an hour. This is due to the high unit cost of providing a pump of sufficient capacity. This can be compensated for by making the tubes narrower and longer (up to 4 m) and arranging the system so that the wort passes through several tubes. This can cause problems with vapor breakout, but this can be solved with the installation of throttling valves that ensure that the wort remains in a liquid form and forced convection occurs. Figure 10.20b illustrates how the return wort line can be diverted to convert the system into a whirlpool at the end of the boil.

The advantage of this system is that it can be started up much earlier than an internally heated system, especially if the wort is preheated. Disadvantages include the undersized pumps, wort damage due to pumping and passing through valves, fouling in the narrow tubes, and the high cost of operating the pump.

An alternative to the aforementioned system is to use a plate and frame heat exchanger in place of the tubes (forced circulation with a plate heat exchanger). Here, the wort is heated from 100 to 105°C with a cross-current of steam. The gaps between the plates should be wide (16 mm) to avoid blockages, and the τ value should be between 50 and 120.[19] Wort is circulated 7–12 times per hour. The system operates at a slight overpressure; so when the boiled wort is returned to the kettle, any volatile is flashed off and the expansion of the liquid creates a vigorous boil. This type of system is flexible, as extra plates can be easily added to the heater to increase capacity. The same heater unit can be used for wort preheating.[19]

The external wort boiler with natural thermosyphon circulation system is similar to the forced circulation system, but uses a thermosyphon for circulation without the need for continuous pumping. Wort is removed from the kettle and passed once through an external boiler. The boiler is positioned near the kettle and below the level of the wort. Tubes are of a similar diameter to those used in internal heaters, but much longer at 2–6 m. Such long tubes are beneficial for nucleate boiling and generating the vapor lift required to form the thermosyphon. A pump is used to start the circulation but is not required once the syphon is established. Again wort preheating is an advantage, as the boil can start as soon as wort is available.[16]

Two configurations of this system are available (Figure 10.20c). One has a venturi tube and a spreader, and the wort is returned to the kettle through a mixing tube. The other has a spreader and the volatiles are removed in the same way as in the internal version. In this case, circulation of the total wort volume occurs six to eight times an hour. Alternatively, the system can be set up as a kettle and whirlpool where the wort is returned to the kettle tangentially just above the surface level, and where the volatiles are removed using the top of the wort as a removal area. The return point must be above the wort level or the resulting vigorous mixing that would occur could damage the vessel. In this system, eight to ten circulations per hour of the total wort volume can occur due to longer tubes being generally used.

The use of a combined boiling and separation system is of interest as it avoids having an additional vessel for solid–liquid separation. Having the heater outside is an advantage as it means that the vessel has no internal parts, which aids cleaning. The energy for the spin of the whirlpool comes from the thermosyphon, so no other input is required and this reduces running costs and shear damage to wort particles.[21]

Advantages of this system include low consumption of energy and flexibility as thermosyphons can be built to any scale. Disadvantages include cost, space occupied in the brewhouse, and length of time required to clean the system.

Dynamic Low-Pressure Boiling

This system is provided by Huppmann (Kitzengen, Germany) and is used in conjunction with an energy storage system.[22] Once the wort has been boiled for a short time to remove any air from the vessel, the vents are closed and pressure is allowed to build up to 150 mbar (103°C), which takes about 4 min. When this point is reached, the steam is cut off and the energy storage system is operated, circulating cooling water until the pressure falls to 50 mbar (101°C). The heater remains on during this period. The small bubbles produced at this point due to the drop in pressure cause stripping of volatiles, known as "flash evaporation." Unlike other boiling systems, stripping occurs in the whole volume of the kettle, not just at the liquid surface. This process is repeated about six times during the boil. The dynamic low-pressure boiling system reduces boiling time as the higher temperature (103°C) causes reactions and volatile stripping to take place faster. Following the last pressure release, the pressure is restored to atmospheric levels and a period of "after boiling" takes place. Boiling takes 45–50 min with an evaporation rate of 3.5–5%.[22]

This system makes use of an internal heater with a low operating pressure (1.2–1.6 bar). The design of the heater outlet cone causes an increase in the level of circulation, so the total kettle volume is circulated 20–30 times per hour. However, this circulation is done at low velocity to avoid shear damage to wort. Mixing and volatile stripping are promoted by a two-level wort spreader that distributes wort horizontally and vertically.

Boiling is described by the Henry–Dalton law which says that the pressure in the gas phase is equal to the pressure of the gases held in the liquid phase. To increase evaporation, the area of gas/liquid interphase must be extended. Such conditions occur at the end of a boil when the pressure is released, and small steam bubbles are created that provide the large surface area. However, in the dynamic system, there are many such releases of pressure. This means that with the other reactions speeded up, the evaporation rate can be reduced to around 4–5%. Compared to normal systems with evaporation rates of 6–8%, this system is claimed to be able to give energy savings of 33–50%.[23]

Merlin System with Wort Stripping

This method is a departure from the aforementioned methods and makes use of a completely different vessel type and arrangement (Figure 10.21). It was introduced by Steinecker in 1998. For heat exchange, the system

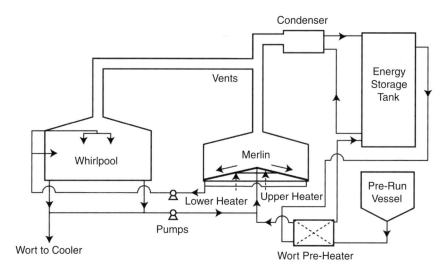

FIGURE 10.21
Merlin wort boiling system with wort stripping.

uses a conical heating surface that uses falling-film and thin film evaporation during preheating, boiling, and an extra stage of wort stripping in place of tubes. The system is designed to be used with an energy storage system and the manufacturers claim an energy saving of 72% over conventional systems.[24]

Wort from the mash separation process is run into a prerun vessel and is preheated using a plate and frame heater. This increases the temperature from 72 to 90°C. Energy is provided by hot water drawn from an energy storage tank. The cooled water from this process (reduced from 96 to around 76°C) is then returned to the tank. The wort is then passed to the Merlin vessel. This is a large circular vessel fitted with a conical heating surface that has two steam heating areas underneath it. The angle of the cone is around 130° distributing 100 hl of wort per 7.5 m^2 area of the conical heating surface.

The wort is preheated to boiling point by being passed over the cone with both heaters on at a temperature of 125–130°C. The wort runs off the cone into a collecting ring and then passes into the second vessel, which is set up as a whirlpool. The wort enters this vessel via an overhead spreader and a tangential wort inlet at the vessel wall, avoiding the formation of a central region of cold wort and to create the whirlpool effect that will aid hot trub settling. Preheating takes around 40 min during which time the wort will pass over the cone about four times. Hops are added during preheating as the boiling stage is too short to allow full isomerization of hop acids.

After preheating, the wort is recirculated into the Merlin vessel for the boiling stage, which makes use of the lower steam heater only at a temperature of around 120°C. This stage takes 35–60 min and the evaporation rate is between 1.5 and 2.5%. During the boil, the wort will pass over the cone four to six times. Speed of circulation, temperature, and time can all be adjusted for optimized wort boiling. After boiling, the liquid falls into the whirlpool vessel and is left to stand for 10–15 min to allow trub sedimentation. The Merlin heaters are switched off at this point. The flow rate of this filling process is 1.0–1.5 m/sec, which is much lower than in a normal whirlpool (usually 4 m/sec). A higher speed is not required as trub formation begins in the collecting ring surrounding the Merlin vessel and little effort is required to make the hot trub flocs settle in the whirlpool. After this, the clear wort is pumped to the Merlin vessel for wort stripping; again, only the lower heater is used at about 125°C. Removal of undesirable volatiles takes place here and a further 1% evaporation occurs. A major advantage of the Merlin system over conventional systems is that any DMS formed during whirlpool separation is removed. Wort is then transferred to the wort cooler. The energy contained in the vapor is recovered using a condenser, and the condensate is used to heat up the water in the energy storage tank to the 96°C required for wort preheating.

This system has the advantage of reducing energy usage. An example provided by Steinecker claimed that the energy required for Merlin wort boiling was 5.537 kW/hl compared to 8.820 kW/hl for a conventional system. With an energy storage system fitted to the Merlin and vapor compression fitted to the other system, the values were 3.093 and 3.424 kW/hl, respectively.[15] Damage to wort is also reduced by the use of the thin film boiling and less fouling takes place. Less protein is coagulated and this can be beneficial to foam stability.

However, the system has several disadvantages. The size of the heating area is difficult to increase in order to provide larger vessels, although more than one cone can be installed in one vessel, with the cones stacked. The system causes increased wort oxidation and this affects color formation. Higher levels of DMS precursor have been reported in Merlin-brewed beers.[15] The evaporation rate is fixed and this makes the system inflexible.

Conventional Boiling with Stripping

Merlin stripping can be combined with conventional boiling systems to obtain the benefits of more DMS removal and reduced boiling time. In this arrangement, the wort is boiled in a conventional kettle, but with the boil time reduced to around 40 min, and is then transferred to the Merlin vessel for stripping. Evaporation is lowered by around 5% and energy usage is reduced by around 25%.[24]

Conventional boiling with Meura stripping has a similar arrangement and aim as conventional boiling with Merlin stripping. Meura stripping has been installed by Interbrew at their Leuven brewery and the technology has been patented by Meura.[25] Wort boiling takes place in a kettle for the same length of time as in a conventional system, but with a significantly reduced evaporation rate. The boiling process is divided into three phases.

This first phase is a vigorous boil for 5 min to solubilize the hops and to promote trub formation, followed by a long phase where the steam is cut off and the vessel is sealed, thus preventing cooling and evaporation. The third phase includes another 5 min vigorous boil in which some volatiles are removed and hot break formation is encouraged. Total evaporation during this stage is below 2%.

From the kettle, the wort is passed to a clarification vessel to remove the hot trub. At the Leuven brewery, this is a settling tank with a centrifuge, but a whirlpool separator could be used instead. After clarification the wort is pumped to a stripping column containing an open packing structure providing a very high surface area (50–500 m^2/m^3). Before entering the column, the wort is heated to boiling point. The wort runs down the column forming a thin film.

Directed against this downward flow is an upward flow of injected live steam vapor (0.5–2.0% of the wort volume), which flows up the column stripping the volatiles from the descending wort. It is most important that the wort is at boiling point, or the energy from the steam will be wasted in heating up the wort rather than stripping it of volatiles. The wort leaving the column has the same or a lower amount of volatiles than found in conventionally boiled wort. The wort is then pumped to the wort cooler.

The process was designed to reduce energy usage and to gain more control of factors such as wort volatile content and color. Interbrew has claimed a 46% reduction in energy consumption, from 83,200 to 45,000 kJ/hl for the whole brewhouse operation. This saving is achieved mainly by the reduction of evaporation. The injection of steam to the stripping column is an extra cost factor, but some of this energy can be recovered.

Energy Recovery Systems in the Brewhouse

All modern plants should be fitted with systems that can recover energy, steam, and condensate for further use. This saves energy and reduces emissions. Modern boiling systems operate as sealed units and vapor produced is no longer released to the atmosphere, but is either removed to an energy storage system or used in the heater unit.

Vapor condensate can be used for a variety of purposes including cleaning or can be used to heat fresh water for brewing. This process will cool the condensate to around 35°C at which temperature it can be discharged. Another option is to filter the water to make it clean enough to be used as boiler feed, in refrigeration or for brewing.[26]

A diagram of a typical brewhouse energy recovery system is shown in Figure 10.22. Here, condensate from the condenser on the kettle, at around 100°C, is either used to top up the energy storage tank, to heat up a mash

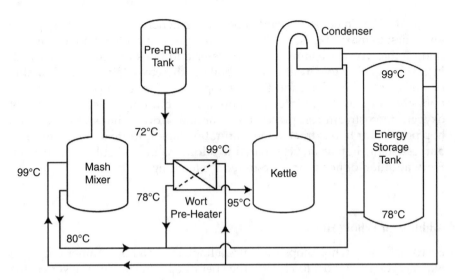

FIGURE 10.22
A brewhouse energy recovery system.

in the mash vessel, or to preheat wort heading for the kettle. The cooled condensate, at 78–80°C, can be collected in the storage tank or used to condense the vapors from the kettle.

Hop Addition

Addition of Hop Products

With regard to hop addition, points to consider are in how many parts the additions should be made, when the partial additions should be made, and the order in which they should be added. These parameters are dependent on beer type and brewing practice. Hops can be added all at once or can be divided over up to four additions. For the production of beer with low bitterness (20–24 BU) and a more subtle hop aroma, hops are usually added all at once at kettle fill up, or up to 20 min after the start of boiling. The latter practice promotes the reaction of malt polyphenols with high molecular weight nitrogen compounds during the boiling period where no hop polyphenols are present, thus minimizing the loss of bittering substances.

If the hops are added in two parts, 70–80% of the total hops should be added after 10–20 min of boiling, and the remaining 20–30% should be added 10–30 min before the end of boiling. In general, the bittering hops are given first to maximize isomerization of their α-acid content and to drive off undesirable volatile compounds. The later additions are aroma hops with increasing quality toward the end of boiling, thereby retaining the desirable hop oils for the beer. Late additions of hops to the kettle or even to the whirlpool give a distinct and intensive hop aroma, but only strict avoidance of oxygen downstream can preserve this aroma. Adding hops in cone form, hop powder, or hop extract to the maturation tank leads to a more intensive, but less stable hop aroma compared to additions in the kettle or whirlpool. Early addition of hops can help to reduce foaming.

Addition of Whole Hops

Breweries using whole hops must use a hopback or a hop strainer (hopjack) when casting the wort to remove the spent hops. In smaller breweries, the wort may be boiled in two vessels, one with the strong worts from the start of the runoff and the other with the weaker worts. Traditionally, the hops are added to the second boil, as extraction is more efficient at lower gravities.

Addition of Hop Pellets and Powders

Processed hop products that simplify handling provide a standardized form of bittering material and reduce the storage space required. In smaller breweries, the hop products are added manually to the copper. The larger the brewery, the greater is the pressure for automated addition. In a 10-t brewhouse producing eight brews per day, 250–300 kg hops may be added.

Addition of Hop Extract

The high viscosity of hop extract must be reduced by heating it to 45–50°C. In some processes, the can of extract is sprayed with a hot liquid (water or last runnings) and continuously turned to homogenize the contents before addition to wort at a predetermined time. In the case of large containers, the contents must be heated, so that it can be emptied into a mixing vessel and added to the wort after mixing. Care has to be taken that the temperature of the hop extract never exceeds 50°C and oxidation has to be avoided. In smaller breweries, cans with hop extract are often pierced, placed in a retaining cage, and directly immersed in the wort kettle. This process has the advantage of easy handling and simple storage of materials that have long shelf lives. A disadvantage is the potential fouling of heating coils.

Wort Clarification

Introduction

Hot wort should be clarified as soon as possible before cooling, removing the hot break (trub) and hop solids that serve no purpose and will cause problems downstream. The amount of trub material present in wort, which consists of complexes of protein, polyphenol, and carbohydrate, ranges from 2 to 8 g/l depending on wort type and gravity. The amount of hop material involved will depend on the type of hop product used. The amount of trub should be reduced to less than 0.1 g/l before cooling, some of the remaining trub material will later come out of suspension as the cold break.[27]

Precipitation of trub is never complete, but the amount passing into the cold wort must be minimized. If wort is allowed to cool before clarification, trub will coagulate and settle out in the vessel. These coagulated trub particles will subsequently be destroyed when the wort is transferred elsewhere, thus resulting in cloudy wort. Cloudy wort can cause cooler blockages, slow fermentations, poor yeast performance, increased risk of contamination, and haze in finished beer.

Separation of trub can be achieved by sedimentation, centrifugation, or filtration. Sedimentation and centrifugation depend on the difference between the densities of wort and trub. Trub has a density of 1.2–2.25 g/ml and 1040 specific gravity (10° Plato) wort has a density of 1.04 g/ml, so trub will settle out.[28] Filtration depends on the size of the particles and these will settle according to Stokes' law. This process can be speeded up by centrifugation, but care must be taken to avoid shear forces that can break flocs into particles too small to be removed.

Stokes' law for gravity:

$$V_g = \frac{d^2(\rho_P - \rho_L)}{18\mu} g$$

Stokes' law for centrifugation:

$$V_c = \frac{d^2(\rho_P - \rho_L)}{18\mu} r\omega^2$$

where: V_g is the terminal velocity of particle under gravitational force (m/sec), V_c is the terminal velocity of particle under centrifugal force (m/sec), d is the particle diameter (m), ρ_P is the particle density (kg/m^3), ρ_L is the liquid density (kg/m^3), μ is the liquid viscosity (N sec/m^2), g is the acceleration due to gravity (9.81 m/sec^2), r is the radius of rotation (m),

ω is the angular velocity of rotation ($=2\pi N/60$ rad/sec), N is the revolutions per minute, $r\omega^2$ is the centrifugal constant (A) (m/sec^2), and Z is the $r\omega^2/g$ (g-value). This specifies the increased settling rate of centrifugation compared to gravity alone.

Separation Systems

A range of systems is available and choice depends on the type of hops used. With whole hops, wort is clarified using hop backs or hop strainers; if pellets are in use, then settling tanks, filters, centrifugation, or whirlpool separators are the methods of choice.

Hop Back

This is a vessel with a false bottom that resembles a mash tun but with a smaller floor area compared to a mash tun due to the smaller quantity of solid material being filtered (Figure 10.23a). The hot wort is run into the vessel with the hops and trub in suspension. The hops settle quickly to form a filter bed on the false bottom. The trub particles also settle out, but according to Stokes' law, these smaller particles settle more slowly and the trub particles will form a layer on top of the hop bed. Thus, the trub will be retained in the spaces between the hop particles, while the clear wort flows through.

As the filter bed is in a loose suspension, it must not be damaged by running off the wort too quickly and creating a high differential pressure. The bed should be between 300 and 600 mm deep and not less than 150 mm or it will be too thin to provide enough filtration.[27] Once the bed is stabilized, the wort can be recirculated until it is bright and can be cast to the cooler. At the end of filtration, the hop bed can be sparged with hot liquor (8 l/kg spent hops) to recover the wort held in the bed. The spent hops can either be left in the hopback to help filter the next brew or can be removed.

This system is flexible when dealing with worts with different amounts of whole hop material, but it is not suitable for worts brewed with hop pellets or extracts as these products do not contain sufficient solid material to form a filter bed.

Hop Strainer

This is a screw conveyor/press arrangement that is surrounded by a fine mesh (Figure 10.23b). Hot wort with hops and trub in suspension is run into the separator chamber. The lower part of this chamber contains an inclined rotary Archimedean screw that draws the wort out of the chamber. This screw is surrounded by a mesh that retains the solid materials and allows the liquid to pass through. At the top of the screw, the hops are

Brewhouse Technology

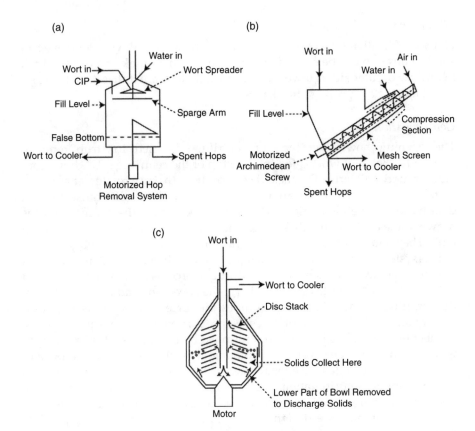

FIGURE 10.23
Older hop separation systems. (a) Hopback, (b) hopstrainer, (c) wort centrifuge.

sparged with hot water and compressed with air to remove wort before being removed.

This process does not give a fully clarified wort, as it removes little trub material and it must be used in conjunction with a settling tank, filter, or whirlpool separator. The system is inflexible, and with declining usage of whole hops it is now rarely used. Compressing hops has a negative impact on beer foam stability because it extracts lipid components from the trub.

Settling Tanks

When boiled wort is allowed to stand in a tank for 20–60 min, solids settle out by gravity according to Stokes' law and the clear wort may be drawn off from the top of the vessel. A simpler version of this is the coolship, which is a shallow open vessel where the wort is left to stand for up to 2 h. Its use is restricted to very small breweries.

Filtration

Hot wort filtration gives excellent quality clear wort with very low levels of solids. Filtration can be carried out by sheet, cloth, or tank and can be aided by the use of filter aids. However, such installations are difficult to clean and maintain and are prone to blockages.

Centrifugation

The opening bowl disk type of centrifuge is used for trub removal (Figure 10.23c). Hot wort is fed into the top of the centrifuge. The bowl contains around 200 vertically stacked conical disks that are rotated at approximately 6000 rpm. The disks are fitted with slots or holes that cause an even distribution of the incoming wort. As the disks rotate, the clarified wort is drawn to the center of the disk stack where it passes up and out to a holding tank. The solids collect at the vessel walls and are ejected as sludge.[27]

This system separates solids from liquids due to their different densities, but the process is accelerated by using centrifugal force, as shown in the modified Stokes' law equation (described earlier). The speed and radius of rotation of the centrifuge creates 6000–12,000 times the force of gravity. As shown in the equation, the amount of force is proportional to the square of the number of revolutions per second multiplied by the radius. Thus, a machine with a 60 mm diameter disk stack rotating at 6000 rpm will exert a force of 11,946 g. This illustrates the effectiveness of such systems to remove trub.[27]

Even though centrifugation is effective, such systems are expensive to operate and maintain. Wort particles occur in a range of masses, so it is difficult to optimize the design so that as little shear damage as possible occurs, although centrifuges are ideal for removing wort from trub cones as long as shear forces are minimized. The advent of whirlpool separators has rendered centrifuges largely redundant.

Whirlpool Separators

Whirlpools can be considered as modified centrifuges (Figure 10.24a). The separation of the solids from liquid is achieved by sedimentation due to gravitational and centrifugal forces and the "teacup" effect. Due to the tangential entry of the wort into the cylindrical vessel, a parabolic flow is established, which causes an increasing pressure gradient across the radius from the middle to the walls of the vessel. Friction between the fluid and the walls and base of the vessel causes the formation of a fluid layer where the centrifugal force on a particle is diminished and the force on the particle directed toward the vessel center exceeds, causing the particle to move there.

In order to aid this process, vessels are designed with large diameters and reduced height to speed up settling. Thus, the height to diameter ratio (H/D) is important. A range of 0.5–1.0 is normal, with a trend toward lower values for new installations. The volume stream of the entering wort should be in

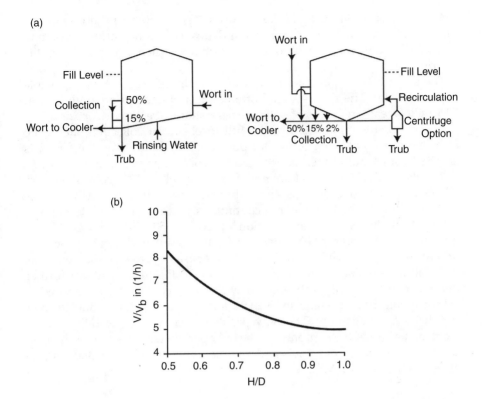

FIGURE 10.24
Whirlpool separators and their operation. (a) Whirlpool types, (b) relationship of wort entry velocity to vessel height/diameter.

relation to the H/D ratio[29] (Figure 10.24b). The wort entry velocity should not be too high because hot trub flocs can be damaged negatively influencing their sedimentation properties.

Wort should be transferred to the whirlpool as soon as boiling is over, at a flow rate of less than 2.4 m/sec. It should enter the whirlpool at a speed of 3.5–5 m/sec, at an angle of 20–30° to the vessel tangent at a height of 1 m from the vessel floor, which gives the maximum efficiency of energy conversion. A vessel with a H/D ratio of 0.6 would fill in about 10 min. The spin time in the vessel is 20–30 min before transfer to the wort cooler begins. To avoid the cone of trub collapsing, the stand time should not exceed the rotation time.[28] Furthermore, the stand time should be minimized to avoid DMS formation from precursors. Whirlpools cannot be used with whole hops unless a hopback or strainer is used initially. High gravity worts do not give such rapid clarification as do worts of lower extract because the difference between the density of the particles and the wort is smaller (Stokes' law). If no hop material is present, separation is poorer and the best results have been claimed with hop powder.[30] Anaerobic conditions are also said to increase the amount of material deposited, particularly by injecting carbon dioxide or nitrogen into the wort line when pumping into the whirlpool.

There are two general types of whirlpool separators available (Figure 10.24a). The first is a flat-bottomed vessel that has an angle of 1 in 50 to allow easy removal of solids. There should be three wort draw-off points at 50, 10–15, and 0% of the full level. Initial collection should be from the top point, then from that positioned at 10–15% as the vessel is emptied. Flow should then be reduced to avoid damage to the trub cone.

An alternative configuration involves a vessel with a conical bottom where the trub collects. This can either be operated as a batch process with recycling to the kettle or as a continuous process with an external centrifuge. With this system, the collection levels are at 50, 10–15, and 2–2.5% of the full level. If a centrifuge is fitted, a collection point at 0% is also provided. Recent development with whirlpools has focused on the production of as dry trub as possible, which can be returned to the mash without further treatment, and thus no centrifuge for wort recovery from trub is needed.[21]

Whirlpools have become the separation system of choice due to their simplicity and reliability. The only problem is a lack of flexibility due to the fixed nature of the equipment. The only parameters that can be changed easily are the filling, spin, and standing times. These can be adjusted to minimize production of DMS.

Whirlpool Kettle

Vessels are available that can act both as kettle and separator. This has the advantage of the economy of having only one vessel, but it must be remembered that the whirlpool acts as a buffer tank increasing flexibility and is relatively inexpensive to install as a separate vessel.

Wort Recovery from Trub

Trub contains about 10–20% solids and it is desirable to recover as much as possible of the wort from this material.[28] This can be achieved by either adding the trub to the lauter tun, using a further settling tank or by centrifugation. Solid trub can be combined with spent grains to produce animal feed.

Wort Cooling and Aeration

Introduction

After whirlpool separation, wort is at a temperature of approximately 95°C and must be cooled to between 8 and 22°C depending on the intended fermentation temperature. The wort must also be aerated to aid yeast growth. Cooling should be carried out as quickly as possible after clarification to reduce formation of DMS and to prevent excess color formation. Care must be taken with the handling of cold wort as it, unlike hot wort, is highly susceptible to microbial contamination. Equipment used to handle cold wort must be kept sterile.

Plate and Frame Heat Exchangers

The method most frequently employed for wort cooling is the plate and frame heat exchanger. The heat exchanger plates are made of stainless steel and are thin (0.5 mm) to allow optimal heat exchange. The surfaces of the plates are embossed to cause turbulent flow and speed up heat transfer. It is difficult to find data for heat transfer, but Alfa Laval (Brussels, Belgium) claim a film coefficient of 3000–6000 W/m°C for clean plates. Some fouling will occur during cooling. Care must be taken with turbulence, as damage to these trub molecules can lead to cloudy wort and, eventually, to hazy finished beer. The τ value should be kept between 40 and 80. At low τ values, more frequent cleaning will be required.[19]

Heat exchangers can be set up to give single-stage or two-stage cooling, with the flows of wort and cooling water arranged to flow in opposite directions (Figure 10.25).

In single stage coolers, wort is cooled with cooled water in counterflow. The total hot liquor requirements of the brewhouse (6.0–7.3 hl water of 82°C for 100 kg of malt) can be provided by the water at 79°C from the wort cooler.

Two-stage coolers have an initial water cooling stage, reducing the wort to around 16°C, followed by an extra circuit using a refrigerant that further cools the wort to 7°C. Two-stage coolers are used when the temperature of the cooling liquor is high or when lower wort temperatures are required.

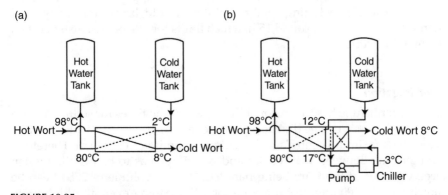

FIGURE 10.25
Wort cooler systems. (a) Single stage, (b) two stage.

It is crucial to optimize the precooling stage because this determines the heat recovery efficiency of the process and the cooling requirements in the subsequent cooling stage. Insufficient sizing leads to higher electricity costs due to a higher requirement of refrigerant in the second cooling stage. Iced water, glycol–water mixtures, and sole or direct evaporation can be used in the second cooling stage. If the cooling water is used elsewhere in the brewing process, this compensates for the evaporation energy losses during boiling.[19]

Removal of Cold Break

As wort is cooled, the trub material that did not precipitate as hot break will come out of solution as cold break and can be removed using several methods.

Cold Sedimentation Tank

In this vessel, the cold trub is removed by sedimentation before the yeast is pitched. Wort is pumped into this closed vessel, and the cold break particles are allowed to settle for 14 h. The maximum fill level should not be higher than 1.2 m so that the cold break particles can settle out quickly. With CIP fitted, the cycle time is about 16.5 h.[9]

Anstellbottich

This is a variation of the sedimentation tank commonly used in Germany. It is an open, rectangular vessel without cooling facilities that can hold one brew. Yeast is added when the wort is pumped into the vessel, which means that the contents have to be transferred a fermentation vessel after 12 h, before the wort starts fermenting. The fill level should not exceed 1.2 m. One cycle including filling, settling of cold break, emptying, and manual cleaning takes around 15 h, which limits the use of the Anstellbottich to small breweries.[9]

Centrifugation

Use of a centrifuge has the major advantage that the times for sedimentation and transfer steps can be drastically shortened. As the viscosity of cold wort is low, the maximum centrifuge throughput is limited to 160 hl/h. Therefore, a buffer tank between wort cooler and centrifuge has to be installed when large volumes of wort are being handled. A wort volume of 278 hl can be centrifuged in 100 min, leaving time for CIP cleaning even at a high brew frequency.[9]

Filtration

Cold wort filtration is carried out with Sieve or Spalt filters, using kieselguhr as a filter aid. Such filters have a throughput of around 10 hl/m^2/h, which means that a filtration area of 27.8 m^2 is needed to filter 278 hl cold wort in 60 min. Including the precoat that must be renewed every four brews, the kieselguhr requirement is around 60 g/hl, which has to be disposed together with the cold trub.[9]

Flotation

Flotation vessels are closed, vertical vessels with an even or slightly angled bottom that are amenable to CIP. Cold break particles adhere to the surface of air bubbles that are introduced from the bottom of the tank. The finely dispersed bubbles float to the wort surface within 2–3 h, where they form a white foam head. Besides cold wort filtration, flotation is the most common system for cold break removal in breweries. In contrast to sedimentation, the wort fill level should be as high as possible in order to provide enough time for the trub particles to adhere to the surface of the air bubbles. Flotation can be carried out with or without the addition of yeast. Flotation vessels must have an extra 40% capacity over the wort volume in order to provide room for the foam head. The ratio between wort fill level and vessel diameter should be around 1.3–1.4.[9]

Aeration

Yeast needs oxygen to multiply and thus wort must be aerated. Aeration can take place either on the hot or cold side of the wort cooler. Aeration on the hot side has the benefit of increased sterility and mixing, but will cause oxidation of polyphenols, which will darken the wort. Thus, air or oxygen is usually added to cooled wort. Some breweries inject sterile air or oxygen between two stages of a plate heat exchanger with the temperature at 10–15°C.

The amount of oxygen required in wort depends on the yeast to be used in the fermentation. The target is 70–90% saturation, which will give an oxygen content of 4–14 mg/l in the fermentation vessel. Wort at 15°C when saturated with air contains 8 mg/l oxygen, but when saturated with oxygen, this gives 38 mg/l; therefore, the choice of gas type is important. Use of air is preferred due to the risk of oversaturation and excessive yeast growth.[27] The use of sterile gases or filtration cartridges to remove contaminants is essential to ensure that the wort remains uncontaminated.

To achieve saturation, the gas being added must be in the form of small bubbles to increase gas diffusion surface area and it must be at a suitable pressure. A variety of sintered injector and mixer systems is available fitted with gas flow measuring devices.

Ceramic or Sintered Metal Candles

With candles, the air is injected through fine pores in the candle into the wort as it flows past. This is a simple and effective method, although cleaning and sterilization of the candles are difficult.

Venturi Pipes

The flow velocity in a pipe is increased if the pipe diameter decreases. In a venturi mixer, air is injected through a jet nozzle at this narrow part of the pipe and intensively mixed with the wort by the turbulent flow in the widened region of the pipe after the constriction (Figure 10.26). A similar device is the two-component jet, where the air is introduced through fine nozzles in the wall instead of through an air jet.

Static Mixers

In a static mixer, the mixing of wort and air is achieved in a reaction section with built-in angled bands that induce turbulent flow.

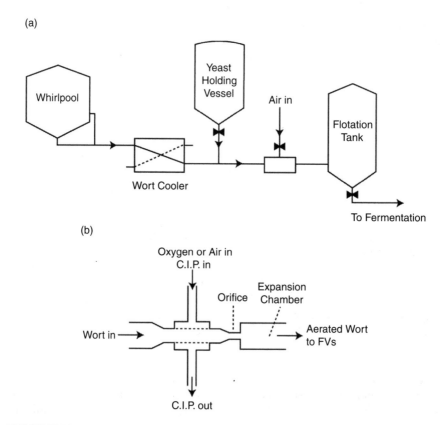

FIGURE 10.26
Wort aeration. (a) Wort aeration and yeast addition system, (b) turbo aerator.

Centrifugal Mixers

These work by forcing air into the wort by centrifugal force, leading to very fine air bubbles and extensive dissolution of the air. A disadvantage is the high energy requirement.

Yeast Addition

Yeast is usually dosed from a buffer tank with a pump or valve. The cell dosage rate is controlled by indirect measurement of cell count by turbidity measurement of the wort/yeast suspension at defined wavelengths. The accuracy of this widely used measurement is $\pm 20\%$ and enables automation of yeast pitching. An alternative method measures the yeast load involving the combination of solids content and flow rate to give a measurement parameter expressing the solids content of a volume in kilogram dry weight. This yeast addition is independent of the consistency of the yeast slurry. Aeration and yeast addition can be performed together in one system, as shown in Figure 10.26.

Wort is fed from the whirlpool, through a plate heat exchanger into a flotation tank. Yeast is added from a storage tank with a frequency-controlled pump. In a second section the wort is aerated as previously described. A fermenter can be used instead of the flotation tank shown here. Yeast addition is spread over the entire duration of wort addition; thus, yeast is distributed uniformly in the fermentation vessel. A weighing cell on the yeast storage tank assesses the total amount of yeast dosed and thus it is possible to add exactly the desired amounts of yeast and air required.

Brewhouse Efficiency

Brewhouse Yield

When measuring the efficiency of a brewhouse, the most important factor is the percentage extract recovery. This is a comparison between the potential extract from the raw materials and the amount of extract actually collected in the fermentation vessel. The calculation is based on laboratory analyses of the raw materials and the volume and gravity of the wort produced. The brewhouse yield depends on the raw material, the brewhouse equipment, the mashing process, the lautering process, and the overall operating methods.[31]

To calculate the brewhouse yield, the following parameters must be known:

- Weight of grist used in kg. This parameter is usually obtained from the automatic grist weighing device.

- Volume of the produced wort in liters.
- The extract content in kg/100 kg. This value is the reading from the hydrometer. For example, 11.4 means there are 11.4 kg extract in 100 kg wort and is the specific gravity of the cast wort. The specific gravity (kg/l) is obtained with the hydrometer reading and a conversion table.
- The contraction or correction factor. This factor compensates for the difference in the volume of wort at 100°C where the wort volume is usually assessed, and at 20°C where the extract content is measured. This factor is 0.96; meaning that from 100 l of casting wort at 100°C, 96 l of cold cast wort at 20°C can be obtained.

$$\text{Brewhouse yield} = \frac{\text{Volume of wort (l)} \times 0.96 \times \text{extract content (kg/100 kg)} \times \text{specific gravity (kg/l)}}{\text{Weight of grist charge (kg)}}$$

For example, if 4600 kg grist was used for a lager brew that produced 331 hl of 10.7°Plato wort, what would the brewhouse yield be?

$$\text{Brewhouse yield} = \frac{33{,}100 \text{ l} \times 0.96 \times 10.7 \text{ kg/100 kg} \times 1.04294 \text{ kg/l}}{4600 \text{ kg}}$$

$$= 77.09\%$$

For calculation of the brewhouse yield, the figures obtained in the laboratory and the brewery must be taken into account along with the extract remaining in the spent grains. The difference between the total extract yield achieved in the brewhouse and the total extract yield obtained in the laboratory should be less than 0.5%. However, for daily use, the difference between the laboratory yield and the brewhouse yield gives a preliminary indication without taking into account the extract losses in the spent grains. Typical yields are in the range of 74–79% and should be as high as possible, but not more than 1% under the air-dry laboratory yield for brewhouses with a lauter tun and not more than 0.7% under the laboratory yield for mash filter brewhouses. The loss of extract in spent grains is considered normal if the soluble extract is under 0.5% for lauter tuns and 0.7–1% for mash filters. The insoluble extract in the spent grains should be in the range of 0.2–0.5% for mash filters.[9]

Brewhouse Capacity

To operate a brewhouse economically, the brewhouse must be used to the greatest possible extent. Depending on the claims of the various manufacturers, a lauter tun brewhouse can produce eight to ten brews per day and the most modern lauter and mash filter brewhouses can produce 12–14 brews. These figures, however, do not take cleaning and other maintenance into account. In this way, lauter tun and mash filter brewhouses can give very similar outputs when performance is measured over longer periods.

The amount of cast wort produced in a year obviously does not equal the output of beer per year as losses occur during fermentation, maturation, filtration, and packaging. This depends on the brewery, but losses are normally in the range of 8–10%. The output capacity of a brewery is quoted in tonnes of grist. As a rule of thumb, 17 kg of malt are used to produce 1 hl of beer. It can be assumed that if eight brews are made per day, 5 days a week, each tonne of grist will produce 100,000 hl of beer in a year.[9]

For 1 tonne of grist, vessels of the following volumes (hl) are required[31]:

Mash converter	60–80
Mash cooker	40–50
Lauter tun	60–80
Wort kettle	80–90
Wort buffer vessel (underback)	30–40

Brewhouse Cleaning

With increasing vessel size, manual cleaning becomes increasingly uneconomic and finally impossible. This has led to the installation of fixed cleaning and sterilizing systems inside the vessels and other pieces of equipment themselves (CIP). In smaller plants, one CIP system is enough to fulfill the cleaning requirements, but due to the risk of passing on contamination from one vessel to the next, larger brewing plants need separate CIP systems for different production areas, thus the brewhouse and the wort path will be cleaned from a separate CIP station.

Automated CIP includes circulation of caustic and hot water in combination with an acidic flush. All liquids should be stored in vessels for reuse, thus the CIP system consists of storage vessels for hot base, hot acid, cold water, and hot water. Sizing of the pipes in the CIP system should be based on the recommended flow velocities for CIP media. On the pressure side of the pipe, the flow should be between 1.6 and 2.0 m/sec and on the return side 1.5–1.8 m/sec.[9]

A possible regime is:

- Hot caustic, brewhouse
- Hot caustic, wort path (subdivision because the brewhouse caustic becomes dirty more rapidly)
- Acid
- Fresh water
- Stored water

Regarding the cleaning and disinfecting agents used, it is very important to examine their composition carefully and their suitability for use with the various materials with which they come into contact. Most brewery vessels and pipes are made of different chrome nickel steels that are not resistant to cleaning and disinfection agents that contain chlorine. So when a hypochloride cleaning solution is used, mixing with acidic cleaning solutions must be avoided, or rapid corrosion will take place.

References

1. Hancock, J.C. and Andrews, J.M.H., Wort boiling, *Ferment*, 9(6):344–351, 1996.
2. Wilkinson, R., Materials handling and milling, *Brew. Guard.*, 132(1):26–31, 2003.
3. Wilkinson, R., Brewhouse plant optimisation. Part I, *Brew. Guard.*, 130(4):29–36, 2001.
4. Huite, N.J. and Westermann, D.H., Effect of malt particle size distribution on mashing and lautering performance, *Tech. Q. Master Brew. Assoc. Am.*, 12:31–40, 1975.
5. Wilkinson, R., Brewhouse plant optimisation. Part II, *Brew. Guard.*, 130(5):22–28, 2001.
6. Wilkinson, R., Mashing/conversion technology, *Brew. Guard.*, 132(2):24–29, 2003.
7. Wilkinson, R., Mash/wort filtration, *Brew. Guard.*, 132(3):20–27, 2003.
8. Wilkinson, N.R. and Andrews, J.M.H., Mashing, cooking and conversion, *Ferment*, 9(4):215–221, 1996.
9. Narziss, L., *Die Technologie der Wuerzebreitung*, 7th ed., Ferdinand Enke, Stuttgart, 1992.
10. Wasmuht, K., Stippler, K., and Weinzierl, M., Pegasus provides rapid mash separation without quality loss, *Brew. Distill. Int.*, 34(2):14–16, 2003.
11. Dillon, W.F., The Strainmaster — 15 years, *Tech. Q. Master Brew. Assoc. Am.*, 12:233–234, 1975.
12. Nguyen, M.T., The universal Nortek mash filter, *Ferment*, 9(6):329–335, 1996.
13. Narziss, L., Krueger, R., and Kraus, T., Technologischer und wirtschaftlicher vergleich von ablautersystemen, *Proc. 18th Congr. Eur. Brew. Conv.*, Copenhagen, 1981, pp. 137–152.
14. Wilkinson, R., Brewhouse plant optimization. Part III, *Brew. Guard.*, 130(6):24–27, 2001.
15. Stippler, K., The Merlin wort boiling system, *Ferment*, 13(1):34–36, 2000.
16. Wilkinson, R., Wort boiling technology, *Brew. Guard.*, 132(4):22–28, 2003.
17. O'Rourke, T., The function of wort boiling, *Brew. Int.*, 2(2):17–19, 2002.
18. O'Rourke, T., Back to basics 9 — wort boiling (part 1), *Brew. Guard.*, 128(8):34–36, 1999.
19. Hyde, A. and Sajland, J., Heat transfer processes in breweries — part 2, *Ferment*, 12(3):26–31, 1999.
20. O'Rourke, T., The process of wort boiling, *Brew. Int.*, 2(6):26–28, 2002.
21. Hancock, J.C., Brewhouse — mashing to cooling, *Brew. Int.*, 2(11):13–16, 2002.
22. Michel, R.A. and Kantelberg, B., Dynamic low-pressure boiling, *Brew. Int.*, 3(11):50–53, 2003.

23. Kantelberg, B., Wiesner, R., John, L., and Breitschopf, J., Advantages of the dynamic low-pressure boiling system, *Brew. Distill. Int.*, 31(3):16–18, 2000.
24. Weinzierl, M., Stippler, K., and Wasmuht, K., Merlin sparkles in terms of beer quality and energy usage, *Brew. Distill. Int.*, 31(4):12–15, 2000.
25. Seldeslachts, D., Wort stripping — the practical experience, *Brew. Guard.*, 128(9):26–30, 1999.
26. Herrmann, H., New developments in brewhouse systems and operations, *Ferment*, 11(1):36–45, 1998.
27. Wilkinson, R., Hot wort separation and cooling, *Brew. Guard.*, 132(5):26–32, 2003.
28. O'Rourke, T., Back to basics 10 — wort boiling (part 2), *Brew. Guard.*, 128(9):38–41, 1999.
29. Denk, V., Whirlpool and trub separation: experimental studies, practical realization and industrial experience, *EBC Monograph XVII*, EBC Symposium, Wort Boiling and Clarification, Strasbourg, 1991, pp. 155–164.
30. Dadic, M. and Van Gheluwe, J.E.A., Experiments with a whirlpool tank, *Brew. Dig.*, 47(9):120–126, 1972.
31. Kunze, W., *Technologie Brauer und Malter*, 8th ed., VLB Verlag, Berlin, 1998.

11

Brewing Process Control

Zane C. Barnes

CONTENTS

Introduction	448
Process Performance	449
Brewhouse Unit Operations	449
Grist Preparation	450
Malt Cleaning	451
Malt Weighing	452
Malt Milling — Grist Particle Size Distribution	452
Grist Storage	454
Dust Extraction and Dust Explosion Prevention	454
Malt Silo Level Detection and Stock Control	455
Grain Flow and Routing	457
Mashing	459
Grist Transfer	460
Water Flow and Temperature	462
Mash Homogeneity	463
Mash Heating	463
Temperature Control	464
Vessel Level Control	464
Conversion Efficiency	466
Wort Separation	466
Mash Transfer	467
Wort Filtration	468
Sparging	470
Effluent	470
Spent Grain	471
Extract Yield to Kettle	471
Wort Boiling	472
Nucleate Boiling	473
Evaporation Rate	474
Level Control	477

Avoidance of Boil-over 477
Liquid Adjunct Metering 479
Hops Addition and Utilization 479
Volatile Stripping 479
Trub Formation 480
Hot Wort Clarification and Cooling 480
Hot Wort Clarification 480
Wort Cooling 481
Wort Flow Rate 482
Wort Temperature 482
Turbidity and Cold Break 484
Conductivity 484
Wort Oxygenation 484
Trub Handling and Wort Recovery 485
Brewhouse Yield 486
Process Control and the Brewer 486

Introduction

The required characteristics of a beer for sale depend upon the consumer being targeted and can be readily defined as in Figure 11.1. However, it is consistency and product stability that will be needed to retain brand loyalty. To be commercially successful, a brewery needs to achieve world-class

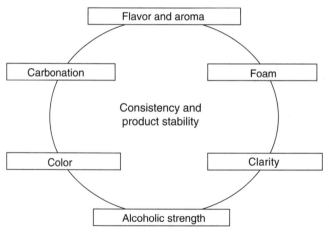

FIGURE 11.1
Defining beer attributes required.

standards of quality and efficiency. Process control applies engineering and technology to harness brewing scientific knowledge in the quest for consistent high-quality beers at competitive cost.

Each of the production operational areas presents some unique and interesting issues, but many of the control requirements are common. The brewhouse is particularly challenging due to the diversity of materials to be processed and the hostile operating environment. This is the production area chosen to demonstrate the wide range of process control techniques used throughout the modern automated brewery.

Process Performance

The key performance indicators to be measured and controlled to ensure that brewhouse quality and efficiency targets are being achieved are shown in Figure 11.2. What is immediately apparent is that many of the wort analysis measurements are off-line, retrospective, and generally cannot be used to correct the current batch before it proceeds to the next process step. In many situations, it is not possible to measure on-line the actual change to be controlled. Instead, past experience is used to predict that by controlling a certain parameter in a certain manner with predefined raw materials, the desired outcome will be achieved.

In this chapter, the term "process control" is used to refer to on-line measurement and real-time process regulation as part of a fully automated system. "Process management," on the other hand, is used to refer to the adjustment of system control values based on off-line measures and also to the management of any interfacing nonautomated activities. Effective process control requires reliable on-line measurement, control technology, and automation. Effective process management requires an experienced and competent brewer. Both are required to achieve excellence in "process performance."

Brewhouse Unit Operations

A review of the required beer characteristics in Figure 11.1 clearly demonstrates the paramount influence of wort on the finished beer. Correct wort composition from the brewhouse is essential for healthy yeast growth and fermentation to provide finished beer alcohol, carbonation, and the fermentation-derived flavors and aromas. The wort additionally provides color, foam, and the flavors derived from malts, adjuncts, hops, and water salts.

Final wort composition has predefined limits set by the specifications for brewing materials and water, but the final composition can be influenced by any of the brewhouse unit operations shown in Figure 11.3. Each of these

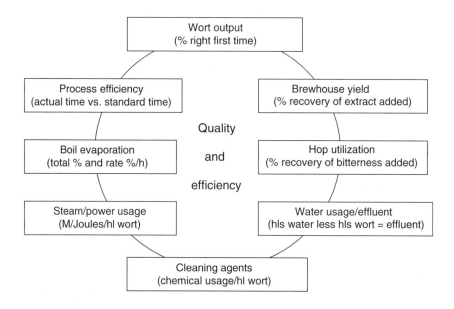

FIGURE 11.2
Brewhouse key performance indicators.

unit operations will now be considered and provided with the process objectives and the key elements for process control and process management. Typical on-line measurement and control techniques are described and, where appropriate, guidelines are offered for correct instrument installation, plant safety, and product protection.

Grist Preparation

The process objective is to mill the correct quantities of malts and adjuncts to produce grist with the optimum particle size distribution required for

Brewing Process Control

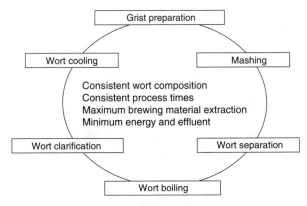

FIGURE 11.3
Brewhouse unit operations — process overview.

efficient wort creation and wort separation. The key activities to be controlled are summarized in Figure 11.4.

Malt Cleaning

If using nonmalted barley, an air/screen cleaner is needed to remove agricultural oversize and undersize foreign bodies (wood, string, seeds, straw, and

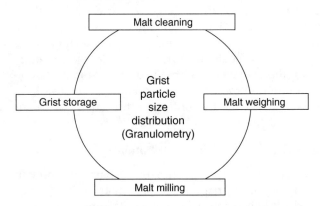

FIGURE 11.4
Grist preparation — key activities to be controlled.

sand). The destoner (gravity selector) removes grain-size objects of higher density than malt, which could damage the mill. Permanent magnets are installed at critical key points to retrieve the metal objects surviving the malting process, which would cause mill damage and any sparking that would risk a dust explosion. These captured ferrous objects must be removed regularly if the magnet is not selfcleaning. Regular removal of separated waste is essential to minimize the risk of fire/explosion and pest attraction.

Malt Weighing

The malt will be weighed either continuously in-line with an electronic tubular or tipping weighing machine, or it will be metered in discrete batches through a weigh bin suspended on either a knife edge weigh beam or load cells.

The load cell is a block of metal to which are attached four electrical resistance strain gauges. Two of the strain gauges are bonded to the metal surface so that they can reproduce any contraction due to deformation under load (the compression gauges). The other two strain gauges do not move and are there to enable compensation to be made for temperature variation that affects the modulus of elasticity and electrical resistance (the passive gauges). The strain gauges are electrically insulated from the metal cube, so that a measurement of the change in their electrical resistance is a measure of the amount of compression of the metal and compression gauges and, therefore, the load applied. The change in resistance is measured by using a Wheatstone bridge-type electrical circuit. The change in dimension of the load cell is normally a maximum of 2 mm.

Regular calibration of the weigher is essential to achieve $\pm 0.1\%$ accuracy and this needs to be constantly confirmed from malt stock control.

Malt Milling — Grist Particle Size Distribution

The gaps on a roller mill or the screen size on a hammer mill will be changed according to the required grist particle size distribution. This will be regularly measured off-line using a laboratory grading sieve (plansifter). The grist profile is critical to achieving maximum conversion of available extract during mashing, but still allowing efficient wort filtration and sparging. A finer grist is better for conversion and sparging efficiency, but a coarser grist is needed to maintain grain bed permeability at lauter tun wort filtration.

Typical plansifter analyses are provided in Table 11.1 for the ASBC and EBC measurement techniques. The fine/coarse balance is achieved by restricting the mill damage to the malt husks to achieve a lauter tun grist with, say, a 20% husk fraction and a resultant 20% flour fraction. However, when using a mash filter, the filtration is achieved with polypropylene

TABLE 11.1

Grist Composition — The Key to Brewhouse Extraction Efficiency

Pfungstadt EBC

	Particle Size (μm)	Screen Wire Thickness (mm)	Screen Mesh Size (mm)	Grist Fraction	Roller Mill Grist	Hammer Mill Grist
Sieve 1	<1250	0.31	1.27	% Husks	18	1
Sieve 2		0.26	1.01	% Coarse grits	7	4
Sieve 3	500–1250	0.15	0.547	% Fine grits 1	35	15
Sieve 4		0.07	0.253	% Fine grits 2	20	20
Sieve 5	125–500	0.04	0.152	% Flour	8	30
Pan	<125			% Fine flour	12	30

ASBC

Screen Number	Screen Wire Thickness (mm)	Screen Mesh Size (mm)	Grist Fraction	Roller Mill Grist
10	0.76	2		13
14	0.61	1.41		20
18	0.48	1		32
30	0.33	0.59		24
60	0.162	0.25		6
100	0.102	0.145		2
Thru				3

	Roller Grist	Hammer Grist
Bulk density (kg/m^3)	350	700
Husk vol/100 g	>550 ml	n/a
Typical roller gaps (mm)	1.6/0.8/0.4	n/a
Typical screen size (mm)	n/a	3.5

cloths (70 μm pore size) and the grist can, therefore, be much finer with a typically 1% husk fraction and 60% flour. It is this finer grist that enables the higher extraction efficiency capability of the mash filter.

Traditionally, North American grists have tended to be coarser with more emphasis on throughput time than extraction efficiency. However, each brewery has to establish the grist profile that best optimizes the interdependent parameters of wort quality, extraction efficiency, and throughput time appropriate to their own brewing plant and management priorities.

In addition to the proportion of husks in the grist, a greater husk particle size will increase the lauter grain bed permeability. It is possible to control the particle size of the husks not only by the mill roller gap settings, but also by the controlled addition of steam/water immediately prior to the mill. This moisture uptake (0.5 to 1% of malt weight) is known as malt conditioning and results in a more pliable husk, reducing its proneness to shattering at the mill. An indication of the coarseness of the husk fraction is given by its husk volume and typically will be >550 ml/100 g of husk.

Grist Storage

A typical 650-hl brewlength requires a 10-ton grist charge to be mashed in 15 min. To concurrently mill and mash would require milling at 40 tons/h. Dry mills are typically limited to 10 tons/h, and in order to eliminate the need for four mills a buffer vessel, the grist case, is used. One mill at 10 tons/h is now adequate for up to 12 brews per day. Grist case discharge control is covered in the section "Mashing."

Dust Extraction and Dust Explosion Prevention

Dust explosion is a constant danger and must never be underestimated. In many countries the malt tower will be classified as a hazardous area for the purposes of electrical regulations. Any incident due to noncompliance will certainly result in prosecution of the brewer. The fulfillment of this onerous duty can be briefly summarized as follows:

1. *Prevent accumulation of combustible materials.* This will be mainly achieved by a well-designed dust extraction system (cyclones/dust filters), removing dust as it is created at each of the processing units and the conveyor and elevator discharges. Malt processing will be interlocked to prevent operation if the dust extraction system is not running and healthy. The building and equipment should be designed with no ledges for dust collection. Rigorous housekeeping is essential to prevent any dust build up or accumulation of other combustible materials.

2. *Avoid any ignition source.* Good design will ensure that machinery and associated electrical equipment will be dust ignition proof rated. However, on-going inspection, preventive maintenance, and statutory records are required. It is essential to remove from the malt any stones and hard objects (screen/destoner) and ferrous objects (in-line magnet) before entry to the mill to avoid sparking. Hot surfaces must be avoided, and high-speed bearings should be monitored for any abnormal temperature increase or excessive vibration. Conveyors and elevators need to be monitored for correct speed of operation and for blockages (see the Section "Malt Flow Control"). There must be a strictly controlled permit procedure for any welding in the area. All equipment must be earthed to avoid static build up and the building itself may need to be fitted with a lightening conductor.
3. *Explosion control.* All explosion risk points should be fitted with explosion vents to direct the explosion to a safe area to avoid any risk to operational staff and secondary explosions within the malt tower. In the event of an explosion panel blowing out, this movement should be sensed and the whole system shut down to fail-safe.

Malt Silo Level Detection and Stock Control

Bulk malt will be generally delivered in 20-ton batches into a silo holding, say, 200 tons (2 days production). As a minimum requirement for automatic operation, sensors are needed to know when the silo goes empty and if it reaches a high level during filling. A continuous indication of the amount of grain in a silo is highly desirable for stock control purposes. Sensors must be compatible with the hazard of dust explosion and, as malt silos are frequently installed outdoors, need to be weatherproof.

The high level/full probe, which is needed to prevent overfilling and avoid a disastrous jamming up of feed conveyors and elevators detects the presence (and absence) of grain at that point. A suitable instrument for fixed point malt detection is the rotary paddle bin monitor. The motion of an electrically driven rotating paddle is resisted by immersion in malt, and the driving motor is caused to rotate within its housing, which triggers a switch to provide an electrical control signal and also cuts power to the motor. When the malt level drops, the loaded tension spring returns the motor to its original position and the unit is reactivated. The position of this sensor should be on top of the silo and set off-center (Figure 11.5). This will allow for the angle of repose of malt (26° from horizontal), avoid damage from the falling grain during filling, and accommodate the quantity of malt still in transit through the intake conveyors and elevators.

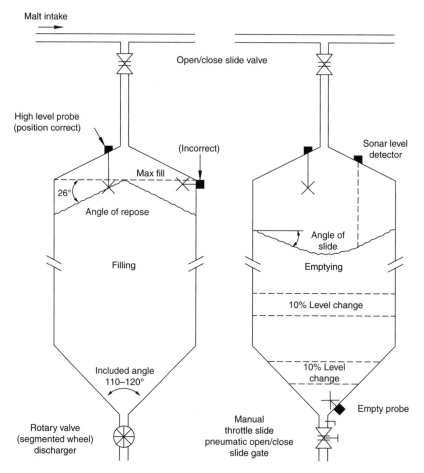

FIGURE 11.5
Malt silo level control.

The low level/empty probe at the silo outlet detects the absence (and presence) of malt and, in an automated system, would initiate the opening of an alternative silo of the same product or close down the malt transfer.

Continuous-level detection between high- and low-level probes presents difficulties, but reasonable accuracy is possible with the sonar-type sensor mounted on top of the silo. This sensor directs a high-intensity sonar pulse down to the grain surface and measures the echo return time. Preknowledge of the sound velocity allows calculation of the distance between the sensor and the surface of the grain, which then enables calculation of the grain level in the silo. Positioning of the sensor requires due consideration for the angle of repose at filling and the angle of slide at emptying (Figure 11.5) and, therefore, should be at midpoint of the radius.

A continuous level detector is calibrated to indicate level from 0% (empty) to 100% (full). Due regard needs to be given to the vessel shape. As demonstrated in Figure 11.5, a change of level in the conical outlet equates to a much smaller quantity of malt than the same change of level in the vertical-walled section.

If the level is to be converted to a kilogram of malt, then the vessel geometry and dimensions are used to create a reference calibration table to provide the volume for any given level in the silo. Knowing the bulk density of malt (510 to 550 kg/m^3) allows conversion of volume to weight, so that the silo level can now be reported as the malt weight.

Because of the difficulties with continuous-level detection, a common practice is to maintain a "paper" stock level for each silo. The quantities transferred into a silo (from weighbridge certificates) and quantities weighed out for brewing are entered to provide an automatic electronic display of the (calculated) malt quantity in the silo. A physical reconciliation is readily achieved when the silo goes empty. In the case of a single silo product being continually topped up and rarely emptying, stock reconciliation is achieved by manually measuring the malt level from the top of the silo (tape measure/dip-stick).

Physical stock reconciliation allows true malt usage to be calculated, which provides an important cross-check for weigher accuracy and brewhouse yield calculations.

Grain Flow and Routing

Malt will be transported by a system of horizontal conveyors and vertical elevators. Due to the fire/explosion risk, it is essential to avoid overloading and overheating caused by blockages. Start up of conveyor/elevators/processing units will be "cascaded" as a strictly controlled sequence, starting with the final destination conveyor (feed into grist case) and finishing with the source conveyor (discharge from the malt silo). Each start will be preconditional on receiving a confirmed running signal from the previous item of the sequence. The shutdown sequence will be "cascaded" in reverse order.

Generally, the conveyors and elevators will be of fixed speed, and the system flow is regulated by the discharge rate of the malt silo. This can be achieved using a manual slide valve positioned to allow the required flow rate, which can result in flow differences depending on how full the silo is. This flow variation is normally acceptable, but a motorized rotary vane discharger (segmented wheel) can be used to ensure a constant flow rate from the silo. Figure 11.6 is a stylized representation of typical conveyors and elevators used. The bucket elevator is a continuous belt with attached buckets. These have been the cause of friction in fires/explosions arising from belt slippage on the top drive roller (insufficient belt tension/blockage resistance) or from rubbing against the casing through misalignment. The

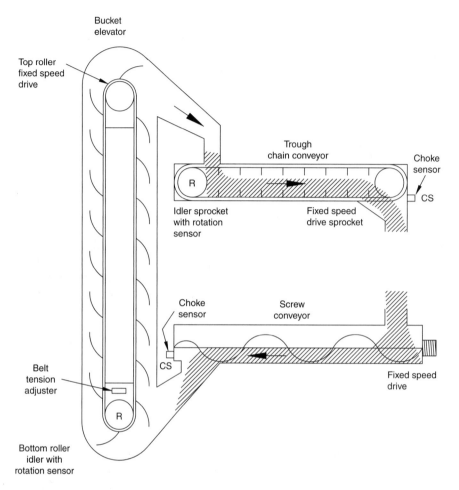

FIGURE 11.6
Malt flow control (styalized).

belt must always be under correct tension, and the bottom idler roller should be fitted with a rotation sensor to confirm the correct belt speed.

In the drag chain conveyor, the chain being pulled discharges malt at the drive motor end. The idler sprocket at the in-feed end should be fitted with a rotation sensor that confirms that the chain conveyor is intact. To detect a blockage at the discharge and drive end of the conveyor, a choke sensor is present to detect any pressure build up.

The screw conveyor has a helicoid flighting that is normally right-handed, with a pitch equal to the diameter. The conveyor is usually driven by a gear motor coupled to sprockets and chains, and the discharge should be fitted with a choke flap or pressure switch.

The rotation sensor (tachometer) is a reed switch that is activated by a magnet on the rotating idler sprocket, so that the revolutions per minute can be determined and an alarm raised if rotation is abnormal. The choke sensor is a diaphragm pressure switch. Pressure creates a deflection of the diaphragm and sufficient movement will activate a microswitch, providing an electrical signal to alarm the control system.

Mashing

The process objective is to slurry the grist with hot water to dissolve the soluble malt components (grist hydration) and then to regulate enzymic activity to achieve the required wort composition (mash conversion). The key activities to be controlled are identified in Figure 11.7. Correct sweet

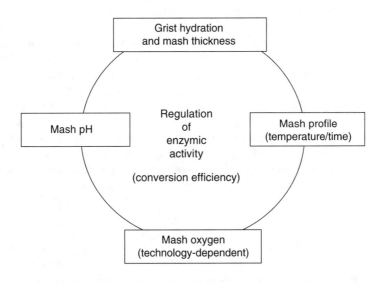

FIGURE 11.7
Mashing — key activities to be controlled.

wort composition is essential for consistent product quality, but process control is limited. The most informative analyses are off-line and retrospective.

Certain attributes of sweet wort composition (color, pH) are predetermined by the specifications chosen for brewing materials and water. However, the enzymic degradation of, particularly, carbohydrates and proteins can be influenced by the mashing conditions.

Ideally, the mash vessel would be fitted with sensors that could directly measure the progress of these enzymic reactions and directly control the final sweet wort composition. Unfortunately, partly due to the abrasive nature of the mash slurry, such probes are not available. Instead, we must rely on off-line measurement.

The simple off-line iodine starch check is used to confirm that starch has successfully gelatinized, liquefied, and saccharified, but sophisticated laboratory analysis is needed to know the ratio of nonfermentable dextrins to fermentable sugars. Similarly, the essential total soluble nitrogen of the wort and the ratio of its subcomponents (proteins, polypeptides, and amino acids) cannot be measured on-line.

Mash thickness, pH, and salt composition do influence enzymic reactions, but, for a given recipe, can be considered as fixed. Therefore, the only on-line process control variable used to influence wort composition is temperature.

A critical responsibility for process management is well specified malt quality and brewing water composition, and vigilant policing by the brewer for conformance; for instance, measurement of wort color immediately after mashing and before grinding the next brew. This provides not only an evaluation of color consistency from the maltster (no color variation from downstream brewhouse processing), but also the opportunity to produce a corrective grist for the succeeding brew. Apart from analysis of salts of the brewing water, regular tasting from the holding tanks will provide an early warning of taints from the water supply or from brewery contamination.

Grist Transfer

If the mashing in cycle time is 15 min, then other activities ancillary to the grist transfer, such as the prior addition of foundation water to the mash vessel, mean that the actual grist transfer needs to be done in 10 min (60 tons/h for 10-ton grist charge). This will typically be achieved with a screw conveyor operating at fixed speed and fitted with a discharge choke sensor.

The grist case will be fitted with an empty probe to signal the completion of grist transfer. Additionally, continuous read out of the grist weight can be achieved by suspending the grist case on load cells. To prevent distorted readings, it is important that the grist case is isolated from the building and surrounding equipment by the use of flexible joints (Figure 11.8). The

Brewing Process Control

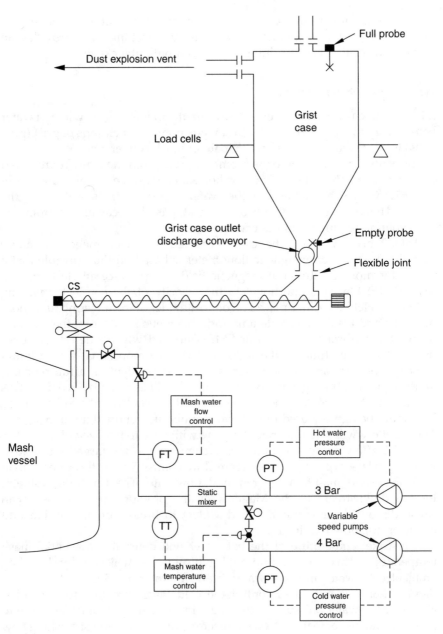

FIGURE 11.8
Grist hydration — mashing in.

grist case weight allows a crosscheck against the malt weigher. It also provides a grist case depletion rate during mashing and provides an alarm to warn that grist bridging has caused outlet blockage.

Water Flow and Temperature

It is normal to use a premasher for the intimate mixing of grist with hot water before entry to the mash vessel. This avoids nonwetted clumps of grist (grist balls) that are then unable to participate in mash conversion.

The control of hot water to the premasher is diagrammatically represented in Figure 11.8. The total volume of hot water required is calculated from the grist weight available and the water to grist ratio specified by the recipe. The required flow rate of hot water is then calculated from the total volume required and the grist transfer time.

Water/wort flow rate is typically measured with a magnetic flow meter (Magflow). The electromagnetic flow meter is based on the principle that a conductor moving through a magnetic field will produce an electromotive force (EMF) that is proportional to the velocity of the conductor passing through it. Field windings placed on opposite sides of the metering tube (nonmagnetic and insulated) create a uniform magnetic field across the pipe. As the wort flows through this field, an EMF is induced, which is detected by electrodes. The magnitude of the induced EMF is proportional to the liquid velocity. The accuracy of the meter is greatly dependent on a full pipeline, and the possibility of gas breakout or part full lines must be avoided. This requires that the meter should be normally installed in a rising main. The liquid being measured must have a minimum conductivity of 5 μS/cm. The Magflow will also integrate flow rate with time to totalize volume.

The Magflow value is continuously compared to the target and adjustment signals sent to the downstream flow control valve (throttling valve) to achieve and maintain the required flow rate. The throttling valve is located downstream of the Magflow as the pressure drop across it can cause dissolved air (from the cold water) to break out of solution and create erratic flow readings.

The strike temperature of the hot water will determine the initial mash temperature. This temperature will typically be achieved by blending ambient cold water into hot water, brewing water at 85°C as depicted in Figure 11.8. It is critical to control the cold water pressure to, say, 1 bar above the hot water pressure, so that the cold water flow is sufficient to attemperate the hot water to the lowest strike temperature and highest mashing flow rate. This pressure difference will be maintained by variable speed pumps responding to direct feedback from diaphragm-type pressure sensors.

The resultant temperature of the mixed hot and cold water is determined after the in-line static mixer by a resistance bulb thermometer. Adjustment signals are continuously sent to the cold water flow regulation valve to achieve and maintain mashing water temperature.

The resistance thermometer uses the principle that the resistance of a pure metallic conductor, such as platinum, varies with temperature. The resistance thermometer is calibrated by measuring resistance at known temperatures. It will provide an accuracy to 0.1°C provided it is immersed in a correct length thermowell. Careful consideration is required for the thermowell location to ensure that this position of measurement is meaningful and representative. For instance, a thermowell inserted into the top of a pipeline that is running part full will provide an incorrect temperature.

Mash Homogeneity

Prior to grist entering the premasher, the mash vessel itself will have received foundation water sufficient to cover the rotating agitator. When the grist slurry enters the body of the mash, it is necessary to keep the nonsoluble grist components (husks) from settling out of suspension and causing outlet blockage. As discussed earlier, the main control over wort composition is temperature, and it is essential that the mash has an evenly distributed temperature, particularly during any heating phase (described later). The achievement of this mash homogeneity will be by use of the mash vessel agitator. To avoid mash shear damage, it is important that the agitator can achieve effective mixing with low tip speed (<2.5 m/sec) and without internal baffles. Shear damage will result in release of high-molecular-weight β-glucan materials and also protein gelation. Both will result in slow mash separation due to reduced grain bed permeability.

Mash Heating

Mash heating can be described by the following formula:

$$Q = U \times A \times \Delta T$$

Q	U	A	ΔT
Mash heat up rate	Heat transfer coefficient	Heater surface area	Temp. difference (steam/wort)

The coefficient of heat transfer for any heater is fixed by the material of construction and its thickness. However, it will be markedly reduced by fouling and needs to be kept clean. The heater surface area will be fixed, but this can be effectively reduced by condensate water logging of the jackets. The steam pressure and its associated temperature are normally given, but the presence of air in the steam will reduce the temperature for a given pressure and will reduce ΔT. Care must be taken with mash heating. Using 3 bar steam at 144°C supplied to heating coils or heating panels would scorch stationary mash onto their surfaces, and any such fouling would dramatically reduce the heat transfer coefficient and heat-up rate. The mash needs sufficient velocity across the heater surface to quickly dissipate the heat away to prevent local overheating, and it also needs to be effectively mixed to ensure a homogeneous temperature distribution throughout the vessel.

In addition to ensuring that the agitator is running correctly, it is also important to confirm that any heating surface is fully immersed in the mash before allowing steam to be supplied. This can either be by level probes located above the heating surface or by a diaphragm-level transmitter installed at the base of the vessel, as shown in Figure 11.9. Both of these present difficulties and are discussed later.

Just before emptying the mash vessel, a rinse is carried out to ensure heater surfaces are cleared of residual products that would otherwise bake on and foul.

In the heating jackets, steam at, say, 144°C is giving up its energy to the mash and condensing to hot water at 100°C. This condensate must be effectively removed to prevent the steam jackets becoming water-logged, which can cause severe vibration (hammer) as well as reducing heating efficiency (steam at 144°C replaced by condensate at 100°C). Moreover, air mixed with steam in the heating jackets will result in a lower temperature at a given steam pressure, resulting in lower heat-up rates. A thermostatic air eliminator is used to vent the jackets until the correct temperature for the given steam pressure is achieved.

Temperature Control

The temperature of the mash is accurately measured with a resistance bulb thermometer in a temperature well of adequate immersion positioned in the lower sidewall of the vessel. When steam heating ceases, there is still residual energy to be dissipated from the heating elements. To avoid mash temperature overshoot, the temperature control system needs to stop heating before the desired set-point has been reached, based either on time or temperature.

Vessel Level Control

A continuous read out of the vessel contents can be achieved using a diaphragm-type pressure sensor located at the base of the vessel. Pressure creates a deflection of the instrument's diaphragm and its fill liquid transfers this pressure to a resistance bridge. The voltage output of the resistance bridge is dependent on pressure.

The level transmitter measures the hydrostatic pressure created by the weight of liquid above it. At constant density, this hydrostatic pressure is a function of the height of the liquid above the sensor, which will be calibrated for 0% (empty) to 100% (the maximum safe working level for the

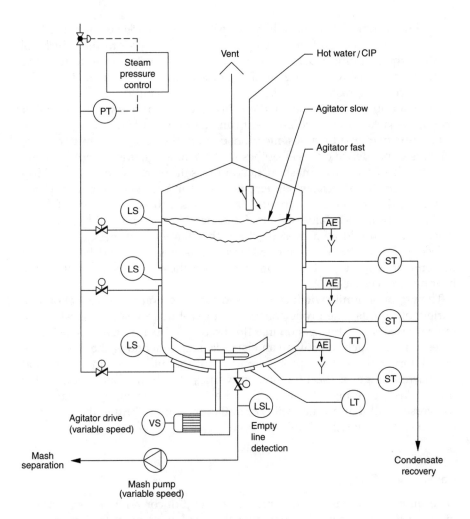

FIGURE 11.9
Mash heating control.

AE Air eliminator – thermostatic vent – steam jackets
ST Strainer/steam trap – condensate removal
LSL Level probe – mash detection
LT Level transmitter
TT Temperature transmitter

vessel). The measured head will be influenced by density and a mash bulk density will need to be assumed, based on the most frequently used water to grist ratio. Any significant variation from this ratio will cause error in level detection. Knowing the geometry of the vessel and its dimensions, it is possible to calculate its volume at any given level and display this.

During heating, when the agitator is used at higher speeds to ensure heat transfer there is a significant change in liquid level due to the whirlpool effect of the spinning mash as shown in Figure 11.9. The level transmitter with stationary/slow-moving mash detects a higher level than when the mash is rotating fast. According to the level transmitter, the level of the mash is too low to allow heating, but in reality, with the agitator at higher speed, the jacket is covered and heat can be safely applied.

Because of the difficulties mentioned with the level transmitter, there remains dependence on level probes to determine whether to apply steam to the jackets, to detect mash at maximum safe working level (high-level alarm), and to detect vessel empty (lack of product in outlet pipe). However, due to the sticky nature of the mash, these probes can foul or bridge to indicate a false signal of the presence of mash. Prior to each mash, there should be a prestart check that the level probes are correctly detecting no mash. As mentioned earlier, the vessel should be rinsed at the end of emptying of each brew to keep these probes as well as the heating surfaces clean.

The vibration limit switch can be used for fixed point level detection in a variety of liquids. The probe consists of a two-tine fork set to vibrate at an intrinsic frequency. Immersion in a fluid reduces this frequency, which activates a limit switch to provide an electrical signal denoting the presence of liquid. Orientation of the probe, if side-wall mounted, should ensure that the narrow edge of the fork tines is vertical to allow easy runoff of the liquid. Also, care must be taken that the tines are far enough away from the vessel wall that normal soil accumulation will not bridge the tines and provide a false signal of liquid present.

Conversion Efficiency

Determination of mash wort density at the end of conversion allows calculation of the efficiency of solubilizing the available extract (as determined in the laboratory). Roller-milled grists should achieve 100% conversion whereas hammer-milled grists should achieve 103% conversion. A roller-milled grist with as is laboratory extract of 76% w/w and a mash thickness of 3.2 l water/kg of malt should yield a mash wort of 20°Plato. Similarly, a hammer-milled grist with as is laboratory extract of 76% w/w and a mash thickness of 2.5 l water/kg of malt should yield a mash wort of 24.5°Plato.

Wort Separation

The process objective is to separate the maximum amount of sweet wort created at mashing from the insoluble grist and coagulated protein (spent

Brewing Process Control

grain). For quality reasons, the recovered wort should be low in turbidity (filtered to a haze <5 EBC or <20 NTU). The mash separation device is usually the rate-limiting step of a brewstream and, therefore, turn around time (TAT) is critical to brewhouse performance. The key activities to be controlled are summarized in Figure 11.10.

Mash Transfer

The avoidance of shear forces by the mash vessel agitator is equally valid at mash transfer. The mash pump should be slow-revving with an open impellor. The mash velocity should be less than 1.5 m/sec both in the pipework and at vessel entry. Shear forces due to severe directional changes or constrictions in the pipework are to be avoided. Bottom entry to the lauter tun is preferred to avoid oxygen pick up, and it is essential to complete transfer with an evenly distributed level grain bed.

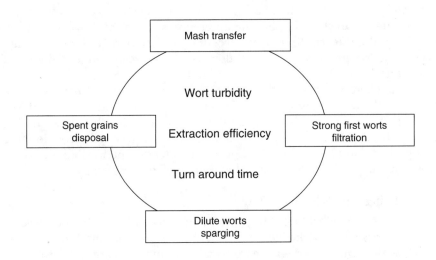

FIGURE 11.10
Wort separation — key activities to be controlled.

Wort Filtration

$$Q = \frac{K \text{ (grain bed porosity)} \times P \text{ (bed differential pressure)} \times A \text{ (surface area)}}{U \text{ (wort viscosity)} \times L \text{ (grain bed depth)}}$$

The above simplified version of the D'Arcy filtration equation was devised using an incompressible sand filter. However, the grain filter bed in the lauter tun is, in fact, compressible. Higher flow rate, viscosity, and differential pressure will compress the bed and reduce grain bed permeability. The D'Arcy equation cannot be used for accurate prediction of mash filtration flux rate, but it provides some useful guidance as to what influences performance and, therefore, which process parameters need to be measured and controlled. The D'Arcy equation identifies that wort filtration will be improved either by an increased value of a top line parameter (K, P, or A) or by a decreased value of a bottom line one (U or L).

A. Surface area of the filter. For a given machine this is fixed.

P. Differential pressure (DP). Wort flow from underneath the false floor is resisted by the permeability of the grain bed and in extreme cases by the free area of the (false) slotted floor. The difference in pressure above and below the false floor is continuously measured by diaphragm-type pressure sensors to ensure that the wort flow does not result in the designed maximum DP being exceeded (Figure 11.11).

K. Grain bed porosity. This is the major parameter that will determine the maximum flow rate of wort that will not exceed the maximum DP permitted. Grain bed permeability is preset by the particle size distribution at milling (see the Section "Grist Preparation") and a coarser grist will provide more bed porosity and faster flow. However, this intended permeability derived from milling can be reduced and the grain bed compressed by:

1. Using too high a wort flow
2. Increasing wort viscosity either by thicker mashes or by increased wort β-glucans (change in malt quality, inappropriate mashing temperature profiles, or release from excessive mash shearing forces)
3. Presence of gelled proteins due to excessive mash-shearing forces

The lauter tun will be fitted with rotating knives/rakes that can be lowered to varying degrees to gently cut open the bed and improve the grain bed permeability. This will reduce DP for a given flow rate but needs to be limited to retain acceptable wort clarity. Careful control of the rake height, and sometimes rotation speed, is needed to avoid any deterioration in wort clarity. Integration of an in-line wort turbidity meter as part of the raking/DP control (Figure 11.11) will ensure this.

U. Wort viscosity. Reduced wort viscosity will in itself improve filtration rate and has the additional bonus of reduced bed compression. However,

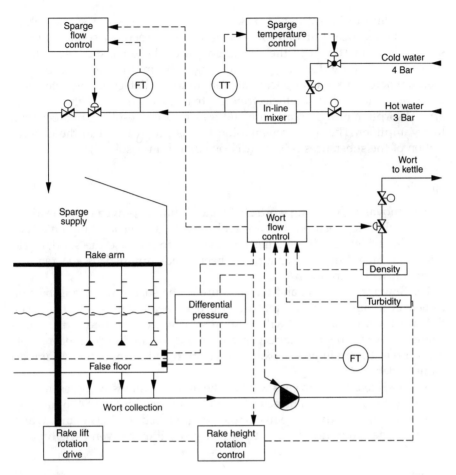

FIGURE 11.11
Wort separation and sparge control (lauter tun).

wort viscosity will be fixed for a given malt composition, mash thickness, mashing program, mash-off temperature, and upstream shear forces.

L. Grain bed depth. A thinner grain bed will increase filtration rate and reduce vessel TAT. However, this faster cycle time is unlikely to compensate for the capacity lost per brew by reducing the grist charge. A minimum bed depth is required by the lauter tun to achieve efficient wort filtration and clear worts. Generally, bed depth is fixed by the machine design.

From the previous discussion, it becomes obvious that efficient wort filtration is very dependent on careful upstream control of malt quality, milling, mashing profile, and avoidance of mash shear forces to achieve the required grain bed porosity. Flow at wort filtration is limited by DP, which can be improved by the use of cutting knives/rakes, but at the expense of wort clarity.

The in-line turbidity meter makes use of Lambert–Beer's law, which relates that the logarithm of the loss of light intensity across a solution of substances is proportional to the concentration of those substances. A lamp directs a precisely focused, constant light beam through the process medium across a known optical path length to the detector. The detector is a hermetically sealed photoelectric cell that measures the light intensity after absorption or light scattering and transmits the resulting photocurrent to an amplifier. The light transmission loss is proportional to the concentration of the substances present (dissolved and undissolved).

Sparging

After removal of the strong worts, hot water is finely sprayed onto the top of the grain bed to wash out the residual extract. Greater leaching efficiency would be achieved with smaller particle size to increase the solid/liquid interfacial area, but this has already been predetermined by the requirements for wort filtration.

The flow rate, volume, and temperature of the sparge water will be controlled in a similar way to that described earlier for mashing. However, the target sparge flow rate will be set to match the flow rate of the worts being removed. This maintains constant hydraulic conditions and prevents exposure of the grain bed to oxidative gelation and formation of an impervious surface layer.

As the eluted wort becomes weaker, the density and viscosity reduce and faster wort collection can be achieved without exceeding the DP limit. This higher flow rate is advantageous for deeper grain beds (>200 kg/m^2), which require larger volumes of water to be sparged within a given cycle time.

Effluent

The final weak worts have increased levels of fatty acids and polyphenols, and sparged worts below 1.0° Plato are generally discarded for this reason. To minimize effluent production, it is important that there should be the minimum possible liquid left in the lauter tun at completion of wort collection, but care is needed to avoid air entrainment and cracks in the grain bed resulting in deterioration of wort quality.

The lauter tun grain bed drainings and under-floor washings are the main source of process waste from each brew (assuming that wort recovery from trub is practiced). This effluent is high in total settleable solids (5000 to 10,000 mg/l) and chemical oxygen demand (10,000 to 20,000 mg/l), which makes it increasingly expensive to dispose of as trade waste.

On-line analysis of chemical oxygen demand and settleable solids is not currently available, and this together with combined drain flows from other areas can make quantitative monitoring difficult. However, this is the major source of effluent in the brewhouse.

A simple self-cleaning screen (30 μm) or a belt press will remove most of the settleable solids and reduce effluent charges. In the future, it is likely to become economic to process this waste further by using ultrafiltration (100 nm) to remove the problematic lipids and polyphenols so that the permeate can be recycled as hot brewing water. If the retentate can be disposed of with the spent grains, then this process waste and its effluent treatment cost can be eliminated. There will also be a reduction in the demand for hot water and its associated water and energy costs.

Spent Grain

Modern lauter tuns can discharge the residual grain bed in 10 min and achieve low levels of retained spent grain on the false floor (DIN Standard <400 g/m^2). Spent grain is sticky by nature and not free flowing, which result in a repose angle and sliding angle approximately twice that of malt. This requires any spent grain holding vessel to have a steep outlet with included cone angle of 70°. It may additionally terminate in a chisel section, and will almost certainly have a rotating Archimedean screw or spiked arm (archbreaker) to prevent the grain from bridging the outlet and preventing discharge. (Bridging can also occur on level probes and result in false detection of grain present at any level in the silo.)

Because spent grain is not free-flowing, any long distance transportation to a storage silo will be by pneumatic conveyor using either pressurized air or steam.

The texture of spent grain and its handleability will vary with moisture content, the presence of high molecular weight β-glucans and pentosans, and residual extract. Regular analysis of spent grain for residual extract will quantify the soluble extract that has not been leached out by sparging and the extractable extract that has not been converted at mashing. Both values should be less than 0.8% of the wet spent grain.

Extract Yield to Kettle

Traditionally, the level in a calibrated kettle would be used to determine the volume of wort, and a saccharometer would be used to determine the specific gravity of the wort. The weight of dry extract can then be calculated and compared to the quantity added with the grist.

There is considerable advantage to be gained by on-line determination of density ex lauter tun. Now that the wort has been separated from the abrasive mash, it becomes practical for the first time to measure density on-line.

The first wort's density from the lauter tun, after allowance for false floor flooding and mash vessel/line rinsing, provides us with the gravity of the mash. Knowing the laboratory extract of the grist and the mash thickness (water:grist ratio) and with 100% mashing conversion, the expected value can be calculated. A mashing inefficiency becomes immediately apparent. During sparging, the decrease in gravity will correspond to decreasing viscosity and this can be factored into wort collection flow rate.

The dilute weak worts are higher in lipids, polyphenols, and silicic acids and worts below 1°Plato are generally discarded for quality reasons. The in-line density meter can be used to terminate wort collection if the wort Plato drops below a preset value.

The in-line density can be continuously integrated with a flow meter volume to calculate the total dry weight extract recovered from the lauter tun or mash filter. Alternatively, a mass flow meter will determine both parameters:

$$\text{kg Dry extract} = \text{hl wort} \times \text{Density} \times °\text{Plato}$$

(All volume and density measurements must be referenced to 20°C. If measuring specific gravity, this needs to be converted to density.)

The densitometer works on the principle that the period of oscillation of a vibrating U-tube is related to the density of the fluid flowing through it. The tube is made to vibrate by the application of a pulsating current and the amplitude of vibration is detected by a sense coil that provides an electrical signal for the frequency of oscillation. In addition, a temperature sensor is mounted on the flow tube, which is used to compensate for the change in the tubes modulus of elasticity as well as referencing the measured density to 20°C.

Wort Boiling

The process objective is to finalize the color and sugar profile by the addition of liquid adjuncts and to create the desired levels of hoppiness and bitterness by the addition of hops. Boiling is needed to isomerize the hop alpha acids, to strip out unwanted malt and hop volatiles, to denature proteins and coagulate proteins/polyphenols as hot break, and to fix the wort composition by terminating all enzymic and microbiological activity surviving the mashing process. As a consequence of boiling and evaporation, there will also be color development and wort density increase that need to be accommodated to achieve finished wort target. Figure 11.12 outlines the key activities to be controlled.

Ideally, the kettle would be fitted with in-line probes to follow the isomerization of alpha acids, the disappearance of dimethyl sulfide, and the coagulation of colloidal size particles (0.1 to 10 µm) to form hot break (30 to 70 µm). This is not yet practical and, therefore, the brewer has to rely on the pragmatic experience to achieve these critical parameters.

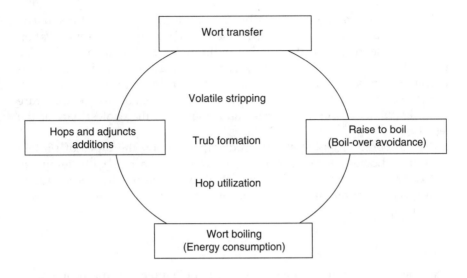

FIGURE 11.12
Wort Boiling — key activities to be controlled.

Nucleate Boiling

The most common boiling technique practiced is nucleate boiling at atmospheric pressure. Steam vapor bubbles nucleate at the surface of the heater to create a two-phase liquid/vapor fluid. This mixture of liquid and steam vapor bubbles is lower in density and becomes displaced by higher-density single-phase wort. It is this density differential between single-phase and two-phase fluids that provides the motive force for the thermosyphon.

The steam vapor bubbles provide a high interfacial surface area contact with the wort. This assists with volatile diffusion, α-acid isomerization, and protein denaturation. Higher vapor bubble intensity in the wort will improve boil effectiveness. Higher heater surface area relative to the volume of wort to be boiled (m^2/hl) will achieve this.

Protein denaturation needs turbulence as well as temperature that is usually referred to as vigor of boil. This tends to be simplified to evaporation rate and vessel turnover (the number of times the kettle contents will come into contact with the heater surface). The technology employed must ensure thorough compound mixing in the body of the vessel and that all wort passes over the heater several times. Measurement of wort temperature in the kettle should be at the base and, therefore, the coolest part of the vessel.

The temperature of the steam used impacts upon the types of protein denatured. Lower steam temperature is considered beneficial to minimize heat damage and improve the foam stability of the finished beer. Again, for a given evaporation rate, a larger heater surface area will use a lower steam temperature.

Evaporation Rate

Heater surface area determines the vapor bubble intensity and the temperature of steam used but is fixed by the technology available. The main control parameter, therefore, becomes evaporation rate:

$$Q = U \times A \times \Delta T$$

Q	U	A	ΔT
Rate of heat transfer (evap. rate)	Overall heat transfer coefficient	Heater surface area	Temp. difference (steam/wort)

Figure 11.13 represents a section across the wall of a heating surface. On the left-hand side of the wall, steam is condensing as a falling film of condensate creating a steam boundary layer. On the right-hand side of the wall, the wort and vapor bubbles form a rising boundary layer of two-phase flow. Figure 11.13a shows the temperature profile across the heater from the steam to the wort. Whilst the evaporation rate depends on ΔT, it is the actual heater surface contact temperature with the wort that determines evaporation rate. This will depend on the overall coefficient of heat transfer (U). Figure 11.13b shows how fouling of the heater surface reduces U and lowers the surface contact temperature, which reduces evaporation rate. In contrast, increasing the steam temperature (pressure) increases U and restores the surface contact temperature for the required evaporation rate. This higher steam temperature to compensate for fouling does not increase the heater surface contact temperature for a given evaporation rate. ΔT as an indicator of contact temperature with wort is only valid for a clean heater surface.

Because heater surface fouling cannot be measured or predicted, control of evaporation by steam temperature (pressure) alone yields variable results. (The heater could be cleaned after every brew to maintain a constant U, but this is time consuming and can reduce capacity). Some improvement to accommodate for fouling can be achieved by monitoring density change

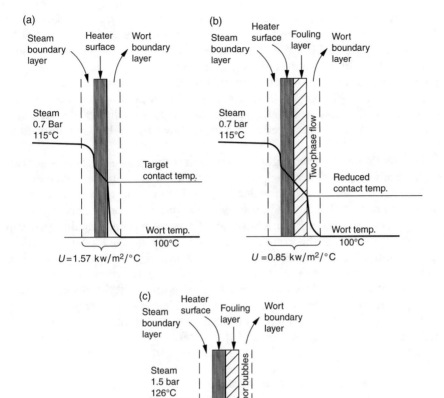

FIGURE 11.13
Temperature profile at heater surface — effect of fouling. (a) Evaporation rate 7%/h clean heater surface, (b) Evaporation rate 4%/h fouled heater surface (no steam compensation), (c) Evaporation rate 7%/h steam temperature increased contact temperature and evaporation rate restored.

during the boil and regulating steam flow to achieve the desired evaporation rate. Obviously, this will be complicated by any extract or hop additions during the boil.

Increasingly, use is being made of controlling the mass flow of the steam to the wort heater. Evaporation of 1 kg of water from the wort requires 1 kg of steam to be condensed (small corrections needed for steam enthalpy, wort density, and system losses). Measuring and controlling the condensation of a constant mass flow of steam at the heater ensures consistent evaporation. As shown in Figure 11.14, if the steam mass flow meter senses a drop in flow due to fouling, it will regulate the steam control valve to restore the mass flow to target. This will result in increased steam pressure and temperature in the heater, which restores the surface contact temperature for the required evaporation rate. Eventually, the steam control valve will be fully open and unable to provide the required steam mass flow and the heater will need to be cleaned.

The steam flow will typically be measured by an orifice meter or a venturi meter. Both work on the principle of a line constriction (orifice or converging cone) creating an increased velocity and kinetic energy and a reduced pressure energy. The differential pressure across this constriction is, therefore, proportional to the steam flow rate. Pressure tappings made upstream and downstream of the constriction at precisely defined positions allow determination of the differential pressure. Knowing the steam supply pipe diameter, the constriction diameter, and the differential pressure allows calculation of the steam flow rate. These meters require that the steam pressure be constant and that the steam flow is laminar. It is critical that the pipework installation achieves specified minimum straight lengths of pipe immediately upstream and downstream of the meter.

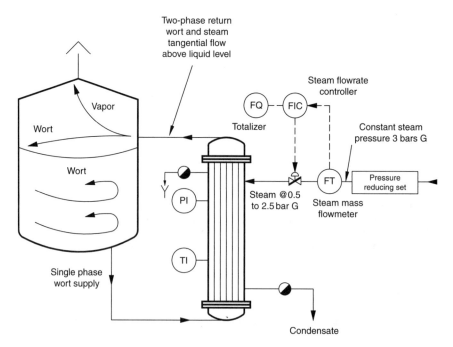

FIGURE 11.14
Steam mass flow control and foam suppression (external thermosyphon heater).

If constant steam pressure and laminar flow cannot be achieved, then a vortex meter should be installed. The steam flow passes around a flow-obstructing element shaped so that vortices are formed alternately on either side of it. The rate of formation and shedding of the vortices is proportional to the rate of flow. A drilled channel connecting both surfaces of the blunt object will experience alternating flow direction caused by the alternating surface eddy formation. This alternating flow in the channel can be sensed by using a thermal or mechanical device.

As discussed in the Section "Mash Heating," it is important that all wort heaters (coils, jackets, percolators, and calandria) are effectively drained of condensate to avoid heating inefficiency and steam hammer.

Level Control

As a minimum, the kettle will be fitted with a high-level probe (maximum safe working level) and an empty probe (no product in emptying pipework). For continuous read out of liquid level, a diaphragm-type pressure sensor can be installed at the base of the vessel. As described for the mash vessel, this can only be a guide if there is dynamic movement and if wort density is varying. Traditionally, the kettle was used to declare the extract recovered from mashing and would be carefully calibrated to know the volume according to the level in the vessel. Because of the shape and internal fixtures of the kettle, the volume of the vessel at a given liquid level is more difficult to calculate, and calibration is more typically achieved by using a certified meter to batch in discrete volumes of water at 20°C and determine the level. It is then possible to use a float connected by pulley to a gauge or a calibrated floating stick to determine the liquid level and, hence, volume in the kettle. At kettle cast, the volume of the wort is measured at 100°C and for extract calculations must be adjusted back to 20°C, the temperature for determining density (wort will contract in volume by 4% from 100 to 4°C).

Avoidance of Boil-over

Boil-over of the kettle is a serious risk to operating personnel and must be avoided at all costs. Traditionally, excessive foaming has been controlled by opening of the kettle manway doors to create a cold down draught. This suppresses the foam, and by visual inspection the operator can regulate the steam supply to prevent boil-over. The practice of open-door boiling is increasingly unacceptable on safety grounds and many brewers now use manway door interlocks to ensure that boiling cannot occur with the manway doors open. Additionally, the kettle will be fitted with a liquid detection probe located just below the manway door level that will automatically close down the steam supply if this risk of boil-over is detected. These techniques improve safety and prevent the kettle spilling over, but stopping the boil results in significant lost time.

The main technique to suppress foam is by use of wort returning from a spreader (Figure 11.15) whether using an internal or external heater. Less frequently used, but very successful, is tangential return above the wort surface from an external thermosyphon, which constantly suppresses the spinning foam (Figure 11.14).

The spreader is very effective once the boil is in full flow, but with an internal percolator this requires the kettle to be fairly full and the body of the wort to be above 96°C. Applying too much steam to the percolator before these conditions have been achieved will result in high levels of foaming, which, without an effective return spreader, will inevitably result in boil-over. Raising the wort temperature ex lauter tun to 100°C (raise to boil) is the most risky time for boil-over. There is an increasing tendency (to reduce construction costs) to eliminate the traditional internal heating coils and bottom heating jackets, which could be used to raise the wort to near to boil before applying full steam pressure to the percolator. To avoid boil-over, the steam needs to be applied carefully to the percolator until vessel contents achieve >96°C, which is likely to incur an unacceptable increase in vessel cycle time. Alternatively, the incoming wort can be

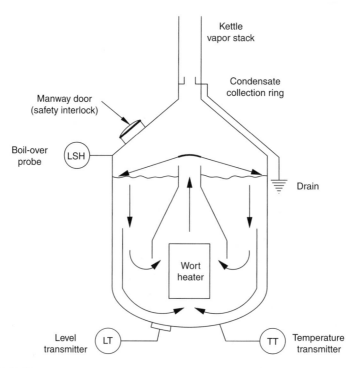

FIGURE 11.15
Wort turnover and foam suppression (internal thermosyphon heater).

preheated with a heat exchanger, so that the percolator steam can be fully applied once the vessel is about two thirds full.

Liquid Adjunct Metering

Liquid cane sugars and corn syrups are metered into the kettle either by a mass flow meter or by a volumetric flow meter. The mass or volume required is calculated from the % (w/w) solids (Brix) of the sugars and the required amount of dry extract required by the recipe.

Hops Addition and Utilization

A 650-hl brewlength using hops with 10% alpha acid and achieving a 65% wort hop utilization would require 35 kg of type 90 pellets to achieve a wort bitterness of 35 EBU. Hop pellets are almost universally purchased in 20/25-kg vacuum-sealed bags to protect the hops from oxidative deterioration and, therefore, need to remain sealed until just before use. For this reason, bulk handling of hop addition has not been developed and the most common practice is manual batch weighing using an industrial balance. The quantity indicated by the recipe needs to compensate for the batch to batch variation in alpha acid content of the hops. Past experience of optimizing hop utilization from the bittering hops and retaining the required hoppiness from aroma hops will decide the timing of different hop additions.

Direct addition of hops to the kettle through the manway door is possible, but safety interlocks can mean disrupting the boil.

Preweighed hops can be placed in small pressure vessels that can then be flushed out with wort from the kettle at the appropriate moment. Manual opening of such a "hop bomb" must be rigorously interlocked to prevent high-pressure boiling wort inadvertently spraying operational staff.

Hop pellets can also be weighed out remotely and blown to the kettle by a pneumatic conveyer, avoiding any possible human contact with boiling wort and any need to disrupt the kettle operations.

Hops are very expensive and any modification to boil procedures (boil time, evaporation rate, wort density, and wort pH) requires a cost–benefit impact on wort hop utilization. Wort hop utilization with type 90 pellets should be 60 to 70% with reference to final batch volume ex brewhouse.

Volatile Stripping

The efficacy of dimethyl sulfide precursor converting to free DMS and then flashing off during boil is a critical parameter, but analytical confirmation can only be retrospective. Experience determines the evaporation rate and boil time that will normally yield the required residual volatiles and it will certainly be technology dependent. Recent developments suggest that large surface area heaters will achieve satisfactory volatile stripping using lower evaporation rates.

A useful assessment of volatile stripping progress is to sample the product condensate from the kettle vapor stack collection ring (Figure 11.15) for flavor and aroma evaluation. The condensate at start of boil will be strong in volatiles, but at the end of boil should be virtually tasteless and odorless.

Trub Formation

At the end of boil, it is useful to carry out a qualitative assessment of how well the protein has been coagulated and how well it is likely to separate at wort clarification.

A sedimentation test at the end of boil is carried out on 1 l of wort in an insulated Imhoff cone. After 5 min, the wort should have a brilliant background clarity and the trub should have compacted to a level below 100 ml/l.

An unsatisfactory sedimentation test is likely to result in poor hot wort clarification and urgent investigation is needed to identify its cause. If boil control is correct, then the most likely cause is cloudy worts from mash separation. This will be due either to inadequate control of upstream operations or an unidentified change in malt quality.

Hot Wort Clarification and Cooling

The process objective is to separate the end of boil hot break (coagulated proteins and undissolved hop residues) to produce a brilliantly clear wort that can then be cooled and oxygenated during transfer to the fermenting vessel. The key parameters to be controlled are identified in Figure 11.16.

Hot Wort Clarification

At the end of boil, the hot break particle size (30 to 80 μm) is sufficient to separate by sedimentation. However, this settling time needs to be restricted due to the continuing formation of aldehydes and the continuing transformation of DMS precursor to DMS, which cannot be flashed off (boil terminated). The sedimentation process can be accelerated by centrifugation, but the whirlpool has become the most widely accepted hot wort clarification technology.

The wort solids at end of boil are approximately 6000 to 8000 mg/l. If this wort has a satisfactory end of boil sedimentation test, then a correctly operating whirlpool will reduce the hot wort suspended solids to <100 mg/l (<0.5 ml/l Imhoff cone).

Brewing Process Control

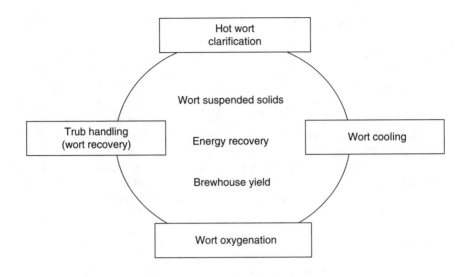

FIGURE 11.16
Wort finishing — key activities to be controlled.

The transfer of wort from the kettle to the whirlpool is critical. To minimize the shear force break up of trub flocs, the pipe-line velocity and whirlpool tangential inlet velocity should be less than 5 m/sec. In order to ensure mass momentum and adequate spin, a short transfer time of 10 min is needed. This low velocity and high mass flow are achieved with large-diameter transfer pipework and care is needed to prevent significant wort losses. Wort transfer from the upper outlet can normally commence within 15 min of finishing wort transfer.

Wort Cooling

Currently, wort cooling is achieved exclusively with a plate heat exchanger using cold brewing water as the coolant, which becomes transformed to hot brewing water, resulting in significant water and energy savings. This process will be either two-stage or single-stage.

In a two-stage process, ambient water at, say, 15°C is used to cool the wort to an intermediate temperature and the ambient cooling water becomes heated to, say, 85°C and is collected for future brewing. In the second stage, a closed loop of iced water at 1 to 2°C is used to reduce the intermediate wort temperature to the required exit temperature of about 10°C.

In single-stage cooling, ambient cold water has been prechilled to about 4°C, so that the outlet wort temperature and recovered brewing water temperature are achieved without using an ice water second stage. Figure 11.17 shows a control system for single-stage wort cooling and its "interconnectedness" with the brewhouse water system and other heat exchangers.

Wort Flow Rate

The hot wort from the whirlpool is pumped at the required speed to achieve a 1-h transfer to the fermenting vessel (FV). The wort flow rate and totalized wort volume are measured by an in-line Magflow meter, which is used to regulate the variable speed drive wort pump. The pump speed is controlled to maintain a constant flow rate, adjusting for decreasing hydrostatic head in the whirlpool and for increasing hydrostatic head in the FV. Also, toward the end of wort transfer, it is essential to ramp down the wort transfer speed to prevent the whirlpool trub cone collapsing and causing an unacceptable carry over of wort solids.

Wort Temperature

The wort temperature at the cooler outlet is measured with an in-line resistance bulb thermometer, and adjustment signals are sent to the chilled water throttle valve to achieve the required wort temperature.

As discussed in the Section "Wort Boiling," the efficiency of the plate heat exchanger will be affected by fouling. As fouling increases the heat transfer coefficient reduces, and a higher flow of chilled water will be needed to achieve the wort outlet temperature. Although the target outlet wort temperature is being achieved, the lower outlet hot water temperature becomes too low for recovery to the brewing hot water tanks. Regular CIP of the plate heat exchanger is essential to maintain the required hot water return temperature and brewhouse water balance.

With all plate heat exchangers, there is a risk of gasket failure resulting in product/coolant interchange. Careful design is required to ensure that the pressure difference between product and coolant cannot exceed the design limit (2 to 4 bar) and cause gasket/plate damage.

A risk analysis is required as to the consequences of such leakage. For instance, it is best when using a refrigerant at −5°C (chilled water

Brewing Process Control

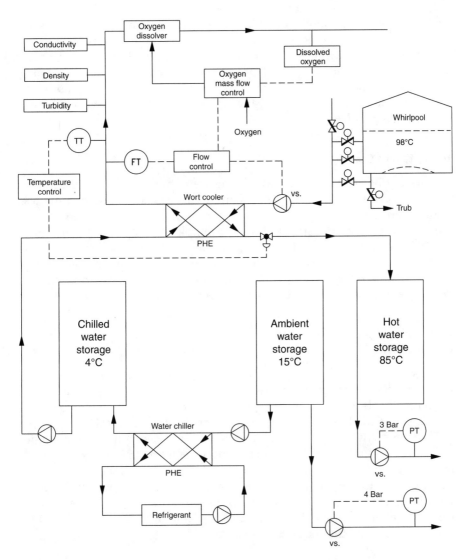

FIGURE 11.17
The plate heat exchanger (PHE) "interconnectedness."

production) to always maintain the product at a higher pressure so that any leakage is from water to the refrigerant. The wort cooler is an interesting dilemma because the "refrigerant," chilled water, is the hot water for future brews.

Other plate heat exchanger duties in the brewhouse will be for heating rather than cooling (hot water tank make up, CIP heating, wort preheating to kettle). Generally, steam will be used as the heating medium and the priority consideration will be to avoid contamination of the condensed steam returning to the boilers. However, if an energy store is in use, this closed loop of 96°C water will become the heating medium for these plate heat exchangers, and the risk of product crossover within the brewhouse is markedly increased. Regular sensory and analytical evaluation of brewing water tanks and the energy store (if applicable) is critical to identifying such a problem.

Turbidity and Cold Break

Despite the hot wort ex whirlpool being visually brilliant, it still contains wort solids of particle size 0.1 to 1.0 mm, which below 60°C become less soluble and creates wort turbidity, which with time will precipitate as cold break (particularly if kettle finings are employed). An in-line turbidity meter is an immediate indication of the level of these solids, providing early warning of an abnormally high value. This will impact negatively on the downstream activities of fermentation and beer clarification and denotes a control problem upstream or an unrecognized change in malt quality.

Conductivity

A consistent conductivity for each brew type indicates that the brewing water composition is consistent and that there has been no gross contamination with cleaning chemicals. An in-line conductivity meter ex wort cooler is an added assurance that processing has been normal.

The conductivity meter measures the ionic concentration of a liquid. Inductive conductivity measurement consists of a transmitting coil generating a magnetic alternating field to induce an electric voltage in the liquid. The ions present in the liquid enable a current to flow, which increases with increasing ionic concentration. The current in the liquid generates a magnetic alternating field in the receiving coil, which is measured and used to determine the conductivity value.

Wort Oxygenation

In order for the yeast to complete fermentation consistently, it needs to be first stimulated into a growth phase. The normal practice is to provide sufficient oxygen in the wort for the yeast to multiply three-to-six-fold at the start of fermentation. A working rule of thumb is 1 mg/l oxygen for each ° Plato to be attenuated.

The pressure of air/oxygen must be sufficient to overcome the maximum back pressure at the wort cooler outlet. The flow of air/oxygen can be manually regulated through a rotameter, but this does not automatically compensate for varying wort flow rates. More accurate air/oxygen flow control is achieved by a mass flow meter that can automatically change its control set-point on the basis of the wort oxygen level required and the varying wort flow rate.

The oxygen thermal mass flow meter works on the principle of measuring the temperature gradient across a sensor that is imparting a known energy transfer into the oxygen flow. The temperature difference across the sensor is proportional to the gas flow.

The air/oxygen will dissolve to varying degrees depending particularly on the method of injection/dispersion into the wort. More finely dispersed bubbles dissolve more efficiently and several alternative techniques are available (sintered candle, wort line venturi, in-line static mixer). In all cases, it is advisable to have an in-line dissolved oxygen (DO) meter located as near to the FV as possible to confirm the actual DO level achieved compared to the level injected at the wort cooler. This will allow for adjustment of injected levels to compensate for dissolving inefficiencies and establish whether an in-line mixer would be beneficial. The in-line DO meter reading at wort line pressure is likely to read higher than that actually staying in solution in the FV at atmospheric pressure. A pragmatic value will be established that achieves consistent fermentations with the required flavor and aroma profiles, and also provides cropped yeast with high viability for pitching future brews.

The in-line DO probe is a teflon membrane-enclosed, electrochemical cell containing a silver anode and gold cathode dipped into a potassium chloride electrolyte solution. An electronic circuit is linked to the anode and cathode, and through an applied voltage a current flows between them. Oxygen penetrates through the membrane and undergoes a reaction at the cathode causing a measurable electric current to flow that is proportional to the number of oxygen molecules per unit of time entering the cell. The permeation rate of oxygen entering the cell is limited by the solubility of oxygen in the membrane and its thickness, and it is proportional to the partial pressure of the oxygen outside the membrane.

Trub Handling and Wort Recovery

The trub recovered from the whirlpool is typically 75% wort and 25% hot break/hop residue and represents 1 to 2% of lost extract if wort recovery is not undertaken. Wort recovery can be achieved by using centrifuges and filters, but is fraught with microbiological risks. A simpler solution is to collect the trub and transfer it into successive brews at wort separation. If a lauter tun is in use, the trub can be introduced at the start of sparging. If a mash filter is in use, the trub can be mixed with the mash in the latter stages of mash transfer to the filter. This way, the extract is recovered but the trub solids are disposed of with the spent grain and significant effluent loadings are avoided.

Brewhouse Yield

An in-line density meter ex the wort cooler can be used to calculate the dry weight extract as described for wort separation (volume and temperarure already being measured to control wort flow rate and temperature). Comparison with extract added to the mash vessel and kettle allows calculation of the percentage of extract recovered from the brewhouse or the brewhouse yield.

Process Control and the Brewer

Over the last 30 years, as with other professions, automation has radically changed the skills required by the operational brewer. The high degree of automation in the modern rapid batch brewhouse has been very successful, and allows simultaneous control of several batches at different process stages by one operator. However, this reduced manpower has also reduced the processed intelligence from highly experienced personnel in direct physical contact with the raw materials, plant, and equipment. Their training, human senses, and experience could instantly recognize change and refer to the brewer for process decision-making.

Today's brewer has to interpret nonevaluated raw data from remote sensors that may be incorrect. The ubiquitous SCADA control screen may not reflect physical reality. Some drive motors and valves may not be fitted with feedback sensors, and when the SCADA screen is displaying their operational status it is in reality only displaying the control request status. It is essential to understand the limitations of the information available. Increasingly, brewing process troubleshooting requires the additional skills of the process engineer and the automation engineer. Sadly, in some breweries, this responsibility is being abrogated to those professions.

Fully automated process control is still futuristic. In the meantime, the challenge for the operational brewer remains as it always has been — to bridge the gap between brewing scientists and engineers so as to ensure world-class brewery performance.

12

Fermentation

James H. Munroe

CONTENTS

Introduction	488
Wort	489
Clarification	489
Aeration	489
Laboratory Analyses	490
Pitching Yeast	491
Microbiological Examination	491
Cell Concentration and Pitching	492
Pitching Process	492
Metabolism and Growth	493
Biochemistry of Fermentation	493
Growth during Fermentation	495
Measurement of Growth	496
Batch Fermentations	496
Lager Fermentation	497
Fermentation Vessels	497
Characteristics of Fermentation	498
Ale Fermentation	501
Laboratory Analyses during Fermentation	502
Factors Affecting Fermentation	503
Yeast Strain and Condition	503
Pitching Rate and Yeast Growth	504
Temperature	504
Oxygen	505
Zinc	505
Trub Carry-Over	505
Fermentor Geometry	505
Interrelationships	506
Related Fermentations	506
High-Gravity Fermentations	506

Accelerated Fermentations	507
High-Pressure Fermentations	508
Continuous Fermentations	509
Low-Calorie Fermentations	510
Definition	510
Production Methods	510
Nonalcoholic and Low-Alcohol Fermentations	511
Definitions	511
Major Deficiencies	511
Production Methods	512
Immobilized Yeast	513
Abnormal Fermentations	514
Symptoms	514
Causes	514
Process Variations	514
Wort Nutrient Deficiencies	515
Yeast Changes	515
Treatments	515
Beer Transfer and Yeast Separation	516
Yeast Cropping Considerations	516
Methods of Cropping	517
Centrifugation	518
Recovery of Carbon Dioxide	520
Purity and Collection Strategies	520
References	521

Introduction

The influence of fermentation conditions on beer flavor cannot be overemphasized. The central role of yeast metabolism in the production of flavor compounds has been well established. Yeast produces these compounds as by-products of the synthesis of compounds necessary for growth and metabolism. The relative concentrations of flavor compounds vary with cell growth patterns, which are dependent on processing conditions. To produce consistent beers, it is imperative to have consistent cell growth in each fermentation. Each yeast strain has its own characteristic response to processing conditions.

One purpose of this chapter is to provide information that will build an understanding of the important metabolic concepts in brewing. This will make it easier to interpret the influence that various changes in processing conditions have on yeast growth and yeast metabolism and their resultant effects on fermentation characteristics and beer quality.

Wort

Clarification

During the kettle boil, a hot break occurs. This is the formation of particulate matter called trub (from the German noun *Trub* meaning sediment or dregs). Hot trub consists of light flocs that contain protein–polyphenol complexes, lipid-rich material, insoluble hop components, and so forth. The formation of particulate matter does not stop at kettle boil, however. As the wort is cooled to reach fermentation temperature, more insoluble matter is formed. Indeed, from kettle boil through fermentation and aging to final polish filtration, precipitates are continually formed.

As described in Chapter 3, before wort is transferred to the fermentor, trub can be collected and the majority removed from wort in a hot wort settling tank or more completely by filtration. Removal of trub from wort will:

- Remove undesirable, bitter substances
- Improve physical stability
- Improve extract efficiency

Many brewers permit some trub carryover into the fermentor, because it has been shown that more vigorous fermentations occur if trub is present. It has been shown that nutrients are concentrated in trub causing the vigor. However, it also seems clear that the particles in trub provide nucleation sites for CO_2 bubble formation, thus improving fermentation vigor.[1]

Aeration

Oxygen is essential for the growth of yeast and therefore for proper fermentation. The amount of oxygen required depends on the yeast strain and its growth requirements for unsaturated fatty acids and sterols. Wort is usually aerated on its way to the fermentor by the injection of sterile air after wort cooling. Wort saturated with air will contain approximately 8 ppm oxygen, depending on the temperature, wort gravity, etc. Because air is injected under pressure, more oxygen may be dissolved in the wort on the way to the fermentor. The concentration of oxygen can be increased, if necessary, by injecting wort with pure oxygen. The length of time the yeast is in contact with oxygen (or, equivalently, the amount of oxygen supplied to the yeast) during the early stages of fermentation greatly affects yeast growth and consequently the flavor of beer. As oxygen is usually the limiting nutrient, it can be used to control yeast growth. Precise and consistent control of aeration is thus quite important for proper fermentation and consistent beer flavor.[2]

Laboratory Analyses

Several organizations publish standard methods for use in brewing laboratories. In North America, the American Society of Brewing Chemists (ASBC) publishes its *Methods of Analysis*[3] and the *Laboratory Methods for Craft Brewers*.[4] In Europe, the European Brewery Convention (EBC) publishes a book of methods, the current version of which is *Analytica-EBC*.[5] In the UK, the Institute of Brewing & Distilling publishes *Methods of Analysis*.[6] In this chapter, references are made to the ASBC *Methods of Analysis*, but in most cases an identical or similar method can be found in the others. The classic work of DeClerk[7] gives methods for many other assays not included in the books mentioned.

A quality wort is important for producing a quality beer, and wort quality should be checked in the laboratory. A wort that is out of specification is generally a result of process deviations or poor-quality raw materials and may portend a fermentation or flavor problem. Depending on any previous history of abnormal fermentations, special analyses may be performed over and above routine work. Only the more important wort analyses are discussed here.

Wort samples can be taken at kettle knockout, before or after the wort cooler, or immediately after introduction into the fermentor. Samples taken from the fermentor will contain yeast that will have already caused changes to the wort. The sample point depends on the information desired. Because worts are contaminated easily, steps should be taken to prevent spoilage.

Worts should be analyzed at least for wort solids and pH. A normal wort pH may range from 5.0 to 5.4, depending on malt/adjunct ratio, mashing conditions, water treatment, etc. Abnormal values may result from contamination by caustic cleaning solutions, contamination by spoilage organisms, improper water treatment, or brewing deviations. Abnormal wort solids could result from poorly modified malt, or other process deviations during mashing.

Total solids concentrations, or wort extract (*Methods of Analysis*: Wort-3), is usually expressed as degrees Plato (°Plato) at 20°C, the percentage of solids based on the specific gravity of sucrose solutions,[8] or as specific gravity. Degrees Balling is sometimes used as a substitute for °Plato. The two tables are not equivalent. In 1900, F. Plato improved on Balling's work and produced a more accurate table. The difference between the tables is about 0.05% w/w units. Because wort contains very little sucrose, the tables are only an approximation of wort sugar concentrations but are useful nonetheless. A very thorough history and treatment of these and other brewing measurements is found in the study by Kämpf.[9]

Another useful assay is wort attenuation limit, which is the final solids concentration after all fermentable extract is utilized. The attenuation limit can be estimated by the technique of rapid fermentation (*Methods of Analysis*: Wort-5,B). Rapid fermentation measurement is used to adjust brewhouse conditions to achieve the desired alcohol content of beer. The attenuation limit can also be used to forecast when the end of fermentation (assimilation of all fermentable carbohydrates) may be expected. The attenuation limit can be expressed as apparent extract in units of °Plato, real extract weight concentration, or specific gravity.

Additional tests can be helpful. For example, dissolved oxygen concentration can indicate whether the wort is aerated sufficiently. Nutritional deficiencies in yeast may occur during fermentation if the concentration of amino acids or zinc is too low. Abnormal protein concentrations may lead to haze or foam problems. Calcium concentration may be important to ensure low oxalate concentrations in beer. If oxalic acid is not removed as insoluble calcium oxalate in the brewhouse, later precipitation as microcrystals can cause filtration problems and appear in beer. Bitterness units (*Methods of Analysis*: Beer-23) can be checked to determine proper hopping and hop utilization. Brewhouse yields and efficiencies can be calculated from wort solids.

Pitching Yeast

Microbiological Examination

Wort is essentially sterile prior to inoculation ("pitching") with yeast. Contamination of pitching yeast with either bacteria or wild yeast must be minimized, as these contaminants will cause undesirable flavors in the beer. Fermentation may even be hindered by high levels of contaminants. Consequently, a microbiological examination of pitching yeast at regular intervals for cellular morphology[10,11] is important to ensure that yeast is in proper condition for fermentation. Samples may be plated on various media (*Methods of Analysis*: Microbiological Control) to look for common brewery bacterial contaminants, such as *Pediococcus* spp. and *Lactobacillus* spp., and wild yeasts such as *Hansenula, Dekkera, Brettanomyces, Candida*, and *Pichia* (see Chapter 16). Other *Saccharomyces* species may also be present. For a more complete discussion of contaminants and their detection, see Priest and Campbell.[12] Contaminants can be controlled to an extent through acid treatment, which is discussed in Chapter 8.

A microscopic examination may also show excess extraneous particles that may interfere with proper yeast performance or cause incorrect pitching concentrations. These particles could be diatomaceous earth, trub, grain particles, etc. Because many of them adsorb onto the yeast, extensive adsorption due to repeated reuse of recovered yeast could lead to fermentation problems.

Cell Concentration and Pitching

Yeast is held in special vessels called brinks dedicated to yeast storage and handling. Yeast brinks are generally jacketed to maintain low temperatures and have agitators to maintain a uniform yeast concentration.

The pitching rate is the amount of yeast added per unit volume of wort. The laboratory usually uses millions of cells per milliliter wort, while the brewer may refer to pounds or volume of yeast/bbl wort. In order to have consistent pitching rates, the yeast concentration in the brink must be known and then metered accurately. Usually, the brewer establishes a relationship for an easily determined quantity, such as total solids in the brink, that has previously been related to the yeast cell concentration.

Total yeast solids can be estimated by an oven-drying method (*Methods of Analysis*: Yeast-5), centrifugation, volumetric or gravimetric wet solids, or turbidity methods; but all these methods also measure solids from trub and other particles, which will result in low and variable yeast concentrations in the fermentor. A better estimate of cell concentration, although more time consuming than total solids methods, is a microscopic-counting method using a hemacytometer (*Methods of Analysis*: Yeast-4). Alternatively, the cell concentration can be estimated from particle-counting instruments.[13]

Pitching rates are ideally based upon the concentration of live yeast cells in the yeast slurry; therefore, the viability of the cell population should be estimated using methylene blue staining (*Methods of Analysis*: Yeast-3,A), by slide culture (*Methods of Analysis*: Yeast-6), or by using plate-count techniques. High dead-cell counts may be due to lengthy storage in the yeast brink, harsh treatment in the brink (pH too low or temperature too high), or lengthy delay in removal of yeast after fermentation.

Pitching Process

Consistent fermentations require consistent pitching rates. Yeast slurry can be metered into the cold, aerated wort stream on the way to the fermentor at the appropriate dilution factor. An alternative is to mix wort and yeast in a starting tank prior to transfer to the fermentor. Yeast concentration can then be measured and adjusted at this point. Another advantage is that holding pitched wort in a starting tank for up to 24 h before transfer to the fermentor allows some undesirable solids to be removed by settling or flotation.[14] Pitching fermentors that are filled with multiple brews require empirical studies to determine the optimum aeration and pitching processes to achieve the desired beer flavor profile.

Equipment based on the electrical capacitance of yeast with intact membranes is also available to meter yeast into wort. This methodology is generally not affected by trub particles or dead cells.

A cell concentration of about 10 million live cells per milliliter in 12°Plato wort is normal and is usually increased in proportion to an increase in wort gravity.[15] A range of 10–25 million cells per milliliter for the pitching rate can be used. The actual pitching rate selected depends on wort gravity, fermentation temperature, yeast strain, proper attenuation, fermentation characteristics, and so forth.

Metabolism and Growth

Aspects of yeast metabolism are included here to explain certain growth and fermentation phenomena that are discussed in this chapter. The importance of controlling yeast growth for producing consistent beers is essential.[16] A more detailed review of yeast metabolism can be found in Chapter 8.

Metabolism refers to the complex biochemical reactions in a living cell. A yeast cell, like other living organisms, must assimilate nutrients from its environment and convert them into energy and molecular building blocks for growth. During this process, the cell excretes by-products of its metabolism. An understanding of the influence of yeast metabolism and growth on beer flavor will aid the brewer in making decisions about process conditions and their influence on beer quality.

Yeast requires carbon and nitrogen sources, vitamins, minerals, and certain growth factors. Brewer's wort supplies fermentable sugars as the carbon source. Amino acids, collectively referred to as free amino nitrogen (FAN), are the principal nitrogen source in wort. However, nitrogen deficiencies may occur if large percentages of adjunct are used, diluting the amino acids derived from malt. The yeast's requirements for specific nutrients depend on the strain. Normal brewer's wort usually supplies sufficient FAN, vitamins, minerals, and other growth factors for adequate yeast growth during fermentation.

Biochemistry of Fermentation

The main biochemical route for a yeast cell to produce energy from sugars in the absence of oxygen is called alcoholic fermentation. The sequence of enzymic reactions that produce pyruvate from glucose is referred to as the Embden–Meyerhof pathway. In the presence of sufficient oxygen, the cell completely oxidizes the resultant pyruvate to CO_2 and water while producing energy for other metabolic processes. In the absence of oxygen, as

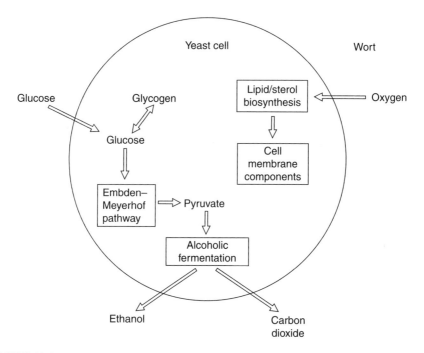

FIGURE 12.1
Beer fermentation. Schematic of a yeast cell in wort assimilating glucose to produce fermentation by-products ethanol and CO_2; oxygen is assimilated to produce membrane components.

with beer fermentations (Figure 12.1), for example, the pyruvate is converted into ethanol and CO_2 primarily by way of acetaldehyde. A summary of biochemical equations expressing these reactions is:

$$\text{Glucose} \xrightarrow{\text{Embden–Meyerhof pathway}} 2 \text{ Pyruvate}$$

$$2 \text{ Pyruvate} \xrightarrow{\text{Alcoholic fermentation}} 2 \text{ Ethanol} + 2 \text{ Carbon dioxide}$$

Fermentable carbohydrates are usually assimilated by brewer's yeast in the following order: fructose, glucose, maltose, and maltotriose. Sucrose is split into fructose and glucose by the yeast-produced enzyme, invertase, and is not assimilated by yeast as sucrose.

Amino acids are assimilated to build proteins within the cell. Proteins are needed for the production of new cells and the synthesis of enzymes for biochemical reactions. Ammonium ions are also used for this purpose, but amino acids predominate in wort. As with sugars, amino acids are assimilated in a specific order which can depend on yeast strain; they fall into four groups distinguished by the order and extent of assimilation.[17]

Besides producing ethanol and CO_2 as its main by-products of metabolism, brewer's yeast produces a large number of other compounds that contribute to the sensory properties of beer. These flavor and aroma compounds can be related to specific metabolic pathways that operate during yeast growth. The production of fusel alcohols has been linked to amino acid biosynthesis. Therefore, the balance of fusel alcohols depends on the balance of amino acids. Volatile esters are formed from an alcohol and a fatty acid and appear in fermenting beer after the appearance of the fusel alcohols. The concentrations of volatile esters, fusel alcohols, and other sensory compounds depend upon wort composition and yeast growth. Chapter 8 offers more details on these compounds and their formation.

An important component of yeast that affects fermentation performance is glycogen. Glycogen is a polymer of glucose produced by the yeast cell as a means for storing energy. Under appropriate conditions, glycogen within the cell is broken down to glucose, which yields biologically useful energy. As discussed in the following text, the cell uses glycogen to initiate certain metabolic functions before assimilation of wort glucose can begin.[18]

Growth during Fermentation

Relating metabolism to yeast growth during fermentation will illustrate the importance of consistent growth in each fermentation. Yeast growth in normal wort is limited by oxygen. Therefore, flavor and other attributes in beer can be altered by using oxygen to control the amount of yeast growth.

In a batch fermentation system, the growth of yeast can be divided roughly into stages or phases: lag phase, logarithmic growth phase, and stationary phase.

The lag phase is the time lag, usually 12–24 h, between inoculation of the wort and the appearance of fermentative activity. However, during the lag phase, important yeast activity occurs. The yeast readjusts to a new environment rich in nutrients. The cells synthesize enzymes needed to utilize nutrients and support growth. Of particular importance is assimilation of oxygen. Molecular oxygen is used by yeast to produce unsaturated fatty acids and sterols,[19] which are essential to cell membrane synthesis. Without sufficient oxygen, yeast growth is restricted and causes abnormal fermentation and flavor changes in the beer. The oxygen is consumed rapidly, usually within 6–10 h. Because wort sugars are not being assimilated early in the lag phase, glycogen is essential as an energy source for cell activity (Figure 12.1).

After the lag phase, there is a short transition into the logarithmic or exponential growth phase. During the logarithmic phase, the cell population increases at a logarithmic rate because of the excess substrate. In practice, strict logarithmic growth is observed for only a short period of time because of the brief period in which sugar is in excess. Cell proliferation occurs by cell budding, in contrast with cell division in other eukaryotes. The individual cells increase in mass and volume to a certain size at which time new buds are formed. The daughter cells (buds) increase in size until, at a critical size, they break from the mother cell, leaving a "bud scar" on the surface of the mother cell.

In the logarithmic growth phase, glucose is in excess and the yeast growth rate reaches a maximum. The maximum fermentation rate, which depends on cell population but not on the cell growth rate, follows closely in time behind the growth maximum.[20] As the cell population increases and the pools of unsaturated fatty acids and sterols are distributed among the increasing cell population, the growth rate slows. By this time, the cells have begun to assimilate maltose. The cells gradually shift into the stationary phase as the sugar concentration falls. The cells begin to settle or flocculate while the last of the sugars are assimilated at a slow rate. During this period, the glycogen concentration has increased and the cells are preparing for an environment nearly devoid of assimilable carbohydrates.[21] The yeast dry-weight total increases throughout fermentation before decreasing slightly at the end of fermentation.[16]

Measurement of Growth

The term yeast growth is ambiguous because it can refer to either the cell population (number of cells) or to the total weight of cells (biomass). Estimation of cell population (cell counts) will vary depending on whether or not one distinguishes between live cells and dead cells or counts buds as cells, etc. The term "cells in suspension" usually is different from total cell population because of yeast settling during fermentation. When growth data are reported, a precise designation of "yeast growth" is always needed. Yeast growth as used in this chapter refers to the increase in cell population.

Cell-counting techniques include hemacytometer counting (*Methods of Analysis*: Yeast-4) and yeast particle counting by instruments configured for that purpose.[13] Microscopic counting is tedious and subject to between-analyst variability. Particle counters cannot distinguish single cells from those with buds or from particulate matter, although most particulate matter is generally smaller in size than yeast and can be excluded electronically by the instrument. Measurement of yeast mass is usually determined by dry weight of a known amount of sample. Almost all methods for biomass estimation include particulate matter in the measurement as a source of error.

Batch Fermentations

Lagers and ales will be discussed separately because of some fundamental differences between them. The emphasis here will be on lagers because they are more prevalent not only in North America, but also in many other parts of the world. Ales still enjoy popularity in Canada and especially in the UK However, lager sales there have increased significantly in recent years.

Because brewing is as much an art as a science, there are many variations in batch-process conditions for beer fermentations practised by different brewers. The term "processing conditions" refers to the physical conditions under which the fermentation is conducted: pitching rate, temperature, pressure, agitation, dissolved oxygen, tank geometry, wort volume, etc. All of these can affect fermentation — simply and interactively. The purpose of this section is to provide a general explanation of events during fermentation, rather than a description of all processing variations. With the general principles in mind, logical choices among the variations can be made.

Lager Fermentation

The production of lagers utilizes a bottom-fermenting strain of yeast. The yeast is called bottom fermenting because the yeast forms flocs or clumps, which being denser than beer, tend to settle to the bottom of the tank towards the end of fermentation. The term flocculation is used to describe the settling of yeast, but yeast may not form true flocs. Brewing strains are selected for the flavor characteristics they impart to beers and for other desirable fermentation characteristics. Some of these characteristics are suitable flocculation properties, genetic stability, suitable attenuation of wort carbohydrates (particularly maltotriose), acceptable foam characteristics in the beer produced, and interaction with fining materials for effective filtration.[22] To an important degree, flavor and aroma qualities in the final product are determined by the yeast strain used.[23,24] Some strains, for example, can produce less diacetyl or fewer sulfur compounds without altering the balance of desired esters and higher alcohols.

Fermentation Vessels

Older fermentors were generally open, relatively shallow tanks. A fermentation cellar often had several floors with 6–12 tanks per floor. This traditional style made CO_2 collection difficult, was prone to contamination, and created dangerous working conditions because of the high CO_2 levels in the air space.

Open tanks were supplanted by closed horizontal and vertical tanks, of which there are many designs[25–27]: the Uni-Tank, Asahi Tanks, and spheroconical tanks. The use of closed vessels gradually gained acceptance through the efforts of Nathan,[26] who held patents from the early 1900s for closed vertical tanks with conical bottoms (cylindroconical tanks). The tanks could be used for fermentation and aging because the yeast can be removed from the cone, leaving the beer undisturbed in the tank. The advantages he claimed for using cylindroconical tanks are still valid today: yeast is kept in suspension better by natural agitation caused by rising CO_2 bubbles resulting in faster fermentations, cooling is more easily controlled, CO_2 collection is facilitated, and removal of yeast from the conical base is more readily done. Clean-in-place systems make tank cleaning easier and eliminate manual cleaning. The tanks are designed to handle modest counterpressure, if required.

Normal tank volumes in larger breweries vary over a wide range, from 500 to 6000 bbl (600–7000 hl). Vertical vessels require from 15 to 25% of tank volume as headspace to allow for foaming during fermentation. The amount of foam generated depends mainly on those process conditions that influence fermentation rate, and on factors that stabilize foam. The concentration of hop constituents in wort is a major factor in the stability of foam during fermentation.

Tanks are filled from the bottom, and several brews are required to fill larger tanks. Because large cylindroconical tanks may take up to 12 h or longer and many brews to fill, aeration and pitching practices will be different from those used with smaller tanks. For example, yeast and air can be metered into all or a portion of the brews needed to fill the tank. When compared to filling smaller tanks, these filling practices may produce different yeast growth patterns in early fermentation leading to differences in the final product. The amount of oxygen available to the yeast will determine yeast growth. Suitable filling procedures are usually established empirically from many possible variations to give the desired flavor. However, consistency of procedures is essential for batch to batch flavor consistency.

Characteristics of Fermentation

Traditional lager fermentations (Figure 12.2) are conducted at temperatures ranging from about 7 to 14°C. The fermentation temperature in a fermentor is usually allowed to rise spontaneously to a few degrees above the pitching temperature. Because of the quantity of heat generated during fermentation (about 140 kcal/kg extract fermented),[20] cooling is needed to maintain a maximum temperature if the brewer wishes to limit the temperature rise. The duration of fermentation traditionally ranged from 8 to 20 days, but with the higher temperatures often used in modern practice, fermentation time may be reduced to about 7 days.

During the first 6–10 h of fermentation, the yeast consumes all of the dissolved oxygen. There is no detectable uptake of glucose during this time. At about 8–16 h, the first signs of active fermentation appear when CO_2 bubbles are formed and a thin foam or head is apparent.

The budding of cells can be observed within 24 h. The temperature, if uncontrolled, begins to rise due to heat generated by fermentation. Within 24–48 h, the rates of yeast growth and carbohydrate assimilation reach

Fermentation

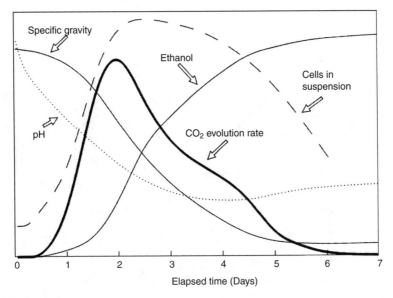

FIGURE 12.2
Fermentation response profiles. Graphical representation showing the relative changes taking place during fermentation.

their maxima. The rate of carbohydrate assimilation and the progress of fermentation can be measured by the CO_2 evolution rate.[16]

The pH falls[28] as organic acids are produced and buffering compounds (basic amino acids and primary phosphates) are consumed. The minimum pH attained during fermentation is a function of three factors: wort pH, wort buffering capacity, and the amount of yeast growth during fermentation. Lower beer pH is associated with a lower wort pH, lower wort buffering capacity, and increased yeast growth.[29] The pH reaches a minimum of 3.8–4.4 before rising slightly toward the end of fermentation. The lowered pH inhibits bacterial spoilage during fermentation.

High kräusen, that is, maximum foam head, occurs at the time of the maximum rate of CO_2 and heat generation. The maximum fermentation rate (activity) also corresponds to the maximum decline in specific gravity. Because foam is stabilized by proteinaceous compounds and isohumulones from hops, a "kräusen ring" of precipitated foam components adheres to the fermentor walls above the liquid as the foam head subsides toward the end of fermentation.

Glucose and fructose are consumed first (Figure 12.3) and as the glucose concentration diminishes, the enzyme systems required for assimilating maltose are synthesized and the yeast begins to utilize maltose and maltotriose. The production of ethanol and other fusel alcohols generally follows the consumption of carbohydrates. There is usually a delay of about 1 day

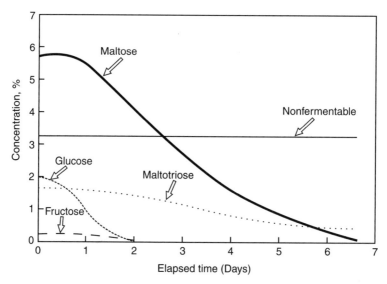

FIGURE 12.3
Carbohydrate assimilation profiles. Graphical representation of fermentable carbohydrates during fermentation. Note that maltose and maltotriose generally are not assimilated in appreciable quantities until most of the glucose is assimilated.

after the beginning of fusel alcohol production before the production of esters is observed. The fermentation rate begins to slow, the kräusen begins to fall, and heat generation diminishes.

During growth, yeast produces two α-acetohydroxy acids, which are excreted into the wort and converted into diacetyl and 2,3-pentanedione. Diacetyl and 2,3-pentanedione are collectively referred to as vicinal diketones or VDKs (see Chapter 13) and have undesirable flavor effects at high concentrations. Yeast assimilates VDKs toward the end of fermentation. When fermentation ceases for lack of fermentable carbohydrates, the amount of VDK precursors may still present a potential flavor defect. It is, therefore, important to allow enough time for the total of the VDKs and their precursors to be reduced below their flavor threshold or to an acceptable concentration before complete removal of yeast. Because the reduction of VDKs has traditionally been treated as a maturation problem, further discussion is left for the next chapter.

When fermentation reaches the attenuation limit and the level of VDKs and their precursors are low enough, the yeast is separated from the beer, which is then transferred to an aging tank. Yeast can be separated by decanting the beer or more effectively by centrifuging or filtering it. Cooling the fermentor before transfer will help the yeast to settle and aid the transfer process by reducing yeast carryover. If ruh storage is desired, some yeast is intentionally carried into aging. In a Uni-Tank operation, the yeast is removed from the bottom of the tank, while the beer remains behind for the aging step.

If a secondary fermentation is desired, the transfer from the fermentor will be done with some fermentable extract still remaining (1–4°Plato). More yeast carryover is permitted so that the cell concentration is $1-4 \times 10^6$ cells per milliliter. Discussion of secondary fermentation and kräusening is left for the next chapter.

Ale Fermentation

In terms of yeast biochemistry during fermentation, there are few differences between ales and lagers. Yet, major differences occur in the processing conditions, traditional fermentation vessels, and methods of yeast recovery. Modern practices, however, are lessening the distinctions of vessel use and yeast recovery.

Ale fermentations traditionally use the top-cropping yeast strain *Saccharomyces cerevisiae*. This strain of yeast rises to the top of the beer toward the end of fermentation because the yeast flocs entrap CO_2, making them buoyant. However, with deep cylindroconical vessels, brewers may use a bottom-cropping *S. cerevisiae* strain to make ale.

The fermentor may be of the open type, which allows easy removal of the yeast crop by skimming or suction. The many fermentors developed in the UK, such as Burton Unions and Yorkshire Squares, were designed with yeast skimming as an important consideration. Closed vessels have become more common, and suction devices are designed to be used in closed tanks. An alternative is to drop the ale out of the bottom of the tank, leaving the yeast crop behind.

Ale wort is pitched at a higher temperature (about 15°C) than in lager fermentations and the temperature is allowed to rise to about 20°C or higher. Early in fermentation while the fermentation is active (24–36 h), the fermenting beer may be transferred to another vessel. This helps mix and aerate the wort. By pump circulation, the mixture may be "roused" instead. The higher temperature and extra aeration causes ale fermentations to be completed in much less time than lagers, as little as 3 days. Secondary fermentation follows a course similar to that used with lagers.

As additional foam head is formed during fermentation, yeast crop removal may be carried out in steps. The yeast retained for repitching must be carefully selected from these separate skimming operations because the yeast collected at different times may have different attenuation characteristics. For example, the yeast crop formed initially may have less ability to fully attenuate the wort.

Short fermentation time, high temperature, and different yeast strain all contribute to a product that has a different balance of various flavor compounds such as fusel alcohols and esters. These give ales a distinctly different flavor from lagers.

Laboratory Analyses during Fermentation

Several laboratory analyses are necessary to follow the progress of fermentation, determine its completion, and monitor it for abnormalities. The simplest measurement, and also one of the most useful, is the specific gravity. After removal of yeast, specific gravity measurement can be estimated, with a hydrometer (*Methods of Analysis*: Wort-4) or more accurately by measurement of mass and volume (*Methods of Analysis*: Wort-2). As sugars are consumed and alcohol produced, the specific gravity falls. One indication of the end of carbohydrate fermentation is that the specific gravity stops declining. In contrast to wort, in which the specific gravity corresponds directly to the solids concentration in wort, the specific gravity does not quantify the true solids (real extract) remaining in beer. Alcohol is less dense than water so the specific gravity is only an indication of the "apparent extract" left in the wort. The apparent extract is always lower than the real extract but is still a useful indicator of fermentation progress and can be used as a beer specification. "Present gravity" is sometimes used to describe the gravity during fermentation. Real extract, which expresses true solids concentration in beer, can be determined from dealcoholized beer (*Methods of Analysis*: Beer-5), or from specific gravity and refractive index (*Methods of Analysis*: Beer-4,C). Both real and apparent extracts are expressed as °Plato. Modern instruments are available to easily estimate the various parameters described here.

A measure of how much carbohydrate was fermented is the real degree of fermentation (RDF), which measures the percent of extract that was fermented (*Methods of Analysis*: Beer-6,B):

$$\text{RDF, \%} = \frac{(O-E)}{O} \times \frac{1}{1-(0.005161 \times E)} \times 100$$

where O is the wort original gravity (°Plato) and E is the real extract (°Plato). The second term accounts for the weight loss due to the increase in yeast mass and loss of CO_2. The second term permits the RDF to be used with high-gravity brewing[30] because it is not affected by dilution. An RDF of 60–70% is normal for lager beers.

Cell counts will indicate abnormal growth patterns during fermentation that may result from improper process conditions, wort composition, or pitching yeast. Dead cells can be determined by staining with methylene blue and counting with a hemacytometer (*Methods of Analysis*: Yeast-3A). The method suffers from subjectivity due to variation in the amount of stain the dead cells absorb. Analysts should standardize their techniques to reduce between-analyst variability. Checks for dead cells can be made during fermentation, at the end of fermentation, or in the yeast brinks after collection. Dead cells tend to remain in suspension and, on a percentage basis, may appear to be increasing toward the end of fermentation when in fact they are not.

More extensive laboratory analyses are possible, which help the brewer diagnose aberrant conditions and maintain quality products. For example, SO_2, alcohol, and VDKs (*Methods of Analysis*: Beer-21, Beer-4, Beer-25, respectively) can be monitored. Estimations of diacetyl and 2,3-pentanedione are commonly used by brewers as a specification before transferring beer or releasing beer from aging. See Chapter 13 for a more complete discussion of VDKs.

Factors Affecting Fermentation

It was emphasized earlier that part of the flavor profile of beer is based on yeast metabolism during fermentation. Consequently, wort composition and process conditions affecting the fermentation performance of yeast will also affect beer quality. These factors interrelate so closely, however, that it is often difficult to clearly single out the influence that any one factor exerts on fermentation or product quality. Attempting to alter one process parameter to influence an outcome almost always causes other, perhaps unwanted, outcomes.

Yeast Strain and Condition

The yeast strain itself is a major contributor to the flavor character of beer and many suitable strains are available to the brewer, while some strains of yeast are unacceptable for brewing because of the poor balance of flavor compounds produced. Each yeast strain may perform differently under a given set of fermentation conditions and the brewer must consider the flavor profile produced by a strain (sulfur compounds, esters, fusel alcohols, etc.). The choice of yeast strain depends on characteristics considered important: the relative importance of oxygen requirements, cropping methods, attenuation limits, fermentation rate, etc.

Yeast handling must be carefully monitored because the condition of the yeast at the time of pitching has been shown to influence fermentation.[31] Yeast stored under extreme conditions may not respond normally, having reduced its glycogen in response to stress. Poor aeration, zinc deficiencies, or residuals from previous fermentations may cause an excessive lag phase or incomplete attenuation.

The flocculation characteristics of a strain are important. The ideal yeast would settle rapidly as the wort reaches limit attenuation. A slowly flocculating strain would have too much yeast in suspension at the end of fermentation and make beer separation more difficult. A rapidly flocculating strain may not ferment fully. On the other hand, if a secondary fermentation is intended, then a more slowly flocculating strain, leaving a greater number of cells in suspension at transfer, would be desirable.

Pitching Rate and Yeast Growth

The pitching rate has an influence on the fermentation rate. The fermentation rate depends on the temperature (see following text) and the cell population. More cells utilizing sugars at a constant rate result in an increased overall rate of sugar utilization. Higher pitching rates give shorter lag phases, higher maximum fermentation rates, and shorter times to complete carbohydrate fermentation. A pitching rate of about $10-15 \times 10^6$ cells per milliliter for a normal gravity lager wort (about 12°Plato) is typical. The precise concentration chosen depends on the yeast strain, fermentation characteristics desired, etc.

Because growth of new cells is limited by the oxygen supply in the wort, the increase in cell population (net cell growth) is nearly independent of initial yeast concentration under fixed, dissolved oxygen conditions. Compared to low pitching rates, higher pitching rates will lead to a lower number of cell doublings. At some point, yeast growth does depend on pitching rate. As the pitching rate becomes too low, the amount of oxygen utilized to make sterols for new cells is lower and overall growth is lowered.[15] Insufficient growth will lead to slower and perhaps stuck fermentations. Poor yeast growth can also lead to high SO_2 concentrations.

Variations in cell growth and therefore variations in fermentation rates, at fixed pitching rates and dissolved oxygen, are possible if other yeast-growth factors in wort vary, such as sterols or zinc. Different yeast-growth patterns will influence the flavor of the beer by changing the proportions of volatile flavor compounds. For this reason, consistent yeast growth in each fermentation should be a primary objective.

Temperature

To minimize the consequences early in fermentation from process variations, fermentation temperature control is necessary for consistent yeast growth. Fermentation rates will increase with temperature by increasing the rate of yeast metabolism giving higher specific fermentation rates. Higher temperatures will give a faster conversion of VDK precursors and shorten the time required to reduce potential VDK off-flavors. Faster fermentation rates raise the peak demand for cooling.[20] If a maximum fermentation temperature setpoint is specified, then the excess heat must be carried away by cooling the fermentor. The quantitative influence of a temperature change will be different for each biochemical reaction, changing the balance of flavor compounds.[2]

Oxygen

The dissolved oxygen concentration in wort will influence yeast activity in fermentation because yeast requires oxygen to produce essential compounds for new yeast cells, as discussed earlier. In normal wort, oxygen is the limiting nutrient of yeast growth. Usually, saturation of wort with air provides enough oxygen for proper yeast growth. However, consistent control over the dissolved oxygen in wort is essential to uniform growth in each fermentation and consistent production of flavor compounds.[2]

Zinc

Zinc, as Zn(II) ion, is required as a cofactor in enzymic reactions within the cell and therefore is a requirement for growth. The addition of zinc has been clearly linked to increased fermentation rates.[32,33] Certain yeast strains may require more zinc than occurs naturally in wort and a supplemental zinc salt may be needed. A zinc concentration of 0.3 ppm should be adequate in most cases. Yeast will utilize nearly all zinc present in wort. It was shown that concentrations of zinc above 1 ppm can inhibit yeast activity.[33]

Trub Carry-Over

As discussed earlier, trub in the fermentor can influence the fermentation rate and filtration performance. Because of the amorphous nature of trub and settling characteristics in the hot wort tank, however, it may be difficult to accurately control the quality and quantity of trub carried into the fermentor. Trub contains a mixture of substances that can be beneficial to yeast performance: zinc, lipids and sterols, and other minerals and nutrients essential for yeast growth. It also provides nucleation sites for bubble formation, increasing the fermentation rate by reducing dissolved CO_2 and encouraging yeast growth, and increasing turbulence from rising bubbles.[1]

Fermentor Geometry

Vessel geometry plays an important role in fermentation. The depth of the tank is important because the hydrostatic head affects CO_2 bubble formation — hence dissolved CO_2 concentration and yeast growth — and mixing. Yeast growth and flavor production have been shown to be reduced under higher dissolved CO_2 concentrations,[34] which would result from greater hydrostatic heads. The differences in fermentation characteristics between horizontal and vertical tanks are a result of natural agitation from convection currents and rising CO_2 bubbles, and a dissolved CO_2 difference. As the height to diameter ratio increases, so does natural agitation and the fermentation rate. Generally, horizontal tanks give slower fermentations and produce more estery beers than vertical tanks.

Yeast-collection procedures for horizontal and vertical tanks differ significantly. Compared to vertical tanks, horizontal tanks generally require manual removal of yeast.

Interrelationships

All of the foregoing factors are interrelated,[35] and a component of the art of brewing is to find the proper combination of process conditions that produce the optimum product. Changing one process can sometimes be compensated for with a change in another. However, in most cases the beer flavor profile will be affected, as the main impact of a change is on yeast growth. For example, additional unsaturated fatty acids and sterols in trub may lessen the requirement for oxygen. A lowered pitch rate may be compensated by a higher temperature in terms of time to limit attenuation.

The requirements of a particular yeast strain for specific nutrients (wort composition) and sensitivity to different environments during processing will dictate process conditions so that the strain performs well.

Related Fermentations

High-Gravity Fermentations

The practice of high-gravity brewing[10] has been well established. Instead of a wort at about 12°Plato, worts up to 18°Plato and perhaps higher are fermented. In principle, brewhouse operations are not affected greatly, although equipment must handle larger grain bills. The fermented beer is subsequently diluted with carbonated water to a prescribed original gravity or to a prescribed alcohol concentration; the water being added as late as practicable in the process.

There are a number of important advantages to high-gravity brewing:

- More consistent beers (uniform alcohol, original gravity, etc.) can be produced because adjustments by dilution can be made more easily at later stages in processing.
- Beers have better physical stability because compounds responsible for haze are more easily precipitated at higher concentrations.

- Better utilization of equipment is achieved because more concentrated worts and beers are processed before dilution. This results in greater finished beer volumes from equivalent wort volumes, thereby increasing plant capacity and brewing flexibility while decreasing unit costs.
- Lower energy costs

There are also some disadvantages:

- Longer fermentation times when compared to normal gravity fermentations under the same process conditions, which is usually addressed by increasing the pitching rate
- Change in flavor characteristics from normal gravity beers, primarily disproportionate increases in volatile esters due to changes in yeast growth patterns
- Reduced foam stability
- Reduced efficiency, as any losses of concentrated materials are more expensive
- Poorer kettle hop utilization results because higher hop concentrations in the fermentor results in greater losses by precipitation in the kräusen ring and adsorption onto the yeast
- Additional equipment for preparation of dilution water
- Yeast recovery systems will receive a greater load because more yeast is grown overall
- Yeast mortality may be greater requiring close monitoring in the yeast room

Brewers who have adopted high-gravity brewing have addressed each disadvantage during the conversion from previous practices, and are successfully using this process.

Accelerated Fermentations

Because of the economic advantage inherent in producing more beer with the same equipment in less time, there have been and continue to be many strategies for accelerating fermentations. Care must be taken because flavor changes will occur when fermentations are accelerated.

Fermentation rates can be accelerated in several ways.[27,36] The lag phase can be shortened by increasing the pitching rate, increasing the initial fermentation temperature, or by blending actively fermenting beer with fresh, aerated wort.[37] A higher temperature will increase the metabolic rate of yeast and speed fermentation.[38] By raising the temperature the profile of aroma compounds may change unfavorably; however, this problem can be offset somewhat by pressure.[34,35] Increased pitching and increased oxygen concentration in wort will also decrease fermentation time. Dissolved oxygen concentration can be increased by injecting oxygen rather than air. Fermentation times can be shortened by keeping yeast in suspension longer. This can be accomplished by using powdery strains of yeast or by mechanically stirring the fermenting beer. The yeast can also be roused toward the end of fermentation by sparging with CO_2 or using a recirculation device.

Flavor changes are to be expected because accelerated fermentation will change yeast growth patterns, which have a major influence on flavor and aroma compounds such as fusel alcohols, esters, and VDKs. The concentration of VDKs are of particular concern. They are related to yeast growth (and other factors as well) and accelerated fermentation may result in higher VDKs than traditional fermentation due to greater precursor production or slower conversion to diacetyl. In any case, the utilization of carbohydrates is completed before the VDKs are reduced to an acceptable level.[16] Therefore, even though the beer may have reached limit attenuation, proper maturation is still required.

High-Pressure Fermentations

Typically only light counterpressure, about 0.5 psi or less, is applied to fermentation vessels. The use of CO_2 overpressure of up to two atmospheres has an important influence on fermentation and beer characteristics. Brewers have used higher temperatures to increase fermentation rate, decreasing overall time, but they have found that excessive volatiles were formed.[39] The use of CO_2 pressure was found to retard yeast growth and volatile formation.[34] Therefore, pressure came to be used to counteract the effects of higher temperatures. It has been shown for many fermentation and beer characteristics that increasing temperature and increasing pressure have opposite effects: fermentation rate, specific fermentation rates, yeast growth, yeast flocculation, pH (minimum during fermentation and beer pH), and so forth.[40] Pressure can be increased in stages during fermentation.

Pressure appears to have an insignificant effect on the maximum concentration of VDK precursors or their conversion rate to VDKs. However, the longer fermentation time under pressure allows more time for the conversion of precursors and subsequent removal of VDKs by yeast.[39]

In order to obtain sufficient yeast growth, pressure is not usually applied at the beginning of fermentation. The high hydrostatic head in tall tanks has an effect on fermentation similar to the use of pressure, although to a lesser extent.

Continuous Fermentations

Continuous fermentations[36,41] have been studied for nearly 100 years with the first patents appearing around the turn of the 20th century. During this time, many different systems have been developed, and research in this area considerably advanced brewing knowledge.

Because the process of continuous fermentation is radically different from conventional fermentation, there are specific requirements to consider,[42] of which the following requirements must be met in designing a workable system:

- A highly flocculent yeast strain capable of producing a satisfactory beer
- Careful control of oxygen
- Correct yeast concentration and level of growth to achieve the proper flavor balance and maintain the culture in an active state
- Steady state conditions with sufficient mixing and a suitable fermentation gradient to achieve consistent product and optimal efficiency
- Beer of consistent flavor must be produced over a range of flowrates to compensate for demand

The advantages of continuous fermentation are greater efficiency in utilization of carbohydrates, better utilization of equipment, a smaller yeast crop for disposal, and better hop utilization as there is less adsorption onto the yeast cells.

There are many disadvantages, however. There is always the danger of microbial contamination arising from the nature of the process. System shutdowns necessitated by contamination are costly. Wort production and beer unit processes (filtration, stabilization, etc.) may not be continuous, and extra tanks are then required for holding wort and beer, which again creates an environment for possible contamination. Additional equipment such as a wort pasteurizer may be required. Running a continuous process does not permit the brewer to easily respond to fluctuations in demand for different products. The beer produced from continuous fermentations will have a different flavor from batch fermented beer. However, acceptable beers from continuous fermentations are being produced and sold commercially.

A system for continuous fermentation was developed using a series or cascade of vessels.[43] In New Zealand, a similar system was integrated with a continuous brewing process.[44] In the Coutts system, beer and recycled yeast are mixed with aerated wort. After yeast growth occurs, the fermenting beer is passed to succeeding vessels. Residence time in the fermentation system is about 30 h. Stirred vessels have also been used.[45]

Low-Calorie Fermentations

Definition

In the U.S., most low-calorie beers contain about 100 cal although superpremium low-calorie beers may be higher. Because the caloric content of beer is largely made up of alcohol and nonfermentable carbohydrates, low-calorie beers are produced by reducing the concentration of both of these components. The original gravity of low-calorie beers, 6–10°Plato, is lower than regular lagers. The alcohol content is also lower: Low-calorie beers contain 2.3–3.4% by weight alcohol compared to lagers with about 3.5–4.0% by weight. The low original gravity and real extract contribute to the sensory sensation of thinness. Low-calorie beers are also called low-carbohydrate beers.

On a equal weight basis, alcohol has more calories than carbohydrates (real extract). The formula for calculating calories in beer demonstrates this relationship (*Methods of Analysis*: Beer-33):

$$\text{Calories (kcal/100 g)} = 6.9\,(A) + 4(E - \text{Ash})$$

where A is the weight percent alcohol, E is the real extract, and Ash is the percent ash (*Methods of Analysis*: Beer-14).

Production Methods

A reduction in calories from residual nonfermentable carbohydrate can be accomplished in a number of ways including:

- Simply diluting regular beer with water
- Extending mash conversion times
- Using glucose as an adjunct rather than rice, corn, or corn syrup that contain nonfermentable carbohydrates; however, there is a danger of glucose repression or fructose block (see Section titled "Abnormal Fermentations")
- Using malt enzymes from either a malt mash or other aqueous extraction process added to fermentation to increase breakdown of dextrins[46]
- Using debranching enzymes to hydrolyze limit dextrins during mashing or fermentation

The advantages of using enzymes are higher alcohol concentrations and lower caloric contents than with the other methods: real degree of fermentation can be as high as 87% with enzymes as compared to about 75% without.

A debranching enzyme can be added to the mash or fermentor to cleave α-1,6-glycosidic linkages in limit dextrins, thus forming fermentable sugars from normally nonfermentable dextrins.[46,47] The yeast utilizes these sugars, and the caloric content from residual carbohydrate is reduced. Enzyme systems are glucoamylase from *Aspergillus niger*, or pullulanase from malted rice or of microbial origin. These enzymes hydrolyze the α-1,6-linkages inefficiently because the temperatures and pH values during fermentation are not those optimal for the enzymes. Sufficient levels of enzyme are used such that limit dextrins are degraded within the time required. Nonetheless, longer fermentation times may still be required to attenuate the wort to a specified level.

If enzymes are added to the fermentor, the enzymic action releases glucose and makes it available to the yeast throughout fermentation, in contrast to conventional fermentations. This alteration in the process can produce a product with a different flavor. Longer fermentation times may produce off-flavors from yeast autolysis due to prolonged starvation of some cells.

If not completely inactivated, these enzymes remain in the beer and continue to produce glucose, thereby sweetening the beer. Accidental mixing of products is another potential problem associated with this process. If enzyme-treated beer becomes mixed with conventional beer, with its greater concentration of limit dextrins, glucose will be produced to a noticeable extent.[48]

Nonalcoholic and Low-Alcohol Fermentations

Definitions

Interest in nonalcoholic and low-alcohol beers has spurred much research into new processes. "Low-alcohol beer" or "reduced-alcohol beer" is defined by the U.S. government as a beer with less than 2.5% alcohol by volume, and in the UK the limit is 1.2% by volume. In the U.S., "near beer" or "nonalcoholic" means less than 0.5% alcohol by volume. "Alcohol-free" means that absolutely no alcohol is present.

There are numerous patents world-wide describing methods for producing nonalcoholic and low-alcohol beers (NAB/LAB). Only the most commonly accepted processes[49,50] will be described. There are two general principles by which an NAB/LAB is produced. Because alcohol is produced by fermentation of carbohydrates, production of an NAB/LAB requires that less carbohydrate be fermented, or that alcohol already formed be removed.

Major Deficiencies

There are two major objections to NAB/LABs. Both are related to sensory qualities. The first pertains to the mouthfeel of such beers. NAB/LABs are "thinner," that is, wanting "palate fullness," a quality of fullness in the mouth.[49] The components of beer that contribute to palate fullness are not well defined, although alcohol concentration has an influence, as do all-malt worts and original gravity.

The second objection to NAB/LABs is that the characteristic aroma of beer is reduced or absent. Because aroma compounds are produced concurrently by yeast with alcohol, one cannot obtain normal concentrations of aroma compounds with reduced alcohol production.

Production Methods

There are several techniques for removing or reducing the produced alcohol, although volatile flavor compounds are also removed with the alcohol. Alcohol can be removed by vacuum or steam distillation and by low-temperature/low-pressure distillation.[49,51] Equipment for this process include thin film evaporators[52,53] and stills.[54] In some processes, to improve product flavor, flavor compounds can be recovered by de-esterification and restored to the treated product. Another method for improving flavor is blending dealcoholized beer with conventional beer to the desired alcohol concentration.

Another technique for removing alcohol is reverse osmosis.[55] In this procedure, a fermented wort flows over a semipermeable membrane under high pressure. The membrane is permeable to water, alcohol, and other small molecules. The permeability of the membrane governs the extent to which alcohol and aroma compounds are removed. The resulting beer is concentrated, and then adjusted to the desired alcohol concentration. Dialysis is another membrane technique in which beer is dialyzed against water. Aroma compounds are reduced along with alcohol.

There are also several techniques for controlling the alcohol concentration by controlling the extent of fermentation.[49] The beers resulting from these incomplete fermentations may contain high concentrations of fermentable sugars that will impart a sweet, worty flavor to the beer. Development of aroma compounds will also be arrested, and the beers will lack a full aroma. The residue of fermentable sugars presents a potential microbiological problem.

A technique for controlling alcohol production in NABs is to use high CO_2 pressure and low temperature during fermentation to reduce yeast activity. Another technique adds harvested yeast from an active fermentation to fresh wort at a rate of $13-30 \times 10^6$ cells/ml at a temperature below 3°C. With a contact time of about 24–48 h, the fermentation rate is practically zero. It was proposed that the flavor compounds developed in yeast during the active fermentation from which yeast is harvested are allowed to diffuse from the yeast cell, imparting beer flavor to the low-alcohol product.[56–58]

By means of incomplete fermentation, a yeast strain that does not utilize maltose can be used for LAB production. *Saccharomycodes ludwigii*[49,50] is closely related to *Saccharomyces uvarum* but will not ferment maltose. With the production of a wort that provides the appropriate fermentable carbohydrate profile, a beer of the desired alcohol concentration can be produced. Although maltose is not as sweet as glucose, the beers will retain some sweetness. The beer flavor will be different because of the new strain; however, the alcohol level is easily controlled.

Another technique for reducing alcohol production for LAB is to use a wort with a high concentration of nonfermentable carbohydrates. Such a wort can be produced in two ways. One is to add a corn syrup or other adjunct with a high concentration of nonfermentables. The other is to use a high-temperature mash cycle.[49,50] A temperature of about 77°C is used for mashing, lautering, and sparging. The elevated temperature inactivates the amylases, preventing formation of normal concentrations of fermentable sugars. Worts produced in this manner may have an RDF of 40–50% at 8°Plato.

Other techniques involve combinations of those mentioned earlier, complicated brewhouse and fermentation cellar procedures, and even simple dilution of normal beer. In all, there are numerous articles and patents that describe many variations on the major ideas presented here.

Several process considerations are unique to NAB/LABs. For alcohol removal methods, high processing temperatures and long processing times are detrimental to product flavor. For alcohol prevention methods, processing to minimize yeast growth and VDK formation is necessary. Because of the absence of alcohol or low alcohol concentration in NAB/LABs, there are two special processing problems. First, the product must be prevented from freezing, which can occur when aging cellar temperatures are near 0°C or below. Second, alcohol and heat have a synergistic effect on microorganisms during pasteurization, so NAB/LABs require higher pasteurization units to ensure microbiological stabilization. Usually, these products have a higher pH than beer, which presents additional microbiological considerations regarding pasteurization.

Immobilized Yeast

In the brewing industry during the 1980s, there was an increase in immobilized yeast technology research. The principle of this technology is to produce a high yeast-cell density in order to rapidly ferment wort and maturate beer. The main advantages are:

- Increased capacity by shortened processing times
- Continuous operation possibility
- Improved efficiency of extract-to-alcohol conversion (i.e., less extract-to-yeast cell production)
- Much improved yeast separation

- Use of yeast strains with other performance or flavor characteristics, without regard for their flocculent characteristics

During early brewing investigations, yeast was immobilized by entrapment, most often in calcium alginate beads or gelatin. Cells can also be attached to the surface of a large number of sintered or porous materials, such as glass, brick, ceramics, stainless steel.[59] Popular reactor designs include packed and fluidized beds. Whether in a gel matrix or on a porous support, it is necessary to obtain sufficient mass transfer of wort nutrients, including oxygen, to the cells and fermentation products, principally CO_2, away from the cells. Therefore, immobilized yeast reactor design centers around maximizing mass transfer. It was shown, however, that sufficient mass transfer was too difficult to achieve, and immobilized yeast reactors were unable to duplicate traditional batch fermentation. Yeast growth is severely limited by design and normal flavor development is not feasible. Most successful immobilized yeast systems were designed for flavor maturation or alcohol-free products.[60] Pilot-scale investigations examined immobilized yeast technology for continuous production.[61]

Abnormal Fermentations

Symptoms

Fermentation is a natural process that occurs freely in nature. As a result, industrial beer fermentations tolerate wide variations in process conditions. Although abnormal fermentations are rare, "stuck" or "hung" fermentations will occur even in the most modern breweries utilizing elaborate process control and monitoring equipment. Symptoms of a stuck fermentation may be a long lag phase accompanied by a very slow fermentation rate, followed by no fermentation activity at all. In other cases, after a normal lag phase active fermentation may simply stop before all fermentable carbohydrates are consumed.

Causes

Process Variations

The pitching rate, yeast viability, level of aeration, and wort fermentability should be the first areas to be investigated; these process conditions are among the easiest to check. Usually, a check of meter calibration and settings, time–temperature records, yeast cell counts (including dead cells), and a microbiological examination will point to the problem. Insufficient wort aeration and low pitching rates are common causes of abnormal fermentations. Successive fermentations in nutrient-deficient worts can gradually lead to deterioration of the yeast and slow fermentations.[35] Checks of yeast cell counts in the fermentor may show early flocculation of yeast or its inability to completely assimilate maltose or maltotriose. A positive starch test will indicate a brewhouse conversion problem.

Wort Nutrient Deficiencies

A more difficult situation occurs if there is a wort nutrient deficiency. If slow or stuck fermentations are a continuing problem, wort deficiencies may be the cause. Under normal conditions, nutrients for yeast growth are in excess. However, if the yeast has been handled poorly, thereby increasing its requirement for a particular nutrient, the wort may not be able to supply it. Also, changes in malt or brewhouse processes or raw materials may change wort composition and produce a nutrient deficiency.

The most common deficiencies are oxygen, zinc, biotin, unsaturated fatty acids and sterols. As discussed earlier, yeast in anaerobic fermentation has a requirement for oxygen, unsaturated fatty acids, and sterols. These requirements are interrelated. Usually, sufficient oxygen will reduce the need for unsaturated fatty acids and sterols in wort and vice versa. Biotin is obtained from malt during mashing and is a growth factor required by most brewing yeasts.

When using high percentages of adjunct, another possible nutrient deficiency is reduced nitrogen content in the wort. Insufficient amino acids will hinder proper yeast growth. In worts with high glucose concentrations relative to the other sugars, "stuck" fermentations may occur from a condition sometimes referred to as glucose repression.[37] In this case, the presence of high concentrations of glucose prevents the yeast from synthesizing the enzymes needed to assimilate the other sugars. A similar condition may arise if a high concentration of fructose is present causing "fructose block." These situations are more likely in worts for low-calorie (low-carbohydrate) beers because the adjunct may be glucose solely.[62]

Yeast Changes

Another possibility is that yeast performance has changed during successive repitchings. Such change may be caused by a build-up in yeast cells of toxic amounts of normal nutrients; adsorption of hop and trub compounds on the surface of the yeast; a mutation that affects sugar utilization; or a contaminating killer-yeast strain, a possibility discovered during the 1970s.[63,64]

Treatments

If a process variation was discovered to cause a stuck fermentation, appropriate action is obvious. Contamination of the yeast requires careful investigation of possible sources and corrective action. Wort nutrients can be increased by adding yeast extract or yeast food. These preparations have a variety of nutrients beneficial to yeast growth.[65] In the case of persistent fermentation problems because of wort nutrient deficiencies, more laboratory analyses and study will be required to pinpoint which compound is needed. Yeast deterioration can be eliminated by using a freshly propagated yeast batch. Increasing trub carryover may have a beneficial effect on fermentation.

If a fermentation is "stuck," one may want to recover the product. The addition of 10–20% volume of actively fermenting beer at high kräusen may help[2]; in severe cases, the blend may need to go up to 50%. Another, although more risky, remedy is to aerate the fermentation. This may activate the yeast, but it introduces the possibility of contamination or it may cause oxidation reactions and off-flavors. If fermentable sugars in the wort are too low, the addition of an amylolytic enzyme to the fermentor may substitute for a poor mashing process.

Beer Transfer and Yeast Separation

Most brewers periodically propagate their yeast from pure cultures. However, generally there is no need to propagate yeast for each fermentation as there is sufficient yeast available from production fermentations to supply nearly all of the needs of a brewery. There is a four- to sixfold increase in yeast concentration during fermentation. Aside from the need to remove most of this yeast from beer prior to aging or secondary fermentation, yeast recovery for reuse in subsequent fermentations is an important process in any brewery.

Yeast Cropping Considerations

Yeast strain may have an important influence on the method chosen for separating yeast from beer. More powdery strains are best removed by centrifugation or filtration, while highly flocculent strains are more efficiently separated by sedimentation. Moderately flocculent strains, which allow more yeast carryover during transfer, are used if a secondary fermentation or ruh storage is desired. The formation of yeast flocs or clumps is usually aided by the presence of calcium ions, low concentrations of fermentable sugars, low pH, and low temperature. These are the conditions at the end of fermentation. Higher amounts of zinc assimilated during fermentation also aid flocculation. Ale strains have been classified into four groups based on the requirement for an inducer for flocculation.[66] It is believed that certain amino acid residues in peptides are the inducers.

The geometry of fermenting vessels also plays a role and, in fact, usually determines the separation method. Shallow vessels of large surface area are conducive to relatively complete sedimentation of highly flocculent yeast. Cylindroconical vessels are better for more efficient yeast separation of less flocculent strains, as the bottom-fermenting yeast collects in the conical bottom. The larger volume cylindrical tanks are even useful for top fermentations.

Methods of Cropping

In traditional top-cropping ale fermentations in open vessels, the yeast crop is skimmed by various devices. A parachute device (inverted funnel) and suction or similar suitable system is common. This technique has also been used successfully with closed fermentors. Other techniques, used principally in the UK, rely on specially designed fermentors that aid in collection of the yeast.[37]

In lager fermentations using horizontal tanks, the beer is simply drawn off for lagering, leaving the yeast behind. The yeast is then collected manually from the vessel floor. This yeast contains other, mostly proteinaceous, sediment from the fermenting beer, and separation of the yeast from the debris is difficult. Modern cylindroconical tanks allow improved separation and collection strategies. Temperature reduction at the end of primary fermentation aids sedimentation of yeast into the cone for easy beer transfer and yeast collection. Most of the yeast can be collected from the conical bottom before the beer, which contains some yeast in suspension, is removed from the fermentor. The beer can then be transferred to ruh storage with the residual yeast present, or the yeast can be removed by passing the beer through a centrifuge. Alternatively, the beer can be transferred first, leaving the yeast layer undisturbed, for more efficient removal. In large vertical fermentors in which bottom-fermenting strains are used, the trub will settle with the early flocculating layer. Removal of this layer to waste will eliminate most trub particles in the remaining yeast crop. The ease of collection in cylindroconical tanks makes it advantageous to use bottom-fermenting strains of *S. cerevisiae* for ale production.

Because yeast will flocculate continually toward the end of fermentation, there will be a somewhat stratified yeast crop with varying flocculation characteristics. Collection of the bottommost layer in bottom-fermenting strains or the topmost layer in top-fermenting strains will harvest the most flocculent yeast, while other layers will contain the least flocculent yeast. Repitching yeast with undesirable flocculation characteristics may lead to fermentation problems.

Separating yeast from green beer by skimming the yeast or decanting the beer will not leave the beer totally free of yeast. If a secondary fermentation or ruh storage is used, some yeast is needed in aging, and yeast carryover is not a problem. If no secondary fermentation is used, then more efficient transfer is advantageous. A better separation of yeast from beer can be accomplished by using centrifuges (discussed next).

Centrifugation

Before centrifuges were used, shallow tanks were needed for proper sedimentation of yeast. With the introduction of centrifuges in the brewery for yeast separation, the need for shallow tanks was eliminated, fostering the use of tall, space-saving tanks. The use of centrifuges for clarification has become more common with the development of equipment that can be used successfully with beer. For example, modern centrifuges have hermetic seals that: (a) exclude air during their operation, (b) minimize temperature increases, and (c) reduce turbulent flow.

There are two principal ways to use centrifuges for beer separation after fermentation. The first method is to use the centrifuge to separate the yeast crop from the entire fermentor. This accomplishes beer transfer and yeast collection in one step. The second is to use a centrifuge to clarify the beer after the yeast has been separated by decanting or skimming for repitching. This two-step procedure is generally used when the collected yeast will be used for repitching, because centrifugation can be stressful to the yeast cell.

Centrifugation can heat the beer and yeast streams. It is necessary to have a cooler for the beer to bring it to aging temperature. If the yeast collected by centrifuge is to be used for repitching, a cooler should be used to reduce the temperature of the yeast to brink temperature as quickly as possible.

The principle of centrifugation is based on Stokes' law governing the rate of settling of particles. A greater settling rate occurs with: (a) a lower viscosity liquid, (b) larger diameter particles, and (c) greater difference in density between the particle and the liquid. None of these factors is usually controllable. However, the settling rate is increased by the considerable force created by the centrifuge (up to 10,000g). A shorter settling distance, which is a common feature of modern centrifuges, also speeds sedimentation.

There are two basic types of centrifuges used in brewery applications. They may be classified according to the solids load they handle. A dewatering or decanting centrifuge is a screw-type conveyor, generally horizontal, that handles larger and more fibrous particles in liquids of up to about 60% solids. The centripetal force moves the particles to the outer surface of the cylindroconical shell, where they are conveyed by the screw to the discharge. It is used mainly for brewery effluents, to recover liquor containing extract from brewer's grains after lautering, and to recover beer from tank bottoms containing yeast. The second and most common type of centrifuge, a clarifying centrifuge, is a disk-stack bowl type in a vertical configuration.

Clarifying centrifuges use either intermittent or continuous solids ejection. Continuous ejection models are referred to as nozzle centrifuges. The disk-stack centrifuges can handle liquids with up to about 30% solids and are ideal for brewery applications, as they are self-cleaning, air tight, and have CIP systems. In a disk-stack bowl centrifuge, the centripetal force moves the particles to the outside of the bowl, where solids are removed intermittently by rapid openings of the bowl, or continuously through a nozzle. Manufacturers produce a large variety of models suitable for a broad range of brewery applications.

Clarifying centrifuges perform best when they receive a feed stream with a uniform concentration of solids. Therefore, they are most efficient for use with: (a) beer that has already been cropped, for example, ruh beer which contains only a low concentration of yeast or (b) fermentations using powdery yeast strains.

The use of centrifuges in the brewery was initially viewed with misgivings, but the advantages and disadvantages are now reasonably clear. Some significant advantages of centrifuges are:

- Rapid and efficient clarification before further filtration steps
- More consistent clarity
- Equipment can be sterilized
- Filter aids are not required
- Space requirements are small
- Most are self-cleaning
- Operate continuously
- Lower beer losses compared to sedimentation
- Minimal oxygen pick-up

Some disadvantages are:

- High maintenance costs
- Beer temperature may increase
- Yeast needed for flavor maturation (VDK reduction) may be too completely removed
- Mechanical break-up of large particles due to shear may increase the concentration of finer haze particles
- Increased mechanical stress or increased yeast temperature may adversely affect yeast being used for repitching
- Improper operation possibly leading to oxygen pick-up, and high noise level

When using centrifuges, probably the three most important factors for maintaining beer quality are oxygen exclusion, a microbiologically clean operation, and a minimal beer temperature rise as a result of processing. If proper cleanliness and strict operating standards are employed, beer separated from yeast by centrifuge will be of excellent quality.

Recovery of Carbon Dioxide

Less than 10% of fermentable sugars are converted into new yeast mass during fermentation.[37] The remaining fermentable sugars are converted into approximately equal quantities of ethanol and CO_2. Depending on wort gravity, about 4 kg of CO_2 are produced/bbl and theoretically collectible; but in practice, about only 80% or less is recoverable[10] depending on fermentor geometry (headspace), fermentation rate, and efficiency of the recovery system.

The uses for CO_2 are for: (a) sparging beer in aging, (b) counterpressure in beer storage tanks, (c) purging filters and transfer lines, and (d) packing operations, which can usually be met with that recovered from fermentation, assuming there are no losses. Major uses are carbonation (1 lb/bbl or about 0.5 kg/hl for each volume CO_2 used), tank counterpressure and transfer (2 lb/bbl), and packaging, especially in can operations (up to 3 lb/bbl).[67] The collection, purification, and liquefaction of CO_2 for reuse thus becomes economically attractive. With careful attention, brewers can meet their CO_2 needs with that recovered from fermentation and have excess to sell.

Purity and Collection Strategies

Because introduction of any oxygen into beer is detrimental to flavor, the CO_2 used in processing must contain as little oxygen as possible. Oxygen in CO_2 is influenced by collection strategies from the fermentor as well as proper design and operation of a purification system. To minimize oxygen impurities in CO_2, the collection timing must be coordinated with the fermentation cycle. In particular, the CO_2 evolved early in fermentation must be vented until its oxygen concentration is below some specified limit, usually 0.01% or less.[68] In-line sensors are available to assist in timing of collection. The duration of venting will depend on the fermentation rate, fermentor geometry, headspace volume, and product type.[67]

Fermentors are disconnected from the collection system toward the end of carbohydrate fermentation when CO_2 production is low. Because fermentors are added to the production stream according to a known brewing schedule, the peak CO_2 load on the collection system can be calculated. From this calculation, the collection system can be properly designed to handle the CO_2 expected from normal production.[69]

References

1. Siebert, K.J., Blum, P.H., Wisk, T.J., Stenroos, L.E., and Anklam, W.J., The effect of trub on fementation, *Tech. Q. Master Brew. Assoc. Am.*, 23:37–43, 1986.
2. Drost, B.W., Fermentation and storage, *Proc. 16th Eur. Brew. Conv. Congr.*, Amsterdam, Elsevier, New York, 1977, pp. 519–532.
3. American Society of Brewing Chemists, *Methods of Analysis*, 9th ed., revised, American Society of Brewing Chemists, St. Paul, MN, 2004.
4. American Society of Brewing Chemists, *Laboratory Methods for Craft Brewers*, American Society of Brewing Chemists, St. Paul, MN, 1997.
5. European Brewing Convention, *Analytica-EBC*, Schweizer Brauerei-Rundschau, Zurich, Switzerland, 2005.
6. Institute of Brewing, *Methods of Analysis*, The Institute of Brewing & Distilling, London, 1998.
7. DeClerk, J., *A Textbook of Brewing*, Vol. 2, Kathleen Barton-Wright, trans., Chapman and Hall, London, Vol. 2, 1958.
8. American Society of Brewing Chemists, *Tables Related to Determinations on Wort, Beer, and Brewing Sugars and Syrups*, American Society of Brewing Chemists, St. Paul, MN, 1975.
9. Kämpf, W., Das Einheiten-und Messwesen im Braugewerbe, *Monatsschr. Brau.*, 21:41–69, 1968.
10. Hardwick, W.A., in *Biotechnology, Vol. 5, Food and Feed Production with Microorganisms*, Reed, G., Ed., Verlag Chemie, Deerfield Beach, FL, 1983.
11. Helbert, J.R., in *Prescott & Dunn's Industrial Microbiology*, 4th ed., Reed, G., Ed., AVI Publishing Co., Inc., Westport, CT, 1982, Chapter 10.
12. Priest, F.G. and Campbell, I., *Brewing Microbiology*, Elsevier Applied Science, New York, 1996.
13. Siebert, K.J. and Wisk, T.H., Yeast counting by microscopy or by electronic particle counting, *J. Am. Soc. Brew. Chem.*, 42:71–79, 1984.
14. Miedaner, H., Optimisation of fermentation and conditioning during the production of lager, *Brewer*, 64:33–39, 1978.
15. Kirsop, B.H., Pitching rate, *Brew. Dig.*, 53:28–32, 1978.
16. Stassi, P., Rice, J.F., Munroe, J.H., and Chicoye, E., Use of CO_2 evolution rate for the study and control of fermentation, *Tech. Q. Master Brew. Assoc. Am.*, 24:44–50, 1987.
17. Jones, M. and Pierce, J.S., Absorption of amino acids from worts by yeast, *J. Inst. Brew.*, 70:307–315, 1964.
18. Quain, D.E. and Tubb, R.S., The importance of glycogen in brewing yeasts, *Tech. Q. Master Brew. Assoc. Am.*, 19:29–33, 1982.
19. Ohno, T. and Takahashi, R., Role of aeration in the brewing process (1). Biosynthesis of lipids in brewer's yeast, Rep. Res. Lab. Kirin Brew. Co., No. 26, pp. 15–23; (2) Behavior of yeast lipids in the brewing process, Rep. Res. Lab. Kirin Brew. Co. No. 26, pp. 25–30, 1983.
20. Fricker, R., The design of large tanks, *Brewers Guardian*, 107:28–37, 1978.

21. Murray, C.R., Barich, T., and Taylor, D., The effect of yeast storage conditions on subsequent fermentation, *Tech. Q. Master Brew. Assoc. Am.*, 21:189–194, 1984.
22. MacLeod, A.M., in *Economic Microbiology, Vol. 1, Alcoholic Beverages*, Rose, A.H., Ed., Academic Press, New York, 1977.
23. Engan, S., Formation of volatile flavour compounds: alcohols, esters, carbonyls, acids, in *E.B.C. Fermentation and Storage Symposium, Monograph V*, European Brewing Convention, Elsevier, New York, 1978, pp. 28–39.
24. Jacobsen, T. and Gunderson, R.W., Cluster analysis of beer flavor components. II. A case study of yeast strain and brewery dependence, *J. Am. Soc. Brew. Chem.*, 41:78–80, 1983.
25. Kleber, W., Systems of fermentation, in *Brewing Science*, Vol. 3, Pollack, J.R.A., Ed., Academic Press, New York, 1987, pp. 329–377.
26. Maule, D.R., A century of fermentor design, *J. Inst. Brew.*, 92:137–145, 1986.
27. Knudsen, F.B., Fermentation — principles and practice, in *The Practical Brewer*, Broderick, H.M., Ed., Master Brewers Association of the Americas, Madison, WI, 1977.
28. Coote, N. and Kirsop, B.H., Factors responsible for the decrease in pH during beer fermentations, *J. Inst. Brew.*, 82:149–153, 1976.
29. Narziss, L., Miedaner, H., and Gresser, A., Heferasse und Bierqualität. Extraktabnahme, pH-Abfall und Farbveränderung während der Gärung, *Brauwelt*, 123:478–486, 1983.
30. Cutaia, A.J., and Munroe, J.H., A method for the consistent estimation of real degree of fermentation, *J. Am. Soc. Brew. Chem.*, 37:188–189, 1979.
31. McCaig, R. and Bendiak, D.S., Yeast handling studies. I. Agitation of stored pitching yeast, *J. Am. Soc. Brew. Chem.*, 43:114–118; Yeast handling studies. II. Temperature of storage of pitching yeast, *J. Am. Soc. Brew. Chem.*, 43:119–122, 1985.
32. Frey, S.W., DeWitt, W.G., and Bellomy, B.R., The effect of several trace metals on fermentation, *Proc. Am. Soc. Brew. Chem.*, 199–205, 1967.
33. Densky, H., Gray, P.J., and Buday, A., Further studies on the determination of zinc and its effects on various yeasts, *Proc. Am. Soc. Brew. Chem.*, 93–100, 1966.
34. Rice, J.F., Chicoye, E., Helbert, J.R., and Garver, J., Inhibition of beer volatiles formation by carbon dioxide pressure, *J. Am. Soc. Brew. Chem.*, 35:35–40, 1977.
35. Drost, B.W., Fermentation and storage, *Proc. 17th Eur. Brew. Conv. Congr.*, Berlin West, Elsevier, New York, 1979, pp. 767–786.
36. Tenney, R.I., Rationale of the brewery fermentation, *J. Am. Soc. Brew. Chem.*, 43:57–60, 1985.
37. Hough, J.S., Briggs, D.E., Stevens, R., and Young, T.W., in *Malting and Brewing Science*, 2nd ed., Chapman and Hall, New York, 1982.
38. Masschelein, C.A., La Levure et Son Environnement, *Proc. 14th Eur. Brew. Conv. Congr.*, Salzburg, Elsevier, New York, 1973, pp. 255–269.
39. Miedaner, H., Narziss, L., and Wörner, G., Über den Einfluss der Gärungsparameter Temperatur und Druck auf die Entwicklung einiger Gärungsnebenproduckte des Bieres, *Brauwissenschaft*, 27:208–213, 1974.
40. Nielsen, H., Hoybye-Hansen, I., Iboek, D., and Kristensen, B.J., Introduction to pressure fermentation, *Brygmesteren*, 43:7–17, 1986.
41. Purssell, A.J.R. and Smith, M.J., Continuous fermentations, *Proc. 11th Eur. Brew. Conv. Congr.*, Madrid, Elsevier, New York, 1967, pp. 155–165.

42. Portno, A.D., Continuous fermentation in the brewing industry — the future outlook, in *E.B.C. Fermentation and Storage Symposium, Monograph V*, European Brewing Convention, Elsevier, New York, 1978, pp. 145–154.
43. Wellhoener, H.J., Ein Kontinuierliches Gär- und Reifungsverfahren für Bier, *Brauwelt*, 94:624–626, 1954.
44. Coutts, M.W., A continuous process for the production of "beer," UK Patents 872,391–400, 1957.
45. Bishop, L.R., A system of continuous fermentations, *J. Inst. Brew.*, 76:172–181, 1970.
46. Pfisterer, E., Enzymes in fermentation, *Tech. Q. Master Brew. Assoc. Am.*, 11:9–16, 1974.
47. Marshall, J.J., Allen, W.G., Denault, L.J., Glenister, P.R., and Power, J., Enzymes in brewing, *Brew. Dig.*, 57:14–22, 1982.
48. Owades, J.L. and Bierman, G.W., A study of the admixture of amyloglucosidase-treated beer with other beers, *Tech. Q. Master Brew. Assoc. Am.*, 15:38–42, 1978.
49. Brenner, M.W., Beers of the future, *Tech. Q. Master Brew. Assoc. Am.*, 17:185–195, 1980.
50. Lieberman, C.E., Low alcohol beers, *Brew. Dig.*, 59:30–31, 1984.
51. Kieninger, H. and Haimerl, J.H., Die Herstellung von Alkoholreduziertem Bier mittels Vakuumdestillation, *Brauwelt*, 121:574–581, 1981.
52. Berger, H.F., Theoretische Grundlagen der Dünnschichtverdampfung zur Alkoholreduktion von Bier am Biespiel Centri-Therm, *Monatsschr. Brau.*, 35:98–101, 1982.
53. Oliver-Dauman, B., Praktische Erfahrungen mit dem Centri-Therm zur Alkoholreduktion von Bier, *Monatsschr. Brau.*, 35:101–106, 1982.
54. Bernstein, K. and Strömsdörfer, F., Gewinnung von technischem Alkohol bei der Diätbierproduktion — Vorläufiger Beitrag, *Monatsschr. Brau.*, 35:107–108, 1982.
55. Anon., New technology improves low-alcohol beer flavor, *Beverage Ind.*, 75(10):9, 1984.
56. Schur, F., Process for the Preparation of Alcohol-Free Drinks with a Yeast Aroma, U.K. Patent GB 2112619, 1982.
57. Schur, F., Ein neues Verfahren zur Herstellung von Alkoholfreien Bier, *Proc. 19th Eur. Brew. Conv. Congr.*, London, Elsevier, New York, 1983, pp. 353–360.
58. Schur, F., Alcohol-free beer — a new method of production, *Brauwelt Int.*, I:87–88, 1984.
59. Ryder, D.S. and Masschelein, C.A., Immobilized yeast in brewing — a current perspective, *Proc. 23rd Eur. Brew. Conv. Congr.*, Lisbon, Elsevier, New York, 1991, pp. 345–352.
60. Pajunen, E., Immobilized yeast applications in the brewing industry, *EBC Monograph*, EBC Symposium, European Brewery Convention, Elsevier, New York, 1995, pp. 24–34, 1995.
61. Pajunen, E., Lommi, H., and Viljava, T., Novel primary fermentation with immobilized yeast, *Tech. Q. Master Brew Assoc. Am.*, 37:483–488, 2000.
62. Bindesboll Nielson, N. and Schmidt, F., The fate of carbohydrates during fermentation of low calorie beer, *Carlsberg Res. Commun.*, 50:325–332, 1985.
63. Maule, A.P. and Thomas, P.D., Strains of yeast lethal to brewery yeasts, *J. Inst. Brew.*, 79:137–141, 1973.
64. Taylor, R. and Kirsop, B.H., Occurrence of a killer strain of *Saccharomyces cerevisiae* in a batch fermentation plant, *J. Inst. Brew.*, 85:325, 1979.

65. Ingledew, M.W., Sosulski, F.W., and Magnus, C.A., An assessment of yeast foods and their utility in brewing and enology, *J. Am. Soc. Brew. Chem.*, 44:166–170, 1986.
66. Stewart, G.G., Russell, I., and Garrison, I.F., Some considerations of the flocculation characteristics of ale and lager yeast strains, *J. Inst. Brew.*, 81:248–257, 1975.
67. Eiker, J., Utilities engineering, in *The Practical Brewer*, Broderick, H.M., Ed., Master Brewers Association of the Americas, Madison, WI, 1977.
68. Huige, N.J., Charter, W.M., and Wendt, K.W., Measurement and control of oxygen in carbon dioxide, *Tech. Q. Master Brew. Assoc. Am.*, 22:92–98, 1985.
69. Collins, T. and Bandy, J., CO_2 generation and harvesting, *Tech. Q. Master Brew. Assoc. Am.*, 37:255–260, 2000.

13

Aging and Finishing

James H. Munroe

CONTENTS

Introduction 526
  Objectives of Aging and Finishing 526
  Component Processes 526
Flavor Maturation 528
  Introduction 528
  Important Flavor Compounds 528
    Diacetyl and 2,3-Pentanedione 528
    Sulfur Compounds 531
    Nonvolatile Flavor Maturation 531
    Yeast Autolysis 532
Lagering and Secondary Fermentation (Kräusening) 532
  Historical Lagering Practice 533
  Kräusening 533
  Lagering without Secondary Fermentation 534
  Addition of Modified Hop Extracts 535
Beer Recovery 535
  Economics 535
  Quality of Recovered Beer 536
Clarification 536
  Gravity Sedimentation 537
  Finings 537
  Filtration 537
  Filters 538
  Sterile Filtration 540
  Transfer to Packaging 541
Stabilization 542
  Flavor Stability 542
  Biological Stability 543
  Physical Stability 543

    Brewhouse Procedures and Filtration 544
    Measurement of Haze 545
Carbonation 546
    Basics of Beer Carbonation 546
    Historical Carbonation 547
    Modern Carbonation 547
Standardization 547
    Blending for Consistency 547
    High-Gravity Brewing 548
References 548

Introduction

Objectives of Aging and Finishing

Aging refers to flavor maturation. At the end of fermentation, many undesirable flavors and aromas of a "green" or immature beer are present. The aging process reduces the levels of these undesirable compounds to produce a mature product.

Finishing refers to the production of a brilliantly clear beverage after aging that remains that way until consumed.

Component Processes

The component processes of aging and finishing are:

1. Lagering or aging
2. Clarification
3. Stabilization
4. Carbonation
5. Blending or standardization

Each process can be accomplished in a variety of ways, but each is independent and can be treated as a unit operation. A brief description of the unit processes is given here and in Table 13.1, with fuller discussions in subsequent sections.

In modern practice, cold aging or lagering is the storage of beer for the purpose of flavor maturation. Because historical practice had additional functions, there are other meanings attached to these terms as explained in a later section.

Aging and Finishing

TABLE 13.1

Unit Operations for Aging and Finishing

Unit Operation	Purpose	Equipment and Methods
Transfer	Yeast separation	Decant beer Centrifuge Filter
Aging (lagering, ruh storage)	Flavor maturation	Some yeast present for VDK reduction CO_2 purge
Stabilization	Protect beer from:	
	1. Oxidized flavor	Keep yeast present Minimize O_2 pick-up Use CO_2 during all transfers of beer Add antioxidants
	2. Biological haze and off-flavors	Pasteurization Sterile filtration
	3. Physical haze (chillproofing)	Tannic acid Fining agents Proteolytic enzymes PVPP Silica gels
Clarification (filtration)	Removal of all suspended particles	Filters
Carbonation	Attain proper CO_2 concentration	Traditional aging Pressurized fermentation CO_2 injection
Standardization (blending)	Uniformity of packaged product	Tankage for transfers Blenders

After aging, clarification is required to remove any remaining yeast and suspended particles formed during cold storage. At least one filtration step is needed before beer is suitable for packaging if a clear, brilliant beer is desired.

Stabilization refers to protecting the finished product from changes that may occur after packaging. These changes are: (a) flavor changes primarily due to oxidation, (b) nonmicrobiological haze caused by the formation of molecular complexes, and (c) haze produced by the growth of bacteria or yeast.

Carbonation is the process of adjusting the carbon dioxide (CO_2) concentration to a specified concentration. Carbonation by injection of CO_2 into beer is done as a replacement for the traditional raising of the CO_2 level by a cold secondary fermentation.

Blending or standardization is the process of mixing batches of beers to achieve uniformity of flavor or analytical characteristics.

Brewers may combine some of these operations, or change the order in which they are carried out. The possible variations are too numerous to detail here, but most brewers, for reasons of economy and product uniformity, attempt to combine some of the unit operations.

Flavor Maturation

Introduction

Flavor maturation is generally considered the most significant outcome of aging. Successful flavor maturation has become more important as beers have become "lighter" in flavor. Taste thresholds of objectionable flavors are lower in lighter beers. In heavier beers, the presence of more flavorful compounds will mask some objectionable flavors and aromas.

Considerable research in the brewing industry has been devoted to understanding flavor maturation. In some cases, it can be described in terms of individual compounds that can be detected in wort and beer permitting the brewer to rely on laboratory tests in addition to taste tests to determine the success of maturation. Taste tests can be unreliable and should be used with the knowledge that tasters vary in their sensitivity to different flavors. Therefore, most brewers supplement tasting with chemical tests and set specification limits on objectionable flavor compounds. In-process beer must meet such specifications and satisfy taste requirements before release to downstream processing.

Because most of the important compounds discussed under flavor maturation are a result of yeast metabolism, the central role of consistent yeast growth during fermentation is again stressed. As in Chapter 12, yeast growth as used in this chapter refers to the increase in cell population during fermentation.

Important Flavor Compounds

Diacetyl and 2,3-Pentanedione

As noted in Chapter 12, diacetyl and the homologous compound 2,3-pentanedione have flavor properties that are important to the brewer. Collectively, diacetyl and 2,3-pentanedione are called vicinal diketones (VDKs). Both compounds have a buttery flavor generally considered objectionable in lighter-bodied lagers but sometimes desirable in ales and more full-bodied beers. Diacetyl has a higher flavor impact than 2,3-pentanedione. The flavor threshold of diacetyl, and other flavor compounds as well, depends on the background flavor intensity of the beer but it is usually detectable at about 0.1 mg/l. It is thought that at subthreshold levels, diacetyl contributes positively to palate fullness. Brewers may speak of diacetyl only, but both VDKs are important to maturation.

Research has elucidated some of the chemical reactions and biochemical pathways of these important compounds.[1-4] Only a summary of the main reactions for diacetyl is given here.

The precursor to diacetyl, α-acetolactate, is produced by the yeast as it synthesizes the amino acids valine and leucine needed for protein synthesis (Figure 13.1). The α-acetolactate is transported out of the cell where it is converted nonenzymically to diacetyl. This step is the slowest or rate-limiting step and is accelerated by a higher temperature and lower pH. The diacetyl is subsequently reassimilated by the yeast and reduced enzymically to butanediol by way of acetoin.[4-7] The importance of this step is that butanediol has virtually no impact on flavor. A similar series of reactions occurs for 2,3-pentanedione, the precursor of which is α-acetohydroxybutyrate.

The brewer should be concerned about the concentration of precursors and whether or not yeast is present to remove VDKs when formed. The important concepts are: (a) the precursors are produced as a result of yeast growth relative to the wort valine and other amino acid concentrations; (b) the precursors are potential flavor-active VDKs; (c) conversion of precursors to VDKs is an extracellular chemical reaction that varies with temperature, pH, etc.; (d) these extracellular reactions are rate limiting in the

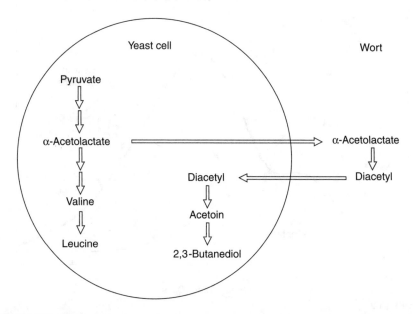

FIGURE 13.1
Mechanism of diacetyl formation. A simplified biochemical production of α-acetolactate by yeast, chemical conversion of precursor to diacetyl, and reassimilation of diacetyl by yeast.

conversion of precursors and removal of VDKs from beer; and (e) yeast assimilates VDKs, and therefore needs to be present to reduce the VDKs as they are formed.

The production of precursors continues throughout carbohydrate fermentation (Figure 13.2). In reality, the profile marked diacetyl in Figure 13.2 includes both diacetyl and its precursor, because the chemical test generally converts the precursor; thus, the graph shows the total potential diacetyl in the final product. Because precursor conversion to diacetyl is the rate-limiting step, with yeast present in fermenting wort, the concentration of diacetyl is small compared to the precursor. Considerable potential for VDK formation remains after active fermentation because of the high concentration of precursors. Regarding VDKs, the maturation process has two objectives: the spontaneous conversion of precursors to VDKs and their removal by yeast. Thus, to hasten the conversion of precursors to VDKs, the temperature in the fermentor can be held higher (e.g., about 15°C) for a period of time after the completion of carbohydrate fermentation. Once the total precursors (potential VDKs) and VDKs fall below a specified level, the temperature can be lowered to aid in yeast sedimentation. However, the higher temperature may lead to other off-flavors from nonvolatile yeast products or yeast autolysis.

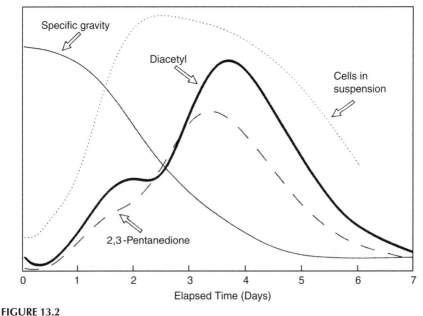

FIGURE 13.2
Production of VDKs during fermentation. The graph shows the approximate relationship of the concentration of diacetyl and 2,3-pentanedione as they relate to yeast cell growth and specific gravity decline during fermentation.

Sulfur Compounds

The subject of sulfur compounds in brewing is broad and complex.[8] They are particularly important because of their very low flavor threshold and a flavor perception that is objectionable. Important sulfur compounds result from yeast metabolism, but many present in beer come from malt and hops. Yeast requires sulfur-containing compounds for synthesis of proteins. Sources of sulfur in wort for yeast metabolism are sulfate ions from water, and thiols and sulfides from raw materials, particularly sulfur-containing amino acids. Sulfur compounds in beer arise through a combination of raw material sources, processing conditions, yeast strain, metabolism and autolysis, and microbial contamination.

Three of the more important volatile compounds are hydrogen sulfide (H_2S), sulfur dioxide (SO_2), and dimethyl sulfide (DMS). Some DMS is formed during fermentation by the action of yeast on dimethyl sulfoxide produced in the kettle, although most DMS comes from the conversion of a precursor in malt during kettle boil.[9] DMS at low concentrations is believed to make a positive contribution to beer flavor.[10] At higher concentrations, DMS can have the objectionable aroma of cooked corn. Chapter 10 discusses the origins and fate of DMS in depth.

Hydrogen sulfide is a product of the transport of sulfide ions by yeast during metabolism of sulfate ions and organic sulfur compounds.[11] Production of H_2S is related to yeast growth. Hydrogen sulfide has the aroma of rotten eggs and the elimination of this compound is usually accomplished by the purging action of CO_2 gas evolution. More H_2S is produced in lager fermentations than ales because of the higher temperatures used.

Sulfur dioxide is also present in beer although usually in concentrations well below 10 ppm, at which level it does not have a flavor impact in most beers. Higher concentrations of SO_2 are produced under conditions of low yeast growth and are beneficial for flavor stability as described later in the section "Flavor Stability."

Nonvolatile Flavor Maturation

Packaged beer contains low concentrations of amino acids, peptides, nucleotides, organic acids, inorganic phosphates, and other ions that contribute to the overall flavor of beer. Some nonvolatile compounds are products of raw materials, normal fermentation, and processing steps. Others are internal components of yeast cells released because of a change in cell permeability following fermentation. Free amino nitrogen, pH, phosphates, color, and invertase activity in beer all increase during storage.[12] It would be reasonable to assume that these increases are dependent on temperature, time, yeast strain, physiological condition of the yeast, fermentor geometry, and so forth.

It is important to note that these changes in nonvolatile compounds are not necessarily undesirable. Nonvolatile compounds can contribute to palate fullness or mouthfeel, and act synergistically with other flavor-active substances and contribute to the overall flavor quality of beer.

Yeast Autolysis

Yeast autolysis is not clearly defined, but the general use of the term refers to the dissolution of dead or moribund cells by their own enzymes. The autolysis products released into the beer result in a sharp, bitter taste and a yeasty aroma. Autolysis occurs under conditions of starvation and high temperature. Holding a fermentation vessel at high temperature (over 15°C) in order to facilitate conversion of the VDK precursors is a condition that can lead to autolysis. It is important therefore not to allow the yeast to remain in beer for long periods at high temperature.

Lagering and Secondary Fermentation (Kräusening)

The term lagering comes from the German verb *lagern*, which means to store, to age, to lay down. The use of this term in the brewing industry is often synonymous with aging and storage, and sometimes other terms that are a consequence of aging such as maturation, conditioning, and secondary fermentation. The term lager beer follows from historical aging practices before refrigeration.

For clarity, the following definitions will be used in this chapter (see Figure 13.1):

> *Primary or main fermentation*: The initial fermentation, during which most of the carbohydrates in the wort are assimilated. If no secondary fermentation is done, then all carbohydrates are assimilated during primary fermentation.
>
> *Secondary fermentation*: Fermentation subsequent to transfer of the beer from primary fermentors with some yeast and fermentable carbohydrates present, during which residual carbohydrates are assimilated. This process is usually done at a reduced temperature. Secondary fermentation is not always used in modern practice.
>
> *Kräusening*: The addition of fermenting wort to the secondary fermentation. Kräusening is not always used in modern practice.
>
> *Lagering*: Nonspecific term applied to aging and other processing following primary fermentation. Historically, lagering included a secondary fermentation followed by a long, cold storage period.

Ruh storage: A practice that refers to cold aging of beer with some yeast present following the completion of fermentation.

Maturation, aging, storage, conditioning: Terms used interchangeably to refer to the maturing of beer flavor. A secondary fermentation may or may not be included.

Historical Lagering Practice

Historically, lagering[13,14] was necessitated by the absence of refrigeration, the need to remove yeast, and the need to control the level of carbonation in beer. Consequently, lager beer was brewed during the colder months and stored in iced caves for long periods of time. Cutting ice for summer storage of beer was well known prior to the advent of refrigeration.

Primary fermentation was usually done at or below 10°C. The resulting beer, containing about 1% fermentable extract, was then transferred to a cold storage cellar along with some suspended yeast.[11] The yeast would assimilate any oxygen picked up during transfer into storage, thus eliminating a potential oxidative flavor problem. Secondary fermentation of the remaining fermentable extract proceeds increasingly slowly while the beer gradually cools over several days. Because CO_2 is more soluble at lower temperatures, the brewer could readily obtain elevated levels of carbonation. Total storage time was up to about 50 days at 0°C. The long, cold storage allowed not only the settling of the remaining yeast, but also the settling of haze-forming material. Extended storage times have been promoted as giving superior flavor maturation. Modern thinking is that long, cold aging is not necessary if the process provides for the elimination of VDKs, their precursors, and other compounds responsible for green beer flavors in immature beer.

Selection of a yeast with the proper flocculation characteristics was obviously important for a long aging process. With a powdery yeast, the transfer to storage and secondary fermentation would carryover too much yeast; the secondary fermentation would occur too quickly, and the yeast would not settle sufficiently at the completion of fermentation. With a very flocculant strain, too little yeast would be carried into secondary fermentation, which would not go to completion unless the yeast were agitated or roused in some way.

Kräusening

Kräusen is a German term meaning "rocky head"; and, in brewing, the word refers to the appearance of the foam head in the primary fermentor. When fermentation is most active, foam formation is greatest and the fermentation is at "high kräusen."

In the practice of kräusening, beer is transferred to the storage cellar after primary fermentation, usually with some residual fermentable extract remaining. Then a volume of high-kräusen beer, about 5–20% of the primary fermented beer volume,[13] is added to the tank. The secondary fermentation continues as in the traditional process, except more rapidly. The degree of secondary fermentation can be controlled by the amount of residual fermentable extract at the transfer and by the amount and fermentable extract of the high kräusen added. In kräusening, a somewhat more flocculant strain can be used because the secondary fermentation is more vigorous than without kräusen, and a flocculant strain is not needed.

Cooling may occur gradually during secondary fermentation, or rapidly at the end to promote yeast settling. The CO_2 produced helps to reach the packaged beer level of carbonation. However, the introduction of high kräusen adds more fermentable extract and produces more flavor compounds. Off-flavors such as H_2S and diacetyl are also increased after being at low concentrations at the end of primary fermentation.[14] Lengthy storage may be required to reduce these undesirable flavor notes to acceptable levels.

Lagering without Secondary Fermentation

Historically, lagering employed shallow open fermentors for primary fermentation, but closed vessels for secondary fermentation in order to maximize carbonation. Modern equipment for refrigeration, carbonation, filtration, etc. obviates the need for secondary fermentation and a long, cold storage period. Modern practice has shortened fermentation and lagering times, and uses rapid cooling after fermentation to aid yeast settling. Techniques for accelerated fermentation, discussed in Chapter 12, accelerate the utilization of carbohydrates. If the wort is fully attenuated during primary fermentation, there is no need for secondary fermentation, and the aging process is principally for flavor maturation.

As described earlier, one of the more important classes of compounds involved in flavor maturation are VDKs. Because the rate of conversion of VDK precursors is temperature dependent, elevating the temperature can be used to hasten the conversion. Short, warm lagering has proven quite effective with minimal deleterious effects on beer quality.[15] This lagering can be accomplished in the presence of yeast by extending the residence time in the fermentor at the upper temperature limit after the wort is fully attenuated. If there is a lack of sufficient suspended yeast, recirculation or "rousing" with a CO_2 purge may help.[15]

With modern equipment, the use of separate vessels is unnecessary; and some unit operations may be combined in a Uni-Tank operation. For example, after a predetermined attenuation limit has been reached, yeast can be removed from the bottom cone and the beer cooled for lagering. Periodic removal of more yeast may be beneficial during the lagering phase to prevent off-flavor development. There are compelling economic advantages for combining fermentation and aging in one tank.[16,17] Other advantages of this concept[16] are: (a) fewer microbiological and foam retention problems because of fewer transfers, (b) more efficient yeast collection, (c) better control of CO_2 levels, with the possibility of eliminating carbonation, and (d) better opportunities for automation.

Addition of Modified Hop Extracts

If modified hop extracts are used wholly or partially to replace the kettle addition of hops, they can be added to beer on transfer to aging. These extracts can be preisomerized hop extracts, which contain iso-α-acids (isohumulone and its homologs), or reduced hop extracts. More details on extracts can be found in Chapter 7. The advantages for adding modified hop extracts to fermented beer compared to adding hops to the kettle include:

- Better utilization (efficiency) of hop-bittering acids (iso-α-acids added); more iso-α-acids surviving in the beer because loss during fermentation is eliminated.
- Kettle-hopping losses are eliminated, producing product with more uniform international bitterness units (IBUs).
- Degradation of iso-α-acids during kettle boil is eliminated.
- Better control of IBU in beers. IBU measurements can be done during aging, and additional extract added if the IBU is below specification.

Beer Recovery

Economics

During normal production, beer is lost in the yeast, in spent filter aids, and in tank bottoms. Yeast cropped by skimming or in connection with beer decantation will have low solids in the slurry. The beer in such slurry, sometimes called barm, may be over 50% beer by weight. It is estimated that up to 2% of total beer output is held up in collected yeast.[18] Tank bottoms may be 2–7% solids. The recovery of beer from these sources may be economically advantageous. The recovery of beer also reduces biological oxygen demand (BOD) and chemical oxygen demand (COD) in brewery effluent, thereby reducing sewer charges, an additional cost saving.

Beer can be recovered from these various sources in several ways. In some cases, yeast strain differences may play a role in the selection of suitable equipment. Methods include centrifuges, membrane or diaphragm filter presses, and other types of filters[19] (discussed later in this chapter).

Quality of Recovered Beer

The microbiological stability of the recovered beer is extremely important. Proper equipment, piping, etc., must be chosen so that acceptable cleanliness can be maintained. Minimum residence time for feedstocks and recovered filtrates is essential for microbiological stability. A flash pasteurization step for the recovered beer is sometimes necessary to obtain a quality filtrate for blending.

A second factor pertinent to the quality of recovered beer is its dissolved oxygen (DO) content. Processing should be carried out under conditions as anaerobic as possible. It is nearly impossible, however, to make transfers without some air pick-up. The introduction of any oxygen will contribute to flavor deterioration. Blending recovered beer at low percentages will help minimize any adverse effects of oxidation.

A third, major consideration is the clarity of the recovered beer. The requirement for clarity after recovery depends upon subsequent processing and blending. If the recovered beer is added to primary production beer during transfer to aging, further clarification occurs downstream. If the recovered beer is added later in the process, it may be necessary to filter it before pasteurization and blending.

Other properties of recovered beer that are likely to vary are the color, pH, and flavor. Color and pH changes generally are not significant because of subsequent blending. Particular attention should be paid to the flavor of any recovered beer. Flavor changes may be caused by yeast autolysis, or the recovery processing steps. Keeping the temperature of the feedstock below 5°C will help minimize flavor changes.

The recovered beer can be blended into normal production beer at any convenient step in the operation. However, the actual choice may depend on the configuration of fermentation and aging equipment, the number and types of beer being brewed, etc. The final extract of the recovered beer may dictate the point of blending; and the recovered beer will have low carbonation. In any case, the brewer must determine a maximum percentage of recovered beer to blend into production beer. Normal practice is to use not more than 10%. Taste testing of blended beers gives more confidence in the use of recovered beer.

Clarification

At the completion of aging, the beer contains some yeast, colloidal particles of protein–polyphenol complexes, and other insoluble material that was driven out of solution by the low pH and the cold temperature during aging. If a brilliant, clear beer is desired, the clarification must remove

Aging and Finishing

these substances before beer packaging can be done. Four basic clarification techniques are used either separately or in combination: (a) sedimentation, (b) use of finings, (c) centrifugation, and (d) filtration.

Gravity Sedimentation

This is surely the simplest method for achieving clarity and was the only method before the development of centrifuges and filters. Historically, the chilling of fermented beer to about 0°C for long periods promoted the sedimentation of yeast and other particles. However, despite its simplicity, caution is needed because yeast autolysis occurs readily, especially if the packed yeast mass begins to heat.[13] With clarification by sedimentation, beer losses are relatively large and clean-up of tank bottoms is costly.

Finings

Although good clarity can be obtained from simple sedimentation, better results can be obtained in less time by using fining agents. Because of their chemical structure, they carry a net positive charge and interact with yeast cells, which are negatively charged, and with negatively charged proteins.[11] Negatively charged proteins have been implicated in haze formation.[20] Consequently, removal of these compounds improves physical stability. Finings increase the volume of tank bottoms and also increase tank clean-up costs and beer losses. The most common fining agent is isinglass, which is made by chemically treating the swim bladders of certain fishes (Chapter 9). Other fining agents are tannic acid, silicates and silica gels, and clays.[13] The use of finings improves subsequent filtration.

Filtration

Filtration generally refers to clarification of beer through several stages to produce a crystal-clear product. The purpose is to remove suspended material and residual yeast, which would otherwise cause the beer to be hazy. The particle size of suspended material in beer is 0.5–4 μm.[21] Particle size information is necessary for the brewer to set filtration parameters.

The mechanisms of filtration can be classified into three types: (1) surface filtration, (2) depth filtration through mechanical entrapment of particles, and (3) depth filtration through adsorption of particles. Surface filtration means that particles are blocked at the surface of the filtration medium because the particles are larger than the pores in the medium. In depth filtration, particles pass into the filtration matrix; the particles are either mechanically trapped in the pores or adsorbed on the surface of the internal pores of the filtration medium.

Filtration may be used at two or more stages after aging, depending on the particulars of cellar operations. The terminology for various filtrations in cellar operations differs from brewery to brewery. The first or primary filtration stage removes the bulk of yeast and suspended material and the second stage produces a brilliantly clear beer. The addition of stabilization agents occurs before primary filtration and they are substantially removed by the filter. Primary filters are almost always powder filters. A turbidity sensor can be installed at the outlet of the filter to monitor filter performance.

As a second stage, polish or final filtration removes any additional suspended solids resulting from lagering at cold temperatures and any adsorbents added for stabilization. These final finishing steps are generally preceded by a final beer-cooling operation to aid precipitation and ensure that the beer reaches the government cellars at the proper temperature. Polish filtration may consist of two separate filters. After a first filter, trap filters may be used as an immediate final stage only to guard against any breakthrough from the upstream filter, not to perform further filtration. Trap filters are usually membrane filters. There should be no further addition of any substance to the beer stream after the last filter, as the introduction of unfiltered liquids may prove harmful to the clarity of packaged beer. Sterile filtration to remove bacteria present in the beer is described in the Section "Sterile Filtration."

Filters

The first category of filters to be discussed use powders or filter aids and are the most popular. The materials for powder filtration include kieselguhr (diatomaceous earth; DE) and perlite (volcanic silicate). Then those that use sheets, cartridges, membranes, etc., will be described. Filters are used not only to clarify beer but also to clarify wort, recover wort from separated trub, and recover beer from tank bottoms.

DE is usually calcined after mining in order to eliminate organic matter. The high porosity of the diatom skeletons is ideal for filter beds, as the liquid passes through the bed while the suspended particles cannot. DE is supplied in a variety of grades from which the brewer chooses to accomplish clarification objectives. The different grades have particle size distributions that affect filter flow rates, filter bed permeability, the degree of filtration (coarse to fine), etc.

Perlite is an ore of volcanic rock containing silica. When crushed and heated, perlite expands to become a light, fluffy powder and is suitable as a filter aid. The expanded perlite is milled and graded, producing filter aids with a range of permeabilities. For any filter aid, the important properties are: good permeability to keep the pressure drop low across the filter and good wetting to ensure uniform dispersion and bed formation. Filters that use powders are sometimes called DE or kieselguhr filters.

Filters that use filter aids (powders) operate on a principle of building a bed or cake of powder on a septum or filter screen. The porous bed creates a surface that traps suspended solids, thus removing them from the beer. Normally, the filter septum is precoated with a filter aid in advance of the beer filtration run. This precoat forms the base layer for the bed. The rough beer to be filtered is dosed with more filter aid, called body feed, at a concentration based on the solids content to be removed. The use of body feed helps to achieve the goal of maximum filter throughput. As beer is run through the filter, the bed increases in thickness because of the body feed, thereby maintaining bed permeability. Various grades of powders are used, depending on the filtration performance desired and beer to be filtered. For example, primary and polish filtrations will use a different grade and thickness of precoat. Body feed may not be required in polish filtration as it is in primary filtration. The different types of DE filters that follow are simply different implementations of these principles.

Filters are operated until the differential pressure rises beyond a designated point, which requires the flow rate to be reduced, or to the point when the bed depth reaches a thickness that bridges the spaces between the septa in the filter. Filter systems are designed to function within a specific range of pressures. An excessive differential pressure can cause: (a) the filter leaves to collapse, (b) the filter shell to burst, or (c) the pumps to fail, as they are not sized to operate with the increased energy needed to maintain the flow rate.

The relationship between filter operational parameters is:

$$\Delta P = \mu V \times L/\beta$$

where ΔP is the differential pressure, V is the specific flow rate (flow rate of beer per unit filter area), μ is the beer viscosity, L is the bed depth, and β is the bed permeability. The specific flow rate will depend on the differential pressure or pressure drop across the filter bed. The differential pressure is directly proportional to the specific flow rate, beer viscosity, and bed depth, and is inversely proportional to the bed permeability. Therefore, at a constant flow rate and viscosity, filter performance depends on the ratio of the bed depth to its permeability. DE is added to the beer (body feed) being filtered in an effort to maintain permeability as filtered particles from the beer reduce the bed permeability. However, this ratio inevitably increases gradually, causing the differential pressure to rise to an unacceptable level resulting in the end of the filter run.

Plate and frame filters are one traditional type used in the industry. They consist of a series of parallel plates covered with filter sheets used to support the filter bed. The frames between the plates control the bed depth. Different numbers of plates can be used, depending on requirements. Most filters allow beer to pass through both sides of the plates, thus doubling the surface area per plate.

Leaf filters consist of a series of circular, stainless steel leaves as perforated support plates. The leaf configuration can be horizontal or vertical. The leaves in horizontal filters have a stainless steel woven septum to support the bed, while vertical filters use the septum on both sides. Their operation is quite similar to plate and frame filters.

The candle filter is of a different design entirely, although the filtration principle is the same. The candles can be porous ceramic but are usually perforated or fluted, stainless steel tubes covered or surrounded by a stainless steel support of various types. This rigid septum is easier to clean than filter leaves used in the powder filters. There is also an operational advantage. The beer is fed to the outside of the candles and the filtrate collected through the inside. The circular design means that the increase in bed thickness during operation is less than other filters, and the pressure drop increase occurs at a slower rate.[22] The ceramic filter can be used for sterile filtration of beer.[23]

Sheet filters are similar in design to plate and frame filters. Whereas the sheet used in powder filters acts as a septum to hold the precoat, in the sheet filter, the sheet acts as the filtration medium. The sheet is usually made of cellulose impregnated with DE. Other materials can be added to achieve both the desired liquid permeability and solids retention. These filters have wide applicability, but are generally used after a primary DE filter because they do not have the capacity of the powder filters. They are also suitable for sterile filtration. Most filters of this type can be easily backwashed and several runs can be made before replacing the sheets. Two disadvantages are the high labor cost of handling the sheets and lack of automation. Sheet filters are often used for keg beer.

A related type of filter is the pulp filter. The cellulose and cotton fibers are formed into circular pads and joined face to face. The filter can be used for primary filtration or later filtration stages and is suitable for sterile filtration.[24] The pads can be reused by washing the material after dispersing the pad fibers in water. Because the pads are reusable, disposal problems are reduced considerably. The use of these filters is very labor intensive.

Cartridge or membrane filters are generally much smaller and serve as sterile filters, and as trap filters that catch breakthroughs of DE occurring upstream. The filter medium is usually a membrane produced from polymeric synthetic materials, for example, Nylon 66 or cellulose esters. With man-made materials, the membrane can be constructed to a desired permeability and mechanical strength with a large surface area. These systems are generally economical and easy to maintain.

Sterile Filtration

Sterile beer filtration is defined as an operation that produces sterile beer ready for packaging with no subsequent pasteurization. As discussed previously, several filter types are suitable for this task: sheet, membrane, ceramic candle, and pulp filters. The type of filter selected for sterile filtration will depend on the brewer's needs and on appropriate features, such as throughput, ease of maintenance, cleaning, and sterilization.

Whichever filter type is used, it must be preceded by at least one other filter that can remove all the colloidal load, including chillproofing agents, and reduce the yeast count, preferably near zero. Typically, the bulk of suspended particles is removed with a DE filter, followed by a DE, sheet, or cartridge filter that will remove residual material sufficiently to reach a haze specification prior to the sterile filter. The sterile filter acts as the final trap of yeast and bacteria.

Sterile filters are not absolute filters. Therefore, the brewer will set a specification for the maximum concentration of bacteria in sterile-filtered beer. As it is possible for a single beer-spoiling bacterium in a bottle or can to spoil the beer, there is a need to balance the risk of spoilage against filter practicality and throughput.

Suitable microbiological sampling and methodology is needed to measure adherence to specifications. Rapid microbiological assay methods are of particular importance to reduce the quantity of product awaiting release to packaging or packaged goods waiting for shipping (see Chapter 16).

To determine if the filtration system will allow microbiological specifications to be met, the brewer must measure the efficiency of the system for removal of microorganisms from beer. This is also called challenge or integrity testing. For example, for a specific yeast/bacterial load in the beer, there is a measured reduction in that load in the beer filtrate. Beer or water is seeded at a known concentration with a beer-spoiling bacterium. Under fixed filtration conditions, beer filtrates are collected and plated. The ratio of colony-forming units before filtration to after filtration, that is, the change in microbiological load across the filter, is called the log reduction value and is expressed as a logarithm. For example, a ratio of 1×10^9 means the system has a log reduction value of 9. A log reduction value of 8–9 is required of a filter system if it is to be useful as a sterile filtration system. Filter media for sterile filtration, particularly sheets and membranes, will have specific log reduction values that help the brewer optimize the system. Filters must be tested with appropriate bacteria. The brewer should select beer-spoiling bacteria common to the brewery to obtain a practical measure of their filter integrity.

Based on the filter system log reduction value and the filtered beer specification, the incoming beer may present a greater microbiological load than that for which the filter is designed. In such cases, sanitation procedures further upstream in the process must be addressed. The sterile filter cannot be expected to remedy poor microbiological practices upstream. In the end, instilling a proper attitude toward sanitation and care with regard to producing sterile beer is invaluable to reducing microbiological problems.

Transfer to Packaging

In the finishing steps of the transfer of aged beer to packaging, a major concern is oxygen pick-up. Chillproofing, dilution, carbonation, and final filtration steps generally occur in a continuous sequence leading to package release or government tanks. These operations are separated by transfer tanks and connected by pipes. Minimizing oxygen pick-up in these vessels and piping/pumping systems is important because it is difficult to correct a package release tank that has a high concentration of dissolved oxygen.

To minimize oxygen pick-up, filter feed, surge, and transfer tanks should be purged with CO_2 or packed with carbonated water prior to use. For filtration, oxygen in filter precoat and body feed in makeup tanks is reduced by CO_2 purging for a sufficient time. Using deaerated water for makeup is also helpful.

A critical area to reduce oxygen pick-up is in the filters themselves; they are usually opened to the atmosphere for cleaning. Even when a closed filter is purged free of the filtration medium, sluicing with water that has not been deaerated presents a risk; the filter may require CO_2 purging. Deaerated water can also be used to purge transfer lines. Account must be made for beer dilution from water left in filters, tanks, and transfer lines.

If packaged beer will not be pasteurized, the transfer of sterile beer to packaging presents additional challenges because the contact of sterile beer with any surface presents an opportunity for contamination. Generally, it is advantageous to dedicate specific transfer and package release tanks for sterile-filtered beer. This reduces the possibility of contamination of tanks and transfer lines from beer which is not sterile. The number of dedicated tanks must be chosen to buffer the sterile filter output with packaging requirements. To prevent contamination of sterile beer further downstream, the use of dedicated bottle and can packaging lines is advantageous.

Stabilization

The stabilization of beers may refer to flavor stability or microbiological stability, although stabilization commonly refers to physical characteristics.

Flavor Stability

Chemical reactions continue to occur after beer is packaged. Many of the changes that lead to stale beer flavor are caused by chemical oxidation. Flavor stabilization, then, generally refers to the protection of beer from oxidative changes. In early research, the cardboard flavor of stale beer was attributed to *trans*-2-nonenal.[25] Furfural and related compounds have also been identified in staled beer.[26] Stale off-flavors are generally attributed to the oxidation of higher alcohols to aldehydes by melanoidins, but there are many more chemical routes participating in the staling of beer. The topic of flavor stability is far ranging and complex,[27] and will not be discussed here in detail. Only two important factors, SO_2 and oxygen, that have a clearly established effect on flavor stability will be discussed.

Sulfur dioxide, in the form of bisulfite ion, protects against oxidative flavor in two ways. It reacts with oxygen, eliminating it from beer and potential oxidation of beer components. It also reacts with aldehydes, which have stale flavors rendering them flavor inactive. While the complex of bisulfite and unsaturated aldehydes is irreversible, the complex with saturated aldehydes is reversible as other chemical species compete for the bisulfite.[28] Sulfur dioxide in beer occurs naturally from fermentation and is increased under conditions of low yeast growth. In addition to naturally occurring SO_2, one can add antioxidants to beer. Potassium metabisulfite or forms of ascorbic acid are sometimes added after fermentation as reducing agents to counteract oxidative changes.

Excluding oxygen from beer is an important step in flavor stability. Because yeast is an oxygen scavenger, once it is removed, any oxygen picked up in processing has the potential to oxidize beer. Therefore, flavor stability is enhanced by excluding oxygen from the beer during aging and finishing operations after the yeast is removed. The use of CO_2 to pack tanks and to transfer beer reduces the possibility of air pick-up. Flavor stability may be enhanced by proper handling of wort in the brewhouse, by reduction of oxygen pick-up during mashing, lautering, and wort cooling, etc.[27]

Biological Stability

Microorganisms can contribute to flavor instability. Certain bacteria (e.g., *Lactobacillus* sp. and *Pediococcus* sp.) and yeasts of the *Saccharomyces* and *Hansenula* genera can spoil beer by producing undesirable flavor compounds, such as VDKs and lactic acid. Generally, brewers conduct microbiological tests specifically for beer spoilage microorganisms. Microorganisms can also grow and form a haze by increasing their number. Proper pasteurization ensures biological stability but requires the heating of beer, which accelerates potential oxidative flavor changes. Biological stability can be achieved by sterile filtration in which microorganisms are removed by special filtration systems. Although sterile beer can be produced by available filtration technology, contamination is still possible during filling and keg operations. In fact, aseptic filling is more difficult than producing sterile beer by filtration.

Physical Stability

Colloidal or nonmicrobiological haze is a result of the precipitation of insoluble complexes formed from beer constituents. The general components are known, but the mechanisms of interaction and complexation are not well understood; see References 29–31 for detailed reviews. There are two types of physical haze: chill haze, which appears when the beer is chilled but redissolves upon warming, and permanent haze, which never fully redissolves under any condition. Beer affected with permanent haze remains cloudy and may even develop a sediment.

Research has shown that several chemical species are present in haze material. The major component appears to be proteinaceous material in the range of 1000–40,000 Da.[32] Other observed components of colloidal haze are polyphenols, and to a lesser extent metal ions and polysaccharides.[33] It is generally believed that complex proteinaceous compounds and polyphenols become associated through hydrophobic and hydrogen bonds involving proline residues of proteinaceous compounds.[30,32] The presence of oxygen may play a role in polymerizing the phenolic constituents. Those portions of the proteins and polyphenols that contribute to haze formation are referred to as the haze-active fractions.

Knowing that haze consists of insoluble protein–polyphenolic complexes, preventative measures can be directed at one or both of these classes of soluble compounds. Three methods are used to "chillproof" beer, as physical stabilization is commonly called: treatment with proteolytic enzymes, use of finings such as tannic acid, and adsorption (Figure 13.3 and see Chapter 9 for details).

Brewhouse Procedures and Filtration

Some remedial measures can be taken during brewing to improve physical stability and to reduce the need for stabilization. However, removal of

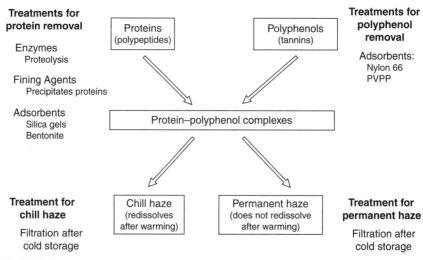

FIGURE 13.3
Chillproofing strategies.

proteins and polyphenols must be done carefully as both contribute to the character of beer — both its flavor and physical characteristics.

Selection of malt with lower soluble nitrogen, good modification, and high diastatic power can lower the proteinaceous content of beer. Proteolysis in the mash tends to reduce protein content, although this reduction of high-molecular-weight protein in the mash may be due to a precipitation mechanism rather than an enzymic one.[34] When adjuncts are used, especially corn or rice, the wort protein level is reduced proportionally. Lower mash pH reduces the solubility of polyphenols. The last runnings from sparging can be high in polyphenol content if the sparge water pH is not carefully controlled.

Proper adjustment of wort boiling helps control the levels of polypeptides and polyphenols. A long and vigorous boil helps coagulate the complexes, and the presence of oxygen will aid the oxidation of polyphenols. However, oxidation of compounds important for flavor stability may also occur. A good, hot break along with efficient wort clarification enhances physical stability.

Measurement of Haze

The measurement of haze or turbidity is based on the principle of nephelometry in which light reflected from particles in solution is measured. The angle of reflection is usually 90°, although smaller forward scattering angles are also useful.[35] The measurement of turbidity depends on the color of the incident light and on the size and shape of the light-scattering particles. Calibration of instruments specifically designed for nephelometry depends greatly on stable particle standards. Formazin is usually used, but more expensive, chemically polymerized spheres can be used.[36] Use of instruments is further complicated by imperfections in the measuring cells. Beer haze determined in bottles introduces large, random errors, whereas the use of optical cells is time consuming. An additional complication arises because different instruments produce different results on the same samples and differ in their responses to particles of different sizes.[36] In-line turbidity meters are often difficult to correlate with laboratory instruments.

It is also possible to rate beer haze visually by comparing the sample with standards usually based on different concentrations of formazin (ASBC *Methods of Analysis*: Beer-27, A).[37] With this method, there is difficulty in obtaining agreement between individuals on the level of turbidity in a sample. At best, this method is qualitative.

Another element of confusion is that different measurement units are used; for example, the American Society of Brewing Chemists (ASBC) and the European Brewery Convention (EBC) use different units of measurement. The major problem, however, lies in trying to quantify the human perception of hazy beer by quantifying particles with a range of sizes and shapes and other light-scattering characteristics; the correlation is not always good.

Carbonation

Basics of Beer Carbonation

Carbon dioxide solubility in beer is usually measured in volumes of CO_2 per volume of beer at standard temperature and pressure. This means that one volume of CO_2 is equal to 0.196% CO_2 by weight or 0.4 kg CO_2/hl (0.92 lb CO_2/bbl).[38] Typical American lagers contain 2.5–2.8 volumes of CO_2. Ales are generally lower. Because beer contains 1.2–1.7 volumes of CO_2 after a normal nonpressurized fermentation,[38] another 1 volume (or about 0.5 kg/hl [1 lb/bbl]) must be added before packaging. Considering that other uses for CO_2 in the process consume CO_2, it is generally economical to recover excess CO_2 from fermentation. In some breweries, losses together with requirements may exceed recovery and CO_2 must be purchased, although under careful conditions breweries can be self-sufficient. The recovery of CO_2 and purity requirements are discussed in Chapter 12.

The amount of CO_2 in solution depends on Henry's law. This law states that the amount of a gas dissolved in a liquid is proportional to the concentration of the gas in the headspace. Therefore, the CO_2 concentration in the beer can be increased by increasing the headpressure of CO_2. Temperature changes the solubility. A temperature increase leads to a decrease in solubility. Therefore, the desired CO_2 concentration can be attained by fixing the temperature and pressure at appropriate settings. Because the solubilities of gases are independent of each other (Henry's law), the level of carbonation has no influence on oxygen pick-up as the product moves through the process.

The time required to reach a desired CO_2 concentration depends on physical factors. Finer bubbles have more surface area per unit weight and dissolve faster than larger bubbles. Moreover, finer bubbles rise more slowly. The longer it takes for bubbles to rise through a tank, the more time there is for solution. Therefore, carbonation stones are designed to form a fine mist of bubbles. If the headspace is filled with CO_2, a larger headspace–liquid interface area will shorten carbonation time. The solution of CO_2 also slows as equilibration is approached.

Pressure and temperature relationships to CO_2 concentrations are used to establish a tank concentration. Measurement of CO_2 in tanks can be done with a sensor separated from the liquid by a membrane permeable to gases. A common alternative to sensors is the Zahm–Hartung method. A metal bottle is filled under controlled temperature and pressure. After establishing equilibrium with the headspace, the temperature and pressure are read and converted by means of a table (*Methods of Analysis*: Beer-13,A)[37] to volumes of CO_2. Corrections can be made for oxygen and nitrogen to improve accuracy. In tall tanks, the CO_2 concentration will be higher at the bottom because of the greater hydrostatic head.

Aging and Finishing

Historical Carbonation

Before there were equipment and methods for recovering, purifying, and reusing CO_2, beer was carbonated by kräusening and low-temperature secondary fermentation, which raised the CO_2 concentration. Thus, "historical carbonation" depended on retaining CO_2 rather than reintroducing it. An alternative approach to retaining natural carbonation is to conduct a secondary fermentation and close the tank at an appropriate time. An overpressure of about 1 atm should yield about 2.7 volumes of CO_2 in beer.[13] These approaches make it difficult to precisely control the CO_2 concentration from batch to batch. Also, allowing the pressure to rise during fermentation may affect yeast growth and change the flavor characteristics of beer (as explained in Chapter 12).

Modern Carbonation

Carbonation can now be done by in-line injection or by in-tank carbonation. In-line injection can be done whenever beer is transferred. However, it cannot be done upstream from DE filtration because CO_2 bubbles would disturb the filter bed. In-tank carbonation usually involves introduction of CO_2 through a carbonation stone in the bottom of the tank. The purpose of the stone is to form fine bubbles of CO_2, which readily dissolve in the beer. Another reason for carbonating in tanks is that oxygen and objectionable aromas can be swept out of the beer if the tank can be open to the atmosphere during the early part of the process. The tank is then closed and carbonation begins.

Standardization

Blending for Consistency

Blending or standardization refers to the mixing of different batches of beer to achieve product uniformity. Generally, blending is done to achieve an exact alcohol concentration, specific gravity, or original gravity. Blending can be done to achieve uniformity in other parameters, for example, bitterness units or color.

Occasionally, blending is used to attenuate an objectionable flavor note. For example, if a fermentation problem led to a high diacetyl concentration or noticeable sulfury character, the beer could be blended with a normal product in an attempt to dilute the objectionable flavor. Blending guidelines are established by brewers to prevent noticeable deviations from flavor uniformity.

High-Gravity Brewing

The need to dilute beer from high-gravity brewing means that additional equipment must be purchased and that the diluent water be treated properly before it can be used to dilute beer. Dilution water must be filtered, sterilized, deaerated, and carbonated prior to use. As already noted, the CO_2 used for carbonation of diluent water must be free of oxygen and other volatile compounds that will alter beer flavor.

Purification of water is usually done with a carbon filter. Deaeration is usually accomplished by spraying water into a closed chamber under vacuum. By adding plate heat exchangers and carbonation equipment, a system can be constructed that produces purified, sterilized, chilled, deaerated, and carbonated water. Finally, accurate metering of diluent water into beer is important for product consistency.

Beer is diluted to the concentration it would normally have if high-gravity brewing were not used. This concentration is called the original gravity (OG), which refers to the solids (°Plato) of the wort from which the beer was produced, whether it was the true wort gravity or not. The OG is related to the "heaviness" of the beer. After dilution, it is useful to know the equivalent wort gravity of the diluted beer in order to maintain product consistency. The OG calculation is given by *Methods of Analysis*: Beer-6,A,[37]

$$OG = \frac{2.0665A + E}{100 + 1.0665A} \times 100$$

where A is the wt% of alcohol and E is the real extract in wt% in the diluted beer.

Dilution calculations can be based on reaching a precise alcohol, OG, etc. To maximize capacity, the dilution step should take place as late as possible in the process. However, this means that beer losses are increased if they occur before dilution. With any blending, whether for dilution of high-gravity brews or for flavor uniformity, sufficient time must be allowed to attain a homogeneous product. Quality checks on the blend are a necessity before release to packaging.

References

1. Haukeli, A.D. and Lie, S., Conversion of α-acetolactate and removal of diacetyl. A kinetic study, *J. Inst. Brew.*, 84:85–89, 1978.
2. Inoue, T., Yamamoto, Y., Kokubo, E., and Kuroiwa, Y., Formation of acetohydroxy acids during wort fermentation by brewers yeast. Rep. Res. Lab. Kirin Brewery Co., Ltd., 1973, No. 16, pp. 11–18.

3. Inoue, T., [Diacetyl formation and amino acid metabolism during beer fermentation]. *Hakko to Taisha*, 36:49–59, 1977.
4. Geiger, E., Diacetyl in beer, *Brauwelt*, 46:1680–1692, 1980.
5. Dellweg, H., Diacetyl und bierreifung, *Monatsschr. Brauwiss.*, 38:262–266, 1985.
6. Godtfredsen, S.E. and Ottesen, M., Maturation of beer with a-acetolactate decarboxylase, *Carlsberg Res. Commun.*, 47:93–102, 1982.
7. van den Berg, R., Harteveld, P.A., and Martens, F.B., Diacetyl reducing activity in brewers' yeast, *Proc. 19th Eur. Brew. Conv. Congr.*, London, Elsevier, New York, 1983, pp. 497–504.
8. Pengelly, B., Sulfurous flavors and aromas. Part I. Organic sulfur and malt, *New Brewer*, 18:17–19, 2001; Part II. Other sources of organic sulfur, *New Brewer*, 18:19–21, 2001; Part III. Inorganic sulfur, *New Brewer*, 18:21–23, 2001.
9. Szlavko, C.M. and Anderson, R.J., Influence of wort processing on beer dimethyl sulfide levels, *J. Am. Soc. Brew. Chem.*, 37:20–25, 1979.
10. Anderson, R.J., Clapperton, J.F., Crabb, D., and Hudson, J.R., Dimethyl sulphide as a feature of lager beer, *J. Inst. Brew.*, 81:208–213, 1975.
11. Hough, J.S., Briggs, D.E., Stevens, R., and Young, T.W., in *Malting and Brewing Science*, 2nd ed., Chapman and Hall, New York, 1982.
12. Masschelein, C.A., The biochemistry of brewing, *J. Inst. Brew.*, 92:213–219, 1986.
13. Coors, J.H., Cellar operations, in *The Practical Brewer*, Broderick, H.M., Ed., Master Brewers Association of the Americas, Madison, WI, 1977.
14. Drost, B.W., Fermentation and storage, *Proc. 16th Eur. Brew. Conv. Congr.*, Amsterdam, Elsevier, London, 1977, pp. 519–532.
15. Pajunen, E. and Enari, T.-M., Accelerated lagering and maturation, in *EBC Fermentation and Storage Symposium, Monograph V*, European Brewing Convention, Elsevier, New York, 1978, pp. 181–191.
16. Haag, O.B., Symposium on 'storage and maturation of beer': (iii) Economics, *Proc. 13th Conv. Inst. Brew.*, (Australia and New Zealand Section), 1974, pp. 81–86.
17. Lindsay, J.H. and Larson, J.W., Large volume tanks — a review, *Tech. Q. Master Brew. Assoc. Am.*, 12:264–272, 1975.
18. Mueller, G., Preparation and processing of brewery yeast sludge using high-pressure chamber-filter presses, *Tech. Q. Master Brew. Assoc. Am.*, 19:57–62, 1982.
19. Young, I.M., The case for beer recovery, *Brew. Dig.*, 60:24–28, 1985.
20. Savage, D.J. and Thompson, C.C., Electrofocusing studies on the formation of beer haze, *J. Inst. Brew.*, 78:472–476, 1972.
21. Morris, T.M., Particle size analysis of beer solids using a Coulter counter, *J. Inst. Brew.*, 90:162–166, 1984.
22. Reed, R.J.R., Beer filtration, *J. Inst. Brew.*, 92:413–419, 1986.
23. Beer, C., Sterile filtration of beer, *Tech. Q. Master Brew. Assoc. Am.*, 26:89–93, 1989.
24. Beckett, M.H., Sterile pulp filtration implementation and optimization at the Adolph Coors Company, *Tech. Q. Master Brew. Assoc. Am.*, 22:53–55, 1985.
25. Jamieson, A.M. and van Gheluwe, J.E.A., Identification of a compound responsible for cardboard flavor in beer, *Proc. Am. Soc. Brew. Chem.*, 192–197, 1970.
26. Hashimoto, N., Oxidation of higher alcohols by melanoidins in beer, *J. Inst. Brew.*, 78:43–51, 1972.
27. Bamforth, C.W., The science and understanding of the flavour stability of beer: a critical assessment, *Brauwelt Int.*, 17:98–110, 1999.

28. Dufour, J.-P., Leus, M., Baxter, A.J., and Hayman, A.R., Characterization of the reaction of bisulfite with unsaturated aldehydes in a beer model system using nuclear magnetic resonance spectroscopy, *J. Am. Soc. Brew. Chem.*, 57:138–144, 1999.
29. Moll, M., Colloidal stability of beer, in *Brewing Science*, Pollack, J.R.A., Ed., Academic Press, New York, 1987, Vol. 3, pp. 1–327.
30. Siebert, K.J., Troukhanova, N.V., and Lynn, P.Y., Nature of polyphenol–protein interactions, *J. Agric. Food Chem.*, 44:80–85, 1996.
31. Siebert, K.J. and Lynn, P.Y., Mechanisms of beer colloidal stabilization, *J. Am. Soc. Brew. Chem.*, 55:73–78, 1997.
32. Asano, K., Shinagawa, K., and Hashimoto, N., Characterization of haze-forming proteins of beer and their roles in chill haze formation, *J. Am. Soc. Brew. Chem.*, 40:147–154, 1982.
33. Dadic, M. and Belleau, G., Beer hazes. I. Isolation and preliminary analysis of phenolic and carbohydrate components, *J. Am. Soc. Brew. Chem.*, 38:154–158, 1980.
34. Lewis, M.J. and Wahnon, N.N., Precipitation of protein during mashing: evaluation of the role of calcium, phosphate and mash pH, *J. Am. Soc. Brew. Chem.*, 42:159–163, 1984.
35. Gales, P.W., A comparison of visual turbidity with turbidity measured by commercially available instruments, *J. Am. Soc. Brew. Chem.*, 58:101–107, 2002.
36. Buckee, G.K., Morris, T.M., and Bailey, T.P., Calibration and evaluation of haze meters, *J. Inst. Brew.*, 92:475–482, 1986.
37. American Society of Brewing Chemists, *Methods of Analysis*, 8th ed., revised, American Society of Brewing Chemists, St. Paul, MN, 1992.
38. Eicker, J., Utilities engineering, in *The Practical Brewer*, Broderick, H.M., Ed., Master Brewers Association of the Americas, Madison, WI, 1977.

14

Packaging: A Historical Perspective

Tom Fetters

CONTENTS

Introduction	551
Packaging in the Early Brewing Industry	552
Kegs	552
Bottles and Stoppers	553
Revolution in the Brewing Industry	554
Crowns	554
Mass-Produced Glass Bottles	555
Glass Treatments	556
Bottle Filling	557
The Expansion of the Brewing Industry	558
Cans	558
The "Tin Can"	559
The Aluminum Can	559
Secondary Packaging	561

Introduction

Packaging has been used by Mother Nature in many ways. Shells, for example, exist in many varieties to protect their tender contents. Nutshells protect nuts, oyster shells protect the oyster, and eggshells protect the egg. Other variations such as the tightly wound leafy husk covering of an ear of corn is nature's protection for the corn kernels; and the heavy peel of the orange, lemon, and grapefruit protects the inner fruit.

    It was natural for man in prehistory to develop jugs and casks to contain liquids as well as straw baskets and woven bags to carry seeds and grain. Glass bottles for wine as well as other ceramic containers were common in the middle ages, but modern packaging dates back only to 1809, to Nicolas Appert's invention of the "tincan" as a solution for Napoleon's need to feed his vast armies. The French Directorate had offered an award of 12,000 francs for a satisfactory method of preserving food which Appert won by closing the can and heat treating the food and can for a specific time and temperature.

Today, packaging ranges from cans for food, beverages, and household and industrial products to plastic bottles and containers for many of the same products; aluminized pouches; fiber boxes and cardboard boxes; and composite containers that blend materials for the best protection at lower prices. Yet, the principles of packaging remain unchanged: to protect the product from shipping damage; to shield the product from spoilage over a reasonable shelf life; and to be economically attractive to use.

The "package" is used primarily to move products through a distribution system to the consumer, but it can also, itself, provide a point of purchase message of "Buy THIS product, please." Through the use of attractive labels and colors, the consumer is drawn to select one container over another when confronted with two or more different choices of the same product. Modern packaging is constantly evolving with lighter weight metal containers, holograms to attract attention, composite materials to gain the best performance from each component, and the attempt to keep up with the ever-changing consumer attitude. As an example, modern soft drink vending machines are being converted from 12 oz aluminum easy open cans to 20 oz polyethylene terephthalate (PET) plastic bottles as manufacturers respond to a younger consumer market that wants more than 12 oz in a serving, and the convenience of resealabiltiy of the bottle. At the same time, adult consumers find that the new PET bottle adds a plastic odor and often contains a flatter product, while giving them more drink than they want at a higher overall cost.

Packaging in the Early Brewing Industry

Kegs

Much of this evolutionary change has been mirrored in the brewing industry where beer packaging traditionally throughout the 1800s was in wooden kegs. The kegs were filled by small, local breweries and delivered by horse-drawn wagons to the taverns in the town and in the surrounding countryside. When the kegs were emptied, they were returned again to the brewery to be cleaned and refilled. The kegs were sealed with a wooden bung that was inserted with a mallet. Later, the wooden bung was replaced with a plastic bung that was mechanically applied.

Prior to the development of the refrigerated railcar, kegs were packed in ice in boxcars for shipment to remote locations and other major cities. Adolphus Busch pioneered the use of a double-sheathed boxcar together with a network of railside icehouses to allow his St. Louis brewed Budweiser brand to become the first national brand in the U.S.

When the first canned beer was released for sale in 1935, 75% of American beer was sold in kegs with the other 25% in glass bottles. Today, cans and bottles account for 90% of beer packaging.

Modern kegs are made of noncorrosive chrome nickel stainless steel. The steel is drawn into two deep drawn, short cylinders with hemispherical ends for greatest strength. The two halves are then welded together to provide a smooth surface that is easily cleanable. A threaded tap-port is sealed with a built-in tap tube, and both ends of the keg are protected from shipping damage by concentric chime rings of rolled stainless steel with integral handholds that are welded to each end. Before shipping, the kegs are pickled, rinsed, and passivated to ensure a hygienically sealed surface. Aluminum kegs have also been used in the past 40 years.

The racking room of a brewery is set aside to clean, fill, and ship kegs. Returned kegs are cleaned with a chlorine solution or a caustic solution and then rinsed thoroughly to remove these cleansers. The keg is then placed at a loading station where it is pressurized with CO_2 before introducing the beer. The beer is forced into the keg against the back-pressure of CO_2, which reduces foaming. When filled, the keg is automatically sealed by the same machine.

Bottles and Stoppers

The first use of cork-sealed glass bottles was in 1602 when Dr. Alexander Nowell discovered that this package could store beer for long periods. As early as 1759, a Virginia brewer wrote to a London ship captain that he wanted 50 gross of "cheap corks for small beer and in case you go to any place where bottles are cheap, buy me 4 groce." Bottles were not cheap at that time and Charles Carroll, who signed the Declaration of Independence, left "3 beer glasses and 18 dozen quarts and 2 dozen Pint bottles" in his will to his heirs.

The breweries of the late 1800s rarely bottled their beer. Instead, they sold beer in kegs to individuals who bottled beer as small businesses. Using hand-blown bottles with a porcelain stopper together with a rubber gasket, these bottlers siphoned the beer from the kegs into the bottles in a one-at-a-time procedure. The beer was not pasteurized and had a very limited shelf life as a consequence. This was bottling for immediate consumption type, and just a step above carrying a tin pail of beer home from the tavern.

The porcelain stopper of the glass bottles was pulled down into the sealed position by a heavy wire bail that leveraged the stopper into sealing the bottle. This bottle, which was very hard to clean for refilling, was never used for carbonated soft drinks in America, and saw only limited use after the crown was introduced. Some imported European beers continued this type of package into the 21st century.

The development of pasteurization by Louis Pasteur in 1876 led to prevention of additional growth of yeast after beer had been packaged. The rapidity with which beer staled before this discovery made mass production of beer of no interest to the brewers. When glass was introduced to the brewing industry, Pabst, a large Milwaukee brewer, converted 10% of its production from kegs to glass in 1893.

Revolution in the Brewing Industry

Crowns

The award of a patent for the crown, or bottle cap, to William Painter in 1892 led to a revolution in both the carbonated soft drink market and the brewing industry. The crown design featured a round disc with a corrugated skirt with 21 pleats that could be crimped into place by a reciprocal crowning movement. Coupled with a natural cork liner, the crown could seal marginal glassware finishes. The crown eliminated the difficulty in washing the bottles that other novel closure systems with restricted openings had required. A crown-finish bottle, with the crown removed, offered the full opening to permit full caustic cleaning of the interior.

Earlier, Painter had invented the Baltimore Loop Seal in 1885, which featured a rubber plug with a wire loop that could be used to pull the plug from the bottle using a hook or nail. Priced at 25 cents per gross, it was a single-use closure. While simple, it nevertheless took some effort to properly seat the plug in the hand-blown bottles of the day. Six years later, Painter submitted the "crown cork" to the patent office in Washington.

Painter formed a new company which grew to be the Crown Cork and Seal Co., Inc. of today. Painter and his successors watched the company grow to include plants around the world and by the 1950s had become the largest crown manufacturer in the world. Other inventors brought out competitive designs and formed their own companies such as Bond Crown, Mundet, Consolidated Cork, and W H Hutchinson and Sons. When Painter's patent expired, the other companies began to manufacture the 21 corrugation seal that had proved to be the best bottle cap.

The crown led directly to a new industry: mass production of beer in glass bottles that allowed even wider distribution. Based on this new way of doing business, the small local brewers were bought out or were forced out of the market by the formation of new regional and national breweries with full distribution across regions of the country.

Natural cork with the holes that were present in the bark as stripped from the trees in Portugal and northern Africa presented a problem in that 1% of the crowns would leak. In 1915, a composition cork made of cork particles that had been glued together and extruded into rods from which the cork liners were sliced was able to provide a leak-free liner. Breweries insisted on an improved crown by having the companies add an aluminum spot, smaller than the cork disk that would protect the beer from the natural cork flavor or odor, yet allowed the cork disk to function as a seal for the bottle.

The use of the crown required an investment in all new bottles with a special crown finish that was useless unless fully uniform. Bottle manufacturers had not been held to high standards before this time and had to develop new machines to provide a seamless finish to accept the new closure. Thus, the glass industry began to modernize as machines were created to mass produce bottles on a scale unheard of before this time.

The next major development was made in the late 1950s with the introduction of the plastic-lined crown with distinct formations that each company touted as superior to all others. Both vinyl plastisol compounds and plastic vinyl pellets were commercially used to manufacture polyvinyl chloride liners of a uniform consistency. While some breweries resisted the change at first, the plastic liner did provide relief from the vagaries of the natural cork production that was subject to weather and climatic conditions.

The twist crown was introduced during the mid-1960s, which allowed the final consumer to easily open a bottle without the use of a separate opener. Made of a lighter gauge steel and using a lubricated plastic liner material, the new crown relied on a threaded glass finish to allow easy removal. Nearly all beer in glass in the U.S. is packed with twist off crowns.

As can production grew dramatically in the brewing industry, the market for glass fell and consequently crown sales fell forcing the revamping of the crown manufacturing market. Major crown suppliers like W. H. Hutchinson and Sons, Consolidated Cork, Mundet Cork, Armstrong Cork, Hoosier Crown, Kerr Packaging, and Bond Crown were absorbed or fell by the wayside with only Crown Cork and Seal and Zapata surviving in the North American steel crown market.

Mass-Produced Glass Bottles

Michael Owens formed the Owens Bottle Company after developing the first completely automatic bottle making machine in 1903. These mass-produced bottles replaced the hand-blown bottles that had been available and provided a new level of accuracy and uniformity in height, weight, and capacity of the bottle as well as providing a perfect crown finish. Owens Bottle purchased the Illinois Glass Company of Alton, Illinois, and the Chicago

Heights Bottle Company to form the new Owens–Illinois Glass Company in April of 1929.

The universally accepted standard beer bottle became the returnable long neck, "export" bottle that is still present in the bar trade in the 21st century. The use of this standard amber beer bottle then led to the formation of other companies that developed new automatic filling and capping machines that relied on the new bottle quality to provide a perfect package. Crown Cork and Seal Co., Inc., was among these companies that included Meyer and some smaller manufacturers.

This 12 fluid ounce "export" bottle weighed about 10 oz. Shortly after the end of prohibition, this bottle served as the model for light-weighting development. Soon new nonreturnable designs evolved like the "Stubbie," the "glass can," and short neck bottles that dropped the glass weight to 6 oz while still containing the 12 fluid ounce standard fill.

Two new varieties of one-way bottles for getting beer to the troops were developed in the Second World War. The "one-way" bottle used 7.5 oz of glass and went to overseas servicemen. The "Throw-away" bottle used only 6.5 oz of glass and was used exclusively for shipments to military camps within the U.S. In 1963, 51% of packaged beer was sold in returnable bottles and 14% in one-way, no deposit bottles.

Glass Treatments

Brewers need to be aware of two treatments that glass bottles may receive during manufacturing. Hot end treatment (HET) is applied just after the bottle is blown and before the stresses are relieved in a lehr, a gas fired oven. The HET is usually a tin or titanium salt spray applied to 540°C bottles in a single-file tunnel to make the bottles resistant to scratching. However, if applied on the finish, the salt residue can severely aggravate the formation of rust on the crimped crown closure. Glass makers are aware of the need to keep the HET away from the finish, but have been known to err on occasion. The American Glass Research Institute has an instrument to measure the HET on glass bottles that can be very useful in problem solving. A perfect fire-polished surface can provide a tensile breaking stress of 100,000 psi, while invisible abrasions can reduce this to 15,000 psi and a diamond scratch will bring the breaking stress down to 5500 psi.

Cold end treatment (CET) is applied at the exit of the lehr where the temperature is now about 40°C. This CET is a stearate solution that is used to provide a slippery surface on the glass bodies to reduce glass binding in the conveying systems. Care should be taken that the CET does not enter the mouth of the bottle.

The "twist-off" crown finish was developed in 1964 to allow a lighter baseweight crown to be crimped over small threads in the finish to permit the

Packaging: A Historical Perspective

consumer to remove the crown by twisting it off. Bottles with the twist-off threads are nearly always designed for the nonreturnable beer bottle and cannot be easily resealed. Brewers should be aware of thread imperfections at the thread split where the body halves meet. High torque issues involving twist-off crowns can often be caused by irregular thread formation caused by glass fillets at the thread split area. Thread mismatch is another type of glass imperfection that can directly lead to high torque complaints. Finally, bottle mold offset can produce bottles that mismatch through the entire finish area and which produce very high removal torques.

The ROPP finish was developed in 1965 for the use of aluminum, roll-on, pilfer-proof closures. This cap was readily adapted to the larger quart bottles where resealability was a consumer preference.

Bottle Filling

Breweries use new one-way and export ware as well as returned bottles from the trade. The new glass can be run directly to the filler, but the returned glass has to be passed through a washer. Here, the bottles enter pockets on a continuous chain and are taken in stages in and out of a caustic solution bath followed by sequential rinse cycles (see Chapter 15 for details). When released from the washer, the bottles go directly to the filler.

At the filler, the bottles receive either a blast of air to remove any dust or debris that might have deposited during shipment or, in some cases, a thorough rinse of fresh water. The bottles then enter the filler where they are pressurized with CO_2 to allow the beer to be filled under backpressure to reduce foaming to a minimum. As the bottle fills, the pressure is slowly released and the beer remains stable as the bottle leaves the filler turret. The bottle is immediately turned into a crowner turret where either a bottle "knocker" or a tiny spritz of high-pressure water impacts the beer surface to produce foam. The foaming beer crests to the top of the bottle, driving out all the air in the headspace, just as the crown or closure is applied.

The bottle is rinsed and then proceeds to a pasteurizer for a hour passage as the temperature is raised to 60°C for 20 min and then cooled again to room temperature. The bottles are then labeled, identified with a packing date and in some cases a "best used by" date and then placed in cases, sixpacks, or hi-cone wrap.

Bottle returns were a problem in many states as bottles were abandoned after use which left parks and beaches littered with empty bottles. Many eastern states adopted a bottle deposit that required the purchaser to pay an additional fee that would be returned when the empty bottle was returned to the store. The deposit was paid to the returning person, which prompted a small business opportunity for kids who scoured the parks for bottles and returned them to collect the deposits.

The Expansion of the Brewing Industry

The impact of the cheap crown cork and the mass-produced, crown finish glass bottle on the brewing industry led to the formation of a large number of regional breweries, and even some breweries with national distribution. The Pabst Brewing Company of Milwaukee became the largest in the nation in 1895 followed by Anheuser–Busch of St. Louis and Schlitz Brewing Co of Milwaukee. Pabst produced more than 1 million barrels of beer annually at this time. By 1900, there were 1816 breweries in operation across the United States Most of these served small local markets.

Beer was now able to be distributed by rail refrigerated freight cars that bore large billboard lettering for the branded beer. Railway trains that included bright yellow URTX (Union Refrigerator) cars marked for Schlitz, Miller, and Pabst as well as white SLRX cars bearing the Anheuser–Busch eagle were popular sights. The problem came when the railroad, with an order for a refrigerator car, would be forced to deliver an empty car emblazoned with another brewery's herald to be filled by the local competitive brew. This led to the banning by the Federal Trade Commission during the late 1930s of brewery advertising on the flanks of refrigerator cars.

By the end of prohibition, only 160 breweries remained in business in the United States By the 1950s, large regional breweries could ship beer in bottles across their home state and into surrounding states. Many of the larger breweries began to build satellite breweries in other states to spread the distribution. Coors, with its single brewery in Golden, Colorado, near Denver, developed a cult following in which college youth would drive long distances across the Midwest to reach a Coors distributor, usually only one or two states away from Colorado, and then return with the beer for a major weekend party.

Cans

The first canned beer was sold in Richmond, Virginia, in January 1935 when G. Krueger Brewing Company of Newark, New Jersey, and American Can cooperatively released this new 12 fluid ounce package. This was a steel, three-piece tin can with protective coatings, on the interior walls, trademarked "Keglined," to protect the beer.

American Can Company licensed the Vaughn Novelty Manufacturing Company to produce the "Quick & Easy" can opener to puncture the end to allow the beer to pour easily for direct consumption. This can was marketed as "un-refillable" and had greater public acceptance over the later cone top can. Vaughn added a crown lifter at the opposite end of their can opener and the "Church Key" became a household necessity.

The wax-lined cone-top beer can was introduced later the same year by Continental Can Company as a tin-plated steel, three-piece can with a cone top that allowed a crown to be crimped in place. This 12 fluid ounce package was released by Schlitz Brewing Company of Milwaukee with much wider distribution across the Midwest and East Coast.

The first nonreturnable bottles came out later the same year. Very quickly, 186 of the 507 licensed brewers of the day adopted the metal can as a means of "no deposit–no return" to compete against the returnable bottle with its deposit.

The "Tin Can"

The coating of steel with tin to eliminate rusting had been first practiced in 1300 in Bohemia by an Englishman who had worked in a tin mine in Cornwall. In 1810, Peter Durand patented the "tin cannister" for preserving foods. By 1853, canned condensed milk was being packaged by Gail Borden. The conventional sanitary can, which was commonly used throughout the 20th century for food packaging, was introduced in 1900.

Beer cans must withstand the maximum internal pressure developed by the product during storage including the high pasteurization temperatures ($60°C$). This internal pressure can reach 100 psi, based on packing 4.3 volume beer and storage at $40°C$.

Continental Can began to produce the flat-top beer can in 1942 and the cone top can for beer was gone by the end of 1956. The Schlitz Brewing Company pioneered the larger, 16 fluid ounce can in 1954 as the public demand for larger servings had an impact on the market.

The Aluminum Can

Aluminum was first produced in 1825 by Hans Christian Oersted. Henri Sainte–Claire Deville first found a way to manufacture aluminum in 1854 and the metal, more precious than gold, sold for $17 a pound. Charles Martin Hall of Ohio found an electrolytic process that he patented in July 1886 and then, with the help of Andrew Mellon, a Pittsburgh banker, started the Pittsburgh Reduction Company, which became The Aluminum Corporation of America, known popularly as Alcoa. Hall's procedure dropped the price per pound to $8 and by 1958 aluminum was a mere $0.26 a pound. In 1957, Continental Can began to work on a draw and wall iron can at their Van Nuys, California, plant.

The aluminum can was first marketed by Coors Brewing of Golden, Colorado, in 1959 via an association with "Primo," a Hawaiian brewery. The aluminum can end appeared in 1960 and was promoted by Schlitz as the "soft top" can.

In 1960, the aluminum producers, Alcoa, Reynolds, and Kaiser, decided to enter the can making business with Reynolds setting up a line in Florida. Alcoa, however, backed out and decided to provide aluminum end stock to Continental Can instead of setting up an entire new business.

Later, in 1962, Iron City Beer, a Pittsburgh Brewing brand, was distributed in Virginia with an Alcoa aluminum tab-opening end that did not require an opener. This blunt ovoid tab lay tight to the end surface and caused some broken fingernails when pulled open. In February 1963, Schlitz introduced the same end nationally and by 1965, 70% of all beer cans were "easy-open."

A variation of the tab, a ring opening for easier grip and safer opening, was introduced to the market in 1965. American Can's "Touch'n Go" and Continental Can's "Ring-pull" were two of the prominent tab improvements that reduced fingernail damage.

Following the conversion to aluminum ends for the three-piece beverage can a two-piece aluminum can with the same aluminum end was soon developed. In 1964, Reynolds Metal Company introduced this new 12 oz container that was used to pack Hamms Beer at St. Paul, Minnesota. Four years later, Kaiser Aluminum was able to bring the price down to become fully competitive with the steel can.

With these innovations, canned beer passed glass bottle sales in 1969 for the first time. The removed tear strip and tab was easily discarded at beaches and parks, and became a hazard to barefoot visitors who complained bitterly about the problem. When the complaints reached the breweries, the pressure was transferred to the can companies to find a solution. After a year or more of development, the can industry introduced the nondetachable tab-opening can.

The can companies had too great an investment in steel cans to transfer to aluminum cans without a struggle. They introduced lighter weight steel cans as well as tin-free steel cans that could be welded (as done by Continental Can) or cemented (as done by American Can). Continental's Conoweld followed American's Miraseam can to market. The production of steel three-piece cans peaked in 1973 when 30 billion cans were produced.

However, aluminum replaced steel as the material of choice for beer cans during the late 1960s and eliminated the use of steel beer cans in America by the 1970s. Some countries, notably Denmark, prohibited the sale of beer cans until 2002 when the law was finally relaxed in that Scandinavian area. The law had been written to protect Denmark's sensitivity to ecology.

Even the aluminum beer can has been changed dramatically in cost, but not in appearance by reducing the metal thickness in stages over the past 20 years. Where a pound of aluminum once yielded 35 cans, this same pound today would produce 44 cans. The original aluminum beer can had a 211 size end that has been reduced down to 204 size through the use of "necking in" engineering to allow the size reduction and the subsequent cost reduction.

Secondary Packaging

Secondary packaging was developed as a means of delivering returnable bottles to and from the retail trade. The 24-cell wooden case was accepted in the years after 1850 as the best means to transport bottles, first by horse drawn wagon and later by truck. The full flap, heavy cardboard reusable carton with 24 cells was later adopted by the brewing industry and is still in use. Smaller, fiber one-way cartons holding 24 or 12 cans became the accepted packaging for groceries and liquor stores, with the six-pack still a viable medium for quick sales. In some cases, an open cardboard tray will hold four six-pack units shrink-wrapped together to provide a unit measure.

Hi-cone multipack plastic collars are used to hold six cans in merchantable units that are much more economical for the brewer than the cardboard wrapper. Small home versions of kegs and "beer spheres" have a small portion of the market aimed at party use. The latest innovations in beer packaging have included can-shaping techniques that give a very individualized look to the package, the use of special textured exterior coatings, the use of holograms on the can exterior, and other ways of attracting consumer attention.

The future holds many new surprises as the market swings back toward a preponderance of smaller microbreweries as the consuming public looks for specialized beers and novelty beers. Malt liquors and fruit-flavored malt liquor drinks such as Mike's Hard Lemonade and Doc Otis as well as the pioneering Zima have shifted the market through a series of brews such as Dry, Dog, Red, and Lite. Packaging developments will no doubt contribute to the technical and marketing success of these new products.

15

Packaging Technology

Alexander R. Dunn

CONTENTS

Introduction	565
Levels of Packaging	566
Packaging Materials	566
Packaging and the Brewing Industry	566
The Cost of Packaging	567
Cans and Bottles	567
Kegs and Casks	568
Capital Cost of Packaging Equipment	569
Packaging as a Marketing Tool	569
Environmental Aspects of Packaging	569
Trends in the Beer Market	570
The British Market	570
The U.S. Market	571
The Canadian Market	571
General Comments	572
Glass Bottles and Bottling	573
Glass	573
Colorless (White Flint) Glass	574
Green Glass	574
Amber Glass	574
Bottle Molding	574
Bottle Treatment	576
Inspection and Dispatch	576
Properties of Bottles	576
Dimensions	576
Bottle Strength	577
Crowns	577
Adhesives	578
Bottling Plant	578

　　　　Bottle Rinsing 579
　　　　Bottle Washing 579
　　　Bottle Filling 581
　　　　Air Evacuation 581
　　　　Filling 582
　　　Crowners 582
　　　Labeling 582
　　　Bottle Fill Height Detection 582
　　　Full Bottle Inspection 583
　PET Bottles 583
　　　PET Bottling Lines 583
　　　Conveying 583
　　　Returnables 583
　　　Rinsing 583
　　　Filling 584
　　　Leak Detection 584
　　　Labeling 584
　Cans 584
　　　Beer Can Materials 585
　　　Can Formation 585
　　　Lids 585
　　　Canning 586
　　　Can Filling 586
　　　Volumetric Fillers 587
　　　Can Seaming 587
　　　The Seamer 588
　　　　Undercover Gassing 588
　　　　Lid Placement 588
　　　　Lid Secured 588
　　　Fill Height Detection in Cans 589
　Kegs and Kegging 590
　　　Kegs 590
　　　Manufacture of Kegs 590
　　　Minikegs 591
　　　Keg Lines 592
　　　Conveyors 592
　　　Offloading/Depalletizing 592
　　　External Washing 593
　　　Spear Check and Torque 593
　　　Internal Keg Cleaning and Sterilizing 593
　　　Keg Filling 595
　　　Types of Keg Racking Machinery 595
　　　　Linear Machines 595
　　　　Rotary Machines 595
　Pasteurization 596

Pasteurization Units 596
Flash Pasteurization 598
Quality Safeguards 598
    Temperature Drop 598
    Gas Breakout 599
    Plate Failure 599
Tunnel Pasteurization 600
    The Water Heating and Spraying System 601
    The Water System 603
Packaging Line Efficiency 604
    Relative Machine Speeds 604

Introduction

Early man used containers for food and other useful items for thousands of years before the introduction of packaging. Animal hides, woven baskets, and pottery were some of the early materials used. There is, however, an important distinction between these early containers and packages. Packaging evolved in order to offer a greater degree of protection to goods as commerce increased. Without packaging, the quality of transported goods would be unacceptable and the end cost would be much greater.

Packaging has a number of definitions in order to cover a wide range of goods and operations. One definition is: "An industrial and marketing technique for containing, protecting, identifying and facilitating the sale and distribution of agricultural, industrial, and consumer products." An EU-derived definition is used with regard to packaging waste and its recovery: "Packaging means all products made of any materials of any nature to be used for the containment, protection, handling, delivery, and presentation of goods."

Beer has some additional requirements when it is packaged, which do not necessarily apply to all foods and drinks. First, pressure; a beer package needs to be able to withstand the pressures generated when a carbonated liquid is heated, either during pasteurization or when transported at high ambient temperatures. Second, light; beer is degraded by sunlight and this must be avoided. Surprising then that glass is so often used for packaging. By using colored glass together with additional external protection and covered transport, this problem can be largely avoided. Finally, impermeability to gases; it is vital that the gas composition of beer is not altered in its package. This means not only preventing CO_2 from getting out, but also preventing access of oxygen.

Levels of Packaging

Packaging comes in essentially three levels. A primary package is one that is in direct contact with the contained product. For beer the material must be inert, pressure resistant, strong, and impermeable. Secondary packaging encloses the primary packages in order to make them easier to handle and store. For bottles and cans, this covers items such as cases, trays, cluster packs, and Hi-Cone. Kegs and barrels go onto pallets or layer boards. For small packages, the next level up, or tertiary package, is usually a pallet, and these are nowadays stretchwrapped for further protection and stability. This produces a pack of around 1 m^3 and weighing about a ton, which is a convenient load for handling, stacking, and transporting.

Packaging Materials

The materials used for packaging beer are widely used elsewhere, but there are certain nuances in the way beer is packaged that make it distinct. None of the primary materials used have ideal properties, but our packages are designed to make the best of what we have.

Glass, for example, has several drawbacks. It does not have an easily printed surface, so we use labels. It is not opaque, so colored glass is generally used to protect the beer. It is also heavy and breakable. However, on the plus side, it is inert, attractive, reusable, and well established.

Steel and aluminum fit the ideal more closely, but need to be coated in various ways to make them inert and to facilitate printing and decorating.

Plastics have had limited success for beer, although polyethylene terephthalate (PET) has been tried. It works for soft drinks but PET cannot be pasteurized as a whole package. Its barrier properties are not quite good enough for a sensitive product such as beer. This is an area where research is ongoing and it may eventually change the market.

Wood is the traditional material for casks but the difficulties with cleaning means that it has given way to metal containers that are lighter, stronger, and easy to clean.

For secondary packaging, there is nothing unique to the brewing industry — although for kegs and casks there are some distinctive shapes for palletizing. The trend in materials for secondary packaging has been to move away from wood and corrugated board. Plastic sheeting and paperboard have come more into use. These materials can enhance the appearance of a pack and assist in marketing. They also improve the strength-to-weight ratio, particularly for plastic shrinkwrap, and this has helped overall in producing more durable and lighter packs and usually at a reduced cost.

Packaging and the Brewing Industry

Packaging is of enormous importance to the brewing industry for a variety of reasons, and this will probably increase. Brewing has moved, over the last century, from being a small-scale industry where the product was consumed from casks to a large-scale industry where beer is transported huge distances. The product is expected to be uniform with a long shelf life rather than being instantly consumed before it can deteriorate. It is, therefore, vital that beer quality is maintained and this has become a huge technical challenge.

The Cost of Packaging

Cans and Bottles

It often comes as a surprise to nonbrewers how cheap the raw materials are in relation to the volume of product obtained and the final price. But it also comes as a surprise that packaging materials cost so much. Indeed, a can lid is probably worth about as much as the raw materials in a can of beer, and the can is worth several times as much. In a brewery with a significant small pack operation, the cost of packaging materials will probably be the biggest single item of spending, followed by employment costs, and then raw materials (Table 15.1). Negotiating a good price for packaging materials is obviously crucial to a successful operation, and it also highlights the need to keep packaging line rejects to a minimum. The quality of these materials and their handling on the line will have a major influence on product cost.

TABLE 15.1

Cost of Packaging in British Breweries

Item	Cost (£/hl)[a]
Total product cost	20–30
Brewing raw materials	3–3.50
Total labor	3.50–4.50
Packaging labor	2–3
Cans	10–15 (size dependent)
Ends	3–4
Cartons	1–1.50
Hi-Cone	0.3–0.4
0.5 l RB	16–20
0.5 l NRB	10–14
Crowns	0.5–1
Labels	1–1.50

[a] These costs are for a midstrength (9°Plato) product and exclude duty, which roughly doubles the product cost exbrewery.

On a unit basis, the cost of a can is in the range of 5–7 pence each and a lid is almost 2 pence. Bottles are comparable in cost to cans (6 pence), but returnable bottles are about 50% heavier and more expensive (around 9 pence) than their nonreturnable counterparts. One can also see from these figures (Table 15.1) that the other items such as cartons, crowns, and labels have a modest cost in comparison with bottles and cans but still significant. Hi-Cone for cans is quite cheap and is popular because of its low cost and effectiveness.

The cost of labor in a packaging plant is significant and it usually comprises more than half the total site workforce. Obviously, this is an area for cost reduction and modern automated lines are designed for minimal manning. Good design of the line is important here so that an operator can monitor more than one machine if they are reasonably close together. Fewer than ten people per shift is now the norm compared with the dozens per shift of the past. Claimed figures, however, need to be scrutinized to see whether they include or exclude peripheral staff such as QC, maintenance, and fork lift operators. Typical comparable costs (in percentage terms) are shown in Table 15.2 for U.S. beer packaging.

Kegs and Casks

The economics of packaging beer in bulk containers are quite different from cans and bottles. Metal containers should last for years and are probably used 10–12 times per year, so the initial outlay in purchasing these containers is well covered by the time the container is taken out of service or lost. The cost of a 50 l keg is £24 and if we assume it is used 100 times, then the cost works out at £0.24/hl, significantly less than buying cans or bottles.

TABLE 15.2

Cost Components of a Pallet of Beer

Component	% Component Cost
Bottled Beer	
Labels	1
Labor	5
Crowns	8
Carriers	9
Beer	28
Bottles and trays	48
Canned Beer	
Labor	2
Carriers	2
Trays	3
Beer	24
Cans	69

The additional costs in the brewery are manpower, which will be similar to bottles and cans, and the utilities and detergents needed to clean and sterilize the containers. These latter costs will probably be around £1/hl. This makes bulk beer a cheap packaging operation, but unfortunately the cost of running a bar or hotel is far greater.

Capital Cost of Packaging Equipment

Most major brewing companies spend very large sums on capital equipment each year as machinery wears out, becomes obsolete, or production requirements change. It is probable that more of this money is spent in the packaging area than in brewing. One of the main reasons for this is that brewing equipment usually has a life span measured in decades, whereas packaging machinery needs replacing or modification on a fairly regular basis. This is particularly so in a dynamic small-pack market where container types, sizes, and secondary packaging are constantly changing. Marketing departments seem to have an unlimited appetite for spending money, especially on new packaging formats.

The cost of a new packaging line is very hard to stipulate because it may or may not include building costs, which on a greenfield site are usually around a third of the total. The size and nature of the line obviously has an effect as well. A typical ballpark figure for a large can line would be £10–15 million and probably £5 million for a new building to house it. If we assume a figure of £15 million then such a line would produce 1.5 million hl per year if operated on three shifts. Over a depreciation period of 10 years this comes to £1.5 million per year or £1/hl, which is not a lot in comparison with an annual bill for cans and ends of about £25 million. However, it does illustrate the need to keep packaging lines busy; a depreciation cost of £3/hl would not be acceptable. Kegging machinery is generally less expensive than a canning or bottling plant and less likely to be altered over its lifetime, so again this is a relatively cheap process.

Packaging as a Marketing Tool

Packaging is a major tool in the marketing of the product. Packaging should sell itself. There are numerous ways in which a package can be changed in shape and color to make it more distinctive and appealing. This not only applies to the bottle or can but also to the tray, shrink, or carton. The liquid is of almost minor importance with some brands where the packaging is of high quality and high price.

Environmental Aspects of Packaging

The environmental aspects of packaging should not be forgotten in the brewing industry. Beer packaging is very visible, before, during, and after its use and excess packaging will only attract adverse reactions. There is a trade-off between minimizing material use and getting the marketing message across at lowest cost. As we shall see subsequently, in many countries the law requires returnable packaging, possibly with a deposit system, to discourage rubbish formation. Where nonreturnable packaging is widely used, collection of the various wastes is to be encouraged.

Trends in the Beer Market

The three basic types of beer package used by brewers — bottles, cans, and kegs — are used all over the world but the proportions vary widely from country to country. There are a number of reasons for this such as social habits, economics, and legislation. The markets in Britain, Ireland, and the Czech Republic are dominated by keg beer. The United States now mainly uses cans and the rest of the world mainly uses bottles. Scrutiny of the changes over the last few decades, in Britain, the United States, and Canada, illustrates how the scene has evolved.

The British Market

Britain is a good example of a society with a tradition of social drinking and the majority of beer is still served from kegs and casks in pubs and bars. Some bottles and cans are consumed on these premises and this is a growing trend partly due to a wider product range. Where sales hardly justify a draught installation, cans and bottles in an illuminated cold

TABLE 15.3

UK Beer Sales by Package Type (%)

Year	Draught	Returnable Bottles	Nonreturnable Bottles	Cans
1960	64	34		2
1965	68	30		2
1970	73	24		3
1975	75.8	16.7	0.6	5.9
1980	78.8	10.3	0.5	10.3
1985	77.3	6.6	2.4	13.8
1990	71.5	5.4	3.4	19.6
1995	65.7	3.0	7.1	24.2
1998	63.9	1.9	10.1	23.6
2000	62.1	1.4	11.0	25.5
2002	58.3	1.0	13.0	27.7
2003	56.8	0.7	13.4	29.1

Packaging Technology

cabinet are an attractive alternative. Pubs and bars are the only places left where returnable bottles are available in any quantity, since supermarkets will not handle returns.

For the take-home trade, supermarkets dominate due to price competition and convenience, and they are a major influence on package styles. Cans have increased in sales enormously and nonreturnable bottles have also grown in recent years, with an amazing variety of decoration and secondary packaging.

The trends over the last four decades are shown in Table 15.3, which illustrates how draught still dominates and how returnable bottles have been replaced by cans and nonreturnable bottles.

The U.S. Market

The U.S. market has shown a trend similar to that in Britain over the years in the way cans have come to dominate the take-home market (Table 15.4). This switch has been even more marked and has been accompanied by rapid market growth. In 1935, 70% of beer was sold on draught, the rest being bottles, with cans about to emerge. By 1950, the whole market had changed and the figures had switched over. This trend continued until about 1980 with draught down to about 10%, cans 65%, and bottles 25%. In recent years, bottles (nonreturnable) have increased at the expense of cans.

The Canadian Market

The changes in the Canadian beer market have been fewer than in Britain or the United States and this is largely due to the culture and legislation around the drinks industry. Canada has a history of using returnable bottles

TABLE 15.4
U.S. Beer Sales by Package Type (%)

Year	Draught	Packaged
1935	70.5	29.5
1940	48.3	51.7
1945	35.7	64.3
1950	28.2	71.8
1955	22.1	77.9
1960	19.3	80.7
1970	14.1	85.9
1980	12.1	87.9
1990	11	89 (cans 59, bottles 30)
1995	11	89 (cans 53, bottles 36)
2000	9	91 (cans 51, bottles 40)

TABLE 15.5

Apportion of Take-Home Packs and Draught in Canada (%)

Year	Bottles	Draught	Cans
1993	73.8	11.4	14.8
1995	69.7	11.5	18.8
1997	68.6	11.9	19.5
1999	69.3	11.7	19.0
2001	69.4	11.2	19.5
2003	67.2	10.4	22.4

and there is strong environmentalist support for continuing with this. Beer for home consumption is only available through controlled liquor stores and the brewers have a deposit system on bottles to increase return rates. This relatively small number of outlets makes control of distribution and the return of empties much simpler than in an open and mixed market. Sales of draught beer in bars have stayed at around 10–11%. Cans have made only modest headway and only in British Columbia and Alberta are their sales ahead of returnable bottles. The recycle/return ethic is very strong with can recycling at 85% and returnable bottles at 97.5% (Table 15.5).

General Comments

A comparison of these three counties emphasizes considerable differences, one dominated by kegs, another by bottles, and the United States by cans. There are no particular technical or economic reasons for these big differences. In Britain, the brewers have substantial control over the retail outlets, so kegs dominate and social patterns have not dramatically changed. In the United States, the market is open and the drinkers have made their choice. Legal and environmental aspects restrict Canada, so returnable bottles dominate.

In the rest of the world, returnable bottles still dominate the market. Part of this is tradition, but simple economics and technical aspects come in. Where breweries are small or medium size, the cost of distribution is not great and the cost of packaging is important. If returnable bottles can be used 20 or 30 times then the whole operation is much cheaper. Handling returnable bottles is always going to be more labor intensive than other forms of packaging, but in many countries this is not a serious problem. Scuffed bottles can be tolerated if the price is right. Glassmaking is also a widely practised industry with easily available raw materials, and the problems of handling returnables are well known. In contrast, kegs are only suitable for high-volume outlets and there is significant expense and expertise needed to maintain quality and the bar equipment. In contrast, no special equipment is needed for dispensing beer from bottles apart from a

TABLE 15.6

A Comparison of Large and Small Pack Pallet Weights (with Product)

Type of Package	Units/Pallet	Pallet wt (tons)	Product wt (tons)	% Product
500 ml can	1920	1.096	0.960	87.6
550 ml RB	1008	1.206	0.554	46
550 ml NRB	1008	0.989	0.554	56
275 ml RB	1512	1.032	0.416	40.3
275 ml NRB	2184	1.113	0.601	54
2 l PET	384	0.851	0.768	90
11 gal (50 l) keg	9	0.603	0.450	75
22 gal keg on locator board	4	0.491	0.400	81.5

decent refrigerator. Cans rely on a sizeable industrial base for their manufacture and distribution, as well as needing a bit more know-how in the packaging plant to maintain quality and efficient running.

The trends in the mature beer markets of the world seem to be toward greater diversity in the types of package available and also toward nonreturnable formats. This may be reversed or modified by future environmental concerns. In the European Union, the Packaging Waste Directive is being implemented with targets being set for recovery and recycling. The impact of new packaging materials is also a factor. PET has been around for about 30 years now and is a major material for water and soft drinks. Numerous announcements have been made over the years claiming that PET has now been modified to improve its barrier properties, making it perfect for beer. We are still not quite there.

Another aspect of packaging and beer distribution is the cost of transport (Table 15.6). If and when oil runs out, transport will get more expensive. It is interesting to compare how much the different package types cost to transport by looking at how much the packaging material itself weighs. In Table 15.6, different beer packages are compared by looking at the weight of beer in a typical pallet on a truck and comparing it to the total weight. It can be seen that PET and cans are very easy to transport, but glass, and particularly returnable glass, is very difficult. This makes the long-distance transport of glass unattractive compared with cans.

Glass Bottles and Bottling

Glass

Glass has been used as a packaging material for centuries and its use for beer goes back a long way. Its use in brewing increased rapidly with the mechanization of the glass molding process about a century ago. Bottles for beer are a major part of the glass market in most countries.

The three main colors for beer bottles are clear (white flint), green, and amber. Color is important from the quality viewpoint with beer and typical absorption curves are shown in Figure 15.1.

Colorless (White Flint) Glass

To obtain completely clear glass, it is necessary to have raw materials with no impurities. The main impurity is usually iron oxide and its levels need to be below 0.04%, otherwise the glass has a blue-green tinge. To a certain extent this can be neutralized by adding "decolorizers," which are traces of cobalt and selenium.

Green Glass

Green glass is obtained by adding small amounts of iron oxide and chromium oxide to the melt. Iron oxide on its own gives a pale green color at levels of about 0.15%.

Amber Glass

Amber is probably the most common color for beer bottles and is obtained by adding carbon as a reducing agent to a glass melt with moderate levels of iron oxide. It also requires a trace of sulfur.

Bottle Molding

The bottle molds are specific to a particular type of bottle and are made from cast iron. Each mold is usually in two halves so that they can be rapidly opened and closed automatically. Molding a bottle takes place

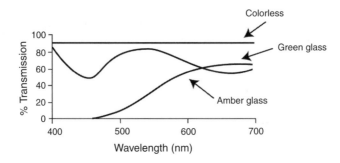

FIGURE 15.1
Typical absorption spectra for different colored glasses used for bottles.

in two stages. First, the gob of molten glass drops into a simple blank mold, which is shaped at the bottom to form the neck of the bottle, that is, upside down. As soon as this is formed, air is blown in from below and molten glass is pushed up into the top of the mold, forming a crude bottle shape. During this brief time, the glass rapidly cools down from about 1000°C as the mold is kept at about 500°C. The sides of the mold are then pulled away and by gripping the neck the part-formed bottle is transferred and inverted into the second mold. Here it is blown again (from the top) and the final bottle shape is made. It is then released and transferred onto a conveyor. The time cycle for a bottle to be formed is only in the region of a few seconds.

The two-stage process described above is known as the blow-and-blow process and is illustrated in Figure 15.2a. Another method of container forming that is gradually becoming more popular is the press-and-blow process. This involves initially inserting a plunger into the mold to press out a blank shape. It is then transferred over to be blown into its final shape as usual (Figure 15.2b). This latter process is increasingly being used for beer bottles and enables lighter bottles to be made with greater uniformity.

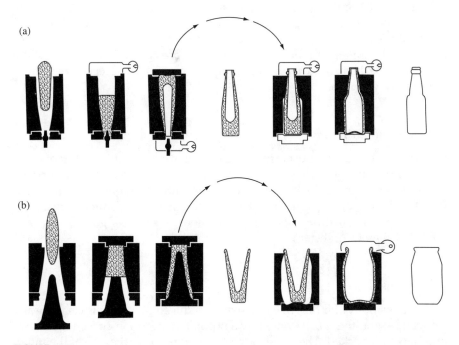

FIGURE 15.2
Bottle forming using (a) the blow-and-blow method (b) the press-and-blow method and method.

Bottle Treatment

When the bottles are transferred from the second mold onto a heated conveyor, the outer skin may be at a temperature of 300°C whereas the interior is at about 500°C. This creates potential stresses in the glass that need to be removed. At these temperatures the glass is almost solid. In order to relieve these stresses, the bottles are reheated to about 550°C and slowly moved through an oven known as the annealing lehr. Gradually, the bottles are cooled down over a period of about 1 h, and they emerge at the back end at a temperature of about 100°C.

Before going into the annealing lehr, the bottles are sprayed with a solution of monobutyl tin chloride. This decomposes and results in tin atoms migrating into the glass. This hardens and strengthens the glass surface.

After emerging from the annealing lehr, the bottles are sprayed yet again on the outside, this time with oleic acid or a similar organic compound. This has the effect of improving the surface lubricity of the glass and improves its resistance to scratching and scuffing, making bulk handling easier.

Inspection and Dispatch

The bottles after annealing and treatment are now ready for despatch, but before this they go through rigorous inspection for flaws and faults. Some of this is still done by skilled operators but increasingly checks are carried out by optical equipment. This involves shining light at all sorts of angles and detecting unusual reflections if there is a breakage. Wall thickness in a bottle can be measured by a radiofrequency machine. Neck size can be measured by a probe.

Properties of Bottles

Quality, dimensional, and performance standards on bottles are frequently agreed in many countries by the members of the industry. In Britain, British Glass is the trade organization representing the glassmakers and it issues booklets on agreed industry standards. In the United States, the Glass Packaging Institute sets overall standards and has good links with the U.S. Brewers Association.

Dimensions

The dimensions of bottles, like any mass-produced article, will vary slightly from target and these usually follow a normal distribution curve when large numbers are checked. There are inevitably variations in the size of the mold, the weight of glass falling into it, and the way the bottle retains its shape when being cooled. The tolerances on height and diameter are, usually for beer bottles, $\pm 1.0-1.5$ mm, depending on size, and a typical range of volume is ± 6 ml or 2%, whichever is the larger.

Bottle Strength

The internal pressure that a bottle can withstand before bursting is of real importance in brewing because this governs the failure rate of bottles when heated in the pasteurizer and their ability to withstand shocks and general rough handling. Surface treatment of bottles before and after annealing has been mentioned already and these treatments have the effect of increasing bottle strength and wear resistance. A bottle during pasteurization may have to withstand 7 bar internal pressure, but in practice it will have an average failure pressure much higher than this. Typical agreed values for beer bottles are 10 bar for nonreturnable bottles and 12 bar for returnables.

It is found that repeated use and abrasion gradually weaken a bottle and consequently returnable bottles need to be substantially heavier (about 50% more) and stronger than nonreturnables. A returnable bottle will probably lose a third of its strength during use before it either fails or is removed due to excessive scuffing. There is also a wide range of bursting pressures because of the variations in the molding conditions. Table 15.7 gives some typical bursting pressures for old and new bottles.

Bottle design is also a factor in determining strength, and sharp angles in a bottle neck or base will weaken it and should be avoided.

Crowns

Metal crowns for bottles were invented in the United States by William Painter in 1892 (see Chapter 14). Since then they have dominated the closure market for glass beer bottles and many soft drinks as well. To fit around the bottle top, the outer edge has 21 serrated "teeth" and the underside is coated with a flowed-in plastisol liner.

Nowadays, crowns are manufactured by the billion all over the world and are regarded as the standard closure for beer and soft drinks bottles. They

TABLE 15.7

Typical Bottle Bursting Pressures

	Pressure (bar)	
	New	Old
High	33	20
Mean	23	15
Low	13	10

have the benefit of low cost, reliability, convenience, and pressure resistance. In addition to the universal prize-off crown, there is also a significant market for twist-off crowns. Crowns are made of low-carbon steel or sometimes stainless steel which is supplied as sheets usually with a thickness of 0.24 mm, although other sizes are also used.

Adhesives

The majority of labels in use are still paper based, and wet adhesives are used to stick them to bottles. These adhesives fall into two main categories: caseins and resins.

Casein is a protein that will dissolve in dilute aqueous ammonia, and when various stabilizers and additives are added, it produces a powerful adhesive for label application. These casein-based adhesives have the advantage of being removable by caustic in a bottle washer. They are fast setting and still work below freezing point.

Resin-based adhesives are derived from starches and dextrins and are also soluble in dilute aqueous ammonia. Self-adhesive labels have grown in popularity in recent years. These labels are mounted on a smooth backing strip and peeled off as they are applied to the bottle. They have the advantage of instant adhesion to the bottle and can be more precisely positioned. They tend to be used with nonreturnable bottles. A problem with returnable bottles is label removal and this is hindered if the adhesive or paper is resistant to wetting as a result of paper coating or lamination.

Bottling Plant

Bottling machinery has evolved over a considerable number of years but continues to change as the demand for increased speed, quality, and level control gets ever greater. Progressive improvements in automation have meant that numbers of operators have steadily fallen from dozens to a mere handful. The stages in the operation of a bottling line are as follows:

- Offload empties
- Wash/rinse empties
- Inspect empties
- Fill and crown
- Pasteurize
- Contents check
- Label and inspect
- Cartons/crates
- Palletize

There are two main types of plant, returnable lines with a bottle washer and nonreturnable lines with a rinser only. In practice, however, lines often have both facilities.

Bottle Rinsing

New nonreturnable bottles have very little in the way of contaminants, but in order to ensure cleanliness they are normally rinsed before filling. Rinsers come generally in two forms, a rotary machine where the bottles are gripped at the neck and then inverted for spraying, and linear machines with an inverted belt where the bottles are gripped at the neck, inverted, and rinsed, and then placed back on a conveyor.

Bottle Washing

The washing of beer bottles on a returnable line is a major component of the operation. There are a number of variables that can affect the operation:

- Temperature
- Detergent strength and composition
- Bottle condition
- Water quality
- Contact time

Bottle washing machines are designed to clean the bottles by a combination of steeping and jetting, so that heat, chemical, and mechanical actions are used to remove labels, glue, foil, dirt, and residual beer. The combination of soaking and jetting is the norm is today's machines.

Bottle washers are large machines and work by loading up rows of dirty bottles into pockets on a continuous carrier chain, where they are held until they are discharged clean at the end of the cycle. There are two basic types of machine:

- Single ended — where the loading and removal of bottles is done at the same end. One operator can easily monitor these machines.
- Double ended — where the discharge end is at the opposite end to loading.

There are numerous varieties of internal layout for bottle washers.

The continuous belt that carries the bottles snakes around inside the machine in a set pattern so that the bottles are put through a series of cycles of jetting, inversion, draining, steeping, rinsing, and so on until they are clean.

A typical sequence of events inside the washer would probably be as follows:

1. *Presoaking and rinsing.* The first requirement once the bottles are loaded up is to drain out any residues and give the bottles a preliminary soak to remove any easily soluble residues. This also serves to warm up the bottles. The water from this presoaking will be high in effluent and should be continuously discharged and replenished.

2. *Immersion.* After rinsing, the bottles are immersed in the main detergent tank where the internal and external surfaces are thoroughly soaked. During this extended steep, the labels, glue, and foil must be loosened as well as any beer residues inside the bottle. When bottles are lifted out to transfer from one soak bath to another, they are jetted with the detergent to help in label and dirt removal.

3. *Rinsing.* The bottles are now conveyed above the soak tanks inverted and subjected to a number of internal and external rinses by jetting to get rid of the detergent residues and any remaining solids. After rinsing, the bottles are drained and then discharged onto the outfeed conveyor to the empty bottle inspector.

4. *Temperature.* Typical temperature ranges from 60 to 85°C for the main detergent tank, but to avoid thermal shock to the bottles, the temperature is ramped up gradually. This is where multiple steep tanks help by having individual temperature control. The presoak tank is usually at 35–40°C and the next at 55–60° with the main soak at 75–85°C. Rinsing is carried out in stages with a drop of 10–15°C between each stage so that the bottles emerge at ambient. A temperature difference of less than 25°C between sections should be enough to avoid breakages in the machine.

5. *Detergent.* The detergent of choice for bottle cleaning is caustic soda (NaOH). A typical strength is 1–3%. Higher strengths should be avoided, otherwise bottle etching and scuffing may increase. A number of formulated bottle washing additives are available, which help to emulsify and disperse the dirt as well as sequester metal ions and suppress foaming. Polyphosphates, EDTA, gluconates, and glucoheptanoates are common additives.

6. *Contact time.* The time needed to soak and clean bottles is temperature and caustic dependent, but a typical total cycle time is 10–15 min. There are two other aspects to bottle washing that deserve mentioning: label removal and ventilation.

    Labels come off in the main caustic steep tank, where the detergent is circulated to maintain its temperature and strength. The suspended labels are removed on a sieve at the infeed to the recirculation pump and then put through a hydraulic press to remove excess caustic. Labels need to be taken out of the soak tank quickly, otherwise they disintegrate.

Ventilation of the bottle washer is needed to remove hydrogen gas, otherwise there is an explosion risk. Hydrogen is generated by the dissolution of aluminum in the foils around bottle necks.

7. *Empty bottle inspection (EBI).* The modern generation of EBIs use solid-state CCD optical inspection systems for checking bottles from as many angles as possible and compare the images with preset values. Any nonconforming bottles are rejected and can be manually inspected when suitable. The machines are relatively compact and do not require manual attention unless reject rates are serious. Good bottles should have reject levels under 1%. The kinds of faults occurring in imperfect bottles are as follows:

- Bottles with residual internal dirt such as dead insects
- External contaminants such as traces of paper or adhesives
- Bottles containing residual caustic
- Defective bottle openings such as a chipped neck or damaged thread
- Cracked sidewalls or inclusions such as gas bubbles or ceramics
- Badly scuffed bottles that have been used too often

Bottle Filling

Large-scale bottle fillers are always rotary machines and the size is related simply to the desired capacity. The largest machines can have about 200 filling heads, be around 5 m in diameter, and produce up to 100,000 bottles per hour. Various ways of filling bottles are used, but in brewing it is always by using gas counterpressure to keep CO_2 in solution and also by the isobarometric method, that is, the bottle pressure is the same as the counterpressure on the beer supply, so beer runs into the bottle effectively by gravity. Bottle contents are controlled by filling to a predetermined height, but recently filling by volume alone has become available. Because of its carbonation, beer is always filled cold between 0 and 3°C.

The main process objectives during filling are:

- No product loss
- Consistent contents
- No microbiological (or chemical) contamination
- No loss of CO_2 or pickup of oxygen

Air Evacuation

Older bottle fillers filled beer against CO_2 counterpressure, but, not surprisingly, it was found that bottle air contents were too high due to the air in the bottle not being removed. Modern fillers have a vacuum system installed that evacuates about 90% of the air in the bottle before counterpressuring, which reduces potential air pickup by a factor of nearly 10. Most recent fillers do this operation twice to get the air contents down to levels of about 20–40 ppb.

Filling

Filling the bottle as it rotates on the filler head takes up about half the available time and there are essentially two different ways of doing this, that is, long tube and short tube.

The older method of filling is with a long tube that descends almost to the bottom of the bottle. Beer is run into the bottle from the filler bowl under the CO_2 counterpressure of about 15–20 psi (1–1.3 bar) and residual gas goes out by way of a vent tube near the top of the bottle. This long tube method is relatively slow because of its length and the diameter being restricted (10–12 mm), but oxygen pickup is quite low because of the quiet filling conditions with the submerged tube.

Fillers with short tubes give a greater throughput because the tube is for venting displaced gas, and beer goes down the outside of the tube. To avoid turbulence, the tube has a conical section on the outside and this deflects beer so that it runs quietly down the bottle walls.

Crowners

Crowns should be applied to bottles as quickly as possible after filling to keep air contents down and prevent loss of beer. As a result, it is usual for crowners to be situated close downstream from the filler, and they are frequently integrated into the filler bloc to get full synchronization of the two operations.

Labeling

The variety of labels available for bottle decoration is enormous, but fortunately the methods for label application are relatively few, and rotary labelers are almost universal in the brewing industry. Water-based wet-glued labels are the most common, especially for returnable bottles, but a higher degree of decoration can be obtained with self-adhesive labels. These are always applied postpasteurization. Labelers are often a major source of stoppages on a packaging line.

Bottle Fill Height Detection

Bottle fill heights can be checked with a gamma source, as with cans, but the transparency of glass means that infrared methods of scanning the neck can also be used. Where the bottle inspector is situated after the application of neck decoration that obscures the product, a gamma source should be used.

Full Bottle Inspection

A full-bottle inspector, which is an image recognition system, rather like the EBI, often checks the position of labels and foil.

PET Bottles

PET Bottling Lines

Bottling lines for PET are similar to those for glass bottles, but a number of variations have evolved since PET was first introduced for soft drinks in 1970. First, the bottles themselves are often delivered new as preforms, solid pieces of PET rather like thick test tubes but with the screw neck and transport flange already in position. The bottles are formed by heating the preforms in an infrared heater and then blowing them out in a mold.

Conveying

Because PET bottles are so light, they have to be conveyed very gently when empty, and many bottling lines have sections where the bottles are conveyed by air blowing.

Returnables

PET bottles are reusable but cleaning has to be modified compared with glass. First, old caps and tamper evident threads have to be removed and the labels taken off. It is common for PET to have large wrap-around labels where there is sufficient overlap for the label to stick to itself like a sleeve, rather than stick to the bottle. This makes label removal much easier as they can be cut with a knife and sucked off.

Bottle washing involves clamping the bottles because they float, and careful temperature control is needed to avoid melting or distortion. After detergent steeping they are rinsed and given a sterile rinse.

Rinsing

A PET rinser usually works by having a rotary machine that inverts the bottles for rinsing and then lowers them onto the conveyor. Bottles must be rinsed sterile.

Filling

Since PET bottles cannot be pasteurized, it is necessary to either sterile filter or flash pasteurize beer just prior to filling and "clean room" technology may be utilized to ensure that microbiological risks are kept to a minimum.

Leak Detection

Because of the flexible nature of PET bottles, it is possible to detect leaking bottles after filling by passing the bottles through a section at the labeler where they are squeezed gently and the internal pressure measured. This test can also be done on empty bottles by sealing the top and pressing the bottle.

Labeling

PET bottles frequently have large wrap-around labels for improved display and these have the added advantage of being easily removable if the bottles are returnable. The labels in this instance would be supplied as a continuous roll and cut as they are applied.

Cans

The first flat-topped beer cans were produced in the United States in 1935, whereas cone-shaped cans were produced in Britain in the same year. In 1955, flat topped, three-piece cans were introduced in Britain; then in 1963, Ernie Fraze of the Dayton Reliable Tool Company invented the aluminum easy open end and a can of 2.11/16 in. diameter (66 mm) known as the 211. The following year the two-piece extruded aluminum can was developed, and over the next 6 years these gradually improved as the techniques for drawing and wall-ironing (DWI) cans were modified. In 1970, these cans were introduced in Britain; then in 1974, the 211 can was replaced by the 209 can with a slightly smaller lid.

At the same time canning speeds were increasing rapidly. By 1980, the three-piece drinks can had gone, and then in 1987 there was a complete change of lid size as the 206 can was introduced. This is still the size used in Britain but in the United States the size of the soft drinks cans have gone further (1991) to the 202 end, 5 mm smaller. There have been other changes to lids apart from size reduction. The detachable ring pull, which came with cans during the 1960s was largely replaced by the "stay on tab" (SOT) in 1989, and in 1995, there was a downgauging of can material.

Beer Can Materials

The two metals of choice for making today's beer cans are steel and aluminum. Aluminum was the first to be used because it is easily drawn and wall-ironed. The metal is not pure aluminum but is an alloy containing small amounts of manganese and magnesium. The lid material is a different alloy containing 4–5% magnesium for greater strength.

The steel used in can making is low carbon steel that is electrolytically coated with tin to a thickness of only about 1 μm. It is then given a passivation treatment and very lightly oiled. These treatments give the steel protection against corrosion and the right surface properties to allow it to be shaped into a can.

The choice of materials for cans differs sharply around the world. In the United States, drinks cans are largely made from aluminum. In Europe, the situation is more evenly balanced because aluminum is more expensive. In Germany, it is steel that is predominantly used, and in Britain it is evenly divided between the two, with can prices being almost identical most of the time.

Can Formation

The metal for can manufacture is supplied as large coils of metal 1.2 m in width weighing 10 tons (steel) and with a thickness of 0.25–0.27 mm. This is uncoiled and lubricated and then fed into a continuous cupping press where circles of metal are cut out and simultaneously formed into shallow cups with sidewalls about 3 cm high. Their diameter is wider than the 66 mm of the final can. The next stage is to redraw the cup to almost the 66 mm diameter and this gives a slightly deeper sidewall. The cup is then forced through a series of tungsten carbide rings, which stretch and iron the sidewall to extend it to just beyond its final size and thin down the walls to 0.09 mm. At the same time, the base, which does not change in thickness, is deformed to give the familiar domed shape. This is frequently profiled on the outer edge to make the cans stackable.

Once this shape has been achieved, the excess metal is trimmed off, the can is thoroughly washed and dried, and then a series of lacquer and decorative coatings are applied. The final stage is to neck in the top of the can by putting it into a die necking machine, which first bends the wall inward and curls the rim outward.

All the cans are then visually inspected for pinholes through a light tester and in more recent times high-speed video cameras have been used.

Lids

The first stage in lid formation is similar to can bodies in that coiled aluminum sheet is the starting point. This sheet is first coated on both sides and then the ends are stamped out by a press that forms the curled edge at the same time. The next stage is to pass the lids through a lining machine, where a precise bead of lining compound is applied around the inside of the curl. This compound will eventually form the seal when the lid is seamed onto the can. The lid is then inspected by a video camera system before being sent forward to fix on the tab.

Beer cans come in a variety of sizes in different countries. The two accepted sizes in mainland Europe are 330 and 500 ml, but in Britain they are 440 and 500 ml. The United States uses imperial units of comparable size to those in Europe. Almost all cans are of 65–66 mm diameter, but lid sizes are variable. The driver for this is reduced weight in the lid, and the latest reduction in size means saving about 0.5 g from the overall weight. The two most common sizes of lid are the 206 (57 mm) and 202 (52 mm).

Wider can diameters are available in the United States and parts of Europe (three-piece) and recently the 65 mm cans have been stretched to pint size (565 ml), which shows that can technology is still developing. Since two-piece cans were introduced in the 1970s, there has been a 30% reduction in the use of materials and developments are continuous. It is probable that further weight reductions can be achieved in the future by using thinner gauge metal and also by altering the base and lid profiles.

Canning

Can lines have a lot in common with bottling lines, especially nonreturnable lines, but a number of features have evolved especially for canning such as seaming, level detectors, Hi-Cone, and final packing. The key sequences in a canning operation are:

- Can depalletizer
- Rinser
- Filler
- Seamer
- Pasteurizer
- Level inspection
- Hi-Cone (or cartons)
- Tray packer
- Palletizer

Can Filling

Can filling has a lot of similarities with bottle filling, and in general appearance the filler has either a central bowl or ring-type beer reservoir. The sequence of operations for cans has to be different because of the low wall strength of cans. They cannot be evacuated to remove air, so CO_2 purging is used to displace the air, either into the filler bowl or to atmosphere. This is followed by CO_2 counterpressuring then filling.

Filling of cans is often by a set of nozzles or annular ring arranged in a circle inside the filling head and the can is filled by beer running down the inside wall. This is very fast, only about 3–4 sec.

Volumetric Fillers

Because of the larger surface area of cans compared with bottles, fill height is much more important in determining can contents. A change of level of only 1 mm means about 3 ml in contents, and the natural variation in conditions gives a noticeable spread of contents. In a move to improve this situation for brewers and other canners, a new type of filler has been developed where a predetermined volume of product is measured out into a chamber above each filling head and then poured into the can once it is purged. The liquid measurement is from the base of the chamber against a back pressure, so it is very quiet. This type of filler dispenses with the conventional beer ring tube and there is a static beer buffer tank outside the filler connected to the individual beer chambers.

Can Seaming

Can seaming is a crucial operation in the production of canned beer with very little margin for error in the way the lids are placed on and mated with the can body. The operation is done at extremely high speeds of well over 2000 cans/min, a long way from the speeds of 40 cans/min attained in 1910, or 500 cans/min in the immediate postwar era. Consistency is absolutely essential, and in addition to this the transfer of cans must be smooth to avoid spillage of beer, and headspace air must be eliminated.

The seaming operation starts back at the filler discharge because the two operations of filling and seaming are fully linked. All three components must be perfectly timed so that can transfer is as smooth as possible to avoid spillage, fallen cans, and jam-ups. In order to keep the exact spacing obtained when the filled cans are discharged from the filler, the transfer conveyor runs at exactly the same speed and there is an indexing conveyor running alongside. This additional conveyor has metal fingers, which stick out across the can conveyor and gently hold the cans in position to stop them sliding backwards or falling. This means the cans are precisely spaced when they enter the seamer.

The cans coming off the filler have a certain amount of foam on them and CO_2 is usually being released as they move. This helps to sweep air away from the top of the can but it is found that some of the bubbles on the beer surface contain a large amount of air. These can be removed by jetting the beer surface with CO_2, water, or steam as the cans move past.

On the transfer conveyor between the filler and seamer, such a "bubble breaker" is located and this simply fires CO_2, etc. downward continuously at the passing cans.

The Seamer

There are several operations taking place inside a seamer in order to produce the finished can:

- Undercover gassing
- Lid placement
- Lid secured (first operation)
- Lid sealed (second operation)
- Can discharge

Undercover Gassing

The bubble breaker upstream on the transfer conveyor removes some of the air trapped in the top of the can but the large surface area of beer means more air is present as the can enters the seamer. Most of this can be removed by having a stream of CO_2 (or nitrogen) moving through the seamer at the point where the lids are being transferred onto the cans. This gas flow rate needs to be controlled and set by trial and error. Too low a flow is not effective; too high a flow starts blowing beer out.

Lid Placement

Lids are supplied stacked together in a paper sleeve about 1 m long and these are fed into an inclined lid feeder chute. This is usually done manually, but can be automated where a line is very fast, or where two or more lines are close to each other. Inside the machine the lids are fed onto a rotating cover feed turret with 6, 12, or 18 pockets. This wheel then passes the lids onto a similar turret that places the lids on top of the cans. In order to avoid lids being fed onto nonexistent cans (i.e., a gap in the filler), there are can sensors on the transfer conveyor that are linked to the lid feed control. In order to bring the cans and lids together, the cans are transferred from the transfer conveyor onto individual lifters rather like the filler. Cans are then raised to meet the lids and held under slight pressure to avoid spillage or, more importantly, the lid slipping off with the considerable centrifugal force.

Lid Secured

The process of fixing a lid onto a can takes place in two stages almost simultaneously. The first stage serves to fix the lid in position by squeezing the outer edge of the lid downward so that it tucks in under the body rim.

Packaging Technology

Spinning the can between two hard pieces of metal, a chuck on the inside edge of the lid, and a roll on the outer edge (Figure 15.3) does this. The outer seaming roll is specially profiled to contact the edge of the lid and curl it under the body hook.

Fill Height Detection in Cans

Metals are completely opaque to most electromagnetic radiation so optical or infrared systems cannot be used. The use of weak gamma ray sources for fill height detection in cans has been around for many years and they use either tritium (H^3) or 241-americium as the radiation source.

These devices work by shining a weak gamma ray across the full can conveyor at a preset but adjustable height. The radiation intensity is measured at the other side of the conveyor. The absorption of radiation will vary because metals, beer, and air all have widely differing coefficients, with metals being the most absorbing, and air the least. Fortunately, the thickness of the can sidewall is so low that it has little effect and the main absorber is the product. By setting the height at a point where the liquid in underfilled cans is just under the beam path, it is possible to detect and reject these cans and divert them for checkweighing. This type of detector is not accurate first, because the radiation source is weak and slightly variable and, second, because the liquid could be slopping about inside the can. The rejected cans should not be dumped, but put through a checkweighing system to pick out accurately those that are genuinely low fill. The acceptable cans should then be fed back into the mainstream.

The usual place to position these detectors is toward the end of the line just prior to packing and after the pasteurizer, so that any leakers can be detected. An additional option is to place one just after the filler and

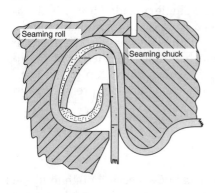 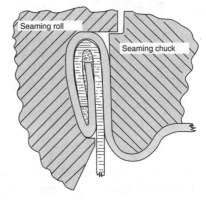

FIGURE 15.3
Detail of can seaming operation.

seamer so that any abnormalities on the filler can be quickly picked up by the filler operator and rectified before a large stock of underfilled cans is produced. This is usually referred to as a gross underfill detector.

Kegs and Kegging

Kegs have emerged over the last 50 years as a major package type, replacing almost all the beer previously dispensed from casks. They dominate the trade in a number of countries, and provide a minor but important part in many others.

Kegs

Metal kegs evolved from wooden casks and are a distinct improvement as a container for dispensing filtered beer in bulk. The advantages of a keg are:

- The container can be pressurized and kept at pressure
- It can be cleaned, sterilized, and refilled without opening
- The extractor allows beer to be removed and displaced with gas with little or no change in carbonation
- The extractor can only be removed at the brewery, safeguarding the product and maintaining sterility and shelf life in trade

In appearance, the larger kegs still look like casks with their barrel shape but kegs of 50 l and under have straight sides. Apart from cask-conditioned beer in Britain, kegs have almost completely taken over the sale of bulk beer. Kegs are simple metal pressure containers with a single threaded fitting in the top center. Both ends are very slightly domed for ease of manufacture and to allow full drainage. At both top and bottom ends of the keg there are metal rims, which are for protection and handling. The height of the top chime or rim is always above that of the extractor. The top of the keg is usually stamped with the owner's name, contents, and pressure rating (usually 60 psi, 4 bar).

Manufacture of Kegs

Kegs are manufactured from aluminum or stainless steel although mild steel (lined) and plastic have been tried. For many years aluminum was favored by numerous brewers because of its low weight, but there has been a move toward stainless steel in recent years for a number of reasons. First, aluminum kegs need to have an internal lacquer to protect the metal from cleaning detergents. There is some unease about the integrity of this lacquer with extended use and the possibility of corrosion damage and metal pickup in beer. Second, there is a significant illegal industry from smelting stolen kegs. This is easy with aluminum, which has a melting point of less than 700°C, whereas stainless steel (melting point 1500°C) is far more difficult. Third, the design and weight of stainless steel kegs has improved over recent years and a 50 l keg is now down to about 11 kg in weight. Stainless steel is also resistant to both acid and alkali cleaning detergents, so it is more versatile in the brewery. Aluminum kegs were probably initially less expensive partly because it is an easy metal to melt and cast, but modern automated techniques of steel forming have eliminated this. Stainless steel kegs at present (50 l) cost around the same or slightly less than aluminum. The worldwide market is about 3.5 million per year.

Steel kegs are manufactured from flat sheet by first stamping out circles of metal and then deep drawing them in presses to make large cups. A hole is formed in the top half to take the threaded neck that is manufactured separately. After trimming, the two halves are welded together. Rolling rings are pressed in as part of the forming process, and these add strength to the container as well as facilitate handling. The protective end chimes are also formed from a flat sheet with a curled in end bead for safe handling and punched out hand holds. These are then welded to the ends of the keg. Normally, the end chimes have the same diameter as the keg but sometimes the bottom chime is of a reduced size that makes the kegs stackable. In this way four pieces of flat sheet end up making a keg, with the neck separately made and fitted.

There are various standard keg sizes; 30 and 50 l are the most common in Europe. Larger kegs of 100 and 164 l (one barrel) are also still used in Britain and there are some in-between sizes both in Britain and the United States. The standard thicknesses for steel are 1.5–2 mm, although lower ratings can be obtained if required.

Minikegs

In order to bridge the size gap between kegs and cans, a number of minikegs are available to consumers as nonreturnable or returnable party kegs. These are typically 5 l capacity and in construction are made like three-piece tinplate cans. The difference, however, is that there is a hole in the top center for a fitting and this allows them to be pressurized, filled, and broached just like beer kegs. The kegs can be supplied sterile with a seal or tap; the taps are available as disposable or reusable units. These taps allow the minikeg to be dispensed over a period of time without greatly affecting beer quality. Shelf life is quite short and obviously depends on storage conditions. Sales of these package types across Europe and in Japan are quite substantial.

Keg Lines

The basic operations in a modern keg filling line are listed as follows:

- Offload/depalletize
- External wash
- Spear check and torque
- Internal clean and sterilize
- Fill
- Checkweigh
- Cap and label
- Palletize

The major differences between a keg line and a smallpack line are the speed of operation, which is measured in containers per hour instead of per minute, and the type of conveying system. At first glance, the cleaning and filling of kegs looks straightforward, but in fact there is quite a lengthy sequence of operations. The other aspects of keg handling are fairly basic and the plant looks simpler but more robust than a bottle or can line. The control of the pasteurizer, however, needs some care and this is dealt with elsewhere.

The layout of a keg line is not uniform and it is usually made to fit an existing building layout. Ideally, a keg line would probably follow the U-shape principle like other packaging lines in order to bring the depalletizer and palletizer together. The most important feature is to ensure that the beer handling components are close together. These are the keg filling machine, pasteurizer, and its buffer tanks. Close control of beer supply and kegs is needed to ensure smooth efficient running so that pasteurizer operation keeps in step with size changes on the line or any stoppages.

Conveyors

Conveyors in a keg plant have to deal with considerable weights (over 100 kg full) with potentially abrasive container rims. Strength and wear resistance is all-important instead of the low friction and smooth running demanded on a bottling line or can line. The conveyors tend to come in two types: chain conveyors and flat conveyors. Chain conveyors are the most widely used (they can turn corners) and consist of linked Y-shaped pieces of metal or plastic with a rectangular cross section running in metal grooves. There are usually two or three strands to the conveyor and they are well spaced to balance the end of a keg.

Offloading/Depalletizing

There is no standard method of handling empty kegs and each brewery tends to have its own particular methods. Manual offloading was normal

Packaging Technology

until relatively recently and is still widely used. Kegs are usually stored in the open on pallets or layer boards and moved to the keg line by a forklift truck. Rather than manhandle them onto a conveyor, kegs are now often gripped by a clamp truck. This is a modified forklift truck with side fittings for squeezing and holding a group of kegs.

External Washing

An external washer consists of a simple tunnel surrounding the empty keg conveyor, where the kegs are fed in at regular intervals and blasted with water and detergent to remove dirt. Some of the jets operate at very high pressures in order to remove the old labels and these jets usually rotate to try and cover the whole of the keg top. The water and detergent can be captured under the conveyors and reused after filtration.

Spear Check and Torque

An additional device often installed nowadays on a keg line is an extractor check. These spears can work loose or get damaged. A torquer in-line will grip the extractor head and tighten it to a preset level to ensure that it does not leak and is also difficult to remove.

Internal Keg Cleaning and Sterilizing

We now move on to the keg racking machine in which there are distinct operations:

- Cleaning
- Sterilizing
- Filling

Usually, cleaning and sterilizing are carried out as one operation but there are minor variations between machines. In the first stage of the cleaning cycle, the keg is located on top of a broaching piece upside down and the extractor depressed to open the keg. Then follows a series of logically programmed events with the keg held in this position. The kegging machine has a whole series of pipelines (Figure 15.4) supplying each station with the following services:

1. Warm water
2. Hot water
3. Steam
4. Detergent

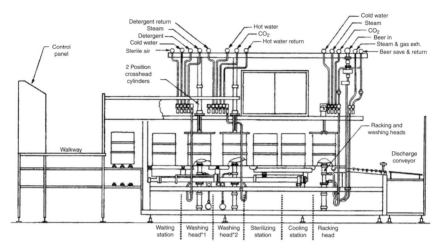

FIGURE 15.4
Side view of linear keg washing and filling machine.

5. Detergent recovery
6. Hot water recovery
7. Condensate return
8. Ullage drain
9. Gas out
10. Air in
11. CO_2/N_2 in

A keg returning from trade is usually dirty inside and may contain significant residual beer and gas at pressure. The first stage in the process is to drain off this residual gas and beer. Once drained, the keg is rinsed out with recovered warm water to remove the last remnants of beer and then this water is purged out with air and put to drain.

The next stage in cleaning is to pump in detergent to wash the internal surfaces thoroughly and remove any scale material or dried-on beer. The single-detergent wash has now been replaced by a double wash using frequently a caustic wash followed by a rinse and then an acid wash. This has the advantage of removing all material, not just that which is susceptible to one detergent.

The next task is to rinse it with hot water and steam sterilize. The steam is used to drive the water out and then build up pressure with the outlet closed. This raises the temperature to well over 100°C and the keg is held at this temperature for several seconds before going on to the filling stage.

Total cleaning times vary widely with keg size, the three main sizes being 30, 50, and 100 l. Some cleaning cycles are as short as 90 sec, others where

Packaging Technology

double-detergent cleaning and steeping are used take up to 5 min. The fashion with new machines is toward longer cycle times — around 4 min.

Keg Filling

The empty, clean keg starts its filling cycle by being full of steam after the sterilization step. This steam is purged out with inert gas and then pressurized. Filling takes place through the gas ports of the inverted keg and gas under pressure is removed down the spear. Accurate dosing of beer into the keg is either by metering or (less frequently) by using a conductivity probe.

Types of Keg Racking Machinery

For very small operations it is possible to purchase manual two- or four-head machines producing just a few kegs per hour. At the other end of the scale speeds can be around 2000 kegs per hour (1000 hl/h). There are two basic designs — linear and rotary.

Linear Machines

Linear machines were invented at quite an early stage in keg development and consist of multiple lanes of conveyors side by side with the keg broaching pieces in the center of the conveyor. The main supply conveyor bringing empty kegs into the machine is at right angles to these rows of conveyors, and the kegs are either manually pushed into the lanes or automatically pulled in by a hook grabbing the rim. They are then moved over the first broaching piece and then clamped and opened to go through the first clean. They then move to the next station and so on.

By having multiple lanes (sometimes over 20) output can be considerable. The disadvantage of these multihead machines is that they are difficult to maintain due to the huge number (hundreds) of valves.

Rotary Machines

More recently, rotary machines have been competing well on new installations with linear kegging machines. These are rather like slow moving bottle fillers in appearance with a number of stations to allow several kegs to be processed at the same time. An extended washing cycle can still be achieved by having a separate washing machine with, say 24 stations, whereas the filling machine has 12 or 16 stations. Another variation is to have two washers and one filler. Typical rotation times are 70–120 sec. These machines have the advantage that the kegs are only broached once on each machine and then they slowly rotate, with the various services switching on and off as the kegs progress around the circle. There is a requirement for all the services (usually around eight) to be piped up the center of the machine and this requires quite an elaborate system of seals to keep various liquids and gases separate. There is potential for space saving by installing such machines, and on a like-for-like output basis, a rotary system has about half the number of mechanical components compared with a linear array.

Pasteurization

The heat treatment of foods and drinks in order to kill off spoilage organisms dates back to the groundbreaking work of Louis Pasteur in the 1870s. He found that heating beer to temperatures between 50 and 55°C was sufficient to preserve it. Within a few years the term "pasteurization" was coined and a number of food and drink industries adopted heat treatment as a means of preserving their products and this is still standard practice today. Not all materials respond to the same temperatures. Milk and foods need to be heated to quite high temperatures. Beer is fortunate in not containing any pathogens and the nature of the common spoilage organisms is such that temperatures of about 60°C will suffice to achieve stability.

In beer, the common contaminants are residual pitching yeast, wild yeast, lactic acid bacteria, acetic acid bacteria, and cocci. Of these, lactic acid bacteria are more heat resistant than most, with acetic acid bacteria the easiest to kill. Occasionally, heat-resistant wild yeasts emerge but this is not a problem in a clean plant with good yeast management. Other factors in the brewers favor are the alcohol and CO_2 content of beer, both of which assist in killing off spoilage organisms. In addition, a well-filtered beer will have a very low microbial loading and this helps in reducing the heat requirements. Lack of oxygen and the presence of hop compounds also helps.

Pasteurization Units

This heat treatment of beer is usually expressed using the term "pasteurization units." A pasteurization unit (PU) is defined, for beer, as the effect achieved by holding beer at 60°C (140°F) for 1 min. In practice, heat treatments aim to give at least 5 PU and 10–30 is normal.

Until the 1950s, the time–temperature relationship was not well understood, and brewers pasteurized their beer largely based on past experience. Around this time, a number of researchers started to look at different spoilage organisms and the effect of various temperatures. It was found that there were differences between organisms, as mentioned previously, and that the temperature effect was exponential, which is not surprising. The effect was not quite the same for all organisms. By plotting a graph of logarithm of time against temperature, a straight line was obtained. As a result of this work two main factors were devised to describe the heat effect:

1. *Decimal reduction time.* This is the time needed to kill 90% of organisms at a given temperature. This varies with the organism and typical D-values at 60°C are 1–5 min, with 2 min being the average.
2. *Temperature dependence value Z (°C).* This is the temperature increase needed to reduce the D-value by 90%. Again this varies with the organism. Values range from 3 to 8°C and the average accepted value is 6.94°C.

From these two factors a formula is derived to describe the PU in which 1 PU is defined as holding beer at 60°C for 1 min and the relationship with temperature is:

$$\text{No. of PU} = 1.393^{(T-60)} \times \text{time (min)}$$

The number of PUs delivered at a range of temperatures over a 1-min period are shown in Table 15.8.

From the figures in Table 15.8, it can be seen that a 20-min hold at 60°C will deliver 20 PUs to a packaged beer, while 20 sec at 72.5°C will do the same for bulk beer going through a flash or plate pasteurizer.

In practice, pasteurization falls into two categories: flash pasteurization and tunnel pasteurization. Flash pasteurization is always used for keg beers (unless sterile filtration is used), as it is not possible to pasteurize such large containers, although it has been tried! Flash pasteurization is also used for some bottle and can filling operations and is needed when

TABLE 15.8

The Number of PUs Delivered at a Range of Temperatures over a 1-min Period

Temp (°C)	PU
60	1
62	1.9
64	3.7
66	7.2
68	14
70	27
72	52
72.5	62

filling pressure-sensitive containers such as PET. This technique has the advantage that the capital cost of installation is not particularly high and the plant does not take up a lot of space.

Tunnel pasteurization is used on bottles and cans and is the most reliable way of producing long shelf life products in these packages as all parts are treated. It relies on a lower temperature than flash pasteurization spread out over a longer time (up to 1 h) because of the time taken for heat to penetrate the package.

The effect of temperature on beer flavor is not entirely clear, apart from the fact that high oxygen levels and pasteurization do not go together. A stale, bread-like flavor is often the result. By keeping dissolved oxygen levels to a minimum, and by using only modest levels of pasteurization, it is possible to produce beers that can stay commercially acceptable for weeks or months. There is also some controversy over whether flash pasteurization is more or less damaging to flavor than tunnel pasteurization. Overpasteurization by either method is deleterious.

Flash Pasteurization

Flash pasteurization involves using a plate heat exchanger to rapidly heat the beer up to a temperature of about 70°C, hold it at this temperature for some seconds, then chill it down again ready for packaging. In practice, the plates in a pasteurizer are sized to give a substantial degree of heat recovery (90–95%). When the incoming beer comes out of the regeneration section, it helps if it is close to its final temperature so that this final heating step can be better controlled. The holding tubes on a pasteurizer usually consist of an elongated spiral of 100- or 150-mm pipe in sections with narrower connections. The most practical holding time is generally 20 sec, so to achieve 20 PU in the beer a target temperature of 72.5°C would be used. This timespan allows good control without the need for huge holding tubes. Brewers seldom use less than 10 PU for flash pasteurization and 20–50 is more common.

Quality Safeguards

There are a number of practical difficulties in operating a pasteurizer that could be serious if allowed to continue. The all-important thing is to ensure that the beer is not underpasteurized and there are various failure modes where this could occur.

Temperature Drop

If the temperature coming out of the heating section of a pasteurizer is not high enough then incomplete pasteurization will result. This is detected by a probe, which should feed a signal back to the controller calling for more heat and should immediately put the pasteurizer onto recycle mode, taking beer from the outlet and feeding it directly into the inlet until temperature conditions are restored. Usually, prolonged recirculation is avoided by shutting down the pasteurizer if the fault is serious, such as a steam supply failure or pump failure on the heating loop.

Gas Breakout

When beer is heated to 72°C, the solubility of CO_2 drops so low that unless a high pressure is maintained in the pasteurizer there will be a breakout of gas, filling the holding tubes with fob. This has a number of unpleasant consequences. First, the expansion of the beer means it flows much faster through the holding tubes in a turbulent state and the beer will be underpasteurized. Second, as the foam collapses on cooling, it will probably form a haze and under these conditions it is usually permanent and visible. There is also the risk that fob will dry and bake onto the holding tubes where it may eventually become infected. Gas breakout can be avoided by good system design but cannot be completely eliminated. It is likely to occur if the beer pump fails or if the beer supply valve fails and shuts, thus starving the system. A pressure monitor should shut the system down and once rectified, the pasteurizer should be cleaned and sterilized before resuming.

Plate Failure

Another occasional cause of pasteurizer failure is leakage through one of the plates. In a simple layout as shown in Figure 15.5, the pressure on the inlet beer side is always going to be higher than that of the pasteurized beer coming back on the other side of the plates. Pasteurizer plates occasionally fail due to stress corrosion. This allows unpasteurized beer to short circuit the pasteurizer, frequently leading to failed kegs in trade. Detecting this problem is not easy. The usual routine is to pressure test the plate pack on a weekly basis.

The solution to the problem is to install an additional pump in the pasteurizer at the end of the holding tubes, the inlet to the regeneration section. This means that the hot pasteurized beer is boosted in pressure as it starts to cool down and exchange its heat with the incoming unpasteurized product. If there is a plate failure in these conditions then it will always be with pasteurized beer getting out, and not raw beer getting in.

Flash pasteurization equipment is mechanically very simple, cheap to buy, and easy to operate (Figure 15.5). It is the ideal system for filling kegs and is becoming more widely used for small packs such as PET and glass bottles. The key to success in this latter application is to ensure that the bottles, caps, and filling equipment are sterile.

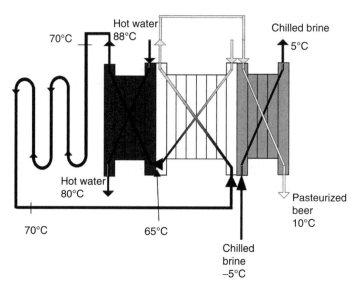

FIGURE 15.5
Schematic representation of a flash pasteurizer.

Flash pasteurizers operate best in steady-state conditions but the downstream packaging line does not always allow this. It is normal to have a downstream buffer tank just prior to the filler to smooth out flow. The solution offered today is to have a variable speed system and to reduce the flow as the sterile buffer starts to fill.

Tunnel Pasteurization

Tunnel pasteurization is similar to flash pasteurization in that it involves heating the package to the correct temperature, holding at that temperature, and then cooling down. The timescale is very much longer, however, up to 1 h, and the peak temperature achieved is lower, at about 60°C. There are a number of reasons why this long timespan is needed. First, the rate at which heat is conducted through a container wall and then through the contents is quite long. There is a "lag" of about 10 min in this heating process. Second, with bottles, a rapid temperature rise would cause thermal stresses that could result in the bottle bursting. Third, there is a steep pressure rise when a highly carbonated package is heated and again there is a risk of bursting (Figure 15.6).

Bottles have a wide range of failure pressures (see the Section "Bottle Strength") and cans are specifically designed and manufactured to withstand only 6 bars. For these reasons, a low temperature–long time profile is the only practical option to achieve the desired PUs (usually about 10) evenly distributed throughout the container.

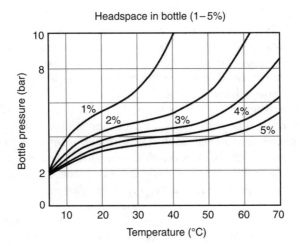

FIGURE 15.6
Relationship between headspace and bottle pressure.

Tunnel pasteurizers are very large pieces of equipment. They consist of a very long enclosed chamber where cans or bottles are fed in at one end on a conveyor, heated and cooled as they travel through, and emerge from the other end. Frequently, they have two decks to save space. The two main components of the pasteurizer are the water spray and circulation systems and the package transport system.

The Water Heating and Spraying System

Cans and bottles are very slowly moved through the pasteurizer, and heating is achieved by spraying warm water onto them. These spray sets are in zones across the length of the machine and are at set temperatures. The first zone, for example, will be at a temperature of, say 20–22°C, to gently warm the container up to about 9–10°C. The water falling past the containers and through the conveyor is collected in a trough underneath for reuse. Water at progressively higher temperatures is sprayed onto the containers to bring them up to 60°C.

The most critical section in the pasteurizer is called the superheat zone, which is the last heating zone of all before the holding section at 60°C. The temperature in the superheat zone must be very accurately controlled at 61–65°C to ensure that the containers are brought up to the correct temperature. Each machine is slightly different and this temperature needs to be established on initial commissioning to give the required PUs. It will also differ between bottles and cans. It is important, if there is a stoppage in the pasteurizer, that the containers do not get overheated if trapped in the superheat zone. This can lead to overpasteurization and burst containers, especially cans, if the temperature goes too high.

The most modern pasteurizers have very sophisticated controls and can compute the number of PUs cumulatively as the containers travel through. This can be used to adjust the temperatures in the event of a stoppage. A more conventional way of checking a machine's performance is to feed through a recording thermometer. There is usually a filled dummy bottle or can on a baseplate with an accurate temperature probe inside, which is connected up to a recorder. It is normally inserted into the pasteurizer, then retrieved at the back end and downloaded. This gives more realistic information as it tells you what is going on inside a container. A typical temperature–time profile for a seven-zone pasteurizer is shown in Table 15.9.

In order to save on water and energy, there is usually a complex system of pipework running backward and forward between the different zones. One can see from the above example that the first and last zones work at the same temperature, as do zones 2 and 6. As the cooling zones will pick up heat from the warm containers, this water is pumped to the front end where it warms up the cold incoming packages. It will lose heat as a result and is then pumped to the back end to cool down more containers again.

Each zone, however, should have facilities for heating up and cooling down its reservoir to cater for start-up and shut-down conditions as well as occasional stoppages. If the pasteurizer is in equilibrium then only the superheat zone needs significant steam input, the others need only a modest top-up. The energy consumption is still quite high with a tunnel pasteurizer, and 50% recovery is the best that can be achieved because the containers go in cold and come out at around 25–30°C. Most modern pasteurizers have more than the seven zones mentioned previously in order

TABLE 15.9

Typical Temperature–Time Profile for a Seven-Zone Pasteurizer

Zone No.	Spray Temp. (°C)	Spray Time (min)	Package Temperature	
			In	Out
1 preheat	22	6	2	9
2 preheat	32	7	9	21
3 (superheat)	65	14	21	60
4 (hold)	60	6	60	60
5 cooldown	40	10	60	43
6 cooldown	32	7	43	36
7 cooldown	22	6	36	28
		56		

to improve temperature control and utilities use. If a pasteurizer does get out of balance, for example, when being emptied, then water consumption will rise due to the heat imbalance and overflow of the troughs as cold water is added. Conversely, steam consumption is high when filling.

A typical time–temperature–PU graph is shown in Figure 15.7. The pressures generated in heated containers were also mentioned and this is illustrated in Figure 15.6. One can see from this that control of CO_2 in package becomes important to keep pressures down and control of contents is another big factor. The only compressible part of a package is the gas headspace and if this is too small the pressure will rise more steeply than in a container with average contents.

The Water System

Pasteurizers are prone to corrosion and slime growth due to the warm damp conditions and it is necessary to add inhibitors to control both of these. Another way to suppress slime is to circulate hot water through all of the sections. The use of inhibitors with cans needs to be done with care since the decoration occasionally suffers if the concentrations are too high, and the cans would be unfit for sale.

Blocked spray jets are a distinct quality hazard in pasteurizers as they will lead to underpasteurization. This occurs due to scale or slime build-up and is best countered by regular inspection of the water tanks underneath and by internal inspection of the machine. Cold spots due to blocked jets or other abnormal flow conditions can be picked up by using a traveling recorder and placing it at different points across the pasteurizer.

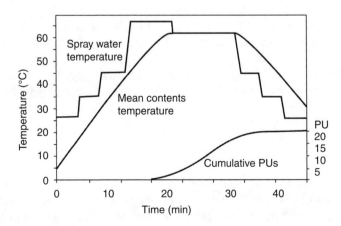

FIGURE 15.7
A time–temperature curve for a three-section machine.

Packaging Line Efficiency

The measurement of a packing line's performance is an essential requirement in any plant and this is usually accompanied by a system of analyzing and rectifying stoppages. This is frequently used to justify capital expenditure for replacing or "beefing up" an item of equipment. There are two components to the overall performance of a line: machine efficiency and machine utilization.

Machine efficiency is usually measured at the slowest component of the line, and in the case of bottling and canning operations, this is almost always the filler. If this machine is rated at, say, 1000 containers per minute and it always runs at this speed, then its efficiency should be 100%. However, if there are, for example, gaps in the supply of empties or some process reason for slow running, then the efficiency will drop.

Machine utilization is the time that the filler is running in relation to the total planned time. If the operation is on three shifts of 40 h each, then the available hours are 120. It may not be busy enough to use the whole 120 h, but say, only 100. The rest is used toward week cleaning, briefing, training, etc. If during the 100 h of planned production there are 23 h of downtime, then the machine utilization is:

$$\frac{100 - 23}{100} = 77\%$$

These two factors of efficiency and utilization are usually combined to give machine effective utilization (MEU):

$$\text{MEU} = \text{ME} \times \text{MU}$$

A typical figure for MEU would range between 65 and 85%. This is often governed not by machine stoppages but factors such as size and quality changes or type of pack.

Some purists would argue that nonprogrammed time should be included in the downtime calculation but this would give a much lower figure and simply cause confusion.

Relative Machine Speeds

For bottling and canning lines the key item of plant is the filler and other machines should be rated relative to it. The best arrangement is where the items immediately before and after the filler are capable of running about 10% faster and the next machines faster still. This is sometimes referred to as the V-graph (Figure 15.8) from the way in which the speeds look when

Equipment	Relative speed (%)
Depalletizer	135–140
Case unpacker	120–125
Bottle washer	110–115
Filler	100
Pasteurizer	110
Labeler	110–120
Bottle packer	120–125
Palletizer	135–140

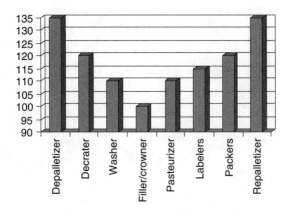

FIGURE 15.8
Relative machine speeds.

plotted on a graph with the order of the machinery. A typical example is given next for a bottling line.

This arrangement of relative speeds means that the line is in a state of compression before the filler with the conveyors usually full, but after the filler bottles are being taken away faster than they can be supplied. This makes the filler the pinch point on the line, but there are good reasons for doing this. First, cost; the filler is probably going to be the most expensive single component of the line and it pays to maximize its throughput. The cost of uprating a filler by 30% would be much higher than for a palletizer. Second, beer quality; the filler performs at its best when there are no interruptions and should always have a supply of containers to fill and no bottlenecks downstream. If a filler is constantly stopping and starting there is going to be more oxygen pickup, variable carbonation, and more variable contents. The filler/closure machine and the pasteurizer are the main machines on the line, where beer quality is directly affected, and they need to be protected.

In summary, it can be seen from the preceding discussion that beer packaging is of great importance and the most expensive part of the whole process. It has changed greatly over the last 50 years, and this will continue. Quality, economics, new technology, and the drive for increased market share and profit will keep this area in a constant state of change.

16

Microbiology and Microbiological Control in the Brewery

Fergus G. Priest

CONTENTS

Introduction	608
Wild Yeasts	608
Molds	610
Bacteria	611
Gram-Positive Bacteria	612
Lactobacillus	612
Pediococcus	613
Gram-Negative Bacteria	613
Acetic Acid Bacteria	613
Enterobacteriaceae	614
Zymomonas	616
Anaerobic Gram-Negative Bacteria	616
Microbiological Quality Assurance and Quality Control	618
Setting Microbiological Standards and Sampling	619
Traditional Microbiological Procedures	621
Rapid Methods	622
ATP Bioluminescence	622
Direct Epifluorescence Filter Technique	623
Antibody DEFT	623
Fluorescence *In Situ* Hybridization	624
Polymerase Chain Reaction	624
The Microbiological Laboratory within the Brewery	625
References	626

Introduction

Beer is microbiologically stable and therefore not subject to the myriad spoilage microorganisms that can colonize most foods or nonalcoholic beverages. It has been subject to exhaustive yeast growth and, therefore, like other fermented foods, it is largely resistant to further microbial development. The reasons for this are several:

1. The low pH (around 4) inhibits most microorganisms
2. The high alcohol concentration is toxic to many microorganisms
3. The antiseptic action of hop α-acids is bacteriostatic to many bacteria, particularly gram-positive types
4. Only residual nutrients (pentose sugars, higher maltooligosaccharides) are available as carbon sources

Despite these factors limiting microbial spoilage, there are various yeasts and bacteria that can flourish in beer, particularly if the storage conditions are poor and oxygen is allowed access. Fortunately, none of these organisms is pathogenic, so the only problem for the brewing microbiologist is consistency of the appearance and organoleptic qualities of the final product.

In this chapter, I will review the predominant spoilage organisms, outline the available technology for detecting and identifying these organisms, and consider the role of the microbiology laboratory in dealing with these problems and assuring consistent end-product quality.

Wild Yeasts

Wild yeasts are generally defined as those "yeasts not deliberately used and not under full control."[1] This definition includes brewing strains that are used for a different style of beer and may have been cross-contaminated in the brewery, as well as nonbrewing yeasts that have gained access from the air or raw materials. It is important to emphasize that there are many genera and species of yeast with diverse physiologies; the only unifying feature is that the organisms are predominantly unicellular. However, many types of yeast have a semifilamentous lifestyle and may form mycelia under various environmental conditions. These hyphae when formed by yeasts contain septa. Most yeasts are members of the ascomycetes in which spores are produced endogenously in an ascus.

The major genera and types of wild yeast encountered in the brewery are listed in Table 16.1[2] and a scheme for their identification is presented in Figure 16.1. Some of these yeasts are strictly aerobic and cannot ferment

TABLE 16.1
Physiological Characteristics of Ascomycete Yeast Genera of the Brewing Industry

Saccharomycetaceae (Vegetative Growth by Multilateral Budding)	
Debaryomyces	Weak or no fermentation
Dekkera	Fermentation (but only under aerobic conditions)
Issatchenkia	Weak fermentation; forms pseudomycelium and surface film (pellicle)
Kluyveromyces	Fermentation, usually vigorous
Pichia	Weak or no fermentation; many form true mycelium or pseudomycelium and surface pellicle
Saccharomyces	Vigorous fermentation, no pellicle
Torulaspora	Vigorous fermentation
Williopsis	Weak or no fermentation
Zygosaccharomyces	Vigorous fermentation
Nadsonioideae (Vegetative Growth by Polar Budding)	
Hanseniaspora	Fermentative
Saccharomycodes	Fermentative
Schizosaccharomycetoideae (Vegetative Growth by Fission or Mycelium)	
Schizosaccharomyces	Fermentative

Source: Adapted from Campbell, I., Wild yeasts in brewing and distilling, in *Brewing Microbiology*, Priest, F.G. and Campbell, I., Eds., Kluwer Academic, New York, 2003, pp. 247–266. With permission.

sugars under anaerobic conditions. *Pichia membranefaciens* is the most common contaminant of beer and wine in this category. The acetic acid-forming *Brettanomyces* and *Dekkera* species, although fermentative, do not usually cause a threat to the brewing process because they cannot flourish under anaerobic conditions. However, they form an important component of the yeast flora of fermenting Belgian lambic beers and can cause problems in ales and lagers if air should gain access. The aerobic yeasts such as *Debbaromyces*, *Pichia*, and *Williopsis* produce yeasty or estery flavors that are most unwelcome.

The fermentative yeasts such as *Kluyveromyces*, *Saccharomyces*, *Torulaspora*, and *Zygosaccharomyces*, on the other hand, can cause serious problems in the fermentation. They are potentially able to compete with the culture yeast, and although they cannot generally kill it, if they grow just a little faster than the culture yeast they will displace the brewing yeast over successive generations. As these wild yeasts neither flocculate well nor interact with finings, they generally pass into conditioning where they can have deleterious organoleptic effects on postfermentation beers, as well as causing haze and turbidity.

Most wild yeasts can cause serious flavor effects; for example, many strains are able to decarboxylate substituted cinnamic acids derived from the barley cell wall. *p*-Coumaric and ferulic acids are decarboxylated into

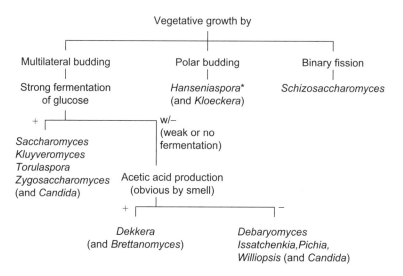

FIGURE 16.1
Simplified identification of common brewing yeasts. (From Campbell, I., in *Brewing Microbiology*, Priest, F.G. and Campbell, I., Eds., Kluwer Academic, New York, 2003, pp. 247–266. With permission.)

their 4-vinyl derivatives, 4-vinylphenol and 4-vinylguaiacol, respectively. Reduction of these molecules produces the 4-ethyl derivatives. These phenolic compounds, which contribute to the characteristic fruity flavor of wheat beers, are most unwelcome in ales and lagers. Other wild yeasts are able to utilize higher maltoologosaccharides and with this super-attenuation of beers.

Generally, wild yeasts are competing with the culture yeast for nutrients, but some yeast possess the "killer" phenotype and actively kill sensitive culture yeast. These strains produce zymocins, proteins that are lethal to sensitive cells. Such killer strains can rapidly displace culture yeasts.

Molds

Airborne microorganisms colonize barley in the field soon after the ears emerge from the leaf sheaths. Climatic conditions are important in determining the types of spore and mold that will contaminate the kernels, but *Alternaria* and *Cladosporium* species are commonly dominant.[3] *Fusarium* head blight has become a major problem, particularly in the wetter regions of North America and northern Europe.[4] Obviously, the growth of head blight fungi occurs at the expense of the grain and in infected crops the grain weight is reduced. Fusaria are also responsible for the synthesis of mycotoxins, small molecules with highly toxic characteristics. The trichothecene, deoxynivalenol (DON), is such a mycotoxin originating from *Fusarium graminarum* and some other species. It has been estimated that the total loss in barley and wheat due to *Fusarium* head blight during 1991 to 1996 in the United States amounted to $3 million. In addition to the reduced yields, the farmer receives a lower price for grain containing DON, even at 1 ppm.

Some of the rarer field fungi, such as the *Aspergillus glaucus* group (which are anamorphs of *Eurotium* species) and some penicillia, can grow under various storage conditions to become dominant as storage fungi.[3] This can give rise to hot spots in the silo as microbial growth ensues, which results in a rapid drop in germinative capacity. Malt typically bears *Eurotium* (*Aspergillus*) species including *Aspergillus fumigatus*, *Rhizopus* species, and penicillia.

Probably, the best-known effect of mold-contaminated barley on beer is the reduced gas stability in conditioned beers known as gushing that results in rapid loss of beer from the bottle on opening. Extensive colonization of barley by *F. graminarum* and other *Fusarium* species has long been associated with the gushing phenomenon that has been attributed to the release of a small peptide-containing substance. Other field fungi, including *Alternaria* species, have been associated with gushing. Mold-contaminated grain can also affect the flavor and color of the finished product, but brewers are unlikely to use such poor-quality raw materials. Although the occurrence of mycotoxins in finished beer is of major health concern, it has been established that mycotoxins are largely degraded during the brewing process and the quality of grain used in brewing is such that mycotoxins are avoided. It is therefore important to determine appropriate microbiological quality parameters for malt and adjuncts.[5]

Bacteria

Of the many thousands of bacteria, few are of concern to the brewer. This is largely due to the reasons mentioned in the "Introduction"; the physicochemical properties of beer are such as to preclude the growth of most bacteria.

Bacteria can be divided into two principal groups depending on the structure of their cell walls. This was first discovered by the Danish microbiologist Gram and in deference to his pioneering work we refer to these as gram-positive and gram-negative. Bacteria of the former group possess a thick cell wall composed almost entirely of a polysaccharide-like material called murein or peptidoglycan. This material complexes the crystal violet/iodine dye used in the Gram stain rather like starch complexes iodine, hence their gram-positive description. Gram-negative cells, on the other hand, contain more lipid in their cell envelopes and do not complex the Gram stain so strongly. This distinction has important implications for brewing microbiology, because gram-positive bacteria are generally sensitive to hop constituents and their growth is inhibited by hop α-acids, whereas gram-negative bacteria are not so seriously affected by these compounds.

Gram-Positive Bacteria

The lactic acid bacteria are the only group of gram-positive bacteria likely to cause a significant threat to beer. These bacteria belong to several genera but share common physiological characteristics (reviewed in Ref. 6). They are fermentative bacteria that do not use oxygen to grow; indeed, many prefer an anaerobic environment whereas others will grow in the presence of air. However, they lack respiratory pathways and none can use oxygen for respiration. Instead, they ferment sugars to predominantly lactic acid as an end-product. Some conduct a homofermentative catabolism of sugars in which lactic acid features as the sole end-product. The heterofermentative lactic acid bacteria, on the other hand, produce lactate, acetate, and carbon dioxide from sugars. Nevertheless, because both types produce lactate, these bacteria have become adapted to an acid environment and grow at pH 3.5 to about 6.0, well fitted to growth in alcoholic beverages. They can generally be recognized by their gram-positive staining properties and lack of the enzyme catalase. The latter can easily be tested for by adding a drop of hydrogen peroxide to a culture. Catalase-positive bacteria reduce the peroxide to water with the rapid evolution of oxygen bubbles; catalase-negative bacteria have no such action.

The lactic acid bacteria are divided into several genera of which members of *Lactobacillus* and *Pediococcus* are the most important to the brewer.

Lactobacillus

The heterofermentative bacterium *Lactobacillus brevis* is the most common beer spoilage bacterium and is detected at high frequency in beer and breweries.[7] Others isolated from spoiled beer include *L. buchneri*, *L. plantarum*, *L. paracasei*, and *L. plantarum*. *L. lindneri* is a close relative of *L. brevis*, which was isolated from lager beers and is also frequently encountered.[8] Other species such as *L. amylolyticus*, which was isolated from malt and wort, has poor beer spoilage properties, probably because of hop sensitivity.[9] It is normally unnecessary to identify lactobacilli to species, a generic identification is sufficient to be aware that problems may exist; nevertheless, species identification will give a clearer indication of spoilage potential.

Not all lactobacilli can grow in, and consequently spoil beer. Like most gram-positive bacteria, lactobacilli are generally sensitive to hop constituents and their growth is impeded in hopped beers. *trans*-Isohumulone and the related colupulone are the major antibacterial components derived from hops and cause leakage of the cytoplasmic membrane and inhibition of amino acid uptake in sensitive bacteria.[10] However, some lactic acid bacteria are hop resistant, especially strains of *L. brevis*, due to the presence of the *horA* gene, which encodes a transporter that expels hop compounds from the cytoplasm.[11] This makes them serious contaminants. They are able to grow in conditioning beer producing a silky turbidity often associated with the buttery flavor of diacetyl. Lactobacilli have also been implicated in the biosynthesis of amines from amino acids in beers.[12] Considerable increases in the concentrations of tyramine and to a lesser extent histamine were found in beers inoculated with mixed cultures of brewery lactic acid bacteria and stored until haze formation. Lactobacilli are much more effective amine producers than pediococci.[13]

Certain thermophilic lactobacilli, especially *L. delbrueckii*, have been noted as contaminants of sweet wort. Indeed, we have detected these bacteria growing during the mash. They are normally killed by the boil, but if the wort is kept sweet for any reason, even stored hot (less than 60°C) it provides an ideal growth medium for this bacterium that produces copious amounts of lactic acid.

Pediococcus

Spherical lactic acid bacteria with a homofermentative mode of metabolism and that divide in two planes to form pairs and tetrads are classified in the genus *Pediococcus*. These bacteria were noted by Pasteur at the end of the 19th century as potent spoilage agents of conditioned beer and, one species in particular, *P. damnosus*, can cause serious spoilage. This bacterium is generally hop tolerant and can grow in finished beer where it is associated with "sarcina sickness" characterized by the production of diacetyl.

Pediococcus inopinatus is also recovered from beer but is less troublesome as a spoilage agent than *P. damnosus*. *P. clausennii* has been isolated recently from spoiled beer.[14] In general, pediococci are serious beer spoilage agents and can be responsible for return of beer from trade. A simple scheme for the identification of gram-positive bacteria associated with beer and breweries is shown in Figure 16.2.

Gram-Negative Bacteria

Acetic Acid Bacteria

These gram-negative, rod-shaped bacteria obtain their energy for growth by oxidizing ethanol into acetic acid. This is a highly aerobic reaction that is used commercially for the production of vinegar, but needless to say is very detrimental to the brewer. Because beer should be stored with limited access of air, spoilage by these ubiquitous bacteria should not occur. However, bacteria of the genus *Acetobacter* are ubiquitous and can cause

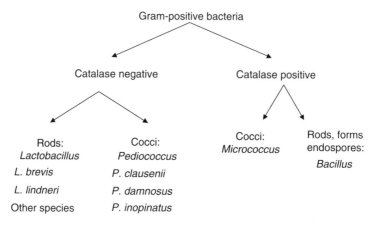

FIGURE 16.2
Simple scheme for the identification of gram-positive bacteria from beer and breweries.

problems in public houses dispensing cask-conditioned beer in which the ale is displaced by air. These bacteria are resistant to hop compounds and are ethanol tolerant and acidophilic, so given air they will grow in beer producing acetic acid, other off-flavors, and turbidity.

Enterobacteriaceae

The family *Enterobacteriaceae* comprises numerous genera of free living and sometimes pathogenic bacteria. Fortunately, none of the pathogenic types, such as *Salmonella* or *Shigella* species, have been found in beer. The enterobacteria are facultative anaerobes able to grow in the presence or absence of air, but they are inhibited by ethanol and low pH so are only responsible for beer spoilage in low alcohol products (<2% by vol) with a relatively high pH (>4.2).

One of the characteristic features of all members of the *Enterobacteriaceae* is the ability to respire under anaerobic conditions with nitrate as an electron acceptor rather than oxygen. In doing so, the nitrate is reduced to nitrite. Unacceptable concentrations of nitrite can be formed in beer if the brewing liquor contains high concentrations of nitrate, as it may do in agriculturally intensive areas. The nitrite reacts with secondary amines of the wort, either chemically or by enzymic catalysis, to form N-nitrosamines (Figure 16.3).[15] These molecules, with their carcinogenic properties, are obviously to be avoided in beer, if necessary by removing nitrate from the water (see Chapter 4) and by controlling the populations of enterobacteria.[16]

Members of the *Enterobacteriaceae* were first noted for their propensity to grow in wort rather than beer. These bacteria gain entry to the wort from water and can grow during the early stages of the fermentation. They will also grow in cooled wort if it is left unpitched for any length of time. Typical

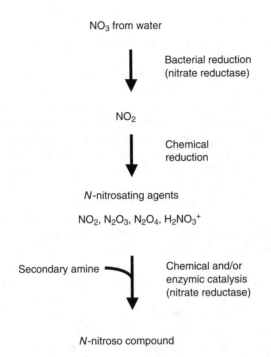

FIGURE 16.3
Pathway for the formation of nitrosamines in fermenting wort. (From van Vuuren, H.J.J. and Priest, F.G., in *Brewing Microbiology*, Priest, F.G. and Campbell, I., Eds., Kluwer Academic, New York, 2003, pp. 219–245. With permission.)

wort bacteria include species of *Citrobacter, Enterobacter, Klebsiella*, and *Rahnella* (reviewed in Ref. 17). Their main effect, if allowed to grow in wort, is the production of dimethyl sulfide (DMS), a sulfur molecule with a very low flavor threshold of around 30 μg/l that contributes a parsnip-like, sulfury flavor to beer. A second common, but not invariable feature of the enterobacteria, is their ability to decarboxylate substituted cinnamic acids to produce the phenolic flavor compounds described in the section "Wild Yeasts."

Obesumbacterium proteus is a unique member of the *Enterobacteriaceae* that is found exclusively in breweries (although it presumably has habitats outside the brewery these have never been described). The bacterium was first noted in ale pitching yeasts as "the short fat rod of pitching yeasts," but has also been described in lager yeasts. Generally considered harmless, the bacterium is responsible for increased levels of DMS and some fusel oils in the finished product. There are indications that there are in fact two genetically different groups within *O. proteus*; one responsible for relatively high levels of DMS and relatively rare, the second more innocuous and more common.[18] Beers brewed with yeast containing the former type of *O. proteus* at about 1% by number will typically contain about 14 to 18 µg/l DMS while the less productive strains produce around 4 µg/l DMS, both below the threshold of about 30 µg/l.

The practice of repitching yeasts, often for many generations in traditional ale breweries, led to yeasts with stable populations of *O. proteus*, generally containing several different strains but typically of the less troublesome type. These bacteria would rise to the top of the fermentation in association with the yeast and be repitched into the next fermentation, thus continuing their succession. They contributed to the flavor of the beer, but often in a characteristic way, which was not considered detrimental. During the early 1970s, *O. proteus* could be isolated from virtually every pitching yeast in use for ale fermentation in the United Kingdom. Today, the situation is different; limited repitching, greater cleanliness, and improved yeast handling have made *O. proteus* relatively rare. Much of this has been driven by the need to reduce N-nitrosamine concentrations.

Enterobacter agglommerans strains found in breweries are now generally classified as *Rahnella aquatalis*.[19] These bacteria grow in either hopped or unhopped wort in the presence or absence of yeast. Like *O. proteus*, they associate with yeast and can be recycled in pitching yeasts collected from lager fermentations.[20] Fermentations contaminated with *R. aquatalis* generally produce beers with a fruity/sulfury aroma and flavor, largely through the increased levels of acetaldehyde, diacetyl, ethyl acetate, and DMS in the beer.

Other enterobacterial contaminants include strains of *Citrobacter, Enterobacter, Hafnia, Klebsiella,* and *Serratia* generally derived from water supplies. They occur most commonly in wort and the early stages of the fermentation and cannot grow in beer. They are associated with sulfury/phenolic off-flavors.

Zymomonas

Bacteria of the genus *Zymomonas* have a unique mode of catabolism among the bacteria in that they conduct an ethanolic fermentation. This is so efficient that it has been seriously considered for the production of fuel ethanol, but it is not used for potable alcohol. Nevertheless, the bacterium is tolerant of ethanol (up to about 10% by volume) and has been associated with spoilage of primed conditioning ales (reviewed in Ref. 17). The organism is very rare in ale breweries and has not been reported in lager breweries where the low conditioning temperatures probably restrict its growth. Beer contaminated with *Zymomonas* has a estery/sulfury flavor due to the production of acetaldehyde and hydrogen sulfide.

Anaerobic Gram-Negative Bacteria

With improved technology resulting in very low oxygen levels in packaged beers, a new range of bacterial contaminants was given the opportunity to proliferate, the strictly anaerobic bacteria that would normally be inhibited by trace amounts of oxygen. First reported by Lee and his colleagues in beers brewed in the United States, similar bacteria were subsequently isolated from spoiled beer in Germany, Scandinavia, and Japan. In a comprehensive taxonomic study, the rod-shaped bacteria were assigned to three genera: *Pectinatus, Selenomonas,* and *Zymophilus*.[21] Gram-negative anaerobic cocci isolated by Weiss in Germany were classified as *Megasphera cerevisiae*.[22] None of these bacteria will grow under normal laboratory conditions even given enhanced CO_2; instead, they must be cultured in a strictly anaerobic environment either by appropriate manipulation of the medium or using anaerobic jars or cabinets.

Pectinatus cerevisiiphilus occurs as slightly curved rods that produce longer helical filaments in older cells. They do not grow in a CO_2-enriched atmosphere but must be cultured under strict anaerobic conditions in a modified MRS agar or similar. However, these conditions also enable lactobacilli to grow, so Lee devised a selective medium that inhibits the growth of lactic acid bacteria with crystal violet and sodium fusidate allowing the recovery of these anaerobes.[23]

Pectinatus frisingensis is morphologically similar to *P. cerevisiiphilus* but can be distinguished by molecular and physiological characters. There is evidence that *P. frisingenisis* is more tolerant of ethanol than *P. cerevisiiphilus* but that the latter grows more quickly in beer.[24] Bacteria of both species

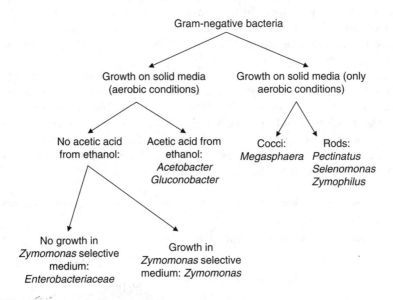

FIGURE 16.4
Simple scheme for the identification of gram-negative bacteria from beer and breweries. (From van Vuuren, H.J.J. and Priest, F.G., in *Brewing Microbiology,* Priest, F.G. and Campbell, I., Eds., Kluwer Academic, New York, 2003, pp. 219–245. With permission.)

can contaminate packaged beer producing considerable amounts of acetic and proprionic acids as well as acetoin and H_2S. The beer becomes turbid with an odor of rotten eggs.

Four strains of *Selenomonas lactiflex* were originally isolated from pitching yeast but these bacteria have not since been reported as spoilage bacteria. They ferment glucose to acetic, lactic, and proprionic acids.[21]

Zymophilus raffinosivorans and *Zymophilus paucivorans* were isolated from pitching yeasts. These curved rods have limited ability to grow in beer and cause spoilage and have seldom been reported as important spoilage agents.[21] Finally, *M. cerevisiae* are gram-negative, slightly elongated, anaerobic cocci that may cause turbidity and off-flavors in bottled beers.

A simplified scheme for the identification of gram-negative bacteria associated with beer and breweries is given in Figure 16.4.

Microbiological Quality Assurance and Quality Control

Quality control (QC) and quality assurance (QA) in the context of brewing distinguish the act of determining the current or very recent microbiological status of the plant and products (QC) from the actions that are put into place to ensure a quality standard (QA). The results of QC are used to provide QA; for example, dirty surfaces in fermenters must be cleaned or poor viability yeast discarded. It is important that the process is not entirely reactive, however, a proactive approach should be adopted to prevent faults occurring.

There are essentially two approaches to microbiological testing, the traditional methods that rely on cultivation of yeasts and bacteria in appropriate media followed by identification of the offending organism, if necessary, and modern rapid approaches that have a minimal reliance on prior cultivation. The move toward rapid methods for microbial detection and identification has been driven by technological developments and also by changes in the industry.[25] Some key motivating factors are:

- Growing market volumes for nonpasteurized beer in cans and bottles
- More low- and nonalcoholic beers
- Increasing variety of flavored sweetened alcopop-type beverages
- Tightened government regulations

Minimizing the time needed to detect a spoilage agent can lead to significant savings through reduced product recalls, extension of shelf life, and consistency of product quality and flavor.

The decision to adopt traditional or rapid methods is to an extent dependent on the critical control point (CCP) under review. For example, processes in the brewhouse, fermenting hall, and storage cellar generally require at least 5 days and so traditional methods are appropriate to provide the green light for the next step. However, other stages such as filtration, bright beer cellar tanks, clean-in-place (CIP), and water services are more constrained by time and rapid tests can provide the necessary information for optimization of the process.

Setting Microbiological Standards and Sampling

Hazard analysis of critical control points (HACCP) has become an essential feature of QC in the food and drink industries including brewing. This involves the systematic assessment of all the steps involved in the brewing process and identification of those steps essential to the hygienic quality of the product. In a large complicated plant, it will be advisable to reduce the operations to a series of connected subroutines.

Many of the CCPs are not microbiological (see Chapter 20); the principal microbiological CCPs are shown in Figure 16.5. At these stages it is essential to monitor for microbiological hazards using either traditional or rapid methods. First, it is necessary to set microbiological standards for each point. What is the maximum allowable level of contamination and by what organisms? Some suggested levels of sensitivity for detection are given in Table 16.2, but these are for guidance only and it is important to establish your own definitive criteria. For example, it may be permissible to use a pitching yeast with limited bacterial contamination by *O. proteus*, but a yeast contaminated with *Pediococcus* should be discarded and the source of contamination determined. Some organizations adopt a "green, amber, red" approach in which green flags adherence to microbiological standards (no action needed), amber indicates minor microbiological concern, the brewing process can continue but some microbiological contamination has occurred, and should be investigated, and, finally, red indicates failure to meet a microbiological standard and production staff must be informed so that corrective action can be taken. It is valuable to prepare trend graphs showing the microbiological status over time for the various stages and products. A trend showing a gradually worsening microbiological situation can give early warning of a problem.

Sampling at the CCPs requires careful consideration so that the results of the microbiological tests are statistically sufficient. Typical samples are liquids; for example, samples of final water rinses from CIP operations or water rinses of containers. Such samples should be of sufficient scope to provide for assurance that the microbiological standard has been achieved. For example, when examining beer, a sufficient quantity should be filtered or forced to allow for detection of contaminants at the required level, usually 100 to 1000 ml for testing of packaged beer depending on the size of the container and the sensitivity required (Table 16.2).

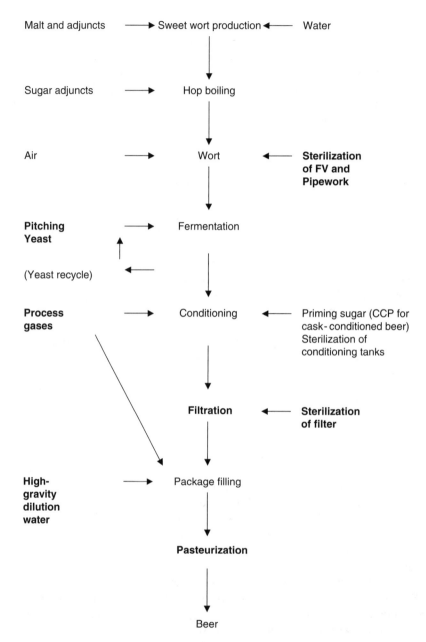

FIGURE 16.5
Flow diagram of beer production showing critical control points (CCPs) for microbiological testing in bold. (From Campbell, I. in *Brewing Microbiology*, Priest, F.G. and Campbell, I., Eds., Kluwer Academic, New York, 2003, pp. 367–392. With permission.)

TABLE 16.2

Suggested Sensitivity Required for Detection of Specific Spoilage Organisms in Brewery Samples

Samples	Sensitivity
Cold aerated wort	1 organism per 25 ml
Pitching yeast	1 bacterium per ml and 1 wild yeast per 10^6 culture yeast
Fermenting wort	1 organism per ml
Tank bottoms	1 organism per ml
Beer in storage	1 organism per ml
Filtered beer	1 beer spoilage organism per 100 ml or 10 to 10^2 nonbeer spoilage organisms per 100 ml
Packaged beer (nonpasteurized or flash pasteurized)	10 to 10^2 nonbeer spoilage organisms per 100 ml
Rinse water (end of cleaning in place)	1 organism per 100 ml

Source: Adapted from Jespersen, L. and Jakobsen, M., *Int. J. Food Microbiol.* 33:139–155, 1996.

Traditional Microbiological Procedures

Culturing of brewery microbes is achieved using a variety of specialist media. These have been discussed in detail elsewhere,[26] here I will describe the principal types. General-purpose media for the cultivation of yeasts and bacteria can be prepared from wort or beer but commercial media provide for consistency and ease of use. Wallerstein Laboratory Nutrient medium (WLN agar or broth) is commercially available and contains sufficient nutrients for the growth of most brewery microorganisms. The inclusion of bromocresol green indicator imparts a variety of colors from yellow/green/blue to bacterial and yeast colonies enabling distinction of different types depending on the pH around the colony.

For distinguishing culture and wild yeasts it is necessary to suppress the growth of the brewing yeast to enable the detection of the relatively low numbers of wild yeasts. Incorporation of the antibiotic cycloheximide (also known as Actidione) at low concentration (10 µg/ml) into WLN agar generally prevents the growth of *S. cerevisiae* while enabling wild, non-*Saccharomyces* yeasts to grow. Inclusion of copper salts is preferred by some for this purpose. *S. cerevisiae* is sensitive to low concentrations of around 6 µg/ml of copper sulfate but this can be modified according to the sensitivity of the brewing yeast to be examined. Addition of copper sulfate at the determined level to WLN agar will suppress the brewing yeast, allowing other yeasts to grow and form colonies. Lysine agar is a synthetic medium available commercially that prevents the growth of the brewing yeast, not by inhibition (as in cycloheximide and copper sulfate media), but by lack of nutrients. Lysine is the sole nitrogen source in this medium. *S. cerevisiae* strains are unable to metabolize lysine as the sole source of nitrogen, preventing their growth and allowing various wild yeasts to grow.

While the media mentioned earlier generally distinguish non-*Saccharomyces* wild yeasts from the culture yeast, devising selective media that will prevent the growth of culture yeast when enabling the growth of wild *Saccharomyces* yeasts is far more challenging. Numerous ingenious media have been formulated for this purpose, but none have been adopted by mainstream brewing laboratories as they fail to work effectively. Instead, the trend has been toward molecular methods for distinguishing brewing strains from wild strains of *S. cerevisiae* and close relatives (see Chapter 8).

Media for the culture of bacteria generally incorporate cycloheximide at about 100 μg/ml to prevent yeast growth. WLN provides a good general-purpose medium for the cultivation of most bacteria including *Enterobacteriaceae*. MacConkey agar is a selective medium for coliforms, which allows the differentiation of the lactose-negative *O. proteus* from various Lac-positive types such as *R. aquatalis*. *O. proteus* grows slowly on MacConkey agar at 30°C, requiring incubation for 36 to 48 h.

Lactic acid bacteria require more nutritious media for optimal growth, Raka Ray No. 3 is a highly nutritious medium that is commercially available and NBB medium has its proponents.[27] The standard medium for lactic acid bacteria is de Man, Rogosa, and Sharpe (MRS) broth. Addition of maltose at about 10 g/l to MRS agar or broth improves this medium for the growth of lactic acid bacteria from beer and breweries. As the bacteria on recovery from the sample may be stressed, it is generally advisable to cultivate them in a CO_2-enriched atmosphere or anaerobically at 30°C for up to 5 days to encourage full recovery.

Acetic acid bacteria and *Zymomonas* are not usually tested for and require specific media that have been described elsewhere.[26]

Finally, the strict anaerobes require culturing in an anaerobic cabinet or similar equipment. SMMP medium has been developed for the selective isolation of *Megasphera* and *Pectinatus*.[23] It contains reducing agents to encourage the growth of anaerobes.

Rapid Methods

Numerous rapid methods for detection and identification of spoilage microorganisms have been developed and examined in the context of brewing microbiology over the past 10 to 20 years and many are now being used routinely.[25] For sensitive detection of trace contaminants, membrane filtration or prior enrichment of samples by growth in a suitable medium are often required for these methods, just as they are for the traditional methods, thus lengthening the time needed to obtain a result.

ATP Bioluminescence

The bioluminescence technique has been available since the 1960s and was applied to food during the early 1970s. It depends on the presence of an elevated concentration of ATP in viable organisms, which is depleted as soon as the cell dies. Thus, detection of ATP correlates with detection of viable cells. The quantification of the ATP is achieved by using it as the energy source in the luciferase enzyme reaction adopted from the firefly, which uses luciferase/luciferin to generate light. The amount of light correlates with the amount of ATP and is detected in a luminometer. Yeasts are more easily detected than bacteria because they contain about 100-fold more ATP than the average bacterium. Kits are available commercially that can detect as few as 100 yeast cells per sample without prior enrichment. Biotrace's Bev-*trace* is specifically designed for detection of viable cells in food and drink products including beer (www.biotrace.com). This approach is used extensively for surface hygiene monitoring and swabbing of process machinery. It can also be used to test the microbiological status of water, which is valuable for assessing effectiveness of CIP. This approach allows real-time estimation of cleanliness, thus making recleaning possible.

Direct Epifluorescence Filter Technique

Direct epifluorescence filter technique (DEFT) combines membrane filtration of samples followed by viability staining using fluorochromes and detection of the cells by epifluorescence microscopy or digital image analysis. Acridine orange has been the favored fluorescent dye but it has its limitations. After excitation, acridine orange fluoresces green in combination with DNA and red with RNA so that viable cells are orange due to the relatively large amounts of RNA and dead cells green. However, in practice it is difficult to distinguish live from dead cells. An alternative combination that is claimed to be more discriminatory is the combination of fluorescein diacetate and propidium iodide, which allows detection of both yeasts and bacteria. Commercial kits are available from Molecular Probes Inc. (www.probes.com) for fluorescent viability staining of both yeasts and bacteria, the LIVE/DEAD kits.

Since microscopical examination of filters is tiring and labor intensive, automated versions of DEFT using computer-assisted scanning of the membrane coupled to computer-enhanced image analysis have been developed. Such systems can be fully automated.

Antibody DEFT

One of the drawbacks of DEFT is that although cells can be distinguished microscopically, they cannot be accurately identified. In the case of lactic acid bacteria this could be very important as only certain species of *Lactobacillus* are serious spoilage organisms. It is therefore useful to incorporate identification into the procedure. A fluorescent antibody stain can be used for highly specific targeting of spoilage microorganisms, although with loss of the viability assessment. Monoclonal antibodies have proven particularly useful in this respect and have been popular for detection and identification of *Pectinatus* and lactic acid bacteria including pediococci.[28] The time-consuming aspect of microscopy was circumvented by Yasui and Yoda,[29] who linked a chemiluminescent immunoassay to a camera for detection of the luminescent spots.

Fluorescence In Situ Hybridization

Fluorescence *in situ* hybridization (FISH) uses a fluorescent nucleic acid probe to specifically hybridize with a nucleic acid target in the cell. These gene probes enter chemically permeabilized cells and bind specifically to nucleic acid targets. A favorite target is the ribosomal(r) RNA because, being a component of the ribosome, it is present in many hundreds of copies. Moreover, when cells die ribosomes are degraded and so this gives an element of specificity toward living cells. Nucleic acid probes can be designed with various levels of specificity enabling groups of bacteria (all lactobacilli, for example) or only certain species to be identified. FISH can be applied to beer samples for the detection of *Pectinatus* without prior cultivation and detection achieved in 5 h rather than several days as required by conventional tests.[30]

This approach has been commercialized as kits by Vermicon (www.vermicon.com). The "VIT-bier plus *L. brevis*" kit comprises fluorescently labeled probes and reagents for detection of a range of beer spoilage lactobacilli and *P. damnosus*. Moreover, as these bacteria fluoresce red, *L. brevis*, the most common beer spoilage bacterium, can be distinguished by green fluorescence. Vermicon also market the "VIT-bier megashpera/pectinatus" kit for FISH analysis of these two anaerobic beer spoilage bacteria.

Polymerase Chain Reaction

Polymerase chain reaction (PCR) is a means of amplifying minute amounts of DNA in a highly specific manner. At the molecular level, the PCR comprises repetitive cycles of DNA denaturation by heat into single strands, highly specific primer annealing and DNA synthesis using a thermostable polymerase. The process is exponential and provides several hundred micrograms of DNA from nanograms of starting material within about 30 cycles. Classical PCR products must be visualized by agarose gel electrophoresis to assess whether material of the correct size has been produced, and on occasion it is necessary to sequence the DNA product to be sure that the correct amplification has taken place. The level of detection is such that prior enrichment by cultivation of the organisms is generally necessary to achieve high sensitivity.

Primers specific for PCR detection and identification of various yeasts and bacteria have been developed over the past 10 years including *Obesumbacterium*, *Megasphera*, *Pectinatus*, *Pediococcus*, and *Lactobacillus* (reviewed in

Ref. 25). In general, these primers target the rRNA gene allowing for varying degrees of specificity according to the region used. Thus, some primers recognize parts of the gene common to all gram-positive bacteria, or even all bacteria whereas other primers are specific at the species level. A particularly innovative approach was to target the hop resistance gene (*horA*) of lactobacilli for PCR detection. In this way, only hop-tolerant (potential beer spoilage) bacteria were detected.[31]

The real-time PCR machine or LightCycler allows simultaneous amplification and detection of the product without recourse to agarose gel electrophoresis. The "foodproof" Beer Screening kit has recently been introduced by BIOTECION Diagnostics GmbH (www.bc-diagnostics.com). With this kit it is possible to test for 14 beer spoilage bacteria in a single test after pre-enrichment to ensure detection of trace contaminants. Beer spoilage lactobacilli (*L. brevis* and *L. lindneri*), *P. damnosus*, *Pectinantus*, *Megasphera*, and *Selenomonas* are all differentiated in the PCR reaction.

The Microbiological Laboratory within the Brewery

Microbiological records of the plant and its products should be stored, preferably in computer-based form, for evaluating any deviations from the norm or indications of adverse hygiene trends. The laboratory is not responsible for hygienic operation of the brewery. The production departments are charged with assuring the quality of the process and the products, but they depend on the data collected by the laboratory and advice of the microbiologist to manage their processes effectively. The modern brewery laboratory is a service provider providing the functions and assistance required by the production staff to assure quality operations.[32] By collecting the necessary data in a reliable and timely fashion, the laboratory provides an essential service to the production departments. To achieve this, the laboratory should choose a combination of both traditional and rapid techniques in a cost-effective manner. Automation and rapid data processing are important to indicate adverse trends and to enable microbiologically informed decisions to be made. Good communication with management is then necessary to achieve efficient and hygienically responsible processing throughout the plant.

The scale of the plant will influence the scope of the microbiological analyses that are undertaken. A microbrewery will have limited resources and probably restrict analyses to determination of yeast quality and perhaps microbiological contamination in the finished product. Large breweries, on the other hand, should monitor progress throughout the process and into the finished product; the problems associated with product recall fully justify a comprehensive microbiological service.

The microbiology laboratory also has a responsibility toward the staff of the brewery providing hygiene training so that the requirements of the health officials are met and the staff operate in a safe environment. Routine contact between laboratory and production staff will allay potential problems and encourage good working practices.

References

1. Gilliland, R.B., Wild yeast spoilage, *Brew. Guard.*, 96:37–45, 1967.
2. Campbell, I., Wild yeasts in brewing and distilling, in *Brewing Microbiology*, Priest, F.G. and Campbell, I., Eds., Kluwer Academic, New York, 2003, pp. 247–266.
3. Flannigan, B., The microbiota of barley and malt, in *Brewing Microbiology*, Priest, F.G. and Campbell, I., Eds., Kluwer Academic, New York, 2003, pp. 113–180.
4. Steffenson, B.J., Combating *Fusarium* head blight: an emerging threat to malting barley quality throughout the world, Proc. 27th Congr. Eur. Brew. Conv., Cannes, 1999, pp. 531–539.
5. Noots, I., Delcour, J.A., and Michiels, C.W., From field barley to malt: detection and specification of microbial activity for quality aspects, *Crit. Rev. Microbiol.*, 25:121–153, 1999.
6. Priest, F.G., Gram-positive brewery bacterial, in *Brewing Microbiology*, Priest, F.G. and Campbell, I., Eds., Kluwer Academic, New York, 2003, pp. 181–217.
7. Hollerova, I. and Kubizniakova, P., Monitoring gram positive bacterial contamination in Czech breweries, *J. Inst. Brew.*, 107:355–358, 2001.
8. Back, W., Bohak, I., Ehrmann, M., Ludwig, W., and Schleiffer, K.-H., Revival of the species *Lactobacillus lindneri* and the design of a species specific oligonucleotide probe, *Syst. Appl. Microbiol.*, 19:322–325, 1996.
9. Bohak, I., Back, W., Richter, L., Ehrmann, M., Ludwig, W., and Schleiffer, K.-H., *Lactobacillus amylolyticus* sp. nov., isolated from beer malt and beer wort, *Syst. Appl. Microbiol.*, 21:360–364, 1998.
10. Simpson, W.J., Cambridge prize lecture, studies on the sensitivity of lactic acid bacteria to hop bitter acids, *J. Inst. Brew.*, 99:405–411, 1993.
11. Sakamoto, K., Margolles, A., van Veen, H.W., and Konings, W.N., Hop resistance in the beer spoilage bacterium *Lactobacillus brevis* is mediated by the ATP-binding cassette multidrug transporter HorA, *J. Bacteriol.*, 183:5371–5375, 2001.
12. Gasarasi, G., Kelgtermans, M., Verstrepen, K.J., Van Roy, J., Delvaux, F.R., and Derdelinckx, G., Occurrence of biogenic amines in beer: causes and proposals of remedies, *Monat. Brauwiss.*, 56:58–63, 2003.
13. Kalac, P. and Krizek, M., A review of biogenic amines and polyamines in beer, *J. Inst. Brew.*, 109:123–128, 2003.
14. Dobson, C.M., Deneer, H., Lee, S., Hemmingsen, S., and Glaze, S., Ziola, B., Phylogenetic analysis of the genus *Pediococcus*, including *Pediococcus claussenii* sp nov., a novel lactic acid bacterium isolated from beer, *Int. J. Syst. Evol. Microbiol.*, 52:2003–2010, 2002.
15. Calmels, S., Ohshima, H., and Bartsch, H., Nitrosamine formation by denitrifying and non-denitrifying bacteria: implication of nitrite reductase and nitrate reductase in nitrosation catalysis, *J. Gen. Microbiol.*, 134:221–226, 1998.
16. Smith, N.A., Nitrate reduction and N-nitrosation in brewing, *J. Inst. Brew.*, 100:347–355, 1994.

17. van Vuuren, H.J.J. and Priest, F.G., Gram-negative brewery bacterial, in *Brewing Microbiology*, Priest, F.G. and Campbell, I., Eds., Kluwer Academic, New York, 2003, pp. 219–245.
18. Prest, A.G., Hammond, J.R.M., and Stewart, G.S.A.B., Biochemical and molecular characterization of *Obesumbacterium proteus*, a common contaminant of brewing yeasts, *Appl. Environ. Microbiol.*, 60:1635–1640, 1994.
19. Hamze, M., Mergaert, J., and van Vuuren, H.J.J., *Rahnella aquatilis*, a potential contaminant in lager beer breweries, *Int. J. Food Microbiol.*, 13:63–68, 1991.
20. Magnus, C.A., Ingledew, W.M., and Casey, C.P., High gravity brewing: influence on the viability of contaminating brewing bacteria, *J. Am. Soc. Brew. Chem.*, 44:158–161, 1986.
21. Schleiffer, K.-H., Leuteritz, M., Weiss, N., Ludwig, W., Kirchhof, G., and Seidel-Rufer, H., Taxonomic study of anaerobic, gram-negative, rod-shaped bacteria from breweries: emended description of *Pectinatus cerevisiiphilus* and description of *Pectinatus frisingensis* sp. nov., *Selenomonas lactiflex* sp. nov., *Zymophilus raffinosivorans* gen. nov., sp. nov., and *Zymophilus paucivorans* sp. nov., *Int. J. Syst. Bacteriol.*, 40:19–27, 1990.
22. Engelmann, U. and Weiss, N., *Megasphaera cerevisiae* sp. nov.: a new gram-negative obligately anaerobic coccus isolated from spoiled beer, *Syst. Appl. Microbiol.*, 6:287–290, 1985.
23. Lee, S.Y., SMMP — a medium for selective isolation of *Megasphaera* and *Pectinatus*, *J. Am. Soc. Brew. Chem.*, 52:115–119, 1994.
24. Tholozan, J.L., Membre, J.M., and Grivet, J.P., Physiology and development of *Pectinatus cerevisiiphilus* and *Pectinatus frisingensis*, two strict anaerobic beer spoilage bacteria, *Int. J. Food Microbiol.*, 35:29–39, 1997.
25. Russell, I. and Stewart, R., Rapid detection and identification of microbial spoilage, in: *Brewing Microbiology*, Priest, F.G. and Campbell, I., Eds., Kluwer Academic, New York, 2003, pp. 267–304.
26. Jespersen, L. and Jakobsen, M., Specific spoilage organisms in breweries and laboratory media for their detection, *Int. J. Food Microbiol.*, 33:139–155, 1996.
27. Kindraka, J.A., Evaluation of NBB anaerobic medium for beer spoilage organisms, *Tech. Q. Master Brew. Assoc. Am.*, 24:146–151, 1987.
28. Ziola, B., Ulmer, M., Bueckert, J., Giesbrecht, D., and Lee, S.Y., Monoclonal antibodies showing surface reactivity with *Lactobacillus* and *Pediococcus* beer spoilage bacterial, *J. Am. Soc. Brew. Chem.*, 58:63–68, 2000.
29. Yasui, T. and Yoda, K., Imaging of *Lactobacillus brevis* single cells and microcolonies without a microscope by an ultrasensitive chemiluminescent enzyme immunoassay with a photon-counting television camera, *Appl. Environ. Microbiol.*, 63:4528–4533, 1997.
30. Yasuhara, T., Yuuki, T., and Kagami, N., Novel quantitative method for detection of *Pectinatus* using rRNA targeted fluorescent probes, *J. Am. Soc. Brew. Chem.*, 59:117–121, 2001.
31. Sami, M., Yamashita, H., Kadokura, H., Kitamoto, K., Yoda, K., and Yamasaki, M., A new and rapid method for determination of beer-spoilage ability of lactobacilli, *J. Am. Soc. Brew. Chem.*, 55:137–140, 1997.
32. Nitzsche, F. and Eggers, G., Hygiene monitoring in breweries — the microbiological laboratory as a service provider, *Brauwelt. Int.*, 20:96–98, 2002.

17

Sanitation and Pest Control

Vernon E. Walter

CONTENTS

Introduction	630
Types of Pests Encountered	630
Integration of Sanitation and Pest Control Methods	630
Location and Environment	631
Landscaping to Minimize Attracting Insects	631
Structural Design to Exclude Pests	632
Perimeter Design	633
Trash and Waste Disposal	634
How to Locate Possible Points of Contamination	635
Equipment Design	635
Grain and Malt Area	635
Brewhouse	636
Can Storage Area	636
Filling Area	637
Insect Control Methods	637
Use of Fumigant Gases	638
Types of Fumigants Used	639
Space Treatments (Fogging)	640
Baits	641
Residual Sprays	641
Residual Dust	642
Heat or Cold Treatments	642
Insect Monitoring Methods	642
Safety	643
Perimeter Rodent Control	646
Regulations Affecting Pest Control	649
Record Keeping	650
Outside Contractors Versus In-House Pest Control	651
Evaluating Pest Control Results	652
Summary	653

Introduction

The modern brewery can be host to microbiological problems, such as stray microorganisms, and to macrobiological problems, such as insects, rodents, and birds. We, as intelligent humans, would never eat from a dirty cracked plate because we would know that bacteria could live in the cracks in the plate and reproduce on the small amount of food contained therein. By the same token, most insect and rodent control depends on similar cleaning and repair to remove the food and shelter for these pests. Sanitation and pest control must go hand in hand and be considered together.

The warm, moist atmosphere of a brewery and the grain ingredients all contribute to the attractiveness of the brewery to pests. It is necessary to examine what can be changed to make the plant less attractive to pests. We also need to have well-sealed buildings and physical barriers to deny entrance to the plant by these pests. These points will be discussed in detail later but it is important to first look at the types of pests that may be encountered in a brewery.

Types of Pests Encountered

Pests can be classified in a number of ways, but in this chapter they will be grouped by the way they can enter the plant or by where they would find harborages in the plant. If we know the source of an infestation, control is easier.

Insects that live and breed outside the plant but enter occasionally to cause problems include ants, crickets, and earwigs. Insects such as flies are attracted to the plant and may breed inside. Numerous beetles such as the red flour beetle are brought into the plant with grains or malt. The psocids such as the silverfish may enter the plant with packaging materials whereas cockroaches enter through or live in floor drains and sewers. Termites can enter through cracks in the floor and may exist almost anywhere in the plant unless control is exercised constantly. Rodents largely come in from outside and birds roost, rest, or feed near the brewery.

Integration of Sanitation and Pest Control Methods

Although most companies will be dealing with existing facilities, it is important to discuss some factors that would enter into selecting a site for a new plant.

Location and Environment

For those deciding on the location of a new facility, a number of factors should be considered in addition to the practical considerations of rail and highway access, quality and quantity of water, and labor potential. For example, locating near a populated area may bring complaints of odors, extra traffic, and so forth. It will also be increasingly difficult to fumigate any part of the facility, as concerns about fumigant gases increase. A heavily populated low-income area may harbor a variety of pests that will then be attracted to the plant.

Rural locations are preferred but there can still be problems. Locations near dairies, stables, kennels, or cattle-feeding lots will mean more flies. A location in a grain area with nearby grain elevators has resulted in flying infestations of red flour beetles and other grain insects. A location near tobacco facilities will often result in flights of cigarette beetles. A location next to a scenic trout stream will run the risk of being blamed for every fish kill.

Obviously, there is no practical location that would be pest-free. It is important, therefore, to choose a location where the pest problems are the easiest to identify and manage. If we know ahead of time that flying insects will be the main problem, we can minimize the number of doors and other plant openings, and provide all appropriate physical barriers.

Landscaping to Minimize Attracting Insects

Landscaping must enhance the image of the plant and its product, but minor changes in the landscaping can make a big difference in controlling the pest problems. It is better to avoid fruit and nut trees. The blossoms and fruit attract honeybees that may become a stinging problem to some workers. The bees will later head for liquid sugar lines or can crushing areas. Rotting fruit that has fallen to the ground attracts flies and birds. Nuts can attract and feed rodents.

Trees or shrubs with tight dense branches will encourage nesting of birds whereas trees with widely spaced branches are less likely to attract birds. In some cases, pruning after planting will be indicated. Similarly, spring-flowering shrubs such as *Spirea* attract the adult dermestid beetles that feed on the pollen. The adult beetles can then fly inside and lay their eggs near a food source such as any type of cereal, or even on the dead insects in a poorly cleaned insect light trap. Dermestid infestations can be difficult to eliminate.

Ponds or any standing water will attract birds as well as some insects. Low spots in parking lots that collect water can be bad, but weedy drainage ditches are the worst. A weedy drainage ditch provides shelter, food, and water for rodents as well as many insects. When the weather turns cold in the fall, the pests will migrate into the plant.

No vegetation should be within 2 ft of the building perimeter as it will provide cover for rodents seeking entrance into the building. Ground covers such as ivy should not be used near any building. Trees should be kept trimmed so that no branch is within 2 ft of the roof edge. Rodents can jump from the tree to a roof!

Most food plants have considerable idle equipment that will probably never be used again. Such storage can be a pest shelter and the equipment should be discarded. Similarly, pallet storage may be a problem. It is better if pallets are stored in a clean dry area. If pallets must be stored outside, they should be on a paved surface and well away from the building. They often serve as shelter for rodents or birds; indeed, an examination of pallets stored outside will often show droppings of rodents and birds as well as a variety of insects. After all of the effort to produce clean products, it makes little sense to ship them on a dirty pallet.

Structural Design to Exclude Pests

Mechanical design to exclude pests starts with recognition of why pests are attracted inside a brewery. The primary points of attractiveness are:

- Physical shelter from unfavorable weather conditions. Pests may seek the warmer building during the winter months or the cooler areas during summer months. Heavy rains will also cause pests to enter a building that would not be their normal breeding site.
- Odors of foods will attract hungry pests.
- Presence of a moisture source. Most pests need water and this is often lacking outside.
- Lights, particularly white lights over a doorway, attract some pests. Sodium vapor lights would be less attractive but mounting lights at least 10 m (30 ft) from a doorway and shining the light back to areas that need illumination is better.

Some pests will enter merely out of curiosity. They live in the grass next to the building and see a small crack and enter. So buildings must be designed to discourage pest entrance and those responsible for construction should recognize these facts and avoid them wherever practical. A well-insulated tight wall will be less attractive than a thin metal wall. More pest invaders are found along a metal wall of a warehouse than along an office wall.

Wall construction can prevent entrance by having fewer cracks and openings. Solid concrete walls such as "precast concrete" slabs are ideal if the joints between panels are kept sealed. Corrugated metal walls are the poorest choice because the sealing strips fall out and the slightest bump on the wall will open a new pest entrance. Insulated metal wall panels can be used for warehouse construction, but will result in a much higher insect population.

Removing grass and other vegetation from a 0.5 m (2 ft) strip around the building provides a barren border between the building and the insects that live in the grass or weeds. This will help discourage many of the pests that more easily enter when the grass touches the wall. When gravel is placed in this strip, regular inspections can be easily undertaken for evidence of either rodents or insects. The barren open area also makes placement of rodent bait stations much more effective. The only hiding place for a rodent that is trying to get into the building is the bait station where it will be poisoned. Weed growth under the gravel can be reduced if heavy roofing paper is placed under the gravel. Newer fiber weed-barrier cloth is sold by nurseries and permits water to flow through rather than collect next to the foundation. It is not quite as effective against weeds but doubling it seems to help.

Perimeter Design

Wall fans with their "self-closing louvers" often give the sanitarian a false sense of confidence. The louvers do not close tight enough to prevent insect entrance and despite the insistence of many people, these fans do get turned off occasionally. A number of pest infestations have been traced to these fans permitting insect entry during weekends or holidays when they were turned off. If wall fans are needed for ventilation, that is, forklift battery charging areas or paint lockers, the wall opening should be screened. Screening reduces airflow by about 50%. If full airflow is needed, it will be necessary to "box out" the opening to assure the total screened surface is double the size of the original opening.

Doorways are needed but doors must not be left open. This is, of course, more common in the tropics but it also happens too often in the United States and Europe. All doorways should be designed so that they can be easily closed and will exclude pests when they are closed.

Metal roll-down doors, when properly installed, are among the best for regularly used doors. However, even these doors will need additional barrier brushes or other materials along some edges. The standard steel door and casing is excellent for pedestrian use, but it often comprises of hollow steel sections that can permit mice to run up inside the doorframe. Proper sealing at the time of installation will only occur if the sanitarian is present at the time.

Air blast fans installed over doorways may help in some situations. They must be properly installed to blow to the outside and must be checked at least semiannually to be sure they are still operating correctly. A 1 m (3 ft) length of ribbon can be held at various points around the door opening to check the efficacy. These devices will not be effective if there is a strong prevailing wind blowing against the opening. Also, because they only knock the flying insect down at best, they are not very effective against some crawling insects. They could be used above a dock door where the blast would blow the insect away and there are times that a dock door is open while a truck is backing in.

Roof openings are the most neglected insect entrance. Unscreened vents and stacks are left next to leaking dust collectors and puddles of rainwater. Do not underestimate the ability of stored product insects to fly from a nearby grain-processing source to the roof vents of a brewery. Roofs near dust collectors should always be smooth rather than gravel to facilitate cleaning.

Trash and Waste Disposal

Trash compactors seem to be designed for two things: as a convenient place to dump trash and also as a pest feeding area. It is almost impossible to clean thoroughly under the compactors as they are usually placed close to the ground. They are rarely curbed and often there is no drain nearby. If possible, the compactor should be elevated on concrete skids to increase the height under the unit. Of course, there must be a comparable raised area for the truck to pick up the disposal unit. The area surrounding the compactor must be smooth concrete and not asphalt or dirt. There must be a slope to a sewer drain with adequate capacity. A good hose of at least 22 mm (1 in.) diameter must be available — steam cleaning can also help. There must also be a regular cleaning schedule of at least once per shift. With some installations, there is a large opening in the wall, which is exposed for 15 min or more when the units are changed. This is the time for a thorough cleaning of the area and placement of a temporary screen over the opening.

Dumpsters are usually placed near a doorway for convenience. This means that the most pest-attractive area is near a door that is often open or partially open. Placing the dumpster 15 m (50 ft) or more away from the building is one solution, but usually a forklift driver will leave another door open when he takes a load out to a remote area. A better solution is an enclosed dock area with good doors between the trash dock and any plant area. If possible, there should be at least one more enclosed area between any process or other critical areas and a trash dock. Electric fly-catching devices should be installed in both areas to intercept flying insects.

Open dumpsters and trash bins placed close to buildings allow pest invasion. Moreover, all trashcans should have self-closing lids, or at least tight-fitting lids, and should be lined with plastic bags to facilitate at least daily disposal and cleaning. A regular cleaning schedule must be followed to assure that all cups and other attractive materials receive proper disposal. Outdoor eating and smoking areas are common at some plants. These areas should be located as far as possible from doorways that lead to production or other critical areas.

Windows that open are unnecessary in a modern brewery. The small amount of light and air movement provided is not worth the potential for insect entry. If light is desired, glass blocks can be used in some areas but any glass presents a risk of breakage and subsequent contamination.

How to Locate Possible Points of Contamination

The modern concept of quality assurance involves a hazard analysis of the potential for contamination at each critical point. Critical points are where contamination may occur if steps are not taken. A trained inspector should do a complete Hazard Analysis Critical Control Point (HACCP) survey. The principles of this program should be kept in mind as each area is discussed.

Equipment Design

Grain and Malt Area

When the malt or grain arrives, the quality should be rechecked for insects or other contamination. However, even clean rice, corn grits, or malt can be contaminated during unloading.

The hatches on hopper cars must be opened for aeration, particularly if the product was fumigated in transit. To protect against insect entry during the aeration period, gauze or screening must be placed over the hatch opening. When this step is skipped, insect contamination has occurred by hairy fungus beetles or other pests.

Unloading at large modern breweries is usually done with a pneumatic system and transfer hoses. When the transfer hoses are left on the ground and not capped, insects and rodents have been known to crawl inside. They will be transferred with the grain. This has happened several times in the past and large amounts of products had to be destroyed. Gravity unloading of railcars into a floor dump cannot be as sanitary as desired. Dirt and insects near the dump may be included with the product. The chance of contamination from stored product insects continues at least through the malt mill and weighing points.

All rail-unloading areas should be paved with concrete to assure that the area can be cleaned. Asphalt, or worse, gravel ballast on the tracks will make proper cleaning difficult or impossible. Any small crack is an ideal breeding area for small beetles. Construction of equipment bases should be engineered to reduce these void areas. The use of tubular steel rather than angle iron is often helpful if the steel units are welded shut.

Silos can develop insect infestations that can be transferred from old stock to new stock if a regular cleaning and fumigation program is not followed. Grain and adjunct cleaning systems utilize equipment that is hard to inspect and clean thoroughly. Spot fumigation is often used for inaccessible areas but a redesign of equipment so that all areas can be inspected and cleaned will eventually be needed.

Brewhouse

At the mash and wort areas, the contamination problem largely concerns cockroaches and flies. Wood should be avoided in production areas as it cannot be totally cleaned and is a preferred resting point for cockroaches. Smooth, well-maintained ceramic walls and floors reduce the chance of problems in this area.

American and oriental cockroaches often use the sewer and drain system as their own private highway. They hide deep in these areas during "fogging" operations with only a few venturing up to be killed. The rest happily feed when the insecticide has dissipated. In some dairy plants, a perforated sleeve is built into the drain. The stainless steel sleeve is about 30 cm (12 in.) long and fits into a special drain. It is removable for cleaning and the perforations allow adequate water flow but cockroaches cannot pass through the small holes. In new plants it is not too expensive whereas retrofitting can be expensive.

Bulletin boards and electrical control panels that are mounted flush to a wall will soon harbor insects behind them. Most of this can be avoided merely by mounting the items out at least 25 mm (1 in.). Cockroaches do not like the light and air movement of a large space.

The floor area around the tubs and kettles often contains loose floor tiles or missing grout. This is an ideal breeding area for phorid flies and drain flies as well as for cockroaches. An inspection with a pyrethrin aerosol should be done when production permits. Proper sealing is a constant operation.

Can Storage Area

Can storage areas have been the source of insect contamination problems in too many cases. The point of contamination is usually a can that sticks slightly out beyond the slip-sheet. The problem is worse when the cans are stored for long periods of time in an outside warehouse. Psocids are the most common problem, but other insects have been found.

Psocids feed on starch and mold. They are, therefore, more common on older cardboard sheets with high starch content and microscopic amounts of mold. The psocid feeds on the starch and the microscopic molds that exist on used sheets and then they may fall into cans. The body fluids of the dying insects can "glue" the insect to the can bottom enough to prevent its removal during the can wash. The psocid will be released later by the beer and can be seen when the beer is poured into a glass. One large brewery had extensive problems with psocids and is no longer in business.

Using slightly larger slip-sheets could prevent this type of infestation. Psocids rarely fall into a can unless part of the lip of the can is exposed (one brewer does utilize these larger sheets). Reducing the starch content in the sheets has been tried and does result in less infestation but the sheets cannot be used as often. Plastic slip-sheets have been used.

In severe cases, the sheets can be fumigated with methyl bromide or phosphine between each use. The newer Profume^R could also be used.

Filling Area

The filler area is, of course, the extreme example of an area where contamination could occur since it is the final point before the product is packaged. The most difficult insect problems involve equipment not currently used. This equipment does not get the same degree of cleaning and inspection, but should receive priority inspection at least monthly.

The filling area presents many contamination possibilities. Part boxes of crown caps are often poorly sealed and can be contaminated with dust or insects. Very few brewers have any cleaning system for crowns. The box is dumped into a funnel-shaped hopper and crowns, bugs, and dusts are funneled to the open containers. It is possible to put an air or water wash at the feed portion of the crown line but this can interfere with the flow of the crowns. The first line of defense must be keeping boxes of crowns sealed.

The filling line is usually covered. At best, the covers merely protect from something falling in from above. They cannot protect from fruit flies or other insects flying in, particularly during line stoppages. Flying insects must be stopped before they get to this area.

There should be at least two closed doorways between packaging and the outside. Can crushing and similar very attractive areas should not be in the same room as packaging. Monitoring with traps for flying insects is important in this area. Any appreciable number of insects caught in this area shows a failure at another point that must be addressed.

Insect Control Methods

I have never seen a brewery that did not occasionally have an insect problem. There are a variety of insect control methods and a good pest control program will choose the options that are safest to the product, the employees, and the environment, and integrate these into one master plan. Among the options are:

- Use of a fumigant gas:
  - Fumigate ingredients before they arrive at the plant
  - Fumigate ingredients in bins or silos
  - Fumigate equipment
  - Fumigate packaging materials
  - Fumigate an entire plant or sections of the plant

- Space spraying of large areas with nonresidual insecticides
- Limited area treatment with nonresidual aerosols
- Baits for ant or cockroach control
- Void treatments with insecticidal dusts
- Crack and crevice treatment with residual insecticides
- Spot treatment with residual insecticides
- Outside bait treatments for flies and other insects
- Nonchemical treatments such as extreme heat or cold

Use of Fumigant Gases

Over the years, fumigant gases have been used frequently in breweries to control insects.

The use of fumigants starts long before ingredients arrive at the brewery. Rice or corn grits and even barley are often fumigated before shipment. In the United States, in-transit fumigation is commonly used to assure that the ingredients arrive pest free. It will become even more important to assure that no pests enter with the ingredients as we lose some of the control methods now used in the brewery. Phosphine products are presently the fumigant of choice for grain.

In the brewery, bins are often fumigated when they are emptied to assure that any insects feeding on the dust clinging to the sides of the bins or at the bottom of the bins will be killed. This must be done very carefully to assure the safety of unprotected workers who often must work nearby during the several days of fumigation. Phosphine products are normally used in this area and they require three of more days to be effective. Methyl bromide can give control in 24 h plus 12 or more hours for aeration, but it is regarded as being more hazardous. Newer fumigants such as Profume[R] may be the answer as they are registered for this use.

Equipment that is difficult to clean thoroughly can become infested in between the fumigations of the entire plant. It is possible to fumigate just the equipment with magnesium phosphide, but this must be handled by specially trained crews.

Fumigation of the entire brewhouse has been utilized by some breweries to assure the lowest possible level of infestation. This is often done several times a year. In some breweries, the cellars are adjacent to the brewhouse and will possibly be contaminated with the fumigant gas. It is very difficult to remove the gas from a cold area and a health hazard could result. This type of fumigation is usually done with methyl bromide so that only 24 h exposure plus 12–24 h aeration time is required. It is difficult to shut down a brewery for longer periods of time. With the projected loss of methyl bromide other materials will be needed. Profume[R], which is discussed later, may be an alternate fumigant for this use.

Fumigation of small amounts of various items such as test samples of different malts or other ingredients can be done in a well-sealed truck trailer or special fumigation chambers.

Types of Fumigants Used

Methyl bromide has been used extensively for many years because of its many advantages:

- Little or no residue problem
- Good kill of all life stages of insects including the egg stage
- Short exposure time of 24 h plus 12–24 h for aeration
- Does not harm electrical equipment
- No significant insect resistance known

Unfortunately, methyl bromide is alleged to destroy the ozone layer and is being banned by international agreement. Developed nations are to stop most uses by 2005. Developing countries will have an additional period of 10 year.

No other existing fumigant has all of the attributes of methyl bromide and no new fumigant has been developed with all of the advantages of methyl bromide. We must look at all possible alternatives including other fumigants.

Phosphine can and should be used on all grain products before they enter the plant area to reduce the chance of introducing new insect infestations. It can also be used to fumigate the brewhouse and other areas but there are problems. Longer fumigations of 3–7 days have been the most successful worldwide but a brewery may not be able to shut down for that length of time. Phosphine, under certain conditions, can corrode copper, gold, silver, and their alloys; so electrical equipment can be damaged. Recent developments of phosphine mixed with carbon dioxide have helped. The gas evolved from pellets or tablets, as has been done in the past, will start at zero gas and peak at 400 ppm or more and then taper off. The mixture with carbon dioxide can maintain a steady 100 ppm or other desired level. This can cut down the exposure time to 2–3 days. The gas never reaches the high peaks that occurred when it was generated from solid formulations and corrosion may be less. Experience will eventually tell us just how safe it is from the corrosive viewpoint, and whether the shorter exposure times will be effective without contributing to a resistance problem. Some companies have used heat with the phosphine and carbon dioxide to permit a lower level of gas and possibly less corrosion.

Dow AgroSciences has been using sulfuryl fluoride for the control of dry wood termites for many years. It is currently being tested as a possible replacement for methyl bromide in food processing plants. It can be used with a

24 h exposure period and handled in a similar manner to the way that methyl bromide has been used. Tests do not show a residue problem and there is no danger of corrosion unless there is an open flame or equivalent. Although the egg kill is not as good as with methyl bromide, the kill of other life stages is comparable. At the time of this writing, it looks like a promising replacement as part of an overall integrated pest management. Dow AgroSciences is working to get the necessary labeling and it will be marketed under the name of Profume^R. Final Environmental Protection Agency (EPA) approval for use in flourmills and other food and beverage plants has now been granted for this product.

Carbonyl sulfide has been developed in Australia and used on grain fumigation and a few other types of fumigation. It sometimes has a strong initial odor that seems to dissipate completely, but many breweries will not want to use this material until more testing has been done. It does give excellent control of all of the insects we would normally encounter in a brewery.

Carbon dioxide alone or combined with heat can control insects, but requires longer exposure than methyl bromide. It could have value in some bin treatments, but probably could not be used to fumigate a brewhouse simply because the building usually cannot be sealed adequately to hold the 60% or higher concentration of carbon dioxide that would be required.

Hydrogen cyanide was used for years, but probably will not be brought back due to its extreme toxicity and other problems. A few other gases are being tested but are not seriously considered at this time.

With any fumigation in a brewhouse or other buildings, safety to employees must be paramount. Even bin fumigations must be done in ways that will protect workers that might travel near the bins.

Space Treatments (Fogging)

Space treatments utilize liquid insecticides dispersed as minute aerosol particles often as small as 5–25 μm in size. When particles are this size, they will float throughout a room for as much as several hours and can contact and kill many different kinds of exposed insects. Space treatments are particularly effective against small flying insects such as fruit flies and other flies, but can kill even large cockroaches when they are exposed.

Space spraying is not the same as fumigation. Fumigant gases move as single molecules and can penetrate cartons, boxes, and even concrete block walls. Fogs can only move between cartons and boxes. The most effective insecticide for space treatments is dichlorvos (DDVP). It formerly was mixed with methyl chloroform (1,1,1-trichlorethylene) that helped it disperse as a very fine vapor. This diluent is no longer available and some of the oil-based diluents have had odor and other problems. There are formulations that use carbon dioxide as an aerosol propellant and have had good success.

The most commonly used insecticide in fogs is synergized pyrethrins and they can be effective but are not as effective as DDVP. Space spraying is commonly used in can warehouses and this can lead to problems if some of the cans stick out beyond the slip sheet when oil-based space sprays are used. Obviously, even a small amount of oil landing in a beer will affect the product. Can warehouses must be inspected prior to any treatment. The use of oversized slip-sheets can also be effective for excluding oil particles and usually results in less infestation by psocids and other insects because there is no any exposed lip allowing them to gain access.

Baits

Baits for insect control are a very old concept, but recent developments have made them the product of choice for most ant and cockroach control situations. Baits are available in liquid, granular, and gel formulations. When properly used, they do not present the hazards to personnel associated with fumigants nor the exposure to contamination problems created by fogging. Although they are initially labor intensive, they are very efficient over a year's time.

For cockroach control, small amounts of a gel or other formulation are injected into the cracks and crevices that are known hiding places for these insects. Control will last for several months or until all of the bait is gone. Baits have always been the first choice for some varieties of ants and baits are now available for most of the types of ants that could be a problem in a brewery. It is usually advisable to use several types of baits as different ants will prefer different foods. Some ants will even feed on sweet baits occasionally, but prefer protein baits at other times. Fortunately, there are many types of bait now available.

Baits are also valuable outside. Granular baits for fly control have been used for many years and can be very useful in trouble spots such as around dumpsters, can crushers, and other areas that are attractive to flies. The flies will normally eat the bait and die in the same area, which seems to attract other flies. Baits should not be used near doorways or other potential pest entrances because you may end up with more flies inside than would normally occur.

Granular-type insecticides are also available to eliminate ants and other insects that can migrate from the outside to the inside. Fertilizer spreaders can be used to apply these materials around the perimeter of the plant. Granular-type materials should never be applied to paved surfaces because birds will eat the baits thinking that they are seeds. Serious bird kills have occurred in this way.

Residual Sprays

We do not have good baits for stored grain insects and they are often the most difficult pests to control in a brewery. Good cleaning and sanitation must be done to aid in the control of stored grain insects. If there is a fine coating of organic dust in many places, there can be lots of stored grain insects and the best pesticides will not work efficiently.

If sanitation is good, residual sprays on the floor or wall areas where stored grain insects have been seen will kill these insects and last for a week or more. These can be applied in spot applications of no more than 0.5 m^2 (2 ft^2) if done carefully. These sprays will be more effective if they are placed back in the narrow cracks that harbor these insects. Each insect will have its own preferred type of crack such as the point where equipment is fastened to the floor. Sealing these areas with caulking after the crack is sprayed will reduce the required labor over a year's time.

Residual Dust

Dust formulations have advantages over spray applications. They will often have a longer residual life because the toxicant is impregnated on a dust surface known to be compatible with the insecticide. Its major, and probably only use in a brewery, is to treat wall voids or other voids. A wet spray injected into a void such as a hollow block wall would hit the rear portion and drip down. Its repellency action would keep the insect away from the spray residue. Dust will float and coat the entire void if properly applied. New dust formulations are even resistant to moisture problems. Small inexpensive hand dusters are adequate for most uses but small electric dusters are now on the market. Dusting should never be done where the dust can drift and cause a contamination problem.

Heat or Cold Treatments

Some flourmills have used "heat treatments" for almost 100 years. When temperatures are held to 135–150°F (57–65°C) for 24 h, lethal temperatures will penetrate to almost all areas of a building. If a building can be heated to 70°F (21°C) when it is 20°F below freezing outside (-29°C), it probably has the capability to use heat treatments with only a limited amount of extra heat during the summer. The cost is usually competitive with normal fumigations. The problem is that breweries are rarely designed to withstand extreme heat. Cellars are often adjacent to the brewhouse and would probably be affected.

Cold storage of sensitive ingredients such as malt samples and hops is a viable practice for controlling pests as well as preserving the quality of the products. The temperatures encountered in cellars would not be lethal but would retard development of most insects.

Insect Monitoring Methods

A thorough flashlight inspection is an indispensable part of any monitoring program but the results can be greatly enhanced with some additional tools.

Pheromones are the natural "perfumes" insects use to attract the opposite sex. They have been synthesized and combined with sticky traps. Pheromones can help discover an infestation in its early stages when control is easier and can help monitor progress in control efforts and suggest areas where more work is needed. The most effective pheromone traps so far are those for the Indian meal moth, the cigarette beetle, the warehouse beetle, and other *Trogoderma* species. At least some of these traps should be placed in any grain storage area where the insects might be found.

There are times when pheromone traps are ineffective. Some areas may have dust conditions that will compromise the sticky trap area. The traps for red flour beetle, confused flour beetle, and saw-toothed grain beetle are not as effective as the others. Newer traps may help with these species.

Glue boards can be placed out of sight in many areas to monitor cockroach and other insect presence. This, coupled with good records, will normally show that over 80% of the problems are in less than 20% of the total area. Priority can then be assigned to the proper areas.

Electric fly grids have been used as control tools for a long time. Their value can be enhanced if the "catch" is carefully examined as it can reveal a great deal about flying insect infestations. In addition to the usual houseflies, phorid flies or drain flies may be found, indicating a drain-cleaning problem nearby. Cigarette beetles or dermestid beetles, or various other strong flyers, may indicate a migration of these pests in the area and need for tighter sealing of the plant.

Samples can be removed from central vacuum systems or just the portable vacuum cleaners and sifted for the presence of insects. Insects that were missed with other techniques can be detected this way. Retained samples of incoming ingredients should be checked after a month for any egg hatch that would not have been visible at the time of arrival.

Pest sighting reports by key employees can be one of the best techniques for locating infestations if properly recorded. Normal pest reporting exaggerates the problem or does not pinpoint the location of the observed pest. A typical remark is that "the bugs are all over." This kind of observation does not permit efficient treatment. More specific detailing is needed. The recording form should tell the name of the observer, description of pests seen, location, date and time, and any remarks that would help. A follow-up section of the report should show the action taken by the technician with documentation of a follow-up inspection no more than 2 weeks later.

Safety

Breweries have always been concerned with the health of their employees, but in today's litigious society and its media-sponsored fear of pesticides, greater care is needed.

Years ago, liquid fumigants were often applied with far fewer safety precautions than are required today and yet few if any injuries were ever reported. This may have made us careless as safety programs sometimes do not get the attention they require. Top priority is naturally assigned to production, and conscious effort from the top executives is necessary to ensure proper precautions are taken. For example, no one should stay in an area where a fumigant or a space treatment is used until the area has been aerated and cleared with proper testing instruments. This may mean that a choice will sometimes be required of whether a routine fumigation will be done or whether an equipment repair will be made. If production will be curtailed or seriously affected without the repair the choice is obvious and the pesticide treatment must be postponed. Good monitoring and record keeping will help decide the necessity of a "routine" fumigation or "routine" space treatment. There are probably too many fumigations or space treatments done that are based on a calendar schedule rather than true necessity.

In-transit fumigated railcars should only be opened by trained and licensed personnel equipped with monitoring tools and safety equipment. Actual tests of arriving hopper cars that had been fumigated ten or more days before showed fumigant levels of over 300 ppm. Although this level would be lethal to a man that entered the hopper car (there should be no reason to enter the car at this point), we did not measure any dangerous level (above 0.3 ppm) in the breathing zone of the workers when the car was opened in the open air and the worker stayed on top of the car. However, it did exceed these levels in a covered, unloading area. Walking on top of a railcar wearing either a gas mask or a self-contained breathing apparatus (SCBA) with their limited visibility cannot be regarded as a safe practice. Large fans were added to assure safety to the workers from either the fumigant or from falls while wearing masks.

The results of tests done one time should never be regarded as "normal" for all conditions at that brewery. It is important to "characterize" the actual exposures at each facility under a variety of conditions so that written guidelines can be furnished to the workers.

Regulations permit the transfer of fumigated bulk material when the gas reading in the headspace is below 0.3 ppm phosphine or above that concentration only with proper monitoring. This does not necessarily mean that there is no significant gas present in the commodity. Much of this gas will be removed and exhausted through the dust collector, but only actual measurements should be trusted. Measurements must be taken in all areas through which the product travels and inside the bins when unloading is completed. No one should ever depend on the odor of phosphine as a warning. Head colds or other factors may interfere with a person's ability to recognize an odor. No unprotected person should be allowed into the area around the bins until all readings are below 0.3 ppm.

Treatment of bins inside the brewhouse should only be done when there is no chance of unprotected persons being exposed to a dangerous level of fumigant. This is very difficult when the gas may be present for three or more days and access to the bins may be in a major travel area. The best way is to rope the area off with "danger tape" and then monitor regularly with instruments to record the actual level. Obviously, written records must be kept and any readings above the threshold limit value (0.3 ppm for phosphine) will require appropriate steps. The use of other gases with shorter exposure times should be considered.

Space spraying presents some hazards. DDVP is considered hazardous and the applicator, if in an exposed area, should wear an SCBA, head covering, and body suit that will protect from any exposure.

Pyrethrin sprays may cause allergic reactions in very few people, but there have been lawsuits by employees against food processors for exposure to pyrethrin "fogs." Cash settlements were made out of court in these cases but it is more important to ensure that employees will not be exposed to either space sprays or fumigants. This will require proper employee notification, placards, and, if possible, locks the keys of which are not generally available. "Clamshell locks" can be obtained that fit over door handles.

Space treatment of can storage areas should not be done if cans are exposed because of short slip-sheets. In some cases, it may be necessary to complete the treatment and then hand remove any cans with exposed lips. These should be discarded anyway because they are more likely to contain dust or insects that may not be removed by the can wash.

Careful crack and crevice treatments are not likely to cause any injury to nearby workers, but it is possible to find very small amounts of the pesticide on untreated surfaces days or weeks later because of pesticide volatilization and migration. This is more severe when the spray is applied as a floor/wall juncture spray, which cannot be recommended as a routine application. The small amounts that can be measured would not normally be considered a contaminant at this time but the brew master should be aware that it could occur.

The primary source of information on the safe application of any pesticide is the pesticide label. The manufacturer should be contacted on any point that is not clear. Material Safety Data Sheets (MSDS) are also valuable, but since they are written for everything from the formulation of the insecticide to its use, they may be overly restrictive for actual use in a brewery.

All technicians or other potentially exposed persons should have a thorough medical examination prior to their first pesticide exposure so that baselines for each individual can be established. Blood cholinesterase tests are needed before exposure to carbamate or phosphate pesticides. No one should be required to wear a respirator until a doctor has certified through examination that the person is capable of working while wearing that type of respirator. A physician must check any symptom of pesticide poisoning.

Perimeter Rodent Control

Careful examination of all incoming products is required under the Good Manufacturing Practices regulations and will have eliminated most cases of pest entry with products but most rodents enter from the outside. They are excellent climbers. If downspouts are not screened, rodents can use them to get to the roof areas. They can also climb rough walls or jump from nearby trees or other buildings.

Excluding rodents is the primary control method. Doorways should be checked at least monthly for any crack that would permit rodent entrance. As a rule of thumb, if a pencil can slide under a door, a small mouse can enter. If your thumb can slide under the door, even a rat can enter. Mice can squeeze through holes as small as 4 mm ($\frac{3}{8}$ in.).

The gravel strip around the building or paved areas next to the building is a very important part of the rodent control measures. Rodents are nearsighted and cannot see a doorway or other opening from a distance. They must creep along a wall until they find an entrance. As they are constantly afraid when they do not have the protection of tall weeds, they will enter any shelter, including rodent bait stations. If there is fresh clean bait, they will usually sample it and can die before they can enter the plant. The best plan for rodent control will have two or more lines of defense. In addition to the bait stations next to the wall of the plant, there should be another row along the perimeter fence line.

There is no magic in the distance between bait stations. No one has ever found a rodent carrying a tape measure. The bait stations along the plant wall should include at least one station within 5 m (20 ft) on either side of any doorway or other potential entrance. More stations should be placed in high-risk areas such as rail doors or forklift ramp areas. Fewer bait stations are needed along long stretches of walls with no openings, such as around cellars.

Experience and good records will determine the appropriate number of bait stations along perimeter fence lines. Although they would normally not be needed along an expressway, there have been cases where this was the primary route for rodent travel. Obviously, if there is a grain-processing plant, neglected buildings, or similar suspected harborage nearby, more bait stations should be installed to intercept this possible source. Bait stations with granular baits or wax blocks are effective for house mice and Norway rats, but are only marginally effective for roof rats, which prefer fruit or meat baits rather than cereal baits.

Roof rats can enter the plant without crawling along the walls. They can jump from a tree limb to a roof when the distance is as much as a meter. It is therefore important to plant trees away from the building or at least keep them pruned to over a meter from a roofline. Roof rats can also crawl along power or telephone wires and enter the plant through the small openings where the wires enter. Ideally, the power lines should be buried so that this will not be a problem. If the rats are using the wires for an entrance, they can still be stopped. A heavy sheet of plastic formed into a tube around each wire can be sealed with tape and will hold for years. If the rodent steps on the tube, it will turn and the rodent will either back off or fall off.

All bait stations should be of the "tamper-resistant" type that has baffles to prevent a child from easily removing the bait. Breweries are very sensitive to adverse public opinion (children have stolen bait out of the bait stations around breweries in the past). The information that children obtained rat poison from your facility is not good public relations. Such tamper-resistant bait stations are expensive and not foolproof, but they are mandated by the EPA on most rodenticide labels. Even these bait stations must be securely closed with a lock or some other method that will make it difficult for even a child with a pocket knife to open the bait station. Special screws are used on the lids of some commercial bait stations. As only a special tool will open them, there is a reduced chance of pilfering.

The bait stations must also be secured to the ground if there is any possibility that a child or nontarget animal could have access to it. "Liquid nail" or similar glues can be used to glue a bait box to a concrete area. On grassy or gravel areas, the bait box can be fastened to a large patio block. Long spikes may be sufficient on asphalt-paved areas. There are also commercial anchoring devices.

All bait stations should be inspected and cleaned at least twice per month and the bait should be replaced at least once per month. Recording stickers are often placed inside the bait station to record each cleaning operation to assure consistent service. Clean stations with fresh bait are far more effective than dirty stations with moldy insect-infested bait.

Multiple catch traps can be used outside as well as in strategic areas inside, but will need protection from damage due to traffic. Traps or stations with glue boards are not subject to the fastening restrictions, but this still may be desirable to prevent stealing. There will rarely be cases where baits will be needed that are stronger than anticoagulants or similar materials. Zinc phosphide-coated baits may be needed as one-time treatments in special situations.

With a well-maintained program, toxic baits will not be needed inside the building. However, multiple catch traps, covered glue boards, and probably some snap traps will be needed to catch the stray rodent that gets inside the building. These traps should be checked at least once per week. Proper maintenance of multiple catch traps is discussed at the end of this chapter.

The perimeter should be inspected at least monthly for any sign of rodent burrowing. This will often occur near drainage ditches or under concrete pads surrounded by dirt. Additional bait stations may be needed when these burrows are found or it may be possible to use phostoxin tablets or wax blocks directly in the burrows if the labels permit and the technician is certified. Use of phostoxin tablets requires a fumigation certification and training.

The numerous reasons why rodent control measures might fail are summarized in Table 17.1.

TABLE 17.1

Some Reasons Why Rodent Control Programs May Fail

Plant environment
  Poor sanitation results in excessive rodent populations
  Weeds provide shelter and food
  Trash and stored equipment provide shelter
  Spills along railroad tracks provide food
  Poor drainage provides water

Outside bait stations
  Not in the travel path of rodents that are entering plant
  No bait or insufficient bait for rodent population
  Old dirty bait that is no longer attractive
  Bait station poorly designed and is not an attractive shelter
  Stored materials and trash provide better shelter
  Bait station so hot or so cold that rodent does not enter
  Rodent resistant to bait used

Rodent proofing
  Some openings not closed
  Openings not sealed with correct materials
  Doors left open during the day
  Rodent proofing at ground level only
  Rodent proofing at doors but not around pipes and other entrances

Inspection of incoming ingredients
  Little or no inspection at the time of arrival
  Open truck not inspected until unloading starts (rodent may have left)
  Shrink wrapped material assumed to be clean — receives no inspection
  "Chimney-packed" pallets not checked in center
  No backlight used even on products preferred by rodents

Multiple catch traps
  Animal (rat) too large for trap and no other traps used
  Trap damaged and not inspected. It cannot catch mice
  Trap wound too tight to catch small mice
  Trap too far from wall to be effective
  Trap left too long in one spot and rodents avoid it
  Trap not cleaned and has bad odor

Baited snap traps
  Poor choice of bait. Undesirable food for rodents
  Bait not tied on. Easily pulled off without triggering trap
  Trigger of trap not against wall
  Trap warped and wobbles when rodent touches, scaring rodent
  Trap sprung by vibration of plant before rodent gets there
  Trap not left out long enough to overcome fear of new objects

(Continued)

TABLE 17.1 *Continued*

Trap not tied down and is dragged off to where rodent can get free
Trap stored with insecticides and is repellent to rodent
Prebaiting not done on "smart old rat"
Trigger angle too high or too low

Glue boards
Poor placement in relation to rodent pathways from shelter to food
Glue layer too thin for size of rodent
Glue has a layer of dirt, making it ineffective
Placed in a moist area where rodents wet feet may make board fail
Board not fastened and is dragged to where rodent can pull it off

Inspection program
Plant inspections infrequent
Inspections only made of inside areas
Inspections made by untrained person
Inspector unwilling to get dirty during inspections
Inspector thinks in terms of chemical control only

Regulations Affecting Pest Control

Laws and regulations concerning the application of pesticides are written at the state and federal levels and may even be written at the city and county levels. The EPA is the governing body at the federal level in the United States. State agencies are usually under a department of agriculture, health, environmental protection agency, or an independent body. Other agencies, such as the Occupational Safety and Health Administration (OSHA) and the Food and Drug Administration (FDA), have regulations that must be followed.

Regulations will differ from state to state but basically pesticides can only be applied in a manner consistent with its label directions and only materials registered with EPA may be used as pesticides. For example, diatomaceous earth used as a filtering medium could not be used as a pesticide as there are no directions on the filtering medium label of how to use it as a pesticide. Diatomaceous earth registered with EPA and packaged in a properly labeled container is a legal and effective pesticide. The requirement is to help to assure proper use.

Fumigants and other "restricted use pesticides" can only be applied by or under the direct supervision of a "certified pesticide applicator." The certified applicator will have passed a state exam to show competence. This certification will need to be renewed on a regular basis by exam or by attendance at special training meetings. In some states, fumigants can be applied and aerated only by a certified individual. Other states have only required that the certified operator be present. Most states will divide the certification into five or more categories for urban pest control. Work in a brewery may require general pest control, termite control, fumigation, bird control, and possibly lawn and ornamental pest control or weed control certifications. This requires a great deal of study and exam time. Some states may permit the use of general use pesticides (pesticides that are not restricted use pesticides) by a noncertified operator if applied on the employer's premises. No commercial brewery should ever consider this action since an accident involving a noncertified person might suggest negligence.

When an outside contractor is used, the brewery should have copies of the certification cards of all personnel who will be working in the plant. Even when an outside contractor does all of the pest control, it is suggested that the master brewer or other responsible person also hold a certification. This will help in administering the program and understanding the implication of all the laws.

Record Keeping

U.S. laws will require that records of all pesticide applications involving the use of restricted use pesticides be kept for at least 2 years (3 years in some states). Some states require records of all pesticides used in a food plant. Breweries should maintain files for at least 10 years after the application of all pesticides.

Even before the 1972 EPA laws, one brewery was accused of permitting a pesticide called lethane to get into a bottle of beer. Ten years of handwritten detailed records of every pesticide application were accepted by the court as evidence that the pesticide in question had never been used at the brewery.

Pesticide records should be kept by both the contracted pest control company and by the brewery. The Pest Control Operator (PCO) records are subject to examination by inspectors and in some cases by attorneys involved in a lawsuit. Pesticide records should not be on the same form that reports pest sightings or sanitation problems. There is no reason to open this kind of information to examination by other people.

Pesticide use reports should contain at least the following information:

- Name of pesticide and registration number
- Amount used
- Location
- Target pest
- Method of application
- Date applied

In addition to the records required by law, there are records that can facilitate a good pest control program. All pest control programs in a brewery should have a written list of what is to be done and when. Good record keeping will track the progress of the program and help avoid any problems. Any change should be done only with written advance permission from the brew master or his designee. Rodent catch records can help evaluate success of the control program and point to any modifications needed.

There are now computerized tracing and recording programs. Maintenance requests that involve sanitation problems should be logged separately and checked to be sure that they are performed in a timely manner. Any remodeling or construction done in or near the plant can cause new pest problems or disclose old obscure ones. Each construction program should result in notification of the person in charge of pest control so that an evaluation can be made of extra steps that may be needed. Poor remodeling practices can create many new entrances for pests. Entrance can be gained from the outside through wall voids left by careless repairs where pipes and electrical wiring were brought into a building.

Outside Contractors Versus In-House Pest Control

The food industry varies in its use of outside firms for pest control service. The cereal industry uses primarily in-house staff at their production facilities and outside contractors at their warehouse and distribution centers. The brewing industry in the United States primarily uses outside contracting firms of pest control operators who have experience in other food plants.

Some states now require one or more years' experience in commercial pest control working for a certified operator before taking an exam as a certified operator. Exceptions are usually made for a person with a college degree in entomology or a related science. As it becomes more difficult to replace certified persons, there will probably be a greater dependence on outside contractors.

Some pest control companies are recognizing this trend and are training people across the country to specialize in work in the more demanding field of pest control in a food plant. A firm that merely treats residential property would not have the expertise or the knowledge of appropriate laws to handle a brewery account.

A qualified outside contractor can offer many advantages:

- A certified staff
- Access to all types of expensive application equipment
- Employees that regularly attend training meetings and know the latest techniques and pesticides
- All pesticides stored off your premise
- Separate insurance and responsibilities in case of an incident
- Not involved in any company politics or rivalry
- Personnel experienced in specialized aspects of pest management such as commodity fumigation

Not all pest control companies are qualified to work in a brewery or any other food plant. Poorly qualified companies can cause the following problems:

- Turnover of technicians can mean workers not familiar with your plant or your safety procedures
- Employees on a commission may try to rush through the large accounts, such as a brewery
- They may not be properly trained in pest control much less pest control in a brewery
- The technician answers to his boss and not directly to the brew master
- The technician may try to hide the extent of infestations so that he will "look good"
- Some companies may have excellent technical staffs but unless they visit you on a regular basis, they are of little value

A successful pest control program starts with detailed specifications usually written by the corporate brewing staff but often with the help of a specialized consultant. The corporate legal staff must then add the clauses that protect the company. The potential for alleged product contamination is always possible regardless of who does the work. Contamination problems or other adverse publicity will fall more heavily on the brewery than on the pest control company. The brewery does not abdicate its responsibilities by hiring an outside contractor. It is vital that the PCO report to one key person, such as the brew master. There must be adequate record keeping to assure the brewery supervisor that the work has been done as planned, but detailed inspections by both in-house and outside inspectors will help assure the quality of the service.

There should be quarterly reviews of the progress of the program between supervisory personnel of the pest control company and the brewery staff. Poor communication is often the reason for problems between the two parties.

Evaluating Pest Control Results

There are various ways to evaluate the results of a pest control program. Obviously, if complaints have dropped to near zero this may be one criterion, but it could give a false sense of security. There should be regular in-house and contract inspections to monitor the progress. The in-house inspection should make a weekly check of key areas and a monthly recorded inspection of the entire plant. Records of pheromone and glue board trap catches should be integrated into the report.

The contract inspection can be made by the corporate staff or one of several excellent outside agencies. It is important when an outside agency

Sanitation and Pest Control

is hired to tell them ahead of time what is expected and any company safety rules that may affect the inspection procedure. The highest corporate officer (the same one who might be named in any FDA action) should see all major reports and at least a summary of the weekly reports.

There are several reasons why inspection programs fail:

- Plant inspections infrequent
- Inspections only made of inside areas
- Inspections made by untrained personnel
- Inspector unwilling to get dirty during inspections
- Inspector thinks in terms of chemical control only

Summary

Insect and rodent control has changed in several ways since the first edition of this book and will change in even more ways in the future. It is important to recognize the principles behind each pest control technique. This will make changes much easier.

The first group of insects that I covered was those that live primarily outside and enter only occasionally. Control of these will always depend on good sanitation and avoiding any vegetation close to the building. This keeps their natural feeding and resting area further away from potential plant entrances. At the present time, a number of insecticides are available to provide residual control of these insects as they try to find entry points. Most of these will need to be applied on a monthly basis during the warm months if a problem exists. It is always better to inspect and monitor with glue boards to determine if control is needed.

The second group comprised flying insects such as flies. Again sanitation is key to their control, but it would be hard to keep the brewry and grounds so clean that a fly would not be attracted. Careful use of space sprays during times when that part of the brewery is not in operation will often be needed. Synergized pyrethrins are currently the first choice. If small flies are kept under control, there will not be a problem with spiders and their webs.

The third group discussed was psocids and those insects that enter with packaging. In the future, more attention will be paid to nonpesticidal controls such as plastic slip-sheets or low-starch slip-sheets. Fumigation will always be an option but there will be a trend away from fumigation whenever possible.

Roaches, ants, and even termites will be controlled with baits. Some of these baits are in use now but even more are being developed. Baits are the least likely to cause contamination and are increasingly cost effective.

Rodent control will probably change less than insect control but will still rely on sealing the buildings so that rodents cannot enter.

18

Brewery By-Products and Effluents

Nick J. Huige

CONTENTS

Introduction	656
Composition and Feed Value of Major Brewery By-Products and Competitive Feeds	658
Brewhouse Effluent, Spent Hops, and Trub	663
Brewhouse Effluents	663
Spent Hops	664
Trub	665
Brewer's Grain	666
Wet Brewer's Grain Handling and Dewatering	666
Brewer's Grain Drying	670
Brewer's Grain Volumes	671
Brewer's Grain Feed Products	672
Value of Brewer's Grain in Ruminant Feeds	674
Value of Brewer's Grain in Nonruminant Feeds	675
Brewer's Grain in Food Products	676
Brewer's Yeast	677
Surplus Yeast Volumes	677
Surplus Yeast Collection and Beer Recovery	678
Surplus Yeast Handling and Storage	680
Inactivation and Preservation of Surplus Yeast	681
Surplus Yeast for Swine	682
Surplus Yeast for Ruminants	683
Surplus Yeast for Poultry	683
Surplus Yeast for Pet Foods	684
Yeast Debittering, Concentrating, and Drying	684
Whole Yeast Food Applications	687
Autolyzed and Hydrolyzed Yeast Products	688
Isolated Yeast Protein, Yeast Extract, and Yeast Glycan	691
Biotechnology Products from Yeast	692

Brewer's Condensed Solubles 692
    Production 692
    Composition and Value 693
Excess Carbon Dioxide 693
Spent Filter Cake 695
Waste Beer 696
    Waste Beer Volumes 696
    Waste Beer By-Products 698
Solid Waste Materials 698
Waste Water Disposal and Treatment 700
    Introduction 700
    Waste Water Volumes and Concentrations 701
    Waste Water Pretreatment 702
    Aerobic Treatment Systems 703
    Anaerobic Treatment Systems 703
    Sludge Treatment, Disposal, and Utilization 705
    Land Application of Brewery Effluents 706
    Production of Single Cell Protein from Brewery Effluents 706
References 707

Introduction

In the process of brewing and packaging beer, the generation of by-products and waste products is unavoidable. Technological advances and improved microbiological control over the past 20 years have enabled the brewer to reduce product losses and to produce valuable by-products from materials that were previously considered waste products. The economic benefits can best be illustrated by Table 18.1.

This table shows the considerable economic advantage derived from minimizing product losses or upgrading waste products to by-products. The following sections discuss the methods and processes used to decrease losses and produce valuable by-products; technological and product quality constraints will be covered as well.

The most common brewery by-products are brewer's grain, surplus yeast, and spent hops. Brewer's who do their own malting also produce malt sprouts. All of these products contain more than 20% protein and are generally sold as protein supplements for animal feeds. The next section compares volume, price, and nutritional value of these brewery by-products to other common protein feeds. The subsequent section discusses brewhouse effluents, spent hops, and trub. Ways to achieve better loss reduction and more efficient by-product recovery are presented. Brewer's grain and various

TABLE 18.1

Opportunities for Upgrading Waste Products

Waste Product	Product Value for Brewing (US $)	By-Product Type	By-Product Net Value (US $)	Waste Product Method of Disposal	Waste Product Cost of Disposal[a] (US $)
1 bbl of 15°Plato wort	7–10	BCS	1–2	Waste treatment	1.50–3.50
1 bbl of beer	6–8	Fuel ethanol	0.50–1.50	Waste treatment	1.00–2.50
1 bbl of surplus yeasts at 15% solids	NA	Feed yeast	0.50–2.00	Waste treatment	2.50–7.00
1 ton of waste treatment sludge	NA	Fertilizer	Give away	Landfill	10–25

[a]Based on 4–10 ¢/lb suspended solids, $5–$15 for sludge hauling, and $5–$15 sludge landfill charges.
Abbreviations: BCS, Brewer's condensed solubles; NA, not applicable.

brewer's grain-related feed products are discussed in the Section "Brewer's Grain." Their processing and uses are discussed.

Surplus yeast, the second largest by-product from breweries, is discussed in the Section "Brewer's Yeast." In a typical lager fermentation, about 0.7 lb of surplus yeast solids are produced per barrel of product. These solids include pure yeast solids, beer solids, and trub solids. Section "Brewer's Yeast" discusses their recovery and market potential. Brewer's condensed solubles are produced by concentrating press liquor and other carbohydrate-rich brewery waste streams; multiple effect evaporators are generally used for concentration. The product is blended with brewer's grain or is sold as a high-energy liquid feed ingredient, as a pellet binder, or as a feedstock for fermentation. The Section "Brewer's Condensed Solubles" gives the particulars.

With good CO_2 recovery and purification equipment, some large breweries have been able to sell excess CO_2 as a by-product, especially when CO_2 from counterpressure systems is recycled (see the Section "Excess Carbon Dioxide").

Spent filter cake slurries are discharged to the sewer or they are concentrated for recycle, for blending with brewer's grain, for landfill, or for processing for agricultural or other industrial use.

Total beer losses in breweries are about 5–7%. Opportunities to reduce losses related to spent filter cake disposal are covered in the Section "Spent Filter Cake."

Breweries and especially packaging plants produce large amounts of solid waste materials, much of which can be recycled or used alternatively. These materials include broken pallets, aluminum, cullet, corrugated paper, fly ash, and spent granular activated carbon. Section "Waster Beer" discusses these.

Even with good waste management, a typical brewery has a waste water volume of 4.5 bbl/bbl of packaged beer. This waste might contain about 2.1 lb of biological oxygen demand (BOD) and 0.9 lb suspended solids. Problems and opportunities are covered in the Section "Solid Waste Materials."

Composition and Feed Value of Major Brewery By-Products and Competitive Feeds

Animal feedstuffs are generally classified as energy feeds or protein feeds depending on their main function in animal nutrition. Protein feeds usually have a crude protein content of more than 20% on a dry solids basis. Brewer's wet grain (BWG), brewer's dried grain (BDG), brewer's spent hops (BSH), brewer's malt sprouts (BMS), and brewer's dried yeast (BDY) are all considered protein feeds. Of the brewery by-products, brewer's grains (wet and dry) constitute by far the largest volume. Table 18.2 shows how the U.S. production and export volumes of dried brewer's grain compare to the annual U.S. production volumes of other competitive protein feeds.

Soybean meal (SBM) is the residue remaining after oil extraction. The protein content is standardized to 50% by dilution with soybean hulls.

TABLE 18.2

U.S. Production Volumes and Export Volumes of Brewer's Dried Grain and Other Protein Feedstuffs

Production Year	U.S. Production			U.S. Export		
	BDG	DDG	SBM	BDG + DDG	SBM	CGF + CGM
1976–77	0.30	0.38	18.5	—	4.6	1.5
1977–78	0.28	0.40	22.4	0.13	6.1	1.8
1978–79	0.31	0.50	24.3	0.18	6.6	2.0
1979–80	0.34	0.50	27.1	0.15	7.9	2.7
1980–81	0.32	0.50	24.3	0.13	6.8	3.0
1981–82	0.26	0.50	24.6	0.16	6.9	3.1
1982–83	0.24	0.75	26.7	0.23	7.1	4.0
1983–84	0.15	0.64	22.8	0.15	5.4	3.9
1984–85	0.16	1.02	24.5	0.13	4.9	3.7
1985–86	0.15	1.28	25.0	0.33	6.0	4.5

Abbreviations: BDG, brewer's dried grain; SBM, soybean meal; CGM, corn gluten meal; DDG, distiller's dried grain; CGF, corn gluten feed.
Note: All annual volumes are given in million tons.
Source: U.S. Department of Agriculture.

Corn gluten meal (CGM) is a corn meal residue that remains after most of the corn starch, germ, and bran have been removed. Corn gluten feed (CGF) is a mixture of corn gluten meal, corn bran, and extractives. Distiller's dried grain (DDG) and distiller's dried grain with solubles (DDGS) are by-products from the manufacturing of whiskey or grain fuel alcohol. They contain the nonextracted portions of corn, possibly also some rye and malted barley, and generally contain yeast.

Table 18.2 shows that BDG is only a minor player among all protein feeds. Since the energy crisis of the 1970s, BDG volumes have decreased in favor of BWG sales. DDG volumes have increased with increased fuel alcohol production. Figures for CGF and CGM production volumes are not available, but their export volumes are substantial. CGF, with volumes of about 96% of the total, is mainly exported to European Economic Community (EEC) countries. Because CGF is a by-product material, there are no tariffs. If the EEC would eliminate the no-tariff status of CGF, this product would flood the D.S. protein feed market and severely depress markets for BDG and BWG.

To assess the value of brewer's by-products in the marketplace, their nutritional quality will be compared with the nutritional quality of competitive feedstuffs. For this purpose, Table 18.3 gives the proximate analysis of various protein feeds. Table 18.4 shows the amino acid profiles. Table 18.5 gives the utilization of the available energy of the feeds for various animals. Table 18.6 and Table 18.7 show mineral composition and vitamin B content in the various feeds. BDG prices are obviously affected by price fluctuations of other feedstuffs. As protein levels vary, price comparisons can best be made on the basis of crude protein content. Average prices

TABLE 18.3

Proximate Analysis of Brewery By-Products and Other Protein and Feedstuffs

	BDG	BWG	GDH	BMS	BDY	DDG	DDGS	CGF	CGM	SBM
DM (%)	92	23	91	92	93	92	92	90	91	89
CP (%)	28	27	23	27.5	48	29.5	29	26	45	50
EE (%)	7.2	6.5	4.5	1.5	1	9	10	3	2.5	1
Ash (%)	4	4.8	6.5	6.8	7	2	5	7	4	7
CF (%)	15	15	26	16	3	14	10	9	5	6
NFE (%)	45.8	46.7	40.0	48.2	41	45.5	46	55	43.5	36

Abbreviations: BDG, brewer's dried grain; BWG, brewer's west grain; BDH; brewer's dried hops; BMS, barley malt sprouts; BDY, brewer's dried yeast; DM, dry matter; CP, crude protein; EE, ether extract (fat); DDG, distiller's dried grain; DDGS, DDG with solubles; CGF, corn gluten feed; CGM, corn gluten meal; SBM, soybean meal; CF, crude fiber; NFE, nitrogen-free extract.
Note: All values except dry matter are shown on a dry matter basis.
Sources: From Crampton, E. W. and Harris, L. E., *Atlas of Nutritional Data on United States and Canadian Feeds*, National Academy of Sciences, Washington, DC, 1972; Hubbell, C. H., *Feedstuffs*, April 22: 14–15, 1985; Preston, D. R., *Feedstuffs*, August 11: 18–22, 1986. With permission.

TABLE 18.4
Amino Acid Composition of Brewery By-Products and Other Protein Feedstuffs

Amino Acid	BDG	BDY	DDG	DDGS	CGF	CGM	SBM	WHO[a]
Alanine	NA	NA	NA	NA	NA	NA	5.3	NA
Arginine	5.1	4.9	3.6	3.8	3.6	3.5	7.4	NA
Aspartic acid	NA	NA	NA	NA	NA	NA	14.0	NA
Cystine	1.2	1.1	1.3	1.3	1.2	1.8	1.5	[b]
Glutamic acid	NA	NA	NA	20.5	17.3	21.3	19.9	NA
Glycine	4.4	3.8	1.9	2.8	5.6	3.8	5.1	NA
Histidine	2.3	2.4	2.2	2.4	2.7	2.5	2.6	NA
Isoleucine	5.6	4.7	3.6	5.5	4.3	5.4	5.3	4.0
Leucine	9.8	7.2	12.0	8.5	9.8	18.0	8.2	7.0
Lysine	3.6	7.0	3.2	2.6	3.1	2.0	6.5	5.4
Methionine	1.8	1.6	1.8	1.8	1.5	2.5	1.4	3.5[c]
Phenylalanine	5.6	4.0	2.2	6.0	3.7	7.2	5.3	6.1[d]
Proline	4.4	NA	NA	10.3	8.2	8.8	6.0	NA
Serine	5.2	NA	NA	5.0	3.6	4.0	5.4	NA
Threonine	3.9	4.7	1.1	3.5	3.6	3.6	4.1	4.0
Tryptophan	1.4	1.1	0.8	0.7	0.7	0.5	1.3	1.0
Tyrosine	4.5	6.3	3.3	2.4	3.1	3.0	3.2	[e]
Valine	6.4	5.2	4.4	5.6	5.1	5.3	5.3	5.0

[a] Amino acid requirements for humans established by World Health Organization in 1973.
[b] Included with methionine.
[c] Cystine + methionine.
[d] Phenylalanine + tyrosine.
[e] Included with phenylalanine.
Notes: All values are expressed as a percentage of the crude protein of the feedstuff. Abbreviations are as given in Table 18.3. NA, not aplicable.
Sources: From Crampton, E.W. and Harris, L.E., *Atlas of Nutritional Data on United States and Canadian Feedstuffs*, National Academy of Sciences, Washington, D.C., 1972; Hubbell, C.H., *Feedstuffs*, April 22:14–15, 1985. With premission.

per ton of crude protein during the 1980s were as follows: BDG, $390; SGM, $410; CGM, $441; CGF, $463; and DDG, $518/ton crude protein. The relatively high price for CGF is probably because a high percentage of CGF is exported. The top relative price for DDG is partly because of the higher energy value for DDG and partly because of good research and marketing efforts by the Distillers Grain Feed Counsel.

The proximate analyses given in Table 18.3 are average compositions for U.S. and Canadian feedstuffs. Actual compositions may vary considerably from these average values. The composition of brewer's grain, for example, varies with the level of added spent hops and trub included, the amount and type of adjunct used, the type of malt, malting and mashing conditions, the degree of dewatering of brewer's grain, and possible inclusions of other by-products such as yeast, brewer's condensed solubles (BCS), and spent filter aid. These factors also make it difficult to obtain a representative sample of brewer's grain.

TABLE 18.5

Utilizable Energy of Brewery By-Products and Other Protein Feedstuffs

	BDG	BWG	BDH	BMS	BDY	DDG	DDGS	CGF	CGM	SBM
DE (cattle)	3.3	3.3	1.7	2.9	3.4	3.7	3.9	3.6	3.8	3.6
NE(L) (cattle)	1.7	1.7	0.7	1.5	1.9	1.9	2.0	1.9	2.0	1.9
NE(M) (cattle)	1.7	1.8	0.7	1.5	1.8	2.0	2.2	2.0	2.1	2.0
NE(G) (cattle)	1.1	1.1	0	0.9	1.2	1.3	1.4	1.3	1.4	1.3
PE (poultry)	2.2	—	—	1.3	1.3	1.6	2.0	1.1	2.0	1.6
ME (poultry)	2.5	2.3	—	1.4	1.9	2.0	2.4	1.7	2.8	2.2
ME (swine)	2.3	—	—	1.7	2.9	2.9	3.0	2.4	3.1	2.8

Abbreviations: DE, digestible energy; NE(L), net energy for lactation; NE(M), net energy for maintenance; NE(G), net energy for growth; PE, productive energy; ME, metabolizable energy. Other abbreviations are as given in Table 18.3.
Note: All values are given in Mcal/kg (kcal/g) on dry matter basis.
Sources: From Hubbell, C.H., *Feedstuffs*, April 22:14–15, 1985; Bath, D., Dunbar, J., King, J., Berry, S., Leonard, R.O., and Olbrich, S. *Feedstuffs*, 32–36, 1986; Preston, D.R., *Feedstuffs*. August 11: 18–22, 1986. With premission.

Crude protein (CP) in Table 18.3 is determined by multiplying the percentage of nitrogen by 6.25. This determination is somewhat misleading because some nitrogen is not associated with protein, as is the case for nucleic acids in yeast. Brewer's yeast, corn gluten meal, and soybean meal are considered high protein products. Brewer's grain, malt sprouts, corn gluten feed, and distiller's grain have similar protein levels whereas the protein level of hops is somewhat lower.

TABLE 18.6

Mineral Composition of Brewery By-Products and Other Protein Feeds

Mineral	BDG	BMS	BDY	DDG	DDGS	CGF	CGM	SBM
Calcium (%)	0.26	0.22	0.12	0.08	0.17	0.37	0.12	0.28
Chlorine (%)	0.15	0.36	0.12	0.06	0.17	0.22	0.15	0.03
Cobalt (ppm)	0.06	—	0.18	0.08	0.11	0.09	0.07	0.09
Copper (ppm)	21.0	—	33.0	38.0	59.0	37.0	30.0	17.0
Iron (%)	0.03	—	0.01	0.02	0.03	0.04	0.03	0.01
Magnesium (%)	0.15	0.18	0.24	0.07	0.21	0.35	0.06	0.27
Manganese (ppm)	37.0	32.0	5.7	16.0	29.0	23.0	13.0	27.0
Phosphorus (%)	0.54	0.73	1.43	0.38	0.84	0.77	0.43	0.63
Potassium (%)	0.09	0.21	1.71	0.16	0.65	0.57	0.16	1.91
Selenium (ppm)	0.70	0.60	1.25	0.35	0.38	0.22	1.0	0.10
Sodium (%)	0.24	1.34	0.09	0.08	0.04	0.52	0.07	0.18
Sulfur (%)	0.30	0.79	0.38	0.43	0.30	0.17	0.46	0.43

Notes: All values are % or ppm (as is basis). Abbreviations are as defined in Table 18.3.
Sources: From Crampton, E.W. and Harris, L.E., *Atlas of Nutritional Data on United States and Canadian Feedstuff*, National Academy of Sciences, Washington, D.C., 1972; Hubbell, C.H., *Feedstuffs*, April 22:14–15, 1985. With premission.

TABLE 18.7

B-Vitamin Content of Brewery By-Products and Other Protein Feeds

	BDG	BMS	BDY	DDG	DDGS	CGF	CGM	SBM
Biotin	0.1	0.4	1.0	0.4	0.6	0.3	0.2	0.3
Choline	1800	1600	3900	1300	3000	2000	330	2800
Folic acid	0.2	0.2	9.7	0.9	0.9	0.3	0.2	0.6
Niacin	44	43	460	37	77	66	53	40
Pantothenic acid	8.5	8.6	110	5.8	13	17	10	15
Riboflavin	1.5	1.4	38	3.0	9.5	2.4	1.5	3.0
Thiamin	0.7	0.7	93	1.7	3.2	2.0	0.6	4.0
Vitamin B-6	0.7	9.4	43	4.0	4.6	15	8.0	6.0

Notes: All values are in ppm (as is basis). Abbreviations are as defined in Table 18.3.
Sources: From Crampton, E.W. and Harris, L.E., *Atlas of Nutritional Data on United States and Canadian Feedstuff*, National Academy of Sciences, Washington, D.C., 1972; Hubbell, C.H., *Feedstuffs*, April 22:14–15, 1985; With premission.

Ether extract (EE) is a measure of the true fat or oil content of the feedstuff, but free fatty acids, glycolipids, phospholipids, fat-soluble vitamins, and hormones are also included. EE values are the highest for distiller's grain products because of the oil content of the corn that is used for mashing.

Carbohydrates in feedstuffs are usually divided into nitrogen-free extractives (NFE) and crude fiber (CF). NFE values in a proximate analysis are determined by difference (100−CP−EE−Ash−CF) and are a measure of soluble or readily extractable carbohydrates. CF is an estimate of carbohydrates resistant to treatment with dilute acid and alkali. For feedstuffs derived from plant residues, CF is mostly cellulose, and to a lesser extent, hemicellulose and lignin. Lignin is a noncarbohydrate, with little or no utilizable energy value. The composition of crude fiber in yeast is quite different; it contains glucans, mannans, and polymeric hexosamines. CF is the principal carbohydrate and digestible energy source in rations for ruminants.[1]

For monogastrics and humans, crude fiber has little energy value, but crude fiber components are nevertheless important dietary constituents. The importance of fiber for human nutrition is discussed in the Section "Brewer's Grain in Food Products." Among the protein feeds, brewer's grain and malt sprouts are especially high in fiber. Spent hops (brewer's dried hops, BDH) are the highest in crude fiber, but the fiber is of low quality and the energy utilization is very low (see Table 18.5). Here, it can be seen that energy utilization of feedstuffs depends on the type of animal and on the function that the energy is used for. Brewer's grain, for example, has a higher energy utilization in poultry than does distiller's grain (DDG), whereas distiller's grain has a higher energy utilization than brewer's grain for cattle, sheep, and swine.

The amino acid profiles of various feedstuffs are compared in Table 18.4. For monogastric animals and humans, the utilization of protein is greatly dependent on its essential amino acids profile. The essential amino acid profile for humans as established by the World Health Organization is also given in Table 18.4. Brewer's and distiller's grain products are limited in methionine and lysine. Brewer's yeast is low in the sulfur amino acids, methionine and cystine, but is higher in lysine.

In general, ruminants have little or no dietary need for amino acids. Microorganisms in the rumen can synthesize their own proteins from simple forms of nitrogen such as ammonium; these microbial proteins are then digested and absorbed in the abomasum and small intestine. Some newer studies[2] suggest that some degree of postrumenal essential amino acid supplementation may be beneficial for dairy cows with high production levels.

The concentration of important minerals in various feedstuffs is given in Table 18.6. Brewer's yeast (BDY) is especially high in potassium, phosphorus and selenium. The importance of these and other trace minerals for various animals is discussed in later sections.

The vitamin content of various by-products is compared in Table 18.7. Brewer's yeast is one of the best sources of B-vitamins. Ruminants have no dietary requirements for B-vitamins since sufficient quantities are produced in the rumen and absorbed from the intestinal tract. B-vitamins are also synthesized by microorganisms in the intestines of monogastric animals, but the ability to absorb these vitamins depends on the animal species. Absorption of certain B-vitamins is especially poor in poultry, a fact that makes poultry very dependent on dietary sources of vitamins.

Brewhouse Effluent, Spent Hops, and Trub

Brewhouse Effluents

Effluents from the brewhouse discharged to the sewer include rinses from the various brewhouse vessels, CIP solutions, brew kettle vapor condensate, and liquor from wort clarification that is too turbid to include with the wort.

After runoff of wort to the kettle is completed, brewer's grain is allowed to drain while the free liquor is collected or sewered. Alternatively, the brewer's grain is immediately conveyed "as is" to the brewer's grain processing area. The exact procedure depends on brewing practice and on the wort clarification device used.

In a lauter tun, which is the most commonly used clarification device, free liquor is usually first drained to a separate holding vessel until this so-called sweetwater becomes too turbid. At this point, it is diverted to the sewer. After it has drained to the sewer, the wet brewer's grain still contains about 77–81% moisture. After discharging the wet brewer's drain to a holding tank, the area under the false bottom is rinsed. The rinse water, which may contain a considerable amount of suspended solids, is flushed to the sewer.

Collection and recycling of sweetwater is done to improve lautering efficiency. If the lautering efficiency is already high, such as in low-gravity brewing or when adequate time is available for lautering, the dissolved solids concentration of sweetwater might be too low (less than 0.8°Plato, for example) to be economically attractive. Other factors that may make recycling of sweetwater unattractive are higher than normal concentrations of suspended solids and soluble β-glucans, which could impede runoff. Proper microbiological control of sweetwater must be done if recycling is employed. If sweetwater is not recycled, all free liquor from the lauter tun is sewered.

Different methods employed at various breweries cause the volume of recycled sweetwater to vary from 0 to 60 bbl for each 1000 bbl of final product. This sweetwater may have an extract concentration ranging up to 3.0°Plato.

It is estimated that the total volume of lauter tun effluent and rinses discharged to the sewer varies between 40 and 120 bbl for each 1000 bbl of final product. The dissolved solids concentration in this effluent may vary from 0.4 to 3.0°Plato whereas suspended solids concentration may range up to 1.0 wt%. For each 1000 bbl of final product, a typical brewery will discharge about 300 lb of dissolved solids and 100 lb of suspended solids to the sewer. The BOD of this effluent is approximately 260 lb.

Other brewhouse effluents are rinses and CIP solutions from brewhouse pipes and vessels. Heating surfaces in the kettle or in external boilers become quickly fouled by a build-up of proteinaceous material. These surfaces may require cleaning after every two brews. Similarly, wort coolers may be CIPed every three to four brews. Wort may also end up in brewhouse effluent through: (a) entrainment by brew kettle vapors; (b) purposely sewering wort to avoid brand mixing; (c) as part of hot wort trub that is not recycled or used as by-product; or (d) by leakage and spillage. Estimated data for BOD and suspended solids of various brewhouse effluents of a typical large North American brewery are summarized in the Section "Waste Water Volumes and Concentrations."

Spent Hops

The average hops usage for U.S. breweries is 0.22–0.35 lb/bbl.[3] As only about 15% of the hop constituents end up in the beer, 85% will become spent hop material requiring disposal at the brewery or the hops processing plant.

When whole hops are used, the spent material is separated from the wort in a hop jack, where the spent hops form a filter bed that traps a large portion of the trub. The remainder of both the hot and cold trubs, which is precipitated as a result of wort cooling, can be removed by settling, if a coolship or starting vessel is used, or can be separated in cold wort diatomaceous earth filters. Some brewers allow a portion of the trub to remain in the wort to increase fermentation vigor.

The composition and feed value of spent hops is compared to the composition of other by-products in Table 18.3 and Table 18.5. The crude protein content of spent hops on a dry matter basis is 23–24%, which is higher than that of hops, probably because of trub inclusion, but it is lower than that of brewer's grain. The protein efficiency ratio (PER) value of the protein is very low.[4] Crude fiber of spent hops is considerably higher than that of brewer's grain, but the net energy that cattle and sheep can derive from spent hops is less than 50% of those obtainable from brewer's grain. The addition of spent hops to brewer's grain therefore compromises quality of the mixture, but it is the most convenient and most prevalent method for disposal of spent hops at the brewery site. If spent hops are discharged to the sewer, a BOD load of 100 lb and a suspended solids load of 200 lb per 1000 bbl of product would result.[5]

O'Rourke[6] discussed a number of ways that spent hops are used in the UK. These include: soil conditioner, fertilizer, and the use of spent hops in chipboard and paper making. Residual hop resins can be used as a binder or extracted with acetone to obtain an unsaturated drying oil for paints.

Spent hops produced at hop processing plants in the United States are used either as a mulch, a component in chipboard, or as fuel. A potential application of spent hops in "kitty litter" has also been suggested. Extracted hop pellets have a fuel value of about 8000 BTU/lb, as the moisture content has been reduced during pelletizing and may therefore be used as boiler fuel. A boiler system for this application requires special biomass burners, biomass storage and delivery systems, and ash removal equipment; such a system is four to five times more expensive than a conventional fuel boiler. Depending on cost and availability of conventional fuels, energy recovery from spent hops might still be economical, in spite of higher capital.

Trub

When hop pellets or hop extracts are used in the brew kettle, a fraction of the hop components will end up in the trub. This fraction may include insoluble hop materials, condensation products of hop polyphenols and wort proteins, and isomerized hop acids adsorbed onto trub solids.

The main component of trub is coagulated proteinaceous material formed in the brew kettle — some also develops during mashing. Most of this precipitate will remain with the brewer's grain in the lauter tun; some is carried over to the brew kettle in the wort along with other fines containing starch, lipids, and plant gums. The amount of trub formed depends on many factors: protein content of the malt, the amount of protolysis during malting, kilning conditions, mashing schedule, polyphenol contents of malts and hops, method of boiling (internal vs. external calandria), length of boil, oxidation during kettle boil, and hopping method.

Trub is separated from wort by sedimentation in a conical bottom hot wort tank or in a whirlpool tank. Even in a whirlpool tank, the sediment still contains considerable wort and recovery of this wort can increase brewhouse yields from 0.6%[7] to 1.5%.[8]

Average values from several large U.S. breweries (expressed per 1000 bbl of packaged beer) are:

- Total trub solids: 1010 lb
- Total suspended solids in this trub: 340 lb
- Total BOD in this trub: 713 lb
- Total recoverable wort extract: 373 lb
- Total BOD after wort recovery: 411 lb

Recovery of wort from trub is accomplished using a filter press, a vibrating screen, a centrifuge, or by recycling the trub to the top of the grains in a lauter tun prior to sparging. The latter method is simple, but has a disadvantage — it can only be employed when the lauter tun is processing the same type of wort at the time of recycle. Recycling trub also has another disadvantage — it slows runoff and decreases the efficiency of wort extraction.[9] Recovery of wort by means of a decanting centrifuge has been successful[8] and has none of these disadvantages. A new method that is currently being developed is recovery of wort from trub by means of cross-flow filtration.

When wort recovered from trub is fed forward, it is important to control the level of residual suspended solids. An increase in wort suspended solids causes a more vigorous fermentation; the solids either serve as nucleation sites for CO_2 bubbles[10] or yeast growth is stimulated by unsaturated lipids and zinc in trub solids.[11]

Trub is generally mixed with brewer's grain. Both have similar amino acid content.[12] Trub contains 30% digestible crude protein, which is about twice the level found in brewer's grain.[13] The addition of trub to brewer's grain will therefore enhance its nutritional value. Trub may also be sold in mixtures with yeast and with centrifuge solids recovered from brewer's grain liquor. This type of mixture can have a protein content close to that of soybean meal and may be an excellent liquid feed for swine.

Brewer's Grain

Wet Brewer's Grain Handling and Dewatering

After wort extraction is complete, the remaining grain solids are discharged to a holding vessel. These remaining solids are referred to as spent grain or preferably as brewer's grain. The wet brewer's grains can be discharged by

TABLE 18.8

Brewer's Grain Processing Options

Wet Grain from Brewhouse
  1. Sell as wet brewer's grain (BWG)
  2. Dry without dewatering.
  3. Dewater by centrifuge or press.

Products from Dewatering Centrifuge or Press

Solids
  1. Sell as pressed brewer's grain (BPG)
  2. Dry in grain dryer "as is"
  3. Dry in grain dryer after blending with other by-product streams

Liquor
  1. Discharge to sewer "as is" or after liquor clarification
  2. Recycle to brewhouse "as is" or after liquor clarification
  3. Concentrate to BCS "as is" or after liquor clarification
  4. Ferment to by-product alcohol "as is" or after liquor clarification

Solids from Liquor Clarifying Centrifuge or Screen
  1. Sell wet or dried as high-protein feed or food
  2. Blend with BWG
  3. Blend with BPG
  4. Blend with dewatered grain and dry

---

screw conveyor, pump, or blower. A rapid discharge will enable the next mash to be clarified without delay. The next process step depends on how the brewer's grain is sold: wet (BWG), dry (BDG), or partially dewatered (BPG). Table 18.8 gives an overview of the various process options.

Wet brewer's grain from a lauter tun contains 77–81% moisture on average, but the moisture level of various portions of the grain may vary considerably. Care must be taken to avoid sanitation problems caused by seepage of free liquor from holding vessels, valves, and pumps. If grains are sold wet, they can be transferred directly into a truck, but for flexibility, outside storage tanks are usually employed. Trucks can drive under the tank and be loaded within 15 min. To avoid freezing problems in the winter, the feed line, the gate valve, and the cone portion of the tank might require heat. Since wet brewer's grain is abrasive and moisture content may vary considerably, a slug flow pump such as that manufactured by Ponndorf is most suited for transfer. Other by-product streams such as grain dust, trub, spent hops, or heat-inactivated surplus yeast may be blended with wet brewer's grain. Care must be taken, however, to avoid free liquor when moisture levels reach 80–81%.

Wet brewer's grain from a Strainmaster has high moisture content (87–90%) and must be dewatered prior to sale. This wet grain is easily pumpable and can be dewatered to 70–72%. This reduces the total weight by approximately 70%.

Brewer's grain from a mash filter or lauter tun may be sold as BWG, or may be partially dewatered and sold as brewer's pressed grain (BPG). If brewer's grain is sold as a dried product (BDG), the wet grain is almost always dewatered prior to drying to reduce drying costs. Mechanical dewatering is done by means of a solid bowl decanting-type centrifuge or by means of continuous presses, such as screw presses or roller presses. In general, wet brewer's grains with moistures of 78–80% can be dewatered to 63–72%. On average, dewatered brewer's grain has 67% moisture.

Solid bowl centrifuges are particularly sensitive to variations in infeed moisture. Slugs of lower moisture grain should be avoided to minimize bearing problems. Even with proper care, maintenance costs can be quite high, partially as a result of the abrasive nature of brewer's grains. An advantage of solid bowl centrifuges is that CIP can be applied. This is important for maintaining sanitary conditions when brewer's grain liquor is recycled as product.

Dewatering is generally more difficult when corn grits are used as adjunct and when milled malt contains excessive amounts of fines. These fines, in the range of $0.2-0.8\,\mu m$, tend to contribute to a dough-like consistency of the brewer's grain, especially after these grains have been pneumatically conveyed — they feel slippery. Such elastic properties are associated with hordein-type proteins that contain disulfide bonds. Such bonds can be randomly oxidized, allowing cross-linking and aggregation to take place with undigested small starch granules, glucans, pentosans, lipids, and proteins. The doughy particles, called "teig" by German brewers, inhibit dewatering. After ending up in the brewer's grain liquor, these particles may become a problem when liquor is concentrated to brewer's condensed solubles.

Brewer's grain liquor from presses or centrifuges can be high in dissolved solids (up to 3.5%) and suspended solids (up to 5%). Brewers have, therefore, looked at opportunities to further clarify this liquor and recycle it to the brewhouse or concentrate it to a valuable feed product or constituent. Various process options are shown schematically in Table 18.8.

The most common are: (a) discharge to sewer; (b) concentrate to byproduct; (c) recycle to the brewhouse; and (d) ferment to produce by-product alcohol.

Discharge of brewer's grain liquor to the municipal sewer or to local streams was formerly the most common way of disposal; however, actions by many local authorities have forced brewers to evaluate the economics of other options. Both dissolved solids and suspended solids contribute to BOD. Dissolved solids (ds) contribute from 0.5 to 1.0 lb BOD/lb ds, whereas suspended solids (ss) contribute to BOD by as much as 0.45–0.75 lb BOD/lb.

Total BOD and suspended solids of press liquor is calculated for the following example: the amount of press liquor is assumed to be 42% of the weight of the wet brewer's grain. For a wet brewer's grain volume of 52 lb/bbl of packaged product, the amount of press liquor would be

22 lb/bbl. If this liquor has a dissolved solids concentration of 2% (wt/wt) and a suspended solids concentration of 1% (wt/wt), the total amount of suspended solids amounts to 0.22 lb/bbl of packaged product and the total BOD is about 0.46 lb/bbl.

Prior to disposal, press liquor is often further clarified. This is done by clarifying centrifuges or screening. Screening equipment includes vibrating screens or hyperbolic screens with screen sizes of 75–100 mesh. Hyperbolic screens can reduce press liquor suspended solids by about 60%. Recovered solids are added back to the wet or pressed grains and become a by-product instead of a pollutant.

If liquor is to be recycled to the brewhouse, a high degree of clarity is required. Suspended solids in recycled liquor tend to interfere with wort clarification and may reduce lautering efficiency. To avoid these problems, suspended solids concentrations should be reduced to less than 0.2% by weight by means of a centrifuge or other clarifying device.

Brewer's grain liquor is an excellent medium for microbiological growth, and good sanitation is therefore required. It is important that dewatering and clarification equipment can be cleaned in place. After clarification, liquor is heated to 165–180°F, and this temperature is maintained in the liquor surge tank prior to brewhouse use.[14,15]

Advantages of recycling liquor are:

1. The value of dissolved solids as extract is about three to five times the value of dissolved solids as a feed material
2. Waste treatment costs are avoided
3. Brewhouse throughput capacity may be increased, as exhaustive wort extraction is not required when a good portion of the extract left in the grain is recycled

Disadvantages of recycling liquor are:

1. Suspended solids and β-glucans may impede wort runoff and cause a slight decrease in lautering efficiency.
2. Recycled liquor has relatively high concentrations of polyphenols, which increase wort color and may contribute to a harsh bitter aftertaste in the resulting beer. The influence of recycled liquor on final product is somewhat controversial as some reports[15] show no flavor changes in resulting beers, while others[9,16] recommend treatment of recycle liquor with activated carbon to avoid flavor problems.
3. The fermentability of recycled liquor solids is usually less than that of wort solids.
4. Careful microbiological control is required.

5. Maintenance costs can be substantial when centrifuges are used for dewatering.

If recycling of liquor as product is undesirable, the use of liquor as by-product may be an option. In this case, liquor can be concentrated to BCS. The concentration process is described in the Section "Brewer's Condensed Solubles."

In those breweries that still produce by-product alcohol from waste beer, it may be beneficial to further ferment waste beer and include brewer's grain liquor in the fermentation medium. Including press grain liquor in by-product fermentations avoids waste treatment costs as discussed in the Section "Waste Beer."

Brewer's Grain Drying

Drying of brewer's grain formerly was commonplace in almost every major U.S. brewery. Drying is energy intensive — an energy equivalent of 1.2–1.5 lb of steam is required to evaporate each pound of water. Since the energy crisis of the late 1970s, the production of dried brewer's grain in the United States has decreased by about 50% (see Table 18.2). Drying is practiced even less in European breweries, where energy costs are higher.

The main problem in operating grain dryers is proper moisture control. At moisture levels of 14% and higher, grains can spoil because of mold growth and the biological action can lead to hot spots in dried grain silos. If grains are over-dried, the grain temperature near the dryer exit rises and the dried grains may be toasted or burned. This may lead to odor pollution problems or smoke emerging from the dryer stacks. Product moisture levels have been controlled indirectly by controlling exhaust air temperatures using feedback control.[17] Direct control of product moisture based on a reliable continuous moisture measurement is anticipated in the near future.

The secret to successful dryer operation is to control the feed rate and avoid slugs of wet material in the feed. High moisture slugs upset outlet moisture control, which can result in wet grain adhesion to the dryer wall or heat transfer surfaces. In direct-fired dryers where high temperatures are employed, adhesion can cause fires.

Another problem that is more prevalent in direct-fired dryers is particulate emissions; some municipalities require scrubbing of the exhaust air.

As already mentioned, other by-products or waste product streams are sometimes mixed with the dryer feed because it is usually more economical to combine these streams into the brewer's grain than to deal with them separately. These streams include malt and adjunct dust, spent hops, hot wort trub, brewer's condensed solubles, surplus yeast, solids from clarifying centrifuges, aging tank sediments, dewatered diatomaceous earth, and chillproofing agents. As most of these streams have more than 70% moisture, recycling and blending of dried grain may be required to keep the moisture of the dryer feed below 65%.

Mixing these by-products and waste product streams into brewer's grain requires a great deal of trial and error as many of the mixtures tend to be "tacky" and cause problems in the mixing conveyor or dryer. Recycling of dried grain helps bring down the overall moisture level, but may not resolve the tackiness problem. Dried grains are often case-hardened during drying and have an outer shell that resists adsorption of moisture. Milling recycled dried grain would reduce the problem, but would require additional equipment.

In those breweries where wet grains cannot be dewatered and marketing of wet grain is difficult, a portion of the dried grain is recycled and mixed with the wet grain for better infeed control. In spite of higher energy costs and increased drying time, drying of wet grain has some advantages over drying of dewatered grain.[18] These are: (a) lower BOD costs; (b) no product lost to the sewer; (c) higher protein (about 2%) in the dried grain; and (d) higher density of the dried grain.

Brewer's Grain Volumes

The amount of brewer's grain that is produced per barrel of beer depends on the amounts of malt and grain adjunct used and on brewhouse efficiencies. The quantity of brewer's grain can best be calculated from a solids material balance made for the brew cycle from mash-in to kettle full.

The solids supplied are malt and grain adjunct; there is an increase in solids when the water of hydrolysis becomes part of the extract solids. The solids produced are extract solids from malt and grits, brewer's grain solids, and solids that are discharged as waste from the mash tun, cereal cooker, and lauter tun. Sweetwater solids are not considered because they are recycled internally, except during the first and last brews of the week.

These contributions to the material balance can be expressed as follows:

- Solids discharged to the sewer: amount depends on operating practice, as discussed in the Section "Brewhouse Effluent, Spent Hops, and Trub"
- Extract solids from malt: lb malt × coarse grain extract × lauter tun efficiency
- Extract solids from adjunct: lb adjunct × extract in adjunct × lauter tun efficiency
- Malt solids: lb malt × (1 − malt moisture fraction)
- Adjunct solids: lb adjunct × (1 − adjunct moisture fraction)
- Hydrolysis solids: extract solids from malt and adjunct × water of hydrolysis fraction

The following example is calculated for a typical brew of a size to produce 1000 bbl of packaged product with an original gravity of 11.5°Plato. It will be assumed that 65% of the extract is from malt and 35% from corn grits. Calculations will be made for a lauter tun efficiency of 96%, a coarse grain extract from malt of 73.6%, an extract yield from grits of 80%, a malt moisture of 4%, a grits moisture of 12%, and water of hydrolysis of 4.3%. Solids discharged to the sewer are assumed to be 500 lb, whereas the total beer loss in brewing and packaging is 7%. For 1000 lb packaged, 1075 bbl of wort is needed at 11.5°Plato. The total extract requirement for this is 33,500 lb.

Malt required: $0.65 \times 33{,}500/(0.96 \times 0.736) = 30{,}818$ lb
Grits required: $0.35 \times 33{,}500/(0.96 \times 0.80) = 15{,}267$ lb
Malt solids: $0.96 \times 30{,}818 = 29{,}585$ lb
Grits solids: $0.88 \times 15{,}267 = 13{,}435$ lb
Water of hydrolysis: $0.043 \times 33{,}500 = 1440$ lb

The material balance thus gives:

- Brewer's grain solids $+ 500 + 33{,}500 = 29{,}585 + 13{,}435 + 1440$
- Brewer's grain solids $= 10{,}460$ lb

For this example, the amount of brewer's grain solids discharged from the brewhouse is about 10.5 lb solids/bbl packaged. At 80% moisture, this will give about 52 lb of wet brewer's grain/bbl packaged.

Brewer's Grain Feed Products

Brewer's grain for animal feed is marketed as BWG, BPG, BDG, or as a component in silage or formulated feeds. Most brewers include trub with their brewer's grain products; trub is hard to market separately and it increases the protein content of the brewer's grain. Malt and grain adjunct dust are also frequently added. Spent filter aid may be added, but it decreases protein content of the brewer's grain. If grains are dewatered, the suspended solids in the liquor are usually recovered and added back to the grain. If the liquor is concentrated to BCS, it is sometimes added back to brewer's grain, especially if the concentration of BCS is too low to be competitive with molasses.

The addition of yeast to brewer's grain for ruminant feed has a number of benefits[5]: (a) it increases palatability, thereby improving voluntary feed intake and rate of gain; (b) it appears to enhance the utilization of other ration components increasing the feed efficiency for steers; and (c) it enhances the milk yield for dairy cows. As surplus yeast slurry can contain as little as 12–15% solids, the addition of yeast slurries to wet brewer's grain may cause problems with free moisture.

For this reason, the addition rate may be limited unless the grain or yeast slurries are partially dewatered.

There is some controversy in the literature about the optimum moisture content of wet brewer's grain. Linton[5] claims that dewatering grain to less than 75% moisture is not advisable because below this moisture level the product becomes difficult to compact and might lead to oxidative degradation during storage. According to Linton, at moisture levels above 80%, free water loss occurs, making storage and handling more difficult. Penrose[19] claims that the ideal moisture is 73%, as liquid separation will occur above this moisture level. Penrose proposes to dewater brewer's grain to 69% moisture and then add back high BOD, high moisture slurries such as surplus yeast to bring the overall moisture back to 73%.

BPG has a moisture level of 65–70%. This product is very palatable and easy to handle and store. Its prime advantage is that transportation costs may be almost 50% lower than for BWG. A disadvantage is that BPG does not store as well: mold growth occurs more rapidly and may work its way through the pile, while the odor might draw flies.

The main drawbacks of BWG compared to BDG are higher transportation costs and the higher tendency for spoilage. The larger the volume of brewer's grain to be marketed from any single brewery, the greater the freight distance and cost. The price per ton of BWG that a brewer receives might therefore go down after a critical volume is exceeded. Dairy businesses are sometimes built around the BWG production level during December and January. During higher production months, an excess of brewer's grain will result. Marketing of brewer's grain is also affected by variability in local crop conditions from year to year as good crops might reduce the perceived need for brewer's grain. Weather conditions might also affect the sale of BWG, as high temperatures increase spoilage and cold weather might cause freezing of product. Even though many farmers have facilities to store BWG during periods when supply exceeds demand, it is beneficial to brewers that have grain dryers to play the market and dry a portion of their grains. These brewers determine the economic optimum ratio of BDG to BWG on a weekly or monthly basis.

Long-term storage of BWG may be done[5] by ensiling the product in a horizontal pit silo, but it is essential that the top be tightly covered with plastic. BWG may also be blended with haylage or corn silage for bunker silo storage. For storage in upright silos, the moisture is usually reduced to prevent excessive seepage. This is done by blending BWG with dry feed such as shelled corn, barley, or mill by-products.

A specialty silage product called Maltlage was developed in the United States by Hunt and Spitzer.[20] The silage is a product of about 50% solids and is prepared by anaerobic fermentation of a mixture of roughage and vitamins and minerals to provide essential elements and to buffer the mixture so that the pH is controlled at 4–4.5.

Value of Brewer's Grain in Ruminant Feeds

Brewer's grain is an excellent feed ingredient for ruminants as it can be combined with inexpensive nitrogen sources, such as urea, to provide all the essential amino acids. A considerable portion of the protein in brewer's grain bypasses the rumen and ends up in the abomasum and in the small intestine of the animal along with microbial protein produced in the rumen.

In the rumen, a portion of the brewer's grain protein is slowly degraded to ammonia and alpha keto acids. Nonprotein nitrogen, such as urea, is utilized in the rumen at a much faster rate. From the ammonia, microorganisms synthesize proteins that the animal can digest and that are needed for maintenance and reproduction. In many feeding situations, rumen microbes cannot synthesize sufficient protein for the animal's needs and a certain amount of bypass protein is required. If too much urea is fed, excess ammonia will result, which is eventually wasted. As soybean meal protein degrades much more readily in the rumen than protein from BDG or DDG, the relative value of SBM is lower; too much goes into microbial protein in the rumen.

Because of its high percentage of bypass protein, the value of BDG protein is rated at 1.8 times the value of SBM protein, whereas DDG has a relative protein value of 2.0 compared to SBM, and DDGS has a value of 1.6.[21] Based on these data, combinations of BDG or DDG can be formulated with corn and urea that have the same crude protein, the same protein efficiency, and the same net energy value for growth as SBM.[22] Table 18.9 shows the results. The costs for these feed combinations is calculated from the average 1986 prices for these feedstuffs for major U.S. markets.

The results in Table 18.9 show that BDG is currently greatly underpriced, especially compared to SBM. The fact that BDG is priced so much lower than DDG is very likely the result of better marketing efforts by the distiller's group. In fact, as seen from the results of this table, BDG is worth $100/ton compared to DDG at $116/ton.

Because of its high crude fiber content, BDG has been found to reduce incidences of rumentitis and liver abscesses, which may severely affect growth

TABLE 18.9

Combinations of Equal Feed Value for Growing Steer

	Price/Ton	SBM (lb)	BDG Combination (lb)	DDG Combination (lb)
SBM	160	2000		
BDG	70		1780	
DDG	116			1508
Corn	74		404	401
Urea	225		141	158
Cost/ton ($)		160.00	93.11	120.08

performance.[23] BDG was also found[24] to reduce skin abnormalities in ruminants.

Drying of brewer's grain slightly increases the percentage of bypass protein, but it appears to have a slightly negative affect on the total feed value, possibly because of a decrease in total digestible nutrients. Linton[25] found that palatability and gain/lb feed (dry basis) for BWG was slightly higher than for BDG.

Milk yields for dairy cows were found to be the same[26] when BDG, DDG, or a mixture of SBM and wheat bran was fed. BWG, however, was more efficiently used for milk production than BDG. This is probably due to the higher solubility of BWG and to its more balanced amino acid pattern. BWG can play a very significant role in dairy cattle nutrition, but because of a large difference in BWG composition between breweries, care has to be taken to use proper nutritional values.[27]

Value of Brewer's Grain in Nonruminant Feeds

Brewer's grain contributes to the protein and energy requirements of swine. The poor balance of essential amino acids for swine in BDG and DDG dictate that these feeds be considered more as energy sources than as protein sources.[28] Digestible energy and metabolizable energy are lower for BDG than for any of the commonly used protein feeds or energy feeds. Supplementation of BDG with brewer's yeast can correct the balance to some extent.

BDG and DDG are beneficial for laying hens, breeder hens, and turkeys. Improvement in egg production rate and hatchability were noted with breeder turkey rations including 40% DDG.[29] For laying hens, up to 20% of BDG could be used without adversely affecting egg production rate.[30] Some researchers found a slight reduction in egg weight, whereas others found no difference. Jensen[31] demonstrated that BDG and DDGS contain an unidentified factor that improves interior egg (albumen) quality. Optimum egg quality was obtained with 20% BDG or DDGS. DDGS at levels of 10–20% was also effective in reducing the incidence of severely twisted legs in broiler breeders and in the prevention of the fatty liver syndrome. As mentioned, this may be associated with selenium, which is also present at high concentrations in BDG. Recent studies show that 20% of BDG in chicken feed decreased low-density lipoprotein cholesterol by 23%.[32]

BDG appears to be an excellent ingredient for inclusion in some types of horse rations.[33] Digestibility studies indicate that its energy content is equal to or greater than oats when fed at levels of 40% or less in pelleted feed. Protein digestibility appears to be comparable to that of a mixture of oats and soybean meal. The use of BDG in rations for growing foals may require some amino acid additions to ensure optimum growth.

In addition to its protein content, BDG is an excellent source of dietary fiber. Fiber fed to monogastric animals may significantly alter their body composition.[34] Fiber-fed pigs, for example, are leaner. Fiber also has important affects in the upper digestive tract as it delays gastric emptying, which causes simple sugars, vitamins, and minerals to be adsorbed more slowly and efficiently. The important health benefits derived from fiber, which is now evident for humans, suggests that research into similar benefits for monogastric animals may be beneficial.

Brewer's Grain in Food Products

As discussed in the previous Section, brewer's grain commands a relatively low price in the feed market. This price may fluctuate considerably depending on prices and on supply and demand of other feed commodities, as well as on tariffs and price support structures instituted by various governments. Therefore, there has been a considerable effort by many brewers to develop food applications for brewer's grain for internal or external use.

During the 1960s and 1970s, processes were developed to isolate protein from brewer's grain for use in human food products. Methods used were enzymic hydrolysis[35] or alkaline extraction followed by precipitation.[36] In both of these processes, protein was considered the most valuable ingredient in brewer's grain and the remaining material, mostly fiber and fat, was to be used as animal feed. Developments during the 1970s and 1980s showed that the fiber in brewer's grain may be an even more valuable ingredient in human nutrition.[37] The role of fiber for human health has created so much interest that the production of food-grade fiber products from brewer's grain must be considered one of the greatest opportunities to upgrade the value of brewer's grain.

The crude fiber content in brewer's grain is about 14–15% on a dry solids basis. Crude fiber is an estimate of lignin and carbohydrates (mostly cellulose) that are resistant to treatment with acid and alkali. Fiber, which is important for human nutrition, includes many other constituents in brewer's grain in addition to the crude fiber. Other components include hemicellulose, pectins, gums, mucilages, and Maillard products. All these components together are referred to as dietary fiber. Dietary fiber is defined as the components of a food that are not broken down by digestive enzymes. Total dietary fiber in brewer's grain is about 56%. Of this, 2.5% is soluble dietary fiber and 53.5% is insoluble. A good review of health benefits of brewer's grain flour is given by Weber and Chaudhary.[38]

Since dietary fiber is not a single entity, different sources of dietary fiber will have different physiological responses. For example, brewer's grain fiber and oat bran have a definite cholesterol-lowering effect not found with pure cellulose. Soluble fiber as well as lignin appear to be important for cholesterol reduction. Some cholesterol-reducing effects have also been found to arise from the presence of a vitamin E-like substance (tocotrienol) in the lipid fraction of brewer's grain. Recently completed studies with brewer's grain flour have demonstrated beneficial effects for accelerating gastrointestinal transit time,[39] cholesterol reduction,[40] and binding of cholesterol oxidation products.[41] Some functional properties of brewer's grain flour in foods are:

- Ease of blending
- Calorie content is about half that of most cereal flours, most comes from fat and protein
- Has high water absorption capacity
- Provides valuable minerals such as Ca, P, Fe, Cu, Zn, and Mg
- Has low-fat absorption, beneficial for batters and coating
- Has uniform tan color, bland flavor, and mildly roasted aroma
- Has 50% more dietary fiber than white wheat bran
- Higher in protein

For product information see Refs. 12, 42, and 43.

Potential applications for brewer's grain flour products are given in Table 18.10. These products can be obtained from brewer's grain by milling followed by screening into a high-fiber product (coarse) and a high-protein product (fine).

Brewer's Yeast

Surplus Yeast Volumes

During fermentation, yeast cell mass increases three- to sixfold. The amount of yeast grown depends on the fermentation conditions of each brewery. The type of yeast as well as the condition of the pitching yeast, such as yeast generation and glycogen content, can also affect yeast growth. Wort constituents that may affect yeast growth include carbohydrates, amino acids, free fatty acids, and trace minerals such as zinc.

TABLE 18.10

Use of Brewers Grain Products in Food Formulations

High Fiber Fraction		High Protein Fraction	
Product Usage	Level of Usage	Product	Level
Breakfast cereal flakes	25–30%	Cookies and brownies	up to 30%
Granola bars	20–25%	Snacks	up to 20%
Muffins	15–20%	Breakfast cereals	up to 20%
Specialty breads and rolls	up to 15%	Pasta	up to 15%
Bagels	up to 15%	Pancakes and waffles	10–15 %
Pizza crust	up to 15%	Doughnuts	up to 10%

After fermentation, most of the yeast is collected as surplus yeast. After beer aging has been completed and the yeast, along with other insoluble material, has settled, so-called tank bottoms are collected as by-product. Typically, the total amount of surplus yeast solids produced in a lager fermentation is about 0.6–0.8 lb/bbl of final product. The solids in surplus yeast include pure yeast solids, beer solids, and trub solids. Trub solids contain approximately the same amount of protein as yeast solids (40–50%) and their removal is not required if yeast is to be sold as animal feed. When yeast is sold for food uses, removal of both trub solids and beer solids is generally necessary.

Surplus Yeast Collection and Beer Recovery

At the end of fermentation, yeast is collected from fermenters by one of the following methods:

1. Pulling (raking) settled yeast from the bottom of the fermenter
2. Separating yeast from the entire fermenter contents using fassing centrifuges
3. Pulling settled yeast, followed by clarification of the rest of the fermenter contents in fassing centrifuges
4. Using various skimming systems when top-fermenting yeasts are employed[44]

Surplus yeast pulled from the bottom of fermenters or aging tanks usually has 10–14% total solids. This surplus yeast may contain as much as 1.5–2.5% of the total beer production. It is usually worthwhile to recover at least a portion of this beer, especially in countries where excise taxes are paid on all beer produced in fermentation, including beer that is wasted.

Many breweries, especially smaller ones, discharge all surplus yeast to the sewer without recovering any entrained beer. This may contribute to considerable sewer loadings. For each 1000 bbl of beer produced, a total of about 900 lb BOD and 600 lb of suspended solids may be discharged in this manner.

Methods to recover entrained beer from surplus yeast have been reviewed in detail for 90 breweries by the Institute of Brewing and the Allied Brewery Traders Association in Great Britain. The review was summarized by Young[45] and Boughton[46]; see also Ref. 44. Their findings are summarized here:

1. *Rotary vacuum filter.* A continuous filter, it requires little operator attention and has fair to good filtrate clarity. Drawbacks include: (a) requires filter aid, (b) the dissolved oxygen of recovered beer is high, and (c) filter cake solids are low (22–25%).

2. *Plate and frame press.* Cake dryness is low (limited low maximum inlet pressure). Filtrate quality is fair to good, but it is labor intensive.
3. *Recessed plate press.* Can recover up to 90% of the entrained beer yielding cake solids of 26–30%. Dissolved oxygen levels are low. Filter aid only required for tank bottoms. Requires operator attention during cake discharge and must operate full.
4. *Pressure leaf filter.* An open mud discharge filter is of this type. Generally more automated, more flexible, but more expensive. Cake moistures range from 26 to 30%. Filtrate quality is good after initial recycle.
5. *Membrane filter press.* After filling, cake can be compressed by expanding a diaphragm. Makes possible processing of variable volume and variable concentration yeast slurries. High capacity (cycle times of 1.5 vs. 2–3 h with others). Cake solids high (30–35%), filtrate quality excellent. Operator attention required during cake discharge only.
6. *Decanter centrifuges.* Can be used to further concentrate yeast slurries from fassing centrifuges or aging tank bottoms to about 25% solids. Filtrate clarity poor, but low labor requirements.

When beer is recovered from yeast by means of filter presses, it is desirable to keep operating pressures below 7 bar to prevent the leaching of undesirable yeast components into the beer. To maximize beer recovery, water washing of the cake may also be applied. Beer and wash water recovered from yeast can be blended into the main product at rates of 1–5%.[47] The beer may be flash pasteurized prior to blending.

Beer recovery from aging or lager tank bottoms presents a real challenge to the brewer. The tank bottoms contain a considerable amount of beer. Suspended solids consist of yeast, which may be partially autolyzed, precipitated proteinaceous material, and finings if these are used. Centrifugal thickening is possible, but very slow. When using conventional vacuum or pressure filters, the filter surfaces tend to blind quickly so filter aid is required. It is sometimes possible to use fermenter yeast as a precoat. Blending of recovered beer from aging tank bottoms into the main product is usually not recommended because of flavor problems. A relatively new technique of crossflow membrane filtration appears to have some promise for handling aging tank bottoms. A clear beer can be obtained[48] as membranes hold back undesirable components, whereas solids can be concentrated to a toothpaste consistency without blinding the membrane filter.

Fassing centrifuges that are used to clarify beer going to aging are usually disk centrifuges with an intermittent bowl opening discharge or a continuous nozzle discharge. To function well they require a reasonably constant feed consistency; this can be accomplished by agitation. When operated well, fassing centrifuges can produce a yeast slurry of more than 20% total solids. In actual practice, an average yeast solids concentration of 17–18% is achievable. Surplus yeast recovered from centrifuges may be as high as 21°C and may require cooling to prevent decomposition.

The high shear forces in fassing centrifuges may damage yeast cells.[49] If the yeast slurry obtained from fassing centrifuges is to be further concentrated by means of yeast presses or decanting centrifuges, use of the beer thus recovered is not recommended as even at a blend ratio of 0.5% flavor differences in the blended product may be noted.

Surplus Yeast Handling and Storage

Yeast cake obtained from presses or rotary vacuum filters can be discharged directly into bins for transport, broken up and bagged or palletized, and stored in refrigerated containers. Larger breweries usually reslurry the yeast cake with water prior to further processing or sale.

The reslurried yeast, as well as the yeast from centrifuges or yeast collected directly from fermenters or aging tank bottoms, is usually pumped to intermediate storage tanks where it awaits in-house processing or shipment to outside processors. Yeast slurries, especially those exceeding 18% solids, are usually moved by a positive displacement pump. A good review of yeast pumps is given by Young.[50] It is useful to have a variable speed drive on the pump and to have oversized inlet and outlet ports.[51] Yeast slurries exhibit a pseudoplastic rheology. Viscosity–shear relationships can vary substantially depending on concentration and temperature.[52]

During storage, yeast can undergo changes that may not be desirable. It is important therefore to control storage and transportation conditions. Total solids may decrease when freshly harvested yeast slurry is stored, depending on storage temperature. Many brewers in the United States sell their surplus yeast as feed yeast to outside processors. Payments are based on the pounds of total solids of the yeast slurry as delivered, but the price per pound of solids is generally a strong function of the total solids concentration. The decrease in solids during ambient storage may be accompanied by considerable foam formation. This foam can be quite stable and may cause yeast loss through the tank overflow pipes. To prevent these problems, chilling of the yeast to below 5°C is recommended. Agitators in the yeast storage tanks are useful to keep contents well mixed and prevent hot spots that will accelerate metabolism. Chilling of the yeast slurry prior to shipment may also prevent overfoaming of tank trucks during transportation in hot climates or over long distances. Table 18.11 lists the compositional changes that can take place when yeast is stored at 21°C.

During fermentation, glycogen content peaks at about 50% of the cell mass — it may still be 30% when the yeast is harvested. As glycogen is metabolized during storage, ethanol and CO_2 are formed. This CO_2 may lead to the excessive foaming. Another change during storage at ambient

TABLE 18.11

Changes in Yeast Slurry Composition During Storage[a]

	Total Slurry		Yeast Fraction		Liquid Fraction	
	0	4	0	4	0	4
Total weight (%)	100	100	52	24	48	76
Total solids (%) (wt/wt)	17.9	13.9	31.3	43.6	3.6	5.6
Ethanol (%) (wt/wt)	4.9	6.9	4.3	6.8	5.6	7.0
Protein (%) (dry basis)	43	53	45	52	24	53
Carbohydrate (%) (dry basis)	43	34	40	37	76	26

[a] 0 = freshly harvested yeast slurry; 4 = after 4 days at 21°C.

temperatures is that some yeast cells autolyze, releasing their cell constituents. This is why a considerable portion of yeast protein ends up in the liquid phase. If the yeast slurry is to be washed prior to further processing, all protein that has leaked into the liquid fraction will be lost. The increase of protein concentration in the total slurry is merely a result of the decrease in carbohydrates. This increase may be of benefit if certain minimum protein specifications are to be met consistently.[53]

Inactivation and Preservation of Surplus Yeast

Yeast for food use or yeast in feed rations for monogastrics is generally inactivated. Feeding of live yeast might cause avitaminosis because of the depletion of B-vitamins in the intestine. If live yeast is fed to suckling sows, there is a danger of diarrhea in piglets.[54] Live yeast can also cause adverse fermentation in the digestive tract of swine leading to bloating, while live yeast cells are utilized less efficiently than dead cells.[55]

Killing of yeast can be done by chemical means or by the application of heat. Thermal death time curves obtained by heating a yeast slurry to various temperatures[53] show that less than 1 min of heating at 60°C is required to reduce viable yeast cells to less than one cell per liter. It should be understood that killing of yeast destroys the yeast membrane, but does not necessarily inactivate all yeast enzymes. Enzymes can be deactivated by heating the yeast slurry to 75°C. After yeast cells are killed, the yeast slurry becomes a very desirable growth medium for other microorganisms. If heat-treated yeast slurry is to be held for more than several days, it is desirable to add preservatives such as organic acids.

If nonviable spray-dried yeast is desired, it is also necessary to pasteurize (heat to 60°C) the yeast slurry prior to drying. Spray drying of fresh yeast was found to give 4–5 log cycles of destruction,[56] which is insufficient to guarantee complete kill when initial cell counts are as high as 10^{10}. Yeast slurries can be heated by the injection of live steam or by passing the slurry through a heat exchanger. Live steam application may be advantageous for breweries that produce a yeast cake, which has to be reslurried anyway. To minimize heating costs, regenerative-type heat exchangers are sometimes employed. Fouling of heat exchange surfaces may occur, especially when some autolysis has already taken place prior to killing the yeast. To minimize fouling problems, scraped surface heat exchangers can be used; it is also beneficial to keep the temperature difference between slurry and heating medium below 10°C.

Chemical treatment to kill live yeast has been done with propionic acid[55] or formic acid.[57] These acids also act as a preservative for yeast. Generally, organic acid treatment is more expensive than heat treatment, but acid-treated yeast is more stable and the organics contribute to the feed value of the yeast. The lower organic acids are skin irritants and they fume. They require appropriate systems of storage, handling, and application to safeguard brewery operators and their working environment.[57] An extensive study was conducted in Canada[58] to determine the effectiveness of various organic acids to kill yeast. A mixture of formaldehyde and propionic acid and a mixture of formaldehyde, formic acid, acetic acid, and propionic acid were much more effective in killing yeast than propionic acid alone.

Surplus Yeast for Swine

Surplus yeast is sold for feed applications as a wet slurry, as dried brewer's yeast, or in mixtures with other brewery by-products. The most common feed application of wet yeast slurry is for feeding swine. Yeast is an excellent source of protein for swine as it contains most of the essential amino acids in adequate quantities; it is somewhat deficient in methionine and cystine. The inclusion of 2 kg of yeast at 15% total solids in the daily diet of a rapidly growing pig will take care of 42% of the lysine requirements and 51, 55, and 49%, respectively, of the threonine, isoleucine, and tryptophan requirements, whereas only 31% of both methionine and cystine requirements are satisfied.[57] In a study in Great Britain,[59] swine were fed up to 3.4 l/day of 13% yeast slurry along with 1–1.5 kg of barley supplemented with a mineral/trace element mix. The results were excellent and yeast fed this way was valued at 200–325 £/ton yeast solids. Part of the value was attributed to a better utilization of barley when yeast was present. Due to the alcohol in the yeast slurry, swine show symptoms of inebriation. The animals become docile, rather than disruptive, which may be advantageous for reducing stress in a controlled environment pighouse.

German studies[60] suggested that swine could be fed an average of 1.9–3 l of yeast slurry/day, completely replacing protein concentrates. Yeast solids were valued 20% above soybean meal.

In addition to providing numerous vitamins, yeast is also important in swine diets because it contributes selenium, copper, and phosphorus.

Selenium concentrations are 12 times higher in yeast than in soybean meal (see Table 18.6). Selenium deficiency has been the cause of higher swine mortality in certain areas of the United States.[61]

Surplus Yeast for Ruminants

Some brewers add surplus yeast at the brewery to brewer's grain destined for ruminant feeding. If dried brewer's grain is produced, the yeast slurry is usually added in a mixing conveyor ahead of the dryer. Sometimes a portion of the dried grain is recycled to adsorb some of the excess moisture. Mixtures of surplus yeast and dewatered grain can be quite sticky and tend to adhere to the conveying equipment. This problem can be minimized by using fresh yeast with high live cell counts that has undergone minimal cell leakage. Coating of the interior surfaces of the dryer can be a problem as well because this may result in dryer fires. Generally, disk dryers are preferred over direct-fired dryers for mixtures of yeast and grain. If surplus yeast slurry is added to wet brewer's grain, it is usually inactivated although this is not required for ruminant feeding.[62] Beneficial effects of the addition of yeast to brewer's grain have been discussed in the Section "Brewer's Grain Food Products."

According to Burgstaller,[54] dairy cows can be fed up to 15 l of fresh yeast slurry per day (1.8 kg solids) together with a high-energy, low-protein fodder such as corn silage. On average, 10 l daily were found to be equivalent to 3 kg of soybean meal, which values yeast solids at 2.5 times SBM solids. Researchers at Michigan State University[63] compared SBM, urea, live brewer's yeast, and inactivated brewer's yeast added to corn silage diets for steers. Crude protein digestability was highest for the live brewer's yeast diet, followed by the inactivated brewer's yeast diet. A recent feeding study at Texas A&M University with inactivated yeast from the Miller Brewing Company showed a 6% increase in milk production compared to a control feed.[64]

Another interesting feed product containing live yeast is yeast culture. This product is prepared by inoculating wet cereal grains or grain byproducts with live yeast, partially fermenting the mash, and then drying the entire medium without killing yeast or destroying vitamins and enzymes. Yeast provides buffering in the rumen and the live yeast is reported to stimulate fermentation in the rumen, which aids in cellulose and fiber digestion.[65]

Surplus Yeast for Poultry

Dried yeast added to poultry rations has been the subject of a number of investigations. It is not known how much, if any, yeast is actually sold for poultry feed. It might be that the price of dried yeast at about 20 ¢/lb is too high to be economical. If wet yeast could be used in poultry feeds, yeast inclusion could be much more economical. In addition to its nutritional quality, wet yeast, properly treated, is believed to be an excellent pellet binder.

Broilers fed diets with up to 20% yeast instead of SBM showed almost identical growth rates as their controls,[66] but the growth was somewhat retarded when yeast replaced part fishmeal and part SBM. Brewer's yeast appears to be especially beneficial for breeder turkeys and laying hens. Brewer's yeast at a 5% level increased egg production rate and hatchability of fertile eggs. Reproductive improvements are attributed to the high level of dietary biotin and dietary selenium in yeast. The selenium in yeast is more beneficial than inorganic selenium added to poultry diets.[67]

Brewer's yeast can help in preventing biotin deficiency in poultry diets, which may result in reduced feed conversion, low egg production, and poor hatchability.[68] Brewer's yeast is also very high in folic acid (see Table 18.7), an important vitamin for turkeys.[69]

Surplus Yeast for Pet Foods

The current usage of yeast in pet foods in the United States is about 15,000–30,000 tons annually. This represents close to 0.6% of the total annual pet food production. It is not known how much of the yeast used for pet food is brewer's yeast.

Dogs depend on dietary proteins to supply a total of ten essential amino acids that cannot be synthesized in the body.[70] Part of these requirements can be supplied by yeast. Yeast might well compare favorably with SBM for dog food, as SBM contains raffmose, which causes flatulence.

There might be a good opportunity to use inactive wet yeast in the mixture prior to extrusion as wet yeast can be provided at reasonable cost and can possibly be used as a binder for the extruded dog food pellets. Some debittering may be required.

Yeast Debittering, Concentrating, and Drying

The hop-derived materials in brewer's yeast reduce palatability for humans or pets. Use of hop extract and isomerized hops produces cleaner, less bitter yeast and postfermentation additions of isomerized hop extract further improves yeast flavor.

When surplus yeast is spun down in a centrifuge, three layers can be seen. The bottom layer consists of coarse trub containing hop resins and grain particles. The middle, light-colored layer contains yeast. The top layer contains fine trub solids and yeast cells. The volume of the top layer varies considerably between breweries and may vary between 5 and 30%. The coarse trub and brown layer are generally removed from the pure yeast layer prior to debittering.

Coarse trub is removed by passing the yeast over a 100–200 mesh vibrating screen. The separated trub particles may be added to brewer's grain for animal feed or sold directly to a hog farm.

Separation of the brown layer, if required, is generally done on a countercurrent multistage washing system; beer solids and the brown layer material are removed. In each washing stage, the heavier white solids are separated by settling in flocculating tanks or by centrifugation.

When beer solids and brown matter are washed out of surplus yeast, it is important that the yeast be as fresh as possible. As described in the Section "Surplus Yeast Handling and Storage," a considerable amount of protein leakage will occur when yeast autolyzes. This soluble protein will then be washed out of the final product.

Effluent from the washing operation is normally sewered, resulting in high BOD and suspended solids loadings. Up to 40% of the BOD of the yeast slurry and 30% of the suspended solids may thus be sewered. If the washing is done at the brewery and a waste alcohol recovery system is available (see the Section "Waste Beer By-Products"), it is useful to include the wash effluent in this alcohol recovery system.

To debitter the pure white yeast, isohumulones adsorbed on the cell wall must be removed. The most common method of debittering is done by dissolving the isohumulones by raising the pH to about 10 using 2% caustic with constant agitation. Sometimes a mixture of caustic soda and sodium tripolyphosphate is used to bring the pH to only 7.[71] The debittered yeast can then be separated by filter or centrifuge. The separated yeast is pumped to a tank where it is rinsed with water. If the proper blandness is not obtained, the process may be repeated.

It is desirable to keep the temperature during the entire debittering process below 10°C and to work with mostly live yeast. Once the osmotic pressure within the yeast cell is impaired, the debittering solution may pass through the cell wall.[72] Prior to drying the debittered yeast, it is usually completely neutralized using phosphoric acid. Washing and debittering involves a loss of 25–33% solids.[72] In the production of autolyzed yeast extract, which will be described in the Section "Autolyzed and Hydrolyzed Yeast Products," some manufacturers choose to remove the bitter constituents from the yeast after the cell contents have been solubilized.[73] Here, the extract is exposed to other materials that remove the isohumulons by adsorption, precipitation, or ion exchange. The bitter solids are generally removed by filtration. Manufacturers are quite secretive about their exact procedures.

After washing and debittering, the yeast suspension might contain 10–15% solids. If this suspension is dried directly, 5–8.5 lb of water per lb of dried product would require removal. As water removal in dryers is usually very inefficient, processors may find it more economical to concentrate the yeast slurry in single stage or multistage evaporators prior to drying. Yeast can be concentrated to 30% solids. Some yeast slurries fed to the evaporator will contain dissolved CO_2 and might require degassing. This is done by preheating the yeast to 83°C and then feeding it through a degasser to the evaporator.

Evaporation usually takes place under vacuum with very short heat contact times. It is important to keep the temperature differences in the heat exchangers to 11°C or less to prevent fouling of the heat exchange surfaces and to avoid the development of burned flavors. Boiling temperatures in a three-effect evaporator, for example, might be 71, 60, and 49°C in the first, second, and third effects, respectively. If the feed slurry contains substantial amounts of ethanol, this ethanol can be conveniently recovered from the first effect.

The principal methods employed for drying yeast are spray drying and drum drying. Drum drying is less capital intensive but it is harder to keep sanitary. Drum dryers generally do not require a preconcentrator but product dryness is harder to control and care must be taken to avoid burned flavors.[51] For this purpose, it is best to use live yeast or yeast that has been inactivated prior to autolysis. Drum dryers can be used, though, to purposely develop nutty or toasty-flavored yeast products. A drum dryer consists of two steam-heated rollers, spaced only 0.01–0.1 in. apart. Yeast slurry is pumped in the trough formed by the two rollers. As the rollers revolve upwards through the pool of yeast they become coated with slurry and yeast is dried immediately. The dried yeast is removed by means of scraping doctor knives. The yeast flakes are then ground in a hammer mill, sifted, and packed.

When spray dryers are used, the yeast slurry is usually concentrated, preheated, and then pumped with a positive displacement pump to the atomizer. Nozzle-type atomizers can be used to create a fine mist of yeast particles, but centrifugal atomizers are more common. Spray dryers can be as much as 80 ft tall. At the top of the dryer, the finely divided yeast slurry is contacted with air that is heated to temperatures of 250–450°C. Due to the latent heat of evaporation, the yeast particles never get too hot during drying. The maximum product temperature is the exit air temperature near the bottom cone of the dryer, which is usually controlled at 70–101°C. Most dried yeast is discharged from the cone, whereas the air is discharged through cyclones to remove entrained yeast. Depending on air pollution regulations, the air may also have to pass through a wet scrubber to remove all particulate matter; in some cases, it might need to be deodorized. The dried yeast may require air cooling prior to bagging.

A new drying system successfully used with yeast is a sonic dehydrator. This type of dryer is more energy efficient than spray dryers and does not require preconcentration. In this system, yeast slurry is sprayed into the exhaust blast of a pulse jet engine. Heat associated with the exhaust blast evaporates water from the yeast and ultrasound enhances the water removal. About 90% of the moisture is removed in the pulse jet itself; the last 10% is evaporated in the collection chamber.

The evaporators or dryers described for concentrating or drying whole yeast can also be used to concentrate or dry yeast autolysates, yeast extract, isolated yeast protein, or yeast cell wall material. For each product, the optimum process parameters must be determined experimentally. Product dryness and product particle size distribution are important parameters. If particles are too fine, separation of product from the exhaust is less efficient and the dried product may also be too dusty. To prevent this, an edible oil is sometimes sprayed onto the product in the spray dryer. Fine particles may also be agglomerated in a fluid bed consisting of a vibrating screen through which dehumidified air is blown. The use of dehumidified air for agglomeration is especially important for spray-dried extracts that can be quite hygroscopic.

Whole Yeast Food Applications

An excellent review of the composition and uses of the main food yeasts is given by Peppler,[74,75] as are the quality guidelines set by various agencies and food associations. In general, brewer's dried yeast for food purposes must be free of fillers and contain more than 45% protein (% nitrogen × 6.25), less than 7% water, and less than 8% ash. Live bacteria should be less than 7500/g, molds less than 50/g, and *Salmonella* should be negative. Minimum limits for vitamins are 120 ppm of thiamine hydrochloride, 40 ppm of riboflavin, and 300 ppm of niacin.

Some dried brewer's yeast is sold in the health food industry as tablets or powders. Of prime importance here are the vitamins and trace minerals. It is possible to add additional minerals that are quickly adsorbed by growing yeast, but this would require further processing. At 75–80 ¢/lb, small margins and low volumes do not make extra processing economically attractive.

The inclusion of yeast in food products is limited by the amount of nucleic acid, primarily ribonucleic acid (RNA), that is present in yeast. In humans, RNA is metabolized to uric acid, which can lead to gout.

A thorough review of food applications of dried, debittered brewer's yeast was presented by Singruen and Ziemba in 1954.[76] It is not known what volumes of yeast are currently used for these applications. Possible uses include:

- As a natural antioxidant in foods of high animal fat content and to preserve vitamins A, E, and D
- In breads, cakes, and certain meat products to increase moisture retention and to prolong freshness
- In doughnuts to retain moisture and freshness, and to curb fat absorption
- As a flavor additive in crackers (1.5–2%) and to prolong freshness and reduce rancidity

- Up to 2% in cake mixes to retain moisture in the baked product and to enhance certain flavors such as spice and chocolate
- In processed cheese to improve flavor
- In peanut butter to curb rancidity
- Up to 3% in baby and infant food for its nutritional properties

Yeast has excellent free-flowing characteristics and is used in seasonings to produce tasty free-flowing products and as nutritional fillers. Dried yeasts are excellent carriers for flavor compounds such as hickory smoke and mesquite flavors.[77] Its low calories (1.69 kcal/g) and nutritional qualities are desirable in low-calorie foods and dietary supplements. Yeast is also used in vegetarian foods as a source of essential amino acids and B-vitamins.

Protein in whole yeast has relatively few functional properties. As noted earlier, yeast has some moisture-binding and flavor-enhancing characteristics. Whole dried yeast was also found to exert a stabilizing effect on water–oil emulsions. The foams obtained were almost comparable to whole egg foams and could be used in meringues to replace up to 25% of egg white on a dry matter basis.[78] Mixtures of debittered dried brewer's yeast and soy protein have also been used in meat analogs, which were used in meatballs (20%), wieners (8%), and hamburgers (8%).[79] To obtain the right texture, the yeast and soy proteins are heat denatured and cross-linked with calcium bridges in a proprietary process.

Another speciality product made from brewer's yeast is a cocoa substitute.[80] The yeast must be debittered and dried to less than 5% moisture and then is roasted.

Autolyzed and Hydrolyzed Yeast Products

In autolysis or self-digestion, the intracellular enzymes break down proteins, glycogen, nucleic acids, and other cell constituents. The autolytic process requires careful application and control of heat to kill cells without inactivating the yeast enzymes. When cells are properly inactivated, autolytic enzymes can break down their specific substrates. As the cells are dead, the cell membrane no longer functions so soluble cell components diffuse freely out of the cell.

Autolysis is usually carried out under moderate agitation at temperatures between 30 and 60°C for 12–24 h. External enzymes such as papain are frequently added to increase yield and rate of hydrolysis. Flavor of the product is governed by temperature and pH. Specific conditions are proprietary.

The onset of enzymic breakdown is sometimes promoted by adding 3–5% sodium chloride or other salts. Salt acts as a plasmolyzing agent by upsetting the osmotic equilibrium between cell contents and surrounding medium. Ethyl acetate (1–2%), amyl acetate, and dextrose or chloroform, where permitted, may also be used as plasmolyzing agents. Cell wall permeability is also increased by the addition of solvents. The use of salt will retard bacterial growth in the final product. Autolysis may be accelerated by the addition of thiamine and/or pyrodoxine.[81] The current trend is away from using solvents and salts in order to capitalize on the natural image of yeast products.

The autolytic process can be shortened to less than 2 h if external enzymes are used[82]: treatment of intact yeast cells with zymolase and lysozyme increases the release of cell constituents during incubation. The addition of pancreatin and promease further increases protein hydrolysis. Use of external enzymes in food ingredients must be approved by regulatory agencies.

Technically, *yeast autolysates* comprise the entire contents of the lysed cell, including water-soluble components, solubilized proteins, and cell wall material.[83] As the entire contents are concentrated or dried, debittering is generally required prior to autolysis. Yeast autolysate or autolyzed yeast is available as a 60–80% solids paste obtained by vacuum evaporation or as a spray-dried or drum-dried product. A 60% solids, low salt paste might require preservatives to prevent microbial spoilage.

If an *autolyzed yeast extract* is desired, the insoluble cell wall material is removed by filtration or by centrifugation. To obtain maximum yield of extract, centrifugation is usually done in a multistage process in which cell wall material is washed with water in a countercurrent process through three or four centrifuges. Debittering might be done by adsorption or ion exchange with other solid materials, which are removed along with the cell wall material. The cell wall material from the last centrifuge is generally used for animal feed. The extract obtained from the first centrifuge might still contain some suspended matter; it is usually filtered. An elegant, less expensive method showing promise uses crossflow membrane filters for this purpose.

Yeast hydrolysates are obtained by breaking down and solubilizing all constituents using chemicals or commercial enzymes. Acid hydrolysis is done on a yeast slurry using hydrochloric acid at elevated temperatures. In one of many possible processes, yeast is boiled at atmospheric pressure in hydrochloric acid for up to 10 h. The product is neutralized by sodium or potassium hydroxide to a pH of 5 or 6 and is subsequently filtered or centrifuged to remove cell wall material. It is decolorized with carbon, and concentrated. Bitter materials and other unwanted constituents may be removed by adsorption, precipitation, or crystallization prior to clarification.[84]

Acid hydrolysis breaks down yeast protein and carbohydrates much more completely than autolysis. Usually, hydrolysis is continued until 90–95% of the potential glutamic acid is released. Glutamates are the most important flavor constituents of hydrolysates. They impart meaty flavor to food products and are flavor enhancers.

Hydrolysates contain 36–40% salt on a dry solids basis. To produce substances with a lower sodium content, up to 40% of the sodium hydroxide may be replaced with potassium hydroxide. More potassium hydroxide would leave an unpleasant metallic taste. Inositates are sometimes added to hydrolysates to increase flavor-enhancing properties. Nucleic acids in yeast are broken down too far during acid hydrolysis to serve as flavor enhancers. Other nutritional compounds such as vitamins are also destroyed during such processing. Process equipment for acid hydrolysis is quite expensive as it requires glass-lined reaction vessels with reflux condensers.

For food applications, autolystates are usually preferred over hydrolysates because they have a cleaner flavor, reminiscent of yeast. They also have less sodium, and are less astringent because of the lower salt content. They are better flavor enhancers as their nucleic acids are not destroyed, and they have better nutritional properties.

Autolystates and whole yeast products made from brewer's yeast must compete with similar products made from primary grown yeasts such as *Saccharomyces cerevisiae* (baker's yeast) or *Candida utilis*. Products made from primary yeasts usually command a premium price. An advantage of primary yeast is that the composition of the yeast cream is generally better controlled. Primary yeast growers also frequently spike their yeast with vitamins and minerals to meet specific customers demands. Minerals such as K, Ca, Zn, Mg, Se, and Cr are best incorporated during the growth phase. Vitamins are added after yeast concentration. Thiamin is adsorbed by yeast up to ten times the normal level, whereas niacin is adsorbed up to four times the normal level. However, riboflavin is not adsorbed.

The International Hydrolyzed Protein Council reports a total U.S. production of yeast autolysates of 10.3 million pounds in 1983 from a total of 13 companies. It is not known what percentage of this is produced from brewer's yeast. The price for autolysates is about $1.70–$2.00/lb. Hydrolysates compete with hydrolyzed vegetable protein (HVP). The 1983, the U.S. production of HVP was 43.7 million pounds.

Yeast autolysates, extracts, and hydrolysates are used in a wide variety of foods as flavors, flavor enhancers, or flavor potentiators.[77,83] Uses include:

- In meat products: meat pies, hamburgers, sausages, etc. (0.5–2%)
- For curing meats (0.5%)
- For retorting meats: to help mask cereal off-flavors
- In chemically leavened baked products: to mask the chemical leavening notes with a yeasty aroma
- In sauces and gravies: meat gravies, cheese sauces (0.4–2.5%)
- In bouillons and soup mixes (up to 10%)
- In soups (0.2–2.0%)
- In seasonings: condiments, marinades, seasoning salts (>95%)
- In cheese and cheese-flavored products (0.5–1.5%)
- In chips and crackers (0.2–2%)

- In batters: for fish products and chicken
- In savory spreads: marmite is 98% yeast extract

Autolyzed yeast extract is often used as a nutrient in industrial fermentations for the production of foods or pharmaceutical products. For example, autolyzed yeast extract is used as a source of nitrogen, vitamins, and minerals in the production of phenylalanine, which is the base material for aspartame (NutraSweet).

Isolated Yeast Protein, Yeast Extract, and Yeast Glycan

In the production of yeast hydrolysates or autolyzed yeast products as described in the previous section, only one food product was obtained from yeast. Undigested cell wall matter and other materials that were not solubilized became part of the final product (autolyzed yeast) or were removed as waste or feed material. In this section, processes are described that obtain multiple food products from yeast.

Cell wall material separated in the production of autolyzed yeast extract can be further treated to yield a purified cell wall material, called yeast glycan.[85] In this process, a slurry of the cell wall material is comminuted by passing it several times through a homogenizer at pressures between 5000 and 15,000 psig. In the processes described in the literature, yeast glycan is produced from baker's yeast. It would appear that brewer's yeast, properly debittered, might also be a good starting material.

Yeast glycan is an excellent thickening agent with a smooth fatlike mouthfeel, even though the product generally contains less than 1% fat.[86] The product may be used to replace all or part of the fat in a food product, which is useful for low-calorie or low-fat products. Viscous suspensions of glycan become thin on heating and thicken on cooling and are relatively independent of pH.[87] Glycan gels on cooling after retorting and has good water-binding characteristics. Potential applications are in salad dressings, retorted foods, sauces, puddings, nondairy shakes and dips, coffee creamers, breading, frosting, and candy. The potential value is $3–4/lb. It is not known whether the process has been commercialized.

In a process developed by Anheuser-Busch, yeast can be fractionated and refined into three useful food products, each of which has better functional properties than the original yeast.[86] In this process, yeast is fractionated into yeast glycan, yeast protein isolate, and yeast extract. Yeast extract has higher flavor-enhancing and flavor-potentiating characteristics than autolyzed yeast extract. Isolated yeast protein disperses quickly and has good water absorption and fat absorption properties. Yeast extract is especially effective in its ability to provide a meat extract-type flavor to various foods. Its flavor-enhancing properties are similar to those of monosodium glutamate and the 5' nucleotides.

During the 1970s, considerable efforts were made to produce single cell protein by fermentation. A part of this effort was to develop equipment that could rapidly rupture cells, making their components available for isolation and purification. As a result, large-scale mechanical disruption of yeast cell walls is currently carried out mainly in high-speed agitator bead mills or in high-pressure industrial homogenizers. An excellent review is given by Kula and Schutte.[88]

Biotechnology Products from Yeast

Brewer's yeast and baker's yeast are the principal sources of a variety of biocatalysts, biochemicals, and metabolic intermediates.[75] Most of these products are isolated and purified in small quantities primarily for use in analytical and biochemical research laboratories. Products include enzymes, such as dehydrogenases; vitamins, such as biotin; amino acids; cytochromes; and the purine components of DNA and RNA.

An interesting opportunity for specialty yeast products might result from the discovery that fractions of surplus yeast exhibit antitumor activity.[67] The fractions were obtained by breaking the yeast cells in a high-pressure homogenizer, separating the cell wall material, and dialyzing the supernatant against water. The highest molecular weight fraction had a higher antitumor activity than a commercial anticancer agent, named Krestin. It is speculated that the activity is due to an increased immunological response.

With the advent of bioengineering, it is possible to modify the genetic make-up of yeast so that the yeast produces greater amounts of useful enzymes or proteins. It is even possible to introduce a factor that causes the yeast to release the useful products extracellularly, negating the need for cell rupture. The challenge is to modify brewer's yeast genetically without altering any of its brewing properties or the quality of the beer. Delta Biotechnology Ltd.[89] has patented a process in which the genes coding for a desirable protein remain silent during a brewery fermentation and are triggered into a second fermentation that is started after harvesting the surplus yeast.

Brewer's Condensed Solubles

Production

Brewer's condensed solubles (BCS) are produced by concentrating brewer's grains liquor by evaporation. The feed to the evaporator may also contain lauter tun drainings, rinse effluents, tank rinses, and waste beer from brewing, packaging, or yeast processing operations. In breweries that recover by-product alcohol, the still bottoms are usually added to the concentrator feed. The final concentration of BCS may vary between 15 and 20% solids for BCS for internal use to between 30 and 60% solids for BCS sold as a liquid feed or as a raw material for industrial fermentations. If

BCS is used internally, the concentrate is usually blended with wet brewer's grain and sold commercially.

Brewer's grain liquor is generally concentrated in a multiple-effect evaporatory system. A mechanical vapor recompression system can also be used to further reduce energy costs. If inexpensive natural gas or fuel oil is available, it might be beneficial to concentrate brewer's grain liquor in a less expensive single-effect evaporator. A submerged combustion evaporator can be used for this purpose.[90] This type of evaporator, which can concentrate liquor up to 20%, does not foul easily and requires only minimal clarification of the feed stream, which can be accomplished by means of a vibrating screen. In the process of concentrating brewer's grain liquor, viscosity of the product can increase considerably — heat transfer surfaces foul easily and high concentrations cannot be achieved. To avoid this, forced circulation-type evaporators are employed in the last stages of concentration. Viscosity of the final product can be decreased substantially by removal of suspended solids or by the use of enzymes.[91] To achieve high concentrations, reduction of suspended solids to below 0.5% (wt/wt) is recommended. This clarification is usually done by means of a centrifuge.

Seven different enzymes were tested[91] for their ability to thin BCS. Cellulase was found to be the most effective enzyme and was also the least expensive. A potential problem with BCS is that unwanted fermentations and mold growth may occur, especially near the top of tank trucks or storage tanks where condensation occurs and an ideal growth medium is created. To prevent this, KMS, propionic acid, or other antimicrobials may be added.

Composition and Value

Brewer's condensed solubles are used as a liquid feed ingredient for ruminants or swine, as a pellet binder, or as a feedstock for fermentation.[92] It is classified as an energy feedstuff and as such it competes with molasses and corn in the finishing rations of steers. BCS is also a promising molasses substitute as it contains more nutrients than molasses. Table 18.12 shows an average proximate analysis for BCS, corn, and cane molasses, in addition to energy values for maintenance, growth, and lactation of ruminants.

BCS is a good-quality nutritional pellet binder at a level of about 3%.[92] This is especially important for layer rations, where crumbling leads to reduced feed intake. BCS also has potential as a feedstock for industrial grade alcohol or the production of organic acids such as citric acid. The addition of enzymes would be required to increase fermentability.

Excess Carbon Dioxide

The total amount of CO_2 that is produced during fermentation is about 9.5 lb/bbl of packaged product. A portion of this CO_2 serves to carbonate

TABLE 18.12

Comparison of BCS and Other Energy Feedstuffs

	BCS	Corn	Cane Molasses
Nitrogen-free extract (% db)	86	82	84
Protein (% db)	10	10	6
Ash (% db)	2.5	2	10
Fat (% db)	1	3.5	—
Crude fiber (% db)	0.5	2.5	—
NE_m (kcal/kg)	2.27	2.18	2.27
NE_g (kcal/kg)	1.48	1.43	1.48
NE_l (kcal/kg)	2.42	2.52	2.60

Abbreviations: NE_m, net energy for maintenance; NE_g, net energy for growth; NE_l, net energy for lactation.

the product in the fermenter. After carbonating the beer, the CO_2 produced is vented until the purity is sufficiently high to start collection. If properly managed, only about 2.0 lb CO_2/bbl of packaged product is needed for saturating the beer and venting of impure product. It is important therefore that CO_2 purity is monitored continuously so that collection can start as soon as the required purity has been reached.[93]

The collected CO_2 is purified, compressed, liquefied, and stored. In the purification system, another 0.5 lb CO_2/bbl packaged beer may be lost. This brings the maximum amount of CO_2 for reuse and sale to about 7.0 lb/bbl packaged beer. CO_2 is used in the brewing department to counterpressure storage and transfer tanks, to carbonate diluent water, to bring carbonation levels of beer up to standard, and to strip dissolved oxygen from diatomaceous earth slurries, diluent water, and beer. Every time a tank is emptied when maintaining a CO_2 counterpressure of 15 psig, about 1 lb CO_2/bbl is used. A similar amount will be used for filling after a tank has been vented. This latter CO_2 usage can be avoided if venting is not required, for example, by using acid cleaning. A reasonable target for total CO_2 usage in brewing is 3.5 lb/bbl packaged.

In the packaging department, CO_2 is used as counterpressure gas during the filling of bottles and cans, in the bubble breakers and rail gassing system, and in the undercover gasser at the can seamer. CO_2 usage is higher for breweries that pre-evacuate prior to filling and that fill a high percentage of cans. For large U.S. breweries, a target usage in packaging of 2–3 lb/bbl appears achievable. Many breweries purchase CO_2, but with proper management and adequate liquid CO_2 storage to accommodate production swings, modern breweries should be able to sell about 0.5 lb/bbl packaged product. Additional CO_2 can be recovered by recycling it from the counterpressure system. This CO_2 is generally higher in oxygen, but if monitored carefully, it can be collected during periods when purity is high. After compressing the recycled gas, it is quite suitable for use in packaging.

Excess CO_2 may be sold to industrial suppliers for carbonation of beverages, refrigeration, or for liquid CO_2 extraction of hops, spices, etc. Excess purified or even impure CO_2 that is vented from fermenters can be used to neutralize alkaline brewery effluents.[94] Vented CO_2 can also be collected and pumped into greenhouses, where higher CO_2 levels can accelerate plant growth by 20%. Another promising use for surplus CO_2 is for fumigating grain silos. It was found to be an effective nontoxic fumigant that does not leave a residue.[95]

Spent Filter Cake

Spent filter cake from primary and polish filtration consists of diatomaceous earth (DE), yeast, trub, and cellulose filter aid and finings, if they are used. Chillproofing materials, such as silica gel or PVPP, may also be present. The filter discharge also contains beer that is left in the filter cake and filter shell.

DE usage for a large U.S. brewery was found to be 0.4 lb/bbl packaged.[96] DE usage and subsequent disposal problems can be decreased by applying automatic body aid feed control. This is done[97] by measuring the turbidity of the beer to be filtered and feed forward control.

The amount of organic suspended matter in the cake was estimated at 0.02 lb/bbl packaged,[96] whereas silica gel and PVPP may contribute 0.1 and 0.05 lb of suspended solids/bbl packaged, respectively. Some brewers use a separate filter to take out PVPP, which is then regenerated. The amount of beer left in the filter cakes and filter shells of the primary and polish filters was found to be 0.0065 bbl/bbl packaged.

Some breweries employ dry discharge filters in which most of the beer is pushed out of the filter shell and the cake by CO_2. The cake is then mechanically removed from the filter elements by spinning, cutting, or vibrating and is generally hauled to a landfill. In most municipalities, the cost for landfill disposal is considerably less than for sewer disposal.

Most breweries discharge spent filter cake and rest beer to the sewer along with a considerable amount of rinse water. Typical sewer loadings based on the above example are 0.52 lb suspended solids/bbl packaged and 0.16 lb BOD/bbl packaged. Alternatives to flushing are recycling[98] and slurry concentration methods.

Spent filter cake slurries can be concentrated by a variety of vacuum or pressure filters[96] to a cake of about 35–39% solids. This cake can be disposed of or utilized in a number of ways, most of which require drying or regeneration to burn off organic matter. Options for cake disposal or utilization include:

- Haul to landfill
- Mix with wet or dried brewer's grain

- Spread on farmland
- Use as soil additive for nursery crops
- Dry and mix with poultry feed for fly control as it destroys larvae
- Dry or regenerate and use as insect control in grains
- Regenerate, reuse in-house
- Regenerate, sell as filter aid for other industries
- Regenerate, use in low-density insulation materials
- Regenerate, use for flow control in powders and fertilizers
- Regenerate, use as filler material in the paint and paper industries
- Regenerate, use as insecticide carrier
- Use in the production of cement or roofing tiles

Waste Beer

Waste Beer Volumes

Waste beer is a major source of BOD in the brewery, but as beer is wasted in so many different areas, substantial reduction of beer losses is difficult. Loss of some beer is unavoidable in both brewing and packaging if product quality is to be maintained and product specifications are to be met consistently. The BOD value of beer depends on the type of beer; for premium U.S. beers, it is approximately 24 lb/bbl. A reasonable correlation seems to exist between alcohol, residual dissolved solids, and the BOD of that beer:

$$\text{BOD (lb/lb beer)} = 0.81 \times \text{dissolved solids (lb/lb beer)} + 1.63 \times \text{alcohol (lb/lb beer)}$$

Beer losses in brewing and by-product processing include:

- Loss with unrecovered fermenter yeast
- Loss in presses and centrifuges during yeast harvesting
- Loss due to overfoaming of surplus yeast storage tanks
- Loss during stages of surplus yeast processing (see the Section "Brewer's Yeast")
- Left in fermenters
- Left in spent filter cake and filter shells (see the Section "Spent Filter Cake")
- Left with tank bottoms in aging
- Left in chillproofing systems
- Left in tanks and lines after a production run

- Leaking from pumps and turnbacks
- Loss as a result of disconnecting hoses and turnbacks
- Left in government tanks and supply lines to packaging
- Out-of-spec beer

There are many opportunities to reduce beer losses as outlined by Bott.[99] Listed are process modifications, operational procedures, operator involvement, and process automation. An important area of improvement has been in new methods to fill a pipeline at the beginning of a run and to displace all the beer at the end of a run.

Beer losses in packaging and distribution include:

- Packaging line start-up losses
- Filler bowl dumps of warm beer
- Filler valve leakage
- Bottle overfoaming at jetter or knocker
- Spillage from cans at bubble breaker and transfer
- Crowner losses, can seamer losses
- Bottle, can, and keg overfills
- Short fills
- Glass breakage on conveyors and in pasteurizers
- Labeller and packer losses
- Snifting losses and beer spillage at the racking head
- Keg short fills and leakers
- Overaged bottles and cans returned from the trade

In most packaging plants, 80% of the beer losses can be accounted for by jetter losses, seamer losses, under- and overfills, and bowl dumps. Goetz[100] discusses a number of methods to reduce packaging beer losses. Some overfoaming because of jetting is necessary to expel the oxygen from the headspace of the bottle, but overfoaming of about 1% of the contents should be more than sufficient to accomplish this.

A newly developed monitoring system should be helpful in reducing beer losses from fillers. This system measures fill volumes of individual packages after closing and relates these to specific positions on the filler head. By observing the digital display of the results, the operator can adjust individual fill valves where necessary or temporarily eliminate certain fill positions.

Beer losses at the beginning of a packaging run can be reduced by using cold diluent water to chill the filler bowl instead of beer. If beer warms up during a packaging run because of unscheduled down time, bowl dumps can be avoided by utilizing a blowback system to return beer to the brewing department.

It is obvious that beer losses vary widely from one brewery to another. For large U.S. breweries, it is estimated that beer losses in brewing vary from 2 to 5% and in packaging from 2 to 6%; average overall beer losses are probably 5–7%.

Waste Beer By-Products

To decrease waste treatment costs, a number of breweries have built elaborate systems to collect waste beer and to concentrate the alcohol and/or dissolved solids to obtain useful by-products. Two basic systems are used: (1) a fermenter/distillation system in which ethanol is the main product; and (2) an evaporator/still system in which both ethanol and BCS are produced.

In the fermenter/distillation system,[101] waste beer is combined with other high-strength brewing wastes in a fermentation system. Fermentation is carried out at 85°F with pitching rates of 200 million cells of surplus yeast. Amyloglucosidase enzyme is added to the fermenter to increase fermentability. The still bottoms are clarified in a decanter centrifuge prior to being sewered. Centrifuge solids may be added to brewer's grain or sold separately. The 190 proof ethanol may be dried to 200 proof in a molecular sieve system.

The concentration of fermentable substrate and/or alcohol in each waste stream should be considered separately to determine if it makes economic sense to include that waste stream in the process. Utility costs, waste treatment costs, enzyme costs, and the price that ethanol can be sold for will be factors in determining which waste streams to include.

Ethanol produced at 190–200 proof is used in the manufacturing of products such as vinegar, beverage alcohol, cosmetics, pharmaceuticals, and tobacco. Up to 10% is frequently used in super unleaded gasoline, previously called gasohol. Because of its use in fuel alcohol, the price of alcohol has fluctuated enormously along with the large fluctuations of crude oil prices. Prices in the United States have also been affected by a large glut of low-priced fuel alcohol from Brazil.

A number of United States breweries have developed food flavorings produced by concentrating and/or spray drying beer.[102] These beer flavor products, which no longer contain ethanol, can be used as flavor potentiators and flavor enhancers in products like soups, sauces, dressings, spice mixtures, batters, and coatings.[103]

Solid Waste Materials

Breweries, and especially their packaging plants, produce large amounts of solid waste materials, much of which can be recycled or used for alternate applications. These materials include broken pallets, aluminum, cullet, corrugated paper, cartons, paper, fly ash, and spent granular activated carbon.

The alternative to recycling is the disposal of solid waste in landfill sites, which might be costly at times. Recycling, on the other hand, requires labor, sorting equipment, and space. Employee involvement is essential for a successful recycling program. A comprehensive review of the available literature on recycling is available from the National Soft Drink Association.[104]

Aluminum waste results from underfilled aluminum cans, cans damaged in the seamer, on conveyors, or in the pasteurizer, cans in damaged packages, and overaged canned products. These cans are collected in dumpsters, brought to the dump area, and flattened in a can crusher. Glass, cartons, and plastic materials should be kept out of the can dumpsters or removed prior to crushing. Recycle rates of about 0.1 lb of aluminum per barrel packaged are not uncommon in large U.S. breweries.

Bottles that are improperly filled or contain product that does not meet specifications may be emptied by hand and recycled back to the rinser or bottle washer. Labor requirements for this are very high and most large breweries resort to crushing the glass and selling the cullet to bottle manufacturing companies along with cullet from damaged bottles. Some color sorting of bottles is required prior to crushing as cullet for amber bottles must consist of at least 90% amber glass, while cullet for green bottles requires a minimum of 80% green glass. Current prices for cullet are $45–55/ton delivered. The cullet may only contain minimal amounts of foreign matter. Large vacuum systems can remove paper, plastics, and aluminum closures, whereas caps are removed by magnet. The total amount of cullet that is collected and recycled may be as high as 1 lb/bbl of packaged product. Besides recycling glass for bottle manufacturing, cullet has also been used in asphalt mixtures to make roads more skid-proof as well as in bricks and glass wool insulation.[105]

Large amounts of corrugated materials and folding cartons, partitions, etc. become available at the uncaser. Here, the boxes must be flattened, put on pallets, and strapped. These materials can be recycled to box manufacturing companies or to local dealers. Slip sheets for cans may be kept separate and shipped back to can manufacturers. Damaged or wet paper goods, old hardshells with staples, cardboard sleeves, paper, etc. can all be collected in dumpsters and compacted in a baler for sale as waste paper. Some waste paper dealers even provide compacter/trailers for this purpose. For each barrel of product, about 0.3–0.4 lb of reusable corrugated materials may be recycled at a value of $50/ton.

If coal is used as fuel in the boiler house, substantial amounts of fly ash may result. Fly ash is used in building blocks or cement mixes. For this application, a constant loss on ignition of 4% or lower is desired. A consistent low loss on ignition requires a consistent boiler load, which is not always easy to achieve. Other uses for fly ash are as a component in road base materials or as an easy-flowing filler material for underground piping systems.

Many breweries use granular activated carbon (GAC) to remove trace quantities of organic impurities from brewing water. GAC needs to be replaced after 1–3 years when its adsorption capacity becomes too low. As an alternative to landfill, spent GAC may be regenerated by carbon manufacturers.

Waste Water Disposal and Treatment

Introduction

Brewery waste water is relatively simple and is highly biodegradable. A complicating factor, however, is that waste water volumes, pH, and concentrations of included solids vary constantly. In order to get a good characterization of the waste water, flow rates and concentrations must be measured simultaneously during an extended period of time. The flow proportional sampling method is most suited for this purpose.

As mentioned, the suspended solids in the effluent contain organic matter such as grain, trub, yeast, and label pulp as well as inorganic materials such as filter aids and silica gel. Dissolved solids are mainly from beer, wort, and cleaning and sanitizing solutions. The BOD is usually used to index the concentration of biodegradable organics in brewery waste streams. BOD determinations are cumbersome and not very accurate. However, they have historically been used to assess the pollution potential of waste waters and have become the basis for design and operation of waste water treatment plants.

Waste water from a brewery may be discharged several ways: (a) directly into a river or ocean; (b) directly into a municipal sewer system; (c) into a river or municipal system after pretreatment; and (d) into the brewery's own waste water treatment plant.

Discharges into public waters are often subject to limitations in organic load, suspended solids, pH, temperature, and chlorine. The maximum allowable BOD limit for discharge of effluents into public waters near densely populated areas can be 20–30 ppm and may be as low as 10 ppm. Surcharges for discharge into municipal systems are based on average volumetric load, average BOD load (5–10 cents/lb BOD is not uncommon), average suspended solids load (5–10 cents/lb suspended solids), and on peak discharge rates. Breweries are sometimes also required to contribute to construction costs of municipal treatment facilities. It is becoming common for large breweries to construct their own complete waste water treatment facility or to pretreat their effluent. The high costs that are often required for waste treatment offer brewers an additional incentive to eliminate unnecessary wastes and to optimize the reuse of effluents.

Waste Water Volumes and Concentrations

As discussed, discharges of BOD and suspended solids from various areas of a brewery may vary considerably, depending on waste management and on the degree of recycling that is applied by a particular brewery. Table 18.13 is therefore only an example of waste discharges in a typical large North American brewery.

Average BOD discharged reported for 12 large North American breweries range from 0.96 to 3.7 lb/bbl with an average of 2.1 lb BOD/bbl packaged. For German breweries with careful waste management, Rosenwinkel and Seyfried[106] report an average of 1.55 lb BOD/bbl packaged, with a range of 0.9–3.03. Suspended solids discharges for seven North American breweries averaged 0.9 lb suspended solids/bbl packaged, with a range of 0.3–2.2.

Concentrations of BOD and suspended solids vary considerably, depending on where samples are taken and how much water is used for dilution. Many brewers have made a concerted effort to reduce water usage in order to lower costs for water, water treatment chemicals, and waste water treatment surcharges based on flow. Not all waste water requires treatment. Noncontact cooling water and rinse water for nonreturnable bottles and cans, for example, is relatively clean and may be discharged directly into a river or storm sewer depending on temperature and chlorine limitations.

Volumes of waste water that are discharged to a sanitary sewer or treatment plant from large modern breweries are in the range of 2.5–6 bbl/bbl packaged. Volumetric discharges are usually higher for smaller breweries, or for older breweries, or breweries located in hot climates.

TABLE 18.13

Typical Sewer Load from a Large Brewery

Brewery Source	BOD (lb/bbl Packaged)	Suspended Solids (lb/bbl Packaged)
Lauter tun rinse and drain	0.26	0.10
Trub and wort losses	0.11	0.06
Other brewhouse losses, rinse, and CIP	0.08	0.05
Press liquor	0.46[a]	0.22[a]
Surplus yeast handling waste	0.08	0.05
Spent filter materials	0.16	0.52[b]
Fermenting and finishing waste	0.44	0.10
Packaging waste	0.72	0.08
Total	2.31	1.18

[a]Can be avoided when brewer's grain is sold wet, or when liquor is used as a by-product.
[b]The majority of this can be disposed of as solid waste, if special equipment is available.

Waste Water Pretreatment

Most large breweries require some degree of waste water treatment, whether their effluent is discharged to public waters, to municipal treatment systems, to their own aerobic or anaerobic systems, or used for land application. In many cases, mere pretreatment might be sufficient to meet local regulations. Pretreatment is done by physical, chemical, or biological methods, or by combinations of these.

Physical pretreatment methods for breweries include coarse suspended solids removal, flotation, and sedimentation. The first process step in pretreatment is usually screening to remove coarse suspended solids such as labels, caps, glass fragments, plastic and other scrap materials, and grain particles.

After screening, the effluents are usually passed through a rectangular grit removal chamber equipped with a continuously operated scraper mechanism. Grit chambers can also serve as a mixing chamber for pH control systems and as a preaeration unit to prevent anaerobic conditions in the primary clarifier.[107]

Major problems associated with treating brewering effluents include the highly variable nature of flow, BOD concentration, and pH. The most efficient action that can be taken is the installation of a buffer or equalization tank that will be most effective when the contents are well mixed.[108] Aeration or oxygenation may be required to prevent microbial production of hydrogen sulfide odors.[109] Some breweries collect their caustic CIP and bottle washer effluents in a separate tank and gradually meter the contents into the treatment plant influent.

Dissolved air flotation is an effective pretreatment method. In this process, air is dissolved in water under pressure and comes out of solution as tiny bubbles to that particles attach. To obtain large flocs, metal coagulants and polyelectrolytes (flocculation agents) are used. Lunney[110] found a removal of 60% suspended solids and 45% COD, whereas Hughes[111] achieved a removal of 95% suspended solids and 6% COD. Acids, caustic, and DE are sometimes used to aid in sludge dewatering in a filter press.

Chemical pretreatment is used in a number of breweries by neutralizing caustic effluents from CIP systems and bottle washers with waste CO_2. Neutralization with sulfuric or hydrochloric acids is usually not recommended because of their corrosive nature and sulfate and chloride discharge limitations.[94] Waste CO_2 sources that may be used were described earlier. Packed trickle flow beds can be efficiently used with good control.[112]

Biological pretreatment systems include aerobic systems with short resident time or anaerobic systems. BOD may be reduced up to 60 or 70%, which may be all the treatment that is required, or it may be the first stage in a full treatment process. Aerobic systems used for pretreatment can be tanks with plastic support media for biological growth such as bio-disks or Flocor systems, or it can be as simple as an aerated equilization tank. Some type of primary clarifier is required for sludge removal.

Aerobic Treatment Systems

Most biological treatment systems used in breweries today are aerobic activated sludge systems, although the trend since the mid-1980s has been toward anaerobic treatment systems. Reviews of aerobic-activated sludge systems are given by Leeder,[113] Miller Brewing,[114] and Barrett and Hayden.[115]

Newer aerobic systems use plastic media that support biological growth. These media can be rotating sheets as in the rotating biological contacter, self-supporting packing materials as in biotowers, or low-density material kept in suspension by fluid motion (fluid bed systems). All require less space than do the conventional activated sludge plants as they provide more efficient contact between waste water and sludge.

In the rotating biological contacter, a series of plastic sheets is mounted on a shaft. The closely spaced sheets, 1 cm apart, rotate in waste water at about 1.5 rpm and are about 40% submerged. The biological growth on the sheets drags waste water along, which is aerated as the sheets emerge from the liquid. Excess build-up of biomass sloughs off and is carried out with the liquor to a clarifier.[116]

A biotower is basically a packed bed with a plastic packing material of high specific surface area and porosity. Biomass grows as a thin film on the packing while waste water trickles over the biomass. The waste water is usually recycled to obtain sufficient flow for uniform wetting. Unpleasant odors may be a problem with these systems.[113]

A fluid bed system is probably the most efficient biological growth system as a very high density of biomass can be kept in suspension. An example of a fluid bed system is the Captor process in which the biomass is held captive within small plastic spongelike blocks that are kept in suspension by circulating currents caused by a submerged aeration system. Another advantage of this system is that a settler is not required; excess sludge is recovered directly by mechanical removal of the sludge-laden sponges from the tank, squeezing the sludge out, and returning them to the tank.[113]

Another novel aerobic system requiring only a small land area is the deep shaft aerator system, which is successfully operated at a Canadian brewery.[117] In this process, waste water is contacted with biomass and compressed air in an underground shaft 152 m deep, which promotes a high rate of biological oxidation.

Anaerobic Treatment Systems

Anaerobic digestion is a complex process in which a large variety of bacteria break down organic materials and convert them into methane and CO_2 in a ratio of about 3:1. The bacteria are generally grouped into three basic types.[118] The first group consists of acid-forming bacteria called acidogens, which provide extracellular enzymes that hydrolyze soluble and insoluble complex organics and convert the hydrolysis products into fatty acids, alcohols, CO_2, ammonia, and hydrogen. The second group, called acetogens, transforms the products from the first process into acetic acid, hydrogen, and CO_2. The third group consists of methanogens, which convert acetic acid, H_2, alcohols, and some of the CO_2 into methane.

As a result of recent advances in the basic biochemical understanding of the process, anaerobic systems have found a wider acceptance as biological treatment systems for brewery effluents. Advantages of the anaerobic system over aerobic treatment mentioned in the literature[119] are: (a) low-energy consumption especially in hot climates; (b) net producer of methane gas, which can be used as a fuel; (c) 70% less space requirements; (d) less nutrient requirements; (e) up to 90% less sludge production; and (f) lower capital cost because of a higher loading rate. A number of potential disadvantages for anaerobic systems have also been discussed[108]; these include: (a) the slow growth rate of methane forming bacteria; (b) anaerobic systems are more sensitive to shock discharges than aerobic systems and generally require a balancing tank; (c) BOD removal rates vary and may not always be sufficient; and (d) besides 70–80% methane, biogas is saturated with water and contains 20–30% CO_2 and 1000–2000 ppm H_2S, which may cause problems in steam boilers.

A typical design for an anaerobic digester is a packed column in which the biomass is attached to plastic media. A system of this type with a downflow mode has been successfully applied to treatment of distillery wastes.[120]

The most commonly employed anaerobic system in breweries is the upflow anaerobic sludge blanket (UASB) system, developed at the University of Agriculture, Wageningen, The Netherlands. In this system, the influent is uniformly distributed over the bottom of the reactor and is then passed through a bed of active anaerobic sludge. Gas, liquid, and sludge are separated from each other above the bed.[121] Most of the sludge returns to the bed, while a portion may be diverted to the surplus sludge tank. Difficulties with this system have been excessive H_2S formation and the loss of solids from the sludge bed as a result of shock loads.

A good solution to these problems has been the design of a two-stage process. The first stage is a stirred tank reactor where acid formation takes place and the second stage is a UASB reactor where methane formation takes place.[122] This two-stage system was found to be quite stable and COD removal rates of 70% and higher were obtained at temperatures as low as 15–20°C. Problems with H_2S have been solved by removal of H_2S from biogas by silica gel adsorption[123] or by the addition of iron to form iron sulfide. The newest design of anaerobic digestor, which appears promising, is the anaerobic fluid bed reactor. In this system, the biomass is attached to an inert carrier that is maintained in a fluidized state by an upward flow of waste water. The main advantage of this system is that a high biomass concentration can be maintained, resulting in a very compact system. A two-stage system has been employed for the treatment of a yeast plant effluent, with a fluidized bed reactor for each stage.[124] As the biogas from this system contains 1–2% H_2S, it is highly corrosive and reactors are made of glass fiber-reinforced polyester with PVC lining.

In general, anaerobic treatment systems work well for effluents with high BOD concentrations. To reduce costs, it is desirable to have a separate collection system for high-strength brewery effluents destined for anaerobic treatment. If complete treatment of all waste is desired, the low strength effluents should be treated aerobically along with the effluent from the anaerobic system.

Sludge Treatment, Disposal, and Utilization

One of the more expensive steps in waste water treatment is the handling and dewatering of excess sludge. This is especially true for aerobic treatment systems where large quantities of sludge are produced. A high level of solids concentration is required to reduce transportation costs or to keep fuel costs down if sludge is dried or incinerated. Landfill requirements sometimes also dictate maximum allowable moisture levels.

Excess sludge that is collected from settling tanks generally contains 1–2% solids. This sludge can be held in consolidation tanks for an additional 2–3 days to increase the solids concentration to about 4%.[113] Solids can also be concentrated to about 4% by means of dissolved air flotation[125] or by means of centrifuges. Sludge squeezed from the plastic spongelike media in the Captor process[113] has solids concentrations of up to 6%.

Further dewatering may be done by means of vacuum or pressure horizontal belt filters, by rotary vacuum filters, or by plate and frame or recessed plate filter presses. A precoat of fly ash or diatomaceous earth is often used in filter presses to prevent cloth blinding. Overall dewatering rates may be enhanced considerably by using chemical coagulants such as alum, lime, ferric chloride, and body feed such as fly ash, diatomaceous earth, or paper fiber.[126]

Lime contributes to the value of the sludge if the product is to be utilized as fertilizer. Fly ash and diatomaceous earth are often brewery by-products that need to be disposed of anyway, whereas ferric chloride may sometimes be obtained inexpensively as pickle liquor, which is a by-product from the metal plating industry.

By adding about 25% lime, 10% ferric chloride, and 10% fly ash based on sludge solids, sludge from an aerobic treatment plant can be consistently dewatered to 35–50% solids in a filter press. Sludge from the deep shaft process, which has 3–5% solids after flotation, can be dewatered to 16–17% solids by using a belt press.[117] Sludge from most breweries is hauled to landfill, but some brewers have found the product to be useful as fertilizer or as animal feed.

At the Coors brewery, sludge from a flotation chamber at about 4% solids may be dried to over 95% solids by means of a Carver–Greenfield drying process.[125] In this process, tallow is added to the sludge at a ratio of

15 parts of oil to 1 part of sludge solids. The oil allows evaporation of the water from the sludge solids in a quadruple-effect evaporator to almost complete dryness. After drying, the oil is removed by centrifugation and expelling, although about 10–20% remains in the dried biomass. The product has about 45–50% protein and is low in metals and high in vitamins. Because of its relatively high content of vitamin B-12 (2–3 mg/lb), the material has been registered as a vitamin B-12 supplement. The product has been successfully tested in various animal feeds, especially for turkeys.[127]

Dewatered sludge (40–45% solids) including lime, ferric chloride, and fly ash from a Miller Brewing Company aerobic treatment plant has been used as a fertilizer for hay and as an alternative liming material in a multiyear field test.[128] On a dry basis, the by-product contains 2.6% N, 0.64% P, 0.11% K, and 16.8% Ca. The by-product has been registered as a fertilizer in one state and as a by-product liming material with nitrogen in another. The material was applied to a hay field test plot at 0, 5, 10, and 20 dry tons/acre. At the highest rate, soil pH increased from 6.1 to 6.7. Two-year total hay yields were 148–212% and crude protein harvested was 171–303% of the unfertilized control. Both immediately available and slowly released nitrogen are provided to crops and soil. On a dry basis, the product is equivalent to a 5–1–0 ($N-P_2O_5-K_2O$) high-lime fertilizer.

Land Application of Brewery Effluents

A number of large Anheuser-Busch breweries in the United States use the system of land application of selected high-strength brewery effluents.[129] The land treatment systems are designed for an annual average fluid loading of 0.12 in./day, a suspended solids loading of 43 lb/acre/day, and a nitrogen loading of 585 lb N/acre/year. The land application system has been combined with a year-round turf operation to assure intensive management and utilization of nutrients.

The high-strength segregated waste streams are collected in above-ground steel tanks with a combined hold-up capacity of 3.6 days. The tanks must be aerated to minimize odor problems. The pH is adjusted to an appropriate level for the particular soil. The effluents have to be passed through a screen to prevent coarse suspended matter from plugging up the rotating irrigation systems. The soil requires extensive preparation to be able to maintain a minimum water table of 3 ft over the entire area. Regulatory monitoring and effluent sampling requirements are extensive. Turf grass may be harvested two to three times per year.

Production of Single Cell Protein from Brewery Effluents

In the aerobic and anaerobic waste treatment systems discussed, the microorganisms that were used to break down organic matter consisted of a wide variety of different bacteria, protozoa, and rotifers. It has been demonstrated that certain yeast species and filamentous fungi can also successfully reduce BOD in brewery effluents, whereas providing a biomass that may be more acceptable to regulatory agencies and to the public at large as a high protein feed or food supplement.

Torula and *Candida* yeasts have been found to degrade brewery effluents successfully. Noel and Bertrand[130] reported on biodegradation of brewer's grain liquor and waste beer by means of a *Candida* yeast isolated from brewer's grain liquor. This yeast is able to break down alcohol and reducing sugars substantially with a biomass yield of 46–47%. After preclarification to remove suspended solids, the BOD of a mixture of 35% waste beer and 65% brewer's grain liquor was reduced by 88% at a temperature of 33°C and a detention time of 10 h. The pilot fermentation was carried out aerobically at a pH of 3.8 with urea as a nitrogen source. No information is available on the ability of full-scale systems to handle shock loads and prevent infections.

It is also possible to use selected fungi species to effectively reduce the BOD of brewery waste systems. Advantages of fungi are that they have a high cellular activity and that the biomass is easily separated using centrifuges, vibrating screens, or belt filters without the need to use coagulants or filter aids. Fungi were successfully used to reduce the BOD of a corn wet milling plant by 90–96% in a 50,000 gallon pilot unit.[131] The aerobic system had a detention time of 16–24 h and a pH range of 3.5–6.

Aspergillus oryzae and *A. bridgeri* were found to be most effective fungi in reducing brewery waste (Church, B.D., Personal Communication, 1984) with BOD reductions of waste beer of over 98%. Fungal biodegradation of brewery waste is a relatively new process that still requires considerable development. More extensive pilot tests are required to determine how dominance of the desired fungi can be maintained and to prevent bacterial infections that apparently killed *A. oryzae* fungi growing on brewery waste.[132] Most importantly, approval and acceptance of fungal mass as an animal feed is still a major area to be addressed.

References

1. Tacon, A.G., Nutritional evaluation of animal and food processing wastes, *Effluent Treat. Biochem. Ind. Conf.*, England, 1979.
2. Bergen, W.G., Amino acid nutrition in ruminants — fact or fiction, *Proc. Distill. Feed Conf.*, 40:91–99, 1985.
3. Grant, H.L., Hops, in *The Practical Brewer*, Broderick, H.M., Ed., MBAA, Impressions, Inc., Madison, WI, 1977.
4. Wallen, S.E. and Marshall, H. F., Protein quality evaluation of spent hops, *J. Agric. Food Chem.*, 27:636, 1979.
5. Linton, J.H., Pollution abatement through utilization of wet brewery by-products in live stock feeding, *Brew. Dig.*, November:42–46, 74, 1973.

6. O'Rourke, T., Efficient handling of brewing by-products, *Brew. Guardian*, September:39–43, 1980.
7. Marx, G., Scopel, S., and Wackerbauer, K., Hot trub separation and removal, and recovery of the entrapped wort, *Brauwelt Int.*, 1:34–40, 1986.
8. Ruggles, R.E. and Hertrich, J.D., Wort recovery from trub with a decanter centrifuge, *Tech. Q. Master Brew. Assoc. Am.*, 22:99–102, 1985.
9. Narziss, L., Reicheneder, E., and Ngo-Da, P., Qualitative Aspekte der Verwendung von Extractresten bei der Bierbereitung, *Brauwelt*, 41:1810–1822, 1982.
10. Siebert, K.J., Blum, P.H., Wisk, T.J., Stenroos, L.E., and Anklam, W.J., The effect of trub on fermentation, *Tech. Q. Master Brew. Assoc. Am.*, 23:37–43, 1986.
11. Schisler, D.O., Ruocco, J.J., and Mabee, M.S., Wort trub content and its effect on fermentation and beer flavor, *J. Am. Soc. Brew. Chem.*, 40:57–61, 1982.
12. Townsley, P.M., Preparation of commercial products from brewer's waste grain and trub, *Tech. Q. Master Brew. Assoc. Am.*, 16:130–134, 1979.
13. O'Rourke, T., Making money out of spent grains and by-products, *Brew. Guardian*, January:31–39, 1984.
14. Chyba, G.W. and Dokos, J.H. Process of brewing beer and treating spent grains, U.S. Patent 4,197,321, 1980.
15. Coors, J.H. and Jangaard, N.O., Recycling of spent grain press liquor, *EBC Proc. 15th Congr.*, Nice, 1975, pp. 311–321.
16. Kieninger, H. and Narziss, L. Die Widerverwendung von Extracthaltigen Nebenprodukten bei der Bierbereitung, *Proc. 17th Eur. Brew. Conv. Congr.*, Berlin, 1979, pp. 213–236.
17. Myron, T.J. and Shinsky, F.G., Product moisture control for steam-tube and direct fired dryers, *Tech. Q. Master Brew. Assoc. Am.*, 12:235–242, 1975.
18. Gilmore, W.H., Treatment of spent grains, hops and yeast, *Tech. Q. Master Brew. Assoc. Am.*, 8:73–78, 1971.
19. Penrose, J.D.F., Upgrading grains, *Brewer*, January:4–7, 1982.
20. Hunt, L.A. and Spitzer, E.H., Livestock feed composition and method of preparing the same, U.S. Patent 3,875,304, 1975.
21. Klopfenstein, T. and Goedeken, F., Animal protein products: bypass potential, *Feed Manag.*, 37:12–22, 1986.
22. Poos, M.I., Ruminant nutritional utilization of alcohol production by-products, *Feedstuffs*, June 29:19–22, 1981.
23. Thompson, G.B., Johnson, R., and Hutcheson, D., Evaluation of brewers dried grains in finishing rations in cattle, *USBA Feed Conf. Proc.*, 49–65, 1977.
24. Oster, A., A note on the use of brewers' dried grains as a protein feedstuff for cattle, *Anim. Prod.*, 24:279–282, 1977.
25. Linton, J.H., Acceptability of mixtures of spent yeast and brewers grains as components of cattle diets, *Brew. Dig.*, March:58–60, 1971.
26. Conrad, H.R. and Rogers, J.A., Comparative nutritive value of brewers wet and dried grain by dairy cattle, *USBA Feed Conf. Proc.*, 26–35, 1977.
27. Chandler, P.T., Wet brewers grain: its role in dairy cattle feeding, *Feedstuffs*, April 27:21, 24, 75, 1987.
28. Pond, W., Limitations and opportunities in the use of fibrous and by-product feeds for swine, *Distill. Feed Conf. Proc.*, 36:71–85, 1981.
29. Sullivan, T., Brewers dried grains in rations for starting, growing, finishing and breeder turkeys, *USBA Feed Conf. Proc.*, 71–83, 1977.

30. Harms, R., Dried grains as an ingredient in the formulation of feeds for laying and breeding hens, *USBA Feed Conf. Proc.*, 66–70, 1977.
31. Jensen, L., Distillers feeds and unidentified nutritional factors for poultry, *Distill. Feed Conf. Proc.*, 31:33–39, 1976.
32. Qureshi, A.A., Chaudhary, V., Weber, F.E., Chicoye, E., and Qureshi, N., Effects of brewers grain and other cereals on lipid metabolism in chickens, *Nutr. Res.*, 11:159–168, 1991.
33. Ott, E.A., Use of brewers dried grains in horse rations, *USBA Feed Conf. Proc.*, 84–97, 1977.
34. Van Soest, P., A nutritional requirement for fiber in non-ruminants and ruminants, *Feed Manag.*, 35:36–39, 1984.
35. Bavisotto, V.S., Recovery of edible products from spent grains and yeasts, U.S. Patent 3,212,902, 1965.
36. Ernster, J.H., Process for utilizing barley malt, U.S. Patent 3,846,397, 1974.
37. Institute of Food Technology, Dietary fiber: scientific status summary of IFT's expert panel on food safety and nutrition, *Food Technol.*, January:35–39, 1979.
38. Weber, F.E. and Chaudhary, V.K., Recovery and nutritional evaluation of dietary fiber ingredients from a barley by-product, *Cereal Foods World*, 32:548–550, 1987.
39. Lupton, J.R., Morin, J.L., and Robinson, M.C., Barley bran flour accelerates gastrointestinal transit time, *J. Am. Diet. Assoc.*, 93:881–885, 1993.
40. Lupton, J.R., Robinson, M.C., and Morin, J.L., Cholesterol-lowering effect of barley bran flour and oil, *J. Am. Diet. Assoc.*, 94:65–70, 1994.
41. Huang, Z., Keenan, A.R., and Addis, P.B., Lowering of plasma cholesterol oxidation products by barley bran: a pilot trial, presented at the Institute of Food Technology Annual Meeting, New Orleans, June, 1992.
42. Chaudhary, V.K., High dietary fiber products, U.S. Patent 4,341,805, 1982.
43. Niefind, H.J., Meuser, F., and Kohl, G., Spent grains for human nourishment, *Brauwelt Int.*, 91–97, 1983.
44. Stead, I., Yeast separation, *Brew. Guardian*, October:31–36, 1986.
45. Young, I.M., The case for beer recovery, *Brew. Dig.*, 60:24–28, 1985.
46. Boughton, R.A., Are you getting the best out of your yeast? *Brewer*, July:2–6, 1983.
47. Mueller, G., Preparation and processing of brewery yeast sludge using high-pressure chamber filter presses, *Tech. Q. Master Brew. Assoc. Am.*, 19:57–62, 1982.
48. Smith, M.G., Ultrafiltration, *J. Inst. Brew.*, 92:315–316, 1986.
49. Siebert, K.J., Stenroos, L.E., and Reid, D.S., Filtration difficulties resulting from breakage of yeast during centrifugation, *Tech. Q. Master Brew. Assoc. Am.*, 24:1–8, 1987.
50. Young, I.M., Some aspects of yeast and tank bottom filtration using filter presses, *IBG/ABTA Seminar*, Bristol, October, 1981.
51. Smith, A.J.T., Recovery and handling of surplus yeast, *Brew. Guardian*, June:35–38, 1979.
52. Labuza, T.P., Barrera Santos, D., and Roop, R.N., Engineering factors in single cell protein production. I. Fluid properties and concentration of yeast by evaporation, *Biotechnol. Bioeng.*, 12:123–134, 1970.
53. Ingledew, W.M., Langille, L.A., Menegazzi, G.S., and Mok, C.H., Spent brewers yeast — analysis, improvement and heat processing considerations, *Tech. Q. Master Brew. Assoc. Am.*, 14:231–237, 1977.

54. Burgstaller, G., Frische Bierhefe fur die Rinder-und Schweinemast, *Brauwelt*, 113:817–818, 1973.
55. Witting, R., Flussige Bierhefe in der Schweinefutterung, neues Behandlungsverfahren, *Brauwelt*, 116:1095–1097, 1976.
56. Labuza, T.P., LeRoux, J.P., Fan, T.S., and Tannenbaurn, S.R., Engineering factors in single-cell protein production. II. Spray drying and cell viability, *Biotechnol. Bioeng.*, 12:135–140, 1970.
57. Lonsdale, C.R., Brewers' yeast as a food for pigs, *Brewer*, March:92–94, 1985.
58. Ingledew, W.M. and Burton, J.D., Chemical preservation of spent brewers yeast, *J. Am. Soc. Brew. Chem.*, 37:140–146, 1979.
59. Neame, C., New use for brewer's yeast, *Food Manuf.*, February:53–61, 1981.
60. Witting, R. and Ellenberger, W., Flussige Bierhefe in der Schweinefutterung — Neues Behandlungsverfahren, *Brauwelt*, February 16:202–209, 1978.
61. Mahan, D., Vitamin E and selenium needs for swine, *Distill. Feed Conf. Proc.*, 33:30–37, 1978.
62. Linton, J.H., Brewery by-product utilization, presented at Midwest Conference, MBAA, July 23–25, LaCrosse, WI, 1982.
63. Trevis, J., Live, killed brewers yeasts make good protein sources, *Feedstuffs*, February 18, 1980.
64. Muirhead, S., Brewers liquid yeast improves milk production in lactating cows, *Feedstuffs*, February 8:10, 1988.
65. Peppler, H.J. and Stone, C.W., Feed yeast products, *Feed Manag.*, August, 1976.
66. Vananuvat, P., Value of protein for poultry feeds, *CRC Crit. Rev. Food Sci. Nutr.*, 9:325–344, 1977.
67. Watanabe, N., Inaida, O., Yamada, K., and Kavokawa, T., New approaches to using spent brewer's yeast, *J. Am. Soc. Brew. Chem.*, 38:5–8, 1980.
68. Scott, M., Importance of biotin for chickens and turkeys, *Feedstuffs*, February 23:59–67, 1981.
69. Froehli, D.M., Importance of folic acid in turkey diets explored, *Feedstuffs*, April 27:26–28, 1987.
70. Corbin, J., Recent advances in nutrition of the dog, *Distill. Feed Conf. Proc.*, 34:21–32, 1979.
71. Nay, E.V., Washing and debittering brewer's yeast, *Brew. Dig.*, 29:52–61, 1954.
72. Singruen, E., Brewer's yeast — its recovery and uses, *Brew. J.*, 108:69–70, 1953.
73. Acraman, A.R. and Smith, A.J.T., Problems encountered in the utilization of brewer's yeast for human food purposes, *Brewer*, October:318–323, 1976.
74. Peppler, H.J., *The Yeasts*, Rose, A.H. and Harrison, J.S., Eds., Academic Press, San Diego, CA, 1970, Vol. 3, Chap. 8.
75. Peppler, H.J., Production of yeast and yeast products, in *Microbial Technology*, 2nd ed., Peppier, H.J. and Perlman, D., Eds., Academic Press, New York, 1979, Vol. 1, Chap. 5.
76. Singruen, E. and Ziemba, J.V., Brewers dried yeast — new uses in foods, *Food Eng.*, October:50–52, 238, 1954.
77. Hobson, J.C., Yeast — a central role, *Food*, January:41–42, 1985.
78. Koivurinta, J., Junnila, M., and Koivistoinen, P., Functional properties of brewer's grain, brewer's yeast and distillers stillage in food systems, *Lebensm-Wiss. Technol.*, 13:118–122, 1980.
79. Gibson, D.L. and Dwivedi, B.K., Production of meat substitutes from spent brewers' yeast and soy protein, *Can. Inst. Food Technol. J.*, 3:113–115, 1970.

80. Liggett, J.J., Preparation of a cocoa substitute, U.S. Patent 4,312,890, 1982.
81. Akin, C. and Murphy, R.M., Method of accelerating autolysis of yeast, U.S. Patent 4,285,976, 1981.
82. Knorr, D., An enzymic method for yeast autolysis, *J. Food Sci.*, 44:1362, 1979.
83. Dziezak, J., Special report on yeast and yeast derivations, *Food Technol.*, February:104–125, 1987.
84. Acraman, A.R., Processing brewers yeast, *Proc. Biochem.*, September:313–317, 1966.
85. Robbins, E.A. and Seeley, R.D., Process for the manufacture of yeast glycan, U.S. Patent 4,122,196, 1978.
86. Seeley, R.D., Fractionation and utilization of bakers yeast, *Tech. Q. Master Brew. Assoc. Am.*, 14:135–139, 1977.
87. McCormick, R.D., Baker's yeast — world's oldest food is newest source of protein and other ingredients, *Food Prod. Dev.*, July–August:17–19, 1973.
88. Kula, M. and Schutte, H., Purification of proteins and the disruption of microbial cells, *Biotechnol. Prog.*, 3:31–42, 1987.
89. Delta Biotechnology Ltd., U.K. Patent Application 2175 590A, 1986.
90. Stein, J.L., Dokos, J.H., Brodeus, T., and Radecki, M.R., Concentration of brewery spent grain liquor using a submerged combustion evaporator, *Proc. Fourth Natl. Symp. Food Proc. Wastes*, Syracuse, NY, March 26–28, 1973.
91. Sebree, B.R., Chung, D.S., and Seib, P.A., Brewers condensed solubles. II. Viscosity and viscosity reduction of brewers condensed solubles by cellulase and beta-glucanase, *Cereal Chem.*, 60:49–52, 1983.
92. Sucher, R.W., Brewery by-products, *AFMA 11th Annual Liquid Feed Symposium*, Omaha, NE, September 15, 1981.
93. Huige, N.J., Charter, W.M., and Wendt, K.W., Measurement and control of oxygen in carbon dioxide, *Tech. Q. Master Brew. Assoc. Am.*, 22:92–98, 1985.
94. Lom, T., A new trend in the treatment of alkaline brewery effluents, *Tech. Q. Master Brew. Assoc. Am.*, 14:50–58, 1977.
95. Hill, P., New way to fumigate grain: CO_2 — It's non-toxic and residue free, *Food Eng.*, July:118, 1982.
96. Wendt, K., Koziboski, E., and Huige, N., Alternatives for disposal of spent filter cake, *Tech. Q. Master Brew. Assoc. Am.*, 18:140–144, 1981.
97. Link, C.E., An automatic filtration system, *Tech. Q. Master Brew. Assoc. Am.*, 22:56–60, 1985.
98. Sommer, G. and Metscher, M., Recycling von Kieselgur aus der Bierfiltration, *Monatschr. Brau.*, July 31:310–312, 1979.
99. Bott, N., Efficient control of brewery process losses, *Brew. Guardian*, August:27–35, 1980.
100. Goetz, J.A., Beer loss in can packaging, *Tech. Q. Master Brew. Assoc. Am.*, 23:111–114, 1974.
101. Sidor, L.L. and Knight, P.D., Converting brewery waste to ethanol, *Tech. Q. Master Brew. Assoc. Am.*, 23:111–114, 1986.
102. Sidoti, D.R., Flavor enhancement and potentiation with beer concentrate, U.S. Patent 4,590,085, 1986.
103. Duxbury, D.D., Beer flavor enhances batters and coatings, *Food Proc.*, January:37, 1987.
104. National Soft Drink Association, *Container Recycling Bibliography*, National Soft Drinks Association, Washington, DC, 1983.

105. Sternberg, R.W., Packaging waste — whose responsibility? *Canner/Packer*, 140:29–31, 1971.
106. Rosenwinkel, K.H. and Seyfried, C.F., Reinigung von Brauereiabwassern, *Brauwelt*, 28:1229–1232; 29:1263–1270, 1984.
107. McWhorter, T.G. and Zielinski, R.J., Waste treatment for the Pabst Brewery in Perry, Georgia, *Brewer*, September:485–490, 1974.
108. Klijnhout, A.F. and van Eerde, P., Centenary review. Some characteristics of brewery effluent, *J. Inst. Brew.*, 92:426–434, 1986.
109. Seddon, A.W. and Woodland, R., Optimization of effluent treatment in a large new brewery, *Tech. Q. Master Brew. Assoc. Am.*, 8:49–52, 1981.
110. Lunney, M.J., Dissolved air pressure flotation treatment of brewery effluent, *Brew. Guardian*, January:13–17, 1981.
111. Hughes, D.A., Dissolved air flotation. An alternative treatment option, *Brewer*, June:266–269, 1987.
112. Lampinen, L. and Quirt, F., Flue gas neutralizing, *Tech. Q. Master Brew. Assoc. Am.*, 24:86–89, 1987.
113. Leeder, G.I., On-site effluent treatment for the brewing industry, *Brewer*, June:213–216, 1986.
114. Miller Brewing Company, Brewery cost-cutting projects: "made the American way," *Energy Manag. Technol.*, October:37–40, 1985.
115. Barrett, P. and Hayden, P.L., Treatment of brewery wastewaters by biological treatment, *FroG. Ind. Waste Conf.*, 853–868, 1978.
116. Scroggins, D., Sewer surcharge reduction — the real cost, *Brew. Dig.*, August:32–34, 1983.
117. LeClair, B., Performance monitoring program — Molson's brewery deep shaft treatment systems, *Proc. 39th Ind. Waste Conf.*, Purdue University, 1985, pp. 257–268.
118. Murray, C.R., Elliott, A., McKee, R., and Scott, R., Anaerobic digestion and its application in the brewery-A bench scale investigation, *Inst. Brew. Central S. African Sect. 1985 Conf. Proc.*, March 9–14, 1986, pp. 358–385.
119. Love, L.S., Anaerobic degradation: more cost effective — more efficient treatment of brewery effluents, *Tech. Q. Master Brew. Assoc. Am.*, 24:51–54, 987.
120. Bolivar, J.A., The Bacardi Corporation digestion process for stabilizing rum distillers wastes and producing methane, *Tech. Q. Master Brew. Assoc. Am.*, 20:119–128, 1983.
121. Sax, R.I., One giant step for Heileman. Advantages of the Biothane system, *Mod. Brew. Age*, October:5–10, 35, 1982.
122. Swinkels, K.Th.M., Vereijken, T.L.F.M., and Hack, P.J.F.M., Anaerobic treatment of wastewater from a combined brewery, malting, and soft-drink plant, *Proc. 20th Eu. Brew. Conv. Cong.*, Helsinki, 1985, 563-570.
123. Chou, T., Lin, T.Y., Hwang, B.J., and Wang, C.C., Selective removal of H_2S from biogas by a packed silica gel adsorber tower, *Biotechnol. Prog.*, 2:203–209, 1986.
124. Heijnen, J.J., Enger, W.A., Mulder, A., Lourens, P.A., Keijzers, A.A., and Hoeks, F.W.J.M.M., Application of anaerobic fluidized bed reactors in biological waste water treatment, *Starch/Stärke*, 12:419–427, 1986.
125. Bays, J., Waste activated sludge. A new brewery by-product, *Tech. Q. Master Brew. Assoc. Am.*, 14:47–49, 1977.

126. Martin, J.L. and Hayden, P.L., Pressure filtration of waste industrial sludges. Modelling design and operational parameters, *31st Annual Purdue Ind. Waste Conf.*, May 4–6, 1976, pp. 1–29.
127. Kienholz, E.W. and Moreny, R.E., A new brewery by-product for turkey diets, *Poult. Sci.*, 60:1879–1883, 1981.
128. Naylor, L.M. and Severson, K.V., *Brewery Wastewater Treatment Residuals as a Fertilizer*, Department of Agricultural Engineering Report, Cornell University, Ithaca, NY, 1984.
129. Keith, L.W., Land application of brewery wastewater, *Tech. Q. Master Brew. Assoc. Am.*, 18:201–204, 1981.
130. Noel, C. and Bertrand, J., Bioconversion des Effluents de Brasserie en Proteines d'Organismes Unicellulaires, *Bios*, 16:40–46, 1985.
131. Church, B.D., Erickson, E.E., and Widmer, C.M., Fungal digestion of food processing wastes, *Food Technol.*, February:36–41, 1973.
132. Irvine, R.L., Fermentation industry waste treatment literature review (24 references), *J. Water Pollut. Control Fed.*, 52:1333–1340, 1980.

19

Beer Stability

Graham G. Stewart

CONTENTS

Introduction 715
Biological and Nonbiological Instability 715
   Biological Instability 716
   Nonbiological Stability 716
      Physical Stability 716
      Flavor Stability 719
      Foam Stability 721
      Gushing 724
      Light Stability 724
Summary 725
References 726

Introduction

To the beer consumer, freshness is key to good drinkability. Compared to most other alcoholic beverages, beer is unique because it is unstable when in the final package. From the day it is bottled, the quality starts to decrease. The rate of decrease depends on time, temperature, and batch variability. In this chapter, those factors that lead to changes in beer quality and how stability can best be achieved will be examined.

Biological and Nonbiological Instability

Beer instability can be divided into biological and nonbiological instability.

Biological Instability

Biological instability involves contamination by bacteria, yeast, and mycelial fungi (see Chapter 16). Fortunately, beer is an inhospitable environment for microbial growth; it has a low pH (less than 4.4), ethanol is present, there is a limited range of usable nutrients, and bacteriostatic hop acids are present. In addition, the environment is anaerobic and the liquid is carbonated. Most potential contaminants originate from the raw materials or from unclean brewing equipment. Barley can contain the *Fusarium* mold, which can release mycotoxins or cause gushing (spontaneous ejection of beer from its container — can or bottle as detailed later). The malt can also carry bacteria that may contribute unpalatable flavors or potentially carcinogenic nitrosamines, and can cause turbidity and filtration problems. It is therefore important to exclude these spoilage organisms from the brewing process. A modern plant and good hygiene help, and many breweries pasteurize their beer to ensure biological stability.

Nonbiological Stability

Nonbiological stability of beer involves a wide range of chemical processes and can be divided into a number of categories: physical, flavor, foam, gushing, and light.

Physical Stability

With a few notable exceptions, consumers prefer their beer to be bright and free of particles. When beer is stored, it has the potential to produce haze and the brightness will be compromised. Beer physical stability (also called colloidal stability or simply haze formation) cannot be ensured by treating beer with one "super-product" that will solve everything. Stability will be affected by the whole brewing process; consequently, care must be taken at every stage. Raw materials are the source of haze precursors. Although there are a number of types of beer haze, the primary reaction is the polymerization of polyphenols and their binding with specific (sensitive) proteins. When beer is cooled below 0°C, chill haze will form, which consists of a reversible association of small polymerized polyphenols and protein. When the beer is restored to room temperature, this haze redissolves and the beer becomes bright again. If beer is chilled and warmed a number of times, or if beer is stored at room temperature for an extended period (6 months or longer), permanent haze will form. This haze does not redissolve even when the beer is warmed to 30°C or higher.

The balance between flavanoid polyphenols (tannoids; Figure 19.1) and sensitive protein largely dictates physical (colloidal) stability. Beers can differ widely in the contents of these species, the relative levels of which depend upon raw materials and the process conditions employed. Haze will not form, or its formation will be slowed, when either of these components is removed, or the factors promoting the interaction are largely excluded.

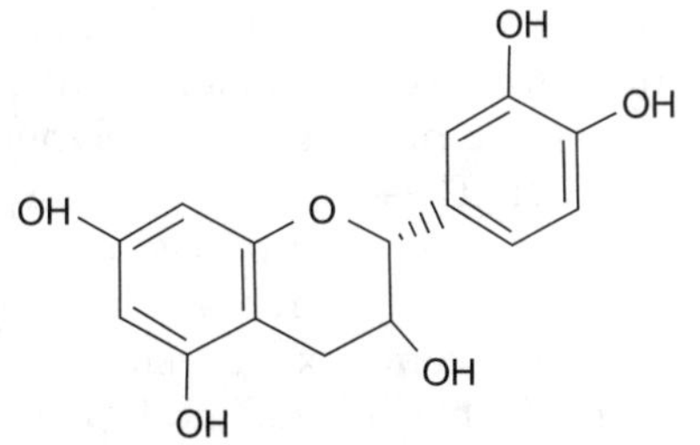

FIGURE 19.1
Catechin — a typical beer polyphenol.

Chill haze is defined as a colloidal haze that appears in beer on cooling to under 0°C and dissolves again on warming to 20°C. With time, chill haze changes to permanent haze that no longer dissolves. Chill and permanent hazes have almost the same composition.

Permanent haze is important in the case of beers that are expected to have a long shelf life (e.g. beers that are to be exported). Consequently, the measures that are taken to delay or prevent the appearance of colloidal hazes are important. The time required for the appearance of permanent haze can vary significantly depending on the beer and the storage conditions. In general, it appears a few months after packaging, but it can occur within a few weeks with substandard raw materials and a poor brewing operation.

Haze formation is increased by a number of factors of which storage temperature has the greatest influence because an increase in temperature raises the rate of the reactions.[1] Pasteurization, in particular, accelerates colloidal haze formation.

Oxidation (the presence of oxygen) has a great effect on beer haze formation. Extensive oxidation can increase the rate of haze appearance manifold. Heavy metal ions (particularly iron) can promote the formation of colloidal haze. Movement of beer accelerates haze formation because of the rapid interaction of colloids. Light encourages oxidation and consequently haze formation.

Beer chill haze consists of a loose bonding of high molecular weight proteins with highly condensed polyphenols (predominantly anthocyanogens). Small amounts of carbohydrates and inorganic materials are included in these loosely bound aggregates. This loose bonding is broken again on warming. Haze formation occurs as a result of dissolved colloidal particles colliding and increasing the formation of hydrogen bonds between them. In course of time, increasingly large aggregates come together until they are visible as haze.

Haze formation corresponds to the presence of "sensitive proteins" (defined as substances precipitated with tannic acid) and "tannoids" (defined as those polyphenols adsorbed by polyvinylpolypyrolidone [PVPP]). The driving force for haze formation is the interaction of hydrophilic groups on these sensitive proteins with polyphenols. There are also hydrophobic proteins in beer. These surface active species are important for foam stability. There are a number of procedures that can be employed to retard or prevent haze formation by removing polyphenols or sensitive proteins from the beer. The complex "sensitive" protein degradation products can be enzymically hydrolyzed and polyphenols or sensitive proteins can be removed during brewing. Maturing beer can be stored cold in order to precipitate the haze precursors and packaged beer can be stored cold to retard haze formation.

Beers with a prolonged shelf life can be produced by employing commercial stabilizers. The main stabilizing agents currently in use (which can be used singly or together) are silica gel preparations. These important stabilizing agents bind the hydrophilic polypeptides, but have little effect on the foam-promoting hydrophobic polypeptides. They are used in amounts of 50–150 g/hl and are usually dosed into the beer before filtration. There are two types of silica gel used in brewing; hydrogels that have a moisture content of more than 30% and xerogels (dry gels) with a 5% water content.

PVPP selectively removes phenol-containing substances. PVPP binds to polyphenols as it has a very similar structure to the amino acid proline.[2] Both have five-membered, saturated, nitrogen-containing rings with amide bonds and no other functional groups. It is not certain whether PVPP binds to the same part of the polyphenol molecule to which polypeptides bind. Selection depends on the pH-sensitive formation of hydrogen bonds, which are split again in alkaline solution with the release of the adsorbed phenol compounds. Regeneration of PVPP with hot caustic is very effective. PVPP and silica gels have been used together with good results because both polyphenol and sensitive protein components are removed.[3]

Proteolytic enzymes are employed as stabilizing agents, but with the advent of silica gels, they are less popular. The enzymes employed include papain (from the latex of papaya), bromelain (from pineapple), and ficin (from figs). These enzyme preparations are not very specific, and as well as hydrolyzing haze-specific proteins, they often hydrolyze the hydrophobic foam-specific polypeptides. Consequently, the use of these enzymes often requires the addition of a foam-enhancing agent such as propylene glycol alginate.

Problems in the brewhouse process (e.g., starch, pentosans, oxalate, β-glucan, carbohydrate and protein from damaged yeast, and solid materials including filter aids) can lead to visible haze. In beers that are shipped over long distances and subject to agitation, "bits" can be formed. These bits contain protein, and perhaps pentosans. Particles can also form from the interaction of incompatible stabilizing agents in the beer (e.g., the crosslinking of papain with propylene glycol alginate during pasteurization). Propylene glycol alginate can also interact with proteins, such as uncomplexed isinglass finings to form larger bits.

Flavor Stability

The flavor stability of a beer depends primarily on the oxygen content of the packaged beer. It is now clear that beer flavor stability is influenced by all stages of the brewing process[4]: preservation of reducing substances by avoidance of oxygen pick-up during mashing, lautering, and wort boiling; elimination of substances that are prone to react with flavor-active compounds such as carbonyl molecules by good mashing and wort separation procedures, and prevention of ion pick-up such as iron and copper.[5] Controlled exposure of the wort to heat is important to limit the formation of Maillard reaction products (produced as a result of heating sugars with amino acids) and related substances. The role of such reaction products in beer flavor staling reactions is ambiguous, and there are reports of their positive and negative influences.[6]

In many foods such as milk, butter, vegetables, vegetable oils, and beverages, staling is caused by the appearance of various unwanted unsaturated carbonyl compounds, and it is now becoming increasingly clear that the same is true of beer staling. As already discussed, packaged beer has a limited shelf life. The phenomenon of beer aging or staling has been intensively investigated with a view to understanding and controlling it. Despite intensive studies over the past 30 years, the mechanisms of staling are not fully understood. The actual compounds responsible for stale flavor vary during prolonged storage, as evidenced by changes in the flavor profile of beer (Figure 19.2), and typical lager staling flavors over time are summarized in Table 19.1.

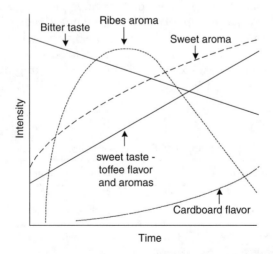

FIGURE 19.2
Sensory changes in beer flavor with time in packaged product. (Adapted from Dalgliesh, C.G., *Proc. 16th Eur. Brew. Conv. Congr.*, Amsterdam, DSW Dordrecht Press, 1977, pp. 623–659. With permission.)

TABLE 19.1

Typical Staling Flavors

Taint	Flavor	Likely Cause
Oxidized	Papery, cardboard, "dull," toffee	Storage/oxygen
Catty/Ribes	Tomcats, black current leaves, tomato plants	High in package oxygen
Aldehyde	Rotting apples	Storage/high oxygen
"Cooked"	Over pasteurized, grainy, "dull," toffee	Storage/high oxygen/high temperature

Source: From Dalgliesh, C.G., *Proc. 16th Eur. Brew. Conv. Congr.*, Amsterdam, DSW Dordrecht Press, 1977, pp. 623–659. With permission.

The compounds causing the sweetish, leathery character of very old beers have not been identified. However, there is evidence that the papery cardboard character of 2–4-month old beer is due to unsaturated aldehydes. The most flavor-active aldehyde, which has been conclusively proven to rise beyond flavor threshold levels, is *trans*-2-nonenal.[7] Other aldehydes such as nonadienal, decadienal, and undecadienal may also exceed threshold levels. Foster et al.,[8] working with light beer, showed that as unpleasant stale compounds increased during staling, simultaneously fresh apple and pear/fruity esters decreased (Table 19.2).[9]

Although there are many factors that will influence the flavor stability of beer, the oxygen level in the final package is of paramount importance. It is critical that this level in beer immediately prior to packaging is as low as possible (less than 100 μg/l) and that oxygen pick-up during filling is minimal. The adverse effects of oxidation on the flavor of finished beer have been known for a considerable time, and some brewers add SO_2 (bisulfite) or other antioxidants, such as ascorbic acid, to beer prior to packaging to provide protection against oxygen pick-up. This can improve flavor stability.[7] The effectiveness of bisulfite, besides its antioxidant properties, is its

TABLE 19.2

Degree of Lager Staling and Flavor Descriptors

Degree of Staling	Flavor Descriptors
Fresh beer	Clean
Slightly stale	Increased sweet, ribes, cardboard, papery, decreased bitterness, estery character
Stale beer	Increased bread-like character, decline in body
Very stale beer	Honey-like flavor
Extremely stale beer	Sherry-like

Source: From Boccorh, R.K. and Paterson, A., The 6th Sensometrics Meeting, Dortmund, Germany, 2002. With permission.

ability to bind carbonyl compounds into flavor-neutral complexes.[10–12] The reaction is reversible, and in the presence of an excess of bisulfite there will be an increase in the flavor-neutral adduct. A decrease in SO_2 in the stored beer shifts the equilibrium toward release of carbonyl substances with the attendant development of aged character.

The addition of bisulfite to fresh beer minimizes increases in free aldehyde concentration during aging. When added to stale beer, bisulfite lowers the concentration of free aldehydes and effects the removal of the cardboard flavor. However, over time, the bisulfite will be oxidized to sulfate, thus increasing the concentration of free aldehydes again.

Sensory analysis experiments by Bushnell et al.[13] suggested that SO_2 is rapidly consumed upon addition to beer and that when added at levels of less than 10 mg/l (a value that does not require labeling in the United States), it has limited ability to restrict aging of beer as detected by a trained sensory panel. They suggest that the impact is due more to SO_2 acting as an antioxidant, than as a carbonyl binder.

Foam Stability

When beer is sold, the stability of the foam in a glass of beer is considered by many consumers to reflect the quality of the product. The increasing use of adjuncts (unmalted sources of carbohydrate) and the associated decrease in malt in the grist today, together with the employment of high-gravity brewing techniques (details later) reduces foam values in many beers.

There are many foam-promoting compounds in beer such as iso-α-acids from hops, protein, metal ions, polysaccharides, and all have an important role to play in foam formation and stability.[14,15] However, the backbone of foam is protein. Many methods have been tried and extolled for their virtues in the isolation and characterization of foam-positive beer polypeptides; for example, separation of foaming proteins by hydrophobic interaction chromatography has long been a standard technique. Once separated, the proteins are investigated to discover if they are related to beer foam potential or stability. On the basis of such experiments, it has been proposed that certain sizes of protein are important in the formation and stabilization of foam. For example, 40-, 10-, and 8-kDa proteins have all been postulated as major foam-stabilizing molecules. The polypeptides of greatest hydrophobic character produce the most stable foam, and it is the hydrophobic property that is more important than size.[16,17]

The use of high-gravity brewing is essential for the future economic viability of the brewing industry. High-gravity brewing is a procedure which employs wort at higher than normal concentration, and therefore requires dilution with water later in the process. This process, by reducing water employed in the brewhouse, increases production capacity without adding to the existing brewing plant. Most major brewing companies worldwide have revised their production processes to accommodate high gravity brewing procedures as a means to reduce capital expenditure. Although this process has many advantages,[18] one of the problems that still exists is that beers brewed at higher gravities exhibit poor foam stability. The effect of high-gravity brewing on head retention with respect to hydrophobic polypeptide levels has been examined throughout the brewing and fermentation of high gravity (20°Plato) and low gravity (10°Plato) worts (Figure 19.3).[19]

Three notable features of the data are worthy of highlighting:

1. At "kettle full," the level of hydrophobic polypeptides was similar in the 20 and 10°Plato worts. This is in spite of the use of twice the amount of malt grist to produce the high-gravity wort. This implies there was a major failure to extract hydrophobic polypeptides during the high-gravity mash.

2. There was much greater loss of hydrophobic polypeptides during fermentation of the high-gravity wort, so that by the end of fermentation the hydrophobic polypeptide content of the high-gravity fermented wort was just over 50 mg/l, markedly lower than that of the low-gravity fermented wort.

3. When the high-gravity beer was diluted to 4.5% alcohol by volume, equivalent to the low-gravity beer, it contained half the hydrophobic polypeptide content of the low-gravity brewed beer.[20]

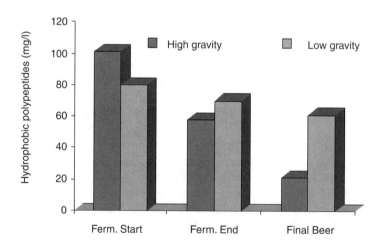

FIGURE 19.3
Changes in the levels of hydrophobic polypeptides during the brewing process (final high-gravity beer diluted to 4.5% alcohol by volume equivalent to the low-gravity beer).

The head retention of the high-gravity brewed beer was less than that of the low-gravity brewed beer. This contrasts with the low-gravity brewed beer where the hydrophobic polypeptide in this foam accounted for over 40% of the total polypeptide. Therefore, not only is the polypeptide of high-gravity brewed beer reduced, but so is the hydrophobic content of its foam, which would adversely influence its stability. The amino acid profile of the hydrophobic polypeptides recovered from beer foam, unlike polypeptides involved in haze formation (where glutamic acid and proline account for 40–50% of the total amino acid composition), showed no amino acid present in a distinctive amount.[21]

It has already been discussed that the fermentation stage is a key step in which hydrophobic polypeptides are lost during the brewing process (Figure 19.3). Two factors could account for the loss of hydrophobic polypeptide during fermentation. First, fermentation is known to be responsible for the loss of a large amount of foam-active substances and this problem is exacerbated during the fermentation of high-gravity worts. Second, yeast secretes proteolytic enzymes into the fermenting wort, and these enzymes will reduce the foam stability of finished beer through protein degradation (hydrolysis) that will occur during fermentation and storage.

Analysis of proteinase A activity (using the fluorimetric method described by Kondo et al.[22]) in wort and beer during the brewing process showed, as would be expected, that freshly boiled wort contained no enzyme activity (Figure 19.4). However, during fermentation, proteinase A was secreted into wort by yeast cells. Proteinase A increased throughout the fermentation with the highest enzyme activity occurring at the end. Considerably higher amounts of proteinase A were released during the 20°Plato fermentations compared to the 10°Plato fermentations.

During high-gravity brewing, increased stress on the yeast, in the form of elevated osmotic pressure and ethanol concentrations, stimulates the secretion

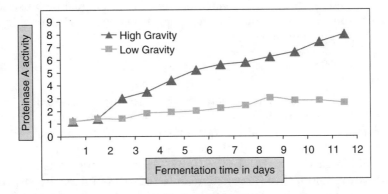

FIGURE 19.4
Yeast proteinase A activity released into wort during fermentation of high (20°Plato) and low-gravity wort (10°Plato).

of proteinase A into the wort during fermentation. Preliminary *in vitro* studies in this laboratory have shown that both ethanol and increased osmotic pressure (simulated using sorbitol that is not metabolized by brewer's yeast) stimulate the secretion of proteinase A by brewer's yeast strains.

Gushing

Excess foaming in a beer is regarded as deleterious and is known as gushing or "wild beer." Gushing is the violent, uncontrolled ejection of beer from the package at the time it is opened and involves the loss of a significant portion of the contents. There are two different classes of gushing: sporadic and epidemic. Sporadic gushing occurs as a result of minor production deviations that are generally difficult to pinpoint. Epidemic or longer term serious gushing may be caused by several factors. Perhaps the most widely discussed cause is the use of weathered (damp) barley. If barley is harvested when wet, fungus infection, particularly *Fusarium*, can develop during the malting process, resulting in beer susceptible to serious gushing. The formation of mycotoxins such as deoxynivalenol (DON) has been paralleled with the development of gushing potential and the screening of barley and malt for these metabolites may offer a means of reducing beer gushing problems.[23] Hydrophobins are small, moderately hydrophobic proteins produced by filamentous fungi. They have been found on the cell walls of hyphae and on spore surfaces, and they can also be secreted into the culture medium. Studies by Sarlin et al.[24] suggest that these proteins, when present, can cause gushing in beer.

Other factors, such as increased levels of carbonation or a carbonating system operating without proper controls, can produce beers that have the potential to gush. Calcium oxalate microcrystals (otherwise known as beerstone) may precipitate in the finished beer and cause gushing. These minor crystals are thought to form nuclei for carbon dioxide gas emissions, but treatment and filtration will overcome this cause of gushing. Excessive levels of iron and other nuclei-forming particles such as sediments also contribute to gushing problems.

Light Stability

Beer is sensitive to light. When exposed to light, especially in the 350–500 nm range, an objectionable aroma known as lightstruck character is produced. The beer is said to be "sunstruck" and the aroma and taste referred to as "skunky". This character is predominantly due to the formation of 3-methyl-2-butene-1-thiol (MBT). This compound has a very low sensory threshold.[25] MBT formation is avoided commercially by using brown glass, which has low light transmittance properties (see Chapter 15), and by replacing natural hops with chemically reduced hop acids that fail to produce MBT (see Chapter 7).

Light instability in beer results from hop components. Hops in brewing have a number of roles: they impart bitterness to beer, provide characteristic hop aromas, suppress growth of certain microorganisms, particularly Gram-positive bacteria, assist in beer foam stability (previously discussed) and contribute polyphenols to the protein/polyphenol complex during wort boiling (previously discussed).

Light of the wavelength of 350–500 nm can penetrate clear and green glass and cause nauseous off-flavors in beers bottled in such containers and in drinking glasses. Researchers have shown, using laser flash photolysis, that the isohumulones, the bitter principles in beer, are decomposed by light-induced reactions and that upon the absorption of visible light, riboflavin is excited and interacts with isohumulones, as well as with oxidized and reduced derivatives thereof.[26]

When the beer is exposed to light, one of the side chains on the hop iso-α-acid is cleaved, and the highly reactive radical that is liberated combines with sulfur-containing compounds to produce MBT (Figure 7.9 of Chapter 7), which has a flavor threshold in the order of parts per trillion, and the skunky aroma is one of the most flavor-active substances in beer.[27] Specialized hop extracts (produced using liquid CO_2 or ethanol as a solvent) have been developed to combat hop sensitivity to light[28] (see Chapter 7). In essence, pairs of hydrogen atoms are catalytically added to the isomerized α-acid. There are three principal types of such extracts (called reduced extracts) currently available on the market: rho iso-α-acid, tetrahydro-iso-α-acid, and hexahydro-iso-α-acid.

All of these materials are bitter to varying degrees, some improve beer foam cling and stability and protect beer against light-struck, sun-induced skunky flavors. Normally, all of these materials are used as a postfermentation addition in order to achieve maximum benefit and optimum utilization. In order to achieve complete light-strike protection, no iso-α-acids can be used in any other part of the process. Even repitched yeast with iso-α-acids absorbed onto their surface will provide sufficient material for photolytic cleavage to occur and the resultant production of MBT.

Summary

There are two major keys to stabilizing beer. The first is to monitor storage temperature, as a relatively small change in storage temperature has a large adverse effect. The second key is to target the lowest practical level of oxygen in the final container, as levels as low as 0.1 mg/l are theoretically sufficient to afford staling potential.[29]

The chemistry of beer instability involves a number of complex reactions involving proteins, carbohydrates, polyphenols, metal ions, thiols, and carbonyls. Our understanding of these reactions has progressed over the past 25 years, but we are far from a complete comprehension of beer instability reaction systems.[30] It is important when evaluating flavor stability to measure the time to appearance of the aged character rather than just the intensity.[31] Research has focused on compounds that develop during aging, and little research has been conducted on compounds that decline in level during this time. Both are very important to the final character of the beer.

References

1. Bamforth, C.W., Beer haze, *J. Am. Soc. Brew. Chem.*, 57:81–90, 1999.
2. Siebert, K.J. and Lynn, P.Y., Mechanisms of beer colloidal stabilization, *J. Am. Soc. Brew. Chem.*, 55:73–78, 1997.
3. McMurrough, I. and O'Rourke, T., New insight into the mechanism of achieving colloidal stability, *Tech. Q. Master Brew. Assoc. Am.*, 34:271–277, 1997.
4. Narziss, L., Centenary review: technological factors of flavor stability, *J. Inst. Brew.*, 92:346–353, 1986.
5. Irwin, A.J., Barker, R.L., and Pipasts, P. The role of copper, oxygen, and polyphenols in beer flavor instability, *J. Am. Soc. Brew. Chem.*, 49:140–149, 1991.
6. Bright, D.R., A Study of the Antioxidant Potential of Speciality Malt, PhD Thesis, Heriot-Watt University, Edinburgh, Scotland, 2001.
7. Dalgliesh, C.G., Flavour stability, *Proc. 16th Eur. Brew. Conv. Congr.*, Amsterdam, DSW Dordrecht Press, 1977, pp. 623–659.
8. Foster, R.T., Samp, E.J., and Patino, H., Multivariate modeling of sensory and chemical data to understand staling in light beer, *J. Am. Soc. Brew. Chem.*, 59:201–210, 2001.
9. Boccorh, R.K. and Paterson, A., Variations in flavour stability in lager beers, Poster 05, The 6th Sensometrics Meeting, Dortmund, Germany, 2002.
10. Barker, R.L., Gracey, D.E.F., Irwin, A.J., Pipasts, P., and Leiska, E., Liberation of staling aldehydes during storage of beer, *J. Inst. Brew.*, 89:411–415, 1983.
11. Lermusieau, G., Noël, S., Liégeois, C., and Collin, S., Nonoxidative mechanism for development of *trans*-2-nonenal in beer, *J. Am. Soc. Brew. Chem.*, 57:29–33, 1999.
12. Nyborg, M., Outtrup, H., and Dreyer, T., Investigations of the protective mechanism of sulfite against beer staling and formation of adducts with *trans*-2-nonenal, *J. Am. Soc. Brew. Chem.*, 57:24–28, 1999.
13. Bushnell, S.E., Guinard, J.-X., and Bamforth, C.W., Effects of sulfur dioxide and polyvinylpolypyrrolidone on the flavor stability of beer as measured by sensory and chemical analysis, *J. Am. Soc. Brew. Chem.*, 61:133–141, 2003.
14. Bamforth, C.W. and Kanauchi, M., Interactions between polypeptides derived from barley and other beer components in model foam systems, *J. Sci. Food Agric.*, 83:1045–1050, 2003.
15. Bamforth, C.W., The relative significance of physics and chemistry for beer foam excellence: theory and practice, *J. Inst. Brew.*, 110:259–266, 2004.
16. Bamforth, C.W., The foaming properties of beer, *J. Inst. Brew.*, 91:370–383, 1985.

17. Lusk, L.T., Goldstein, H., and Ryder, D., Independent role of beer proteins, melanoidins, and polysaccharides in foam formation, *J. Am. Soc. Brew. Chem.*, 53:93–103, 1995.
18. Stewart, G.G., Bothwick, R., Bryce, J., Cooper, D., Cunningham, S., Hart, C., and Rees, E., Recent developments in high gravity brewing, *Tech. Q. Master Brew. Assoc. Am.*, 34:264–270, 1997.
19. Cooper, D.J., Stewart, G.G., and Bryce, J.H., Some reasons why high gravity brewing has a negative effect on head retention, *J. Inst. Brew.*, 104:83–87, 1998.
20. Cooper, D.J., Stewart, G.G., and Bryce, J.H., Hydrophobic polypeptide extraction during high gravity mashing — experimental approaches for its improvement, *J. Inst. Brew.*, 104:283–287, 1998.
21. Leiper, K.A., Beer Polypeptides and Their Selective Removal with Silica Gels, PhD Thesis, Heriot-Watt University, Edinburgh, Scotland, 2002.
22. Kondo, H., Yomo, H., Furukubo, S., Fukiu, N., Nakatani, K., and Kawasaki, Y., Advanced method for measuring proteinase A in beer and application to brewing, *J. Inst. Brew.*, 105:293–300, 1999.
23. Haikara, A., Gushing induced by fungi, *Proc. 6th Eur. Brew. Conv. Symp.*, Symposium on the relationship between malt and beer, Helsinki, Brauwelt-Verlag Nurnberg, 1980, pp. 251–259.
24. Sarlin, T., Linder, M., Nakari-Setälä, T., Penttilä, M., Kotaviita, E., Olkku, J., and Haikara, A., Fungal hydrophobins as predictors of the gushing activity of barley and malt, Am. Soc. Brew. Chem. Annu. Meeting, Santa Ana Pueblo, NM. *Am. Soc. Brew. Chem. Newslett.*, 62:PB15, 2003.
25. Stephenson, W.H. and Bamforth, C.W., The impact of lightstruck and stale character in beers on their perceived quality: a consumer study, *J. Inst. Brew.*, 108:406–409, 2002.
26. Huvaere, K., Olsen, K., Andersen, M.L., Skibsted, L.H., Heyerick, A., and De Keukeleire, D., Riboflavin-sensitized photooxidation of isohumulones and derivatives, *Photochem Photobiol Sci.*, 3:337–340, 2004.
27. Irwin, A.J., Bordeleau, L., and Baker, R.L., Model studies and flavour threshold determination of 3-methyl-2-butene thiol in beer, *J. Am. Soc. Brew. Chem.*, 51:1–3, 1993.
28. Guzinski, J.A. and Stegink, L.J. Stable aqueous solutions of tetrahydro and hexahydro iso-alpha acids, U.S. Patent 5,200,227, 1993.
29. Bamforth, C.W., Enzymic and non-enzymic oxidation in the brewhouse. A theoretical consideration, *J. Inst. Brew.*, 105:237–242, 1999.
30. Bamforth, C.W., Making sense of flavor change in beer, *Tech. Q. Master Brew. Assoc. Am.*, 37:165–171, 2000.
31. Bamforth, C.W., A critical control point analysis for flavor stability of beer, *Tech. Q. Master Brew. Assoc. Am.*, 41:97–103, 2004.

20
Quality

George Philliskirk

CONTENTS

Introduction	730
Quality Management Systems	730
ISO 9000	731
Hazard Analysis Critical Control Point	733
Customer Requirements	737
Systems Integration	738
Materials Control	740
Supplier Relationships	740
Specifications	742
Auditing	743
Raw Materials	747
Other Materials	749
Analytical Control	750
Repeatability and Reproducibility	752
Collaborative Trials	752
Analytical Methods	753
Microbiological Methods	754
Sensory Analysis	757
Difference Tests	758
Descriptive Tests	758
Preference Tests	759
Scaling Tests	759
Drinkability Tests	759
Trained Tasting Panel	760
Customer and Consumer Feedback	761
Goodwill	761
Safety	762
Improvement	762

Process Control 764
   Variation 764
   Beer Quality at the Point of Sale 769
Acknowledgments 769
References 769

Introduction

We can be sure that in every chapter in this handbook the word "quality" will feature prominently at some stage, although the context may appear to be different in many cases. So what, exactly, do we mean by "quality?" Learned texts on the subject, of which there are many, will quote a range of definitions with varying degrees of sophistication and elaboration. My favorite definition (because it is the simplest) is that "quality is meeting the customer requirements." This chapter will focus on meeting the requirements of the customer (or consumer) for beer, be it dispensed into the glass, delivered via a can or bottle in a variety of packaging formats. The emphasis will be on the practical issues related to managing the quality of beer throughout the supply chain, from the purchase of raw materials to presenting the finished product to the customer. As mentioned earlier, there are several excellent books, which will give the reader a detailed insight into the many facets of quality management (*Total Quality Management* by John S Oakland[1] is particularly recommended). However, the scope of this chapter and the nature of this handbook are such that the focus of the content will be on what we actually do in the brewery and packaging hall to deliver quality through the supply chain.

Quality does not happen by chance or because we talk about it. To achieve quality we must work at it by understanding our processes — the work we do every day — and actually improving them. In other words, quality needs to be managed effectively at all levels of organization to meet the requirements of the customer. The customer in this context is not only the end user but also the recipient of a process, product or service throughout the supply chain. This chapter then, will guide the reader through managing quality in the beer supply chain.

Quality Management Systems

A quality management system (QMS) refers to the activities that are carried out within an organization to satisfy the quality-related expectations of its customers. The scope of a QMS may vary considerably from an unwritten, word-of-mouth set of procedures operated in a very small microbrewery to a sophisticated, documented series of manuals, procedures, and interactions. For most larger breweries, the core of most QMSs is the ISO 9000 Quality Systems Standard.

ISO 9000

ISO is recognized as the short name for the International Organization of Standardization, an international agency consisting of almost 100 member countries. The origins of the ISO 9000 standards stem from military specification standards established during the 1940s (US MIL SPEC), and subsequently developed into BS 5750 in the United Kingdom and EN 9000 in the rest of Europe. These latter standards were incorporated into ISO 9000 with the latest revision ISO 9001: 2000 replacing the 1994 versions (ISO 9000: 1994). Organizations adopt ISO 9000 standards for different reasons, but the main objectives are aimed at providing:

- Control of company operations to achieve, sustain and improve the quality of the products
- Assurance to company management that internal controls are effective
- Assurance to the customer that the product conforms to requirements

However, the organization also benefits because use of ISO 9000 serves as a basis to:

- Achieve better understanding and consistency of all quality practices throughout the organization
- Ensure that continued use of the required quality system year after year
- Improve documentation
- Improve quality awareness
- Strengthen organization/customer confidence and relationships
- Yield cost savings and improve profitability
- Form a foundation and discipline for improvement activities within the QMS

Note the emphasis on the word "improvement" in these perceived benefits. A significant shift in the emphasis of the latest ISO 9000 standard has been to move to reduced bureaucracy (documentation) and to focusing on customer needs/satisfaction as a basis for driving improvement activities.

1. Customer Focus
   - Understanding current customer needs
   - Understanding future customer needs
   - Meeting customer requirements
   - Striving to exceed customer expectations

2. Leadership
   - Establishing unity of purpose and direction for the organization
   - Establishing the organization's internal environment

3. Involvement of People
   - Developing abilities fully
   - Using abilities to maximum benefit

4. Process Approach
   - Managing resources as a process
   - Achieving desired results more efficiently

5. Systems Approach to Management
   - Identifying
   - Understanding, and
   - Managing the interrelated processed of a system to effectively and efficiently attain objectives

6. Continual Improvement
   - Making improvements a permanent objective

7. Factual Approach to Decision Making
   - Analysing data and information logically

8. Mutually Beneficial Supplier Relationships
   - Creating value through mutually beneficial, interdependent relationships

FIGURE 20.1
ISO 9000:2000 quality management principles.

Figure 20.1 lists the eight quality management principles identified in ISO 9000:2000.[2] These management principles are reflected in the five clauses of the ISO 9001:2000 Quality Systems Standard. Organizations seeking to demonstrate their conformance to the standard are audited by an accredited third party and if successful, are "registered" to ISO 9001:2000 and receive a certificate that is accepted by customers. However, complying with the ISO 9001 standard does not indicate that every product or service meets the customer's requirements, only that the quality system in use is capable of meeting them.

Although ISO 9001 is usually at the core of a brewery's quality management system, whether it is registered or not, most breweries need to comply with other specific requirements either from regulatory bodies or customers. The legal requirements imposed by government (in the United Kingdom from both the UK government and the European Union) in the area of food safety have increased significantly over the last decade. This legislation has reflected increasing customer awareness of aspects of quality and safely relating to food and drink, from raw materials through production and processing up to and including packaging and point of sale.

Hazard Analysis Critical Control Point

Brewers have addressed these requirements primarily through the application of Hazard Analysis Critical Control Point (HACCP) methods. The key elements of a HACCP system can be listed as follows[3]:

1. Analysis of the potential hazards in a food/drink business operation
2. Identification of the points in those operations where hazards may occur
3. Decisions as to which points are critical to ensure food/drink safety (critical points)
4. Identification and implementation of effective control and monitoring procedures at the critical points
5. Review of the analysis of food/drink hazards, the critical points, and the control and monitoring procedures, periodically and whenever the business operations change

In HACCP terms, hazards are defined as any microbiological, physical, or chemical contaminant which may potentially gain access to the food or drink product and cause harm to the consumer. Although hazards clearly relate to safety, in the brewing context any failure to ensure safety is also a failure in quality and the application of HACCP is therefore seen as an integral part of a QMS.[4]

When carrying out a HACCP program, the principles of HACCP are implemented in the series of stages outlined in Figure 20.2. A detailed review of the application of HACCP in brewing is beyond the scope of this chapter, but several key points are worth emphasizing:

1. *Prepare a Flow Diagram.* The purpose of the flow diagram is to provide a detailed description of the process to help the HACCP team carry out the hazard analysis. The flow diagram is essential to the HACCP team when identifying hazards in the process. The flow diagram should be an activities diagram, showing each process step in the order in which it is carried out, including rework routes. All material additions and services should be shown in the diagram.
2. *Identify the CCPs.* A critical control point is a step or procedure in the brewing process where control is essential to prevent, eliminate, or reduce a hazard to an acceptable level. The World Health Organization (WHO) recommends that CCPs should be determined using the HACCP decision tree (see Figure 20.3).

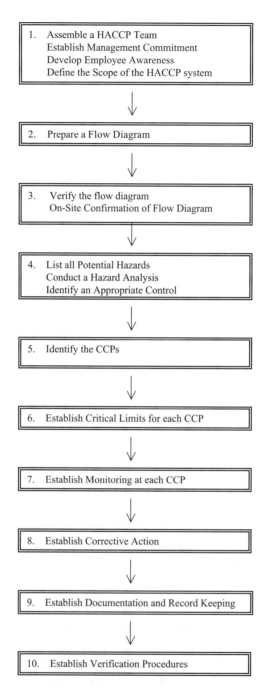

FIGURE 20.2
Stages of HACCP.

Quality

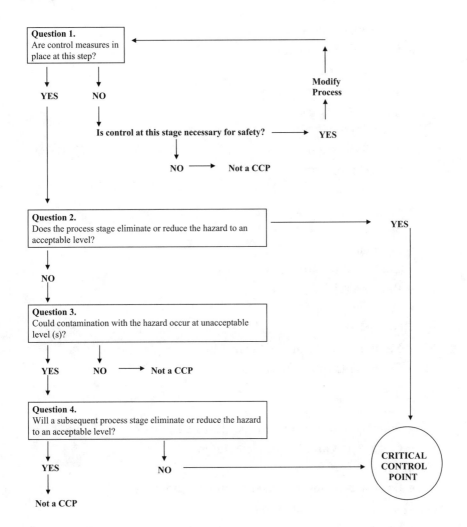

FIGURE 20.3
HACCP decision tree.

Before embarking on a HACCP implementation program, a number of issues need to be addressed as prerequisites for a successful outcome.

1. Plant and building design
   - Site location
   - Site perimeter and grounds
   - Site layout/product flow
   - Building fabric
   - Categorization of risk area

2. Process plant and equipment
   - Plant specification and layout
   - Maintenance
   - Calibration
3. Supplier quality assurance
   - Specifications
   - Meet legal requirements
4. Housekeeping and hygiene
   - Cleaning of process plant
   - Housekeeping (cleaning)
   - Staff facilities
   - Personal hygiene
   - Control of hazardous chemicals
   - Pest control
   - Waste disposal
5. Glass policy
   - Minimize use to prevent contamination
   - Complaints fully investigated
6. Transport
   - Suitable for purpose
   - Loading and unloading procedures
   - Good repair and hygiene condition
7. Training
   - Emphasis on tasks related to quality and safety
8. Quality management system
   - Document control
   - Retention of appropriate records
   - Training
   - Instrument calibration
9. Product recall
   - Batch coding and identification
   - Retain distribution records beyond shelf life
   - Record all health and safety complaints
   - Recall team established
   - Recall procedures in place
   - Communication channels defined

Whilst the issues listed may seem formidable, most breweries will already have in place policies and procedures which cover most if not all of these requirements.

Customer Requirements

It is not uncommon for some customers to impose their own specific requirements for quality and/or safety inspections. Until fairly recently, in the United Kingdom, each major retail customer (particularly the larger supermarket groups) would audit a supplying brewery on a regular basis. This practice was both time-consuming and expensive for both parties involved, and inevitably involved much unnecessary duplication of effort on behalf of the quality assurance (QA) department of the brewery. In the United Kingdom, and elsewhere in Europe, the major retail groups in each country decided to establish a set of common standards against which suppliers would be audited. This abolished the need for auditing by the individual retailers and instead required suppliers to meet the standards demanded by third party auditing of accredited bodies. In the United Kingdom, this has become known as "BRC Accreditation" — this is an abbreviation for the "British Retail Consortium Technical Standard and Protocol for Companies Supplying Retailer Branded Food Products." Note the emphasis on the "retailer branded," that is, those products which carry the name of the retailer, known in the United Kingdom as "own-label." Retailer branded products represent over 50% of all food in the United Kingdom, and although "own-label" beer is a substantially lower proportion of all beer sold, it is still significant.

Under the terms of the UK's Food Safety Act (1990), retailers have an obligation to take all "reasonable precautions" and exercise "due diligence" in the avoidance of failure, whether in the development, manufacture, distribution, advertising, or sale of food products to the consumer. That obligation in the context of retailer branded products includes the verification of technical performance of food (or beer) production sites. To address this obligation, the British Retail Consortium (BRC) developed a technical standard[5] for those companies supplying retailer branded food products.

The standard requires:

- The adoption of HACCP
- A documented quality management system
- Control of factory environment standards, product, process, and personal

The technical standard provides for a certificate of inspection to be awarded at one of two levels: foundation level and high level, the latter demanding additional requirements, particularly in the quality management system. For the inspection process to have credibility, it must be conducted by bodies that are both independent and competent. It is a requirement that inspections using the standard are carried out by bodies formally accredited to the European standard EN 45004 (General criteria for the operation of various types of bodies performing inspection). There are a number of benefits arising from the introduction of the BRC Technical Standard:

- A single standard and associated protocol, allowing inspection to be carried out by inspection bodies, who are accredited against a European standard
- Single verification commissioned by the supplier in line with an agreed inspection frequency, allows suppliers to report upon the current status to those customers recognizing the standard
- The standard is comprehensive in scope covering all areas of product safety and legality
- Within the associated inspection protocol, there is a requirement for ongoing surveillance and confirmation of follow up of corrective actions on nonconformance
- As inspection bodies are accredited against a European standard, there will be future recognition of inspection bodies in countries where product is sourced

Unfortunately, to date, there has been a failure to agree a common international standard for inspection.[6] Despite several years of work to establish a single standard, in Europe four separate standards have been established that are deemed to have the key elements against which all other food safety standards are to be benchmarked: the Efsis Standard; the BRC Technical Standard; the Dutch HACCP audit and the German-based IFS (International Food Standard). To complicate things further, Australia and the United States are working on a standard called SQF2000. Additionally, ISO is working on a food safety standard called ISO 22000, which is due to be ready for 2005. Clearly, with the continuing growth of truly global brewing companies where movement of beer across international borders is an essential part of optimizing the supply chain, it is important that common international standards are established and recognized to avoid unnecessary (and costly) duplication and bureaucracy.

Systems Integration

The ISO 9001:2000, HACCP, and BRC Technical Standard systems exist as stand-alone requirements, despite the fact that all are very closely interrelated. Most breweries are committed to a whole range of standards covering a range of activities deemed necessary to meet legislative, consumer, environmental, and good management practices. For example, at Carlsberg UK's Northampton brewery, the following standards are maintained:

- Quality — ISO 9001
- Environmental — ISO 14001
- Health and Safety — OHSAS 18001
- Climate Change Levy (CCL)
- British Retail Consortium (BRC)
- Investors in People (IIP)
- Integrated Pollution Prevention and Control (IPPC)
- HACCP
- Feed Materials Assurance Scheme (FEMAS)

There is clearly much scope for duplication within these various standards, leading to inefficiencies in utilizing management time in maintaining these standards and in confusion for plant operators when confronted by a whole range of procedures and work instructions allocated to the different standards. Many breweries have addressed these problems by developing integrated management systems (IMS). The key elements of the approach to integration are as follows:

- Look for requirements common to each standard
- Apply a consistent approach
- Avoid duplication
- Adapt existing reports where possible
- Keep it simple!

An IMS is structured along the lines of the ISO standards format with three sections: an operations manual, integrated procedures, and integrated training/operating methods.

The operations manual will encompass:
- Policy statements
- Commitment to the standards
- Outline system description
- Process definitions
- Organizational plan
- Relationship to other parts of the business

The integrated procedures will cover:
- Management of the business
- Defines responsibilities

- Process map
- How the standards are met
- Action summaries from studies
- Supply chain and support processes
- General system procedures

The integrated training/operating methods will ensure:

- One set of instructions containing only key operating requirements
- Multipurpose records
- Training support
- No separate instructions for quality, environment, health and safety, etc.

The structure of an IMS used by Carlsberg UK at its Northampton Brewery is shown in Figure 20.4. Although the scope of the IMS embraces various operational activities, the IMS is often referred to as "The Quality System," in part reflecting how the structures and procedures defined under a quality management system serve as a template for the effective management of the operational activities in a brewery. This systems integration enables ownership and involvement by the teams operating in a particular area of the brewery, allowing easy access for operators through local PCs, minimizing paper proliferation and facilitating operator training. Auditing say, a brewhouse area, covers the full range of activities within that area rather than a specific focus on quality, environmental, or safety procedures.

The QMS establishes the framework on which the process of meeting the requirements of the customer is built. A process is the transformation of a set of inputs, which can include actions, methods, and operations into desired outputs, in the form of the products, information, or services. Clearly, to produce an output that meets the requirements of the customer, it is necessary to define, monitor, and control the inputs to the process. Achieving control of the inputs should result in the desired outputs without recourse to detailed monitoring and inspection of these outputs. This is the goal of modern quality control operations in brewing. We will now look at how this can be achieved.

Materials Control

Supplier Relationships

The materials used in brewing and packaging beer are diverse, from the basic brewing materials of malt, hops, yeast, and water through to the chemicals used in plant cleaning, process gases, plant items, packaging, and

Quality

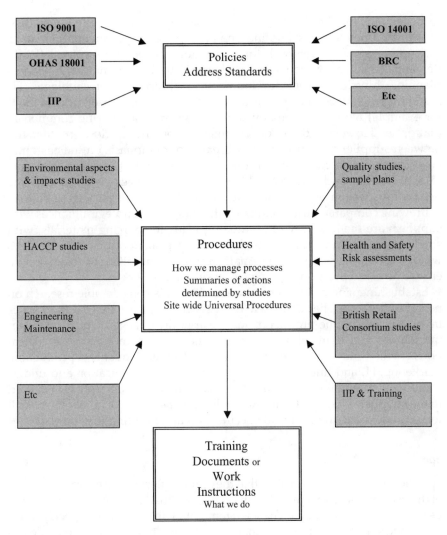

FIGURE 20.4
Structure of an integrated management system.

dispense equipment at the bar. It is essential that in all materials or services purchased, the supplier is capable of meeting the requirements demanded by the user. These requirements do not merely relate to the conformance to specification or price. Factors such as deliveries on time (with the correct quantity at each time), rapid response to programme changes, technical troubleshooting, and product improvement are also important considerations in assessing the worth of a supplier. As brewers increasingly seek to reduce inventories of material stocks, the onus on the supplier increases. In many instances, the supplier (vendor) is given sole responsibility for the delivery or stockholding of materials (malt, bottles, etc.) to the brewhouse or bottling line in response to production demands. This process, often referred to as vendor managed inventory (VMI), demands close liaison between supplier and brewer, particularly between planning functions. The development of information technology (IT) systems greatly facilitates this process, particularly in those systems (SAP, for example) with intercompany communication links sharing common IT platforms. However, these developments demand a close partnership between supplier and customer, with particular emphasis on the customer enjoying complete confidence in the supplier's ability to deliver a product to specification and on time. This confidence reduces or eliminates the need for inspection of materials entering the brewery or packaging hall.

Brewing companies are recognizing that long-term strategic alliances with suppliers are the best way of optimizing value in sourcing materials, with single-sourcing arrangements increasingly prevalent. Gone are the days when the brewer sourced his malt supply from different maltsters to ensure he could even out differences in quality!

Establishing single-sourcing contracts demands considerable research on behalf of both supplier and user. The user must ensure that all the parties involved in using the product or service are fully engaged whilst establishing the contract terms and drawing up the product and service specifications. This may involve purchasing, technical/quality, production, marketing, IT, and finance functions. Clarity of communication and understanding is essential if the needs of the user are to be accurately transmitted to the supplier. From the technical/quality perspective, the involvement will focus on devising the technical specification and auditing the supplier.

Specifications

Specifications must be for real, in that they must reflect the actual requirements of the user and not a theoretical wish list of items that do not relate to the actual use of the product supplied. There is a temptation in drawing up specifications to include specific requirements that are not really required, or to include tolerances on certain parameters that are either out with the process capability of the supplier, or the analytical capability of the measurement (see later sections on process control and analysis), or both.

The simpler the specification the more likely the supplier can deliver the product consistently within specification. More complex specifications usually result in added cost and more disagreements between supplier and user. It is important to involve suppliers in the drafting of specifications. The supplier must be aware of the constraints on the use of the material by the customer, and work with the customer to minimize any possibility of downstream problems caused by material not being fit for the purpose. Regular reviews of material performance are essential between supplier and customer. These reviews not only focus on the results of the analysis of incoming materials and delivery conformance, but also on the downstream process performance influenced by the material. For example, in reviewing malt performance with a malt supplier, the brewery manager will look at brewhouse throughput rates, runoff performance, wort-free amino nitrogen levels, fermentation, and beer filtration performance. The advantage of a single-sourcing agreement is that in the case of malt performance, it leaves no scope for the maltster to blame another supplier!

It is unusual for breweries to routinely monitor the quality of incoming materials. It is a considerable cost to the brewery in both sampling and analysis and, given a close relationship with the supplier, the customer should have confidence in the ability of the supplier to deliver in specification and on time. Occasionally, a supplier may notify the customer of a parameter that may be slightly outside specification. The customer may choose to reject this delivery or accept the delivery as a concession. However, these should be rare occurrences and certainly recorded for review at the supplier meetings. A typical lager malt specification is shown in Table 20.1. The analytical specification lists the parameters to be measured together with the acceptable tolerances permitted. The certificate of analyses section defines the requirements for predelivery analysis by the customer. Section 3, the notes and additional specification requirements, lists further specific requirements from the customer, most of which will be applicable to all breweries. Analysis refers to the standard analytical methods used, whether European Brewing Convention (EBC), Institute of Brewing & Distilling (IBO), or American Society of Brewing Chemists (ASBC) etc., and is important in cases of dispute between maltster and brewer. The section on barley varieties needs to be updated on a fairly regular basis given the turnover of barley varieties available to growers. Note that in this specification, the purchaser will not permit blending of batches made from different varieties — some purchasers are also reluctant to accept blends of the same variety, and this may also be stipulated in the specification. Finally, the age of malt off-kiln is restricted to not less than two weeks — some brewers insist on not less than four weeks.

Whilst the specification is important for both supplier and purchaser in that it defines those attributes of the material that are relevant to the purchaser and capable of definitions, it is of limited value in establishing confidence in a new supplier, particularly a new supplier with no proven track record of supply. A key element in establishing confidence is through the process of auditing.

Auditing

Auditing is an essential part of maintaining and developing a quality management system and in evaluating the worth of suppliers. Audits are mandatory requirements of many certification schemes, such as ISO 9000, and are a way of obtaining objective feedback on how effective the quality system is working, what is working well, what can be improved, and what is not fulfilling the planned levels of performance.

TABLE 20.1
A Typical Lager Malt Specification

Item	Parameter	Unit	Lower Limit	Target	Upper Limit
1.1	Moisture	%		4.5	5
1.2	Extract, fine (dry)	%	81		
1.3	Saccharification time	min		10	15
1.4	Color	EBC		3.2	3.7
1.5	Protein (equivalent to total nitrogen) (equivalent to total nitrogen)	%	9 1.4		
1.6	Total soluble nitrogen (dry)	%	0.56	0.62	0.68
1.7	Kolbach index	%	Not specified		
1.8	Diastatic power (dry)	WK units	250		
1.9	Wort pH	%	5.8		
1.10	Beta-glucan in wort	mg/l			200
1.11	Friability	%	80		95
1.12	Partly unmodified grains	%			4.5
1.13	Glassy grains	%			2.5
1.14	Modification (Carlsberg)	%	85		
1.15	Homogeneity (Carlsberg)	%	65		
1.16	Dust (<0.25 mm)	%			0.5
1.17	Dust and extraneous material (2.2 mm screen)	%			2
1.18	DMS precursors	mg/kg			4.5
1.19	Arsenic	mg/kg			0.5
1.20	Lead	mg/kg			1
1.21	Nitrosamines	ug/kg			2.5
1.22	Ochratoxin A	ug/kg			3
1.23	Gushing	g/bottle			0
1.24	Foreign seeds				None

Certificates of Analysis

    Parameters 1.1 to 1.7 — to be included on certificate for each shipment prior to delivery.
    Parameters 1.18 to 1.23 — to be included on certificate for each shipment prior to delivery, or, for regular suppliers:
        Nitroamines — quarterly (direct fired kilns only)
        DMS precursors — quarterly
        Ochratoxin A — monthly
        Gushing — monthly or as agreed (if low risk of Fusarium in growing area, otherwise each batch).
        Pesticides and heavy metals in barley — monthly.

Notes and Additional Specification Requirements

    The use of insecticide in direct contact with barley is not permitted.
    The malt must be free from any genetically modified material.
    Neither the barley used for malting nor the malt is to be exposed to attack form molds, insects, or other pests.
    All analyses are on samples as delivered and (except for moisture) on the basis of dry matter.
    Analysis of malt batches for delivery are to be sent to the receiving brewery prior to delivery with any results outside specification highlighted. In addition, details (including the proportion and relevant analysis) are to be included of any component

(Continued)

TABLE 20.1 *Continued*

batch in a blend which does not comply with the specification, even though the blended batch itself complies with the specifications. Each batch, or component batch of a blend, is to be identified by a batch code and off-kiln date.

Malt batches that do not comply with the limits of the specification are not to be delivered without the approval of the receiving brewery.

Analysis
Use EBC methods, except where otherwise stated, for check analysis. In case of dispute with a supplier of purchased malt, use EBC methods or other such methods as may have been agreed with the supplier. In case of legal dispute, use the legally approved method.

Barley Varieties
Permitted varieties: Halcyon, Optic, Fanfare, Gleam, Chariot, Regina, and Pearl
Other varieties may be tested for approval by breweries as they become approved for malting through the NIAB/IOB procedures.
The variety of barley malted, its crop year and place of origin are to be declared on the analysis certificates.
Blending of malt batches made from different varieties is not permitted.

Storage of Malt After Kilning
Malt is planned to be used not less than 2 weeks off kiln.
The age of malt off kiln is to be declared on the analysis certificate for any batch or component of a batch to be offered for delivery at less than 14 days off kiln.

Any audit must be measured against the requirements of the system. In the case of ISO 9000, the five clauses of the standard define the requirements of the system, and regular structured internal and external audits are required to test compliance to the standard.

The auditing of a supplier is usually on a fundamentally different basis. Instead of a formal audit of the QMS, the auditor will usually focus on those elements of the supplier's activities that relate directly to the quality of the product or service supplied. As mentioned earlier, compliance to the ISO 9000 standard does not necessarily guarantee that the product supplied will meet the requirements of customer (although it is a good starting point!). The requirements of a supplier audit are usually defined by the purchaser, and the scope and frequency of the audits will normally be based on the risk elements to the purchaser. For a new supplier, the initial audit is very rigorous, involving a thorough examination of the supplier's processes, the quality management system in general, and the state of housekeeping and hygiene. The result of the audit will usually be either:

- Approved
- Approved with comments
- Not approved

This first audit, or approval audit, should normally be carried out ahead of the first purchase from the specific plant. A number of stages are usually involved in preparing for this first audit:

1. Completion of a technical questionnaire that will usually focus on questions on:
   - The quality management system, including registration and certification to ISO 9000, HACCP, BRC, etc.
   - Process plant description and flow diagram
   - Quality control (QC) procedures and analytical checks and facilities
   - Specific food safety related issues such as containment checks and product recall procedures
   - Management responsibility
2. Submission by the supplier of representative samples of the product to be supplied together with the suppliers own analytical results. These will be checked against the purchaser's analyses of the samples.
3. The lead auditor (there will usually be at least two auditors for an approved audit) will agree on the audit date, inform the supplier of the scope of the audit, and issue an agenda.
4. Auditors should be chosen with (i) appropriate experience of the material or process to be purchased, (ii) detailed knowledge of quality management systems and auditing, and (iii) practical brewing experience. The audit team should communicate ahead of the audit by meeting or corresponding on how to proceed at the audit and discuss the information already available. The lead auditor will be responsible for bringing along relevant documentation and checklists to be used.

At the audit, the agreed agenda is followed. The audit tour of the plant and laboratories is the most important part of the audit, and time is allocated accordingly. The auditors will meet after the tour and the findings from the audit are discussed, including the necessary action points. On the basis of the findings, it is judged whether the plant can be approved, approved with comments, or not approved. The lead auditor will present the main findings and conclusions at the closing meeting with the suppliers. All required action points must be addressed and agreed with the supplier.

Audit criteria for the assessment of the audit findings need to be defined; typically, the findings will fall into three categories: satisfactory, not quite satisfactory, and not satisfactory.

If any of the activities is judged to be "not satisfactory" the plant is "not approved." If any of the areas (process, QMS, or housekeeping/hygiene) are deemed "not quite satisfactory" the plant is "approved with comments," and the not quite satisfactory issues need to be addressed. Plants that are "approved" without any comments may have some minor issues raised.

Any action points arising from the audit should be addressed by the supplier and followed up by the auditors within a reasonable timescale. Once a new supplier or plant is approved, it is normal to submit the supplier initially to higher levels of surveillance on incoming materials and product performance than for well-established suppliers with a proven track record. It is again important to stress the value of regular review meetings and customer feedback to the supplier. Thereafter, suppliers will be subjected to routine audits on an annual or biannual basis, depending on the importance of the material purchased and the degree of risk of failure. The routine audit will follow the same format as the approval audit, excepting that the audit team will often be fewer in number and the areas covered will be less extensive.

The results of the audits will be disseminated to all interested parties. In larger, often multinational brewing companies with international sourcing of materials and services, the data from these audits are often collated and the database shared between breweries. Carlsberg breweries have such a database known as Carls Audit, which provides information on actual or potential suppliers for materials for Carlsberg and Tuborg production. It is primarily used for entry of supplier data and audit reports, and it is the responsibility of the users, that is, the breweries, to keep the database updated with their current suppliers. Audit reports on the suppliers are filed in Carls Audit and can be shared with other Carlsberg brewery users, often preventing unnecessary multiple auditing of the same supplier.

In summary, auditing of suppliers is a key aspect of establishing confidence in the supplier and minimizing the need for inspection at the point of supply or use. Additionally, feedback through the audit process can help the suppliers to improve their own performance — although auditing can sometimes be painful to the supplier, it is also a potential source of free consultancy!

Raw Materials

The basic raw materials used in brewing are malted barley, water, hops, and yeast. Additional sources of fermentable extract may be used to augment or replace the malted barley, depending on beer style, tradition, or economics. The malted barley and hops are, in particular, subject to biological variability created by changes to the growing season and the ongoing development of new varieties designed to deliver improved agronomic performance to the grower, whilst ensuring that the requirements of the processor and end-user (brewer) are met. This biological variability is an important factor to be considered in drafting specifications, particularly for malted barley. Despite the best endeavors of maltsters and brewers, different malt batches that conform to the agreed specification may perform differently in the brewhouse. New varieties and new season barleys should be assessed carefully before introduction, and both the malting process and brewhouse procedures may need to be manipulated to deliver a consistent output. It is important that maltster and brewer work closely together to minimize the impact of any changes, and a review of the malt specifications is needed from time to time to ensure that it is relevant to the brewer's needs.

Seasonal and varietal changes in hops also need to be taken into account, particularly where the contribution of hops to the aroma characteristics of a beer is considered important. Measurements of α-acid content of hops from harvesting through storage (particularly in the presence of air) show an almost linear decline in alpha levels, but also increases in the so-called hard resins, which can lead to the development of unacceptable "cheesy" aromas. The hop storage index (HSI) is a useful measurement of these changes in storage. Processed hops in which the α-acids are isomerized and extracted or the essential oils extracted tend to enjoy greater stability with storage time. Although specifications for hop products used for

TABLE 20.2

Typical Specifications for Hop Brewing Pellets

Specification	Limits	Method of Analysis
Moisture, %	Max. 9.0	EBC 7.2 (most recent)
α-Acid, % w/w		EBC 7.7 (most recent)
Cohumulone, %	Max. 35 of total α-acids	EBC 7.7 (most recent)
Pesticides	EC directives (90/642/EEC) and (2001/48/EC)	
— for U.S. grown hops	U.S. regulations	
Heavy metals, mg/kg	Lead max. 2.0	
HIS before shipment	Declared for each batch	ASBC Hops-12

	Growing Area		
Variety	United States (Washington, Oregon, Idaho)	Germany (all areas)	Other
Approved varieties (× = Approved)			
Brewers Gold			Turkey
Magnum		×	
Millenium	×		
Northern Brewer		×	Belgium
Nugget	×	×	Spain
Pheonix			England
Pilgrim			England
Super Styrian			Slovenia
Taurus		×	
Warrior	×		
Zues	×		

Delivery
Packaging: Alufoil soft packs, air and aroma tight. Packed under inert atmosphere.

Quality

bittering purposes can be reasonably well defined in terms of variety, source, and α-acid content (Table 20.2), assessing the quality of hops used in late-hopping or dry-hopping of cask beers is more problematic in the absence of sound analytical measurements. Again, close liaison between hop supplier and brewer is demanded with auditing of processing plants and hop stores needed to monitor the changes in quality.

The presence of unauthorized pesticides and herbicides possibly used in the growing of hops should be monitored carefully, as also the possibility of unacceptably high levels of approved substances. Whilst the brewer will define these limitations in the specification, and the hop supplier will issue the appropriate certificate of analysis confirming conformance at the time of delivery, it is incumbent on the brewer to periodically send samples of the delivered hops or hop products to what is usually a specialist laboratory for independent verification. These checks would normally form part of a HACCP programe.

Water comprises about 94% of the content of beer and can exert a significant influence on the quality of beer (see Chapter 6). In addition to the effects of brewing liquor on brewing performance and flavor, the quality of the water is a significant contributor to the safety and wholesomeness of beer. Water quality is usually carefully regulated through national legislative controls, and the water used directly in brewing (mashing, sparging, dilution) should conform to these regulations. Where breweries purchase water, from either municipal supplies or utilities companies, certificates of analysis are provided on a regular basis to confirm compliance to the agreed standards. Where breweries utilize their own sources of water (often boreholes), the quality of the water must be monitored rigorosly and, if necessary, treated to attain the required standards. There are potentially four major classes of contamination found in water[7]; microbiological, inorganic, trace organic substances, and pesticides/agrochemicals. The implications of contamination by these four groups of materials are discussed in detail in Chapter 6.

An effective yeast management system is a major requirement for brewing beer of consistently high quality. Chapter 8 covers this topic comprehensively, but it is worth emphasizing the importance of a structured yeast management programme that accommodates strain maintenance, propagation techniques, contamination checks, viability and vitality, DNA fingerprinting techniques (if available), and procedures for harvesting, storing and repitching the yeast in process. The avoidance of stress factors (high-gravity fermentations, shear forces, high storage temperatures, nutrient deficiencies, and high generation numbers) needs to be carefully managed.

Other Materials

Although malt, hops, water, and yeast can be described as the basic raw materials of brewing, control of which is essential to delivering beer of a suitable quality, many other materials need to be purchased and controlled, applying the same principles outlined for malt and hops. Detergents, process gases, process aids, additives, packaging materials, etc. must be specified, suppliers audited and approved, and performance monitored in brewery and packaging hall. Packaging materials, (bottles, cans, crates, kegs, cartons, etc.) constitute a significant cost element in beer and impact not only on production efficiencies but also on final presentation to the customer or consumer. The need for well-defined technical specifications, agreed with the supplier, is essential.

Analytical Control

The emphasis on the role of analytical control has shifted significantly over the past 25 years or so in brewing operations. Traditionally, brewers employed large numbers of chemists and microbiologists to sample and measure at every stage of the brewing process, from raw materials at intake through to inspection of finished product in keg, bottle, or can. These inspections were borne out of a lack of confidence — in the supplier to deliver materials to specification, and in the ability of the process to control the operation. Despite these inspections, which spawned large QC departments (at large expense), the costs of internal failure (rework, disposal) and external failure (returns, customer complaints, lost business) were high.

The development of QMSs, with an increasing emphasis on controlling the inputs to a process rather than inspecting the outputs, have reduced considerably the costs of both the inspection (QC departments) and the internal and external failures. However, there is still a need for measurements at various stages of the brewing and packaging processes to confirm that the process is still in control. Ideally, these measurements or analyses should be made in-line, with suitable feedback control for process adjustment, or on-line, where the operator can directly intervene to adjust the process if necessary. For more sophisticated measurements or nonroutine checks, a central laboratory function needs to be available. This can be either an in-house facility or, increasingly, outsourced to an accredited external laboratory service provider. Confidence in the accuracy and reliability of the measurements, be they at the time or in a central laboratory, is essential. Rigorous equipment calibration procedures need to be in place (a key element in the ISO 9000 standard) supported by regular monitoring between reference laboratories. This regular monitoring and critical appreciation of laboratory performance enables a realistic assessment of marginal results to be made and the source of variation can then be ascertained — process error, sampling error, or laboratory error. Obtaining a representative sample needs careful thought and consideration, particularly for large bulk solids such as barley, malt, and hops. For example, when sampling a wagon containing malt or barley, the number of sampling points recommended can vary depending on the size of the wagon.[8]

Sampling of liquids (beer) is generally much easier, but factors such as carbonation, layering, and the position of the sample point need to be considered. Laboratory errors can arise from inherent errors in the methods, equipment, and personnel conducting the test. The methods used should be both accurate and precise, where accuracy is defined as the closeness to a true or conventionally accepted value, and precision describes the amount of variation in a method. Classically, this relationship between accuracy and precision is best demonstrated by considering a number of arrows aimed at a target. The degree of scatter of the arrows in the target is referred to as precision, whilst their overall closeness to the centre is known as accuracy. Figure 20.5 illustrates this point.

Any analytical method should be capable of giving an accurate measure of the true value every time a measurement is made. Checks on the accuracy and precision of analytical methods are made by comparing the results of the analysis of a sample both within and between laboratories to establish the repeatability ($r95$) and reproducibility ($R95$) of the method.

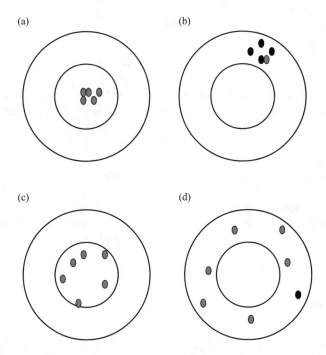

FIGURE 20.5
Accuracy and precision. (a) Accurate and precise, (b) precise but inaccurate, (c) accurate but imprecise, (d) inaccurate and imprecise.

Repeatability and Reproducibility

Checks within the same laboratory, that is, the repeated analyses by the same analyst, on the same equipment, on the same sample, at the same time, give a measure of the repeatability of that method. The $r95$ value relates to a range of results that fall within the stated range 95% of the time, that is, 19 out of 20 analyses should lie within this range.

Reproducibility ($R95$) relates to checks between laboratories, that is, the repeated analysis of same sample by different analysts using different equipment, at different times, gives a measure of the reproducibility of that method. Again, the 95% rule applies.

Inevitably, the values of $R95$ will always exceed those of $r95$. For example, mean precision values for bitterness measurements (BU) were determined by 16 laboratories in a major UK brewing company using 33 sample pairs over a 3-year period. Within a range of bitterness values from 18 to 32 BU, $r95$ was 1.0 and $R95$ 4.1. Knowledge of the $R95$ values of an analyte is particularly important in establishing specification limits. The specification must take into account this potential interlaboratory variation as well as sampling errors and the process capability (see later) of the supplier.

Collaborative Trials

Collaborative trials between two or more laboratories are frequently held to check on the reliability of methods between different laboratories and to identify any problem area that can be resolved by further development. The protocols for these trials are established as an international standard (ISO 5725), and the guidelines are incorporated into the recommended methods for brewing analysis. Many large brewing companies maintain their own ongoing collaborative schemes (Carlsberg Breweries have a scheme called CILAS — Carlsberg Inter Laboratory Analysis Scheme), but one of the major open schemes is run jointly from the United Kingdom by Brewing Research International (BRI) and the Laboratory of the Government Chemist (LGC). This is known as the Brewing Analytes Proficiency Scheme (BAPS) and participants include breweries from around the world. Samples of a single beer are dispatched every four weeks to the participants and a range of analytes are covered, all on an optional basis. These include:

- Alcohol, original gravity, present gravity
- Bitterness
- Color
- pH
- Haze
- Carbon dioxide
- Diacetyl (2,3-pentanedione)

- Dimethyl sulfide
- Fusel alcohols
- Total soluble nitrogen and free amino nitrogen
- Head retention value (Rudin, Nibem)
- Sulfur dioxide
- Chloride, sulfate, nitrate, phosphate

Results from each "round" of samples are collated and analyzed statistically and each participant is given a coded number to assess their performance against the group as a whole. Breweries are often obliged to notify their customers of their "BAPs number" to enable the customer to check on the analytical capability of the supplier. Confidence in the supplier established through this scheme will often enable the customer to minimize or obviate totally any analysis on incoming product.

Analytical Methods

It is beyond the scope of this chapter to detail all the methods available for the analysis of raw materials, beer-in-process, coproducts, finished beer, and packaging used in brewing. Brewers have been fortunate in the availability of methods used in brewing, with methods from:

- The Institute of Brewing and Distilling (IBD — formerly known as the IOB methods)
- The European Brewery Convention (EBC)
- The American Society of Brewing Chemists (ASBC) and
- The Methodensammlung der Mitteleuropaischen Brautechnischen Analysen Kommission (MEBAK)

However, with the growth of major international brewers over the past 10 years, represented in areas of the world traditionally using the "local" methods, there is clearly a requirement to develop a single international methodology for brewing analysis. In recent years, the IBD and EBC have been collaborating on developing a single set of methods that hopefully will ultimately embrace ASBC and MEBAK methods.[9]

Certain beer quality parameters are common to almost all breweries,[10] and are as follows:

- Original gravity (OG)
- Present gravity (PG)
- Alcohol
- Color
- Bitterness

- Haze
- pH
- Head retention
- Carbon dioxide
- Dissolved oxygen

Microbiological and flavor evaluation checks are considered in the following text.

The extent to which these basic quality parameters and other checks are employed in the brewing process will depend on the availability of resources (equipment, trained staff) and the confidence in the materials supplied and in the control of the process. One suggested brewing process analysis program is as shown in Table 20.3. The frequency of sampling and testing will vary, depending on the need for immediate, in-process corrective action, or as indicator of trends. For example, of the parameters listed in Table 20.3, some may need checking on every batch (alcohol, color, bitterness, carbon dioxide, dissolved oxygen, pH) whereas others may require less frequent (often weekly or monthly) checks (free amino nitrogen, volatiles, malt extract).

Microbiological Methods

A detailed account of the microorganisms of relevance to brewing is given in Chapter 16. Brewers are fortunate in that beer provides a hostile

TABLE 20.3

Brewing Process Analysis Program

Character	Materials	Wort	Fermentation	Maturation	Bright Beer
Alcohol	No	No	Yes	Yes	Yes
Original gravity	(Malt, extract)	Yes	Yes	(Possible)	(Possible)
Present gravity	(Malt, extract)	Yes	Yes	(Possible)	(Possible)
Color	Yes	Yes	Yes	Yes	Yes
Bitterness	(Hops)	(Possible)	(Possible)	Yes	Yes
Haze/clarity	(Water)	(Possible)	No	No	Yes
pH	(Water)	Yes	Yes	Yes	Yes
Head retention	No	No	(Possible)	Yes	Yes
CO_2	No	No	No	No	Yes
Dissolved oxygen	No	(Cold wort)	No	Yes	Yes
TSN/FAN	Malt	Yes	No	No	No
Attenuation limit	No	Yes	Yes	No	No
Yeast count	No	No	Yes	(Possible)	No
Microbiology	(Water)	(Cold wort)	Yes	Yes	Yes
Taste	Yes	(Possible)	Yes	Yes	Yes
Volatiles	No	No	Yes	Yes	(Possible)
Diacetyl	No	No	Yes	(Possible)	(Possible)

Source: Adapted from O' Rourke, T., *Brewers' Guardian*, 129:21–23, 2000.

environment for the growth or survival of potentially pathogenic organisms. However, microbial contamination of beer, or at any stage of the brewing and packaging processes, can have serious consequences for beer quality and cost.

The raw materials used in brewing (malt, cereals, sugars, hops, and water) are potential sources of microbes, although they are usually only relevant through to the stage of wort boiling which effectively sterilizes the wort. However, the presence of some fungal contaminants on malt and some cereals, notably *Fusarium* species, can lead to the development of "gushing" in finished beer and the formation of mycotoxins, compounds considered to be potentially carcinogenic. Although some mycotoxins can survive the brewing process, the concentrations are greatly reduced. "Gushing" tests on malt are available and provide a reasonable indicator of potential problems. It is usual for malt specifications to include reference to this test (Table 20.1). Table 20.4 summarizes the possible occurrence of microbial spoilage organisms at different stages of the brewing process and their possible consequences.

The elimination of microbiological spoilage potential is achieved by the rigorous adherence to several key principles:

- Plant and equipment design and layout to facilitate ease of cleaning
- Processes for effective plant cleaning and sterilization

TABLE 20.4

Microbial Spoilage Organisms in the Brewing Process

Stage	Organisms Found	Possible Effects
Mashing	Heat-tolerant lactic acid bacteria	ATNC[a] development
Cooled wort	*Obesumbacterium proteus*, Enterobacteria	Off-flavors ("parsnips"), ATNC
Pitching yeast	Wild yeasts, *O. proteus*, *Lactobacillus*, *Pediococcus*, acetic acid bacteria	Off-flavors, diacetyl, abnormal fermentations
Fermentation	Wild yeasts, *Lactobacillus*, *Pediococcus*	Off-flavors, diacetyl, abnormal fermentations
Maturation/storage	*Pediococcus*, *Lactobacillus*, *Zymomonas*, *Pectinatus*, Wild yeasts	Sour taste and off-flavors with sulfury aromas, Diacetyl, Turbid beers
Bright/packaged beer	*Pediococcus*, *Lactobacillus*, *Zymomonas*, *Pectinatus*, Wild yeasts	Sour taste and off-flavors with sulfury aromas, Diacetyl, Haze
Cask-conditioned beer	Acetic acid bacteria; Wild Yeast; *Lactobacillus*; *Pediococcus*; *Zymomonas*	Aldehydic and acid beers Sourness. Haze, Off-flavors, and aromas

[a] ATNC, apparent total nitroso compounds.

- Trained operatives and sound operational procedures
- Effective pasteurization or sterile filtration techniques
- Effective monitoring of microbial quality

The principles applied in the use of HACCP techniques are equally useful in identifying and nullifying the risks of microbial contamination (Chapter 16). In addition to the potential risks from brewhouse materials mentioned earlier, a number of other additions to the process can serve as sources of microbial contamination after wort has been boiled and cooled, namely:

- Recovered beer
- Dilution liquor
- Process gases (oxygen, nitrogen, and carbon dioxide)
- Finings
- Priming sugars
- Filter aids
- Foam enhancers

Each must be assessed, and the microbiological risks eliminated. Inadequate pasteurization regimes are often the cause of microbial problems in packaged beer, particularly in keg beers. Although these problems can be related to the incorrect application of the appropriate pasteurization units (PUs), the most usual cause is leakage in the cooling side plates of the pasteurizer. Modern plate-heat exchangers usually incorporate pressure differential systems to prevent leaks (usually caused by pinholes in the plates) that lead to contamination of the beer stream with unsterilized water, beer, or coolant.

Beer packaging, notably into kegs and casks should be carefully monitored to ensure that both the containers and filling heads are properly cleaned and sterilized. Similarly, filling heads on bottles and can lines must be maintained and cleaned thoroughly; although correct pasteurization of the filled bottle or can will usually prevent microbiological problems.

Where microbiological problems do occur in brewing and packaging, it is usually worthwhile to complete a thorough microbiological audit of the process. The general aims of the audit can be summarized as follows:

- To improve the quality of the beer
- To identify (and remove) potential sources of contamination
- To check on the adequacy of cleaning and sterilizing regimes
- To assess the procedure for sampling and testing

This last part is important in that much routine microbiological testing is either worthless, unnecessary, or inadequate. An audit is an opportunity to revisit current procedures and determine if they are really necessary. Many tests have been introduced on an *ad hoc* basis as a result of a specific problem that may have occurred years earlier and is no longer relevant.

The methods used for identifying and enumerating the microorganisms found in brewing are well documented in the various recommended methods of analysis (IBD, ASBC, EBC, and MEBAK). Of particular interest in recent years has been the development of "rapid methods" designed to produce results either instantly or within several hours rather than the more traditional methods that can take several days or even weeks. These rapid methods include the use of ATP bioluminescence for detection of microbial contamination and various molecular biology techniques, largely founded on the polymerase chain reaction (PCR), which are used to amplify specific DNA sequences representative of the various spoilage organisms (see Chapter 16).

In addition to the use of PCR in microbiological monitoring, it can also be applied to the detection of genetically modified (GM) raw materials that could be used in brewing. In the EU, foods that contain or are made from GM organisms are required to be labeled. Using Q-PCR it is possible to measure accurately the levels of, for example, genetically modified maize in a consignment used for brewing.

Monitoring microbiological contamination by these rapid methods will develop extensively over the next few years as techniques are refined and simplified, and costs of reagents and equipment reduced to economic levels. Although larger breweries may well embrace this technology, smaller breweries may take advantage of the accredited laboratories that are able to offer this service.

Sensory Analysis

The consumer's perception of beer quality is usually based on a reaction to a complex mix of expectations, which are associated with the effects of:

- Beer style (ale, lager, stout, wheat beer, etc.)
- Branding/advertising
- Color
- Clarity
- Foam
- Flavor and aroma
- Temperature
- Beer glass
- Carbonation/nitrogenation
- Mouthfeel

Many of these perceptions are outside the control of the brewer, but for those factors directly influenced by the brewing and packaging processes, the control of beer flavor and aroma are the most significant. Whilst the various analytical and microbiological methods referred to earlier provide fairly objective tools for identification and quantification, the very nature of the differing responses by the senses to different flavors and aromas makes this identification and quantification difficult.

Despite this fundamental limitation, brewers and flavor analysts have developed robust procedures that enable sensory analysis to be a valuable tool in the monitoring and control of beer quality. There are five basic types of flavor evaluation methods[11]:

- Difference tests
- Descriptive tests
- Preference tests
- Sealing tests
- Drinkability tests

Difference Tests

These are used by trained tasters to establish if there is a difference between one or more samples. In the triangular taste test, three beers are presented, two of which are identical and the third is from a different batch. In the absence of a difference, 33% of the answers will be correct — statistically significant deviation from this percentage indicates a difference between the beers. Similarly, the duo–tri test also uses three samples but here a control beer is used as a reference and is tested against the same control beer and a test beer; the taster must match the reference. In the absence of a difference, 50% of the answers will be correct. Difference tests are useful for monitoring consistency in a beer and if changes to the process or raw materials (new hop, for example) have affected beer quality. Where the same brand is produced at more than one brewery, this test is used to check for differences in the beer.

Descriptive Tests

These tests rely on highly trained assessors to estimate the flavor of beers using an established vocabulary of flavor terms. These terms were agreed internationally during the 1970s to describe all the characteristic flavors found in beer. Each flavor was given a specific name and a chemical was assigned to each character to act as a standard reference. For example, a floral (rose) aroma is referenced to 2-phenylethanol; a peardrop or banana aroma is referenced to isoamyl acetate. This vocabulary is characterized by the Beer Flavor Wheel.[11] Descriptive tests can be used in establishing the flavor profile of a beer or in trueness-to-type tests.

Preference Tests

The need to understand the likes and dislikes of consumers is critical. Preference tests, in which consumers are asked to compare two beers, express a preference, and sometimes comment on any perceived character ("too bitter," "too gassy"), are used to monitor a beer's performance in the market against its competitors.

Scaling Tests

These tests are designed to quantify or rank particular flavor attributes (or defects) in a beer. Trained tasters are used.

Drinkability Tests

Most sensory analysis involves the smelling and drinking of relatively small quantities of beer. However, consumers will usually drink significantly greater volumes of beer and this can be used as a measure of acceptability or "drinkability" of a particular beer. In consumer tests where beers are being compared and tasted "blind" over a protracted session, the actual volume of beer consumed is a good indicator of the preference of the consumer.

Consumer tests are an important aspect of determining the customers' requirements and, as such, are an integral element of the ISO 9000 standard. These tests can be used to monitor a beer's performance in the market against its competitors, assess the changes in recipe formulation, or in the evaluation of a new beer. Consumer research is expensive — large numbers (over 300) of consumers are usually required, often of defined gender and age profile and in distinct regions. Those selected and willing to participate (the rewards usually include free beer, a taxi home, and a cash payment) are invited to a location and are given a selection of beers to drink. The participants are asked to fill in a questionnaire at varying stages of the session to assess the initial reactions to the beer and the reactions after consuming several glasses. Preferences are sought and the reasons for the preferences are elucidated through the questionnaire, which will usually also include other relevant demographic and buying habit questions.

A key element in all comparative tests is the establishment of trueness-to-type of the test beers. This is particularly relevant in large-scale consumer tests where competitors' beers are being assessed. Whilst familiarity with one's own beer facilitates ready identification of trueness-to-type, with competitors' beers this is considerably more difficult. As well as analytical compliance to specification, a trained panel of tasters must be used to ensure that the beers are "normal" and that no unusual defects are present.

Often sensory data from a trained panel is superimposed on consumer research results in what is termed a preference map (Figure 20.6). Information is obtained from this map on the relationship of a beer's current

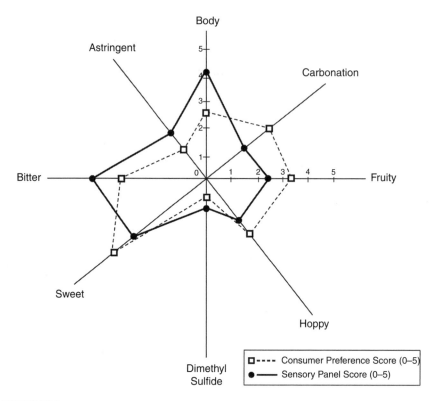

FIGURE 20.6
"Spider diagram" showing the results of a sensory analysis of a sample of beer against a consumer research preference model with a 0–5 intensity scale.

sensory characteristics to the consumer palate. This map could be used to align the beer with consumer preferences by modifying the beer (ingredients, processes, and specifications) to maximize customer acceptability.

Trained Tasting Panel

Sensory analysis depends on the sensitivity of the sensory panel. Tasters are selected on the basis of their ability to discriminate between certain flavors and aromas. Not everyone can demonstrate the required level of sensitivity, and those that can, require training in identifying and naming the range of characteristics found in beer. This process can take several months before a taster qualifies for membership of the taste panel. Subsequently, taste panelists should be periodically exposed to standard flavors and smells to test acuity. The role of the taste panel should be compared to that of an analytical instrument and, as such, it should be subject to calibration exercises and collaborative testing.

Quality

Trained tasters are used in in-process quality assurance and in testing the final product. The former may involve a simple go/no-go for a beer movement and is often carried out on the line. Final product tasting needs to be more rigorously controlled. Tasting rooms should be specifically designed for the purpose, with the following basic conditions:

- Quiet
- Dim light
- Separation of tasters, preferably in booths
- Good ventilation and air conditioning
- Odor free (no smoking or extraneous smells)
- Tasters not to use perfumes
- Tasters not to have consumed food or drink immediately prior to the session

The final product tasting is an essential part of the product release. An unsatisfactory beer should not be released to trade. Data from taste panels can be used to establish trends in the changes in beer flavor and aroma. Software is now available to handle large amounts of sensory data and interpret and present the results in a meaningful format.

Customer and Consumer Feedback

Whilst one of our goals in meeting the customers' requirements is a total absence of defects, it is unreasonable to pretend that this is achieved. Errors and mistakes do occur in servicing the needs of the customer and consumer, and where these errors do occur, it is important that we have systems in place that can react appropriately to the complaint or feedback generated. These systems have three key objectives:

1. To maintain the goodwill and loyalty of the customer/consumer
2. To use the information gathered to quickly identify product faults that may be potentially injurious to the consumer
3. To use the information gathered and subsequent investigations to generate improvements in the product or service

Goodwill

A complaint that is handled promptly and efficiently is usually appreciated by the customer. Although fraudulent claims do sometimes occur, the majority of complainants do have issues that they perceive as not delivering the quality promise of the supplier. The advent of telephone "helpline" or "careline" numbers on the package facilitates ready access to suppliers by the consumers, who are learning to complain more vociferously and effectively. It is essential that in responding to these complaints, either by telephone, letter, or e-mail, the business has procedures in place that address the needs of the customer. Of particular importance is the selection and training of the staff who will be communicating directly with the complainant.

Safety

Brewers are obliged to deliver to customers and consumers products that are safe and wholesome. Failure to do so could result in prosecution by regulatory authorities and by the customer for damages as a result of injury or ill-health. Additionally, the publicity generated by such incidents could seriously damage the brand and manufacturer concerned. It is incumbent on suppliers to monitor complaints by customers and consumers, to identify in the market the product that may be deemed to be unsafe, and to take appropriate, rapid action to remove that product from sale and notify customers of the risk. Suppliers must have policies and procedures in place, which address these potential incidents. Crisis management teams should be constituted and rehearsed to ensure that if and when these incidents occur the appropriate mechanisms are in place to effectively manage the process. Problems with glass in bottles constitutes the greatest proportion of these incidents, and it is important that in monitoring complaints that staff are fully alert to the technical issues involved in identifying these problems.

Improvement

Although we may have confidence in the effectiveness of our quality management systems to meet our customers' requirements, nevertheless there may be instances where our controls and checks in the process are unable to detect or prevent faults in the product. The consumer will inspect and drink each pack or glass of beer we offer for sale. As such, our consumers can give us the benefit of a 100% inspection system. This information is extremely useful in highlighting and rectifying problem areas. In draft beers, when trade outlets complain of faults, it is important to categorize the fault. Typically, the supplying brewery will seek information on:

- Product
- Container type
- Batch code
- Fault
  - Flat
  - Fobbing (excessive foam)
  - Hazy/cloudy

- Flavor
- Aroma
- Shortfill

This information will be collated and used to identify and track particular problems. It is useful to benchmark the data collected with other brewers to highlight issues of concern. For example, two breweries in the same group producing the same beer in the same container size and in the same market, showed significant ($\times 2$) differences in return levels. On investigating the returns data, the major difference was related to a high level of "flat" complaints in one brewery, which was traced to a breakdown in the plastic seals in the keg extractors caused by high sterilization temperatures on keg cleaning.

Similar considerations apply in the analysis of complaints from beers supplied in cans or bottles. Complaints are again categorized into defined faults, typically:

- Taste
- Flat
- Foreign objects
- Ringpull/crown
- Damaged/leaking
- Shortfill
- Packaging

Many of these issues are not readily identified by periodic inspection of the end-product, although this is not surprising given the statistical probability involved. Additionally, damage during the distribution chain from brewery to retail outlet or consumer's house can often not be readily anticipated. This is particularly true of damage to cartons of beer, sometimes, the result of poor basic design or failures in the gluing mechanisms on the line.

Effective customer and consumer feedback is essential in driving improvement, and it is important that the complaint handling operation has the appropriate procedures and systems in place to report quickly on problems, both to activate a withdrawal of beer from trade if necessary and to stimulate corrective action in the process.[12]

Procedures are included as part of the quality management system, and set out the scope of the complaints process, define roles and responsibilities, authority levels, escalation procedures, define handling standards, and provide guidance on reimbursement levels. It is important that procedures for referring complaints to the complaints function are clearly understood across the business.

Systems are required to log, track, analyze and report on complaint data. Although simple paper systems can be used, computer-based systems are now readily available to facilitate this process. Typically, the following information needs to be captured:

- Complaint identification reference
- Name, etc. of the consumer
- Receipt data
- Product details (pack size, type, best before, production codes, etc.)
- Fault description
- Date and place of purchase
- Priority status/seriousness
- Correspondence and e-mail and telephone contacts
- Reports form investigations
- Complaint status
- Reimbursement details
- Date of resolution
- Follow-up corrective measures

This information should be readily available and presented in the appropriate format.

Process Control

Having identified the factors that should be measured, monitored and controlled in the brewing and packaging of beer, how do we establish that we have control of the process? We have seen that the innate variability of the raw materials used in brewing and the complex physical, chemical, biochemical, and biological transformations in the brewing process comprise a formidable challenge to the brewer to produce a beer that consistently meets the requirements of the customer. Controlling the inputs to the process should ensure that the outputs conform to the stated requirements and without recourse to unnecessary inspection of the outputs. The emphasis in process control is to ensure that by using data as feedback on the performance of a process, we can identify sources of variation and then work to reduce or eliminate this variation.

Variation

Variation or variability appears in two significantly different forms: common or natural variability and special or nonrandom variability. Natural variability occurs when a process is functioning normally and is invariably present. Because it is normal, this natural variability can only be significantly reduced by changing one or more inputs to the process, for example, better plant or raw materials, procedural changes and improved methods of analysis. Nonrandom variability is the contribution to the overall variability that can be attributable to "assignable causes," that is, some specific, unplanned occurrence such as plant malfunction; analytical or operator error or instrument calibration fault. These are features of a system that is not behaving normally and, unlike natural variability, nonrandom variability is amenable to being significantly reduced or eliminated entirely by the application of appropriate control procedures and remedial action.

A frequency distribution due to natural variability is called the normal distribution. This is represented by the classical bell-shaped curve (Figure 20.7), where the data can be defined in terms of the mean or average, with the standard deviation as the best measure of the spread of data around the mean. With data that conform to such a normal distribution, 99.7% of the data will fall within ± 3 standard deviations of the mean value. It is not normally possible to fully characterize the total data population of interest. However, it is possible to gain a meaningful picture of that total population with estimates of the true mean and standard deviation by taking samples. In simple terms, with more samples, their means approach

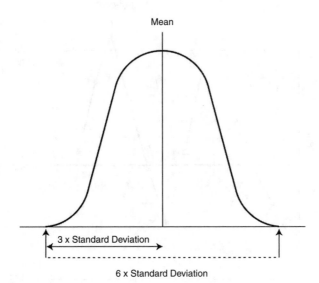

FIGURE 20.7
The normal distribution.

a normal distribution, even when the number of samples is relatively small (say 3 or 4). If a process is operating within so-called statistical control, that is, not subject to the influence of assignable causes, it is to be expected that the data points from the samples will be randomly distributed around the mean line, and within ± 3 standard deviations of the process means. This is the primary criterion for identifying a process being within or without statistical control. These data are used to construct control charts (Figure 20.8) that have the following basic features:

- A line denoting the overall mean
- Lines either side of the mean line that are called upper and lower control limits

These are usually located at distances representing ± 3 standard deviations from the mean line.

Values plotted inside the control limits show that the process is stable and operating under the influence of random variability. Values outside the control limits indicate that the process is unstable and is subject to the influence of one or more assignable causes, which can be investigated and (hopefully) eliminated to bring the process back under control.

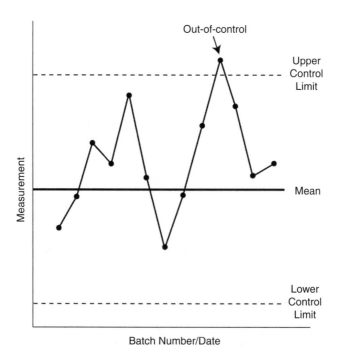

FIGURE 20.8
Control chart for variables.

Quality

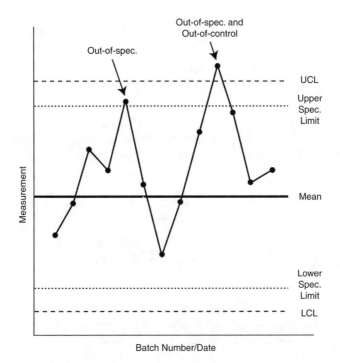

FIGURE 20.9
Control chart with control limits and specification limits.

It is possible to have a process that is in statistical control but is delivering a product that is outside specification! This is illustrated in Figure 20.9. If the specification has been set too tight, then the process may not be capable of meeting that specification. This may demand either a change of specification or changes to the process to improve its capability to reduce the normal variation.

Process capability can be measured relative to the control limits and the specification limits. Where the control limits and specification limits coincide, the process capability, Cp, is said to have the value of 1.0. Cp values of 1.0 or greater mean that the process is capable, that is, can generate product that is all within specification. Cp values less than 1.0 mean that the process is not capable — the process will generate product that is out of specification. In practice, it is always advisable to target Cp values greater than 1.0 to allow for any slight drift in the average of the product and still produce in specification.

Although Cp values indicate whether or not the process is capable of producing within specification, it does not actually tell us that we are producing in specification. A further measurement, the capability index, CpK, is used to determine whether we are actually producing in specification.

The CpK can be calculated from the relationship:

$$\text{CpK} = \frac{\text{Specification Range}}{\text{Upper Control Limit} - \text{Lower Control Limit}}$$

However, if the process mean is not centrally located, upper (C_U) and lower (C_L) capability indices are calculated:

$$C_U = \frac{(\text{Upper Specification Limit} - \text{Process Mean})}{(3 \times \text{Standard Deviation})}$$

$$C_L = \frac{(\text{Process Mean} - \text{Lower Specification Limit})}{(3 \times \text{Standard Deviation})}$$

The smaller of C_U or C_L is the capability index, CpK.

In situations of one-sided specifications, for example, maximum haze or dissolved oxygen level or minimum head retention value, only the relevant of the two measures, C_U or C_L, will apply. Like Cp, to allow for slight shifts in process, a minimum CpK value of 1.3 is usually sought. As the CpK increases, the need for routine measurement decreases. Periodic checks on CpK are required to ensure that CpK is not decreasing, thereby increasing the risk of product out of specification.

The applications of statistical process control (SPC)[13] are increasing in the brewing industry. In the United Kingdom, volume control of container contents and the control of alcohol levels in finished products is rigorously monitored by government authorities using statistical control methods that have been agreed with the brewing industry. Codes of Practice have been published detailing the approved methods of calculation. However, in addition to these controls on finished product, SPC is being applied throughout the brewing and packaging processes to monitor and control, minimizing the risks and costs associated with quality failure (rework, replacement raw materials, scrapping of product, reinspection, rescheduling of production, etc.).

The application of SPC techniques has led to the development of the Six Sigma approach to business improvement and has been adopted by many organizations throughout the world. Six Sigma equates to the six standard deviations referred to in the normal distribution curve and provides a focus for reducing the variation in all work processes, by developing a structured approach for improvements in process capability. A detailed review of Six Sigma techniques is beyond the scope of this chapter, but for more information refer to *Six Sigma for Managers* by Greg Brue.[14]

Beer Quality at the Point of Sale

It could be argued that the responsibility of the brewer for the quality of beer ends at the brewery gate. However, "meeting the customers' requirements" extends beyond the brewery gates, particularly at the point of presentation of draft beer. Much of the good work done in the brewery and packaging hall can be undone by a lack of care and attention in the cellar and bar. A detailed review of cellar management and bar practice is beyond the scope of this chapter, but the reader is recommended to Ref. 15 which highlights the important features of ensuring good-quality beer at the point of sale. Effective stock control, correct storage conditions, good hygiene, appropriate dispense equipment, clean glassware, and critically, staff training, are all issues that need to be addressed.

Brewers have an important role in educating their customers to ensure that the consumers' requirements and expectations are at least met and preferably exceeded. Beer is arguably "the best long drink in the world," and it is incumbent on those involved in the brewing industry supply chain — suppliers, brewers, packers, distributors, and retailers — to deliver products and services to the highest standards of quality.

Acknowledgments

The author would like to thank his former colleagues in Carlsberg UK for their support in preparing this chapter and in particular; Dr. Chris Dickenson, Tony Trusler, Laurence May, Lynda McGroggan, and Neil Burton.

References

1. Oakland, J.S., *Total Quality Management*, 3rd ed., Butterworth–Heinemann, Boston, MA, 2003.
2. Peach, R.W., Peach, B., and Ritter, D.S., *The Memory Jogger 9000/2000*, 1st ed., GOAL/QPC, 2003.
3. Hargreaves, L., Microbiology, C.I.P and HACCP, *The Brewer*, 86:251–256, 2000.
4. British Beer and Pub Association, *Guide to Managing Food Safety in the Brewing Industry Using the HACCP Approach*, British Beer and Pub Association, 2003.
5. British Retail Consortium, *Technical Standard and Protocol for Companies Supplying Retailer Branded Food Products*, British Retail Consortium, 2000.
6. Birks, S., Standards under inspection. *Food Manufacturer*, November:47–48, 2003.
7. Baxter, D., The influence of brewing liquor on beer safety and quality, *Ferment*, 12:13–18, 1999.
8. O'Rourke, T., Quality management, *Brewers' Guardian*, 130:20–22, 2001.

9. Sharpe, R., Amalgamation of IGB and EBC methods, *Brewer Int.*, 2:39, 2002.
10. O'Rourke, T., Analytical measurement, *Brewers' Guardian*, 129:21–23, 2000.
11. Simpson, W.J. and Canteranne, E., Sensory analysis in modern brewery operations, *Brauwelt Int.*, IV:280–286, 2001.
12. Powesland, T., Consumer feedback, in *Excellence in Packaging of Beverages*, Browne, J. and Candy, E., Eds., Binsted Group Plc., 2000, pp. 703–707.
13. Oakland, J.S., *Statistical Process Control*, 4th ed., Butterworth–Heinemann, Boston, MA, 2002.
14. Brue, G., *Six Sigma for Managers*, McGraw-Hill Education, New York, 2003.
15. Buggey, L.A., Bennett, S.J.E., and Pain, M., *Brewer Int.*, 2:15–19, 2002.

21

Microbrewing

Johannes Braun and Brian H. Dishman

CONTENTS

Introduction	773
Definition of Pub Breweries	773
Definition of Microbreweries	774
Differences	774
Location Requirements	775
Choosing the Right Products	776
How to Choose the Right Equipment	778
Kind of Equipment	778
Sizing the Equipment	781
Quality Control in Microbreweries	783
Operating Methods — An Introduction	785
Pub Breweries	785
Microbreweries	785
Legal Points	786
Marketing	787
Customer Targeting	787
Pub Brewery	787
Microbrewery	788
Concept! Concept! Concept!	790
Pub Brewery Concept	791
Microbrewery Concept	791
Beer Theme	792
Food Theme	792
Hospitality Theme	793
Entertainment Theme	794
Event Theme	794
Mission Statement	795
Sales Plans	795
Pub Brewery Sales Plans	795

The Microbrewery Sales Plan 796
  Cost Control 798
Staff Organization and Training (Owner's Standpoint) 799
  Pub Brewery Management Selection 799
  Microbrewery Management 800
  Chain of Command 800
  Employee Handbooks 801
  Operations 801
  General Manager 802
  Brewery Manager 802
  Sales Management 803
  Advertisement and Marketing 803
  Information Management 803
  Kitchen Manager 804
  Hospitality Manager 804
  Operation Reporting and Filing System 804
  Customer Interaction Planning — Pub Brewery 805
  Customer Interaction Planning — Microbrewery 805
  Staff Training System 807
  Hiring Staff 807
  Staff Training 807
  Staff Motivation 808
Claim Management 808
  Customer Expectations 808
  Understanding! 809
  Probable Causes 809
  Planning for Claims 809
  Health and Sanitation Checks 810
  Drink Driving 810
Sales Methods 811
  Point of Sales 811
  Point of Presence 812
  Events 812
Advertising and Promotion Plans 813
  Planning 813
  Product Ambassadors 814
  Feedback 814
  Professional Assistance 815
References 815

Introduction

There is a long history of microbreweries in many countries of the world. Initially, brewing was carried out as home brewing by women for domestic use only. It was a part of the daily housework next to cooking and baking bread. Later on, some of these home brewers started to offer their beer to travelers. The beer was sold in the pub only. Bamforth[1] writes "Out of the domestic brewing scene came the development of breweries each selling its own beer in a room out front." These were the first pub breweries. Nowadays, the term microbrewery is used for many different styles of breweries and the definition varies from country to country. Generally, the term microbrewery describes a brewery that is smaller than the average sized brewery in the area. Therefore pub breweries, boutique breweries, craft breweries, and cottage breweries are also referred to as microbreweries from time to time.

Definition of Pub Breweries

A pub brewery is an installation that brews its own beer on premise in the tavern for consumption on-premise at that tavern only.[2] Eschenbach[3,4] similarly defines a pub brewery as a small brewery, which produces beer only for one's own pub, and the brewery is situated in, or directly next to this pub. It generally does not produce bottled beer. Marchbanks,[5] however, gives a slightly broader definition. He describes a pub brewery as a licensed premise with a small brewery installed in it. It usually produces only draft beer for sale on the premise with perhaps some local free trade support. Output varies from 2 to 80 hl per week. Braitinger[6] considers breweries with an output up to 1500 bbl/year as pub breweries, while Piendl[7] describes a pub brewery as a brewery where beer is brewed directly in the pub and most of the production is also sold there. Beer for take out in siphons is normally included, but bottling of filtered beer is not part of the operation of a brewpub. Radtke[8] mentions as the characteristic points for a pub brewery that 95% of the beer is sold in its own restaurant, the beer is an unfiltered product and it is not bottled. However, Caspary[9] does not specify a pub brewery by output or sales routes but by the atmosphere it creates. For him the determining points are that the brewhouse is situated in the pub, without a glass wall around, and that the customer can see the brewing equipment. The first pub breweries of the modern times where established in England during the 1970s. Five to ten brewpubs existed in Britain at this time.[5] By the end of the decade, the first brewpubs in the United States and Canada were established[6] and in 1981 the first pub breweries in Germany were developed.[10] Now the German Brauereiadressbuch 2002[11] lists 336 pub breweries in Germany. In Japan, the first pub brewery was established in 1995.

Definition of Microbreweries

Microbreweries have a much longer history in Europe. In Germany, there are still 400 breweries that produce less than 4000 hl per year[12] and, consequently, very few new microbreweries have been established. In other countries like Britain, United States, and Canada, very few traditional microbreweries existed during the 1970s. Therefore, numerous new microbreweries have been developed in the last 30 years. The 2002 Company Directory of the Institute of Brewing & Distilling[13] lists 239 Smaller/Microbreweries in the United Kingdom and Eire, and in Germany 833 breweries with an output of less than 10,000 hl/year were registered in 2000.[14] This represents 65.6% of the breweries registered in Germany.

Microbreweries are very often defined by their annual output. Braitlinger[6] defines microbreweries to have an output between 1500 and 15,000 bbl/year. Marchbanks[5] writes small independent breweries have a capacity of 20–300 bbl/week. They are set up on industrial trading estates or old brewery sites and can produce both draft and packaged beers. The Texas State Legislature characterizes a microbrewery as follows, "A small brewery that produces beer and packages it primarily for sale at retail outlets." The brewing industry defines microbreweries as those producing less than 15,000 bbl/year. Coulter[2] states: a microbrewery is a small independent free standing brewing operation that sells in one way or another to taverns and/or the public at arm's length, that is, other than size, a business like any of the major breweries. A microbrewery is generally understood to have an annual output of less than 10,000 bbl/year.

Differences

From these various definitions it becomes clear that pub breweries and microbreweries are two very different forms of business (Table 21.1). They have

TABLE 21.1

Differences between Pub Breweries and Microbreweries

	Pub Brewery	Microbrewery
Size	Less than 3000 hl/year	Less than 15,000 hl/year
Sales route	At least 95% of production is sold only in the pub	Beer is primarily sold for retail outlets
Processing	Unfiltered beer	Filtered and bottled beer
Type of business	Restaurant business with brewing show	Brewery business
Target customer	Local	Individualists
Location	High customer frequency, high image	Cheaper ground and easy access
Shelf life	Unnecessary	Minimum 1 month
Competitors	Other restaurants	Brewing companies

different needs for location of equipment, products, sales routes, quality control, and target customers. For the success of an operation these points have to be considered during planning of a new brewery.

Location Requirements

As seen in the previous section, pub breweries and microbreweries are two different types of business. Therefore, the right choice of location (Table 21.2) is crucial for success of the whole operation. Pub breweries are a kind of restaurant business; therefore, the requirements regarding location are similar to those for restaurants.

A heavily populated area with a high percentage of wealthy people and only a few big breweries in the locality is ideal, with big cities, university towns, or tourist areas topping the list.[15,16] Similarly, towns where small breweries used to be situated and have closed are suitable for pub breweries. The pub brewery should have a hinterland of about 30 km with at least 150,000 people resident.[17] Tourists are not included in this figure because numbers can vary considerably over longer time periods. Radtke[18] recommends locating the pub brewery in towns with an historical center where cultural and sports events are held frequently. The best location is a historical building in the city center that used to be a brewery in former times. Others do not think that the exact location is very important for the success of a pub brewery, but two key criteria should be considered[19]: (i) The pub brewery should be in a location where customers pass by throughout the day. Locations that are only busy in the evenings are not suitable. (ii) The location should be easy to find, accessible with public transport, provide sufficient parking space, should include a beer garden and no other pub brewery should be in close proximity. Some additional points to be evaluated before choosing a location include[20]: (i) population growth in the area, (ii) long-term employment growth, (iii) growth of retail sales, (iv) tourism, (v) sports and cultural events in the area, (vi) types of transportation, (vii) number and type of brewpubs, and (viii) restaurants and bars and their pricing. Finally, Schoolmann[21] emphasizes the following points when

TABLE 21.2

Location Requirements for Pub Breweries and Microbreweries

Pub Brewery	Microbrewery
Highly populated area	Easy to access (motorway)
Public transport	No legal restrictions
All day customer frequency	Sufficient supply of electricity water
Beer garden	Effluent discharge

choosing a location for a restaurant: (i) neighboring buildings, (ii) population density, (iii) growth potential, (iv) public financing programs, (v) parking space, (vi) public transport, (vii) distance to hotels, (viii) cultural events, (ix) competition from other restaurants, and (x) planned road developments.

For microbreweries where the main business is the sale of kegged and bottled beer requirements regarding the location are quite different. As the product has to compete with other bottled beer products, the price becomes more important, and production and investment costs have to be minimized. So it is necessary to find a cheap location, one that is suitable for building a brewery. Peterson[22] lists the following criteria. The location should be easy to access: railway lines, pedestrian areas, big shopping centers, or schools impede access. The neighboring location should not have a negative affect on the brewery operation: legal restrictions have to be researched in advance. What are costs the involved for sufficient supply of electricity and water and disposal of effluent? Also, the water quality, depending on the yearly seasons, has to be researched. The location should be large enough to provide space for loading and unloading of trucks, for transport of raw materials or the final product (trucks need at least 20 m turning space).

Choosing the Right Products

Many authors have given suggestions on what they think microbrewery beer should be, but the most important characteristic for microbrewery or pub brewery beer is that it should be different from the existing beers available in the area.[16] Particular recommendations include: (1) The product should be produced by a natural process and not like a mass market beer.[23] (2) The beers from pub breweries should be unfiltered in order to maintain the best flavor and drinkability.[18] (3) Two or three different beer types should be offered in order to keep the novelty of the product.[18] (4) Beer types, which are typical for the area the pub breweries are situated in, should be considered, because a beer produced locally will always mean more to the consumer than one from the neighboring region.[24] (5) As the products have to be tailored to the market, beers from pub breweries and from microbreweries need to be designed differently (Table 21.3).

One of the big differences lies in the shelf life of the products. A pub brewery that sells only in its own restaurant can serve beer directly from the lager tank, unfiltered and not stabilized. Because the customer comes to the place of production, he or she can enjoy the freshest beer available in its purest stage. Production costs can be kept low because there is no filtration, pasteurization, bottling or kegging equipment, and processing costs. The brewmaster on a daily basis can easily monitor the quality of

TABLE 21.3
Product Differences between Pub Breweries and Microbreweries

	Pub Breweries	Microbreweries
No. of items	High	Low
Taste	Different from others in area	Unique
Filtration	Unfiltered	Filtered
Shelf Life	Less than a week	More than 4 weeks
Pasteurization	None	Might be necessary
Stabilization	None	Finings
Packaging	Siphon for take out only	Bottling and kegging
Quality control	Simple	Sophisticated

the final beer, and beer that does not meet the standards can be readily removed from the market.

A microbrewery, on the other hand, that has to bottle the beer and sell it through wholesalers and retailers has to ensure a longer shelf life. Generally, the longer the shelf life required the greater the processing of the beer after maturation. Therefore, cost-intensive processes that help to extend the shelf life like filtration, stabilization, sterile filtration, pasteurization or cool delivery have to be carried out. All this processing has an effect on the taste of the beer. Especially in small breweries, filtration is a hazardous process because of difficulties with sufficient throughput and the avoidance of oxygen pick-up.[5] Much effort has to be made to keep the taste of the beer as close to that coming directly from the lager tank and, as every brewmaster can confirm, this is where you can enjoy beer in its freshest, purest, and most delicious state. In order to ensure the stability of the bottled product, a serious quality control system must be installed. The quality assurance has to be carried out more seriously in a microbrewery than in a pub brewery, and this adds to the cost of the final product. On the other hand, as the distribution area is expanded, the brewery output can be extended along with the extension of the shelf life requirement.

Another advantage of a pub brewery is the possibility to offer many different types of beer at different seasons of the year, because the production facility is very flexible and the marketing costs relatively low. For microbreweries, the cost of marketing many different beer types and explaining the differences to their customers are much higher, and it can be difficult to find the proper shelf space in the shops.

Other criteria used to determine the product range are the investment and production costs. Many pub and microbreweries in Britain and the United States prefer top fermenting beer types because of their tradition but also because of the shorter production time and, therefore, reduced tank capacity requirements.

In Germany and Japan, where mainly bottom fermenting lager type beer is produced, this increases investment and production cost. In order to choose the most suitable products, decisions on sales routes, target customers, sales areas, sales price, packaging styles, and distribution systems have to be made.

How to Choose the Right Equipment

Kind of Equipment

As previously discussed, different sales routes require different products and, therefore, different equipment. Four different brewery designs are shown in Figure 21.1. The easiest way to produce beer for one's own restaurant is by extract brewing. Advantages are the low investment cost and easy operation, as the wort production process is not carried out in the pub brewery. Further, no malt storage area is required and no spent grains have to be removed. However, there are a number of disadvantages. One of the main attractions for the consumer to drink beer in the pub brewery is that the whole brewing process is visible and one can easily understand how beer is produced. The consumer desires beers that are different from those produced by big breweries and is attracted by the natural and traditional production process.[25] Using extract that is produced on a large scale in factories has the possibility of conflicting with the image of the hand-crafted pub brewery beer. Moreover, because of the standardized extracts, it may be difficult to create the unique beer types that are an important attraction for each pub brewery.

An alternative route to reduce investment cost is the Swiss-based Back and Brau system where wort is produced in a central brewery and then shipped by tank lorry to the pub breweries where it is fermented and matured. Advantages are low investment cost and lower personnel cost as no qualified brewer has to be on site. The disadvantages are similar to these of extract brewing and, as no brew kettle is necessary, another attraction for the customer is removed.

Another way of keeping investment cost low is to limit production only to ale type beers. In this case, malt is milled in a two-roller mill or can be bought readily milled from the maltser. Mashing takes place in the lauter tun and a single step infusion mashing is carried out. First runnings are drawn off into the wort kettle and sparging is carried out. The wort is boiled in the wort kettle, and spent hops are removed in the whirlpool which can be also, combined with the kettle. Beer is cooled in a plate cooler and fermented in jacketed and insulated fermenters. After fermentation, the beer can be racked in casks or kegs for secondary fermentation.

Advantages are low investment costs as no maturation tanks are necessary. Further, the number of fermenters can be reduced due to the short fermentation time of 2 or 3 days. Costs for tank cooling equipment, such as ice banks or glycol coolers can be kept low due to higher fermentation

Microbrewing

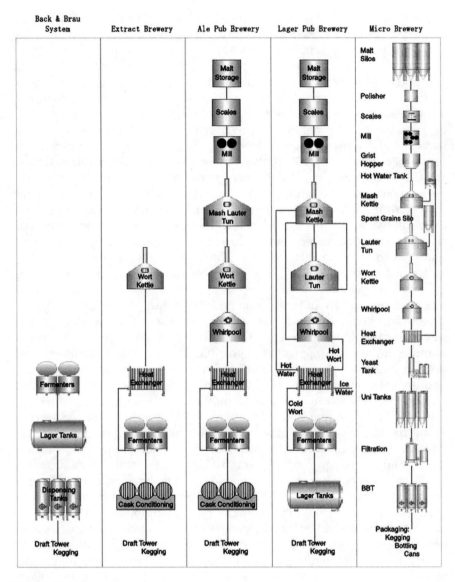

FIGURE 21.1
Different types of pub breweries and the microbreweries.

temperatures. As a disadvantage, the lack of flexibility has to be mentioned. As customer preferences change over time, different types of beer become the trend, requiring major changes in plant design. This may be especially true in markets where no long brewing tradition exists.

Therefore, a pub brewery design that enables production of lager or ale type beers should be preferred even though the initial cost will be higher.

In a lager type pub brewery, mashing-in takes place in the combined mash and wort kettle. In addition to infusion mashing, decoction mashing should be possible to allow for poor malt quality and to provide the different characters of the different types of beers. The lauter tun should be equipped with knives in order to ensure short lautering times. The wort is drawn off from the lauter tun and boiled in the combined mash and wort kettle that can be heated by steam, electricity, or directly by an oil or a gas burner. The hot break is removed in the whirlpool and the wort is cooled in the plate cooler to low temperature for bottom fermentation. As the beer ferments for 7–10 days, more fermenters are required than in an ale beer brewery. In order to ensure constant beer quality, the fermentation vessels must be equipped with cooling jackets and an automated temperature control. Maturation takes place in maturation tanks for a couple of weeks at temperatures below freezing. Lager tanks do not necessarily need an automated temperature control. Therefore, sufficient tank space has to be made available and cooling of these tanks by room cooling or glycol has to be ensured. The final beer can be stocked in dispensing tanks or kegs before serving.

After the flexibility of production and the quality issues, the equipment should be chosen by its appeal to the customer. The brewing process should be staged like a show, and the customer should be able to experience as much of the brewing process as possible. Therefore, it is an advantage if the customer can watch, smell, and hear the process when beer is produced. In order to be able to understand the whole process, even malt storage and milling should be made visible to the public. The brewhouse should be situated within the seating area and easily visible from most of the seats. It should not be covered by glass so that the customers can also enjoy the smell of the boiling wort or the hot spent grains being removed. The more the customer can experience the more authentic the story of natural pure hand-crafted beer becomes. Fermenters should be open with a lid that can be closed during cleaning. The fermenting and maturation room also should be made visible to the customer by big windows so that an atmosphere of beer production can be conveyed even on days when the brew kettle is not used. Automation of the brewing equipment should be limited to the level that one person can operate all the brewing equipment in order to run the brewery cost efficiently. Highly automated plants are not only expensive but also reduce the amount of brewing work that can be shown to the customer. In order to create an authentic atmosphere, the customer should be able to see the brewer working. Removing spent grains or cleaning tanks by hand stresses the character of the hand-made beer.

Microbreweries that produce bottled or kegged beer for wholesale do not need to show the brewery character. The brewery design should concentrate on being able to produce a high-quality product at low cost. In order to ensure sufficient shelf life, filtration of the beer is necessary. In order to keep costs low, fermentation and maturation can be carried out in unitanks. To ensure beer quality and independence of operation, a yeast propagation and storage system should be available. The bottling line should be able to provide low oxygen concentrations (less than 0.1 mg/l in the final product) in order to ensure flavor stability. Depending on the filtration system and the distribution method, pasteurization of bottled and kegged beer might be necessary.

Sizing the Equipment

One of the most important factors for the success of a pub brewery is the quality of the beer. The shelf life of the beer produced in pub breweries is short because of the simple and natural production method, and a key element is to serve the beer fresh to the customer. The fresher the beer the better the taste and the lower the production costs. Consequently, it is important not to oversize the capacity of the brewing vessels. If the customers demand for the beer increases, more tanks can be added and the brew kettles used more frequently. Therefore, room for extensions should be considered at the outset.

In order to run a pub brewery with a capacity of 1000 hl/year profitably, 750 hl beer/year should be sold.[15] The capacity of the brewhouse has to be calculated considering the three consecutive months with the highest output. For pub breweries, this is usually during the summer months, when the beer garden can be operated and about 15% of the annual output per month is necessary. Microbreweries with bottled and kegged beer usually need 10% of the annual output per month. A formula to calculate the necessary wort volume per week is as follows:

$$W_{week} = \frac{B_{year} * f_{max}}{(1 - L) * f_{w/m}}$$

where W_{week} is the wort necessary per week in hectoliters (hl), B_{year} is the yearly planned beer production, f_{max} is the beer necessary during the month with the highest output in percent of the yearly output, and $f_{w/m}$ is the number of weeks per month.

For an annual beer production of 1000 hl, 39.47 hl of wort have to be produced per week

$$W_{week} = \frac{1000 * 0.15}{(1 - 0.05) * 4} = 39.47 \, hl.$$

Realistically, one person, producing four brews each week, should operate the pub brewery. One day should be reserved for cleaning purposes. Therefore, a brewhouse with a capacity of 10 hl should be designed for the aforementioned example. For the right choice of fermenters and maturation tanks, many points that are important need to be considered. First of all, the number and the types of beers should be determined. Then, the fermentation and maturation times and the technology have to be considered.[16]

The fermenter size should be designed for one brew to enable production of one batch per day and a filling time of less than 12 h. Usually a headspace of 25% is required; therefore, four fermenters with a capacity of 12.5 hl would be necessary if the average fermentation time is 7 days. As mentioned earlier, open fermenters have the advantage of attracting the customers' attention but are more labor intensive. Closed fermenters are easier to clean and fermentation and maturation can be carried out in the same vessel. The size and the number of maturation tanks depend on the number of beer types to be sold. For up to three beer types, double batch maturation tanks can be used, but if more than three types are to be sold at the same time, single batch maturation tanks will be necessary in order to ensure the freshness of the beer. If an average maturation time of 4 weeks is assumed and a headspace of 10% is provided, a total maturation tank volume of 176 hl is needed.

The space requirements for different sizes of pub breweries are listed in Table 21.4.[26] For a brewery with an annual output of 1000 hl, space requirements of 35 m^2 for the brewhouse, 20 m^2 for the maturation area, and 12 m^2 for utilities are typical.[27]

Radtke[18] suggests 100 m^2 for the brewing equipment of a brewery with an annual output of 1000 hl; the seating requirements are given in Table 21.5.

TABLE 21.4

Space Requirements for Different Sizes of Pub Breweries

Yearly Output (hl)	1000	1500	2000	3000
Brewhouse size (hl)	10	10	15	20
Minimum area requirement brewery (m^2)	130	130	180	220
Malt storage (m^2)	15	15	20	40
Hop storage	4	4	4.5	5
Yeast	1	1	1	1
Water supply	4	4	4	4
Malt mill	2	2	2	3
Brewhouse	15	15	20	20
Fermenters	15	15	20	25
Maturation tanks	50	50	80	100
Cleaning systems	4	4	4.5	4.5
Cooling system	5	5	5	6
Bottling kegging	10	10	15	20
Customer seats	100	160	220	250

Source: Adapted from Flad, W. Brauwelt, 27:1111–1114, 1990. With permission.

TABLE 21.5

Brewery Capacity Depending on Consumption Per Seat

hl/Seat and yr	100 Seats	150 Seats	200 Seats	300 Seats	400 Seats
6	600 hl/yr	900	1200	1800	2400
7	700 hl/yr	1050	1400	2100	2800
8	800 hl/yr	1200	1600	2400	3200
9	900 hl/yr	1350	1800	2700	3600
10	1000 hl/yr	1500	2000	3000	4000

Source: Radtke, K.H. Brauwelt, 18:893–895, 1995. With permission.

Quality Control in Microbreweries

In pub breweries where the beer is consumed only in the restaurant and no long shelf life has to be ensured quality control can be limited, but nevertheless some checks are necessary and cannot be neglected (Table 21.6). Most of checks are simple, but help the brewmaster to keep the beer taste constant. Malt should be checked visually before each brew and assessed by taste and smell, as off-flavors can develop during transportation or storage or by spoilage due to grain insects or molds. A complete malt analysis should be ordered annually from an independent lab in order to check the malt analysis of the malt supplier. Hops should be checked by smell before each brew. Water taste and pH should be checked before every brew, and once a month the hardness should be determined in order to monitor the water supply.

The mash should be analyzed: each brew for pH and saccharification using iodine check. During lautering, wort turbidity should be checked visually and the gravity of the first runnings, kettle full, and last runnings should be recorded for each brew. The attenuation limit, pH, gravity, and the color of the cast wort should be checked for every brew and the brewhouse yield and evaporation should be calculated as follows:

$$Y = \frac{V[hl] * OG[P] * 0.96}{Malt[100 \text{ kg}]}, \quad E = \frac{(Vkf - Vcw) * 60}{Vcw * tb}$$

A complete wort analysis should be ordered annually from an independent laboratory.

The yeast should be checked visually by color and smell. Also, the yeast count should be determined by counting chamber, and the microbiological status by cultivation in NBB broth or similar (see Chapter 16) should be checked before pitching. During fermentation, daily records of extract and temperature must be taken. Yeast count should be recorded during the

TABLE 21.6

Recommended Checks for Quality Assurance

Medium	Frequency	Analyses	Method
Malt	Every brew	Taste, looks, smell	Visual
	Every year	Malt analysis	External laboratory
Hop	Every brew	Smell, looks	Olfactory, visual
Water	Every brew	Taste, smell, pH	pH meter
Yeast	Every brew	Smell, color	Olfactory, visually
	Every brew	Cell count	Microscope, cell chamber
		Microbiology	NBB broth
Mash	Every brew	pH, saccharification	pH meter, iodine test
Lauter wort	Every brew	Turbidity, pH, extract of first runnings, last runnings and kettle full	Visual, pH meter, Saccharometer
Cast wort	Every brew	Color, pH, OG attenuation Limit	Visual, pH meter,
	Every year	Full wort analysis	External laboratory
Fermenting wort	Everyday	Temperature, pH, extract	
	First 72 h	Yeast count	Counting chamber
Green beer		Extract, pH, yeast count, attenuation	Counting chamber
Maturing beer	Everyday	Pressure, temperature	
	Every week	Extract, color, yeast count, pH	Counting chamber, pH meter
		Attenuation, foam stability, smell, taste	Saccharometer, comparator
		Microbiological	NBB broth
Final beer		Extract, color, yeast count, pH	Counting chamber, pH meter
		Attenuation, foam stability, smell, taste	Saccharometer, comparator
	Every year	Bitter units, CO_2 content, alcohol content	External laboratory

first 72 h. Color, pH, yeast count, and attenuation of the green beer should be recorded. During maturation, temperature, pH, pressure, yeast count, extract, color, and foam stability should be checked once a week. The final beer should be thoroughly assessed for pH, attenuation, remaining extract, taste, and foam stability. An external lab should be utilized to check the beer every 6 months for alcohol level and bitter units as well as CO_2 content.

Further, all brewing equipment such as fermenters, lager tanks, dispensing tanks, hoses, pipelines, and kegging equipment should be checked visually and microbiologically by swabbing and cultivating in NBB broth. These simple quality checks can be easily carried out by a trained brewer with little equipment but are very important to ensure a constant beer quality that is one of the main factors of a successful pub brewery.

For microbreweries that produce both kegged and bottled beer for wholesale, providing an extended shelf life becomes much more labor and

TABLE 21.7

Microbiological Checks to Ensure Shelf Life

	Method/Media	Incubation Time at 27°C
Yeast	NBB broth	5–6 days anaerobic
Pitching wort	Membrane filtration on NBB agar	5–6 days anaerobic
Fermenting beer	NBB broth concentrate	6–7 days
Maturing beer	Membrane filtration on NBB agar	5–6 days anaerobic
Filtered beer	Membrane filtration on NBB agar	5–6 days anaerobic
Water	Membrane filtration on NBB agar	5–6 days anaerobic
Kieselgur	NBB broth	5–6 days anaerobic
Compressed air	NBB agar	5–6 days anaerobic
Cleaned bottles	Rinse with NaCl on NBB agar	5–6 days anaerobic
Cleaned kegs	Rinse water on NBB agar	
Bottled beer	Storage at 20°C	2 months

cost intensive because of the requirements of the quality control system. Microbiological analysis of the beer at different points along the production path must be carried out in order to spot contamination at an early stage. Therefore, careful consideration should be given to whether the quality control system should be outsourced or carried out in-house. Schmidt[28] recommends checks given in Table 21.7.

Operating Methods — An Introduction

Pub Breweries

As mentioned earlier, a pub brewery is not simply a beer production facility. It is a much more complex business unit. Not only does one have to contend with the production of high-quality beer, but also with the extremely stressful operation of a high-quality restaurant. A successful pub brewery requires consistent management of three separate business operations at the same time.

These three operations include the brewery, the kitchen, and the dining room service system, and each has to be properly managed so that the customer receives an exceptional experience each time he or she visits the pub brewery. Exceptional experience translates into customer satisfaction, without which there will be no repeat business. Once delivery of the exceptional experience has been mastered, then cost control must also figure into the overall picture such that the repeat business translates into profitability for the pub brewery.

Microbreweries

While pub breweries are indeed complex business units, microbreweries should not be considered easier to operate. Actually the opposite is probably true. There are many more things within the brewing system that must be considered. Microbreweries are producing a product with a longer shelf life and therefore require much higher-level equipment. The average pub brewery simply serves the beer they have produced on-premise, whereas a microbrewery must package the beer using multiple packaging systems, such as kegging, bottling, and possibly even canning lines. Moreover, as discussed previously, the packaged beer must also have a guaranteed shelf life requiring a filtration system as well as a more complex quality control mechanism.

Microbreweries also require a sales team that replaces the hospitality team in the pub brewery. The target customers are different, as are the methods needed to accomplish the sales. However, as with pub breweries, an exceptional product experience must be delivered to the various targeted customers in order to build repeat business that translates into profits for the microbrewery.

Legal Points

A microbrewery is a business that from its inception is planning to produce a fairly large volume of beer packaged for sale in various locations, possibly even crossing national boundaries. Therefore, legal issues that could arise from its sale in other locations must be taken into account. No matter which type of brewery is being planned, it would be a good idea to consult with a lawyer who could explain the legalities for sales of the product in the various markets that are planned. Product liability laws should be well understood in all the planned sales areas.

While it is important to look at product liability laws, it is also a good idea to have a thorough understanding of local, state, and national laws governing all aspects of brewing and the property where the brewery is being considered. It is also smart to have more than one location in the research pipeline. One location may seem like the perfect spot, but zoning laws as well as many other problems could surface once more specific research begins. Consulting with a real estate appraiser may be very helpful. The real estate appraiser will be able to provide a very large amount of information about proposed sites that are very important for the decision making process. A real estate appraiser should be able to provide proper demographic data that can be invaluable in the selection of a site that matches up with the previously researched target customer. Do not become overly attached to any one spot before researching all the facts of all the possible sites. With the international problems associated with effluent and waste disposal, the local utilities should be well understood, especially laws governing the disposal of brewery by-products and effluent. One would be well advised to keep track of legislation in this specific area.

Last, but not least, the taxation laws with regard to beer production should be carefully checked and understood in the location where the brewery will be operating. Understanding what reports are necessary and when the reports must be filed are two very important points. All legal issues should be researched early in the development of the brewery plans.

Marketing

Customer Targeting

Customer targeting has been heavily covered in the technical discussion previous to this section. It is indeed the most important concept and must be determined before any equipment is selected, a location is purchased, or a building is constructed. It is necessary, at the outset, that one has already determined what is to be accomplished with the business being planned. Many people have made the mistake of starting a pub brewery or microbrewery simply because they thought it would be an interesting business or that they could brew fantastic beer at home. However, it is very important to know what type of customer is to be targeted, where this customer exists, and how to get this targeted customer to a location where the products of the pub or microbrewery will be available.

Pub Brewery

For the pub brewery deciding whom to target as the customer is a fairly complex task. It is necessary to understand the demographics of the targeted sales area as a first step, and then asking the next set of questions is critical. Once those questions have been posed, researching the market to find the answers is a big job. Below are some sample questions that could assist with the pub brewery; other questions of relevance to the special circumstance of the business being planned should also be considered.

1. What type of customer is being targeted?
   a. Age
   b. Gender
   c. Marital status
   d. Income level
   e. Proximity to the pub brewery
   f. Tourist
   g. Others
2. What type of image should the pub brewery project to the customer?
3. Why should the customer come to the pub brewery?

4. What unique services or products will entice the targeted customers to come to the pub brewery?
5. Should specific beer types be produced to attract the customers targeted?
6. Should specific foods be produced to attract the targeted customer?
7. Is a certain level of service needed to attract the targets?
8. (Continuation of other relevant questions for the business.)

Microbrewery

For a microbrewery, customer targeting in many respects, is a much more difficult problem than for the pub brewery. The customer is no longer simply the end-user who happens to come along to drink the beer and have a bite to eat. The management team must think in terms of on-premise sales, meaning those shops which elect to sell the beer to their patrons as well as off-premise sales, meaning those who purchase the beer from a retail store for private consumption. Therefore, it is necessary to consider who the target customer really is.

1. What type of end-user is being targeted?
    a. Demographics
        (i) Age
        (ii) Gender
        (iii) Marital status
        (iv) Income level
    b. Everyday consumption
    c. Special occasion consumption
    d. Planned purchase
    e. Purchase location
        (i) Liquor store
        (ii) Convenience store
        (iii) Grocery store
        (iv) Specialty store
        (v) On-line shopping
        (vi) Catalog shopping
        (vii) Home delivery
        (viii) Others
    f. Spontaneous purchase
    g. Tourist

h. Consumption habits
     (i) On-premise consumption
        - Restaurant
        - Pub
        - Others (continuing list as required)
     (ii) Off-premise consumption
        - Home
        - Other private location
  i. Others
2. What type of on-premise customer is being targeted?
   a. Restaurants
      (i) Areas of availability
         - Local
            o Independent shops
            o Chain shops
         - Regional
            o Independent shops
            o Chain shops
         - National chain
            o Independent shops
            o Chain shops
         - International areas
            o Independent shops
            o Chain shops
      (ii) What type of restaurant?
         - Theme restaurant
         - Ethnic restaurant
         - Steak house
         - Seafood
         - Others (continuing list as required)
      (iii) Shop requirements
         - Package type (kegs, bottles, cans, other)
         - Dispensing equipment
         - Glassware
         - Advertisement materials
         - Other requirements
   b. Pubs and Bars
      (i) What areas of availability
         - Local metro area

- Localized region
- National
- International

(ii) What type of pub?
- Special theme pub
- Irish pub
- Beer pub
- Sports bar
- Any other type

(iii) Shop requirements
- Package type (kegs, bottles, cans, other)
- Dispensing equipment
- Glassware
- Advertisement materials
- Other requirements

3. What type of image should the microbrewery project to the on-premise customer?
4. What type of image should the microbrewery project to the off-premise customer, alternatively called the end-user?
5. Are there any purchase price optimums?
   a. Price sensitivity
   b. Discount pricing
   c. Premium pricing
6. Why should any of the targeted customers purchase the micro-brewed beer?
7. What unique services or products will entice the target customers to purchase from the microbrewery either as an end-user or an on-premise customer?
8. What level of service is needed to attract or maintain sales from the targets?
9. Should specific beer types be produced to attract the customers targeted?
10. Should specific events or sponsorships be considered in order to attract the target customers?
11. Continuation of other relevant questions for the business

Concept! Concept! Concept!

It goes without saying that the reader of this book is interested in producing the finest beer possible, but for a pub brewery to properly function, it is necessary to create a concept which will attract enough customers to provide the proper turnover such that the business is profitable. There are many types of concepts one could choose and it will also in some way dictate the kind of beer produced at the pub as well. Some may very well choose to a make a British ale type of beer for one reason or another, while others may choose to produce Belgian-style beers or German-style lagers; whatever the choice, it is important that the other themes of the business integrate with the beer theme to tie it all together.

Pub Brewery Concept

The themes of the beer, food, and hospitality are all very important to attract and keep the targeted customer. If the beers were British ales and the food Korean-style vegetarian, the operators of this establishment might very well confuse its customers or worse, put them off in such a way that they would never come back to the pub brewery. The beer, the food, the dining room hospitality and the atmosphere or as they say in Germany, "Gemuetlichkeit" are all very important pieces of the puzzle. Therefore, one must be sure that all the pieces have been identified and that they fit together appropriately in order to have a properly functioning business unit.

Microbrewery Concept

Unlike a pub brewery that has food and atmosphere and possibly a small concert stage to attract customers, the microbrewery must stand on the merits of the product and the service that it provides to the targeted customers. Therefore, *brand image* must be very carefully considered, beginning with the *beer theme*. The beer theme must be well considered. Once again, it is necessary to look at the market that is to be targeted and then make the beer theme decisions. There are many important decisions to be made; however, this is a fundamental one that will affect all things done at the brewery itself. If it is assumed that entertainment plays a major role in the attraction of a major segment of the target customers, then an *entertainment theme* should also be considered. Another method employed by many breweries to attract targeted customers, as well as to build brand image is to participate in various events. Local as well as other events in the areas where the beer is marketed can be a large part of the brand building. Events give the brewery the chance to expand market share by giving out samples as well as selling single portions of their beer to people who might otherwise never be exposed to the product.

In these ways, brand image must be developed. This task is more easily accomplished if a concept, which is well planned and understood by all who participate in the business, is employed; therefore, the use of themes can assist with the understanding by the brewery staff. In order that all members may understand why the business exists, it is critical to have a *mission statement* that is well known and appreciated by everyone.

Beer Theme

Once again it is time to look at the demographics and the brainstorm, which were conducted earlier, and make some decisions on beer type. Various types of beers are produced worldwide. However, one must look at the local market and make determinations based on the information that has been collected. It may very well be that a decision has been made to produce something totally different from what is available in the local market. There are many different decisions possible, which, however, should be sustainable by a perceived demand in the market. The decision regarding which beer to produce will impact on all decisions made with regards to the other themes, as well as the brand image of the brewery, whether it is a pub or microbrewery.

Having a proper understanding of the history and culture associated with each beer lends credibility to those whose job it is to sell the product. It is imperative the beer that is produced and sold has a culture built around it that can be sensed at all levels by the target customers. This must be effected by producing consistent messages that communicate the theme, history, and culture of the beer to the customers. For example, if the microbrewery is producing German beer, then the Reinheitsgebot should be explained to the customer as well as why using only German raw materials is necessary for producing beers that adhere to the German traditions. Perhaps the beer has a very special history. If so, this history should be thoroughly communicated to the targeted customers. Proper advertisement and point of sale materials should be produced that allow for simple display of this information in order to reinforce brand image with the end-user. If this is properly done, then a consistent theme will be accomplished that is easier to communicate to the customers and, in general, is much more believable.

For the pub brewery, everything possible should be done that can set the brewpub apart from other culinary or drinking destinations. One very strong possibility is the use of the brewing facility as a big show, the centerpiece of the business, not in terms of just the product, but also in terms of how it is made. Emphasize the production methods, by making visible each and every step of the process. Think of it as dinner theater! For instance, one might make an announcement that a new beer is born at some point during the brewing process such as when the beer is transferred to the whirlpool or to the fermentation vessels. Announcements of each step when it is being performed keeps the excitement going. One could even involve the brewery pub guests by giving out samples of first wort or allowing them to see and touch the steaming spent grains while they are being removed.

Food Theme

Many pubs have made wonderful beers and then because of a lack of good food have not been able to attract a large enough clientele to remain in business. It is also possible that the inverse is true; however the point is, all parts must come together to create a whole much larger than any of the other pieces by themselves. The food should be able to stand upon its own merits as well as consistently follow the main theme of the pub brewery. For example, if the pub is producing cask-conditioned ales, it might be a good idea to produce excellent British food items. One possible menu might include "fish and chips" that could be seasoned at the table with malt vinegar produced at the pub brewery itself.

In general, a food theme for a microbrewery is not considered important, but it can be utilized in many ways. One possibility would be to provide certain types of food that cater to brewery tours. Lunch or dinner tours that have been booked in advance create a great atmosphere for building personal relationships with end-users. Another possibility would be to include special beer food that can be made on-site at special events. This could possibly entice people who might not normally be interested in the microbrewed beer, to have a taste of the beer as a side to the food, creating the possibility to entice this person into the microbrewery's customer base.

Hospitality Theme

The hospitality theme follows the beer and the food themes. It includes not only the decoration of the pub brewery, but also the uniforms of the hospitality staff, kitchen staff, and the brewery staff. The hospitality theme also affects the dishes and cutlery used and even if the tables are covered. It is very important that the central theme is followed. The way that the staff speak and interact with the target customers should also be affected. It should go as far as treating the target customers as if they were the center of the universe.

The hospitality staff should be required to recognize and remember the names of their repeat guests. There are restaurant companies in the U.S. who give special rewards to their service staff who are able to recognize 1000 guests on sight. Plaques on the walls of the pub brewery for everyone to see is a strong motivational tool for the members of the pub brewery staff (management included!).

What cannot be emphasized sufficiently is the development of a customer database. It is necessary to become intimately familiar with the pub's guests. Databases allow the hospitality staff to store and easily retrieve birthdays or other special days of the repeat customers and send out invitations to them for a complimentary drink, meal, or a one-time discount. Emphasize parties and special events. Remember, frequent repeaters are what push the business forward and these customers are the ones who advertise the pub brewery the best, utilizing word of mouth.

For the microbrewery, once again a hospitality theme is not generally considered. However, if a brewery tour is to be developed as a way to educate the end users or even the on-premise owners and their staff, then a hospitality theme is indeed warranted. It could be that the microbrewery is allowed to sell beer directly to the public. If this is the case, then it is important that those who are in contact with the end users know how to treat the customer and present the customer with a proper image of the business itself.

Entertainment Theme

The brewery may elect to provide entertainment at certain events throughout the year. If undertaken, this, should be considered important for the development of brand image. In order that consistency from event to event is maintained, an entertainment theme should be established. The theme should specify the types of entertainment utilized at predetermined events, where the inclusion of sponsored entertainment has been determined to have a positive impact on the brand image perceived by the customers who come to the event. The theme should also have a positive effect on the number of customers who participate in the event as well as on sales figures in some predefined manner.

For example, a microbrewery or brewpub has determined that a significant proportion of their end-users are fans of jazz music. Therefore, in order to reinforce the brand image and enlarge the end-user base, jazz bands could be hired to provide entertainment for specific events throughout the year. It might also be that a specific band holds sway over a group of potential targets, therefore, this band might be linked to perform at certain events throughout the year in which the microbrewery is the exclusive beer.

In general, a pub brewery's entertainment theme is a part of the hospitality theme. However, it may be the case that the amount of entertainment held at the pub brewery is such that a separate policy needs to be established. Perhaps specialist entertainment staff are needed to help manage stage timing and equipment for use during specific events.

Event Theme

The *event theme* is tied directly to the previously mentioned entertainment theme for the microbrewery and with the hospitality theme for the pub brewery. However the main point here is "How" the event participation will be executed. There are many types of events in which the brewery can participate. Potential events could include beer tastings, beer festivals, various music concerts or festivals, arts festivals — the list is endless. However, the main factors that should be examined to determine event participation are:

1. What is the cost of the event and is the cost justified?
2. Are target customers going to participate in the event?

3. Can demographic data collection be implemented at the event?
4. Will the event contribute in a meaningful way to brand image development or the overall image of the brewery itself?
5. Can the customer base be expanded by this event?
6. Is the event well advertised and will this advertisement reach a broad segment of potential target customers?

Before any event participation is planned, a thorough examination of the event and the aforementioned factors should be carefully considered.

Mission Statement

Does the brewery have a mission statement? If so, has the brewery management actually considered the content of its mission statement? Does it incorporate the basic themes of the business in a way that is easily understood? It is much easier to develop the strategies and the themes of the brewery if the management as well as all members of the staff of the microbrewery understand why it is in business. It is the base upon which all decisions should be made about who is targeted as a customer, what beer is produced, and the culture that will be communicated to those targeted during sales activities, whether the business is a brewpub or a microbrewery.

Sales Plans

An annual sales plan is essential. A sales plan may be easily accomplished; however, annual goals have to be established and then consistently used as benchmarks year after year as the way to measure the growth of the brand over time. The evaluation will then allow the management to understand if the brewpub or microbrewery is achieving its goals in terms of beer sales and other products sold. Once the annual goals are determined, then monthly and weekly goals should also be selected and tracked. This type of analysis will also assist the management at a more detailed level, as fluctuations based on seasonal conditions will invariably occur. If these data are properly collected and analyzed, then trends and seasonal variations in products sold can eventually be predicted and managed to the company's advantage.

Pub Brewery Sales Plans

Every manager has his or her own ideas on what data should be analyzed, so the discussion below is simply one opinion. First, it is considered a good idea to determine the *required annual sales*. Next, it is possible to break the annual sales figure into the following: (a) beer sales, (b) food sales, (c) ancillary products, and (d) offsite events and catering. Each of these four main categories may be subdivided such that a basic analysis may be performed.

1. Beer Sales
   a. Required annual beer turnover/restaurant seat
   b. Percent on-site sales
      (i) Percent on-premise consumption
      (ii) Percent take-home bottles or other packaged beer
   c. Percent off-site sales
      (i) Percent bottle beer
      (ii) Percent kegged beer
2. Food sales per year
   a. Required annual food turnover/restaurant seat
   b. Percent from main menu items
   c. Percent from à la carte items
   d. Percent from side dishes
   e. Percent from snack items
   f. Percent from dessert items
   g. Percent from "other" drink items
3. Annual sales from ancillary products
   a. Percent from promotional merchandise
   b. Percent from products made from beer
   c. Percent from products made from wort
4. Annual sales from off-site events and catering

The Microbrewery Sales Plan

It is extremely important for the microbrewery to clearly establish what goals should be analyzed and then maintain the analysis on a yearly, monthly, weekly, and even, perhaps, a daily basis. It is necessary for a business of this size to be able to see the trends of product sales as a whole and sales broken down between segments and various shops. This allows adjustments in advertising as well as marketing to be made so that advantage may be taken of the trends that are discovered.

1. Total annual beer sales plan
2. Sales by beer type (analyze on a yearly, monthly, and weekly basis)
   a. Main beer types
      (i) Total sales
      (ii) Percent sales by package type

b. Seasonal beers each type
      (i) Total sales
      (ii) Percent sales by package type
      (iii) Analyze on a yearly and monthly basis
3. Sales by package type (analyze on a yearly and monthly basis)
   a. Percent kegged beer sales
      (i) Total kegged beer sales
      (ii) Broken down by keg sizes
   b. Percent bottled beer sales
      (i) Total kegged beer sales
      (ii) Broken down by bottle sizes
   c. Percent canned beer sales
      (i) Total canned beer sales
      (ii) Broken down by can sizes
   d. Percent other
4. Sales by area (analyze on a yearly, monthly, and weekly basis)
   a. Percent local trade
   b. Percent regional trade
   c. Percent national trade
   d. Percent international trade
5. Percent tourist trade (analyze on a yearly and monthly basis)
6. Sales by distribution method (analyze on a yearly and monthly basis)
   a. Percent restaurants,
   b. Percent pubs
   c. Percent bars
   d. Percent liquor shops
   e. Percent super markets
   f. Percent convenience stores
   g. Percent department stores
   h. Percent tourist gift shops
   i. Percent catalog shopping
   j. Percent on-line shopping
   k. Percent home delivery
   l. Percent other
7. Percent event sales (analyze on a yearly and monthly basis)

A basic analysis of the sales of each on-premise shop would be wise. It would allow the sales staff to make proposals that could be utilized to increase sales of the shop. It is important to understand each shop's base clientele as well as the ebb and flow of the customers. If the basics are analyzed in a consistent manner, much assistance can be offered to the shops and their owners and staff. A basic analysis should, at a minimum, contain the following data, and the analysis should be carried out on monthly and annual bases. If the shop is large with extensive sales, a weekly analysis of some of the data might be a good idea.

Basic sales analysis data:

1. Total beer sales
2. Total volume beer sold by types (if more than one)
3. Total volume beer sold by package types (if more than one)
4. If available, average customer counts
   a. By time (when are the peak and idle times?)
   b. By weekdays
   c. By months
   d. Annually
5. Know the number of seats in each shop and the average beer consumption per guest
   a. If the owner will give the average seat turnover, a rough sales guess is possible
   b. This information could be very useful if the shop owner is purchasing beer indirectly from a liquor shop and no direct information is forthcoming
6. Check normal beer sales against those same sales during promotions and events

Cost Control

Once the sales plan is established, it is necessary to set the stage for the inevitable accumulation of costs of operations. It is important to go through the operation, conducting a review of monthly receipts as well as an annual review to identify any changes in the major cost centers. These could be anything from variable to fixed costs; however, some may be more easily adjusted that others. No matter how difficult or painful, all avenues should be explored and a list of these cost centers properly maintained. These cost centers, which may be adjusted, may even change from year to year depending on many different economic variables.

Once the cost centers have been established, it is necessary to look at individually at them to determine the potential for reduction for each respective item. This is a process that could take a long period of time depending on what type of cost is being considered. It is essential, however, to thoroughly explore the potentials for cost reduction and to realize that cost reduction opportunities are always becoming available. Therefore, the management, armed with the analysis tools provided by the financial data and knowledge of their respective operation processes, should always be vigilant for these opportunities.

Examples of these opportunities could be the raw materials for brewing and the currency changes endemic to the financial markets. In many countries, the brewing tax or the tax on raw materials, and even on some of the packaging materials such as glass bottles, is a very large expense of the production. In some countries, the addition of a certain amount of raw materials not considered to be "normal" brewing ingredients may be used to bring the tax bill down. Japan, for instance, allows that if the amount of malt used is under 25%, the tax rate will be cut by a little more than half. This specific product is called "Happoshu" and has become very popular in Japan recently. Happoshu is a good example of analyzing the current situation of laws and then developing a product that reduces costs and dramatically increases profits (see Chapter 22).

Beer transportation can be a large expense for many microbreweries. It might be of interest to a smaller microbrewery to offer its customers special price incentives to pick their beer orders up directly from the brewery itself, if this is legally allowed.

For the pub brewery, other possibilities could be the establishment of a flexible menu that allows for the change of menu items based upon seasonal produce or surpluses in the food market. Another very strong possibility involves menu items such as soups, salads, breads, casseroles, desserts, etc. that utilize food left over from the production of other main menu items.

Staff Organization and Training (Owner's Standpoint)

Pub Brewery Management Selection

Before the selection of the management team of the pub brewery commences, it is essential to know the needs of the pub brewery. These needs will change depending on the target customer and the theme of the overall pub business that translates into what type of beer, food, and decoration as well as the equipment that will be utilized by the pub brewery. Therefore, the following questions need to be answered:

1. What are the needs of the brewery management and staff?
2. What are the needs of the kitchen management/chef and staff?
3. What are the needs of the hospitality management and staff?
4. What are the needs of the general manager (if he or she exits)?

Once these questions are answered, then selection of the managers may proceed. The selection process itself requires that the expectations of each position are already known. While each person selected for the position will have his or her own ideas about how to operate their respective businesses, the owner should already understand what he wants from the management staff in order that all parties have similar expectations from the employment contracts. With this in mind, an interview strategy should be well planned, and all laws regarding what can and cannot be asked during a job interview be well understood, so that no embarrassing situations occur during the delicate process of job interviews.

Microbrewery Management

Basic management selection for a microbrewery follows the same basic rules laid out in the pub brewery section. However, there are major differences in that there are no kitchen and hospitality staff. Instead, the microbrewery will have a large sales staff and a sizable administrative staff. It is possible someone will be in charge of the brewery information systems as well as marketing and advertising. Once again, it is necessary to emphasize that before the management team is selected it is important that the overall needs of the business are known. Understanding the expectations of the management and the staff will help them all work together in a more efficient manner.

Chain of Command

Once the managers have been selected, conversations with each manager individually and together in a group should be conducted so that all operations of the brewery are understood by the management personnel. This is the first step in establishing common ground for management of all parts of the brewpub and microbrewery.

After these conversations have been completed, the owner and general manager may very well decide that new information should be integrated into the master plan for the brewery. Once the new managers have their feet on the ground, they should come up with a plan for each of their respective sections, which can be integrated into an overall plan of execution. Once the owner and general manager have understood the plans received and made any adjustments required, it is time to make the decisions on the chain of command.

This is a very important step. Occasionally, there will be times when one of the managers is not present, and it is important that all people know exactly how the operation is to continue. If this point is not made clear at the beginning of the business, problems can easily result when a situation occurs for which no one was prepared. These decisions need to be made and written down so that they may become a major section of the employee handbook that should be present in all businesses.

Employee Handbooks

Employee handbooks are often overlooked as being unnecessary or even an item which can be done without "until there is enough time" to think about such luxuries. The handbook or a written set of rules for everyday conduct of work is absolutely mandatory. The management should know how the work and information is to flow throughout the organization, as should all members of the business.

All employees should understand where the dangers in the business are and how to be safe when working in those areas where potential danger exists. Everyone should be able to have a basic understanding of the brewery operation and be able to explain to anyone who asks. In a like manner, all members should be able to explain the hospitality requirements of the pub brewery to anyone at anytime. The pub brewery's food is also a major part of the business. All aspects of the kitchen operation and all items on the menu should be understood by everyone, including the brewmaster and brewery staff. For the microbrewery, all administrative staff and sales staff should have a basic understanding about what everyone else is doing within the organization.

Black boxes are simply tinderboxes for dissent that is never welcome in the workplace. Therefore, the employee handbook should cover all the aforementioned parameters, including when employees should be trained and how the training process works. It should also discuss the information system and the flow of the information on a daily, weekly, monthly, and yearly basis. All employees should have a basic understanding about how this information is used and why it is important, or else they will not properly function within the information reporting system. Understanding is the key motivating factor for each member of the pub brewery operation to function properly.

Operations

The management positions needed are of course up to the owner; however, a list of management positions to be filled might include:

Possible management positions — pub brewery

1. General Manager — Manage all sections and set the overall goals
2. Brewery Manager — Manage the brewery including staff and maintain proper production
3. Kitchen Manager — Manage the kitchen including staff and maintain quality control of the food
4. Hospitality Manager — Manage the hospitality staff and maximize customer satisfaction

Possible management positions — microbrewery

1. General Manager — Manage all sections and set the overall goals
2. Brewery Manager — Manage the brewery and maintain proper production
3. Sales Manager — Manage the sales staff and maximize customer satisfaction
4. Information Manager — Manage all data and insure the flow of information
5. Marketing and Advertising Manager — Manage and expand the customer base

It is possible that the one person could fill two or more management positions. However, job focus is a very important element in any of the management positions described in the list above. In many microbreweries, only a brewery manager and a sales manager are present to operate the business. A member of the administration staff operates the information system, while an outside company reporting to either the sales manager or the brewery manager handles the marketing and advertising functions.

General Manager

As described previously, the general manager, or the manager executing the day-to-day operation of the microbrewery is in charge of developing the overall plan for the entire business. Each section shall receive from the general manager its respective job requirements and budgets within which to perform. It is the general manager's job to see that proper goals are set for each of the section managers, and that these are carried out with periodic evaluations.

Brewery Manager

The brewing of beer is the specific job required of the brewery manager or the brewmaster. It is his or her job to see that the planned output of the brewery meets the needs of the sales department. This job requires that raw materials arrive at the brewery in a timely fashion, that the beer moves through the tanks in such a way that the proper amount of beer is always available, and that the beer is always properly produced meeting the required specifications in a consistent manner. Quality control is another part of the brewmaster or managers job. The quality of the packaged beer when it leaves the brewery must be guaranteed. A minimum shelf life must be assured as well. Consequently, any claim handling ultimately comes back to the brewery manager.

Sales Management

The output of the microbrewery becomes the responsibility of the sales manager. In general, it is the sales manager's job to recruit on-premise customers. Part of the job of recruiting these shops is to service them. The sales team must provide the customer with exceptional service, including information supply and market analysis. The sales team must also provide product handling and, on rare occasions, claim handling. It is also necessary for the sales staff to work closely with the marketing and advertising staff to provide sales promotions and other special events for the shops, to assist in sales of the products. If these jobs are handled correctly then the shops should do well with the beer supplied by the microbrewery.

Advertisement and Marketing

The marketing and advertising manager has the responsibility for recruiting end-users to the beer. An analogy could be made to hunting. The sales manager is the hunter with the gun. The marketing team is the hunting dog that can help the hunter find the location of the target. The advertising team is like a decoy that entices the prey to a specific location where the hunter lies in wait. The marketing team should determine where the target customers are likely to be, and help direct the sales team members in that direction. The advertisement team utilizes various forms of promotions and promotional items to lure the end users to the beer or the shops where the beer is available. By utilizing various advertising tools and events, the marketing and advertising manager polishes the brand image and works hard to attract as many end-users to the microbrewery's beers as possible.

Information Management

In any brewery, information management is the key controlling factor. A database system should be set up that can maintain records on all key data. This type of system could be managed by either a turnkey database system, or database built to specification using a number of database engines available on the market. A rough sampling of what could be considered important data might look like the following.

Key database files

1. Raw materials inventory
2. Packaging material inventory
3. Beer production volumes
4. Packaged beer inventory
5. Quality assurance tracking
6. Claim management

7. Client contact data
8. Client contract data
9. Client sales and invoicing (on-premise customers)
10. End-user sales invoicing (whenever possible)
11. Transportation data
12. Overall sales data (broken down for analysis in accordance with the sales plan)
13. Financial statements
14. Personnel matters

Once the system is set so that data similar to the above are being tracked, then it is easy for the management team to see what is happening inside the brewery as well as in the various markets where the beer is being sold. This data is vital to the sales manager and the advertising and marketing manager for maximizing the sales and expanding the customer base.

Kitchen Manager

The kitchen manager or if preferred, the head chef, manages all aspects of the kitchen operations. The chef is in charge of analyzing the customers' tastes and maintaining a menu that consistently exceeds the pub brewery guest's expectations. This is done by seeing that all the special details are looked after, such as raw material inventories, cook training, and recipe development. One very important point is that those preparing the food under the chef's supervision follow recipes developed by him or her.

Hospitality Manager

The hospitality manager is the one who pulls all the sections of the brewpub together. This manager brings the guests together to eat and drink. The staff, under the direct management of the hospitality manager, serve both the beer and food produced in their respective areas in the dining room in an atmosphere that complements the food and makes the guests comfortable and interested in repeat patronage. It is also the job of the hospitality manager to keep track of the guests and develop systems in which guests can be lured back to the pub brewery; for example, birthday cards with complimentary coupons for food and beverage.

Operation Reporting and Filing System

Another name for the operation reporting and filing system is the management information system or MIS. It is essential for the business to keep a record of what is going on at the brewery. There are various turnkey systems available on the market with the price being dependent on what types of information management are required.

Some type of information system is generally included with a cash register point-of-sales system (for the pub brewery). The financial information is necessary for tax reporting purposes, but also so that one can see the financial and sales trends over a period of time, and forecast what the needs will be in the future. This information will also assist with the evaluation of the business. The data can be compared to the sales plan that should have been made at the beginning of the year, and then decisions can be made regarding food production, beer production, and even, depending on how far one goes, what types of advertising should be considered or perhaps how it should be adjusted for the upcoming year.

A customer database that tracks how many times each end user makes purchases is very important. Possible data to collect could be: (1) addresses, (2) telephone numbers, (3) e-mail addresses, (4) birthdays, (5) anniversaries, (6) when newsletters have been sent, (7) attendance at beer club functions, or (8) other special events. These types of data allow the hospitality or sales team to make predictions on who is going to come to the pub, at what times, and then be prepared in advance to welcome these guests personally. Perhaps the pub brewery has a special beer mug cabinet for certain members — it might be possible to have the beer mug polished just for the special occasion. This is an example of how to "wow" a guest and to attempt to exceed their expectations at the pub brewery.

Customer Interaction Planning — Pub Brewery

Many businesses simply begin operations without planning the employee–customer interaction. The relationship with the customer should be planned and managed at all times. Even if the overall business plan is to simply produce food and brew beer for a small-scale pub brewery, it is still important to think through all steps of the brewery, the kitchen, and the hospitality functions. It is also important to turn around and think through the process from the prospect of the guest or customer as well. It should be considered whether the management or the staff have the correct point of view when performing the services of the pub brewery. In the end, the question is, "Will the customer ultimately be pleased with the service or the products purchased?"

Customer Interaction Planning — Microbrewery

In order to fully utilize the staff of the brewery, especially the sales staff and others such as the managers of the various sections that come into direct, daily contact with the public customer, interaction should be managed. Those who become the public face of the business must speak with one voice on all things about the microbrewery. Therefore, it is important that these people fully understand the beer, its culture, heritage, and traditions.

They must also fully understand the brewery's customers, both on- and off-premise.

Therefore, in order to provide a common framework for these people who become the public face or brand ambassadors of the brewery, it is important that all aspects of when, where, why, and how they meet the public are properly understood. Basic knowledge of how the customers interact with the microbrewery and what their expectations are is a necessity. Once these interaction points are planned, the microbrewery can better anticipate the outcome of these customer interactions.

Example questions to assist with customer interaction planning:

1. At what point do people in any of these three areas of operation come into contact with guests at the pub brewery?
2. How do they come into contact with the guests?
3. What is the desired outcome of this contact?
4. How should the contact be conducted?
5. Who should make the contact?
6. How should the contact be terminated?
7. How should the contact be evaluated?

First determine the contact type: for example, a Beer Fan Club Event. Determine which personnel members are to be present and with which guest type each staff member will probably have contact. A table, such as the example shown here in Table 21.8, should be made, possibly with more space dedicated to job explanation as well as a place for the location and the time the job should be executed. Rather than just generic titles, actual names should be used, where possible. It is important to include as much detail as possible. The goal here is to be as specific as possible so that there are no questions regarding what should be done during the function or the specific job listed at the top of the table.

TABLE 21.8

Customer Interaction Planning

Beer Fan Club Event (Example)				
Staff Members	Event Job	Event Guest	Pub Guests	Shop Guests
Hospitality manager	Master of ceremonies	Event direction	As needed	Problem assistance
Hospitality staff	Guest assistance	F&B services	F&B services	Sales service
Kitchen Manager	Menu explanation	Food Q&A	As needed	Emergency only
Kitchen staff	Food production	Food service	Food service	Emergency only
Brewmaster	Beer explanation	Beer Q&A	As needed	Emergency only
Draft staff	Beer drafting	Beverage service	Beverage service	Emergency only

All staff should understand what their roles are in each *customer interaction plan*. There should be enough of these charts in order to help all the members of staff to understand their functions. The previous example is also quite useful for the microbrewery in planning events or concerts. There are many different types of customer interactions, and it is imperative that they are properly planned so that the best results can be produced in brand-image building.

Staff Training System

Once management understands the proposed flow and operation of the business, it is time to plan the *training system*. Training is more difficult at the opening of a business than during the actual operation. After the brewery is open, there are employees who have an extended amount of experience who can either assist with or perform the actual training of new staff. Preparation is a necessity for the training program. If it is possible to plan and prepare a script, this is by far the best method for training new members of the staff. Every job at the brewpub or microbrewery should have a training plan and someone who is officially in charge of the training process. Once the training processes have been finalized, hiring may begin in the case of a new brewery.

Hiring Staff

Recruiting new staff is one of the most difficult jobs. It is made even more difficult because of legislation in some countries regulating the questions that can be asked during an interview. As mentioned previously, find out what they are in the country or location where the pub brewery is being operated. Make sure that the job or position that is being filled is understood, perhaps a copy of the *customer interaction plan* as well as a flow of the jobs within the section where the interviewee will be potentially working could be helpful.

One key point to remember is that questions based on gender, religion, age, racial background, or medical issues can all be considered taboo depending on where the interview is taking place. It is best to focus on the job description and what qualifications and experience the applicant has, which will assist them in fully performing the job for which they are interviewing.

Staff Training

The goal for the business is to exceed the expectations of the guest; therefore, a properly trained employee is a requirement. Well-trained employees are better suited to do their jobs, are happier, and remain longer with the business. A long-term employee who understands his or her job and performs it properly, helps with the goal of exceeding customer expectations, because a level of consistency can be expected. These are the people who in the end make the business a success. They are the ones who make sure it all works.

The management team can be the best at their particular jobs, but if they do not get the training done and instill an understanding of what is needed from each individual in their own job (job ownership) then the business cannot succeed. It is the troops, after all, which fight a war, not the planners back at the base. Therefore, it is imperative to have a system in place which determines "Who" needs to be trained and "When" training should be convened. If the system is mature enough, it might even consider "How" the person should be trained.

Staff Motivation

Staff motivation can come in many forms, but communication is the grease that keeps the wheels spinning. It is necessary to make sure that all members of staff understand what all other members of staff are doing. Rewarding the members of the various teams for jobs well done is a great method of motivation. If the brewery produces new products, it can be very useful to give out samples of the new product to all members of the staff first so that they can feel valued. They also have a way of feeding the public rumor mill about new products and can help build customer anticipation. Finding ways to publicly praise members of staff is also very good. Putting plaques with names on the walls for public viewing is also quite useful.

Operation handbooks can be useful by making it difficult for people to deny that they never heard mention of specific rules. Providing the handbooks, and requiring each member of staff to sign a special acknowledgment section in their handbook as proof that the contents have been read and fully understood, protect the management and the brewery against claims of ignorance in the case of an accident. It also gives everyone a point of reference when they have a question about certain procedures. Referring back to the handbook during staff meetings also helps to reinforce the message management might be trying to give out at various meetings.

Claim Management

Customer Expectations

Claims are a part of business. People are human and somewhere along the way of purchasing a product, some part of the experience may not occur as was initially expected by the consumer. The failure to meet a consumer's expectations has the potential for becoming a customer claim. This is why the term "exceeding the customer's expectations" has been used so often in this text. If the customer's expectations are always surpassed, then a claim is less likely to occur. However, this is not possible 100% of the time. With claim management, the key is to understand how the microbrewery and its products interact with the consumer. It would also be helpful if the expectations of the consumer were understood. This refers back to *customer interaction plan*. One more point to remember: consumers are not infallible, they make mistakes just like everyone else. The trick is to help the consumer realize this without telling them they are wrong.

Understanding!

The key word for claim management is "understanding." As has been described throughout this text, one must know the brewpub or microbrewery. There are many interactions occurring within the various organizations operating within the business and understanding these are the key. Having said this, in general, claims originate with a customer against something that has happened to them as a consumer. This is another reason to have completed numbers of *customer interaction plan* tables. If the whole staff and particularly the management understand the various interactions between the staff and the various sections of the brewpub or microbrewery, then it is easier to perhaps come to a happy conclusion with a customer claim.

Probable Causes

Evaluate the business for areas that have a potential for claims. This might require that a frequent repeat customer be asked a favor and assist you with their unique perspective in the business. This person could help the business understand which unique areas could be sources of a claim. Once those areas are understood, action must be taken in advance to reduce the possibility of a claim. Proactive management that pushes for removal of probable sources of claims is just good business and helps maintain the good reputation of the facility and the products produced.

After the most likely areas of claims have been determined, management should consider what are the most likely reasons a claim could be made. Generally, big claims that tend to get blown out of proportion usually occur when a customer has more than one reason to make a claim, but has been "patient" with the business. This is a situation to be avoided. It is the job of the hospitality staff or the sales staff to give the customers ample opportunity to make suggestions as well as voice grievances when things are not perfect. It is the hospitality or sales staff's job to facilitate communication at the pub brewery or microbrewery between the guests and those members of the staff that need to know, be they other members of the staff or management. A line of communication must exist at all times during the operation of the business.

Planning for Claims

It may sound very silly to plan for claims, but actually claims are one of the best friends of business. Guests often tell us what the business does right, but generally, they do not tell what has gone wrong. The average customer, when feeling they have been wronged will simply not go back to the business, and then go on to tell all their friends and family what a bad experience they had while patronizing that business. Because of this scenario, one must look at claims as a possibility to improve the business and find ways to get customers to report problems with the service or the product. Therefore, "How" the business responds to a claim is just as important as providing the customer an avenue to voice a claim.

Therefore, plan for handling claims! Finding the probable causes of claims was discussed earlier, so take that one step further and create various scenarios to cover what to do when those claims or even more importantly, the unplanned claim occurs. One more thought, please remember that everyone is human, even our best customers do make mistakes! It is in how we listen and then explain the businesses situation to the guest, and handle the claim itself, that makes the difference between a customer going home unhappy or satisfied.

Health and Sanitation Checks

It cannot be impressed upon the management and staff enough, that maintaining the sanitary conditions and cleanliness of the work facility at all times are of utmost importance. The brewer is already aware of this fact, for without proper sanitary conditions, beer production would not be possible. However, even though managers of other sections of the business also intrinsically know this, it can happen that this very important point can be either forgotten or simply brushed off to the side due to being "too busy." It would be a very good idea to establish sanitation monitors in all sections of the facility so that a bad situation can never occur. Realize that the local health officials will be making their own checks from time to time. The management should, therefore, take a proactive approach and ask for a voluntary sanitation check by the local health officials to see if any problems exist. In order to become acquainted with the local health regulations, request a meeting with them and discuss those issues of concern. Find out their information needs in advance, and then be sure that the brewery is fulfilling those needs in a timely manner.

Drink Driving

"Drink driving" has become a major topic worldwide in the last decade. It is a foolish pub business that has no plan to assist those guests who have overindulged in alcohol and plan to drive home. In some areas of the world, the most probable end result of an automobile accident by the aforementioned guest would be the arrest of the pub manager. While this is possibly the worst-case scenario, this is what management is all about — managing the business so that such accidents do not occur. Many pubs now have designated driver programs available at their businesses. When a group of people come into the pub, the person who is designated the driver may simply tell the hospitality staff and then receive alcohol-free drinks at either a substantial discount or even for free. Education is the key to maintaining customer safety. All members of the business staff should be educated and creative ways should be developed to educate the guests as well.

While the microbrewery itself is most likely not selling beer for consumption, a "drink driving" program or initiative should be in place. The program could be used as an example for the on-premise shops that are visited by the microbrewery's sales staff. The brewery wants to maintain a well-polished public image, and having such an education program and encouraging the shops that sell the microbrewery's beer to adopt a drink drive policy could be very helpful from a public relations point of view.

Sales Methods

Point of Sales

Now that many of the management issues of the business have been covered, it is time to look at some of the methods that may be employed to push sales. While there may be a lot of books available which explain the way to sell products, it is very important to come up with methods that are specific to the services and products for sale at the pub brewery or the on-premise shops supplied by the microbrewery. First of all, think through all the possible locations within the facility where purchase decisions are made. Now, consider what type of advertisement might fit unobtrusively into that specific area. It could be anything. Creativity is the most important point when thinking of point of sales (POS) advertisements.

Point of Sales Locations

1. Front of the business (signage)
2. Entry way
3. Waiting area
4. Customer menu
5. Walls of the facility
6. Bar counter
7. Areas around the kitchen
8. Areas around the brewery
9. Customer tables
10. Customer toilets
11. Concert stage (If available)
12. Others

Point of Presence

The microbrewery wants its beer to be present at all locations where purchase decisions are made. This may be referred to as point of presence (POP). Items that remind the customer of the business and the products available are very important to the overall sales at the business. Merchandise items are a wonderful point of presence that the customer themselves pay to use. The best example of POP advertising is merchandise that has been so well designed that the customer uses it as a part of their lifestyle, reinforcing a subliminal message about the business at various times throughout the day or week. If the items may be worn in public, then the business has a "point of presence" that is moving throughout the community. Nothing can be better than advertising that is capable of being seen in various locations over a period of time from which the business has received a fee. Other POP may be pamphlets that have been left for potential customers to pick up at local tourist locations such as local hotels or other locations friendly to the brewery. Special events that have been sponsored by the business or are being directly managed by the brewery are great POP opportunities.

Events

Promotion of the business is essential; therefore, various events should be planned throughout the year that will attract a diverse group of customers. These events should showcase various strengths of the brewpub or microbrewery. Not all people are simply interested in the food, the beer, or even the hospitality and atmosphere. Perhaps their interest can be piqued when other things are combined with the beer, food, service, and atmosphere of the pub brewery. One such possibility could be concerts showcasing local musicians or even well-known musicians or stage actors if the budget allows. When planning such an event, it is best to think about the tastes of the majority of your clientele and make decisions on the type of music that will not alienate them. If the Monday or Tuesday night customer base is very weak, the inclusion of music or other stage performance that does not match the main clientele maybe of benefit because it would attract a different group of customers that diversifies the customer portfolio.

Participation in local community events is a very important way of polishing the image of the brewery. Most large cities have a tourist board or a committee that oversees such events. Join the local Chamber of Commerce to stay on top of all the newest events. If the city has no beer events, such as an October Fest, one could be started by the brewery first, and then if it grows in size, ask the city for assistance in growing the event into something for the community at large. There are no guarantees, but the worst they can do is say "No."

Other events that may be helpful are tours, which can actually be a method of consumer education about, not only the beer, but also what the business as a whole is all about. Brewing seminars are also very popular. These can be used as a method of attracting to the facility home brewers who might not otherwise come. They give the business a chance of expanding and diversifying its consumer base and starting new lines of communication within the community. Special beer-themed events are a way to share beer culture with the community at large. When producing seasonal beers, it is a plus if the sales of the beer corresponds with an event which prominently exposes the special points of the beer from a native cultural prospective. Food-themed events allow the chef and the kitchen team to shine. They also attract a group of people who like beer, but are more into food. A "Beer and Dine" event could be operated in a way that only a limited number of participants is selected in some way that makes them feel special. Recruiting to this type of event allows both the food production and brewery staff to shine at the same time. This type of event definitely requires the use of a database with comments that allow the management to determine those customers that would best fit this type of event. All of the events discussed have one thing in common: customer base building and diversification through either an event that brings in new types of customers at different operation times or reinforces their desire to spend more time at the business itself.

Sponsored events:

1. City sponsored events
2. Local events
3. Brewery tours
4. Brewing seminars
5. Beer-themed events
6. Food with beer-themed events
7. Fan club events
8. Beer and dine events
9. Beer tasting events
10. Sponsored concert events

Advertising and Promotion Plans

Planning

As always, planning is the first step in the process of management. It is necessary to look at the annual sales plan in order to develop a plan for the promotions, advertisements, and the key events of the year. As was said previously, establishing events that correspond to seasonal beers as well as national and local holidays is a must. Performing a media audit indicates that medium which is most attractive to those consumers who have been targeted by the brewery. Once the best medium has been selected, advertisements that are directly targeted at the brewery's customer base and potential customers are devised and published or transmitted by the appropriate route. Possible media examples are newspapers, magazines, radio, television, flyers, direct mail, newsletters, Internet homepage and email, as well as news lists on the Internet.

Another major part of the planning process is the utilization of demographic data that have been collected during promotions and advertisements from previous periods for market analysis. The data collected should give the marketing and advertising manager some idea about who belongs to the brewery's customer base. This goes back to the analogy used earlier about the hunter. The advertisements and promotions should be used as a way to lure the members of the hunted customer base, as well as those who have been targeted as potential customers. While performing the advertisements and promotions, it is essential that further demographic data collection be planned for analysis. These data can tell the marketer if the ads are hitting their mark, if the mark has been missed, or even perhaps has changed.

Product Ambassadors

Another method in which the business can promote its products is by the use of *product ambassadors*. The owner and all managers of the various parts of the business should be considered ambassadors. Any time these people go out in public in the name of the business, it is essential that they promote the product in a planned manner. It is even possible that such situations could come about on a private basis, and if it is possible and not a problem for the manager, then again a promotional plug should be considered. If other members of the staff also feel so inclined, they should be provided with instructions that would allow them to become a proper ambassador for the business. This assists with another very important form of promotion, word of mouth.

Feedback

Feedback, especially that which is negative, can be of very good use to the brewery. It helps expose the weak points of the business. Therefore, not only should the customers be given this opportunity but so should the staff as well. Feedback may be in the form of comment cards as well as encouraging email from a portion of the Internet homepage that would not require the input of a name or any other identifying information. As discussed earlier, it is the job of the hospitality staff to provide ample opportunity to the guests to give feedback. A forum for the reporting of the verbal feedback should be built into the management system.

Professional Assistance

It may be that within the business there are in sufficient ideas or hardware for carrying out advertisement or promotion in a manner that properly supports the business. Photography as well as design and desktop publishing are all very complex jobs that many professionals spend their lives mastering. Depending on the individual needs of the brewery, it may be suggested that a local advertising and/or design company be solicited to assist with the advertisement and promotional activities. Even if someone at the brewery has the skills for developing materials used for promotions and advertising, contacts within the media and other mass communication companies are more easily found when such a company is on retainer. These types of contacts can lead to lower costs when contracting with advertising companies. Promotions and advertisements are a very difficult area but are required for continuing the expansion and diversification of the customer base of the business, not to mention sales.

References

1. Bamforth, C.W., A brief history of beer, Proceedings of the 26th Convention of the Institute of Brewing — Asia Pacific Section, 2000, p. 6.
2. Coulter, M., The step-by-step approach to planning, building and running a small brewery is the only way to fly, *New Brewer*, 4:27–33, 1987.
3. Eschenbach, R., Gasthausbrauerei, *Brauwelt*, 43:1916–1918, 1982.
4. Eschenbach, R., *Gasthausbrauereien*, Hans Carl Verlag, Nuernberg, 1993.
5. Marchbanks, C., Avoiding the pitfalls when brewing on a small scale, *Brew. Distill. Int.*, March, 15–17, 1987.
6. Braitinger, M., General aspects in microbrewing, *Tech. Q. Master Brew. Assoc. Am.*, 23:100–103, 1986.
7. Piendl, J.S. and Schuergraf, W., Ueber die Bierherstellung in deutschen Gasthausbrauereien, *Brauindustrie*, 4:404–410, 1990.
8. Radtke, K.H., Die Gasthausbrauerei, *Brauwelt*, 1/2:35–36, 1995.
9. Caspary, R., Braeuhaus — System Caspary, *Brauwelt*, 30/31:1222–1225, 2000.
10. Radtke, K.H., Die Gasthausbrauerei, *Brauwelt*, 5/6:229–230, 1995.
11. Anon., *Brauerei Adressbuch*, 22nd ed., Fachverlag Hans Carl, Nuernberg, 2002.
12. Reischl, G., Gasthausbrauerei Kleinbrauerei, *Brauindustrie*, 6:630–637, 1989.
13. Anon., *The Institute and Guild of Brewing Company Directory 2002*, The Institute of Brewing and The Institute and Guild of Brewing, London, 2002.
14. Heyse, K.U., *Brauwelt Brevier 2002*, Fachverlag Hans Carl, Nuernberg, 2002.
15. Eschenbach, R., Gasthausbrauerei Technik Technologie Konzeption, *Brauwelt*, 48:2218–2233, 1983.
16. Eschenbach, R., Die betriebswirtschaftlichen Grundlagen der Gasthausbrauerei, *Brauwelt*, 44:2110–2116, 1986.
17. Kraemer, T. Gasthausbrauereien von der Idee ueber das Konzept zur Realisierung, *Brauindustrie*, 73:823–827, 1988.

18. Radtke, K.H., Planungsanlagen einer Gasthausbrauerei, *Brauwelt*, 18:893–895, 1995.
19. Schulz-Hess, G., Sind Gasthausbrauereien noch im Trend, *Brauwelt*, 3:61–66, 2000.
20. Admire, M., Creating a sound feasibility study, *Brew. Pub Mag.*, November/December:1–4, 2000.
21. Schoolmann, G., Abseits, Standortkriterien im Gastgewerbe (November 30, 2001) Available at: http://web.archive.org/web/20011205094510/http://abseits.de/standort.htm (November 25, 2002).
22. Peterson, H., *Brauereianlagen*, 2nd ed., Carl Hans Verlag, Nuernberg, 1993.
23. Zentgraf, G., Das Erlebnis de Bierbrauens, *Brauwelt*, 3:62, 1983.
24. Hook, A., Local small scale brewing: re-inventing the wheel? *Brew. Distill. Int.*, February, 26–34, 1993.
25. Kallinowski, G. Gasthofbrauerei ein gastronomischer Trendsetter, *Brauwelt*, 47:2305, 1987.
26. Flad, W., Hausbrauerei mit 10 hl Ausschlagmenge, *Brauwelt*, **27**:1111–1114, 1990.
27. Muenker, R., Hausbrauerei mit 1000 hl Jahresausstoss, *Brauwelt*, 44:2105–2107, 1991.
28. Schmidt, H.J., Mikrobiologische Qualitaetssicherung in der Brauerei, *Brauwelt*, 44:81–85, 1991.

22

Innovation and Novel Products

Inge Russell

CONTENTS

Introduction	817
Taxes and Innovation in New Beverages	818
Happoshu	818
Japanese — Third Category Beers	818
Malternatives in North America and Flavored Alcoholic Beverages in Europe	819
The Blurring of the Lines between Spirits and Beer	819
Health and Innovation	820
Nutraceuticals, Functional Beverages, and Health Claims	820
Low-Carbohydrate Beers	821
Widgets	823
Colder Beers	824
Ice Beers	824
Superchilled Beer	825
Immobilized Cell Technology — Progress	825
The Aluminum Bottle	826
The Plastic Beer Glass	827
Neuroeconomics	827
References	828

Introduction

Innovation is critical to all industries and the brewing industry is no exception. The customer is always looking for something new in the marketplace. A new taste, new technology or a convenience, or quality enhancement. Some of the more recent innovations in the industry are described in the following text.

Taxes and Innovation in New Beverages

The tax structures in countries are often the drivers behind innovation and new beer styles. This can be clearly seen in the "malternative" beverage section in North America and in the "happoshu" market in Japan.

Happoshu

Happoshu, which means "sparkling liquor," is a beer-like low-malt alcoholic beverage that is extremely popular with consumers in Japan because of its lower price relative to regular beer. This price difference arose due to a loophole in the tax law. In Japan, the Liquor Tax Law defines beer to be an alcoholic beverage with 66.7% malt content and Japanese beer was traditionally taxed on malt content. Happoshu contains less than 25% malt, giving it an alcohol tax less than the amount imposed on standard beer.

Happoshu, at a lower price than beer, and marketed as a sparking malt beverage, was launched in October 1994 by Suntory Breweries. Happoshu cost two thirds of the price of regular beers and was a surprise hit. Sales rose quickly, and each of Japan's leading breweries quickly launched their own brands. By 2003, the Japanese beer market was segmented into 61% beer and 39% happoshu. In May 2003, there was a tax increase on happoshu, which reduced sales, but happoshu still commands a significant portion of the market and offers a beer-like beverage at a competitive price. Happoshu is a beer made with less than 25% malt and as much as 75% corn or rice adjunct. Although the alcohol content is not lower than regular beer, it has tax and production advantages due to the low malt content.

In 1994, brewing regulations in Japan were changed and the minimum annual volume of output required for a beer-brewing license was lowered from 2000 to 60 kl. For happoshu, the requirement was set at 6 kl. This easing of the regulations made it easier for smaller breweries to start producing low volume, hand-crafted products. This category also allowed for the production of brews made with ingredients not approved for use in beer. Hence, the Japanese craft brewers have had an opportunity to experiment with unusual ingredients including vegetables and fruits.

In the ever-changing market, today happoshu is competing against lower-priced beverages such as "chuhai" a drink based on Japanese shochu liquor and soda water.

Japanese — Third Category Beers

These new beer products are alcoholic beverages with a beer taste that are in a lower tax class due to ingredients. The pea-based "Draft One," manufactured by Sapporo Breweries Ltd and which made a nationwide debut in February 2004, is an example. Similar in taste to beer, "third beer" is cheaper than beer, because it sidesteps taxes levied on beverages that contain more malt as it obtains its starch from legumes such as peas. It is expected that happoshu will be the product to lose the most market share to this lower-priced rival.

Malternatives in North America and Flavored Alcoholic Beverages in Europe

The market for flavored alcoholic beverages (FABs) in the United Kingdom has been in existence for over a decade and is now mature. It is another example of beverage producers responding in creative, if unanticipated, ways to government regulation. This market is now dominated by large spirit companies such as Diageo and Bacardi — Martini. A major increase in excise duty in the United Kingdom in 2002 slowed growth, but there is still a rapid introduction of new products into this market. Six brands now account for three quarters of the volume.

In North America, there is also a preponderance of these specialty malt beverages. One of the first was ZIMA Clear. It was introduced in 1992 by Coors and was available nationally in th the USA by 1994. It was marketed as a refreshing, lightly carbonated alcohol beverage, an alternative to traditional beer. It distanced itself from beer in terms of appearance, flavor, and marketing. This was in contrast to the "clear beer" that Miller tested in the same time period. The next wave in this category involved variations on hard lemonade, with products such as Hooper's Hooch during the mid-1990s followed by Mike's Hard Lemonade and numerous similar products. These products were characterized by their sweet taste. In European countries with different tax arrangements, the malternative niche is occupied by the so-called "alcopops." These are diluted, flavored spirit-based drinks — with no pretensions to beer.

The Blurring of the Lines between Spirits and Beer

Smirnoff Ice and Smirnoff Ice line extensions, leaders in the malternative category, are examples of products that illustrate how the distinction between spirits and malt beverages is becoming blurred. The regulations and definitions of what is a "malternative," what is the alcohol source, and how is it taxed are complex and vary from country to country.

Tequiza was launched in 1999 by Anheuser-Busch. This successful new-product introduction was marketed as an American-made malt beverage, combining lager beer with blue agave nectar and a natural flavor of lime and real Mexican tequila. Tequiza surpassed established brands to become one of the four best-selling high-end beers in supermarkets, where micro, import, and specialty segments compete.

In 2004, a drop in the malternative market was observed, but a new market for "low carbohydrate" beverages was developing, and products such as Miller Brewing's Skyy Sport were the first malternatives to highlight low-carb positioning. This particular beverage is an ultra-premium, low-carb, citrus-flavored malt beverage with a splash of cranberry.

In 2005, Anheuser-Busch was the first major brewer to infuse beer with caffeine, ginseng, and guarana, the latter a caffeine-bearing herb used in Brazilian soft drinks. Each 10 Oz can of the product named "B-to-the-E" contains 54 mg of caffeine. The product is fruit-flavored with aromas of hops, blackberry, raspberry, and cherry and is served over ice.

Health and Innovation

Nutraceuticals, Functional Beverages, and Health Claims

There have been numerous publications in recent years that have linked moderate drinking and health claims. The red wine industry has been very successful at leveraging sales due to the publicity surrounding some of the findings relating to the ingestion of resveratol and drinking red wine. The scientific literature has seen numerous health studies in recent years that suggest that there is a positive link between moderate beer consumption and good health, and there are a number of websites devoted to updating the public on this area. One such site is http://www.beer-and-health.com.

Most research indicates that moderate consumption of alcohol provides significant health benefits, primarily through reducing the risk of cardiovascular disease, but also with respect to reduction in risks associated in other health areas. With few exceptions, the epidemiological data from over 20 countries in North America, Europe, Asia, and Australia demonstrate a 20–40% lower coronary heart disease (CHD) incidence among moderate drinkers. Moderate drinkers exhibit lower rates of CHD-related mortality than both heavy drinkers and abstainers.[1-4]

In Japan, in response to consumer demand, Japanese beer companies have been developing a new genre of beers with physiological functions — beers that are effective against obesity, beers that are beneficial for gout, etc. Low-calorie beers are enjoying good sales, and beer containing dietary fiber and beer with a lower purine base are also becoming popular. The prediction is that the market for beer with physiological functions will continue to grow in Japan for as long as the healthcare product sector continues to boom.

A beer marketed as a functional beer was launched in the United States in January 2005. This Swedish beer "Aventure Functional Beer" is made by a process developed by Lund University in Sweden. The brewing process employs oats instead of barley and uses enzyme technology to create a beer containing high levels of active β-glucan.[5] Research studies have shown that natural grain fibers can lower cholesterol. Clinical studies are underway in Sweden to evaluate the effect of a beer containing high levels of β-glucan on cholesterol levels.

Low-Carbohydrate Beers

Low-carbohydrate beers, commonly referred to as "low-carb" beers were the "hot new product" in North America in 2004. This was a very successful new category launch. The industry generally recognizes light beers as having low calorie counts; but this craze was not about calories but rather focused on low-carb counts, with the beers being advertised as containing fewer carbohydrates rather than fewer calories. Why all the interest in low carbohydrate products? The main driver was the Atkins diet craze, which focused not on calories, but on limiting carbohydrates in the diet. Someone starting on the Atkins diet is allowed a very low intake of carbohydrates per day during the induction phase (but calories are not limited) — the amount of carbohydrates is normally less than 20 g in the induction phase with a higher allowance once weight loss is achieved, but the total consumption of daily carbohydrates is still very low compared to a normal North American diet. The Atkins diet books have been on the bestseller list for years and over 15 million copies have been sold. These books[6] are now complemented by books such as the South Beach Diet,[7] which preach very similar principles. Reduce carbohydrate intake and slimness will follow is the promise. This low-carb revolution now includes over 30 million Americans avoiding starchy, sugary foods and drinks and has triggered the creation of at least 13 low-carb beers worldwide, and the number is growing steadily.[8]

Anheuser-Busch sold nationally one of the first "low-carb beers" with the highly successful launch of Michelob Ultra — marketed as a low-carb product with 2.8 g of carbohydrate per 12 oz bottle. They used an extended mashing process to reduce the carbohydrate content. By 2004, Michelob Ultra had captured over 3% of the U.S. market (its debut was in September 2002). Although low-carb beer drinkers are often shifting from light beer products, the low-carb beers tend to be priced as premium products thus having a positive impact on revenue. Are low-carb beers just a short-term fad? Only time will tell, but it should be noted that when light beers were first introduced there were similar doubts — it took 30 years, but light beers now outpace regular beers in sales, and four out of the ten top-selling brands today are light beers.

Table 22.1 illustrates the levels of carbohydrates in some typical North American beers. As can be seen, the Miller Lite product was already relatively low in carbohydrate but was traditionally marketed in terms of

TABLE 22.1

Carbohydrate, Calorie, and Alcohol Content Comparisons

Beer Category	Carbs (g) per 12 Oz	Calories per 12 Oz	Alcohol by Volume
Low-carb Beers			
DAB Low Carb	2	92	Not listed
Michelob Ultra	2.6	95	4.2
Rolling Rock Green Light	2.6	91.4	Not listed
Coors Aspen Edge	2.6	94	4.1
Light Beers			
Miller Lite	3.2	96	4.5
Coors Light	5.0	102	4.2
Bud Light	6.6	110	4.2
Regular Beers			
Budweiser	10.6	145	5.0
Heineken (U.S. Lager)	11.5	150	5.0
Miller Genuine Draft	13.1	143	5.0

calories rather than its low carbohydrate content. Another product that has been on the market for a long time is the German Dortmunder's DAB Diat Pils, developed many years ago for diabetics and, hence, very low in carbohydrate and now marketed in the United States as DAB low carb.

Most light beers are made from highly fermentable worts that leave very little carbohydrate in the finished beer. A long mash rest at 65.5–68.3°C (150–155°F), the addition of "debranching enzymes" (specifically glucoamylase) and high adjunct ratios are some of the common methods used by large brewers to produce light beer.

What brewing processes are used to make low carb beers? Several methods can be used individually or in combination, and lessons from light beer production are helpful. The goal is to further reduce the level of carbohydrates (dextrins) that are normally present in a beer at the end of fermentation. Tools include adjusting the amounts of grains and sources of carbohydrate, adjusting the mashing temperature and conversion rest to ensure that the complex carbohydrates are broken down into sugars that the yeast will metabolize to alcohol, and ensuring that there is time for the yeast to utilize all of the sugar. Some brewers employ a krausening process where freshly fermenting wort is added to restart the fermentation. Selection of yeast strains can also influence the amount of the carbohydrates that are utilized, as some strains can utilize smaller dextrins. The option to add commercial enzymes to break down the carbohydrates (i.e., dextrins) is also there if the brewer is willing to go this route. The problem with removing all of the dextrins that are normally present at the end of regular beer fermentation to get to an extremely low carbohydrate count is that a beer with a thinner texture and little mouthfeel or color often results. To compensate for this, brewers use caramel malt for color, specialty malts for flavor, and careful attention to hopping when making these low-carb products.

Is the low-carb trend in North America a fad only related to the Atkins diet, and as a category will it quickly fade with the next diet trend, or will it morph into something much bigger? Industry analysts are divided about the staying power of low-carb beer, and it appears that this sector is growing at the expense of the light beer sector. Speculation is that as it is rooted in the much bigger concern of the awareness of the dangers of obesity in America, and there is so much concern today regarding nutrition, obesity, and healthy beverages, low-carb will probably remain as a viable category for the foreseeable future.

Widgets

The widget is a device inserted into a can to introduce nitrogen into canned beer to produce a firm head similar to draft beer when it is poured. The widget technology is of particular interest for all types of beer with a low CO_2 content, such as ales. In 1984, Guinness developed the first "widget" for this purpose and successfully patented it.[9] They had developed a system to imitate the surge of bubbles created by tiny holes in the tap used to pour normal draft Guinness. Many other widgets have since been developed and patented.

The Guinness widget is a small plastic nitrogen-filled sphere with a tiny hole. The widget is added to the can before sealing and it floats in the beer with the hole just slightly below the surface of the beer. Before the can is sealed, a small amount of liquid nitrogen is added to the beer to pressurize the can, and as the pressure increases the beer is forced into the sphere through the hole compressing the nitrogen inside the sphere. When the can is opened, the pressure drops inside the can, the beer inside the sphere rushes out through the tiny hole and this causes the CO_2 dissolved in the beer to form many tiny bubbles. These bubbles, when they rise to the surface of the beer, form a distinctive creamy head. Other brewers soon developed their own "widgets" and the widget was introduced to draft beer in 1992, lager in 1994, and cider in 1997.

In 1999, Guinness introduced its "rocket widget" enabling drinkers to drink the beer straight from the bottle. Once the bottle is opened, the rocket "takes off," releasing a stream of bubbles into the beer, creating the same creamy head as if it had been poured from a tap. This ensures that the stout keeps its head and that the beer remains smooth by "recharging" the drink at each tip of the bottle.

Whitbread and Heineken jointly filed for a widget in a bottle, a patent that was granted in 1996.[10,11] This widget consists of a hollow plastic insert, which is squeezed in through the neck of the bottle and floats on the beer surface. It has a hole or valve on its top and a one-way valve on its underside. As the bottle is sealed, pressure inside is increased and gas enters the widget through the top hole or valve, increasing the widget's internal pressure. On opening the bottle and releasing its pressure, gas from the widget jets into the beer from the underside hole giving it a creamy texture.

In some widgets, the nitrogen is stored under high pressure inside a plastic casing and in others the nitrogen is absorbed onto the surface of a folded sheet of polymer. The nitrogen is released into the beer when the inside pressure drops after opening the container. The innovative work of Guinness was rewarded in 1991 with the Queen's Award for Technological Achievement. Widget technology is today accepted as a standard system for producing canned or bottled beverages where a smooth and creamy nitrogen head is desired, and the technology continues to evolve and improve as companies continuously tweak the technology.

Colder Beers

Ice Beers

Modern ice beers, now a common brand in most product portfolios, were first introduced in Canada during the early 1990s.[12] References to ice beer in Germany can be found that go back more than a 100 years. Brewers observed that when beer in kegs was allowed to freeze, a concentrated high-alcohol liquid separated from the ice. Eisbock, a beer style with German roots, is made by lowering the temperature of a Dopelbock until ice crystals form, then removing the crystals through filtering. Traditionally, this method was used to produce very strong alcoholic beers. Modern ice beers, however, are not usually strongly alcoholic and are made with a number of processes, basically all involving the superchilling of beer prior to the final filtration step. The beer is cooled below freezing, causing the formation of ice crystals. It is filtered or subjected to other processes, which remove a portion of the ice crystals from the beer. The resultant beer contains slightly less volume than the beer which entered the process. After this freezing process, brewers often restore some of the volume of water lost when ice crystals are removed depending on the final alcohol desired. The ice process results in a smoother beer and often slightly higher alcoholic strength. In addition, the process removes some of the unwanted haze-forming proteins and tannins resulting in a beer with better stability.

Superchilled Beer

Following after ice beers was the concept of "superchilled" when serving draft lager beers. This technology offers the consumer a colder drink than that available from a typical draft system.

One system that has been developed delivers a beer with a layer of soft crystals of frozen beer beneath the head. An integrated cooling and dispense system was designed to accomplish this. The lager is dispensed at subzero temperature and ice formation in the glass at the end of dispense is triggered via an ultrasonic pulse. This technology, called ARC, is well described in the 2003 European Brewery Convention Congress Proceedings.[13]

Another way that beer is offered as superchilled is through the use of superchilled draft tap that delivers the beer into the glass at 2–4°C. This innovation was launched in 1998 by Guinness. Scot-Co's method to deliver extra-cold draft beer is a head injection tap with o-ring cooling, resulting in the carbon dioxide being locked in and a superior quality head. Carlsberg developed a new vortex tap, in conjunction with engineers at the University of Birmingham, which in addition to delivering a perfect head dispenses 30% faster.

Immobilized Cell Technology — Progress

There are three main applications for immobilized cell technology in the brewing industry:

1. The production of alcohol-free and low-alcohol beers
2. Use of the technology to reduce time for secondary fermentation/aging
3. Use of the technology to reduce time for primary fermentation

Immobilized cell technology has great promise, and has already found application for rapid aging of beer and for the production of low-alcohol beer. There are a number of systems currently in commercial use. Finland has been a leader in implementing rapid aging and Bavaria Brewing in the Netherlands has for many years been using immobilized cell technology for producing low-alcohol beers.[14,15] The goal of using immobilized cell technology for main fermentation, as well as for aging, is being pursued by a number of companies using various immobilization matrices, such as wood chips, spent grains, ceramic beads, and cellulose carriers.[14-21]

Why the interest in this technology and exactly what does it involve? Immobilized cells are defined as yeast cells physically confined or localized in a certain defined region of space, with retention of their catalytic activity and viability — these yeast cells can be used repeatedly and continuously.

The parameters that are looked for include the following:

Low capital cost
- High productivity
- Mechanically simple

Low operating cost
- Continuous operation
- Simple operation
- Low energy input

Operational control and flexibility
- Controlled oxygenation
- Controlled yeast growth
- Rapid start-up and shut-down
- Control of contamination

To be commercially viable, the system must be optimized for the interrelated factors of cell physiology, mass transfer, immobilization procedures, and reactor design to ensure high specific rates of fermentation, independent of yeast growth. The system must be able to consistently produce beer with the desired sensory and analytical profile. Many of the original systems, such as those developed by Kirin brewing company, were found to work and produce an acceptable beer, but the complexity of the system was found to be a barrier to large-scale implementation. Ongoing research is required, and indeed is in progress, to manufacture a system that will address all of the parameters listed above on an industrial scale in a consistent and cost-effective manner. Currently, the performance of immobilized technology in brewing still lies below that of traditional batch fermentation, and there is a lack of worldwide acceptance due to unproven long-term performance. However, its potential to revolutionize the industry is a distinct possibility.

The Aluminum Bottle

The aluminum bottle offers advantages over glass bottles in terms of ease of recycling, lower cost of raw materials, weight, and removal of dangers associated with broken glass in venues such as sport events and dance floors of bars and clubs. The bottles have three times the aluminum of a typical beer can, giving them superior insulation and keeping the beer cold longer (up to 50 min longer than a glass bottle). An estimated 2 billion aluminum bottles were sold in Japan in 2003. In 2003, Heineken released a limited edition aluminum bottle to select markets. In North America, Pittsburgh Brewing Company introduced the aluminum bottle nationwide in 2004 and Anheuser Busch used the aluminum beer bottle for a number of its brands in selected markets the same year.

The Plastic Beer Glass

A plastic glass that is nearly unbreakable and hence offers safety considerations, as well as has the ability to produce an excellent head on the beer, that is the promise of a new plastic technology.[22] Invicta Ltd, a plastics engineering firm in the United Kingdom, together with Coors, have developed a new plastic beer glass designed to make the head on a beer last longer. Etching nucleation points on the bottom of the glass to produce bubbles is not new, but this is the first time that it has been attempted with plastic beer glasses made of a high quality polycarbonate, a new material of superior quality, able to withstand high temperatures in washes and which does not discolor. A laser etching process, similar to one used by glass manufacturers, creates an uneven surface on the bottom of the glass. This allows nucleation to take place when beer hits the bottom of the glass, with bubbles rising to the top to create the head. Launched in March of 2005, there are no data yet as to customer acceptance of these glasses, but manufacturing costs are stated to be lower than for existing PET glassware.

Neuroeconomics

Neuroeconomics is a newly emerging science, the intersection of economics and neuroscience, and it has the potential to offer tremendous insight into why customers buy certain products. Neuroeconomics uses the technology of brain scans such as functional magnetic resonance imaging or positron emission topography to literally view neurological reaction to stimuli. The stimuli can be taste, a product photo, or an advertising message. There has always been an element of doubt with consumer focus groups and other research methods. Because brain activity does not lie, there is no longer the question of "is the consumer just telling you what you want to hear?".

While duplicating some classic tests in the laboratory, for example, the Coke versus Pepsi taste challenge, researchers discovered that blind tastings yielded one set of results and stimulated the right side of the brain, while brand-revealed tastings generated an entirely different set of results and stimulated the top of the brain.[23] Their conclusion was that brand attributes and associations demonstrated a marked ability to override taste-based preferences. In other words, brands are so powerful that we are sometimes more likely to buy something we identify with than something we like better, based on taste. How this will influence what new innovations come to market in the brewing industry will be interesting to monitor as this new tool of neuroeconomics comes into more widespread usage.

References

1. Rimm, E.B., Williams, P., Fosher, K., Criqui, M., and Stampfer, M.J., Moderate alcohol intake and lower risk of coronary heart disease: meta-analysis of effects on lipids and haemostatic factors, *Br. Med. J.*, 319:1523–1528, 1999.
2. Mitchell, M.C., Update on alcohol research in 2001, *Proc. 28th Eur. Brew. Congr.*, Budapest, 2001, Paper 2 CD ROM.
3. Baxter, E.D. and Walker, C.J., Beer and health research, *Proc. 28th Eur. Brew. Congr.*, Budapest, 2001, Paper 9 CD ROM.
4. Mukamal, K.J., Conigrave, K.M., Mittleman, M.A., Camargo, C.A., Jr., Stampfer, M.J., Willett, W.C., and Rimm, E.M., Roles of drinking pattern and type of alcohol consumed in coronary heart disease in men, *N. Engl. J. Med.*, 348:109–118, 2003.
5. Triantafyllou, A.O., Preparation of wort and beer of high nutritional value and corresponding products, U.S. Patent Application 20040170726, September 2, 2004.
6. Atkins, R.C., *Dr. Atkin's Diet Revolution*, Bantam Publishing, New York, 1972.
7. Agatson, A., *The South Beach Diet*, Random House, New York, 2003.
8. Bradford, J.J., Drink beer, lose weight? The low carb phenomenon, *Beer Mag.*, 25:18–21, 2004.
9. Arthur Guinness Son & Company, A beverage package and a method of packaging a beverage containing gas in solution, GB Patent 2183592, October 4, 1989.
10. Whitbread & Co Ltd, Heineken Tech Services (NL), Carbonated beverage container, GB Patent 2280886, October 23, 1996.
11. Whitbread & Co Ltd, Heineken Tech Services (NL), Beverage container, GB Patent 2280887, October 23, 1996.
12. Labatt Brewing Company Limited, Improvements in production of fermented malt beverages, U.S. Patent 5,304,384, April 19, 1994.
13. Smith, S. and Quain, D., The world first "super chilled" draught lager — challenges and opportunities, *Proc. 29th Eur. Brew. Congr.*, Dublin, 2003, Paper 141, CD ROM.
14. Mensour, N.A., Margaritis, A., Briens, C.L., Pilkington, H., and Russell, I., Application of immobilized yeast cells in the brewing industry, in *Immobilized Cells: Basics and Applications*, Wijffels, R.H., Buitelaar, R.M., Bucke, C., and Tramper, J., Eds., Elsevier, Amsterdam, 1996, pp. 661–671.
15. Mensour, N.A., Margaritis, A., Briens, C.L., Pilkington, H., and Russell, I., New developments in the brewing industry using immobilized yeast cell bioreactor systems, *J. Inst. Brew.*, 103:363–370, 1997.
16. Onaka, T., Nakanishi, K., Inoue, T., and Kubo, S., Beer brewing with immobilized yeast, *Nat. Biotechnol.*, 3:467–470, 1985.

17. Pajunen, E., Tapani, K., Berg, H., Ranta, B., Bergin, J., Lommi, H., and Viljava T., Controlled beer fermentations with continuous one-stage immobilized yeast reactor, *Proc. 28th Eur. Brew. Congr.*, Budapest 2001, Paper 49, CD ROM.
18. Virkajärvi, I., Vainikka, M., Virtanen, H., and Home, S., Productivity of immobilized yeast reactors with very-high-gravity worts, *J. Am. Soc. Brew. Chem.*, 60:188–197, 2002.
19. Shen, H.-Y., Moonjai, N., Verstrepen, K.J., and Delvaux F.R., Impact of attachment immobilization on yeast physiology and fermentation performance. *J. Am. Soc. Brew. Chem.*, 61:79–87, 2003.
20. Brányik, T. and Vicente, A.A., Continuous primary fermentation of beer with yeast immobilized on spent grains — the effect of operational conditions, *J. Am. Soc. Brew. Chem.*, 62:29–34, 2004.
21. Virkajärvi, I. and Pohjala, N., Primary fermentation with immobilized yeast; some effects of carrier materials on the flavour of the beer, *J. Inst. Brew.*, 106:311–318, 2000.
22. Anon., Invicta get a head in plastic glasses, *Materials World*, March 2005.
23. McClure, S.M., Li, J., Tomlin, D., Cypert, K.S., Montague, L.M., and Montague, R.M., Neural correlates of behavioral preference for culturally familiar drinks, *Neuron*, 44:279–287, 2004.

Index

A

Abnormal fermentations, 514–516
  causes, 514–515
  symptoms, 514
  treatments, 515–516
Abstraction
  ground water and, 99–100
  surface water and, 99
ABV. *See* alcohol by volume.
Accelerated fermentations, 507–508
Acetic acid bacteria, 613–614
Acetobacter. See acetic acid bacteria
Acid addition, decarbonation and, 126
Acids, brewing process supplements and, 337–339
Adhesives, glass bottles and, 578
Adjunct cookers, mashing-in systems and, 403–404
Adjuncts, 161–174, 335
  barley, 164–166
  corn grits, 163
  definition of, 161–163
  liquid form, 169–172
  refined corn starch, 167–168
  rice, 163–164
  sorghum, 166–167
  torrified cereals, 169
  wheat starch, 168–169
Adsorbent fiber technique, hop oils and, 265
Aeration
  centrifugal mixers, 441
  ceramic metal candles, 440
  sintered metal candles, 440
  static mixers, 440
  venturi pipes, 440
  wort and, 489–490
  wort cooling and, 439–441
Aeration systems, 130–131
Aging, brewing and, 86–87
Aging, finishing and, 525–548
  beer carbonation, 546–547

beer recovery, 535–536
clarification, 536–542
component processes, 526–528
flavor maturation, 528–532
lagering, 532–535
secondary fermentation, 532–535
stabilization, 542–545
standardization, 547–548
unit operations of, 527
Air evacuation, bottle filling and, 581–582
Alcohol by volume (ABV), 7
Alcohols, yeast excretions and, 306–307
Ale vs. porter, 8–9
Ales, 56–68
  Belgian origin, 65–68
  British origin, 56–61
  fermentation and, 501–502
  French origin, 65–68
  German origin, 64–65
  India pale ale, 8–9
  Irish origin, 61–62
  North American origin, 62–64
Aleurone layer, 144–146
Aluminum bottle, 826–827
Amber glass, 574
Anaerobic gram-negative bacteria, 616–618
  treatment systems, waste water and, 703–705
Anaerobic treatment systems, waste water and, 703
Analytical control, 750–761
  collaborative trials, 752–753
  microbiology methods, 754–757
  repeatability and reproducibility, 752
  sensory analysis, 757–760
  trained tasting panel, 760–761
Anstellbottich, cold break and, 438
Antibody DEFT, 623–624
Aphids, 252
ATP bioluminescence, 622–623

Auditing, materials control and, 743–747
Autolyzed yeast, 688–691

B

Bacteria, 611–618
  gram-negative, 613
  gram-positive, 612
Baits, insect and control methods and, 641
Barley, 139–158
  as adjunct, 164–166
  growing of, 139–142
  structure and function, 142–150
    aleurone layer, 144–146
    embryo, 149–150
    husk, 142–143
    pericarp, 143
    starchy endosperm, 146–149
    testa, 143–144
Barley malt types, beer style creation and, 44
Batch processing, fermentation and, 496–503
BCS. *See* brewer's condensed solubles
Beer
  ales, 56–68
  barley malt types, 44
  bock beer, 48–49
  boiling time, 46
  canned, 18
  carbonation of, 546–547
  chloride and, 111
  classifications, 39–73
  creation of, 42–48
  defined, 41
  differentiation of, 50–52
  drinking, violence and, 23
  enzymes and storage, 347–348
  equipment configuration, 46
  evaluation, 51–52
  fermentation temperature, 47
  filtration, 47
  globalization, 23–27
  grains, 45
  guidelines, Brewers Association, 52–54
  hops, 43–45
  hops analysis and, 266–268
  hydrogen and, 110
  ice, 21
  India pale ale, 49
  inorganic ions and, 108–109
  instability, 715–726
  iron and, 111
  Japan, 20–21
  lagers, 68–73
  lautering of, 46
  listing of, 55–73
  lite, 21
  magnesium and, 110
  marketing, 47–48
  mashing, 46
  maturation time, 47
  milling, 46
  multiplant brewing, 24
  origins, 39–73
  packaging, 47, 541–542
  pilsener lager, 49
  potassium and, 109–110
  processing, 45–46
  quality, 118–121
  recovery, 535–536
  sodium and, 109
  stability, 715–716
  stout, 49
  sulfate and, 111
  taxation, 23
  transfer, fermentations and, 516–520
  United Kingdom, 20
  United States, 21
  waste, 696–698
  yeast, 45
Belgian origin ales, 65–68
Beta extract, 239
Bicarbonate, inorganic ions and, 120
Bitter flavor
  hops chemistry and, 212–217
  iso-α-acids, 212–213
Bittering
  cost of, 246–248
  units (BU), 266
  value, hops storage issues and, 209
Blending consistency, 547
Bock beer, 48–49
Boil-over, wort boiling and, 477–479
Boiling time, beer style creation and, 46

Index

Bottles
  air evacuation, 581–582
  bottom fermentation, 10–12
  fill height detection, 582–583
  filling of, 581–582
  industrial brewing and, 13
  molding, 574–575
  packaging
    cost of, 567–568
    technology and, 573
  plant, 578–581
  rinsing, 579
  stoppers and, early history, 553–554
  washing, 579–581
Brackish water, 101
Breeding, hops and, 190–191
Brettanomyces, 5
Brewer's condensed solubles (BCS), 692–693
  composition of, 693
  production of, 692–693
  value of, 693
Brewer's by-products and effluents
  brewer's condensed solubles, 692–693
  excess carbon dioxide, 693–695
  solid waste materials, 698–700
  spent filter cake, 695–696
  waste beer, 696–698
  waste water, 700–707
Brewer's grain, 666–670
  drying of, 670–671
  feed products, 672–675
  liquor, 668
  nonruminant feeds, 675–676
  ruminant feeds, 674–675
Brewer's yeast
  brewery by-products and effluents, 677–692
  inactivation and preservation of, 681–682
  surplus
    collection and beer recovery, 678–680
    storage handling, 680–681
    swine, 682–683
    volumes, 677–678
    use in human foods, 676–677
  volumes, 671–672

Brewers Association style guidelines, 52–55
  active beer brewing, 53–54
  analysis, 53
  industry feedback, 54
  tasting and recording, 53
Brewery by-products, 655–707
  brewer's grain, 666–670
  economic uses of, 656–658
    feedstuffs, 658–663
  effluents, 655–707
    economic uses of, 656–658
  spent hops, 663–664
  trub, 665–666
Brewhouse
  capacity of, 442–443
  cleaning of, 443–444
  efficiency of, 441–444
    yield calculations, 441–442
  technology, 383–444
    efficiency and, 441–444
    heat transfer, 388–390
    hop additions, 429–430
    layout, 386–388
    milling, 392–397
    raw materials intake, 390–391
      removal of foreign objects, 391
      storage, 390–391
    wort
      boiling, 414–429
      classification, 431–437
      cooling and aeration, 436–441
  unit operations, 449–450
  yield, wort temperature and, 486
Brewing
  companies, consolidation of, 24–26
  consumption of water, 95–96
  facilities
    sanitation and pest control integration, 630–637
  history of, 1–32
  industrial. *See* Industrial brewing.
  miscellaneous ingredients, 333–378
    acids, 336–337
    biological stabilizers, 377–378
    chillproofing, 358–371
    clarifiers, 352–358

Brewing (*Continued*)
   defoaming agents, 350–352
   enzymes, 340–349
   fining agents, 352–358
   flavor stabilizers, 371–374
   foam stabilizers, 374–378
   overview, 78–79
   salts, 339
   water treatment, 336–340
   yeast nutrients, 349–340
  overview, 77–89
  process control, 447–486
   dust explosion prevention, 454–455
   dust extraction, 454–455
   grain flow and routing, 457–459
   grist
    preparation, 450–459
    storage, 454
   hot wort clarification and cooling, 480–486
   malt
    silo level detection, 455–457
    stock control, 455–457
   mashing, 459–466
   performance indicators, 449
   unit operations, 449
   wort
    boiling, 472–480
    separation, 466–472
  processing steps, 79–89
   aging, 86–87
   clarification, 87
   fermentation, 84
   malting, 79–80
   mashing, 81–82
   milling and adjunct use, 80–81
   packaging, 87–88
   trub removal, 82–83
   warehousing and distribution, 89
   wort
    boiling, 82
    cooling, 83
    separation, 82
   yeast
    handling, 83–84
    pitching, 84
    removal, 85
  quality
   clean-in-place, 619
   critical control point, 619; 733–737
   hazard analysis of critical control points, 619; 733–737
   microbiology and, 618–625
    procedures, traditional, 621–622
    standards, 619–621
   rapid methods, 622–624
   sampling, 619–621
   water and, 94–98
British market, packaging trends and, 570–571
British origin ales, 56–61
Bromine, 133
BU. *See* bittering units
Buffer systems, pH control and, 114
By-products, brewing and, 655–707

C

Calcium, inorganic ions and, 118
Calcium salts, 130
Canadian market, packaging trends and, 571–572
Cans, 584–590
  fill height detection, 589–590
  filling of, 586–587
  formation of, 585
  lids, 585–586
  materials, 585
  packaging and, 558–561
   cost of, 567–568
  seamer, 588–590
  seaming, 587–588
  volumetric fillers, 587
Capacity, brewhouse and, 442–443
Capillary electrophoresis, 260
Capital cost, packaging equipment and, 569
Carbon dioxide excess, 693–695
Carbon dioxide recovery, during fermentation, 520–521
Carbonate, inorganic ions and, 120

Index

Carbonation
　beer and, 546–547
　historical approach, 547
　modern approach, 547
Carbonyl compounds, yeast excretions and, 308–309
Carrageenans, 354–355
Casks, packaging costs of, 568–569
CCP. *See* critical control point
Cell concentration, pitching yeast and, 492
Cell viability, pure yeast cultures and, 319
Cell wall, yeast structure and, 284–285
Centrifugal mixers, aeration and, 441
Centrifugation
　cold break and, 438
　wort classification and, 434
　yeast separation, 518–520
Ceramic metal candles, aeration and, 440
Chillproofing
　brewing process supplements, 358–371
　enzymes usage, 360–362
　insoluble absorbents, 364–370
　oxidation reactions, 370
　precipitants usage, 363–364
Chips, as clarifiers, 358
Chloride
　effect on flavor, 111
　inorganic ions and, 120
Chloride:sulfate balance, effect on flavor, 111
Chlorination, 131
Chlorine dioxide, 132–133
CIP. *See* clean-in-place
Cis:trans ratio, 211–212
Claims management, microbrewery and, 808–811
Clarification, 87, 536–542
　aging and finishing and, 536–542
　filters and filtration, 538–540
　finings, 537
　gravity sedimentation, 537
　hot wort, 480–481
　packaging of beer, 541–542
　sterile filtration, 540–541

Clarifiers
　as brewing process supplements, 352–358
　chips, 358
　gelatin, 355–358
　isinglass, 355–358
Cleaning, brewhouse and, 443–444
Clean-in-place (CIP), 619
CO_2 extracts, 225–228
Coagulation, 123–124
Cockroaches, brewhouse and, 636
Cohumulone ratio, 211
　foam stabilization and, 238–239
Cold break
　Anstellbottich, 438
　centrifugation, 438
　cold sedimentation tank, 438
　filtration, 439
　flotation, 439
　wort
Cold sedimentation tank, 438
Cold treatments, insect control methods and, 642
Colder beers
　ice beers, 824–825
　superchilled beer, 825
Collaborative trials, analytical control and, 752–753
Colloidal stabilizers. *See* chillproofing
Colloidal systems, inorganic ions and, 113
Colorless glass, 574
Concentrating yeast, 684–687
Conductivity, wort temperature and, 484
Consolidation, brewing companies and, 24–26
Contamination
　organic, 121–122
　pest control and sanitation integration, 635
　pure yeast cultures and, 317–318
Continuous fermentations, 509
Continuous process, failure of, 31
Conveying, PET bottles and, 583
Conveyors, kegs and, 592
Cooling, hot wort and, 481–482
Copper, inorganic ions and, 119

Corn grits, 163
Corn starch, refined, 167–168
Critical control point (CCP), 619; 733–736
Crowners, glass bottles and, 582
Crowns
  glass bottles and, 577–578
  packaging and, 554–555
Cultural issues, beer style creation and, 48
Customer and consumer feedback, 761–764
Customer targeting, microbrewery marketing and, 787–790
Cytoduction, yeast genetics and, 294–295
Cytoplasmic structures
  cytoskeleton, 289
  cytosol, 288
  endoplasmic reticulum, 288
  Golgi complex, 288
  peroxisomes, 288
  vacuoles, 287–288
  yeast and, 287
Cytoskeleton, 289
Cytosol, 288

D

Debittering yeast, 684–687
Decarbonation
  acid addition, 126
  heating, 124
  lime treatment, 125–126
Defoaming agents
  brewing process supplements, 359–352
  general classes of, 352
  natural origin, 351
  synthetic origin, 351–352
DEFT. See direct epifluorescence filter technique
Deoxygenated water, 134–135
Diacetyl
  flavor compounds and, 309; 528–530
Direct epifluorescence filter technique (DEFT), 623

Direct fired kettles, wort boiling systems and, 419–420
Disease, hops and, 187
Dosing systems, Hop products benefits and, 219
Drying
  brewer's grain, 670–671
  malt, 150–151
  yeast, 684–687
Dust
  explosion, prevention of, 454–455
  extraction, brewing process control, 454–455
Dynamic low-pressure boiling, wort boiling systems and, 425

E

Effluent
  brewing and, 655–707
  wort separation and, 470–471
Electrodialysis, 129–130
Embryo, barley and, 149–150
Endoplasmic reticulum, 288
Energy feeds. See feedstuffs
Energy recovery systems, 428–429
English ales. See British origin ales
Enterobacteriaceae, 614–616
Environment, packaging and the, 569–570
Enzymes
  brewing process supplements, 340–349
  chillproofing and, 360–362
  fermentation supplementation, 345–347
  used in beer storage, 347–348
Enzymic deoxidation, 373–374
Equipment
  configuration, beer style creation and, 46
  microbreweries and, 778–783
Esters, yeast excretions and, 307
Ethanol, kettle extracts and, 224–225, 227–228
European-Germanic origin lagers, 68–69

Index 837

Evaporation rate, wort boiling and, 474–477
Excess carbon dioxide, 693–695
External heaters
  jackets, wort boiling kettles and, 421
  Kalandria systems, 423
  wort boiling systems and, 423–425
External washing, kegs and, 593
Extract, yeast and, 691–692
Extraction, grist composition and, 453

F

FABs. *See* flavored alcoholic beverages, 819
Facility design, pest control and sanitation
  perimeter, 633–634
  structural, 632–633
Fats and waxes, hops and, 205
Feed products, brewer's grain and, 672–675
Feedstuffs, brewery by-product economic uses and, 658–663
Fermentable carbohydrates, beer styles and, 42
Fermentation, 84–85, 487–521
  abnormal, 514–516
  accelerated, 507–508
  ales, 501–502
  as enzymes supplements, 345–347
  batch processing, 496–503
  beer transfer, yeast separation, 516–520
  bottom, 10–12
  carbon dioxide recovery, 520–521
  characteristics of, 498–501
  continuous, 509
  factors affecting, 503–506
  fermentor geometry, 505–506
  high-gravity, 506–507
  high-pressure, 508
  immobilized yeast, 513–514
  laboratory analyses, 502–503
  lager fermentation, 497
  low-alcohol, 511–513
  low-calorie, 510–511
  nonalcoholic, 511–513

  oxygen and, 504–505
  pH control and, 117–118
  pitching
    rate, 504
    yeast and, 491–495
  related fermentations, 506–514
  temperature, 504–505
  trub carry-over, 505
  vessels used, 497–498
  vicinal diketones, 309; 500
  wort, 489–491
  yeast
    growth during, 495–496, 504
    strain and condition, 503–504
  zinc, 504–505
Fermentor geometry, 505–506
Fill height detection, cans and, 589–590
Filling area, brewery equipment design and, 637
Filling kegs and, 595
Filling PET bottles and, 584
Filter cake, spent, 695–696
Filters, clarification and, 538–540
Filtration
  beer style creation and, 47
  clarification and, 537–538
  cold break and, 439
  wort classification and, 434
Fining agents
  as brewing process supplements, 352–358
  carrageenans, 354–355
  characteristics of, 353
  furcellaran, 354–355
Finings, 537
Finishing. *See* aging and finishing
FISH. *See* fluorescence *in situ* hybridization
Flash pasteurization, 598
Flavonoids, 201–204
  hops analysis and, 265, 268
Flavor
  maturation, 528–532
  nonbiological stability and, 719–721
  stabilizers
    as brewing process supplements, 371–374
    enzymic deoxidation, 373–374

Flavor (*Continued*)
  reducing agents, 372–373
  stabilization and, 542–543
Flavor compounds
  2,3-pentanedione, 309; 528–530
  diacetyl, 528–530
  flavor maturation and, 528–530
  nonvolatile maturation, 531–532
  sulfur, 531
  yeast autolysis, 532
Flavored alcoholic beverages (FABs), 819
Flocculation, 123–124
  yeast and, 310–312
Flotation, cold break and, 439
Fluorescence *in situ* hybridization (FISH), 624
Fluoride, inorganic ions and, 121
Foam
  hops chemistry and, 212–217
  iso-α-acids, 212–213
  nonbiological stability and, 721–724
Foam stabilization, 238–239
  cohumulone ratio, 238–239
Foam stabilizers,
  as brewing process supplements, 374–378
  metal salts, 376
  proplylene glycol alginate, 376
  quillaia extract, 376
  yucca extract, 376
Fogging treatment. *See* space treatments
Food Safety Act, water quality and, 104
Fractionated hop oils, 240
French origin ales, 65–68
Full bottle inspection, glass bottles and, 583
Fumigants, 639–640
Furcellaran, 354–355

G

Gas breakout, pasteurization safeguards and, 599
Gas chromatography (GC), hop oils and, 263–265

Gas removal, 134–135
GC. *See* gas chromatography
Gelatin, clarifiers and, 355–358
Gene mapping, hops breeding and, 191
Genetic manipulation, yeast and, 296–298
  recombinant DNA, 297–298
  spheroplast fusion, 296–297
Genetically modified (GM) yeast strains, 31–32
Genetics, killer yeast and, 295–296
Genetics, yeast and, 289–296
  cytoduction and, 294–295
  hybridization and, 292–294
  mutation and selection, 290–292
  rare mating and, 294
  strain improvement and, 290–295
Germanic origin ales, 64–65
Germanic origin lagers. *See* European-Germanic origin lagers
Germination, malt and, 151–157
Glass, 573–574
Glass bottles
  adhesives, 578
  amber glass, 574
  colorless glass, 574
  crowners, 582
  crowns, 577–578
  dimensions, 576–577
  fill height detection, 582–583
  filling of, 557–558, 581–582
  full bottle inspection, 583
  glass treatments, 556–557
  green glass, 574
  inspection and dispatch, 576
  labeling, 582
  molding, 574–575
  packaging and, 555–557
  plant, 578–581
  properties of, 576
  strength, 577
Globalization, beers and, 23–27
Glycan yeast, 691–692
Glycogen, yeast and, 312–313
Glycosides, 200
GM. *See* genetically modified
Golgi complex, 288

Index

Grain areas, brewery equipment design and, 635
Grain flow and routing, brewing process control and, 457–459
Grains, beer style creation and, 45
Gram-positive bacteria, 612–613
  Lactobacillus, 612–613
  Pediococcus, 613
Gram-negative bacteria, 613–618
  acetic acid bacteria, 613–614
  anaerobic, 616–618
  Enterobacteriaceae, 614–616
  Zymomonas, 616
Granular activated carbon filtration, 131
Gravity sedimentation, 537
Green glass, 574
Grist composition, extraction efficiency, 453
Grist preparation, 450–459
  cleaning, 451–452
  milling, 452, 453
  storage, 453
Ground water, 99–100
Guidelines, beer styles, 52–54
Gushing, nonbiological stability and, 724

H

HACC. *See* hazard analysis of critical control points
Hammer milling, 397
Happoshu, 81
Hard resins, 197
Hard water, 100
Hardness, water and, 102–103
Hazard analysis of critical control points (HACC), 619, 733–737
Haze, stabilization and, 545
Health, innovative brewing and, 820–823
Heat transfer
  brewhouse technology and, 388–390
  materials, 390
Heat treatments, insect control methods and, 642

Heating
  decarbonation and, 124
  mashing and, 463–464
Heavy metals, inorganic ions and, 119
Heptylparaben, as biological stabilizer, 377
Hexa. *See* Hexahydroiso-α-acid
Hexahydroiso-α-acid (Hexa), 237
High-gravity
  brewing, 548
  fermentations, 506–508
High pressure liquid chromatography (HPLC)
  analysis, resin acids and, 266–268
  methods, 257–260
History, of brewing, 1–32
Homogeneity
  hop product benefits and, 219
  mashing and, 463
Hop products
  additions, 429–430
  back, wort classification and, 432
  benefits of, 218
    heavy metal residue reduction, 219
    homogeneity, 219
    improved efficiency, 220
    increased stability, 218
    reduced extract losses, 219
    reduction in chemicals, 219
    use of automated dosing systems, 219
    volume reduction, 218
  classification of, 221
    isomerized, 228–233
    nonisomerized, 221–228
    reduced isomerized, 233–239
  development of, 217
  essences, 240
  extract, 430
    lagering and, 535
  miscellaneous
    beta extract, 239
    fractionated hop oils, 240
    hop essences, 240
    oil-enriched extracts, 240
    pure hop oils, 240
    type 100 pellets, 239
  oils, analysis of, 262–269
    adsorbent fiber technique, 265

Hop products (*Continued*)
   gas chromatography, 263–265
   steam distillation method, 262–263
   sulfur compounds, 263
  other uses, 241–242
  pellets, 430
  powders, 430
  storage of, 241
  resins
   biological stabilizer, 377–378
   soft resins, 192–196
  strainer, wort classification and, 432–433
Hops, 177–269
  analysis of, 248–269
   beers, 266–268
   bittering units, 266
   flavonoids, 266, 268
   oils, 262–269
   polyphenols, 265
   resin, 252–262
   resin acids, 266–268
   worts, 266–268
  beer style creation and, 43, 45
  breeding of, 189–191
   gene mapping, 191
   new variety development, 190–191
  chemistry of, 191–192
   bitter flavor, 212–217
   fats and waxes, 205
   foam, 212–217
   glycosides, 200
   isomerization, 209–212
   oils, 197–200
   pectins, 204–205
   polyphenols, 200–204
   resins, 192–197
   storage, 205–209
   whole hops, 191–192
  classification of, 182–184
  contemporary farming of, 181
  cultivation of, 184–186
  drying of, 187–188
  examination of, 248–252
   aphids, 252
   by hand, 248–251
   leaf and stem, 251–252
  harvesting of, 186–187
  history of, 181
  irrigation, 185–186
  major producing countries, 182
  oils, 197–200
   oxygentated compounds, 198–200
   sulfur compounds, 200
   terpenes, 197–198
   wort and beer analysis, 268–269
  packing of, 187–188
  pests, diseases, 187
  plant description, 182–184
  plant protection, 186
  products, 217–242
  resins, 191–192, 197
  spent, 663–664
  storage issues
   bittering value, 209
   general, 205
   kilning, 206
   moisture, 206
   oils, 206
   resin acids, 207–209
   sulfur compounds, 207
   variety of hops, 205–220
  types, 182–184, 188–189
  usage of, 242–248
   bittering cost, 246–248
   mixing calculations, 243
   product choice, 242
   utilization of, 243
  wort boiling and, 479
Hot water conditioning, 394
Hot wort clarification and cooling, 480–486
  cooling, 481–482
  wort flow rate, 482
  wort temperature, 482–484
HPLC. *See* high pressure liquid chromatography
Human food applications
  brewers grain and, 676–677
  whole yeast and, 687–688
Husk, 142–143
  damage to, 142
  microorganisms and, 142–143
Hybridization, yeast genetics and, 292–294

Index

Hydrogen
 effect on flavor, 110
 inorganic ions and, 118
Hydrolyzed yeast, 688–691

I

Ice beers, 21, 824–825
IKE. *See* magnesium-salt isomerized kettle extract
Immobilized cell technology, 825–826
Immobilized yeast, fermentation and, 513–514
India pale ale, 8–9, 49
Industrial brewing
 bottling, 13
 bottom fermentation, 10–12
 continuous process, 31
 fermenter composition, 30
 genetically modified (GM) yeast strains, 31–32
 history of, 1–32
 malting process, 29–30
 microbrewers, 22
 pasteurization, 13
 porter beer, 4–6
 porter vs. ale, 8–9
 process, mechanization and measurement, 6–7
 scientific approach, 12–14
 sparging, 13
 steam boiling, 30
 twentieth century, 15–32
Ingredients, beer styles and, 42
Innovative brewing, 817–828
 aluminum bottle, 826–827
 colder beers, 824–825
 flavored alcoholic beverages, 819
 happoshu, 818
 health and innovation, 820–823
 immobilized cell technology, 825–826
 low-carbohydrate beers, 821–823
 malternatives, 819
 neuroeconomics, 827–828
 plastic beer glass, 827
 spirits and beer, 819–820
 third category beers, 818–819
 widgets, 823–824

Inorganic constituents, water and, 101–103
Inorganic ions
 beer flavor, 109
 beer quality and, 118–121
 bicarbonate, 120
 calcium, 118
 carbonate, 120
 chloride, 120
 colloidal systems, 113
 copper, 119
 effect in brewing, 107
 effect on flavor, 109–111
 effects of beer, 108–109
 effects of wort, 108–109
 effects on malt enzymes, 112–113
 fluoride, 121
 heavy metals, 119
 hydrogen, 118
 indirect effect in brewing, 107–121
 iron, 119
 magnesium, 118
 manganese, 119
 nitrate, nitrites, 120
 pH control, 113
 phosphate, 120
 potassium, 119
 silicate, 120
 sodium, 118
 sources of, 108
 sulfate, 120
 water and, 107–121
 yeast requirements, 112
 zinc, 119
Insect control methods, 637–642
 baits, 641
 fumigant gases, 638–640
 heat or cold treatments, 642
 residual sprays, 641–642
 space treatments, 640–641
Insect monitoring methods, 642–643
Insoluble absorbents, chillproofing and, 364–370
Inspection, glass bottles and, 574
Instability, beer, 715–726
Internal heater with thermosyphon, wort boiling systems, 421–423
Internal steam coils, wort boiling kettles and, 420–421

Ion exchange system, 126–128
Ionic concentrations, brewing water composition and, 43
Irish origin ales, 61–62
Iron
  effect on flavor, 111
  inorganic ions and, 119
Irrigation, hops and, 185–186
Isinglass, 355–358
ISO 9000, quality management systems and, 731–733
Iso-α-acids, 212–213
  reduced, 213–216
Iso-extract, 231–233
  typical analysis, 232
Isomerization, 209–212
  cis:trans ratio, 211–212
  cohumulone ratio, 211
  late hopping, 212
Isomerized
  hop products, 228–233
    isomerized pellets, 228–230
    kettle extracts, 230–233
    stabilized pellets, 228–230
  kettle extracts, 230–233
    iso-extract, 231–233
    magnesium-salt, 230–231
    potassium-form, 231
  pellets, 228–230
  reduced, 233–239

J

Japan, beer and, 20–21

K

Kalandria systems, 423
Keg lines, kegs and 592
Kegs
  conveyors, 592
  external washing, 593
  filling, 595
  internal cleaning and sterilizing, 593–595
  keg lines, 592
  manufacture of, 590–591
  minikegs, 591
  offloading/depalletizing, 592–593
  packaging
    costs of, 568–569
    early history of, 552–553
    packaging technology and, 590–595
    racking machinery, 595
    spear check and torque, 593
Kettle extracts
  CO_2 extracts, 225–228
  ethanol, 224–225, 227–228
  isomerized, 230–233
  nonisomerized hop products and, 224–228
Kettles, wort boiling systems and
  external heating jackets, 421
  internal steam coils, 420–421
Killer yeast, 295–296
Kilning
  hops storage and, 206
  malt and, 151–157
Kräusening, secondary fermentation and, 533–534

L

Labeling
  glass bottles and, 582
  PET bottles and, 584
Laboratory analyses, wort and, 490–491
Lactobacillus, 612–613
Lagering
  aging and finishing and, 532–535
  historical practices, 533
  hop extracts, 535
  secondary fermentation, 534–535
Lagers
  beer styles and, 68–73
  European-Germanic origin, 68–69
  fermentation of, 497
  North American origin, 69–71
  pilsener, 49
Land application, waste water and, 706
Landscaping, pest control and sanitation integration, 631–632
Late hopping, isomerization and, 212

Index

Lauter tuns, 406–408
Lautering, beer style creation and, 46
Lead conductometric methods, 254–255
Leaf and stem, examination of, 251–252
Leak detection, PET bottles and, 584
Legal points, microbreweries operating methods and, 786–787
Level control, wort boiling and, 477
Lid placement, 588
Lid security, 588–589
Lids, cans and, 585–586
Life cycle, yeast and, 289–296
Light, nonbiological stability, 724–725
Lime, decarbonation and, 125–126
Linear racking machines, kegs and, 595
Lipid metabolism, yeast nutritional requirements and, 299–301
Liquid adjunct, 169–172
  metering, wort boiling and, 479
Liquor, recycling of, 669–670
Lite beers, 21
Low-alcohol fermentations, 511–513
Low-calorie fermentations, 510–511
  definition, 510
  production methods, 510–511
Low-carbohydrate beers, innovative brewing and, 821–823
Low-pressure steam conditioning, 393–394
  hot water conditioning, 394

M

Magnesium
  effect on flavor, 109–110
  inorganic ions and, 118
Magnesium-salt isomerized kettle extract (IKE), 230–231
Malt, 139–158
  areas, brewery equipment design and, 635
  cleaning, grist preparation and, 451–452
  drying of, 150–151
  enzymes, inorganic ions and, 112–113
  germination of, 151–157
  handling of, 150–151
  kilning and, 151–157
  milling, grist preparation and, 452, 453
  production, 150–158
  quality of, 151–157
  silo level detection, 455–457
  steeping of, 151–157
  stock control, 455–457
  storage of, 150–151
  varieties of, 157–185
  weighing, grist preparation and, 452
Malternatives, 819
Malting, 79–80
Malting process, 29–30
Maltose, yeast nutritional requirements and, 302–304
Malts
  average analysis, 156
  cereals, other, 172–174
  malted sorghum, 173–174
  oats, 172
  wheat, 172
Manganese, inorganic ions and, 119
Marketing, microbreweries, 787–799
  beer style creation and, 47–48
  concepts, 791
  cost control, 798–799
  customer targeting, 787–790
    event themes, 792–795
    missions statement, 795
  packaging and, 569
  pub brewery concept and, 790–793
  sales plan, 795–798
Mash
  acidification, 404–405
  brewery equipment design and, 636
  conversion, 398–413
  filters, 409–410
  kettle, 404
  separation, 398–413
Mash transfer, wort separation and, 467
Mash tuns, 399–401
Mashing, 81–82, 459–466
  beer style creation and, 46
  brewing process control and, 459–466
  conversion efficiency, 466
  grist transfer, 460–462
  heating, 463–464

Mashing (*Continued*)
  homogeneity of, 463
  mashing-in systems, 401–402
  purpose of, 398
  temperature control, 464
  vessel level control, 464–466
  water flow and temperature, 462–463
Mashing-in systems, 401–405
  adjunct cookers, 403–404
  comparison of, 413
  lauter tuns, 406–408
  mash acidification, 404–405
  mash conversion vessel, 402–403
  mash filters, 409–410
  mash kettle, 404
  mash separation systems, 405–406
  membrane mash filters, 411–412
  Nortek mash filter, 412–413
  strainmaster, 408–409
Materials control, 740–750
  auditing, 743–747
  other materials, 749–750
  quality and, 740–764
  raw materials, 747–749
  specifications, 742–743
  supplier relationships, 740–742
Maturation time, beer styles and, 47
  flavor and, 528–532
Mechanization and measurement, brewing process and, 6–7
Membrane mash filters, 411–412
Merlin system, wort boiling systems and, 425–427
Metabolism, pitching yeast and, 493
Metal salts, foam stabilizers and, 376
Microbe culturing, 621–622
Microbiology, 607–626
  bacteria, 611–618
  brewery and, 607–626
  constituents, water and, 103–104
  content, water and, 122
  laboratory record keeping, 625
  methods, analytical control and, 754–757
  molds, 610–611
  procedures, microbe culturing, 621–622

  quality assurance and control, 618–625
  standards, 619–621
  wild yeasts, 608–610
Microbreweries, 771–828
  choosing right products, 776–778
  claims management, 808–811
  definition of, 774
  equipment choices, 778–783
  industrial brewing and, 22
  location requirements, 775
  marketing, 787–799
    beer theme, 792
    concept of, 791
    cost control, 798–799
    customer targeting, 787
    entertainment theme, 794
    event theme, 794–795
    food theme, 792–793
    hospitality theme, 793–794
    mission statement, 795
    sales plans, 795–798
  operating methods, 785–787
    legal points, 786–787
  pub breweries, 773
    differences between microbreweries, 774, 777
  quality control, 783–785
  sales methods, 811–815
  staff organization and training, 799–808
Microorganisms, production of, in waste water, 104, 706–707
Milling
  adjunct use and, 80–81
  beer style creation and, 46
  hammer milling, 397
  mash
    conversion, 398–413
    separation, 398–413
  reasons for, 392
  roll mills, 392
  wet milling, 394–395
Mineral content
  calcium salts, 130
  decarbonation by heating, 124
  electrodialysis, 129–130
  ion exchange system, 126–128

Index

reverse osmosis, 128–129
water and adjustment of, 125–134
Minikegs, 591
Mitochondria, yeast structure and, 287
Moisture, hops storage and, 206
Molds, microbiology and, 610–611
Multiplant brewing, beers and, 24
Mutation, yeast genetics and, 290–292

N

Near infrared analysis (NIR) 260–262
Neuroeconomics, 827–828
NIR. *See* near infrared analysis, 260–262
Nisin, as biological stabilizer, 377
Nitrate, inorganic ions and, 120
Nonalcoholic fermentations, 511–513
Nonisomerized hop products
 kettle extracts, 224–228
 pellets, 221–224
Nonruminant feeds, brewer's grain and, 675–676
Nonvolatile maturation, flavor compounds, 531–532
Nortek mash filter, 412–413
North American ales, 62–64
North American origin lagers, 69–71
Nucleate boiling, 473–474
Nucleus, yeast structure and, 287
Nutritional requirements, yeast and, 298–309

O

Oats, malts and, 172
Offloading/depalletizing, kegs and, 592–593
Oil-enriched extracts, 240
Oils, hops
 storage and, 206
 See also Hops, oils
Organic compounds, removal from water, 130–131
Organic constituents, water and, 103
Organic contamination, 121–122
Organohalides, 121

Oxidation, 124–125
 reactions, chillproofing and, 370
Oxygen
 fermentation and, 504–505
 yeast nutritional requirements and, 298–299
Oxygenated compounds, 198–200
Ozonation, 133

P

Packaging
 beer and, 541–542
 beer style creation and, 47
 brewing and, 87–88
 cans, 558–561
 costs of, 567–570
 crowns, 554–555
 early history, 552–554
 environmental aspects, 569–570
 equipment, capital cost, 569
 glass bottles, 555–557
 history, 551–561
 line efficiency, 604–606
 machine speeds, 604–606
 market trends, 570
 marketing tool, 569
 review of current technologies, 565–567
 secondary, 561
 technology, 563–606
 trends
  British market, 570–571
  Canadian market, 571–572
  rest of world, 572–573
  United States market, 571
Pasteurization
 flash, 598
 industrial brewing and, 13
 packaging technology and, 596–606
 quality safeguards, 598–600
 safeguards, 599–600
 tunnel, 600–603
 units (PU), 596–598
PCR. *See* polymerase chain reaction
Peaty water, 101
Pectins, 204–205

Pediococcus, 613
Pellets, nonisomerized hop products and, 221–224
2,3-Pentanedione, 309; 528–530
Pericarp, 143
Periplasmic space, yeast structure and, 287
Peroxisomes, 288
Pest control, 629–653
  equipment design,
    can storage areas, 636–637
    filling area, 637
    grain and malt areas, 635
    mash areas, 636
    pest control and sanitation integration, 635–637
  brewing facilities new construction, 630–637
  brewing facility location, 631
  facility structural design, 632–633
  factors in brewing facility construction and, 630–637
  landscaping, 631–632
  outside contractors vs. in-house, 651–652
  perimeter design, 633–634
  points of contamination analysis, 635
  regulations, 649–651
    record keeping, 650–651
  trash and waste disposal, 634–635
Pests, hops and, 187
Pests, types, 630
PET bottles, 583–584
  bottling lines, 583
  conveying, 583
  filling, 584
  leak detection, 584
  returnables, 583
  rinsing, 583
Pet foods, surplus yeast and, 684
PGA. *See* propylene glycol alginate
pH control
  beer quality and, 113–118
  buffer systems, 114
  fermentation, 117–118
  inorganic ions and, 113
  key reactions, 114
  wort compostion, 116–117
  wort production, 115–116

Phenols, 121
Phosphate, inorganic ions and, 120
Physical stabilization, 543–544
Physical, nonbiological stability and, 716–721
PIKE. *See* potassium-form isomerized kettle extracts
Pilsener lager, 49
Pitching, pure yeast cultures and, 319–320
Pitching process, pitching yeast and, 492–493
Pitching rate, fermentation and, 504
Pitching yeast, 491–492
  cell concentration, 492
  fermentation, 493–495
  growth during fermentation, 495–496
  measuring growth, 496
  metabolism and growth, 493
  microbiological examination, 491–492
  pitching process, 492–493
Plant protection, hops and, 186
Plasma membrane, yeast structure and, 286–287
Plastic beer glass, 827
Plate failure, pasteurization safeguards and, 599–600
Plate and frame heat exchangers, wort cooling and aeration, 437–438
Plate heat exchangers, wort temperature and, 482–484
Polarographic methods, 253–254
Polymerase chain reaction (PCR), 624–625; 757
Polymers, yeast and, 312–313
Polyphenols, 200–204
  flavonoids, 201–204
  hops analysis and, 265
  proanthocyanidins, 201
Porter, disappearance of, 19
Porter beer, 4–6
Porter beer, *Brettanomyces*, 5
Porter vs. ale, 8–9
Potassium
  effect on flavor, 109–110
  inorganic ions and, 119

Potassium-form isomerized kettle
  extracts (PIKE), 231
Poultry, surplus yeast and, 683–684
Precipitants, chillproofing and,
  363–364
Press liquor, 668–669
Pretreatment, waste water and, 702
Proanthocyanidins, 201
Process quality, 764–769
  point of sale, 769
  variation, 764–768
Process water, 97
Processing aids, brewing process and,
  335–335
Prohibition, 16–18
  Volstead Act, 17
Propagation, pure yeast cultures and,
  316–317
Propylene glycol alginate (PGA),
Protein feeds. *See* feedstuffs
Protein yeast, 691–692
PU. *See* pasteurization units
Pub breweries
  definition of, 773
  location requirements, 775
  marketing, 787–788, 791
  microbreweries and, differences,
    774, 777
  operating methods, 785
  space requirements, 782
Pure hop oils, 240
Pure yeast cultures, 313–322
  cell viability, 319
  collection techniques, 320–321
  contamination, 317–318
  pitching, 319–320
  propagation, 316–317
  scale-up, 316–317
  storage, 314–316, 321
  strain selection, 314
  transportation of, 321–322
  yeast washing, 318–319

Q

QMS. *See* quality management systems
Quality, 730–769
  materials control, 740–764
  process control, 764–769

  quality management systems,
    730–740
  Also see brewing quality
Quality control, microbreweries
  and, 783–785
Quality management systems
  (QMS), 730–740
  customer requirements, 737–738
  hazard analysis critical control
    point, 733–737
  ISO 9000, 731–733
  systems integration, 738–740
Quillaia extract, foam stabilizers
  and, 376

R

Racking machinery
  kegs and, 595
  linear, 595
  rotary, 595–596
Rapid methods
  antibody DEFT, 623–624
  ATP bioluminescence, 622–623
  brewing quality and, 622–624
  direct epifluorescence filter
    technique, 623
  fluorescence *in situ* hybridization,
    624
  polymerase chain reaction, 624–625
  spoilage detection, 622–624
Rare mating, yeast genetics and, 294
Raw materials, materials control and,
  747–749
Recombinant DNA, yeast genetic
  manipulation and, 297–298
Record keeping, pest control
  regulations and, 650–651
Reduced iso-α-acids, 213–216
Reduced isomerized products,
  233–239
  characteristics of, 238
  foam stabilization, 238–239
  hexahydroiso-α-acid (Hexa), 237
  rho-iso-α-acid (RHO), 233–234
Reducing agents, flavor stabilizers
  and, 372–373
Refined corn starch, 167–168
Reinheitsgebot, 3

Repeatability and reproducibility, analytical control and, 752
Residual sprays, insect control methods, 641–642
Resin, analysis of, 252–262
    capillary electrophoresis, 260
    HPLC methods, 257–260
    lead conductometric methods, 254–255
    NIR reflectance mode analysis, 260–262
    polarographic methods, 253–254
    spectrophotometric methods, 256–257
    types of, 253
Resin acids
    analysis by HPLC, 266–268
    hops storage and, 207–209
Returnables, PET bottles and, 583
Reverse osmosis, 128–129
RHO. *See* rho-iso-α-acid
rho-iso-α-acid (RHO), 233–234
Rice, 163–164
Rinsing, PET bottles, 583
Rodent control, 646–649
Roll milling, 392–393
Rotary racking machines, kegs and, 595–596
Ruminant feeds, brewer's grain, 674–675
Ruminants, surplus yeast and, 683

S

Safety, sanitation and pest control and, 643–645
Sales methods, microbrewery and, 811–815
Salt water, 101
Salts, as brewing process supplements, 339
Sampling, microbiological standards and, 619–621
Sampling sensitivity, spoilage, 621
Sand filtration, 124
Sanitation, 629–653
    brewing facilities construction, 630–637

    insect control methods, 637–643
    perimeter rodent control, 646–649
    pest control regulations, 649–650
    safety, 643–645
Scale-up, pure yeast cultures and, 316–317
Science, modern industrial brewing and, 27–32
Scientific approach, industrial brewing and, 12–14
Scottish ales. *See* British origin ales
Seamer
    cans and, 588–590
    lid placement, 588
    lid secured, 588–589
    undercover gassing, 588
Seaming, cans and, 587–588
Seasonal influences, water and, 100
Secondary fermentation
    aging and finishing and, 532–535
    Kräusening, 533–534
    lagering and, 534–535
Semidissolved solids, water treatment and, 123–125
Sensory analysis, analytical control and, 757–760
Separation systems, wort classification and, 432
Service water, 98
Settling tanks, wort classification and, 433
Silicate, inorganic ions and, 120
Silver ions, 133
Sintered metal candles, aeration and, 440
Sludge treatment, waste water and, 705–706
Society, brewing and, 16–23
Sodium
    effect on flavor and, 109
    inorganic ions and, 118
Soft resins, 192–196
Soft water, 100–101
Solid phase micro-extraction, hop oils and, 265
Solid waste materials, brewer by-products and effluents and, 698–700

Sorghum, 166–167
  malted, 173–174
Space treatments, insect control
    methods and, 640–641
Sparging
  industrial brewing and, 13
  wort separation and, 401; 470
Spear check and torque, kegs and,
    593
Specifications, material controls and,
    742–745, 748
Spectrophotometric methods,
    256–257
Spent grain, wort separation and,
    471
Spent hops, brewery effluents and,
    663–664
Spheroplast fusion, yeast genetic
    manipulation and, 296–297
Spirits and beer, innovative brewing
    and, 819–820
Spoilage
  detection of, 622–624
  sampling sensitivity and, 621
Stability, 218, 715–716
Stabilization
  aging and finishing and, 542–545
  biological, 543
  flavor, 542–543
  haze, 545
  physical, 543–544
  procedures and filtration,
    544–545
Stabilized pellets, 228–230
Staff organization, microbrewery and,
    799–808
Staff training, microbrewery and,
    799–808
Standardization
  aging and finishing, 547–548
  blending consistency, 547
  high-gravity brewing, 548
Starchy endosperm, barley and,
    146–149
Static mixers, aeration and, 440
Steam boiling, 30–31
Steam distillation method, 262–263
Steep conditioning, 395
Steeping, malt and, 151–157

Sterile filtration, clarification and,
    540–541
Sterile microfiltration, 134
Sterilization
  bromine, 133
  chlorination, 131
  chlorine dioxide, 132–133
  ozonation, 133
  silver ions, 133
  sterile microfiltration, 134
  ultraviolet irradiation, 134
  water and, 131–134
Storage
  hops and, 205–209
  malt and, 150–151
  surplus brewer's yeast and,
    680–681
  yeast and, 321
Stout, 49
Strainmaster, 408–409
Sugar, beer style creation and, 45
Sulfate
  effect on flavor, 111
  inorganic ions and, 120
Sulfur
  compounds, 200
    hop oils and, 263
    hops storage and, 207
  flavor compounds and, 531
  yeast excretions and, 307–308
Superchilled beer, 825
Suppliers, materials control and,
    740–742
Surface water, 98–99
Surplus yeast
  collection, 678–680
  handling and storage, 680–681
  pet foods, 684
  poultry, 683–684
  ruminants, 683
  swine, 682–683
  volumes, 677–678
Suspended solids, removal of
  coagulation, 123–124
  flocculation, 123–124
  oxidation, 124–125
  sand filtration, 124
  water treatment and, 123–125
Swine, yeast surplus and, 682–683

T

Tasting panel, analytical control and, 760–761
Taxation, beers and, 23
Temperance movement, 16–18
Temperature control, mashing and, 464
Temperature drop, pasteurization safeguards and, 598–599
Terpenes, 197–198
Testa, barley and, 143–144
Tetra. *See* tetrahydroiso-α-acid
Tetrahydroiso-α-acid (tetra), 234–237
Thermosyphon, internal heater with, 421–423
Third category beers, 818–819
Torrified cereals, 169
Transportation, pure yeast cultures and, 321–322
Trash disposal, pest control and sanitation integration, 634–635
Trehalose, yeast and, 312–313
Trub
 brewery effluents and, 665–666
 carry-over, fermentation and, 505
 formation, wort boiling and, 480
 handling, wort temperature and, 485
 recovery, wort classification and, 436
 removal, 82–83
Tunnel pasteurization, 600–603
 water heating and spraying system, 601–603
 water system, 603
Turbidity, wort temperature and, 484
Type 100 pellets, 239

U

Ultraviolet irradiation, 134
Undercover gassing, seamer and, 588
United Kingdom, beer and, 20
United States, beer and, 21
United States market, packaging trends and, 571

V

Vacuoles, 287–288
VDK. *See* vicinal diketones
Vegetative reproduction, yeast and, 289
Venturi pipes, aeration and, 440
Vessel level control, mashing and, 464–466
Vessels, fermentation and, 497–498
Vicinal diketones (VDKs), 309; 500
Volatile stripping, wort boiling and, 479–480
Volstead Act, 17
Volume, brewer's grain and, 671–672
Volume reduction, hop product benefits and, 218
Volumetric fillers, cans and, 587

W

Warehousing and distribution, brewing and, 89
Waste beer, 696–698
Waste disposal, pest control and sanitation integration, 634–635
Waste water
 aerobic treatment systems, 703
 anaerobic treatment systems, 703–705
 concentrations, 701
 disposal and treatment, 700–707
 land application, 706
 microorganism production, 706–707
 pretreatment, 702
 sludge treatment, 705–706
 volumes, 701
Water, 93–135
 brewery usage, 94–98
  consumption of, 95–96
  functions, 95
 chemical characterization, 100
  brackish, 101
  hard water, 100
  peaty water, 101
  salt water, 101
  soft water, 100–101

Index

composition, 42
general-purpose, 97
ground water, 99
hardness of, 102–103
heating and spraying system, tunnel pasteurization and, 601–603
inorganic constituents, 101–103
inorganic ions, 107–121
microbiological constituents, 103–104, 122
microorganisms, 104
minor constituents, 102
organic constituents, 103
organic contamination, 121–122
process water, 97
quality of, 93–94, 104–106
   Food Safety Act, 104
   parametric values, 105–106
   World Health Organization, 104
sample compositions of, 107
seasonal influences, 100
service water, 98
sources, 98
standards for ingredient usage, 98–106
surface water, 98–99
system of, tunnel pasteurization and, 603
temperature of, mashing and, 462–463
trace constituents, 102
treatment of, 122–135
   aeration systems, 130–131
   brewing process supplements and, 336–340
   gas removal, 134–135
   granular activated carbon filtration, 131
   mineral contents, 125–134
   organic compound removal, 130–131
   semidissolved solids, 123–125
   sterilization, 131–134
   suspended solids, 123–125
Water flow, mashing and, 462–463
Water sterilization. *See* sterilization
Wet milling, 394–395
   steep conditioning, 395
Wheat malts, 172

Wheat starch, 168–169
Whirlpool kettle, wort classification and, 436
Whirlpool separators, wort classification and, 434–436
White flint glass. *See* colorless glass
Whole hops, 191–192
   hop additions and, 430
Whole yeast, food applications of, 687–688
Widgets, innovative brewing and, 823–824
Wild yeasts, 608–610
   physiological characteristics of, 609
World Health Organization, water quality and, 104
Wort
   cooling and aeration, 83, 436–441
   filtration, 468–470
   flow rate, 482
   hops analysis and, 266–268
   inorganic ions and, 108–109
   laboratory analyses, 490–491
   metabolism, yeast nutritional requirements and, 301–302
   nitrogen, yeast nutritional requirements and, 304–306
   oxygenation, wort temperature and, 484–485
   preheating, 419
   production, pH control and, 115–116
   recovery, wort temperature and, 485
   separation, 82, 466–472
   stripping, marlin system and, 425–427
   temperature, 482–484
      brewhouse yield, 486
      cold break, 484
      conductivity, 484
      plate heat exchangers, 482–484
      trub handling, 485
      turbidity, 484
      wort oxygenation, 484–485
      wort recovery, 485
Wort aeration, 489–490
Wort boiling, 82
   boil-over, 477–479
   brewhouse technology and, 414–429

Wort boiling (*Continued*)
  brewing process control and, 472–480
  evaporation rate, 474–477
  hops addition and utilization, 479
  level control, 477
  liquid adjunct metering, 479
  nucleate boiling, 473–474
  objectives of, 417–419
  preheating, 419
  principles of, 414–416
  systems, types, 419–428
    conventional boiling with stripping, 427–428
    direct fired kettles, 419–420
    dynamic low-pressure boiling, 425
    energy recovery systems, 428–429
    external heaters, 423–425
    internal heater with thermosyphon, 421–423
    kettles, 420–421
    merlin system with wort stripping, 425–427
  trub formation, 480
  types, 416
  volatile stripping, 479–480
Wort clarification and cooling. *See* hot wort clarification and cooling
Wort classification
  brewhouse technology and, 431–437
  centrifugation, 434
  filtration, 434
  hop back, 432
  hop strainer, 432–433
  separation systems, 432
  settling tanks, 433
  trub recovery, 436
  whirlpool kettle, 436
  whirlpool separators, 434–436
Wort composition, pH control and, 116–117

Y

Yeast, 281–323
  addition, wort cooling and aeration, 441
  autolysis, flavor compounds and, 532
  autolyzed, 688–691
  beer style creation and, 45
  biotechnology, 692
  brewers, inactivation and preservation of, 681–682
  collection techniques, 320–321
  concentrating, 684–687
  cropping, yeast separation and, 516–518
  debittering, 684–687
  drying, 684–687
  excretions, 306–309
    alcohols, 306–307
    carbonyl compounds, 308–309
    diacetyl and pentane-2-3-dione, 309
    esters, 307
    sulfur compounds, 307–308
  extract, 691–692
  flocculation, 310–312
  genetics, 289–296
    cytoduction, 294–295
    genome, 290
    hybridization, 292–294
    killer yeast, 295–296
    manipulation, 296–298
    mutation and selection, 290–292
    rare mating, 294
    strain improvement, 290–295
  glycan, 691–692
  glycogen, 312–313
  growth during fermentation, 495–496, 504
  handling, 83–84
  hydrolyzed, 688–691
  life cycle, 289–296
    vegetative reproduction, 289
  nutrients, as brewing process supplements, 349–340
  nutritional requirements, 298–309
    excretions, 306–309
    lipid metabolism, 299–301
    maltose, 302–304
    oxygen, 298–299
    wort, 301–302, 304–306
  pitching, 84–85, 491–492
  protein, 691–692
  pure cultures, 313–322
  removal, 85

Index

requirements, 112
separation, 516–520
storage polymers, 312–313
strain and condition, fermentation and, 503–504
strains, genetically modified, 31–32
structure of, 283–289
  cell wall, 284–285
  cytoplasmic structures, 287
  mitochondria, 287
  nucleus, 287
  periplasmic space, 287
  plasma membrane, 286–287
surplus, collection and beer recovery, 678–680
taxonomy of, 282–283
trehalose, 312–313
washing, 318–319
whole, food applications of, 687–688
Yeast, wild, 608–610
Yield calculations, brewhouse efficiency and, 441–442
Yucca extract, foam stabilizers and, 376

Z

Zinc
  fermentation and, 504–505
  inorganic ions and, 119
Zymomonas, 616